Coherent States and Applications in Mathematical Physics

Theoretical and Mathematical Physics

This series, founded in 1975 and formerly entitled (until 2005) *Texts and Monographs in Physics* (TMP), publishes high-level monographs in theoretical and mathematical physics. The change of title to *Theoretical and Mathematical Physics* (TMP) signals that the series is a suitable publication platform for both the mathematical and the theoretical physicist. The wider scope of the series is reflected by the composition of the editorial board, comprising both physicists and mathematicians.

The books, written in a didactic style and containing a certain amount of elementary background material, bridge the gap between advanced textbooks and research monographs. They can thus serve as a basis for advanced studies, not only for lectures and seminars at graduate level, but also for scientists entering a field of research.

More information about this series at http://www.springer.com/series/720

Didier Robert · Monique Combescure

Coherent States and Applications in Mathematical Physics

Second Edition

Springer

Didier Robert
Departement de mathematiques
Laboratoire Jean-Leray
Nantes University
Nantes, France

Monique Combescure
Batiment Paul Dirac
IPNL
Villeurbanne, France

ISSN 1864-5879 ISSN 1864-5887 (electronic)
Theoretical and Mathematical Physics
ISBN 978-3-030-70844-3 ISBN 978-3-030-70845-0 (eBook)
https://doi.org/10.1007/978-3-030-70845-0

This Springer imprint is published by the registered company Springer Nature Switzerland AG
The registered company address is: Gewerbestrasse 11, 6330 Cham, Switzerland

Preface to the Second Edition

The co-author of the first edition, M. Combescure, agrees that D. Robert would take care of the revisions and extension of the first edition.

In this second edition we have revisited the 12 chapters of the first edition and added some minor corrections. These chapters are now gathered together in three parts: I, II and IV.

In Part I the basic mathematical structures of coherent states are explained for the harmonic oscillator coherent states. In part II we consider coherent states in various non-Euclidean settings. The 2-torus coherent states are explained with application to quantization of cat maps and quantum ergodicity. Then we consider coherent states in the sense of Gilmore-Perelomov for the groups $SU(2)$ and $SU(1,1)$ and we introduce coherent states for the hydrogen atom which are related to the group $SO(4)$ and the unit sphere $\mathbb{S}^3$.

Part III is new. Here we present more advanced results involving coherent states, in five chapters.

In Chaps. 10 and 11 we go more deeply into the connections between coherent states, Weyl quantization and Fourier integral operators. In particular we explain in a general setting the Herman-Kluk formula for the time-dependent Schrödinger propagator. Chapter 12 is concerned with semi-classical measures and quantum chaos: a quantum ergodic theorem is explained. In Chap. 13 we consider open quantum systems. Exact computations are given for quadratic Hamiltonians which are valid for the quantum Brownian motion, by solving a master equation. As a consequence we can estimate the decoherence phenomenon of a coherent state model for the Schrödinger cat.

In Chap. 14 we present an adiabatic decoupling result for multi-component systems without crossing eigenvalues, like the Dirac Hamiltonian. As a consequence we compute the time evolution of wave packets for such systems in the semi-classical regime.

The content of the last Part IV is not new. Coherent states are introduced for infinite systems of bosons, with application of the Hepp's method to the classical limit of large systems. Then we consider finite systems of fermions and we introduce supersymmetric systems with explicit examples using coherent states.

Now, I would like to thank my collaborators during these last years for the numerous exchanges with them concerning the matter studied in this book: Clotilde Fermanian-Kammerer, Caroline Lasser, Alberto Maspero and Laurent Thomann. In particular special thanks to Clotilde and Alberto for their careful readings of the new chapters of Part III.

This book would not exist without the essential contributions of Monique Combescure to the first edition.

At the end, I thank my wife Marie-France for her understanding and patience during the preparation of this new edition and my son Julien for his help for programming pictures with Julia.

Nantes, France Didier Robert
August 2020

Preface to the First Edition

The main goal of this book is to give a presentation of various types of coherent states introduced and studied in the physics and mathematics literature during almost a century. We describe their mathematical properties together with application to quantum physics problems. It is intended to serve as a compendium on coherent states and their applications for physicists and mathematicians, stretching from the basic mathematical structures of generalized coherent states in the sense of Gilmore and Perelomov[1] via the semi-classical evolution of coherent states to various specific examples of coherent states (hydrogen atom, torus quantization, quantum oscillator).

We have tried to show how the field of applications of coherent states is wide, diversified and still alive. Because of our own ability limitations we have not covered all the field. Besides this would be impossible in one book. We have chosen some parts of the subject which are significant for us. Other colleagues may have different opinions.

There exist several definitions of coherent states which are not equivalent. Nowadays the most well known is the Gilmore-Perelomov [112, 113, 209] definition: a coherent state system is an orbit for an irreducible group action in an Hilbert space. From a mathematical point of view coherent states appear like a part of group representation theory.

In particular canonical coherent states are obtained with the Weyl-Heisenberg group action in $L^2(\mathbb{R})$ and the standard Gaussian $\varphi_0(x) = \pi^{1/4}e^{-x^2/2}$. Modulo multiplication by a complex number, the orbit of φ_0 is described by two parameters $(q,p) \in \mathbb{R}^2$ and the L^2-normalized canonical coherent states are

$$\varphi_{q,p}(x) = \pi^{-1/4}e^{-(x-q)^2/2}e^{i((x-q)p+qp/2)}.$$

[1]They have discovered independently the relationship with group theory in 1972.

Wavelets are included in the group definition of coherent states: they are obtained from the action of the affine group of $\mathbb{R}(x \mapsto ax+b)$ on a "mother function" $\psi \in L^2(\mathbb{R})$. The wavelet system has two parameters: $\psi_{a,b}(x) = \frac{1}{\sqrt{a}}\psi\left(\frac{x-b}{a}\right)$.

One of the most useful property of coherent system ψ_z is that they are an "overcomplete" system in the Hilbert space in the sense that we can analyze any $\eta \in H$ with its coefficient $\langle\psi_z, \eta\rangle$ and we have a reconstruction formula of η like

$$\eta = \int dz \tilde{\eta}(z)\psi_z,$$

where $\tilde{\eta}$ is a complex valued function depending on $\langle\psi_z, \eta\rangle$.

Coherent states (with no name) were discovered by Schrödinger (1926) when he searched solutions of the quantum harmonic oscillator being the closest possible to the classical state or minimizing the uncertainty principle. He found that the solutions are exactly the canonical coherent states φ_z.

Glauber (1963) has extended the Schrödinder approach to quantum electro-dynamic and he called these states *coherent states* because he succeeded to explain coherence phenomena in light propagation using them. After the works of Glauber, coherent states became a very popular subject of research in physics and in mathematics.

After the Glauber concept of coherent states several generalizations were elaborated: Gazeau-Klauder coherent states, squeezed coherent states, Perelomov coherent states, wave-packets transform, and may be others, with many applications in different fields in mathematics and physics; some of one is detailed in this book.

There exist several other books discussing coherent states. Perelomov's book [208] plaid an important role in the development of the group aspect of the subject and in its applications in mathematical physics. Several other books brought contributions to the theory of coherent states and worked out their applications in several fields of physics; among them we have [3, 108, 167] but many others could be quoted as well. There is a huge number of original papers and review papers on the subject, we have quoted some of them in the bibliography. We apologize to the authors for forgotten references.

In this book we put emphasis on applications of coherent states to semi-classical analysis of Schrödinger type equation (time dependent or time independent). Semi-classical analysis means that we try to understand how solutions of the Schrödinger equations behave as the Planck constant is negligible and how classical mechanics is a limit of quantum mechanics. It is not surprising that semi-classical analysis and coherent states are closely related because coherent states (which are particular quantum states) will be chosen localized close to classical states. Nevertheless we think that in this book we have given more mathematical details concerning these connections than in the other monographs on these subjects.

Let us give now a quick overview of the content of the book.

The first half of the book (Chapter 1 to Chapter 5) is concerned with the canonical (standard) Gaussian Coherent States and their applications in semi-classical analysis of the time-dependent and the time-independent Schrödinger equation.

The basic ingredient here is the Weyl-Heisenberg algebra and its irreducible representations. Relationship between coherent states and Weyl quantization is explained in Chapters 2 and 3. In Chapter 4 we compute the quantum time evolution of coherent states in the semi-classical régime; the result is a squeezed coherent state whose shape is deformed, depending on the classical evolution of the system. The main outcome is a proof of the Gutzwiller trace formula given in Chapter 5.

The second half of the book (Chapter 6 to Chapter 12) is concerned with extensions to coherent states systems related to other geometry settings. In Chapter 6 we consider quantization of the 2-torus with application to the cat map and an example of "quantum chaos".

Chapter 7 and 8 explain the first examples of non-canonical coherent states where the Weyl-Heisenberg group is replaced successively by the compact group $SU(2)$ and the non-compact group $SU(1,1)$. We shall see that some representations of $SU(1,1)$ are related to squeezed canonical coherent states, with quantum dynamics for singular potentials and with wavelets.

We show in Chapter 9 how it is possible to study the hydrogen atom with coherent states related to the group $SO(4)$.

In Chapter 10 we considered infinite systems of bosons for which it is possible to extend the definition of canonical coherent states. This is used to prove mean-field limit result for two-body interactions: the linear field equation can be approximated by a non-linear Schrödinger equation in $\mathbb{R}^3$ in the semi-classical limit (large number of particles or small Planck constant are mathematically equivalent problems).

Chapters 11 and 12 are concerned with extension of coherent states for fermions with applications to supersymmetric systems.

Finally in Appendices we have a technical section A around the stationary phase theorem and in section B we recall some basic facts concerning Lie algebras, Lie groups and their representations. We explain how this is used to build generalized coherent systems in the sense of Gilmore-Perelomov.

The material covered in these books is designed for an advanced graduate student, or researcher, who wishes to acquaint himself with applications of coherent states in mathematics or in theoretical physics. We have assumed that the reader has a good founding in linear algebra and classical analysis and some familiarity with functional analysis, group theory, linear partial differential equations and quantum mechanics.

We would like to thank our colleagues of Lyon, Nantes, and elsewhere, for discussions concerning coherent states. In particular we thank our collaborator Jim Ralston with whom we have given a proof of the trace formula, Stephan Debièvre and Alain Joye.

To conclude we wish to express our gratitude to our spouses Alain and Marie-France whose understanding and support have permitted us to spend many hours for the writing of this book.

Lyon and Nantes
October 2011

Monique Combescure
Didier Robert

Contents

Part II Coherent States in Non Euclidian Geometries

Part I
Basic Euclidian Coherent States and Applications

Chapter 1
Introduction to Coherent States

Abstract In this Chapter we study the Weyl–Heisenberg group in the Schrödinger representation in arbitrary dimension n. One shows that it operates in the Hilbert space of quantum states (and on quantum operators) as a phase-space translation. Then applying it to a Schwartz class state of arbitrary profile we get a set of generalized coherent states. When we apply the Weyl–Heisenberg translation operator to the ground state of the n-dimensional Harmonic Oscillator, one gets the standard coherent states introduced by Schrödinger [233] (Naturwissenshaften 14:664–666, 1926) in the early days of quantum mechanics (1926). Later the coherent states have been extensively studied by Glauber [118, 119] (Phys. Rev. 131:2766–2788, 1963; Phys. Rev. 130:2529–2539, 1963) for the purpose of quantum optics and it seems that their name comes back to this work. The standard coherent states have been generalized by Perelomov [208] (Generalised Coherent States and Their Applications, 1986) to more general Lie groups than the Weyl–Heisenberg group. We also introduce the usual creation and annihilation operators in dimension n which are very convenient for the study of coherent states. We show that coherent states constitute a non-orthogonal over-complete system which yields a resolution of the identity operator in the Hilbert space and which allows a computation of the Hilbert–Schmidt norm and of the trace of respectively Hilbert–Schmidt class and trace-class operators. We study their time-evolution for the quantum Harmonic Oscillator hamiltonian and show that a time evolved coherent state located around phase-space point z is up to a phase a coherent state located around the phase-space point z_t, where z_t is the phase-space point of the classical flow governed by the Harmonic Oscillator. This property was described by Schrödinger as the non-spreading of the time evolution of coherent states under the quantum Harmonic Oscillator dynamics. We also show how to go from the Schrödinger to the Fock–Bargmann representation using the standard coherent states.

D. Robert and M. Combescure, *Coherent States and Applications in Mathematical Physics*, Theoretical and Mathematical Physics,
https://doi.org/10.1007/978-3-030-70845-0_1

1.1 A Very Brief Introduction to the Heisenberg Uncertainty Principle

One of the main motivation to introduce and study coherent states comes from their close relationship with quantum mechanics. It is well known that a remarkable property is that they are the only states in the Hilbert space $L^2(\mathbb{R}^n)$ which minimize the uncertainty Heisenberg principle. To fix our notations we start by recalling a mathematical formulation of the uncertainty principle due to H. Weyl.

Let $\mathscr{H}$ be any infinite dimensional Hilbert space and $\hat{A}$, $\hat{B}$ two self-adjoint operators in $\mathscr{H}$. Recall that when theses operators are not bounded, self-adjointness requires a condition on their domains $D(\hat{A})$, $D(\hat{B})$ (see [217]).

The meaning of the *hat* on operators is to remember that very often an operator $\hat{A}$ is obtained after *quantization* of a classical observable A.

Let be $\psi \in \mathscr{H}$, $\|\psi\| = 1$ and define for $\psi \in D(\hat{A})$:

$$\langle \hat{A} \rangle_\psi := \langle \psi, \hat{A}\psi \rangle, \quad \text{(average)}. \tag{1.1}$$

$$\Delta_\psi \hat{A} := \langle (\hat{A} - \langle \hat{A} \rangle_\psi \mathbb{1})^2 \rangle_\psi, \quad \text{(variance)}. \tag{1.2}$$

Then the following Heisenberg inequality holds true for any $\psi \in D(\hat{A}) \cap D(\hat{B})$, $\|\psi\| = 1$

$$\boxed{\Delta_\psi \hat{A} \Delta_\psi \hat{B} \geq \frac{1}{4} |\langle [\hat{A}, \hat{B}]\psi, \psi \rangle|,}$$

where $[\hat{A}, \hat{B}] = \hat{A}\hat{B} - \hat{B}\hat{A}$ is the commutator. The proof is similar to the usual proof of the Cauchy–Schwarz inequality [217].

Application: $\hat{A} = \hat{P}$, $\hat{B} = \hat{Q}$, then if $\psi \in D(\hat{Q}) \cap D(\hat{P})$, $\|\psi\| = 1$, we have

$$\boxed{\Delta_\psi \hat{P} \Delta_\psi \hat{Q} \geq \frac{\hbar}{4}.}$$

Moreover

$$\Delta_\psi \hat{P} \Delta_\psi \hat{Q} = \frac{\hbar}{4} \tag{1.3}$$

if and only if there exist $q, p \in \mathbb{R}$, $\lambda > 0$, $C \in \mathbb{C}$, such that

$$\psi(x) = C \exp\left(\frac{ip \cdot x}{\hbar} - \frac{\lambda(x-q)^2}{2\hbar} \right). \tag{1.4}$$

We shall see that the r.h.s in (1.4) defines a coherent state localized in the phase space $\mathbb{R}_x \times \mathbb{R}_\xi$ close to (q, p). Recall that $\hbar = \frac{h}{2\pi}$ where h is the Planck constant, $h \approx 6,626 \times 10^{-34}$ m^{2}kg/s, small at the macroscopic scale. Notice the Fourier transform formula:

$$\int_{-\infty}^{+\infty} e^{-ix\cdot\xi} e^{-\frac{x^2}{2}} dx = \sqrt{2\pi} e^{-\frac{\xi^2}{2}},$$

so it results that the Fourier transform $\tilde{\psi}(\xi)$ is localized in p.

1.2 The Weyl–Heisenberg Group and the Canonical Coherent States

1.2.1 The Weyl–Heisenberg Translation Operator

Consider quantum mechanics in dimension n (n degrees of freedom). Then the position operator $\hat{Q}$ has n components $\hat{Q}_1, \ldots \hat{Q}_n$ where $\hat{Q}_j$ is the multiplication operator in $L^2(\mathbb{R}^n)$ by the coordinate x_j. Similarly the momentum operator $\hat{P}$ has n components $\hat{P}_j$ where

$$\hat{P}_j = -i\hbar \frac{\partial}{\partial x_j}, \tag{1.5}$$

$\hbar$ is the Planck constant divided dy 2π. $\hat{Q}$ and $\hat{P}$ are selfadjoint operators with suitable domains $\mathscr{D}(\hat{Q})$ and $\mathscr{D}(\hat{P})$.

$$\mathscr{D}(\hat{Q}) = \left\{u \in L^2(\mathbb{R}^n) \mid x_j u(x) \in L^2(\mathbb{R}^n), \quad \forall j = 1, \ldots, n\right\}.$$

$$\mathscr{D}(\hat{P}) = \left\{u \in L^2(\mathbb{R}^n) \mid \frac{\partial u}{\partial x_j} \in L^2(\mathbb{R}^n), \quad \forall j = 1, \ldots, n\right\}.$$

The operators $\hat{Q}$ and $\hat{P}$ obey the famous Heisenberg commutation relation

$$[\hat{P}_j, \hat{Q}_k] = -\delta_{j,k} i\hbar \tag{1.6}$$

on the domain of $\hat{Q}\cdot\hat{P} - \hat{P}\cdot\hat{Q}$. The bracket $[\hat{A}, \hat{B}]$ is the commutator:

$$[\hat{A}, \hat{B}] = \hat{A}\hat{B} - \hat{B}\hat{A}.$$

On the intersection of the domains $\mathscr{D}(\hat{Q}) \cap \mathscr{D}(\hat{P})$ the operator $p\cdot\hat{Q} - q\cdot\hat{P}$ is well defined for $z = (q, p) \in \mathbb{R}^{2n}$, where the dot represents the scalar product:

$$p\cdot\hat{Q} = \sum_1^n p_j \hat{Q}_j.$$

It is selfadjoint so it is the generator of a unitary operator $\widehat{T}(z)$ called the Weyl–Heisenberg translation operator:

$$\widehat{T}(z) = \exp\left(\frac{i}{\hbar}(p \cdot \hat{Q} - q \cdot \hat{P})\right). \tag{1.7}$$

Now we use the following Baker–Campbell–Hausdorff formula

Lemma 1 *Consider two anti-selfadjoint operators $\hat{A}$, $\hat{B}$ in the Hilbert space $\mathscr{H}$, with domains $D(\hat{A})$, $D(\hat{B})$. We assume the following conditions are satisfied:*
(i) There exists a linear subspace space $\mathscr{H}_0$ dense in $\mathscr{H}$, which is a core for $\hat{A}$ and $\hat{B}$.
(ii) $\mathscr{H}_0$ is invariant for $\hat{A}$, $\hat{B}$, $e^{t\hat{A}}$, $e^{t\hat{B}}$, $\forall t \in \mathbb{R}$.
(iii) $\hat{A}$ and $\hat{B}$ commute with $[\hat{A}, \hat{B}]$ in $\mathscr{H}_0$ and $i[\hat{A}, \hat{B}]$, well defined in $\mathscr{H}_0$, has a self-adjoint extension in $\mathscr{H}$.
Then we have

$$\exp(\hat{A} + \hat{B}) = \exp\left(-\frac{1}{2}[\hat{A}, \hat{B}]\right) \exp(\hat{A}) \exp(\hat{B}). \tag{1.8}$$

Proof Let us introduce

$$F(t)u = e^{-t^2/2[\hat{A},\hat{B}]} e^{t\hat{A}} e^{t\hat{B}} u,$$

where $u \in \mathscr{H}_0$ is fixed. Let us compute the time derivative

$$\begin{aligned} F'(t)u &= -t[\hat{A}, \hat{B}] e^{-t^2/2[\hat{A},\hat{B}]} e^{t\hat{A}} e^{t\hat{B}} u \\ &+ e^{-t^2/2[\hat{A},\hat{B}]} e^{t\hat{A}} (\hat{A} + \hat{B}) e^{t\hat{B}} u. \end{aligned} \tag{1.9}$$

The only difficulty is to commute $\hat{B}$ with $e^{t\hat{A}}$. But we have, using the commutations assumptions,

$$\frac{d}{dt}(e^{t\hat{A}} \hat{B} e^{-t\hat{A}}) = e^{t\hat{A}} [\hat{A}, \hat{B}] e^{-t\hat{A}} = [\hat{A}, \hat{B}].$$

So we get that

$$F'(t) = (\hat{A} + \hat{B}) F(t) \tag{1.10}$$

and the formula (1.8) follows. □

Using this formula, one deduces the multiplication law for the operators $\widehat{T}(z)$:

$$\widehat{T}(z)\widehat{T}(z') = \exp\left(-\frac{i}{2\hbar}\sigma(z, z')\right) \widehat{T}(z + z'), \tag{1.11}$$

where for $z = (q, p)$, $z' = (q', p')$, $\sigma(z, z')$ is the symplectic product:

$$\sigma(z, z') = q \cdot p' - p \cdot q', \tag{1.12}$$

and

$$\hat{T}(z)\hat{T}(z') = \exp\left(-\frac{i}{\hbar}\sigma(z,z')\right)\hat{T}(z')\hat{T}(z),$$

which is the integral form of the Heisenberg commutation relation. In particular we have:

$$(\hat{T}(z))^{-1} = (\hat{T}(z))^* = \hat{T}(-z),$$

since the symplectic product of z by itself is zero.

The fact that the Weyl–Heisenberg unitary operator is a translation operator can be seen in the following lemma:

Lemma 2 *For any $z = (q, p) \in \mathbb{R}^{2n}$ one has*

$$\hat{T}(z)\begin{pmatrix}\hat{Q}\\ \hat{P}\end{pmatrix}\hat{T}(z)^{-1} = \begin{pmatrix}\hat{Q}-q\\ \hat{P}-p\end{pmatrix}. \tag{1.13}$$

Proof Let us denote $\hat{L}(z) = p\cdot\hat{Q} - q\cdot\hat{P}$ if $z = (q, p)$. We have easily

$$i\hbar\frac{d}{dt}\left(e^{\frac{it}{\hbar}\hat{L}(z)}\hat{Q}e^{-\frac{it}{\hbar}\hat{L}(z)}\right) = e^{\frac{it}{\hbar}\hat{L}(z)}[\hat{Q},\hat{L}(z)]e^{-\frac{it}{\hbar}\hat{L}(z)}.$$

But we have $[\hat{Q},\hat{L}(z)] = -i\hbar q$. So we get the formula for $\hat{Q}$. With the same proof we get the formula for $\hat{P}$.

Corollary 1

$$\hat{T}(z) = e^{-iq\cdot p/2\hbar}e^{\frac{i}{\hbar}p\cdot\hat{Q}}e^{-\frac{i}{\hbar}q\cdot\hat{P}}. \tag{1.14}$$

Proof Let

$$\hat{U}(t) = e^{-it^2q\cdot p/2\hbar}e^{\frac{it}{\hbar}p\cdot\hat{Q}}e^{-\frac{it}{\hbar}q\cdot\hat{P}}.$$

Using Lemma 1 we get

$$\frac{d}{dt}\hat{T}(tz) = \frac{d}{dt}\hat{U}(t) = \frac{i}{\hbar}\hat{L}(tz)\hat{U}(t).$$

Hence the corollary follows. □

Let us specify the situation in dimension 1. We introduce:

$$e_1 = \frac{i}{\sqrt{\hbar}}\hat{P}$$

$$e_2 = \frac{i}{\sqrt{\hbar}}\hat{Q}$$

$$e_3 = i\mathbb{1}$$

We easily check that

$$[e_1, e_2] = e_3, \qquad [e_1, e_3] = [e_2, e_3] = 0.$$

This means that the operators $\hat{Q},\ \hat{P},\ \mathbb{1}$ generate a Lie algebra denoted by $\mathfrak{h}_1$ which is the Weyl–Heisenberg algebra. The elements of this algebra are defined using triplets of coordinates $(s; x, y) \in \mathbb{R}^3$ by:

$$W = xe_1 + ye_2 + se_3. \tag{1.15}$$

In quantum mechanics it is more convenient to use the following coordinates:

$$W = -\frac{it}{2\hbar}\mathbb{1} + \frac{i}{\hbar}(p\hat{Q} - q\hat{P}),$$

where the real numbers $t, q,\ p$ are defined as

$$q = -\sqrt{\hbar}x, \qquad p = \sqrt{\hbar}y, \qquad t = -2\hbar s.$$

Then we can calculate the commutator of two elements $W,\ W'$ of the Lie algebra $\mathfrak{h}_1$:

Lemma 3

$$[W, W'] = (xy' - yx')e_3. \tag{1.16}$$

Notice here that $\sigma((x, y), (x', y')) = xy' - x'y$ *is the symplectic product of* (x, y) *and* (x', y').

Proof We simply use the Lemma 1. □

For any W in $\mathfrak{h}_1$ we can define the unitary operator e^W and we get a group using (1.8). This group is denoted $\mathbf{H}_1$. It is a Lie group and its Lie algebra is $\mathfrak{h}_1$. The Lie group $\mathbf{H}_1$ is simply $\mathbb{R}^3$ with the non commutative multiplication

$$(t, z)(t', z') = (t + t' + \sigma(z, z'), z + z') \text{ where } t \in \mathbb{R}, z \in \mathbb{R}^2. \tag{1.17}$$

We deduce (1.17) from an elementary computation. If $W, W' \in \mathfrak{h}_1$ using (1.8) we have

$$\mathrm{e}^W \mathrm{e}^{W'} = \mathrm{e}^{W''} \text{ where } W'' = \frac{1}{2}[W, W'] + W + W'.$$

Using the (t, q, p) and (t', q', p') coordinates for W and W' respectively, we get the corresponding coordinates (t'', q'', p'') for W'' such that

$$t'' = t + t' + \sigma(z, z'), \quad z'' = z + z',$$

which is the Weyl–Heisenberg group multiplication (1.17).

In the same way we define the Weyl–Heisenberg algebra $\mathfrak{h}_n$ and its Lie Weyl–Heisenberg group $\mathbf{H}_n$.

The Weyl–Heisenberg group $\mathbf{H_n}$ and Schrödinger representation, $n \geq 1$. The Weyl–Heisenberg Lie algebra $\mathfrak{h}_n$ is a real linear space of dimension $2n + 1$. Any $W \in \mathfrak{h}_n$ has the decomposition

$$W = -\frac{it}{2\hbar}\mathbb{1} + \frac{i}{\hbar}(p \cdot \hat{Q} - q \cdot \hat{P}), \text{ where } \hat{Q} = (\hat{Q}_1, \ldots, \hat{Q}_n), \quad \hat{P} = (\hat{P}_1, \ldots, \hat{P}_n),$$

$(t; q, p) = (t; z) \in \mathbb{R} \times \mathbb{R}^{2n}$ is a coordinates system for W. The Lie bracket of W and W', in these coordinates, is

$$[W, W'] = \frac{i}{\hbar}\sigma(z, z')\mathbb{1}.$$

This reflects the Heisenberg commutation relations (1.6).

As for $n = 1$ a group multiplication is introduced in $\mathbb{R} \times \mathbb{R}^{2n}$ to reflect multiplication between operators e^W. So, $\mathbf{H_n}$ is the set $\mathbb{R} \times \mathbb{R}^{2n}$ with the group multiplication

$$(t, z)(t', z') = (t + t' + \sigma(z, z'), z + z'), \tag{1.18}$$

where σ is the symplectic bilinear form in $\mathbb{R}^{2n}$:

$$\sigma(z, z') = q \cdot p' - q' \cdot p, \text{ if } z = (q, p), \ z' = (q', p').$$

$\mathbf{H_n}$ is a Lie group of dimension $2n + 1$.

The Schrödinger representation is defined as the following unitary representation of $\mathbf{H_n}$ in $L^2(\mathbb{R}^n)$:

$$\rho(t, z) = \mathrm{e}^{-it/2\hbar}\hat{T}(z), \quad (t, z) \in \mathbf{H_n}.$$

In other words the map $(t, z) \mapsto \rho(t, z)$ is a group homomorphism from the Weyl–Heisenberg group $\mathbf{H_n}$ into the group of unitary operators in the Hilbert space $L^2(\mathbb{R}^n)$.

By taking the exponential of W one recovers the Weyl–Heisenberg Lie group defined above:

$$e^W = e^{-it/2\hbar} \exp\left(\frac{i}{\hbar}(p\hat{Q} - q\hat{P})\right) = e^{-it/2\hbar}\hat{T}(z).$$

Recall that $z = (q, p)$.

Remark 1 The Schrödinger representation is irreducible, this will be a consequence of the Schur Lemma 10. According to the celebrated Stone–von Neumann theorem (see [104, 241]) the Schrödinger representation is the unique irreducible representation of $\mathbf{H_n}$, up to conjugation with a unitary operator, for every $\hbar > 0$.

1.2.2 The Coherent States of Arbitrary Profile

The action of the Weyl–Heisenberg translation operator on a state $u \in L^2(\mathbb{R}^n)$ is the following:

$$(\widehat{T}(z)u)(x) = \exp\left(-\frac{i}{2\hbar}q \cdot p\right) \exp\left(\frac{i}{\hbar}x \cdot p\right) u(x-q). \tag{1.19}$$

Physically it translates a state by $z = (q, p)$ in phase space. One has a similar formula for the Fourier transform that we denote $\mathcal{F}$ defined as follows:

$$\mathcal{F}u(\xi) = (2\pi\hbar)^{-n/2} \int_{\mathbb{R}^n} \mathrm{e}^{-\frac{i}{\hbar}x\cdot\xi} u(x)dx,$$

$$(\mathcal{F}(\widehat{T}(z)u)(\xi) = \exp\left(\frac{i}{2\hbar}q \cdot p\right) \exp\left(-\frac{i}{\hbar}q \cdot \xi\right) \mathcal{F}(u)(\xi - p),$$

which says that the state is translated both in position and momentum by respectively q and p. Now taking any function u_0 in the Schwartz class $\mathcal{S}(\mathbb{R}^n)$ the coherent state associated to it will be simply

$$u_z(x) = (\widehat{T}(z)u_0)(x). \tag{1.20}$$

A useful example for applications is the following generalized Gaussian function. Let us denote $\Gamma \in \mathfrak{S}^+(n)$ the set of symmetric complex $n \times n$ matrices Γ such that its imaginary part $\Im\Gamma$ is positive-definite. Then we can take $u_0 = \varphi^{(\Gamma)}$, with $\Gamma \in \mathfrak{S}^+(n)$ and

$$\varphi^{(\Gamma)}(x) = (\pi\hbar)^{-n/4}\det{}^{1/4}(\Im\Gamma)\mathrm{e}^{\frac{i}{2\hbar}\Gamma x\cdot x}. \tag{1.21}$$

Coherent states are closely related with the wave-packet transform.

Definition 1 Let $z = (q, p) \in \mathbb{R}^{2n}$ and $u \in \mathcal{S}(\mathbb{R}^n, \mathbb{C})$ (a profile) . We set

$$\mathcal{WP}_z^\hbar u(x) = \hbar^{-n/4}\,\mathrm{e}^{\frac{i}{\hbar}p\cdot(x-q)} u\left(\frac{x-q}{\sqrt{\hbar}}\right). \tag{1.22}$$

The operator

$$u \mapsto \mathcal{WP}_z^\hbar u$$

is a unitary map in $L^2(\mathbb{R}^n, \mathbb{C})$ which maps continuously $\Sigma_1^k(\mathbb{R}^n)$ into $\Sigma_\hbar^k(\mathbb{R}^n)$, where

$$\Sigma_\hbar^k(\mathbb{R}^n) = \{f \in L^2(\mathbb{R}^n, \mathbb{C}),\ (-\hbar^2\Delta + |x|^2)^{k/2} f \in L^2(\mathbb{R}^n, \mathbb{C})\},$$

equipped with the $\hbar$-dependent graph-norm. We also denote $\Sigma_\hbar^k = \Sigma_1^\hbar(\mathbb{R}^n)$.

Gausssian coherent states are the Wave packet transform of Gaussian profile, up to a phase factor:

$$\varphi_z^\Gamma = e^{-\frac{i}{2\hbar}q\cdot p}\mathcal{WP}_z^\hbar(g^\Gamma),\ \Gamma \in \mathfrak{S}^+(n). \tag{1.23}$$

Moreover we have the following properties whose proofs are left to the reader

- $\mathcal{WP}_z^\hbar = e^{-\frac{i}{2\hbar\hbar}p\cdot q}\,\hat{T}(z)\,\Lambda_\hbar$.
- $\varphi_z = e^{\frac{i}{2\hbar}p\cdot q}\mathcal{WP}_z^\hbar g$ where g is the standard Gaussian.
- For all z, z' in $\mathbb{R}^{2n}$,

$$\mathcal{WP}_{z+z'}^\hbar = e^{-\frac{i}{\hbar}p\cdot q'}\,\mathcal{WP}_z^\hbar\,\Lambda_\hbar^{-1}\mathcal{WP}_{z'}^\hbar$$

where $\Lambda_\hbar u(x) = \hbar^{-n/4}u(x\hbar^{-1/2})$.

1.3 The Coherent States of the Harmonic Oscillator

1.3.1 Definition and Properties

They have been introduced by Schrödinger and have been extensively studied and used. They are obtained by taking as reference state u_0 the ground state of the harmonic oscillator

$$u_0(x) = \varphi_0(x) = (\pi\hbar)^{-n/4}\exp\left(-\frac{x^2}{2\hbar}\right) \tag{1.24}$$

Thus $\varphi_z := \hat{T}(z)\varphi_0$ is simply a Gaussian state of the form

$$\varphi_z(x) = (\pi\hbar)^{-n/4}\exp\left(-\frac{i}{2\hbar}q\cdot p\right)\exp\left(\frac{i}{\hbar}x\cdot p\right)\exp\left(-\frac{(x-q)^2}{2\hbar}\right). \tag{1.25}$$

$$(\mathcal{F}\varphi_z)(\xi) = (\pi\hbar)^{-n/4}\exp\left(\frac{iq\cdot p}{2\hbar}\right)\exp\left(-i\frac{q\cdot\xi}{\hbar} - \frac{(\xi-p)^2}{2\hbar}\right). \tag{1.26}$$

φ_z is a state localized in the neighborhood of a phase-space point $z = (q, p) \in \mathbb{R}^{2n}$ of size $\sqrt{\hbar}$ in all the position and momentum coordinates. Then it is a quantum state which is the analog of a classical state z obtained by the action of the Weyl–Heisenberg group $\mathbf{H}_n$ on φ_0. They are also called canonical coherent states. They have many interesting and useful properties that we consider now.

It is useful to use the standard creation and annihilation operators :

$$\mathbf{a} = \frac{1}{\sqrt{2\hbar}}(\hat{Q} + i\hat{P}), \tag{1.27}$$

$$\mathbf{a}^\dagger = \frac{1}{\sqrt{2\hbar}}(\hat{Q} - i\hat{P}), \tag{1.28}$$

$\mathbf{a}^\dagger$ is simply the adjoint of $\mathbf{a}$ defined on $\mathscr{D}(\hat{Q}) \cap \mathscr{D}(\hat{P})$. Furthermore a simple consequence of the Heisenberg commutation relation is that:

$$[\mathbf{a}_j, \mathbf{a}_k^\dagger] = \delta_{j,k}. \tag{1.29}$$

Then the Hamiltonian of the n-dimensional harmonic oscillator of frequency 1 is

$$\hat{H}_{\text{os}} = \frac{1}{2}(\hat{P}^2 + \hat{Q}^2) = \hbar \sum_{j=1}^{n} (\mathbf{a}_j^\dagger \mathbf{a}_j + \frac{n}{2}) = \frac{\hbar}{2}\left(\mathbf{a}^\dagger \cdot \mathbf{a} + \mathbf{a} \cdot \mathbf{a}^\dagger\right) \tag{1.30}$$

It is trivial to check that the ground state φ_0 of $\hat{H}_{\text{os}}$ is an eigenstate of $\mathbf{a}$ with eigenvalue 0. A question is: are the coherent states φ_z also eigenstates of $\mathbf{a}$? The answer is yes and is contained in the following proposition:

Proposition 1 *Let $z = (q, p) \in \mathbb{R}^{2n}$. We define the number $\alpha \in \mathbb{C}^n$ as*

$$\alpha = \frac{1}{\sqrt{2\hbar}}(q + ip). \tag{1.31}$$

Then the following holds

$$\hat{T}(z)\mathbf{a}\hat{T}(z)^{-1} = \mathbf{a} - \alpha. \tag{1.32}$$

Moreover

$$\mathbf{a}\varphi_z = \alpha\varphi_z. \tag{1.33}$$

Proof We simply use Lemma (2) to prove (1.32). Then we remark that:

$$\hat{T}(z)\mathbf{a}\hat{T}(z)^{-1}\varphi_z = \hat{T}(z)\mathbf{a}\varphi_0 = 0 = (\mathbf{a} - \alpha)\varphi_z.$$

□

The Baker–Campbell–Hausforff formula (1.8) is still true for annihilation-creation operators but we need to adapt the proof with the following modifications.

Let $\mathscr{H}_0$ be the linear space spanned by the products $\phi_\alpha(x)e^{\eta \cdot x}$ where $\alpha \in \mathbb{N}^n$ and $\eta \in \mathbb{C}^n$. We can extend the definition of $\hat{T}(z)u$ for every $u \in \mathscr{H}_0$ and $z \in \mathbb{C}^{2n}$.

Lemma 4 *For every $u \in \mathscr{H}_0$, $z \mapsto \hat{T}(z)u$ can be extended analytically to $\mathbb{C}^{2n}$. Moreover $\hat{T}(z)u \in \mathscr{H}_0$ and we have for every $z, z' \in \mathbb{C}^{2n}$ and every $u \in \mathscr{H}_0$,*

$$\widehat{T}(z)\widehat{T}(z')u = \exp\left(-\frac{i}{2\hbar}\sigma(z, z')\right)\widehat{T}(z + z')u, \tag{1.34}$$

where σ is extended as a bilinear form to $\mathbb{C}^{2n} \times \mathbb{C}^{2n}$.

Proof Using formula (1.19), we can extend $\hat{T}(z)u$ analytically to $\mathbb{C}^{2n}$. So we can define

$$\exp\left(\frac{i}{\hbar}(p\cdot\hat{Q}-q\cdot\hat{P})\right)u := \hat{T}(z)u, \quad \text{for } z=(q,p)\in\mathbb{C}^{2n}$$

Now with the same proof as for (1.8), we get that (1.34) is still true for every $z, z' \in \mathbb{C}^n$. □

Using (1.34), in the creation and annihilation operators representation we have that

$$\hat{T}(z) = \exp(\alpha\cdot\mathbf{a}^\dagger - \bar{\alpha}\cdot\mathbf{a}) = \exp\left(-\frac{|\alpha|^2}{2}\right)\exp(\alpha\cdot\mathbf{a}^\dagger)\exp(-\bar{\alpha}\cdot\mathbf{a}). \tag{1.35}$$

Recall that by convention of the scalar product $\cdot$ we have:

$$\bar{\alpha}\cdot\mathbf{a} = \sum_{j=1}^{n}\bar{\alpha}_j\mathbf{a}_j, \quad \alpha\cdot\mathbf{a}^\dagger = \sum_{j=1}^{n}\alpha_j\mathbf{a}_j^\dagger.$$

Using (1.35) we have, since $\exp(-\bar{\alpha}\cdot\mathbf{a})\varphi_0 = \varphi_0$:

$$\varphi_z = \exp\left(-\frac{|\alpha|^2}{2}\right)\exp(\alpha\cdot\mathbf{a}^\dagger)\varphi_0. \tag{1.36}$$

Two different coherent states overlap. Their overlapping is given by the scalar product in $L^2(\mathbb{R}^n)$. We have the following result:

Proposition 2

$$\langle\varphi_z,\varphi_{z'}\rangle = \exp\left(i\frac{\sigma(z,z')}{2\hbar}\right)\exp\left(-\frac{|z-z'|^2}{4\hbar}\right). \tag{1.37}$$

Proof: We first establish a useful lemma:

Lemma 5

$$\langle\varphi_0,\hat{T}(z)\varphi_0\rangle = \exp\left(-\frac{|z|^2}{4\hbar}\right). \tag{1.38}$$

Proof We use (1.35). So we get

$$\langle\varphi_0,\hat{T}(z)\varphi_0\rangle = \exp\left(-\frac{|\alpha|^2}{2}\right)\langle\varphi_0, e^{\alpha\cdot\mathbf{a}^\dagger}e^{-\bar{\alpha}\cdot\mathbf{a}}\varphi_0\rangle = \exp\left(-\frac{|z|^2}{4\hbar}\right)\|e^{-\bar{\alpha}\cdot\mathbf{a}}\varphi_0\|^2$$

But since φ_0 is an eigenstate of $\mathbf{a}$ with eigenvalue 0, we simply have

$$\|e^{-\bar{\alpha}\cdot\mathbf{a}}\varphi_0\| = 1.$$

□

The operator $\hat{T}(z)$ transforms any coherent state in another coherent state up to a phase:

Lemma 6

$$\hat{T}(z)\varphi_{z'} = \exp\left(-\frac{i}{2\hbar}\sigma(z,z')\right)\varphi_{z+z'}.$$

Proof The proof is immediate using (1.11).

The overlap between φ_z and $\varphi_{z'}$ is given by:

$$\langle\varphi_z,\varphi_{z'}\rangle = \langle\hat{T}(z)\varphi_0,\hat{T}(z')\varphi_0\rangle = \exp\left(\frac{i}{2\hbar}\sigma(z,z')\right)\langle\varphi_0,\hat{T}(z'-z)\varphi_0\rangle,$$

where we have used (1.11). Now using the lemma for the last factor we get the result. □

In the particular case of the dimension n equals one, the kth eigenstate ϕ_k of the harmonic oscillator (the Hermite function, normalized to unity) is generated by $(\mathbf{a}^\dagger)^k$:

$$\phi_k = (k!)^{-1/2}(\mathbf{a}^\dagger)^k\varphi_0,$$

so that expanding the exponential, formula (1.36) gives rise to the following well-known identity:

$$\varphi_z = \exp(-|\alpha|^2/2)\sum_{k=0}^{\infty}\frac{\alpha^k}{\sqrt{k!}}\phi_k.$$

In arbitrary dimension n, the operator $(\mathbf{a}_j^\dagger)^k$ excites the ground state of the harmonic oscillator to the kth excited state of the jth degree of freedom. More precisely let $\mathbf{k} = (k_1,\ldots,k_n) \in \mathbb{N}^n$ be a multiindex. The corresponding eigenstate of $\hat{H}_{\mathrm{os}}$ is:

$$\phi_{\mathbf{k}}(x) = \phi_{k_1}(x_1)\ldots\phi_{k_n}(x_n) \tag{1.39}$$

and it has eigenvalue $E_{\mathbf{k}} = (k_1 + k_2 + \cdots + k_n + n/2)\hbar$. Note that this eigenvalue is highly degenerate, except E_0. We have

Lemma 7

$$\phi_{\mathbf{k}} = \prod_{j=1}^{n}\frac{(\mathbf{a}_j^\dagger)^{k_j}}{\sqrt{k_j!}}\varphi_0. \tag{1.40}$$

The physicists often use the ket notation for the quantum states. Let us define it for completeness:

$$|0\rangle = \varphi_0,$$

$$|\mathbf{k}\rangle = \phi_{\mathbf{k}}$$

and they also designate the coherent state with the ket notation:

$$|z\rangle = \varphi_z.$$

Then we have:

Lemma 8

$$|z\rangle = \exp\left(-\frac{|z|^2}{4\hbar}\right)\sum_{k_j=1}^{\infty}\frac{\alpha^{\mathbf{k}}}{\mathbf{k}!}|\mathbf{k}\rangle, \tag{1.41}$$

where $\alpha_j = \frac{q_j+ip_j}{\sqrt{2\hbar}}$ *and*

$$\alpha^{\mathbf{k}} = \alpha_1^{k_1}\alpha_2^{k_2}\dots\alpha_n^{k_n},$$

$$\mathbf{k}! = k_1!k_2!\dots k_n!.$$

1.3.2 The Time Evolution of the Coherent State for the Harmonic Oscillator Hamiltonian

A remarkable property of the coherent states is that the Harmonic Oscillator dynamics transforms them into other coherent states up to a phase. This property was anticipated by Schrödinger himself [233] who describes it as the non-spreading of the coherent states wavepackets under the Harmonic Oscillator dynamics. Furthermore the time-evolved coherent state is located around the classical phase-space point of the harmonic oscillator classical dynamics.

Let $z := (q, p) \in \mathbb{R}^{2n}$ be the classical phase-space point at time 0. Then it is trivial to show that the phase-space point at time t is just $z_t := (q_t, p_t)$ given by

$$z_t = F_t z,$$

where F_t is the rotation matrix

$$F_t = \begin{pmatrix} \cos t & \sin t \\ -\sin t & \cos t \end{pmatrix}$$

We have the following property:

Lemma 9 *Define*

$$\begin{pmatrix} \hat{Q}(t) \\ \hat{P}(t) \end{pmatrix} = e^{-it\hat{H}_{os}/\hbar}\begin{pmatrix} \hat{Q} \\ \hat{P} \end{pmatrix} e^{it\hat{H}_{os}/\hbar}.$$

Note that $\hat{Q}(-t)$, $\hat{P}(-t)$ *are the so-called Heisenberg observables associated to* $\hat{Q}$, $\hat{P}$. *Then:*

(i)

$$\begin{pmatrix} \hat{Q}(t) \\ \hat{P}(t) \end{pmatrix} = F_{-t} \begin{pmatrix} \hat{Q} \\ \hat{P} \end{pmatrix}. \tag{1.42}$$

(ii)

$$e^{-it\hat{H}_{os}/\hbar}\hat{T}(z)e^{it\hat{H}_{os}/\hbar} = \hat{T}(z_t).$$

Proof (i) One has, using the Schrödinger equation and the commutation property of $\hat{Q}$, $\hat{P}$ that

$$\frac{d}{dt}\begin{pmatrix} \hat{Q}(t) \\ \hat{P}(t) \end{pmatrix} = \begin{pmatrix} -\hat{P}(t) \\ \hat{Q}(t) \end{pmatrix}.$$

Then the solution is (1.42).

(ii) Then

$$e^{-it\hat{H}_{os}/\hbar}(p \cdot \hat{Q} - q \cdot \hat{P})e^{it\hat{H}_{os}/\hbar} = p \cdot \hat{Q}(t) - q \cdot \hat{P}(t) = p_t \cdot \hat{Q} - q_t \cdot \hat{P}.$$

By exponentiation one gets the result. □

Proposition 3 *The quantum evolution for the harmonic oscillator dynamics of a coherent state φ_z is given by*

$$\boxed{e^{-it\hat{H}_{os}/\hbar}\varphi_z = e^{-itn/2}\varphi_{z_t}}.$$

See the picture (1.1), end of this chapter.

Proof

$$e^{-it\hat{H}_{os}/\hbar}\varphi_z = e^{-it\hat{H}_{os}/\hbar}\hat{T}(z)e^{it\hat{H}_{os}/\hbar} \times e^{-it\hat{H}_{os}/\hbar}\varphi_0$$

$$= \hat{T}(z_t)e^{-itn/2}\varphi_0 = e^{-itn/2}\varphi_{z_t}.$$

where we have used that φ_0 is an eigenstate of $\hat{H}_{os}$ with eigenvalue $n\frac{\hbar}{2}$. □

In Chap. 3 we shall see a similar property for any quadratic hamiltonian with possible time-dependent coefficients. Then the quantum time evolution of a coherent state will be a squeezed state instead of a coherent state, located around the phase-space point z_t for the associated classical flow which is linear (since the Hamiltonian is quadratic).

1.3.3 An Overcomplete System

We have seen that the coherent states are not orthogonal. So they cannot be considered as a basis of the Hilbert space $L^2(\mathbb{R}^n)$ of the quantum states, in the naïve sense. Instead they will constitute an overcomplete set of continuous states over which the states and operators of quantum mechanics can be expanded.

We can now introduce the Fourier–Bargmann transform that will be studied in more details in Sect. 1.3. We start from $u \in L^2(\mathbb{R}^n)$, $\|u\|^2 := \int_{\mathbb{R}^n} |u_0(x)|^2 dx = 1$. Let us define the Fourier–Bargmann transform by the following formula

$$\mathscr{F}_u^{\mathscr{B}} v(z) =: v^\sharp(z) = (2\pi\hbar)^{-n/2} \langle u_z, v\rangle, \; z = (q, p) \in \mathbb{R}^{2n}. \tag{1.43}$$

If u is the standard Gaussian φ_0, the associated Fourier–Bargmann transform will be denoted $\mathscr{F}^{\mathscr{B}}$.

Proposition 4 *$\mathscr{F}_u^{\mathscr{B}}$ is an isometry from $L^2(\mathbb{R}^n)$ into $L^2(\mathbb{R}^{2n})$.*
In particular the orbit $\mathbb{R}^{2n} \ni z \mapsto \hat{T}(z)u$ is dense in $L^2(\mathbb{R}^n)$.

Proof We have

$$\langle u_z, v\rangle = e^{\frac{i}{2\hbar} p\cdot q} \int_{\mathbb{R}^n} v(x)\overline{u_0(x-q)} e^{-ix\cdot p/\hbar} dx. \tag{1.44}$$

From Plancherel theorem we get

$$(2\pi\hbar)^{-n} \int_{\mathbb{R}^n} |\langle u_{(q,p)}, v\rangle|^2 dp = \int_{\mathbb{R}^n} |v(x)u_0(x-q)|^2 dx. \tag{1.45}$$

Then we integrate in q variable and change the variables: $q' = x - q$, $x' = x$, so we get that $\mathscr{F}_u$ is an isometry. It results easily that the orbit of u is dense in $L^2(\mathbb{R}^n)$. □

Le be $\mathbf{U}(L^2(\mathbb{R}^n))$ the unitary group of the Hilbert space $L^2(\mathbb{R}^n)$.

Corollary 2 *The (projective) group representation: $\mathbb{R}^{2n} \ni z \mapsto \hat{T}(z) \in \mathbf{U}(L^2(\mathbb{R}^n))$ is irreducible.*

Proof Let V be a non trivial subspace of $L^2(\mathbb{R}^n)$ such that $\hat{T}(z)V \subseteq V$ for any $z \in \mathbb{R}^{2n}$. Choose $u \in V$, normalized. We know that the orbit of u is dense so V is dense in $L^2(\mathbb{R}^n)$ which entails that the Schrödinger representation is irreducible. □

By polarization we get that the scalar product of two states $\psi, \quad \psi' \in L^2(\mathbb{R}^n)$ can be expressed in terms of $\psi^\sharp$, $(\psi')^\sharp$:

$$\langle \psi', \psi\rangle = \int dz \overline{(\psi')^\sharp(z)} \psi^\sharp(z). \tag{1.46}$$

We deduce, using Fubini theorem that the function $\psi^\sharp(z)$ determines the state ψ completely. We have the reconstruction formula:

$$\psi = (2\pi\hbar)^{-n/2} \int_{\mathbb{R}^{2n}} dz \psi^{\sharp}(z) \varphi_z. \tag{1.47}$$

Another way to prove that the coherent states gives us a resolution of identity is to apply the following Schur's lemma (see also Appendix). A direct proof uses Corrollary 2:

Lemma 10 *If $\hat{A}$ is a bounded operator in $L^2(\mathbb{R}^n)$ such that*

$$\hat{A}\hat{T}(z) = \hat{T}(z)\hat{A}, \quad \forall z \in \mathbb{R}^{2n}$$

then

$$\hat{A} = C\hat{\mathbb{1}},$$

for some $C \in \mathbb{C}$.

We deduce that the coherent states provide a resolution of unity as follows. Define the measure:

$$d\mu(z) = Cdz = Cdq_1dq_2\ldots dq_ndp_1dp_2\ldots dp_n, \tag{1.48}$$

where and $C \in \mathbb{C}$ is a constant to be determined later. Let $|z\rangle\langle z|$ be the projection operator on the state $|z\rangle$. We consider the operator

$$\hat{A} = \int d\mu(z)|z\rangle\langle z|.$$

We have the following result:

Proposition 5 *$\hat{A}$ commutes with all the operators $\hat{T}(z)$.*

Proof Using (1.11) we get:

$$[\hat{A}, \hat{T}(z)] = \int d\mu(z') \left(\exp\left(-\frac{i}{2\hbar}\sigma(z, z')\right) |z'\rangle\langle z' - z| \right. \tag{1.49}$$

$$\left. - \exp(\left(-\frac{i}{2\hbar}\sigma(z, z')\right) |z + z'\rangle\langle z'| \right). \tag{1.50}$$

Now using the change of variable $z'' = z + z'$ in the last term we get zero.

Therefore in view of the Schur's lemma $\hat{A}$ must be a multiple of the identity operator:

$$\hat{A} = d^{-1}\mathbb{1}.$$

We determine the constant d by calculating the average of the operator $\hat{A}$ in the coherent state $|z\rangle$:

$$d^{-1} = \langle z|\hat{A}|z\rangle = \int d\mu(z')|\langle z|z'\rangle|^2 = \int d\mu(z') \exp\left(-\frac{|z'|^2}{2\hbar}\right).$$

The constant C can be chosen so that $d = 1$. Therefore the resolution of the identity takes the form:

$$\int d\mu(z)|z\rangle\langle z| = \mathbb{1}, \tag{1.51}$$

where $d\mu(z)$ is given by (1.48) and the constant C is such that

$$C \int_{\mathbb{R}^{2n}} dz \exp\left(-\frac{|z|^2}{2\hbar}\right) = 1.$$

This gives

$$C = (2\pi\hbar)^{-n}. \tag{1.52}$$

□

The resolution of identity (1.51) allows to compute the trace of an operator in terms of its expectation value in the coherent states. Let us recall what the trace of an operator is when it exists.

Definition 2 An operator $\hat{B}$ is said to be of trace class when for some (and then any) eigenbasis e_k of the Hilbert space one has that the series $\langle e_k, (\hat{B}^*\hat{B})^{1/2}e_k\rangle$ is convergent. Then the trace of $\hat{B}$ is defined as

$$\mathrm{Tr}(\hat{B}) = \sum_{k\in\mathbb{N}} \langle e_k, \hat{B}e_k\rangle. \tag{1.53}$$

An operator $\hat{B}$ is said to be of Hilbert–Schmidt class if $\hat{B}^*\hat{B}$ is of trace class.

Notice that the r.h.s in (1.53) is always defined with values in $[0, +\infty]$ if B is any everywhere defined non negative operator.

Notations: The space of trace class operators in $\mathscr{H}$ is denoted $\mathfrak{C}_1(\mathscr{H})$ and the Hilbert–Schmidt class is denoted $\mathfrak{C}_2(\mathscr{H})$. They are Banach spaces for their corresponding norms: $\|B\|_{\mathrm{Tr}} = \mathrm{Tr}((B^*B)^{1/2})$ and $\|B\|_{\mathrm{HS}} = (\mathrm{Tr}(B^*B))^{1/2})$. For more details see [217].

Proposition 6 *Let $\hat{B}$ be an Hilbert–Schmidt operator in $L^2(\mathbb{R}^n)$ then we have*

$$\|\hat{B}\|^2_{HS} = (2\pi\hbar)^{-n} \int_{\mathbb{R}^{2n}} \|\hat{B}u_z\|^2 dz. \tag{1.54}$$

If $\hat{B}$ is a trace-class operator in $L^2(\mathbb{R}^n)$ then we have

$$\mathrm{Tr}(\hat{B}) = (2\pi\hbar)^{-n} \int_{\mathbb{R}^{2n}} \langle u_z, \hat{B}u_z\rangle dz. \tag{1.55}$$

Proof Let $\{e_j\}$ be an orthonormal basis for $L^2(\mathbb{R}^n)$ (for example the Hermite basis ϕ_j). We have

$$\begin{aligned}\|\hat{B}\|_{HS}^2 &= \sum_j \|\hat{B}e_j\|^2 \\ &= \sum_j \|(\hat{B}e_j)^\sharp\|^2. \end{aligned} \tag{1.56}$$

But we have

$$(\hat{B}e_j)^\sharp(z) = \langle \hat{B}e_j, u_z\rangle = \langle e_j, \hat{B}^* u_z\rangle. \tag{1.57}$$

Using Parseval formula for the basis $\{e_j\}$ we get

$$\sum_{j\geq 0} \|\hat{B}e_j\|^2 = (2\pi\hbar)^{-n} \int_{\mathbb{R}^{2n}} \|\hat{B}^* u_z\|^2 dz. \tag{1.58}$$

Using that $\|\hat{B}\|_{HS}^2 = \|\hat{B}^*\|_{HS}^2$ we get the first part of the corollary.

For the second part we use that every class trace operator can be writen as $\hat{B} = \hat{B}_2^* \hat{B}_1$ where $\hat{B}_1$, $\hat{B}_2$ are Hilbert–Schmidt. Moreover the Hilbert–Schmidt norm is associated with the scalar product $\langle \hat{B}_2, \hat{B}_1\rangle = \mathrm{Tr}(\hat{B}_2^* \hat{B}_1)$. So we get

$$\begin{aligned}\mathrm{Tr}(\hat{B}) = \mathrm{Tr}(\hat{B}_2^* \hat{B}_1) &= (2\pi\hbar)^{-n} \int_{\mathbb{R}^{2n}} \langle \hat{B}_2 u_z, \hat{B}_1 u_z\rangle dz \\ &= (2\pi\hbar)^{-n} \int_{\mathbb{R}^{2n}} \langle u_z, \hat{B} u_z\rangle dz. \end{aligned} \tag{1.59}$$

□

These formulas will appear to be very useful in the sequel.

1.4 From Schrödinger to Bargmann–Fock Representation

This representation is well adapted to the creation-annihilation operators and to the Harmonic oscillator. It was introduced by Bargmann [20]. In this representation the real phase space $\mathbb{R}^{2n}$ is identified to $\mathbb{C}^n$ as real spaces:

$$(q, p) \mapsto \zeta = \frac{q - ip}{\sqrt{2}}$$

and a state ψ is represented by the following entire function on $\mathbb{C}^n$,

$$\psi_{\mathrm{Hol}}^\sharp(\zeta) = \psi^\sharp(q, p) \mathrm{e}^{\frac{p^2+q^2}{4\hbar}}.$$

Recall that $\psi^\sharp(z) = (2\pi\hbar)^{-n/2}\langle\varphi_z, \psi\rangle$, $z = (q, p)$.

Proposition 7 *The map $\psi \mapsto \psi^\sharp_{\mathrm{Hol}}$ is an isometry from $L^2(\mathbb{R}^n)$ into the Fock space $\mathscr{F}(\mathbb{C}^n)$ of entire functions f on $\mathbb{C}^n$ such that*

$$\int_{\mathbb{C}^n} |f(\zeta)|^2 \mathrm{e}^{-\frac{\zeta\cdot\bar\zeta}{\hbar}}|d\zeta\wedge d\bar\zeta| < +\infty.$$

$\mathscr{F}(\mathbb{C}^n)$ is an Hilbert space for the scalar product

$$\langle f_2, f_1\rangle = \int_{\mathbb{C}^n} f_1(\zeta)\overline{f_2(\zeta)}\mathrm{e}^{-\frac{\zeta\cdot\bar\zeta}{\hbar}}|d\zeta\wedge d\bar\zeta|. \tag{1.60}$$

Proof A direct computation shows that $\psi^\sharp_{\mathrm{Hol}}$ is holomorphic : $\partial_{\bar\zeta}\psi^\sharp_{\mathrm{Hol}} = 0$. Recall that the holomorphic and antiholomorphic derivatives are defined as follows.

$$\partial_\zeta = \frac{1}{\sqrt{2}}(\partial_q + i\partial_p),\quad \partial_{\bar\zeta} = \frac{1}{\sqrt{2}}(\partial_q - i\partial_p).$$

We can easily get the following explicit formula for $\psi^\sharp_{\mathrm{Hol}}$.

$$\psi^\sharp_{\mathrm{Hol}}(\zeta) = (\pi\hbar)^{-3n/4}2^{-n/2}\int_{\mathbb{R}^n}\psi(x)\exp\left[-\frac{1}{\hbar}\left(\frac{x^2}{2} - \sqrt{2}x\cdot\zeta + \frac{\zeta^2}{2}\right)\right]dx. \tag{1.61}$$

The transformation $\psi \mapsto \psi^\sharp_{\mathrm{Hol}}$ is called the Bargmann transform and is denoted by $\mathscr{B}$. Its kernel is the Bargmann kernel:

$$\mathscr{B}(x, \zeta) = (\pi\hbar)^{-3n/4}2^{-n/2}\exp\left[-\frac{1}{\hbar}\left(\frac{x^2}{2} - \sqrt{2}x\cdot\zeta + \frac{\zeta^2}{2}\right)\right], \tag{1.62}$$
$$\text{where } x \in \mathbb{R}^n,\quad \zeta \in \mathbb{C}^n.$$

Recall the notations $x^2 = x\cdot x$, $\zeta^2 = \zeta\cdot\zeta$, $\zeta\cdot\bar\zeta = |\zeta|^2$.

Using that $\psi \mapsto \psi^\sharp$ is an isometry from $L^2(\mathbb{R}^n)$ into $L^2(\mathbb{R}^{2n})$, we easily get that

$$\int_{\mathbb{C}^n} |\psi^\sharp_{\mathrm{Hol}}(\zeta)|^2 \mathrm{e}^{-\frac{\zeta\cdot\bar\zeta}{\hbar}}|d\zeta\wedge d\bar\zeta| = \|\psi\|_2^2. \tag{1.63}$$

Hence $\mathscr{B}$ is an isometry form $L^2(\mathbb{R}^n)$ into $\mathscr{F}(\mathbb{C}^n)$.

For convenience let us introduce the Gaussian measure on $\mathbb{C}^n$, $d\mu_{\mathscr{B}} = \mathrm{e}^{-\frac{\zeta\cdot\bar\zeta}{\hbar}}|d\zeta\wedge d\bar\zeta|$.

It is not difficult to see that $\mathscr{F}(\mathbb{C}^n)$ is a complete space.

If $\{f_k\}$ is a Cauchy sequence in $\mathscr{F}(\mathbb{C}^n)$ then $\{f_k\}$ converges to f in $L^2(\mathbb{C}^n, d\mu_{\mathscr{B}})$. So we get in a weak sense that $\partial_{\bar\zeta} f = 0$ so f is holomorphic hence $f \in \mathscr{F}(\mathbb{C}^n)$. □

Let us now compute the standard harmonic oscillator in the Bargmann representation. We first get the following formula

$$\int_{\mathbb{R}^n} \partial_x \psi(x)\mathscr{B}(x,\zeta)dx = \int_{\mathbb{R}^n} \psi(x)\left(\frac{x}{\hbar} - \frac{\sqrt{2}}{\hbar}\zeta\right)\mathscr{B}(x,\zeta)dx \tag{1.64}$$

$$\partial_\zeta \int_{\mathbb{R}^n} \psi(x)\mathscr{B}(x,\zeta)dx = \int_{\mathbb{R}^n} \psi(x)\left(\frac{\sqrt{2}}{\hbar}x - \frac{\zeta}{\hbar}\right)\mathscr{B}(x,\zeta)dx. \tag{1.65}$$

Hence

$$\mathscr{B}(x\psi)(\zeta) = \frac{1}{\sqrt{2}}(\hbar\partial_\zeta + \zeta)\mathscr{B}\psi(\zeta) \tag{1.66}$$

$$\mathscr{B}(\hbar\partial_x\psi)(\zeta) = \frac{1}{\sqrt{2}}(\hbar\partial_\zeta - \zeta)\mathscr{B}\psi(\zeta). \tag{1.67}$$

Then we get the Bargmann representation for the creation and annihilation operators

$$\mathscr{B}[\mathbf{a}^\dagger\psi](\zeta) = \zeta\mathscr{B}[\psi](\zeta). \tag{1.68}$$

$$\mathscr{B}[\mathbf{a}\psi](\zeta) = \partial_\zeta\mathscr{B}[\psi](\zeta). \tag{1.69}$$

So the standard harmonic oscillator $\hat{H}_{\mathrm{os}} = \hbar(\mathbf{a}^\dagger\mathbf{a} + \frac{n}{2})$, has the following Bargmann representation

$$\hat{H}^\sharp_{\mathrm{os}} = \hbar\zeta\cdot\partial_\zeta + n\frac{\hbar}{2}. \tag{1.70}$$

Remark 2 It is very easy to solve the time dependent Schrödinger equation for $\hat{H}^\sharp_{\mathrm{os}}$. If $F \in \mathscr{F}(\mathbb{C}^n)$ such that $\zeta\,\partial_\zeta F \in \mathscr{F}(\mathbb{C}^n)$, then $F(t,\zeta) = \mathrm{e}^{-\frac{it}{2\hbar}}F(\mathrm{e}^{-\frac{it}{2\hbar}}\zeta)$ satisfies

$$i\hbar\partial_t F(t) = \hat{H}^\sharp_{\mathrm{os}}F(t),\ \ F(0,\zeta) = F(\zeta). \tag{1.71}$$

Moreover if $\hbar = 1$ and if we put $\mathrm{e}^{-t/2}$ in place of $\mathrm{e}^{-it/2}$ we solve the heat equation $\partial_t F(t) = \hat{H}^\sharp_{\mathrm{os}}F(t)$.

We shall see now that the Hermite functions ϕ_α have a very simple shape in the Bargmann representation. Let us denote $\phi^\#_\alpha(\zeta) = \mathscr{B}\phi_\alpha$. Then we have

Proposition 8 *For every $\alpha \in \mathbb{N}^n$, $\zeta \in \mathbb{C}^n$,*

$$\phi^\#_\alpha(\zeta) = (2\pi\hbar)^{-n/2}(\alpha!)^{-1/2}\zeta^\alpha. \tag{1.72}$$

Moreover $\{\phi^\#_\alpha(\zeta)\}_{\alpha\in\mathbb{N}^n}$ is an orthonormal basis in $\mathscr{F}(\mathbb{C}^n)$.

Proof Let us first recall the notations in dimension n. For $\alpha = (\alpha_1\cdots\alpha_n) \in \mathbb{N}$, $\alpha! = \alpha_1!\cdots\alpha_n!$ and for $\zeta = (\zeta_1,\cdots,\zeta_n) \in \mathbb{C}^n$, $\zeta^\alpha = \zeta_1^{\alpha_1}\cdots\zeta_n^{\alpha_n}$.

We get easily that $\langle\zeta^\alpha,\zeta^\beta\rangle = 0$ if $\alpha \neq \beta$.

It is enough to compute, for $n = 1$, $\|\zeta^k\|^2_{\mathscr{F}(\mathbb{C})}$ and this is an easy computation with the Gamma function.

Let us prove now that the system $\{\zeta^\alpha\}_{\alpha\in\mathbb{N}^n}$ is total in $\mathscr{F}(\mathbb{C}^n)$.

Let $f \in \mathscr{F}(\mathbb{C}^n)$ be such that $\langle \zeta^\alpha, f\rangle = 0$ for all $\alpha \in \mathbb{N}^n$. f is entire so we have

$$f(\zeta) = \sum_\alpha f_\alpha \zeta^\alpha,$$

where f_α are the Taylor coefficient of f at 0. The sum is uniformly convergent on every ball of $\mathbb{C}^n$. On the other side from Bessel inequality, we know that the Taylor series $\sum_\alpha f_\alpha \zeta^\alpha$ converges in $\mathscr{F}(\mathbb{C}^n)$. But we can see that $\{\zeta^\alpha\}_{\alpha\in\mathbb{N}^n}$ is also an orthogonal system in each ball with center at 0. Then we get that $f_\alpha = 0$ for every α hence $f = 0$.

Let us remark here that we could also prove that the system $\{\phi^\#_\alpha(\zeta)\}_{\alpha\in\mathbb{N}^n}$ is orthogonal using that Hermite functions is an orthonormal system and $\mathscr{B}$ is an isometry.

Finally, let us prove formula (1.72). It is enough to assume $n = 1$. We get easily that $\phi^\#_0 = \frac{1}{\sqrt{2\pi}}$. So for every $k \geq 1$, we have, using (1.68),

$$\phi^\#_k(\zeta) = \mathscr{B}\left[\frac{(a^\dagger)^k}{\sqrt{k!}}\phi_0\right](\zeta) = \frac{\zeta^k}{\sqrt{2\pi k!}}.$$

□

Then we get the following interesting result.

Corollary 3 *The Bargmann transform $\mathscr{B}$ is an isometry from $L^2(\mathbb{R}^n)$ onto $\mathscr{F}(\mathbb{C}^n)$. The integral kernel of $\mathscr{B}^{-1}$ is*

$$\mathscr{B}^{-1}(\zeta, x) = (\pi\hbar)^{-3n/4}2^{-n/2}\exp\left[-\frac{1}{\hbar}\left(\frac{x^2}{2} - \sqrt{2}x\cdot\bar{\zeta} + \frac{\bar{\zeta}^2}{2}\right)\right], \quad (1.73)$$

where $x \in \mathbb{R}^n$, $\zeta \in \mathbb{C}^n$.

We also get that the Bargmann kernel is a generating function for the Hermite functions.

Corollary 4 *For every $x \in \mathbb{R}^n$ and $\zeta \in \mathbb{C}^n$ we have*

$$\mathscr{B}(x, \zeta) = \sum_{\alpha\in\mathbb{N}^n} \frac{\zeta^\alpha}{((2\pi)^n\alpha!)^{1/2}}\phi_\alpha(x). \quad (1.74)$$

Proof Compute the Fourier coefficient in the Hermite basis of $x \mapsto \mathscr{B}(x, \zeta)$. □

The standard coherent states also have a simple expression in the Bargmann–Fock space.

Let φ_X be the normalized coherent state at $X = (x, \xi)$.

Proposition 9 *We have the following Bargmann representation for the coherent state φ_X*

$$\mathscr{B}[\varphi_X](\zeta) = (2\pi\hbar)^{-n/2} \mathrm{e}^{\frac{\bar{\eta}}{\hbar}(\zeta - \frac{\eta}{2})}, \tag{1.75}$$

where $\eta = \frac{x - i\xi}{\sqrt{2}}$.

Proof A direct computation gives

$$\mathscr{B}[\varphi_X](\zeta) = (\sqrt{2}\pi\hbar)^{-n} \int_{\mathbb{R}^n} dy \exp\left[-\frac{1}{\hbar}\left(y^2 - y(x + i\xi + \sqrt{2}\zeta) - \frac{\zeta^2}{2}\right)\right]. \tag{1.76}$$

Then we get the result by Fourier transform of the Gaussian e^{-y^2}. □

One of the nice properties of the space $\mathscr{F}(\mathbb{C}^n)$ is existence of a reproducing kernel.

Proposition 10 *For every $f \in \mathscr{F}(\mathbb{C}^n)$ we have*

$$f(\zeta) = (2\pi\hbar)^{-n} \int_{\mathbb{C}^n} \mathrm{e}^{\frac{\bar{\eta}\cdot\zeta}{\hbar}} f(\eta) d\mu_{\mathscr{B}}(\eta), \quad \forall \zeta \in \mathbb{C}^n. \tag{1.77}$$

Proof It is enough to assume that f is a polynomial in ζ and that $\hbar = 1$. So we have

$$f(\zeta) = \sum_{\alpha} c_\alpha \frac{\zeta^\alpha}{(2\pi)^{n/2}(\alpha!)^{1/2}}, \quad \text{with} \quad c_\alpha = \int_{\mathbb{C}^n} \frac{\bar{\eta}^\alpha}{(2\pi)^{n/2}(\alpha!)^{1/2}} f(\eta) d\mu_{\mathscr{B}}(\eta). \tag{1.78}$$

Hence we get

$$f(\zeta) = (2\pi)^{-n} \int_{\mathbb{C}^n} \left(\sum_{\alpha} \frac{\zeta^\alpha \bar{\eta}^\alpha}{\alpha!}\right) f(\eta) d\mu_{\mathscr{B}}(\eta) = (2\pi)^{-n} \int_{\mathbb{C}^n} \mathrm{e}^{\bar{\eta}\cdot\zeta} f(\eta) d\mu_{\mathscr{B}}(\eta). \tag{1.79}$$

□

Remark 3 The function $e_\zeta(\eta) = (2\pi\hbar)^{-n} \mathrm{e}^{\frac{\bar{\eta}\cdot\zeta}{\hbar}}$ is a representation of the Dirac delta function in the point ζ. Note that e_ζ is not in $\mathscr{F}(\mathbb{C}^n)$. Moreover we have $f(\zeta) = \langle e_\zeta, f \rangle$ and $|f(\zeta)| \le (2\pi\hbar)^{-n} \|f\|_{\mathscr{F}(\mathbb{C}^n)}$.

Using the Bargmann representation we can give a proof of the well known Mehler formula concerning the Hermite orthonormal basis $\{\phi_k\}$ in $L^2(\mathbb{R})$. It is sufficient to assume that $\hbar = 1$.

Theorem 1 *For every $w \in \mathbb{C}$ such that $|w| < 1$ we have*

$$\sum_{\mathbf{k} \in \mathbb{N}^n} \phi_{\mathbf{k}}(x)\phi_{\mathbf{k}}(y) w^k = \pi^{-n/2}(1 - w^2)^{-n/2} \exp\left(-\frac{1 + w^2}{2(1 - w^2)}(x^2 + y^2) + \frac{2w}{1 - w^2} x \cdot y\right), \tag{1.80}$$

where $k = |\mathbf{k}| = k_1 + \cdots k_k$, $x, y \in \mathbb{R}^n$.

Proof The case $n \geq 2$ can be easily deduced from the case $n = 1$. So let $n = 1$.

The left and right side of (1.80) are holomorphic in w in the unit disc $\{w \in \mathbb{C},\ |w| < 1\}$. So by analytic continuation principle it is enough to prove it for $w = e^{-t/2}$ for every $t > 0$. Hence the right side of (1.80) is the heat kernel denoted $K_{\rm os}(t; x, y)$ of the harmonic oscillator $\hat{H}_{\rm os}$.

Using Remark 2 and inverse Bargmann transform we get easily the following integral expression for $K_{os}(t; x, y)$

$$K_{\rm os}(t; x, y) = 2^{-1}\pi^{-3/2} e^{-\frac{x^2+y^2}{2}}$$

$$\int_{\mathbb{C}} \exp\left(\sqrt{2}(x\cdot\bar{\zeta} + wy\cdot\zeta) - \frac{1}{2}(w^2\zeta^2 + \bar{\zeta}^2) - \zeta\bar{\zeta}\right) |d\zeta \wedge d\bar{\zeta}|. \tag{1.81}$$

The last integral is a Fourier transform of a Gaussian function as it is seen using real coordinates $\zeta = \frac{q-ip}{\sqrt{2}}$, $z = (q, p)$. We have

$$K_{\rm os}(t; x, y) = 2^{-1}\pi^{-3/2} \int_{\mathbb{R}^2} e^{-\frac{1}{2}Az\cdot z - iz\cdot Y} dY, \tag{1.82}$$

where

$$Y = \begin{pmatrix} i(x + wy) \\ wy - x \end{pmatrix}, \quad A = \frac{1}{2}\begin{pmatrix} 3 + w^2 & i(1 - w^2) \\ i(1 - w^2) & 1 - w^2 \end{pmatrix}, \quad w = e^{-t/2}. \tag{1.83}$$

A is a symmetric matrix, its real part is positive definite and $\det(A) = 1 - w^2$. So we have (see [153] or Appendices A, B and C)

$$K_{\rm os}(t; x, y) = \pi^{-1/2}(1 - w^2)^{-1/2} e^{-\frac{x^2+y^2}{2}} e^{-\frac{1}{2}A^{-1}Y\cdot Y}. \tag{1.84}$$

The Mehler formula follows. □

We can deduce in particular the very classical formula for the heat kernel of the harmonic oscillator ($n = 1$, $\hbar = 1$).

Recall $\hat{H}_{\rm os} = \frac{1}{2}(-\frac{d^2}{dx^2} + x^2)$.

Corollary 5 (Mehler formula) *Let* $K_\tau(x, y)$ *the integral kernel for the operator* $e^{-\tau\hat{H}_{\rm os}}$ *for* $\Re\tau > 0$.

K_τ *is in the Schwartz space* $\mathcal{S}(\mathbb{R}_x \times R_y)$ *and has the following expression*

$$\boxed{K_\tau(x, y) = \frac{1}{\sqrt{2\pi\sinh\tau}} \exp\left(-\frac{x^2 + y^2}{2\tanh\tau} + \frac{xy}{\sinh\tau}\right)}. \tag{1.85}$$

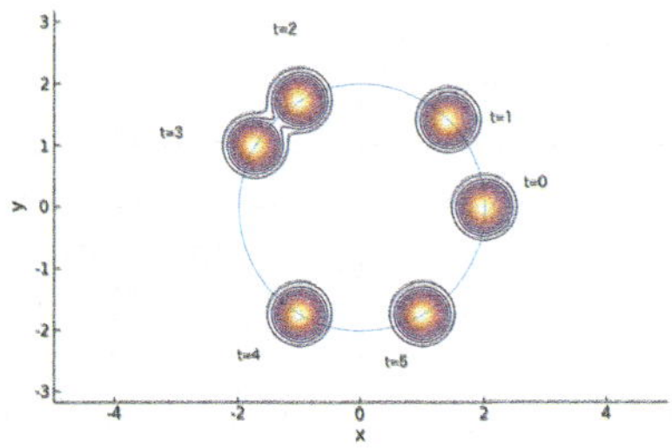

Motion of Coherent States: Motion of the Schrödinger coherent states driven by the harmonic oscillator. The initial data is $\psi_0 = \varphi_{2,0}$ and at time t, $\psi_t = \mathrm{e}^{-it/2}\varphi_{z_t}$, z_t is moving on the circle with center at $(0,0)$ and radius 2. The localization of the coherent state ψ_t is given by the Bargmann transform of $\varphi_z(t)$ which is a Gaussian centered at z_t.

Chapter 2
Weyl Quantization and Coherent States

Abstract It is well known from the work of Berezin [24] (Commun. Math. Phys. 40:153–174, 1975) in 1975 that the quantization problem of a classical mechanical system is closely related with coherent states. In particular coherent states help to understand the limiting behaviour of a quantum system when the Planck constant $\hbar$ becomes negligible in macroscopic units. This problem is called the semi-classical limit problem. In this chapter we discuss properties of quantum systems when the configuration space is the Euclidean space $\mathbb{R}^n$. So that in the Hamiltonian formalism, the phase space is $\mathbb{R}^n \times \mathbb{R}^n$ with its canonical symplectic form σ. The quantization problem has many solutions, so we choose a convient one, introduced by Weyl [259] (The Classical Groups, 1997) and Wigner [261] (Group Theory and Its Applications to Quantum Mechanics of Atomic Spectra, 1959). We study the symmetries of Weyl quantization, the operationnal calculus and applications to propagation of observables. We show that Wick quantization is a natural bridge between Weyl quantization and coherent states. Applications are given to the semi-classical limit after introducing an efficient modern tool: the semi-classical measures. We illustrate the general results proved in this chapter by explicit computations for the harmonic oscillator. More applications will be given in the following chapters, in particular concerning propagators and trace formulas for a large class of quantum systems.

2.1 Classical and Quantum Observables

The quantization problem comes from quantum mechanics and is a mathematical setting for the Bohr correspondence principle between the classical world and the quantum world.

Let us consider a system with n degrees of freedom. According the Bohr correspondence principle, it is natural to check a way to associate to every real function A on the phase space $\mathbb{R}^{2n}$ (classical observable) a self adjoint operator $\hat{A}$ in the Hilbert space $L^2(\mathbb{R}^n)$ (quantum observable). According the quantum mechanical principles, the map $A \to \hat{A}$ has to satisfy some properties.

D. Robert and M. Combescure, *Coherent States and Applications in Mathematical Physics*, Theoretical and Mathematical Physics,
https://doi.org/10.1007/978-3-030-70845-0_2

(1) $A \to \hat{A}$ is *linear*, $\hat{A}$ is self-adjoint if A is real and $\hat{1} = \mathbb{1}_{L^2(\mathbb{R}^n)}$.
(2) *position observables*: $x_j \to \hat{x}_j := \hat{Q}_j$ where $\hat{Q}_j$ is the multiplication operator by x_j
(3) *momentum observables* : $\xi_j \to \hat{\xi}_j := \hat{P}_j$ where $\hat{P}_j$ is the differential operator $\frac{\hbar}{i}\frac{\partial}{\partial x_j}$.
(4) *commutation rule and classical limit*: for every classical observables A, B we have

$$\lim_{\hbar \to 0}\left(\frac{i}{\hbar}[\hat{A}, \hat{B}] - \widehat{\{A, B\}}\right) = 0.$$

Let us recall that $[\hat{A}, \hat{B}] = \hat{A}\hat{B} - \hat{B}\hat{A}$ is the commutator of $\hat{A}$ and $\hat{B}$, $\{A, B\}$ is the Poisson bracket defined as follows

$$\{A, B\}(x, \xi) = (\partial_x A \cdot \partial_\xi B - \partial_x B \cdot \partial_\xi A)(x, \xi), \; x, \xi \in \mathbb{R}^n.$$

Let us remark that if we introduce $\nabla A = (\partial_x A, \partial_\xi A)$ then we have $\{A, B\}(x, \xi) = \sigma(\nabla A(x, \xi), \nabla B(x, \xi))$ (σ is the symplectic bilinear form).

If the observables A, B depend only on the position variable (or on the momentum variables) then $\hat{A} \cdot \hat{B} = \widehat{A.B}$, but this is no longer true for a mixed observable. This is related to the non-commutativity for product of quantum observables and the identity: $[\hat{x}_j, \hat{\xi}_j] = i\hbar$ so, the quantum observable corresponding to $x_1\xi_1$ is not determined by the rules (1)–(4).

We do not want to discuss here the quantization problem in its full generality (see for example [104]). One way to choose a reasonable and convenient quantization procedure is the following, which is called Weyl quantization (see [153] for more details). Let L_z be a real linear form on the phase space $\mathbb{R}^{2n}$, where $z = (p, q)$, $L_z(x, \xi) = \sigma(z, (x, \xi))$ (every linear form on $\mathbb{R}^{2n}$ is like this). It is not difficult to see that $\hat{L}_z$ is a well defined quantum Hamiltonian (i.e. an essentially self-adjoint operator in $L^2(\mathbb{R}^n)$). Its propagator $\mathrm{e}^{\frac{-it}{\hbar}\widehat{L_z}}$ has been studied in Chap. 1.

For $\psi \in \mathscr{S}(\mathbb{R}^n)$, we have explicitly (see Chap. 1)

$$\mathrm{e}^{\frac{-it}{\hbar}\widehat{L_z}}\psi(x) = \mathrm{e}^{-\frac{i}{2\hbar}t^2 q\cdot p}\mathrm{e}^{\frac{it}{\hbar}x\cdot p}\psi(x - tq). \tag{2.1}$$

So, the Weyl prescription is defined by the conditions (1)–(4) and the following

(5)

$$\mathrm{e}^{-iL_z(x,\xi)} \to \widehat{\mathrm{e}^{-iL_z}} = \hat{T}(z).$$

In other words Weyl quantization and the exponential commutes on linear forms on the phase space.

We shall use freely the Schwartz space $\mathscr{S}(\mathbb{R}^n)$[1] and its dual $\mathscr{S}'(\mathbb{R}^n)$ (temperate distributions space).

Proposition 11 *There exists a unique continuous map $A \to \hat{A}$ from $\mathscr{S}'(\mathbb{R}^{2n})$ into $\mathscr{L}\left(\mathscr{S}(\mathbb{R}^n), \mathscr{S}'(\mathbb{R}^n)\right)$ satisfying conditions (1)–(5).*

Moreover if $A \in \mathscr{S}(\mathbb{R}^{2n})$ and $\psi \in \mathscr{S}(\mathbb{R}^n)$ we have the familiar formula

$$(\mathrm{Op}_\hbar^w)\psi(x) := \hat{A}\psi(x) = (2\pi\hbar)^{-n} \iint_{\mathbb{R}^{2n}} A\left(\frac{x+y}{2}, \xi\right) \mathrm{e}^{i\hbar^{-1}(x-y)\cdot\xi}\psi(y)dyd\xi, \tag{2.2}$$

and $\hat{A}$ is a continuous map from $\mathscr{S}(\mathbb{R}^n)$ to $\mathscr{S}(\mathbb{R}^n)$.

The hermitian conjugate of $\hat{A}$ is the quantization of the complex conjugate of A i.e. $(\hat{A})^* = \hat{\bar{A}}$. In particular $\hat{A}$ is Hermitian if and only if A is real.

Proof Here it is enough to assume that $\hbar = 1$.

Let us consider the symplectic Fourier transform in $\mathscr{S}'(\mathbb{R}^{2n})$. Assume first that $A \in \mathscr{S}(\mathbb{R}^{2n})$ and define

$$\tilde{A}_\sigma(z) = \int_{\mathbb{R}^{2n}} A(\zeta)\mathrm{e}^{-i\sigma(z,\zeta)}d\zeta. \tag{2.3}$$

We have the inverse formula

$$A(X) = (2\pi)^{-n} \int_{\mathbb{R}^{2n}} \tilde{A}_\sigma(z)\mathrm{e}^{i\sigma(z,X)}dz. \tag{2.4}$$

For $\psi, \eta \in \mathscr{S}(\mathbb{R}^n)$ we have

$$\langle\psi, \hat{A}\eta\rangle = (2\pi)^{-n} \int_{\mathbb{R}^{2n}} \tilde{A}_\sigma(z)\langle \mathrm{e}^{i\hat{L}_z}\psi, \eta\rangle dz. \tag{2.5}$$

In other words we must have:

$$\hat{A}\psi = (2\pi)^{-n} \int_{\mathbb{R}^{2n}} \tilde{A}_\sigma(z)\hat{T}(z)\psi dz. \tag{2.6}$$

□

Definition 3 For a given operator $\hat{A}$, the function A is called the contravariant symbol of $\hat{A}$ and the function $\tilde{A}_\sigma$ is the covariant symbol of $\hat{A}$.

Let us remark that we have the inverse formula

---

[1]Recall that $f \in \mathscr{S}(\mathbb{R}^n)$ means that f is a smooth function in $\mathbb{R}^n$ and for every multiindices α, β, $x^\alpha \partial_x^\beta f$ is bounded in $\mathbb{R}^n$. It has a natural topology. $\mathscr{S}'(\mathbb{R}^n)$ is the linear space of continuous linear forms on $\mathscr{S}(\mathbb{R}^n)$.

Proposition 12 *If $\hat{A}$ is a continuous map from $\mathscr{S}'(\mathbb{R}^n)$ to $\mathscr{S}(\mathbb{R}^n)$ then we have for every $X \in \mathbb{R}^{2n}$,*

$$\tilde{A}_\sigma(X) = \mathrm{Tr}(\hat{A}\hat{T}(-X)). \tag{2.7}$$

Proof For $X = 0$ the formula is a consequence of the Fourier inversion formula.

For any X we use that the Weyl symbol of $\hat{T}(-X)$ is $z \mapsto \mathrm{e}^{-i\sigma(z,X)}$ □

As a consequence we have a first norm operator estimate. If $\tilde{A}_\sigma \in L^1(\mathbb{R}^{2n})$ we have

$$\|\hat{A}\| \leq (2\pi)^{-n} \int_{\mathbb{R}^{2n}} |\tilde{A}_\sigma(z)| dz. \tag{2.8}$$

The r.h.s in formula (2.2) can be extended by continuity in A to the distribution space $\mathscr{S}'(\mathbb{R}^{2n})$.

Let us compute now the Schwartz kernel K_A of the operator $\hat{A}$ defined in formula (2.6). We have:

$$K_A(x, y) = \int_{\mathbb{R}^n} \tilde{A}_\sigma(x - y, p)\mathrm{e}^{ip\cdot(x+y)/2} dp. \tag{2.9}$$

Using inverse Fourier transform in p variables, we get

$$K_A(x, y) = (2\pi)^{-n} \int_{\mathbb{R}^n} A\left(\frac{x+y}{2}, \xi\right) \mathrm{e}^{i(x-y)\cdot\xi} d\xi. \tag{2.10}$$

and this gives (2.2). The other properties are easy to prove and left to the reader. □

Let us first remark that from (2.10) we get a formula to compute the $\hbar$-Weyl symbol of $\hat{A}$ if we know its Schwartz kernel K

$$A(x, \xi) = \int_{\mathbb{R}^n} \mathrm{e}^{-\frac{i}{\hbar}u\cdot\xi} K\left(x + \frac{u}{2}, x - \frac{u}{2}\right) du. \tag{2.11}$$

Sometimes, we shall use also the notation $\hat{A} = \mathrm{Op}_\hbar^w A$ ($\hbar$-Weyl quantization of A). Hence we shall say that $\hat{A}$ is an $\hbar$-pseudodifferential operators and that A is its Weyl symbol. For applications it is useful to be able to read properties of the operator $\hat{A}$ on its Weyl symbol A. A first example is the Hilbert-Schmidt property.

Proposition 13 *Let $\hat{A} \in \mathscr{L}\left(\mathscr{S}(\mathbb{R}^n), \mathscr{S}'(\mathbb{R}^n)\right)$. Then $\hat{A}$ is Hilbert-Schmidt in $L^2(\mathbb{R}^n)$ if and only if $A \in L^2(\mathbb{R}^{2n})$ and we have*

$$\|\hat{A}\|_{HS}^2 = (2\pi\hbar)^{-n} \iint_{\mathbb{R}^{2n}} |A(x, \xi)|^2 dx d\xi. \tag{2.12}$$

In particular if $\hat{A}$ and $\hat{B}$ are two Hilbert-Schmidt operators then $\hat{A}.\hat{B}$ is a trace operator and we have

$$\mathrm{Tr}(\hat{A}.\hat{B}) = (2\pi\hbar)^{-n} \iint_{\mathbb{R}^{2n}} A(x, \xi) B(x, \xi) dx d\xi. \tag{2.13}$$

Proof We know that

$$\|\hat{A}\|_{HS}^2 = \iint_{\mathbb{R}^{2n}} |K_A(x,y)|^2 dxdy.$$

Then we get the proposition using formula (2.10) and Plancherel theorem. □

We shall see later many other properties concerning Weyl quantization but most of time we only have sufficient conditions on A to have some property of $\hat{A}$, like for example L^2 continuity or trace-class property.

Let us give a first example of computation of a Weyl symbol starting from an integral kernel. We consider the heat semi-group $e^{-t\hat{H}_{os}}$, of the harmonic oscillator $\hat{H}_{os}$. Let us denote $K_w(t;x,\xi)$ the Weyl symbol of $e^{-t\hat{H}_{os}}$ and $K(t;x,y)$ its integral kernel. From formula (2.11) we get

$$K_w(t;x,\xi) = \int_{\mathbb{R}^n} e^{-\frac{i}{\hbar}u\cdot\xi} K\left(t; x+\frac{u}{2}, x-\frac{u}{2}\right) du. \tag{2.14}$$

Using Mehler formula (Chap. 1) we have to compute the Fourier transform of a generalized Gaussian function, so after some computations, we get the following nice formula

$$K_w(t;x,\xi) = (\cosh(t/2))^{-n/2} e^{-\tanh(t/2)(x^2+\xi^2)}. \tag{2.15}$$

Recall that $x^2 = x\cdot x = |x|^2$.

2.1.1 *Group Invariance of Weyl Quantization*

Let us first remark that an easy consequence of the definition of Weyl quantization is the invariance by the group of translations in the phase space. More precisely we have, for any classical observable A and any $z \in \mathbb{R}^{2n}$,

$$\hat{T}(z)^{-1}\hat{A}\hat{T}(z) = \widehat{\tau_z A}, \text{ where } \tau_z A(X) = A(X-z),\ X \in \mathbb{R}^{2n}. \tag{2.16}$$

Hamiltonian classical mechanics is invariant by the action of the group $\mathrm{Sp}(n)$ of symplectic transformations of the phase space $\mathbb{R}^{2n}$. A natural question to ask is to quantize linear symplectic transformations. We shall see later how it is possible. In this section we state the main results.

Recall that the symplectic group $\mathrm{Sp}(n)$ is the group of linear transformations of $\mathbb{R}^{2n}$ which preserves the symplectic form σ. So $F \in \mathrm{Sp}(n)$ means that $\sigma(FX,FY) = \sigma(X,Y)$ for all $X,Y \in \mathbb{R}^{2n}$. If we introduce the matrix

$$J = \begin{pmatrix} 0 & \mathbb{1} \\ -\mathbb{1} & 0 \end{pmatrix}$$

then

$$F \in \mathrm{Sp}(n) \iff F^t J F = J \tag{2.17}$$

where F^t is the transposed matrix of F.

If $n = 1$ then F is symplectic if and only if $\det(F) = 1$.
Linear symplectic transformations can be quantized as unitary operators in $L^2(\mathbb{R}^n)$:

Theorem 2 *For every linear symplectic transformation $F \in \mathrm{Sp}(n)$ and every $A \in \mathscr{S}(\mathbb{R}^{2n})$ we have*

$$\hat{R}(F)^{-1}\hat{A}\hat{R}(F) = \widehat{A \cdot F}. \tag{2.18}$$

Moreover $\hat{R}(F)$ is unique up to multiplication by a complex number of modulus 1.

Definition 4 The **metaplectic group** is the group $\mathrm{Met}(n)$ generated by $\hat{R}(F)$ and $\lambda \mathbb{1}$, $\lambda \in \mathbb{C}$, $|\lambda| = 1$.

Remark 4 A consequence of Theorem 2 is that $\hat{R}$ is a projective representation of the symplectic group $\mathrm{Sp}(n)$ in the Hilbert space $L^2(\mathbb{R}^n)$. It is a particular case of a more general setting [255].

More properties of the metaplectic group will be studied in the next chapter. Let us give here some examples of metaplectic transform

- The Fourier transform $\mathscr{F}$ is associated with the symplectic transformation $(x, \xi) \mapsto (\xi, -x)$.
- The partial Fourier transform $\mathscr{F}_j$, in variable x_j, is associated with the symplectic transform:

$$(x_j, \xi_j) \mapsto (\xi_j, -x_j), \quad (x_k, \xi_k) \mapsto (x_k, \xi_k), \text{ if } k \neq j.$$

- Let A be a linear transformation on $\mathbb{R}^n$, the transformation $\psi \mapsto |\det(A)|^{1/2}\psi(Ax)$ is associated with the symplectic transform

$$F_A\begin{pmatrix} x \\ \xi \end{pmatrix} = \begin{pmatrix} Ax \\ (A^t)^{-1}\xi \end{pmatrix}.$$

- Let A be a real symmetric matrix, the transformation $\psi \mapsto \mathrm{e}^{iAx\cdot x/2}\psi$ is associated with the symplectic transform

$$F = \begin{pmatrix} \mathbb{1} & 0 \\ A & \mathbb{1} \end{pmatrix}.$$

2.2 Wigner Functions

Let $\varphi, \psi \in L^2(\mathbb{R}^n)$. They define a rank one operator $\Pi_{\psi,\varphi}\eta = \langle \psi, \eta \rangle \varphi$. Its Weyl symbol can be computed using (2.11).

Definition 5 The Wigner function of the pair (ψ, φ) is the Weyl symbol of the rank one operator $\Pi_{\psi,\varphi}$. It will be denoted $\mathscr{W}_{\varphi,\psi}$. More explicitly we have

$$\mathscr{W}_{\varphi,\psi}(x,\xi) = \int_{\mathbb{R}^n} \mathrm{e}^{-\frac{i}{\hbar}u\cdot\xi}\varphi\left(x+\frac{u}{2}\right)\overline{\psi\left(x-\frac{u}{2}\right)}du. \tag{2.19}$$

An equivalent definition of the Wigner function is the following:

$$\mathscr{W}_{\varphi,\psi}(z) = (2\pi\hbar)^{-n}\int_{\mathbb{R}^{2n}} \langle\varphi, \hat{T}(z')\psi\rangle e^{-i\sigma(z,z')/\hbar}dz', \tag{2.20}$$

where $\hat{T}(z) = \mathrm{e}^{-i\hat{L}_z}$.

We can easily see that (2.19) and (2.20) are equivalent using formula (2.6) and Plancherel formula for symplectic Fourier transform.

The Wigner functions are very convenient to use. In particular we have the following nice property:

Proposition 14 *Let us assume that* $\hat{A}$ *is Hilbert-Schmidt and* $\psi, \varphi \in L^2(\mathbb{R}^n)$. *Then we have*

$$\boxed{\langle\psi, \hat{A}\varphi\rangle = (2\pi\hbar)^{-n}\int_{\mathbb{R}^{2n}} A(X)\mathscr{W}_{\psi,\varphi}(X)dX}. \tag{2.21}$$

If $A \in \mathscr{S}'(\mathbb{R}^{2n})$ *and if* $\psi, \varphi \in \mathscr{S}(\mathbb{R}^n)$, *the formula (2.21) is still true in the weak sense of temperate distributions.*

Proof Let us first remark that $\langle\psi, \hat{A}\varphi\rangle = \mathrm{Tr}(\hat{A}\Pi_{\psi,\varphi})$. Hence the first part of the proposition comes from (2.13).

Now if $\psi, \varphi \in \mathscr{S}(\mathbb{R}^n)$ then we get easily that $\mathscr{W}_{\psi,\varphi} \in \mathscr{S}(\mathbb{R}^{2n})$. On the other side there exists $A_j \in \mathscr{S}(\mathbb{R}^{2n})$ such that $A_j \to A$ in $\mathscr{S}'(\mathbb{R}^{2n})$. So we apply (2.21) to A_j and we go to the limit in j. □

What Wigner was looking for was a quantum equivalent of the classical probability distribution in the phase space $\mathbb{R}^{2n}$. That is, associated to any quantum state a distribution function in phase space that imitates a classical distribution probability in phase space. Recall that a classical probability distribution is a nonnegative Borel function $\rho\ ; Z \to \mathbb{R}^+$, $Z := \mathbb{R}^{2n}$, normalized to unity:

$$\int_Z \rho(z)dz = 1,$$

and such that the average of any observable $A \in L^\infty(\mathbb{R}^{2n})$ is simply given by

$$\rho(A) = \int_Z A(z)\rho(z)dz.$$

From Proposition 14 we see that a possible candidate is

$$\rho(z) = (2\pi\hbar)^{-n}\mathscr{W}_{\varphi,\varphi}.$$

Actually in the physical literature the expression above (with the factor $(2\pi\hbar)^{-n}$) is taken as the definition of the Wigner function but we do not take this convention.

In the following we denote by $\mathscr{W}_\varphi$ the Wigner transform for φ, φ.

What about the expected properties of $(2\pi\hbar)^{-n}\mathscr{W}_\varphi$ as a possible probability distribution in phase space? Namely:

- positivity
- normalization to 1
- correct marginal distributions

Proposition 15 *Let $z = (x, \xi) \in \mathbb{R}^{2n}$ and $\varphi \in L^2(\mathbb{R}^n)$ with $\|\varphi\| = 1$. We have:*
(i)

$$(2\pi\hbar)^{-n}\int_{\mathbb{R}^n} \mathscr{W}_\varphi(x, \xi)d\xi = |\varphi(x)|^2,$$

which is the probability amplitude to find the quantum particle at position x.
(ii)

$$(2\pi\hbar)^{-n}\int_{\mathbb{R}^n} \mathscr{W}_\varphi(x, \xi)dx = |\tilde{\varphi}(\xi)|^2,$$

which is the probability amplitude to find the quantum particle at momentum ξ.
(iii)

$$(2\pi\hbar)^{-n}\int_{\mathbb{R}^{2n}} \mathscr{W}_\varphi(x, \xi)dxd\xi = 1.$$

(iv) $\mathscr{W}_\varphi(x, \xi) \in \mathbb{R}$.

Proof (i) Let $f \in \mathscr{S}$ be an arbitrary test function. We have:

$$\begin{aligned}\int_{\mathbb{R}^n} \mathscr{W}_\varphi(x, \xi)f(\xi)d\xi &= \int dy\bar{\varphi}(x + \frac{y}{2})\varphi(x - \frac{y}{2})\int d\xi e^{-i\xi\cdot y/\hbar}f(\xi)\\ &= (2\pi\hbar)^n\int_{\mathbb{R}^n} dy\bar{\varphi}(x + \frac{y}{2})\varphi(x - \frac{y}{2})(\mathscr{F}f)(y).\end{aligned} \quad (2.22)$$

By taking for the usual Fourier transform $\mathscr{F}f$ an approximation of the Dirac distribution at $y = 0$ we get the result.

(ii) is proven similarly.

(iii) follows from the normalization to unity of the state φ.

(iv) We have:

$$\mathscr{W}_\varphi(z)^* = (2\pi\hbar)^{-n}\int dz'\langle\varphi, \hat{T}(-z')\varphi\rangle e^{i\sigma(z,z')/\hbar}.$$

and the result follows by change of the integration variable $z' \to -z'$ and by noting that $\sigma(z, -z') = -\sigma(z, z')$. □

Let us now compute the Wigner function $\mathscr{W}_{z,z'}$ for a pair $(\varphi_z, \varphi_{z'})$ of coherent states.

Proposition 16 *For every $X, z, z' \in \mathbb{R}^{2n}$ we have*

$$\mathscr{W}_{z,z'}(X) = 2^n \hbar^{-n} \exp\left(-\frac{1}{\hbar}\left|X - \frac{z+z'}{2}\right|^2 - \frac{i}{\hbar}\sigma(X - \frac{1}{2}z', z - z')\right). \tag{2.23}$$

Proof It is enough to consider the case $\hbar = 1$. Let us apply formula (2.20).

$$\mathscr{W}_{z,z'}(X) = (2\pi)^{-n} \int_{\mathbb{R}^{2n}} \langle \varphi_z, \hat{T}(z'')\varphi_{z'}\rangle e^{-i\sigma(X,z'')} dz''. \tag{2.24}$$

Using a formula from Chap. 1, we have

$$\begin{aligned}\langle \varphi_z, \hat{T}(z'')\varphi_{z'}\rangle &= \langle \varphi_z, \varphi_{z'+z''}\rangle e^{\frac{i}{2}\sigma(z',z'')} \\ &= e^{-\frac{1}{4}|z-z'-z''|^2} e^{\frac{i}{2}\sigma(z,z'+z'')+\sigma(z',z'')}.\end{aligned} \tag{2.25}$$

Using the change of variables $z'' = z - z' + u$, we have to compute the Fourier transform of the standard Gaussian $e^{-|u|^2/4}$ and (2.23) follows. □

We have the following properties of the Wigner transform:

Proposition 17 *Let $\varphi, \psi \in L^2(\mathbb{R}^n)$ be two quantum states. Then $\mathscr{W}_{\varphi,\psi} \in L^2(\mathbb{R}^{2n}) \cap L^\infty(\mathbb{R}^{2n})$ and we have:*

(i)

$$\|\mathscr{W}_{\varphi,\psi}\|_{L^\infty} \le 2^n \|\varphi\|_2 \|\psi\|_2.$$

(ii)

$$\|\mathscr{W}_{\varphi,\psi}\|_{L^2} \le (2\pi\hbar)^{n/2} \|\varphi\|_2 \|\psi\|_2.$$

(iii) Let $\varphi, \psi \in L^2(\mathbb{R}^n)$. Then we have:

$$|\langle \varphi, \psi\rangle|^2 = (2\pi\hbar)^{-n} \langle \mathscr{W}_\varphi, \mathscr{W}_\psi\rangle_{L^2(\mathbb{R}^{2n})}.$$

Proof (i) is a simple consequence of the definition of the Wigner transform and of the Cauchy–Schwarz inequality. For the proof of (ii) we note that

$$\int dz |\mathscr{W}_{\varphi,\psi}(z)|^2 = \int dx d\xi \left| \int dy e^{i\xi\cdot y/\hbar} \bar{\varphi}(x + \frac{y}{2}) \psi(x - \frac{y}{2})\right|^2.$$

Using an approximation argument, we can assume that $\varphi, \psi \in L^1(\mathbb{R}^n) \cap L^\infty(\mathbb{R}^n)$. So we have that $\bar{\varphi}(x + \frac{y}{2})\psi(x - \frac{y}{2}) \in L^2(\mathbb{R}^n, dy)$. According to Plancherel theorem we have:

$$(2\pi\hbar)^{-n} \int d\xi \left| \int dy e^{i\xi\cdot y/\hbar} \bar{\varphi}(x + \frac{y}{2}) \psi(x - \frac{y}{2})\right|^2 = \int dy |\bar{\varphi}(x + \frac{y}{2})\psi(x - \frac{y}{2})|^2,$$

so that:

$$\int dz|\mathscr{W}_{\varphi,\psi}(z)|^2 = (2\pi\hbar)^n \int dx \int dy|\bar{\varphi}(x+\tfrac{y}{2})\psi(x-\tfrac{y}{2})|^2$$
$$= (2\pi\hbar)^n \|\varphi\|^2\|\psi\|^2 \tag{2.26}$$

□

The Wigner transform operate "as one wishes" in phase space, namely according to the scheme of classical mechanics:

Proposition 18 *Let $\varphi, \psi \in L^2(\mathbb{R}^n)$ and $\hat{T}(z)$, $\hat{R}(F)$ be respectively operators of the Weyl–Heisenberg and metaplectic groups, corresponding respectively to*
– a phase-space translation by vector $z \in \mathbb{R}^{2n}$
– a symplectic transformation in phase-space
We have:

$$\mathscr{W}_{\hat{T}(z')\varphi,\hat{T}(z')\psi}(z) = \mathscr{W}_{\varphi,\psi}(z-z') \tag{2.27}$$
$$\mathscr{W}_{\hat{R}(F)\varphi,\hat{R}(F)\psi}(z) = \mathscr{W}_{\varphi,\psi}(F^{-1}z). \tag{2.28}$$

Proof We have the nice group property of the Weyl–Heisenberg translation operator:

$$\hat{T}(-z')\hat{T}(X)\hat{T}(z') = \exp\left(-\frac{i}{\hbar}\sigma(X,z')\right)\hat{T}(X),$$

so that

$$\mathscr{W}_{\hat{T}(z')\varphi,\hat{T}(z')\psi}(z) = (2\pi\hbar)^{-n}\int dX \exp\left(-\frac{i}{\hbar}\sigma(z-z',X)\right)\langle\varphi,\hat{T}(X)\psi\rangle \quad = \mathscr{W}_{\varphi,\psi}(z-z').$$

As a result of the property of the metaplectic transformation we have:

$$\hat{R}(F)^{-1}\hat{T}(z')\hat{R}(F) = \hat{T}(F^{-1}z').$$

Therefore

$$\mathscr{W}_{\hat{R}(F)\varphi,\hat{R}(F)\psi}(z) = (2\pi\hbar)^{-n}\int dz'\langle\varphi,\hat{T}(Fz')\psi\rangle e^{-i\sigma(z,z')/\hbar} =$$

$$(2\pi\hbar)^{-n}\int dz''\langle\varphi,\hat{T}(z'')\psi\rangle e^{-i\sigma(z,Fz'')/\hbar} = (2\pi\hbar)^{-n}\int dz'\langle\varphi,\hat{T}(z')\psi\rangle e^{-i\sigma(F^{-1}z,z')/\hbar},$$

where we have used the change of variable $Fz' = z''$ and the fact that a symplectic matrix has determinant one. □

Now we get a formula to recover the Weyl symbol of any operator $\hat{A}$ in $\mathcal{L}\left(\mathcal{S}(\mathbb{R}^n), \mathcal{S}'(\mathbb{R}^n)\right)$:

Proposition 19 *Every operator $\hat{A} \in \mathcal{L}\left(\mathcal{S}(\mathbb{R}^n), \mathcal{S}'(\mathbb{R}^n)\right)$ has a contravariant Weyl symbol A and a covariant Weyl symbol $\tilde{A}$ in $\mathcal{S}'(\mathbb{R}^{2n})$.*

We have, in the distribution sense in general, in the usual sense if $\hat{A}$ is bounded in $L^2(\mathbb{R}^n)$,

$$A(X) = (2\pi\hbar)^{-2n} \iint_{\mathbb{R}^{4n}} \langle \varphi_{z'}, \hat{A}\varphi_z \rangle \mathcal{W}_{z',z}(X) dz dz'. \tag{2.29}$$

$$\tilde{A}_\sigma(X) = (2\pi\hbar)^{-n} \int_{\mathbb{R}^{2n}} \langle \varphi_{z+X}, \hat{A}\varphi_z \rangle \mathrm{e}^{-\frac{i}{\hbar}\sigma(X,z)} dz. \tag{2.30}$$

Proof We compute formally. It is not very difficult to give all the details for a rigorous proof.

We apply inverse formula for the Fourier–Bargmann transform (see Chap. 1). So for any $\psi \in \mathcal{S}(\mathbb{R}^n)$, we have

$$\hat{A}\psi(x) = (2\pi\hbar)^{-2n} \iint_{\mathbb{R}^{4n}} \langle \varphi_{z'}, \hat{A}\varphi_z \rangle \langle \varphi_z, \psi \rangle \varphi_{z'}(x) dz dz'. \tag{2.31}$$

So we get a formula for the Schwartz kernel K_A for $\hat{A}$,

$$K_A(x, x') = (2\pi\hbar)^{-2n} \iint_{\mathbb{R}^{4n}} \langle \varphi_{z'}, \hat{A}\varphi_z \rangle \overline{\varphi_z(x')} \varphi_{z'}(x) dz dz'. \tag{2.32}$$

Then we apply formula (2.11) to get the contravariant symbol A.

The formula for the covariant symbol follows from (2.7) and trace computation with coherent states. □

The only but important missing property to have a nice probabilistic setting with the Wigner functions is *positivity* which is unfortunately not satisfied because we have the following result, proved by Hudson [159] for $n = 1$ then extended to $n \geq 2$ by Soto-Claverie [240].

Theorem 3 *$\mathcal{W}_\psi(X) \geq 0$ on $\mathbb{R}^{2n}$ if and only if $\psi = C\varphi_z^{(\Gamma)}$ where C is a complex number, Γ a complex, symmetric $n \times n$ matrix with a positive non degenerate imaginary part $\Im\Gamma$, $z \in \mathbb{R}^{2n}$, where we define the Gaussian*

$$\varphi^{(\Gamma)}(x) = (\pi\hbar)^{-n/4} \det{}^{1/4}\Im\Gamma \exp\left(\frac{i}{2\hbar}\Gamma x \cdot x\right). \tag{2.33}$$

Proof We more or less follow the paper of Soto-Claverie [240].

We can check by direct computation that the Wigner density of $\varphi_z{}^{(\Gamma)}$ is positive (according the definition we have to compute the Fourier transform of the exponent of a quadratic form). We can also give the following more elegant proof. First, it is enough to consider the case $z = (0, 0)$. Second, it is possible to find a metaplectic transformation F such that $\varphi_0{}^{(\Gamma)} = \hat{R}(F)\varphi_0$ (see the section on symplectic invariance and Chap. 3 for more properties on the metaplectic group). Hence we get $\mathscr{W}_{\hat{R}(F)\varphi_0}(X) = \mathscr{W}_{\varphi_0}(F^{-1}(X))$. But we have computed above $\mathscr{W}_{\varphi_0}$, which is a standard Gaussian, so it is positive.

Conversely, assume now that $\mathscr{W}_\psi(X) \geq 0$ on $\mathbb{R}^{2n}$. We shall prove that the Fourier–Bargmann transform $\psi^\#(z)$ is a Gaussian function on the phase space. Hence using the inverse Bargmann transform formula, we shall get that ψ is a Gaussian.

Let us first prove the two following properties

$$\psi^\#(z) \neq 0, \quad \forall z \in \mathbb{R}^{2n}, \text{ and} \tag{2.34}$$

$$|\psi^\#(z)| \leq C\mathrm{e}^{\delta|z|^2}, \quad \forall z \in \mathbb{R}^{2n}, \quad \text{for some } C, \delta > 0. \tag{2.35}$$

We have seen that

$$\begin{aligned} |\langle \psi, \varphi_z \rangle|^2 &= (2\pi\hbar)^{-n} \int_{\mathbb{R}^{2n}} \mathscr{W}_\psi(X)\mathscr{W}_{\varphi_z}(X)dX \\ &= 2^n \int_{\mathbb{R}^{2n}} \mathscr{W}_\psi(X)\mathrm{e}^{-\frac{1}{\hbar}|X-z|^2}dX. \end{aligned} \tag{2.36}$$

The last integral is positive because by assumption $\mathscr{W}_\psi(X) \geq 0$ and $\int \mathscr{W}_\psi(X)dX = 1$.

Using again (2.36) we get easily (2.34). The second step is to use a property of entire functions in $\mathbb{C}^n$. Let us recall that in Chap. 1, we have seen that the function

$$\psi_a^\#(\zeta) := \exp\left(\frac{p^2 + ip\cdot q}{2\hbar}\right)\psi^\#(q, p) \tag{2.37}$$

is an entire function in the variable $\zeta = q - ip \in \mathbb{C}^n$. Moreover we get easily that $\psi_a^\#(\zeta)$ satisfies properties (2.34). To achieve the proof of Theorem 3 we apply the following lemma, which is a particular case of Hadamard factorization theorem for $n = 1$, extended for $n \geq 2$ in [240].

Lemma 11 *Let f be an entire function in $\mathbb{C}^n$ such that $f(\zeta) \neq 0$ for all $\zeta \in \mathbb{C}^n$ and for some $C > 0$, $\delta > 0$,*

$$|f(\zeta)| \leq C\mathrm{e}^{\delta|\zeta|^m}, \quad \forall \zeta \in \mathbb{C}^n. \tag{2.38}$$

Then $f(\zeta) = \mathrm{e}^{P(\zeta)}$, where P is a polynomial of degree $\leq m$.

□

2.3 Coherent States and Operator Norms Estimates

Let us give now a first application of coherent states to Weyl quantization. We assume first that $\hbar = 1$.

Theorem 4 (Calderon–Vaillancourt) *There exists a universal constant C_n such that for every symbol $A \in C^\infty(\mathbb{R}^{2n})$ we have*

$$\|\hat{A}\|_{\mathscr{L}(L^2,L^2)} \le C_n \sup_{|\gamma|\le 2n+1, X\in\mathbb{R}^{2n}} |\partial_X^\gamma A(X)|. \tag{2.39}$$

Beginning of the Proof. From (2.32) we get the formula

$$\langle \psi, \hat{A}\eta\rangle = (2\pi)^{-n} \iint_{\mathbb{R}^{4n}} \langle \varphi_{z'}, \hat{A}\varphi_z\rangle \psi^\#(z')\overline{\eta^\#(z)}dzdz'. \tag{2.40}$$

We shall get (2.39) by proving that the Bargmann kernel $K_A^{\mathscr{B}}(z,z') := \langle \varphi_{z'}, \hat{A}\varphi_z\rangle$ is the kernel of a bounded operator in $L^2(\mathbb{R}^{2n})$. Let us first recall a classical lemma:

Lemma 12 *Let (Ω, μ) be a measured (σ-finite) space, K a measurable function on $\Omega \times \Omega$ such that*

$$m_K := \max\{\sup_{z\in\Omega} \int_\Omega |K(z,z')|dz', \sup_{z'\in\Omega} \int_\Omega |K(z,z')|dz\}.$$

Then K is the integral kernel of a bounded operator T_K on $L^2(\Omega)$ and we have

$$\|T_K\| \le m_K.$$

So the Calderon–Vaillancourt will be a consequence of the following

Lemma 13 *There exists a universal constant C_n such that for every symbol $A \in C^\infty(\mathbb{R}^{2n})$ we have*

$$|K_A^{\mathscr{B}}(z,z')| \le C_n(1+|z-z'|)^{-2n-1} \sup_{|\gamma|\le 2n+1, X\in\mathbb{R}^{2n}} |\partial_X^\gamma A(X)|. \tag{2.41}$$

Proof We have already seen that

$$K_A^{\mathscr{B}}(z,z') = \int_{\mathbb{R}^n} A(X)\mathscr{W}_{z',z}(X)dX =$$

$$2^n \int_{\mathbb{R}^{2n}} A(X)\exp\left(-\left|X - \frac{z+z'}{2}\right|^2 - i\sigma(X - \frac{1}{2}z', z-z')\right)dX. \tag{2.42}$$

First remark that we have

$$|\langle \varphi_{z'}, \hat{A}\varphi_z\rangle| \le \sup_{X\in\mathbb{R}^{2n}} |A(X)|. \tag{2.43}$$

So we only have to consider the case $|z'-z| \ge 1$. The estimate is proved by integration by parts (as it is usual for oscillating integral).

Let us introduce the phase function

$$\Phi = -\left|X - \frac{z+z'}{2}\right|^2 - i\sigma(X - \frac{1}{2}z', z - z'). \tag{2.44}$$

Computing $\partial_W \Phi$, we have $|\partial_X \Phi| \geq |z - z'|$ hence

$$\frac{\overline{\partial_X \Phi} \cdot \partial_X}{|\partial_X \Phi|^2} e^{\Phi} = e^{\Phi}. \tag{2.45}$$

So we get the wanted estimates performing $2n + 1$ integrations by parts in the integral (2.42) using formula (2.45).

This achieves the proof of Calderon–Vaillancourt theorem. □

Corollary 6 *$\hat{A}$ is a compact operator in $L^2(\mathbb{R}^n)$ if A is C^∞ on $\mathbb{R}^{2n}$ and satisfies the following condition:*

$$\lim_{|z|\to+\infty} |\partial_z^\gamma A(z)| = 0, \quad \forall \gamma \in \mathbb{N}^{2d}, \ |\gamma| \leq 2n+1. \tag{2.46}$$

Proof Let us introduce $\chi \in C^\infty(\mathbb{R}^{2n})$ such that $\chi(X) = 1$ if $|X| \leq \frac{1}{2}$ and $\chi(X) = 0$ if $|X| \geq 1$. Let us define $A_R(X) = \chi(X/R)A(X)$. For every $R > 0$, $\hat{A}_R$ is Hilbert-Schmidt hence compact. Using the Calderon–Vaillancourt estimate, we get

$$\lim_{|R|\to+\infty} \|\hat{A} - \widehat{A_R}\| = 0.$$

So $\hat{A}$ is compact. □

Using the same idea as for proving Calderon–Vaillancourt theorem, we get now a sufficient trace class condition.

Theorem 5 *There exists a universal constant τ_n such that for every $A \in C^\infty(\mathbb{R}^{2n})$ we have*

$$\|\hat{A}\|_{\mathrm{Tr}} \leq \tau_n \sum_{|\gamma|\leq 2n+1} \int_{\mathbb{R}^{2n}} |\partial_X^\gamma A(X)| dX. \tag{2.47}$$

In particular if the r.h.s is finite then $\hat{A}$ is in the trace class and we have

$$\mathrm{Tr}\hat{A} = (2\pi)^{-n} \int_{\mathbb{R}^{2n}} A(X) dX. \tag{2.48}$$

Proof Recall that $\hbar = 1$. From (2.29) we know that $\hat{A}$ has the following decomposition into rank one operators

$$\hat{A} = (2\pi)^{-n} \iint_{\mathbb{R}^{4n}} \langle \varphi_{z'}, \hat{A}\varphi_z \rangle \Pi_{z,z'} dz dz'. \tag{2.49}$$

But we know that $\|\Pi_{z,z'}\|_{TR} = 1$. So we have

$$\|\hat{A}\|_{\mathrm{Tr}} \leq (2\pi)^{-n} \iint_{\mathbb{R}^{4n}} |\langle \varphi_{z'}, \hat{A}\varphi_z\rangle| dz dz'. \tag{2.50}$$

Using integration by parts as in the proof of Calderon–Vaillancourt, we have

$$|\langle \varphi_{z'}, \hat{A}\varphi_z\rangle| \leq C_N(1+|z-z'|)^{-N} \sum_{|\gamma|\leq N} \int_{\mathbb{R}^{2n}} \mathrm{e}^{-|X-(z+z')/2|^2} |\partial_X^\gamma A(X)| dX, \tag{2.51}$$

with $N = 2n+1$. Now perform the change of variables $u = (z+z')/2$, $v = z - z'$ and using Young inequality we get

$$\iint_{\mathbb{R}^{4n}} |\langle \varphi_{z'}, \hat{A}\varphi_z\rangle| dz dz' \leq \tau_n \sum_{|\gamma|\leq N} \int_{\mathbb{R}^{2n}} |\partial_X^\gamma A(X)| dX. \tag{2.52}$$

Hence (2.47) follows.

We can get (2.48) by using approximations with compact support A_R like in the proof of Corollary 6.

Remark 5 Using interpolation results it is possible to get similar estimates for the Schatten norm $\|\hat{A}\|_p$ for $1 < p < +\infty$ (see [217] for the Schatten classes).

Let us now compute the action of Weyl quantization on Gaussian coherent states.

Lemma 14 *Assume that $A \in S_0(w)$ (w is a temperate weight, see definition 6). Then for every $N \geq 1$, we have*

$$\hat{A}\varphi_z = \sum_{|\gamma|\leq N} \hbar^{\frac{|\gamma|}{2}} \frac{\partial^\gamma A(z)}{\gamma!} \Psi_{\gamma,z} + \mathcal{O}(\hbar^{(N+1)/2}), \tag{2.53}$$

the estimate of the remainder is uniform in $L^2(\mathbb{R}^n)$ for z in every bounded set of the phase space and

$$\Psi_{\gamma,z} = \hat{T}(z)\Lambda_\hbar \mathrm{Op}_1^w(z^\gamma)g. \tag{2.54}$$

where $g(x) = \pi^{-n/4}\mathrm{e}^{-|x|^2/2}$, $\mathrm{Op}_1^w(z^\gamma)$ is the 1-Weyl quantization of the monomial: $(x,\xi)^\gamma = x^{\gamma'}\xi^{\gamma''}$, $\gamma = (\gamma', \gamma'') \in \mathbb{N}^{2d}$. In particular $\mathrm{Op}_1^w(z^\gamma)g = P_\gamma g$ where P_γ is a polynomial of the same parity as $|\gamma|$.

Proof Let us write

$$\hat{A}\varphi_z = \hat{A}\Lambda_\hbar \hat{T}_1(z)g = \Lambda_\hbar \hat{T}_1(z)(\Lambda_\hbar \hat{T}_1(z))^{-1}\hat{A}\Lambda_\hbar \hat{T}_1(z)g,$$

where $\Lambda_\hbar$ is the dilation : $\Lambda_\hbar\psi = \hbar^{-n/4}\psi(\hbar^{-1/2}x)$ and $\hat{T}_1$ is $\hat{T}$ for $\hbar = 1$.

Let us remark that $(\Lambda_\hbar \hat{T}(z))^{-1}\hat{A}\Lambda_\hbar \hat{T}(z) = \mathrm{Op}_1^w[A_{\hbar,z}]$ where $A_{\hbar,z}(X) = A(\sqrt{\hbar}X + z)$. So we prove the lemma by expanding $A_{\hbar,z}$ in X, around z, with the Taylor formula with integral remainder term to estimate the error term. □

The following Lemma allows to localized observables acting on coherent states.

Lemma 15 *Let A be a smooth observable with compact support in the ball $B(X_0, r_0)$ of the phase space. Then there exists $R > 0$ and for all $N \geq 1$ there exists C_N such that for $|z - X_0| \geq 2r_0$ we have*

$$\|\hat{A}\varphi_z\| \leq C_N \hbar^N \langle z \rangle^{-N}, \quad \text{for } |z| \geq R. \tag{2.55}$$

Proof It is convenient here to work on Fourier–Bargmannn side. So we estimate

$$\langle \varphi_z, \hat{A}\varphi_{z'} \rangle = (2\pi\hbar)^{-n} \int_{\mathbb{R}^{2n}} A(Y)\mathscr{W}_{z,z'}(Y)dY \tag{2.56}$$

As we have already seen, we have

$$\int_{\mathbb{R}^{2n}} A(Y)\mathscr{W}_{z,z'}(Y)dY = $$
$$2^n \int_{\mathbb{R}^{2n}} \exp\left(-\frac{1}{\hbar}\left|Y - \frac{z+z'}{2}\right|^2 - \frac{i}{\hbar}\sigma(Y - \frac{1}{2}z', z - z')\right) A(Y)dY. \tag{2.57}$$

Using integrations by parts as above, considering the phase function $\Psi(Y) = -|Y - \frac{z+X}{2}|^2 - i\sigma(Y - \frac{1}{2}X, z - X)$ and the differential operator $\frac{\overline{\partial_Y \Psi}}{|\partial_Y \Psi|^2}\partial_Y$, we get for every M, M' large enough,

$$|\langle \hat{A}\varphi_z, \varphi_{z'} \rangle| \leq C_{M,M'} \int_{[|Y| \leq r_0]} \left(1 + \frac{|Y - z|}{\sqrt{\hbar}}\right)^{-M} \left(1 + \frac{|z - z'|}{\sqrt{\hbar}}\right)^{M-M'} dY. \tag{2.58}$$

Thefore we easily get the estimate choosing M, M' conveniently and using that the Fourier–Bargmannn transform is an isometry. □

We need to introduce some notations for classes of Weyl symbols controlling their smoothness and their growth.

Definition 6 A positive function w on $\mathbb{R}^m$ is a temperate weight if it satisfies the following property. There exist N, C such that

$$w(X + Y) \leq Cw(X)(1 + |X - Y|)^N, \quad \forall X, Y \in \mathbb{R}^m. \tag{2.59}$$

A symbol $A \in C^\infty(\mathbb{R}^{2n})$ is a classical observable of weight w and gain $\rho \in [0, 1]$ if for every multiindex α there exists C_α such that

$$|\partial_X^\alpha A(X)| \leq C_\alpha w(X)\langle X \rangle^{-\rho|\alpha|}, \ \forall X \in \mathbb{R}^{2n}$$

The space of symbols of weight w and gain ρ is denoted $\mathrm{S}_\rho(w)$.

A basic example of temperate weight is $w_\mu(X) = (1 + |X|)^\mu$, $\mu \in \mathbb{R}$. We shall denote $\mathrm{S}_\rho^\mu = \mathrm{S}_\rho(w_\mu)$.

Remark 6 The product of two temperate weights is a temperate weight and if w is a temperate weight then w^{-1} is likewise a temperate weight.

As proved by Unterberger [245] and rediscovered by Tataru [242], it is possible to characterize the operator class $\widehat{S_0^0}$ on the matrix element $\langle \varphi_{z'}, \hat{A}\varphi_z \rangle$. We state now a semi-classical version of Unterberger result.

Theorem 6 *Let $\hat{A}_\hbar$ be a $\hbar$-dependent family of operators from $\mathscr{S}(\mathbb{R}^n)$ to $\mathscr{S}'(\mathbb{R}^n)$. Then $\hat{A} = \mathrm{Op}_\hbar^w(A_\hbar)$ with $A_\hbar \in S_0^0$ with uniform estimate*[2] *if and only if for every N there exists C_N such that we have*

$$|\langle \varphi_{z'}, \hat{A}\varphi_z \rangle| \le C_N \left(1 + \frac{|z - z'|}{\sqrt{\hbar}}\right)^{-N}, \quad \forall \hbar \in]0,1), \; z, z' \in \mathbb{R}^{2n}. \tag{2.60}$$

Proof Suppose that $\hat{A} = \mathrm{Op}_\hbar^w(A_\hbar)$, with $A_\hbar \in S_0^0$ is a bounded family. We get estimate (2.60) by integrations by parts as above.

Conversely if we have estimates (2.60), using (2.23) and (2.29) we have

$$A_\hbar(X) = (\pi\hbar)^{-n} \iint_{\mathbb{R}^{4n}} \langle \varphi_{z'}, \hat{A}_\hbar \varphi_z \rangle \exp\left(-\frac{1}{\hbar}\left| X - \frac{z+z'}{2} \right|^2 \right.$$
$$\left. + iJ(X - \frac{z'}{2}) \cdot (z - z') \right) dz dz'. \tag{2.61}$$

Using the change of variables $\frac{z+z'}{2} = u$ and $z - z' = \sqrt{\hbar} v$ we get easily that there exists $C > 0$ such that

$$|A_\hbar(X)| \le C, \quad \forall X \in \mathbb{R}^{2n}, \; \hbar \in]0,1]. \tag{2.62}$$

In the same way we can estimate every derivatives of $A_\hbar$, after derivation in X in the integral (2.61). □

The other main fact in Weyl quantization is existence of an operational calculus. We shall recall its properties in the next section.

2.4 Product Rule and Applications

2.4.1 The Moyal Product

One of the most useful properties of Weyl quantization is that we have an operational calculus defined by:

---

[2]This means that for every γ, $\sup_{\hbar \in]0,1]} \|\partial^\gamma A\|_\infty < +\infty$.

The product rule for quantum observables Let us start with $A, B \in \mathscr{S}(\mathbb{R}^{2n})$. We look for a classical observable C such that $\hat{A} \cdot \hat{B} = \hat{C}$. Let us first remark that the integral kernel of $\hat{C}$ is

$$K_C(x, y) = \int_{\mathbb{R}^n} K_A(x, s) K_B(s, y) ds. \tag{2.63}$$

Using relationship between integral kernels and Weyl symbols, we get

$$C(X) = (\pi\hbar)^{-2n} \iint_{\mathbb{R}^{4n}} e^{2i\hbar\sigma(Y,Z)} A(X+Z) B(X+Y) dY dZ \tag{2.64}$$

where σ is the symplectic bilinear form introduced above.

Now let us apply Plancherel formula in $\mathbb{R}^{4n}$ and the following Fourier transform formula

Lemma 16 *Let $f(T) = e^{\frac{i}{2}\langle B.T,T\rangle}$, for $T \in \mathbb{R}^m$ where B is a non degenerate symmetric $m \times m$ matrix. Then the Fourier transform $\tilde{f}$ is*

$$\tilde{f}(\zeta) = (2\pi)^{m/2} |\det B|^{-1/2} e^{i\pi \operatorname{sgn} B} e^{-\frac{i}{2}\langle B^{-1}\zeta,\zeta\rangle}, \tag{2.65}$$

where $\operatorname{sgn} B$ *is the signature*[3] *of the matrix* B.

Proof see [153, 217] □

Hence we get

$$C(x, \xi) = \exp\left(\frac{i\hbar}{2}\sigma(D_x, D_\xi; D_y, D_\eta)\right) A(x, \xi) B(y, \eta)|_{(x,\xi)=(y,\eta)}, \tag{2.66}$$

We can see easily on formula (B.2) that $C \in \mathscr{S}(\mathbb{R}^{2n})$. So that (2.64) defines a non-commutative product on classical observables. We shall denote this product as follows $\boxed{C = A \circledast B}$ (Moyal product).

In semi-classical analysis, it is useful to expand the exponent in (B.2), so we get the formal series in $\hbar$:

$$C(x, \xi) = \sum_{j \geq 0} C_j(x, \xi) \hbar^j, \quad \text{where}$$

$$C_j(x, \xi) = \frac{1}{j!}\left(\frac{i}{2}\sigma(D_x, D_\xi; D_y, D_\eta)\right)^j A(x, \xi) B(y, \eta)|_{(x,\xi)=(y,\eta)}. \tag{2.67}$$

We can easily see that in general C is not a classical observable because of the $\hbar$ dependence. It can be proved that it is a *semi-classical observable* in the following sense.

[3]Let $\{b_j\}$ be the eigenvalues of B, with their multiplicities. Then $\operatorname{sgn} B := \sum_j \operatorname{sgn} b_j$ where $\operatorname{sgn} b_j$ is the sign of b_j.

Definition 7 We say that A is a semi-classical observable of weight w, where w is temperate weight on $\mathbb{R}^{2n}$, if there exist $\hbar_0 > 0$ and a sequence $A_j \in \mathrm{S}_0(w)$, $j \in \mathbb{N}$, so that A is a map from $]0, \hbar_0]$ into $\mathrm{S}_0(w))$ satisfying the following asymptotic condition : for every $N \in \mathbb{N}$ and every $\gamma \in \mathbb{N}^{2n}$ there exists $C_N > 0$ such that for all $\hbar \in]0, 1[$ we have

$$\sup_{\mathbb{R}^{2n}} w^{-1}(z) \left| \frac{\partial^\gamma}{\partial z^\gamma} \left(A(\hbar, z) - \sum_{0 \le j \le N} \hbar^j A_j(z) \right) \right| \le C_N \hbar^{N+1}, \tag{2.68}$$

A_0 is called the principal symbol, A_1 the sub-principal symbol of $\hat{A}$.

The set of semi-classical observables of weight w is denoted by $\mathrm{S}_{sc}(w)$. Its range in $\mathscr{L}(\mathscr{S}(\mathbb{R}^n), \mathscr{S}'(\mathbb{R}^n))$ is denoted $\widehat{\mathrm{S}_{sc}(w)}$.

We may use the notation $\mathrm{S}^{\mu}_{sc} = \mathrm{S}_{sc}(w_\mu)$.

Now we state the product rule for Weyl quantization.

Theorem 7 *Let w, w' be two temperate weights in $\mathbb{R}^{2n}$. For every $A \in \mathrm{S}_0(w)$ and $B \in \mathrm{S}_0(w')$, there exists a unique $C \in \mathrm{S}_{sc}(ww'))$ such that $\hat{A} \cdot \hat{B} = \hat{C}$ with $C \asymp \sum_{j \ge 0} \hbar^j C_j$. The C_j are given by*

$$C_j(x, \xi) = \frac{1}{2^j} \sum_{|\alpha+\beta|=j} \frac{(-1)^{|\beta|}}{\alpha!\beta!} (D^\beta_x \partial^\alpha_\xi A).(D^\alpha_x \partial^\beta_\xi B)(x, \xi).$$

Proof The main technical point is to control the remainder terms uniformly in the semi-classical parameter $\hbar$. This is detailed in the Appendix of the paper [40]. □

Corollary 7 *Under the assumption of the theorem, we have the well known correspondence between the commutator for quantum observables and the Poisson bracket for classical observables, $\frac{i}{\hbar}[\hat{A}, \hat{B}] \in \widehat{\mathrm{S}_{sc}(ww')}$ and its principal symbol is the Poisson bracket $\{A, B\}$.*

A very useful application of the Moyal product is the possibility to get semiclassical approximations for inverse of elliptic symbol.

Definition 8 Let $A(\hbar)$ be a semiclassical observable in $\mathrm{S}_{sc}(w)$ and $X_0 \in \mathbb{R}^{2n}$. We shall say that A is elliptic at X_0 if $A_0(X_0) \neq 0$.

We shall say that A is uniformly elliptic if there exists $c > 0$ such that

$$|A(X)| \ge c w(X), \ \forall X \in \mathbb{R}^{2n}. \tag{2.69}$$

Theorem 8 *Let $A \in \mathrm{S}_{sc}(w)$ be an uniformly elliptic semi-classical symbol. Then there exists $B \in \mathrm{S}_{sc}(w^{-1})$ such that $B \circledast A = 1$ (in the sense of asymptotic expansion in S_{sc}). Moreover we have*

$$\hat{B} \cdot \hat{A} = \mathbb{1} + \mathscr{O}(\hbar^\infty). \tag{2.70}$$

where the remainder is estimated in the L^2norm of operators.

Moreover the semi-classical symbol B of $\hat{B}$ is $B = \sum_{j\geq 0} \hbar^j B_j$ with

$$B_0 = A_0^{-1}, \quad B_1 = -A_1 A_0^{-2}. \tag{2.71}$$

Proof Let us denote $C_j(E, F)$ the j term in the Moyal product $E \star F$. The method consists to compute by induction $B_0, \ldots, B_N$ such that

$$\left(\sum_{0\leq j\leq N} \hbar^j B_j\right) \circledast A(h) = 1 + \mathcal{O}(\hbar^{N+1}). \tag{2.72}$$

We start with $B_0 = \frac{1}{A_0}$. The next step is to compute B_1 such that $B_1 A_0 + A_1 B_0 = 0$. Then to compute B_2 such that

$$C_2(A_0, B_0) + C_1(A_1, B_1) + B_2 A_0 = 0.$$

So we get all the B_j by induction using the asymptotic expansion for the Moyal product.

The remainder term in (2.70) is estimated using Calderon Vaillancourt theorem. □

We give now a local version of the above theorem, which can be proved by the same method.

Theorem 9 *Let $A \in S_{sc}(m)$ be an elliptic symbol in an open bounded set Ω of $\mathbb{R}^{2n}$. Then for every $\chi \in C_0^\infty(\Omega)$ there exists $B_\chi \in S_{sc}^{-\infty}$ such that*

$$\hat{B}_\chi \hat{A} = \hat{\chi} + \mathcal{O}(\hbar^\infty). \tag{2.73}$$

Remark 7 For application it is useful to note that if A depends in a uniform way of some parameter $\varepsilon \in [0, 1]$ then B also depends uniformly in ε. In particular ε may depend on $\hbar$.

2.4.2 Functional Calculus

An useful consequence of the algebraic properties of symbolic quantization is a functional calculus: under suitable assumptions if $\hat{H}$ is an Hermitian semi-classical observable then for every smooth function f, $f(\hat{H})$ is also a semi-classical observable. The technical statement is

Theorem 10 *Let $\hat{H}$ be a uniformly elliptic semiclassical Hamiltonian. Let f be a smooth complex valued function such that, for some $r \in \mathbb{R}$, we have*

$$\forall k \in \mathbb{N},\ \exists C_k,\ |f^{(k)}(s)| \le C_k \langle s\rangle^{r-k},\quad \forall s \in \mathbb{R}.$$

Then $f(\hat{H})$ is a semiclassical observable with a semiclassical symbol $H_f(\hbar, x, \xi)$ such that $H_f(\hbar, \cdot, \cdot) \in \mathrm{S}_{sc}(w^r)$ given by the asymptotic expansion:

$$H_f(\hbar, z) \asymp \sum_{j\ge 0} \hbar^j H_{f,j}(z). \tag{2.74}$$

In particular we have

$$H_{f,0}(z) = f(H_0(z)), \tag{2.75}$$

$$H_{f,1}(z) = H_1(z) f'(H_0(z)), \tag{2.76}$$

$$\text{and for, } j \ge 2,\quad H_{f,j} = \sum_{1\le k\le 2j-1} d_{j,k}(H) f^{(k)}(H_0), \tag{2.77}$$

where $d_{j,k}(H)$ are universal polynomials in $\partial_z^\gamma H_\ell(z)$ with $|\gamma| + \ell \le j$.

A proof of this theorem can be found in [141, 217] and in [83]. We give a sketch of proof in the Appendix C. In particular we can take $f(s) = (\lambda + s)^{-1}$ for $\Im\lambda \ne 0$ (the proof begins with this case) or f with a compact support.

From this theorem we can get the following consequences on the spectrum of $\hat{H}$ (see [141]).

Theorem 11 *Let $\hat{H}$ be like in Theorem 10. Assume that $H_0^{-1}[E_-, E_+]$ is a compact set in $\mathbb{R}^n \times \mathbb{R}^n$. Consider a closed interval $I \subset [E_-, E_+]$. Then we have the following properties.*
(i) $\forall \hbar \in]0, \hbar_0]$, $\hbar_0 > 0$, the spectrum of $\hat{H}$ is discrete and is a finite sequence of eigenvalues $E_1(\hbar) \le E_2(\hbar) \le \cdots \le E_{N_I}(\hbar)$ where each eigenvalue is repeated according its multiplicity.
Moreover $N_I = O(\hbar^{-n})$ as $\hbar \searrow 0$.
(ii) For all $f \in C_0^\infty(I)$, $f(\hat{H})$ is a trace class operator and we have

$$\boxed{\mathrm{Tr}[f(\hat{H})] \asymp \sum_{j\ge 0} \hbar^{j-d} \tau_j(f)}, \tag{2.78}$$

where τ_j are distributions supported in $H_0^{-1}(I)$. In particular we have

$$\tau_0(f) = (2\pi)^{-d} \int_{\mathbb{R}^{2n}} f(H_0(z))dz, \tag{2.79}$$

$$\tau_1(f) = (2\pi)^{-d} \int_{\mathbb{R}^{2n}} f'(H_0(z))H_1(z)dz. \tag{2.80}$$

An easy consequence of this is the following Weyl asymptotic formula

Corollary 8 *If $I = [\lambda_-, \lambda_+]$ such that $\lambda_\pm$ are non critical values for H_0*[4] *then we have*

$$\lim_{\hbar \to 0} (2\pi\hbar)^n N_I = \int_{[H_0(q,p) \in I]} dq dp. \tag{2.81}$$

Remark 8 Formula (2.81) is very well known and can be proved in many ways, under much weaker assumptions.

For a proof using the functional calculus see [217], p. 283–287.

Under our assumptions we shall see in a next Chapter that we have a Weyl asymptotic with an accurate remainder estimate:

$$\boxed{N_I = (2\pi\hbar)^{-n} \int_{[H(q,p) \in I]} dq dp + O(\hbar^{1-n})},$$

using a time dependent method due to Hörmander and Levitan ([152] and its bibliography). For more accurate results about spectral asymptotics see [161].

2.4.3 *Propagation of Observables*

Now we come to the main application of the results of this section. We shall give a proof of the correspondence (in the sense of Bohr) between quantum and classical dynamics. As we shall see this theorem is a useful tool for semi-classical analysis although its proof is an easy application of Weyl calculus rules stated above. The microlocal version of the following result is originally due to Egorov [90]. Beals [22] found a nice simple proof.

For simplicity we shall assume that the semiclassical Hamiltonian $H(t)$ is sub-quadratic, $H(t) \in S^2_{sc}$, in the following sense

$$\begin{aligned} |\partial_z^\gamma H_j(t,z)| &\le C_\gamma, \ \text{ for } |\gamma| + j \ge 2; \\ \hbar^{-2}(H(t) - H_0(t) - \hbar H_1(t)) &\in S^0_{sc}. \end{aligned} \tag{2.82}$$

We also assume continuity and uniformity in time $t \in [-T, T]$.

Theorem 12 (The Semiclassical Propagation Theorem) *Let us consider a time dependent Hamiltonian satisfying 2.82. Let $A \in S^1_0$ be an observable such that $\partial_X^\gamma A \in S^0_0$ if $|\gamma| \ge 1$. Then we have:*

(a) For any $\hbar > 0$ and for every $\psi \in \mathscr{S}(\mathbb{R}^n)$, the Schrödinger equation

$$i\hbar\partial_t\psi_t = \hat{H}(t)\psi_t, \ \ \psi_{t=s} = \psi. \tag{2.83}$$

[4] λ is a non critical value for H means that $\nabla H(z) \neq 0$ if $H(z) = \lambda$.

has a unique solution which we denote $\psi_t = \mathcal{U}(t,s)\psi$. Moreover $\mathcal{U}(t,s)$ can be extended as a unitary operator in $L^2(\mathbb{R}^n)$.

(b) *The time evolution $\hat{A}(t,s) = \mathcal{U}(s,t)\hat{A}\mathcal{U}(t,s)$ of $\hat{A}$, $\hat{A}(s,s) = \hat{A}$, has a semi-classical Weyl symbol $A_\hbar(t,s)$ such that $A_\hbar(t,s) \in S^1_{sc}$. More precisely we have $A(t,s) \asymp \sum_{j\geq 0} \hbar^j A_j(t,s)$, in S^0_{sc}, which is uniform in t, s, for $t, s \in [-T, T]$. Moreover $A_j(t,s)$ can be computed by the following formulas:*

$$A_0(t,s;z) = A(\Phi^{t,s}(z)), \tag{2.84}$$

$$A_1(t,s;,z) = \int_s^t \{A(\Phi^{\tau,t}), H_1(\tau\})(\Phi^{t,\tau}(z))d\tau. \tag{2.85}$$

and for $j \geq 2$, $A_j(t,s;z)$ can be computed by induction on j.

Proof (A sketch). Property (a) will be proved later in this book (see also [191] for $\hbar = 1$). It is easier to prove it if H is time independent because we can prove in this case that $\hat{H}$ is essentially self-adjoint (for a proof see [217]). Then we have

$$\mathcal{U}(t) := \mathcal{U}(t,0) = \exp(-\frac{it}{\hbar}\hat{H}).$$

Let us remark that, under the assumption of the theorem, the classical flow for H_0 exists globally. Indeed, the Hamiltonian vector field $(\partial_\xi H_0, -\partial_x H_0)$ has a sublinear growing at infinity so, no classical trajectory can blow up in a finite time. Moreover, using usual methods in non linear O.D.E (variation equation) we can prove that $A(\Phi^{t,s}) \in S^0_0$ with semi-norm uniformly bounded for t, s bounded.

Now, from the Heisenberg equation and the classical equations of motion we get

$$\frac{\partial}{\partial\tau}\mathcal{U}(s,\tau)\widehat{A_0}(t,\tau)\mathcal{U}(\tau,s) =$$

$$\mathcal{U}(s;\tau)\left\{\frac{i}{\hbar}[\hat{H}(\tau), \widehat{A_0}(t,\tau)] - \{\widehat{H(\tau), A_0\}(\Phi^{t,\tau})}\right\}\mathcal{U}(\tau,s), \tag{2.86}$$

where $A_0(t,s) = A(\Phi^{t,s})$. But, from the corollary of the product rule, the principal symbol of

$$\frac{i}{\hbar}\left([\hat{H}(\tau), \widehat{A_0(t,\tau)}] - \{\widehat{H(\tau), A_0}\}\right)(\Phi^{t,\tau})$$

vanishes. So, in the first step, using the product rule formula, we get the approximation

$$\mathcal{U}(s,t)\hat{A}\mathcal{U}(t,s) - \widehat{A_0(t,s)} =$$

$$\int_s^t \mathcal{U}(s,\tau)\left(\frac{i}{\hbar}[\hat{H}(\tau), \widehat{A_0(t,\tau)}] - \{\widehat{\hat{H}(\tau), A_0}\}\Phi^{t,\tau}\right)\mathcal{U}(\tau,s)d\tau. \tag{2.87}$$

Now, it is not difficult to obtain, by induction, the full asymptotics in $\hbar$. For $j \geq 2$,

$$A_j(t,s;z) = \sum_{\substack{|(\alpha,\beta)|+k=j+1 \\ 0\le\ell\le j-1}} \Gamma(\alpha,\beta)\int_s^t [(\partial_\xi^\alpha \partial_x^\beta H_k(\tau))\cdot(\partial_\xi{}^\alpha \partial_x{}^\beta A_\ell)](\Phi^{t,\tau}(z))d\tau, \tag{2.88}$$

with

$$\Gamma(\alpha,\beta) = \frac{(-1)^{|\beta|}-(-1)^{|\alpha|}}{\alpha!\beta!2^{|\alpha|+|\beta|}} i^{-1-|(\alpha,\beta)|}.$$

The main technical point is to estimate the remainder terms. For a proof with more details see [40] where the authors get a uniform estimate up to Ehrenfest time (of order $\log \hbar^{-1}$). □

Remark 9 If $H(t) = H_0(t)$ is a polynomial function of degree ≤ 2 in z on the phase space $\mathbb{R}^{2n}$ then the propagation theorem assumes a simpler form : $A(t,s) = A(\Phi^{t,s})$ and the remainder term is null. This is a consequence of the following exact formula.

$$\frac{i}{\hbar}[\hat{H},\hat{B}] = \widehat{\{H,B\}}, \tag{2.89}$$

where $B \in \mathrm{S}_0^0$.

Now we give an application of the propagation theorem and coherent states in semiclassical analysis: we recover the classical evolution from the quantum evolution, in the classical limit $\hbar \searrow 0$.

Corollary 9 *For every observable $A \in \Sigma^0$ and every $z \in \mathbb{R}^{2n}$, we have*

$$\lim_{\hbar\searrow 0}\langle \mathscr{U}(t,s)\varphi_z, \hat{A}\,\mathscr{U}(t,s)\varphi_z\rangle = A(\Phi^{t,s}(z)) \tag{2.90}$$

and the limit is uniform in $(t,s;z)$ on every bounded set of $\mathbb{R}_t \times \mathbb{R}_s \times \mathbb{R}^{2n}_z$.

Proof

$$\begin{aligned}\langle \mathscr{U}(t,s)\varphi_z, \hat{A}\,\mathscr{U}(t,s)\varphi_z\rangle &= \langle \varphi_z, \mathscr{U}(s,t)\hat{A}\,\mathscr{U}(t,s)\varphi_z\rangle \\ &= \int_{\mathbb{R}^{2n}} A(t,s;X)\mathscr{W}_{z,z}(X)dX \\ &= (\pi\hbar)^{-n}\int_{\mathbb{R}^{2n}} A(t,s;X)\mathrm{e}^{-\frac{|X-z|^2}{\hbar}}dX.\end{aligned} \tag{2.91}$$

So by the propagation theorem we know that $A(t,s;X) = A(\Phi^{t,s}(X)) + \mathscr{O}(\hbar)$. Hence the corollary follows. □

Remark 10 The last result has a long history beginning with Ehrenfest (1930) and continuing with Hepp (1974), Bouzouina–Robert [40]. In this last paper it is proved that the corollary is still valid for times smaller than the Ehrenfest time $T_E := \gamma_E|\log\hbar|$, for some constant $\gamma_E > 0$.

2.4.4 Return to Symplectic Invariance of Weyl Quantization

Let us give now a first construction of metaplectic transformations. Other equivalent constructions and more properties will be given later (chapter on quadratic hamiltonians).

Lemma 17 *For every $F \in \mathrm{Sp}(n)$ we can find a C^1- smooth curve F_t, $t \in [0, 1]$, in $\mathrm{Sp}(n)$, such that $F_0 = \mathbb{1}$ and $F_1 = F$.*

Proof An explicit way to do that is to use the polar decomposition of F, $F = V|F|$ where V is a symplectic orthogonal matrix and $|F| = \sqrt{F^t F}$ is positive symplectic matrix. Each of these matrices have a logarithm, so $F = \mathrm{e}^K \mathrm{e}^L$ with K, L Hamiltonian matrices, and we can choose $F_t = \mathrm{e}^{tK}\mathrm{e}^{tL}$. F_t is clearly the linear flow defined by the quadratic Hamiltonian $H_t(z) = \frac{1}{2}S_t z \cdot z$ where $S_t = -J\dot{F}_t F_t^{-1}$. □

Now we use the (exact) propagation theorem. $\mathscr{U}(t, s)$ denotes the propagator defined by the quadratic Hamiltonian built in the proof of Lemma 17 and Theorem 12. Then **we define** $\hat{R}(F) = \mathscr{U}(1, 0)$. Recall that $\mathscr{U}(t, 0)$ is the solution of the Schrödinger equation

$$i\hbar\frac{d}{dt}\mathscr{U}(t,0) = \hat{H}(t)\mathscr{U}(t,0), \quad \mathscr{U}(0,0) = \mathbb{1}. \tag{2.92}$$

The following theorem translates the symplectic invariance of the Weyl quantization

Theorem 13 *For every linear symplectic transformation $F \in \mathrm{Sp}(n)$ and every symbol $A \in \mathrm{S}_0^0$ we have*

$$\hat{R}(F)^{-1}\hat{A}\hat{R}(F) = \widehat{A \cdot F} \tag{2.93}$$

Proof This is a direct consequence of the exact propagation formula for quadratic Hamiltonians

$$\mathscr{U}(0,t)\hat{A}\mathscr{U}(t,0) = \widehat{A\Phi^{t,0}} \tag{2.94}$$

□

We can get another proof of the following result (see formulas (2.27)).

Corollary 10 *Let $\psi, \eta \in L^2(\mathbb{R}^n)$. For every linear symplectic transformation $F \in \mathrm{Sp}(n)$, we have the following transformation formula for the Wigner function.*

$$\mathscr{W}_{\hat{R}(F)\psi,\hat{R}(F)\eta}(z) = \mathscr{W}_{\psi,\eta}(F^{-1}(z)), \quad \forall z \in \mathbb{R}^{2n}. \tag{2.95}$$

Proof For every $A \in \mathscr{S}(\mathbb{R}^{2n})$, we have

$$\begin{aligned}\langle \hat{R}(F)\eta, \hat{A}\hat{R}(F)\psi\rangle &= \int_{\mathbb{R}^{2n}} A(z)\mathscr{W}(\hat{R}(F)\psi, \hat{R}(F)\eta)(z)dz \\ &= \langle \eta, \hat{R}(F)^{-1}\hat{A}\hat{R}(F)\psi\rangle \\ &= \int_{\mathbb{R}^{2n}} A(F\cdot z)\mathscr{W}_{\psi,\eta}(z)dz. \end{aligned} \tag{2.96}$$

The corollary follows. □

We have the following uniqueness result.

Proposition 20 *Given the linear symplectic transformation $F \in \mathrm{Sp}(n)$, there exists a unique transformation $\hat{R}(F)$, up to a complex number of modulus 1, satisfying (2.18).*

Proof If $\hat{V}$ satisfies $\hat{V}^{-1}\hat{A}\hat{V} = \widehat{A\cdot F}$ then if $\hat{B} = \hat{V}^{-1}\cdot\hat{R}(F)$, we have that $\hat{B}$ commutes with every $\hat{A}$, $A \in \mathrm{S}^0_0$. In particular $\hat{B}$ commutes with the Heisenberg-Weyl translations $\hat{T}(z)$, hence $\hat{T}(z)^{-1}\hat{B}\hat{T}(z) = \hat{B}$. But we knows that $\hat{T}(z)^{-1}\hat{B}\hat{T}(z) = \widehat{B(\cdot + z)}$. So the Weyl symbol of $\hat{B}$ (it is a temperate distribution) is a constant complex number λ. But here $\hat{B}$ is unitary, so $|\lambda| = 1$. □

2.5 Husimi Functions, Frequency Sets and Propagation

2.5.1 *Frequency Sets*

The Husimi transform of some temperate distributions $u \in \mathscr{S}'(\mathbb{R}^n)$ is defined as follows

Definition 9 The Husimi transform of $u \in \mathscr{S}'(\mathbb{R}^n)$ is the function $\mathscr{H}_u(z)$ defined on the phase space $\mathbb{R}^{2n}$ by

$$\mathscr{H}_u(z) = (2\pi\hbar)^{-n}|\langle u, \varphi_z\rangle|^2, \quad z \in \mathbb{R}^{2n}, \tag{2.97}$$

where φ_z is the standard coherent state defined in Chap. 1.

The Husimi transform in contrast with the Wigner transform is always nonnegative. We shall see below that the Husimi distribution is a "regularization" of the Wigner distribution.

Proposition 21 *For every $\varphi \in L^2(\mathbb{R}^n)$ we have*

$$\mathscr{H}_\varphi = \mathscr{W}_\varphi * G_0$$

where G_0 is a gaussian function in phase space namely

$$G_0(z) = (\pi\hbar)^{-n}e^{-|z|^2/\hbar}.$$

One has $\int_{\mathbb{R}^{2n}} G_0(z)dz = 1$. This means that the Husimi distribution is a "regularization" of the Wigner distribution.

Proof According to the Proposition (17) (iii) we have:

$$\mathcal{H}_\varphi(z) = (2\pi\hbar)^{-n} \langle \mathcal{W}_{\varphi_z}, \mathcal{W}_\varphi \rangle_{L^2(\mathbb{R}^{2n})}.$$

But we know that

$$\mathcal{W}_{\varphi_z}(X) = \mathcal{W}_{\varphi_0}(X - z).$$

We use Proposition (16):

$$\langle \mathcal{W}_{\varphi_z}, \mathcal{W}_\varphi \rangle = 2^n \int_{\mathbb{R}^{2n}} \exp\left(-\frac{|X - z|^2}{\hbar}\right) \mathcal{W}_\varphi(X)dX.$$

This yields the result. □

In semi-classical analysis (or in high frequency analysis) it is important to understand what is the region of the phase space $\mathbb{R}^{2n}$ where some states $\psi \in L^2(\mathbb{R}^n)$ depending on $\hbar$, essentially lives when $\hbar$ is small. For that purpose let us introduce the **frequency set** of ψ

Definition 10 Let $\psi_\hbar \in L^2(\mathbb{R}^n)$, depending on $\hbar$, such that $\|\psi_\hbar\| \leq 1$. We say that $\psi_\hbar$ is negligible near a point $X_0 \in \mathbb{R}^{2n}$, if there exists a neighborhood $\mathcal{V}_{X_0}$ such that

$$\mathcal{H}_{\psi_\hbar}(z) = \mathcal{O}(\hbar^\infty), \ \forall z \in \mathcal{V}_{X_0}. \tag{2.98}$$

Let us denote $\mathcal{N}[\psi_\hbar]$ the set $\{X \in \mathbb{R}^{2n}, \ \psi_\hbar \text{ is negligible near } X\}$. The frequency set $\mathrm{FS}[\psi_\hbar]$ is defined as the complement of $\mathcal{N}[\psi_\hbar]$ in $\mathbb{R}^{2n}$.

Example 1 • If $\psi_\hbar = \varphi_z$ then $\mathrm{FS}[\varphi_z] = \{z\}$

• Let $\psi = a(x)e^{\frac{i}{\hbar}S(x)}$ where a and S are smooth functions, $a \in \mathcal{S}(\mathbb{R}^n)$, S real. Then we have the inclusion

$$\mathrm{FS}[\psi] \subseteq \{(x, \xi) | \xi = \nabla S(x)\}. \tag{2.99}$$

This is called a Lagrangian or WKB state.

There are several equivalent definitions of the frequency set that we now give.

Proposition 22 *Let $\psi_\hbar$ be such that $\|\psi_\hbar\| \leq 1$ and $X_0 = (x_0, \xi_0) \in \mathbb{R}^{2n}$. The following properties are equivalent*

(i)

$$\mathcal{H}_{\psi_\hbar}(X) = \mathcal{O}(\hbar^{+\infty}), \ \forall X \in \mathcal{V}_{X_0}.$$

(ii) There exists $A \in \mathcal{S}(\mathbb{R}^{2n})$, such that $A(X_0) = 1$ and

$$\|\hat{A}\psi_\hbar\| = \mathcal{O}(\hbar^{+\infty}). \tag{2.100}$$

(iii) There exists a neighborhood $\mathcal{V}_{X_0}$ of X_0 such that for all $A \in C_0^\infty(\mathcal{V}_{X_0})$,

$$\|\hat{A}\psi_\hbar\| = \mathcal{O}(\hbar^{+\infty}). \tag{2.101}$$

(iv) There exist $\chi \in C_0^\infty(\mathbb{R}^n)$ such that $\chi(x_0) = 1$ and a neighborhood V_{ξ_0} of ξ_0 such that

$$\langle \chi(x) e^{\frac{i}{\hbar}x\cdot\xi}, \psi_\hbar \rangle = \mathcal{O}(\hbar^{+\infty}) \tag{2.102}$$

for all $\xi \in V_{\xi_0}$.

Proof Let us assume (i). Then we have

$$\mathcal{H}_{\psi_\hbar}(z) = \mathcal{O}(\hbar^{\infty}), \quad |z - X_0| < r_0. \tag{2.103}$$

Using Lemma 15 we have

$$\|\hat{A}\varphi_z\| \le C_N \hbar^{N/2} < z >^{-N}, \quad \text{if} \quad |z - X_0| > r_0/2. \tag{2.104}$$

We have, using linearity of integration,

$$\hat{A}\psi_\hbar = (2\pi\hbar)^{-n} \int dz \langle \varphi_z, \psi_\hbar \rangle \hat{A}\varphi_z.$$

From triangle inequality, we have

$$\begin{aligned} \|\hat{A}\psi_\hbar\| &\le (2\pi\hbar)^{-n} \int dz |\langle \psi_\hbar, \varphi_z \rangle| \|\hat{A}\varphi_z\| \\ &\le (2\pi\hbar)^{-n} \left(\int_{[|z-X_0|<r_0]} dz + \int_{[|z-X_0|\ge r_0]} dz \right). \end{aligned} \tag{2.105}$$

Then we get (iii):

$$\|\hat{A}\psi_\hbar\|^2 = \mathcal{O}(\hbar^{+\infty}).$$

Let us now assume (iii), we want to prove (i).

Let us introduce $\chi \in C_0^\infty(B(X_0, r_0))$, $\chi(X) = 1$ if $|X - X_0| \le r_0/2$. Using Theorem 9 we have $\hat{B}\hat{A} = \hat{\chi} + \mathcal{O}(\hbar^{+\infty})$. Hence $\hat{\chi}\psi_\hbar = \mathcal{O}(\hbar^{+\infty})$. But using Lemma 15 we have $\langle (1 - \hat{\chi})\psi, \varphi_z \rangle = \mathcal{O}(\hbar^{+\infty})$ for $|z - X_0| \le r_0/4$. So we have proved $\langle \psi_\hbar, \varphi_z \rangle = \mathcal{O}(\hbar^{+\infty})$ for $|z - X_0| \le r_0/4$. □

A consequence of this proposition is that Weyl quantization does not increase the frequency set.

Corollary 11 *Let $\psi_\hbar$ be such that $\|\psi_\hbar\| \le 1$, $A \in S_0^0$, then we have*

$$\mathrm{FS}[\hat{A}(\psi_\hbar)] \subseteq \mathrm{FS}[\psi_\hbar]. \tag{2.106}$$

Moreover if A is elliptic at X_0 then we have

$$X_0 \in \mathrm{FS}[\hat{A}(\psi_\hbar)] \iff X_0 \in \mathrm{FS}[\psi_\hbar]. \tag{2.107}$$

Proof Let us assume that $X_0 \notin \mathrm{FS}[\psi_\hbar]$. If χ is like in the proof of the proposition, we have $\widehat{\chi A}\psi_\hbar = \mathcal{O}(\hbar^\infty)$. Applying Lemma 15 we have, for z near X_0,

$$\langle \varphi_z, \widehat{(1-\chi)A}\psi_\hbar\rangle = \mathcal{O}(\hbar^\infty).$$

so we get, z near X_0

$$\langle \varphi_z, \widehat{\chi A}\psi_\hbar\rangle = \mathcal{O}(\hbar^\infty)$$

□

2.5.2 *About Frequency Sets of Eigenstates*

Let us consider a quantum Hamiltonian $\hat{H}$. Assume that $H \in S(w)$ (for some weight w). Let us consider the stationnary Schrödinger equation

$$\hat{H}\psi_\hbar = E_\hbar\psi_\hbar \tag{2.108}$$

where $\|\Psi_\hbar\| = 1$, $\lim_{\hbar\to 0} E_\hbar = E$.

Proposition 23 *The frequency set of $\psi_\hbar$ is in the energy level set $S_E = \{X \in \mathbb{R}^{2n},\ H(X) = E\}$.*

Proof Let $X_0 \in \mathbb{R}^{2n}$ such that $H(X_0) \neq E$. There exist $\delta > 0$, $r_0 > 0$ such that $|H(X) - E| \ge \delta$, for every $X \in B(X_0, r_0)$. Let us choose some $\chi \in C_0^\infty(B(X_0, r_0))$, $\chi(X_0) = 1$. Using Theorem 9 and the remark following this theorem (here at the end $\varepsilon = \hbar$), we can find B such that

$$\hat{B}(\hat{H} - E_\hbar) = \hat{\chi} + \mathcal{O}(\hbar^{+\infty}), \tag{2.109}$$

so we get $\hat{\chi}\psi_\hbar = \mathcal{O}(\hbar^{+\infty})$ hence $X_0 \notin \mathrm{FS}[\psi_\hbar]$. □

Assume now that $\hat{H}$ satisfies the assumptions of the Propagation theorem and $\psi_\hbar$ satisfies the Schrödinger equation (2.108).

Proposition 24 *The frequency set $\mathrm{FS}[\psi_\hbar]$ is invariant under the classical flow Φ^t, for every $t \in \mathbb{R}$.*

Proof Let $X_0 \notin \mathrm{FS}[\psi_\hbar]$. There exists a compact support symbol A elliptic at X_0 such that $\hat{A}\psi_\hbar = \mathscr{O}(\hbar^{+\infty})$.

For every t we have

$$\mathscr{U}(-t)\hat{A}\psi_\hbar = \mathscr{O}(\hbar^{+\infty}) = \mathrm{e}^{\frac{itE_\hbar}{\hbar}}\hat{A}(t)\psi_\hbar.$$

Recall that the principal symbol of $\hat{A}(t)$ is $A \cdot \Phi^t$. So we get that if z is near $\Phi^{-t}(X_0)$, then $\hat{A}(t)\psi_\hbar = \mathscr{O}(\hbar^{+\infty})$, hence $\Phi^{-t}X_0 \notin \mathrm{FS}[\psi_\hbar]$. So we get that $\mathrm{FS}[\psi_\hbar]$ is invariant. □

Remark 11 There are several other names in the literature for the frequency set: "semiclassical wave front set", "microsupport". There are also different variants like "analytic-microsupport" [188].

2.6 Wick Quantization

2.6.1 General Properties

Following Berezin–Shubin [27] we start with the following general setting.

Let M be a locally compact metric space, with a positive Radon measure μ and $\mathscr{H}$ an Hilbert space. For each $m \in M$ we associate a unit vector $e_m \in \mathscr{H}$ such that the map $m \mapsto e_m$ is strongly continuous from M into $\mathscr{H}$. Moreover we assume that the following Plancherel formula is satisfied, for all $\psi \in \mathscr{H}$,

$$\|\psi\|^2 = \int_M |\langle e_m, \psi\rangle|^2 d\mu(m). \tag{2.110}$$

Let us denote $\psi^\#(m) = \langle e_m, \psi\rangle$. The map $\psi \mapsto \psi^\#(m) := \mathscr{I}\psi(m)$ is an isometry from $\mathscr{H}$ into $L^2(M)$. The canonical coherent states introduced in Chap.1 are examples of this setting where $M = \mathbb{R}^{2n}$, $\mathscr{H} = L^2(\mathbb{R}^n)$, $z \mapsto \varphi_z$, with the measure $d\mu(z) = (2\pi\hbar)^{-n}dqdp$, $z = (q, p) \in \mathbb{R}^{2n}$.

Definition 11 Let $\hat{A} \in \mathscr{L}(\mathscr{H})$.
(i) The Wick-covariant symbol of $\hat{A}$ is the function on M defined by $A_c(m) = \langle e_m, \hat{A}e_m\rangle$.
(ii) The contravariant symbol of $\hat{A}$ is the function on M, if it exists, such that

$$\hat{A}\psi = \int_M A^c(m)\Pi_m\psi d\mu(m), \quad \psi \in \mathscr{H}. \tag{2.111}$$

For the standard coherent states example, the covariant symbol is called Wick symbol and the contravariant symbol the anti-Wick symbol.

The covariant symbol satisfies the equality $A_c(m) = \mathrm{Tr}(\hat{A}\Pi_m)$.

Let us compute the anti-Wick symbol of some operator $\hat{A}$ with Weyl symbol A.

We know that the $\hbar$-Weyl symbol of the projector Π_z is the Gaussian $(\pi\hbar)^{-n}\mathrm{e}^{-\frac{|X-z|^2}{\hbar}}$. So we get that the Weyl symbol of $\hat{A}$ is the convolution of its anti-Wick symbol and a standard Gaussian function:

$$A(X) = (\pi\hbar)^{-n}\int_{\mathbb{R}^{2n}} A^c(X)\mathrm{e}^{-\frac{|X-z|^2}{\hbar}}dz. \tag{2.112}$$

This formula shows that if $\hat{A}$ has a bounded anti-Wick symbol ($A^c \in L^\infty(\mathbb{R}^{2n})$) then its Weyl symbol is an entire function in $\mathbb{C}^{2n}$, which is a restriction for a given operator to have an anti-Wick symbol.

Let us remark that the Wick symbol is an inverse formula associated with (2.112):

$$A_c(z) = 2^n\int_{\mathbb{R}^{2n}} A(X)\mathrm{e}^{-\frac{|X-z|^2}{\hbar}}dX. \tag{2.113}$$

Now we give another interpretation of the contravariant symbol. Let us first remark that we have

$$\mathscr{I}^*\cdot\mathscr{I} = \mathbb{1}_{\mathscr{H}}. \tag{2.114}$$

$$\mathscr{I}\cdot\mathscr{I}^* = \Pi_{\mathscr{H}}. \tag{2.115}$$

where $\Pi_{\mathscr{H}}$ is the orthogonal projector in $L^2(M)$ on $\mathscr{H}$ identified with $\mathscr{I}(\mathscr{H})$.

Proposition 25 *Let us assume that $\hat{A}$ has a contravariant symbol A^c such that $A^c \in L^\infty(M)$. Then we have*

$$\hat{A} = \mathscr{I}^*\cdot A^c\cdot\mathscr{I}, \tag{2.116}$$

where A^c is here the multiplication operator in $L^2(M)$.

Proof For every $\psi, \eta \in \mathscr{H}$ we have

$$\langle\eta, \hat{A}\psi\rangle = \int_M \langle\eta, e_m\rangle\langle e_m, \hat{A}\psi\rangle d\mu(m) \tag{2.117}$$

and

$$\begin{aligned}\langle e_m, \hat{A}\psi\rangle &= \textstyle\int_M A^c(m')\langle e_m, \Pi_{m'}\psi\rangle d\mu(m') \\ &= \textstyle\int_M A^c(m')\langle \Pi_{m'}e_m, \Pi_{m'}\psi\rangle d\mu(m').\end{aligned} \tag{2.118}$$

So we get

$$\langle\eta, \hat{A}\psi\rangle = \iint_{M\times M} A^c(m')\langle\Pi_{m'}e_m, \Pi_{m'}\psi\rangle\langle\eta, e_m\rangle d\mu(m')d\mu(m) \tag{2.119}$$

We get the conclusion using the equality

$$\langle \eta, e_{m'} \rangle = \int_M \langle e_{m'}, e_m \rangle \langle \eta, e_m \rangle d\mu(m) \tag{2.120}$$

□

Estimates on operators with covariant and contravariant symbols are easier to prove than for Weyl symbols. Moreover they can be used as a first step to get estimates in the setting of Weyl quantization as we shall see for positivity. The following proposition is easy to prove.

Proposition 26 *Let $\hat{A}$ be an operator in $\mathcal{H}$ with a contravariant symbol A^c. Suppose that $A^c \in L^\infty(M)$. Then $\hat{A}$ is bounded in $\mathcal{H}$ and we have*

$$\|A_c\|_\infty \leq \|\hat{A}\| \leq \|A^c\|_\infty. \tag{2.121}$$

Moreover $\hat{A}$ is self-adjoint if and only if A^c is real and $\hat{A}$ is non-negative if A^c is μ-almost everywhere non negative on M.

For our basic example $\mathcal{H} = L^2(\mathbb{R}^n)$, it is convenient to use the following notation. If A is a classical observable, $A \in S_0^0$, $\mathrm{Op}_\hbar^w(A)$ denotes the Weyl quantization of A and $\mathrm{Op}_\hbar^{\mathrm{aw}}(A)$ denotes the anti-Wick quantization of A. We shall also use the notation $A^{\mathrm{aw}} = \mathrm{Op}_\hbar^{\mathrm{aw}}(A)$.. In other words $\mathrm{Op}_\hbar^{\mathrm{aw}}(A)$ admits A as an anti-Wick symbol. The following proposition is an easy consequence of above results

Proposition 27 *Let $A \in S_0^0$ (more general symbols could be considered). Then we have*

$$\mathrm{Op}_\hbar^{\mathrm{aw}}(A) = \mathrm{Op}_\hbar^w(A * G), \quad \text{where } G(X) = (\pi\hbar)^{-n} \mathrm{e}^{-\frac{|X|^2}{\hbar}}. \tag{2.122}$$

$$\langle \psi, \mathrm{Op}_\hbar^{\mathrm{aw}}(A)\psi \rangle = (2\pi\hbar)^{-n} \int_{\mathbb{R}^{2n}} A(z)\mathcal{H}_\psi(z)dz. \tag{2.123}$$

where $\mathcal{H}_\psi(z)$ is the Husimi function of ψ.

We get now the following useful consequence for Weyl quantization.

Proposition 28 (semiclassical Gårding inequality) *Let $A \in S_0^0$, $A \geq 0$ on $\mathbb{R}^{2n}$. Then there exists $C \in \mathbb{R}$ such that for every $\hbar \in]0, 1]$ we have*

$$\langle \psi, \hat{A}\psi \rangle \geq C\hbar, \quad \forall \psi \in L^2(\mathbb{R}^n). \tag{2.124}$$

Proof We know that $\mathrm{Op}_\hbar^w(A * G)$ is a non negative bounded operator. So the proposition will be proved if

$$\|\mathrm{Op}_\hbar^w(A * G - A)\| = \mathcal{O}(\hbar). \tag{2.125}$$

Using a standard argument for smoothing with convolution, we get that $\hbar^{-1}(A * G - A) \in S_0^0$, with uniform estimates in $\hbar \in]0, 1]$. Hence we get (2.125) as a consequence of Calderon–Vaillancourt theorem. □

These results are useful to study the matrix elements $\langle\psi_\hbar, \hat{A}\psi_\hbar\rangle$, for a family $\{\psi_\hbar\}_\hbar$ in the semi-classical regime [140]. This subject is related with an efficient tool introduced by Lions–Paul [181] and Gérard [110] (see also [45]): the semi-classical measures. This is an application of anti-Wick quantization as we shall see now.

2.6.2 Application to Semi-classical Measures

Semi-classical measures were introduced to describe localization and oscillations of families of states $\{\psi_\hbar\}_\hbar$, $\|\psi_\hbar\| = 1$ (or at least bounded in $L^2(\mathbb{R}^n)$).

Let us first remark that

$$A \mapsto \langle\psi, \mathrm{Op}_\hbar^{\mathrm{aw}} A\psi\rangle$$

is a probability measure $\mu^\hbar$ in $\mathbb{R}^{2n}$. Moreover this probability measure has a density given by the Husimi function of $\psi_\hbar$,

$$d\mu^\hbar = (2\pi\hbar)^{-n}\mathcal{H}_{\psi_\hbar}(z)dz.$$

In particular we have

$$\boxed{|\langle\psi, \mathrm{Op}_\hbar^{\mathrm{aw}} A\psi\rangle| \le \|A\|_\infty}.$$

for every $A \in C_b(\mathbb{R}^{2n})$ (space of continuous, bounded functions on $\mathbb{R}^{2n}$).

Definition 12 A semi-classical measure for the family of normalized states $\{\psi_\hbar\}_\hbar$ is a probability measure μ on the phase space $\mathbb{R}^{2n}$ for which there exists at least one sequence $\{\hbar_k\}$, $\lim\limits_{k\to+\infty} \hbar_k = 0$ such that for every $A \in \mathrm{S}_0^0$, we have

$$\lim_{k\to+\infty}\langle\psi_{\hbar_k}\mathrm{Op}_{\hbar_k}{}^{aw}A\psi_{\hbar_k}\rangle = \int_{\mathbb{R}^{2n}} Ad\mu. \tag{2.126}$$

In other words, the measure sequence $\mu^{\hbar_k}$ weakly converges toward the measure μ.

Remark 12 Semi-classical measures can also be defined for states $\psi_\hbar \in L^2(\mathbb{R}^n, \mathcal{K})$ where $\mathcal{K}$ is an Hilbert space. By the way in this setting Weyl symbols and anti-Wick symbols are operators in $\mathcal{K}$.

We can also define semi-classical measures for statistical mixed states $\hat{\rho}$, where $\hat{\rho}$ is a non negative operator such that $\mathrm{Tr}\hat{\rho} = 1$.

For more applications and properties of these extensions see the huge litterature on this subject; for example see [178].

The following proposition is a straightforward application of properties of Husimi function

Proposition 29 *Let μ be a semi-classical measure for $\{\psi_\hbar\}_\hbar$. Then the support* $\mathrm{supp}(\mu)$ *of the measure μ is included in the frequency set* $\mathrm{FS}[\psi_\hbar]$, $\mathrm{supp}(\mu) \subseteq \mathrm{FS}[\psi_\hbar]$.

Example 2 (i) Let $\psi_\hbar = \varphi_z$, a standard coherent state. Then this family has one semi-classical measure, $\mu = \delta_z$ (Dirac probability)

(ii) Let us assume that the states family $\{\psi_\hbar\}_\hbar$ is tight in the following sense.

$$\lim_{R\to+\infty} \limsup_{\hbar\to 0} \left(\int_{|X|\geq R} |\langle \varphi_X, \psi^\hbar\rangle|^2 dX \right) = 0. \tag{2.127}$$

Then applying the Prokhorov compacity theorem [31] , there exists at least one semi-classical measure for $\psi^\hbar$. One of a challenging problem in quantum mechanics is to compute these semi-classical measures for family of bound states satisfying (2.108). If for some $\varepsilon > 0$, $H^{-1}[E-\varepsilon, E+\varepsilon]$ is a bounded set, this family is tight. For classically ergodic systems it is conjectured that there exists only one semi-classical measure which is the Liouville measure [140].

One important property of semi-classical measures is the following propagation result.

Let us consider the time dependent Schrödinger equation

$$i\hbar\partial_t\psi_\hbar(t) = \hat{H}\psi_\hbar(t), \quad \psi_\hbar(0) = \psi_\hbar, \tag{2.128}$$

where H is a time independent Hamiltonian. We assume that H is real, subquadratic and $\hbar$ independent (for simplicity).

$$\partial_X^\gamma H \in L^\infty(\mathbb{R}^{2n}), \text{ for all } \gamma \text{ such that } |\gamma| \geq 2. \tag{2.129}$$

Let μ be a semi-classical measure for $\{\psi_\hbar\}$.

Theorem 14 *For every $t \in \mathbb{R}$, $\{\psi_\hbar(t)\}$ has a semi-classical measure $d\mu_t$ for the same subsequence $\hbar_k$ given by the transport of $d\mu$ by the classical flow Φ^t:* $\mu(t) = (\Phi^t)^*\mu$.

Proof For every $A \in C_0{}^\infty(\mathbb{R}^{2n})$, the semiclassical Egorov theorem and comparaison between anti-Wick and Weyl quantization, give

$$\langle \psi_\hbar(t), \mathrm{Op}_\hbar^{aw}(A)\psi_\hbar(t)\rangle = \int_{\mathbb{R}^{2n}} A\cdot\Phi^t d\mu_{\psi_\hbar} + \mathscr{O}(\hbar). \tag{2.130}$$

Hence we get the result going to the limit for the sequence $\hbar_k$. □

We have the following consequence for the stationary Schrödinger equation.

Corollary 12 *Let μ be semi-classical measure for a family of bound states $\{\psi_\hbar\}$, satisfying $\hat{H}\psi_\hbar = E_\hbar\psi_\hbar$. Then μ is invariant by the classical flow Φ^t for every $t \in \mathbb{R}$.*

Proof $\psi_\hbar(t) = \mathrm{e}^{-\frac{it}{\hbar}E_\hbar}\psi_\hbar$ satisfies the time dependent Schrödinger equation so using the Theorem we get $(\Phi^t)^*\mu = \mu$. □

Now we illustrate Corollary 2.6.2 on Hermite bound states of the harmonic oscillator.

We assume $n = 1$. We can easily compute the Husimi function $\mathscr{H}_j$ of the Hermite function function ϕ_j.

$$\mathscr{H}_j(q, p) = |\langle\varphi_X, \phi_j\rangle|^2 = \frac{(q^2 + p^2)^j}{2^j j!}\mathrm{e}^{-\frac{1}{2\hbar}(q^2+p^2)}. \tag{2.131}$$

We want to study the quantum measures $d\mu_j = (2\pi\hbar)^{-1}\mathscr{H}_j(q, p)dqdp$ when the energies $E_j = (j + \frac{1}{2})\hbar$ have a limit $E > 0$. So we have $\hbar \to 0$ and $j \to +\infty$. For simplicity we fix $E > 0$ and choose $\hbar = \hbar_j = \frac{E}{j}$.

Let f be in the Schwartz class $\mathscr{S}(\mathbb{R}^2)$. We have to compute the limit of $\int f(X)d\mu_j(X)$ for $j \to +\infty$. Using polar coordinates and a change of variables we have to study the large k limit for the Laplace integral

$$I(j) := \frac{1}{(j+1)!}\int_0^\infty u^j\mathrm{e}^{-\frac{j}{E}u}f(\sqrt{2u}\cos\theta, \sqrt{2u}\sin\theta)du, \quad \theta \in [0, 2\pi[.$$

We can asume that f has a bounded support and $(0, 0)$ is not in the support of f.

Using the Laplace method we get

$$\lim_{j\to-\infty} I(j) = f(\sqrt{2E}(\cos\theta, \sin\theta)). \tag{2.132}$$

So, we have

$$\lim_{j\to+\infty}\int f(X)d\mu_j(X) = \frac{1}{2\pi\sqrt{2E}}\int_0^{2\pi} f(\sqrt{2E}(\cos\theta, \sin\theta))d\theta. \tag{2.133}$$

On the r.h.s of (2.133) we recognize the uniform probability measure on the circle of radius $\sqrt{2E}$. This measure is a semi-classical measure for the quantum harmonic oscillator. Let us remark that the classical oscillator of energy $\sqrt{2E}$ moves on the circle of radius $\sqrt{2E}$ in the phase space.

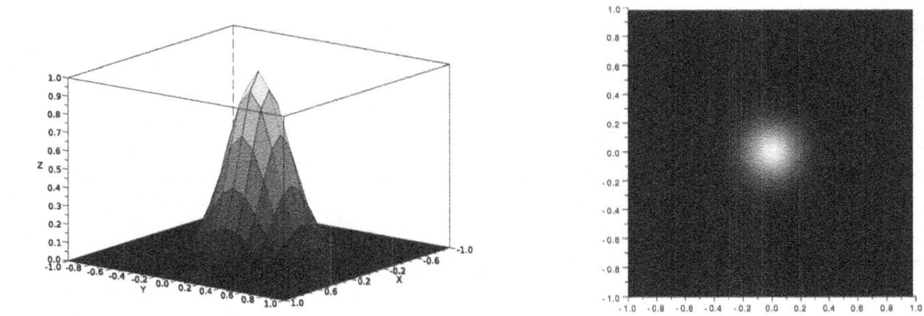

Wigner function of Schrödinger coherent states; on right: horizontal section

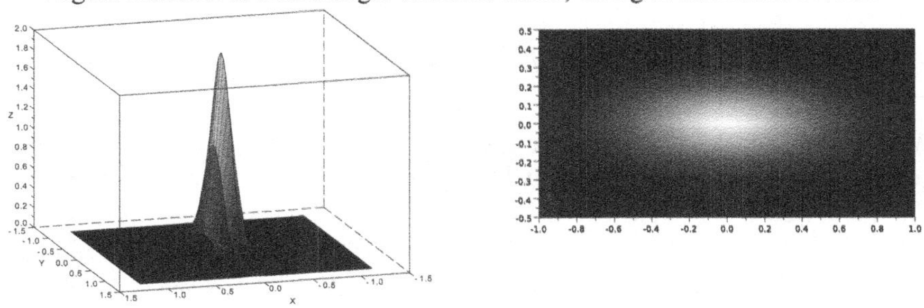

Wigner function of squeezed coherent states; on right: horizontal section

Chapter 3
Quadratic Hamiltonians

Abstract The aim of this chapter is to construct the quantum unitary propagator for Hamiltonians which are quadratic in position and momentum with time-dependent coefficients. We show that the quantum evolution is exactly solvable in terms of the classical flow which is linear. This allows to construct the metaplectic transformations which are unitary operators in $L^2(\mathbb{R}^n)$ corresponding to symplectic transformations. Simple examples of such metaplectic transformations are the Fourier transform, which corresponds to the symplectic matrix J defined in (3.4) and the propagator of the harmonic oscillator, corresponding to rotations in the phase space. The main results of this chapter are computations of the quantum evolution operators for quadratic Hamiltonians acting on coherent states. We show that the time evolved coherent states are still Gaussian states which are recognized to be squeezed states centered at the classical phase space point (see Chap. 8). From these computations we can deduce most of properties concerning quantum quadratic Hamiltonians. In particular we get the explicit form of the Weyl symbols of the metaplectic transformations. These formulae are generalizations of the Mehler formula for the harmonic oscillator. Quadratic Hamiltonians are very important in quantum mechanics because more general Hamiltonians can be considered as non trivial perturbations of time dependent quadratic ones as we shall see in Chap. 5.

3.1 The Propagator of Quadratic Quantum Hamiltonians

A classical quadratic Hamiltonian H is a quadratic form defined in the phase space $\mathbb{R}^{2n}$. We assume that this quadratic form is time dependent, so we have

$$H(t, z) = \sum_{1 \le j,k \le n} c_{j,k}(t) z_j z_k,$$

where $z = (q, p) \in \mathbb{R}^{2n}$ is the phase space variable and the real coefficients $c_{j,k}(t)$ are continuous functions of time $t \in \mathbb{R}$. It can be rewritten as

D. Robert and M. Combescure, *Coherent States and Applications in Mathematical Physics*, Theoretical and Mathematical Physics,
https://doi.org/10.1007/978-3-030-70845-0_3

$$H(t,q,p) = \frac{1}{2}(q,p)S(t)\begin{pmatrix} q \\ p \end{pmatrix}, \tag{3.1}$$

where $S(t)$ is a $2n \times 2n$ real symmetric matrix of the block form

$$S(t) = \begin{pmatrix} G_t & L_t^T \\ L_t & K_t \end{pmatrix} \tag{3.2}$$

G_t, L_t, K_t are $n \times n$ real matrices with G_t, K_t being symmetric, and L_t^T denotes the transpose of L_t. The classical equations of motion for this Hamiltonian are linear and can be written as

$$\begin{pmatrix} \dot{q} \\ \dot{p} \end{pmatrix} = JS(t)\begin{pmatrix} q \\ p \end{pmatrix}, \tag{3.3}$$

where J is the symplectic matrix

$$J = \begin{pmatrix} 0 & \mathbb{1}_n \\ -\mathbb{1}_n & 0 \end{pmatrix}. \tag{3.4}$$

Ler $F(t)$ be the classical flow for the Hamiltonian $H(t)$. It means that it is a symplectic $2n \times 2n$ matrix obeying

$$\dot{F}(t) = JS(t)F(t). \tag{3.5}$$

with $F(0) = \mathbb{1}$. Then the solution of (3.3) with $q(0) = q, p(0) = p$ is simply

$$\begin{pmatrix} q(t) \\ p(t) \end{pmatrix} = F(t)\begin{pmatrix} q \\ p \end{pmatrix}$$

We now consider the quantum Hamiltonian

$$\widehat{H}(t) = (\hat{Q},\hat{P}) \cdot S(t)\begin{pmatrix} \hat{Q} \\ \hat{P} \end{pmatrix}. \tag{3.6}$$

The quantum evolution operator $\mathcal{U}(t)$[1] is solution of the Schrödinger equation

$$i\hbar\frac{d}{dt}\mathcal{U}(t) = \widehat{H}(t)\mathcal{U}(t), \tag{3.7}$$

with $\mathcal{U}(0) = \mathbb{1}$. The following result was already proved in Chap. 2 as a particular case of a more general result. We shall give here a simple direct proof.

[1] We shall explain later why this quantum propagator is a well defined unitary operator in $L^2(\mathbb{R}^n)$.

Theorem 15 *One has for all times t*

$$\mathcal{U}(t)^* \begin{pmatrix} \hat{Q} \\ \hat{P} \end{pmatrix} \mathcal{U}(t) = F(t) \begin{pmatrix} \hat{Q} \\ \hat{P} \end{pmatrix}. \tag{3.8}$$

Proof Define $\widehat{Q}_t = \mathcal{U}(t)^* \hat{Q} \mathcal{U}(t)$, and similarly for $\hat{P}$ (Heisenberg observables). Using the Schrödinger equation one has

$$-i\hbar \frac{d}{dt} \begin{pmatrix} \widehat{Q}_t \\ \widehat{P}_t \end{pmatrix} = \mathcal{U}(t)^* \left[\widehat{H}(t), \begin{pmatrix} \hat{Q} \\ \hat{P} \end{pmatrix} \right] \mathcal{U}(t).$$

But

$$\left[\widehat{H}(t), \begin{pmatrix} \hat{Q} \\ \hat{P} \end{pmatrix} \right] = -i\hbar JS(t) \begin{pmatrix} \hat{Q} \\ \hat{P} \end{pmatrix}.$$

This means that $\widehat{Q}_t, \widehat{P}_t$ must satisfy the linear equation

$$\frac{d}{dt} \begin{pmatrix} \widehat{Q}_t \\ \widehat{P}_t \end{pmatrix} = JS(t) \begin{pmatrix} \widehat{Q}_t \\ \widehat{P}_t \end{pmatrix}$$

which is trivially solved by (3.8). □

3.2 The Propagation of Coherent States

In this section we give the explicit form of the time evolved coherent states in terms of the classical flow $F(t)$ given by the $2n \times 2n$ block matrix form:

$$F(t) = \begin{pmatrix} A_t & B_t \\ C_t & D_t \end{pmatrix}. \tag{3.9}$$

It will be shown that the complex $n \times n$ matrix $A_t + iB_t$ is always non singular. Then we establish the following result:

$$\boxed{\mathcal{U}_t \varphi_z = (\pi\hbar)^{-n/4} \widehat{T}(z_t) (\det(A_t + iB_t))^{-1/2} \exp\left(\frac{i}{2\hbar} (C_t + iD_t)(A_t + iB_t)^{-1} x \cdot x \right)} \tag{3.10}$$

where $z_t = F(t)z$ is the phase space point of the classical trajectory and $\widehat{T}(z)$ is the Weyl–Heisenberg translation operator by the vector $z = (x, \xi) \in \mathbb{R}^{2n}$:

$$\widehat{T}(z) = \exp\left[\frac{i}{\hbar} (\xi \cdot \hat{Q} - x \cdot \hat{P}) \right]. \tag{3.11}$$

This means that $\mathcal{U}_t\varphi_z$ is a squeezed state centered at the phase space point z_t, so the squeezed state moves on the classical trajectory.

We take $\hbar = 1$ for simplicity.

A simple example is the harmonic oscillator:

$$\hat{H}_{\text{os}} = -\frac{1}{2}\frac{d^2}{dx^2} + \frac{1}{2}x^2. \tag{3.12}$$

It is well known that for $t \neq k\pi,\ k \in \mathbb{Z}$ the quantum propagator $\mathrm{e}^{-it\hat{H}_{\text{os}}}$ has an explicit Schwartz kernel $K_{\text{os}}(t; x, y)$ (Mehler formula, Chap. 1).

It is easier to compute directly with the coherent states φ_z. φ_0 is the ground state of $\hat{H}_{\text{os}}$, so we have

$$\mathrm{e}^{-it\hat{H}_{\text{os}}}\varphi_0 = \mathrm{e}^{-it/2}\varphi_0. \tag{3.13}$$

Let us compute $\mathrm{e}^{-it\hat{H}_{\text{os}}}\varphi_z$, $\forall z \in \mathbb{R}^2$, with the following ansatz

$$\mathrm{e}^{-it\hat{H}_{\text{os}}}\varphi_z = \mathrm{e}^{i\delta_t(z)}\hat{T}(z_t)e^{-it/2}\varphi_0, \tag{3.14}$$

where $z_t = (q_t, p_t)$ is the generic point on the classical trajectory (a circle here), coming from z at time $t = 0$. Let $\psi_{t,z}$ be the state equal to the r.h.s in (3.14), and let us compute $\delta_t(z)$ such that $\psi_{t,z}$ satisfies the equation $i\frac{d}{dt}\varphi = \hat{H}\varphi \quad \varphi|_{t=0} = \psi_{0,z}$. We have

$$\hat{T}(z_t)u(x) = e^{i(p_t x - q_t p_t/2)}u(x - q_t)$$

and

$$\psi_{t,z}(x) = \mathrm{e}^{i(\delta_t(z) - t/2 + p_t x - q_t p_t/2)}\varphi_0(x - q_t). \tag{3.15}$$

So, after some computations left to the reader, using properties of the classical trajectories

$$\dot{q}_t = p_t \quad \dot{p}_t = -q_t, \quad p_t^2 + q_t^2 = p^2 + q^2,$$

the equation

$$i\frac{d}{dt}\psi_{t,z}(x) = \frac{1}{2}(D_x^2 + x^2)\psi_{t,z}(x) \tag{3.16}$$

is satisfied if and only if

$$\delta_t(z) = \frac{1}{2}(p_t q_t - pq). \tag{3.17}$$

□

Let us now introduce the following general notations for later use.

F_t is the classical flow with initial time $t_0 = 0$ and final time t. It is represented as a $2n \times 2n$ matrix which can be written as four $n \times n$ blocks :

$$F_t = \begin{pmatrix} A_t & B_t \\ C_t & D_t \end{pmatrix}. \tag{3.18}$$

Let us introduce the following squeezed states: φ^{Γ} defined as follows.

$$\varphi^{\Gamma}(x) = a_{\Gamma} \exp\left(\frac{i}{2\hbar}\Gamma x \cdot x\right) \tag{3.19}$$

where $\Gamma \in \mathfrak{S}^+(n)$. Recall that $\mathfrak{S}^+(n)$ is the Siegel space of complex, symmetric matrices Γ such that $\Im(\Gamma)$ is positive and non degenerate and $a_{\Gamma} \in \mathbb{C}$ is such that the L^2-norm of φ^{Γ} is one.

We also denote $\varphi_z^{\Gamma} = \widehat{T}(z)\varphi^{\Gamma}$.

For $\Gamma = i\mathbb{1}$, we denote $\varphi = \varphi^{i\mathbb{1}}$.

Theorem 16 *We have the following formulae, for every $x \in \mathbb{R}^n$ and $z \in \mathbb{R}^{2n}$,*

$$\mathcal{U}_t\varphi^{\Gamma}(x) = \varphi^{\Gamma_t}(x), \tag{3.20}$$
$$\mathcal{U}_t\varphi_z^{\Gamma}(x) = \widehat{T}(F_t z)\varphi^{\Gamma_t}(x), \tag{3.21}$$

where $\Gamma_t = (C_t + D_t\Gamma)(A_t + B_t\Gamma)^{-1}$ and $a_{\Gamma_t} = a_{\Gamma}\,(\det(A_t + B_t\Gamma))^{-1/2}$.

Beginning of the proof
The first formula can be proven by the ansatz

$$\mathcal{U}_t\varphi_0(x) = a(t)\exp\left(\frac{i}{2\hbar}\Gamma_t x \cdot x\right).$$

where $\Gamma_t \in \mathfrak{S}^+(n)$ and $a(t)$ is a complex values time dependent function. We get first a Riccati equation to compute Γ_t and a linear equation to compute $a(t)$.

The second formula is easy to prove from the first, using the Weyl translation operators and the following known property

$$\mathcal{U}_t\widehat{T}(z)\mathcal{U}_t^* = \widehat{T}(F_t z).$$

Let us now give the details of the proof for $z = 0$.

We begin by computing the action of a quadratic Hamiltonian on a Gaussian ($\hbar = 1$).

Lemma 18

$$Lx \cdot D_x e^{\frac{i}{2}\Gamma x \cdot x} = \left(L^T x \cdot \Gamma x - \frac{i}{2}\mathrm{Tr}L\right) e^{\frac{i}{2}\Gamma x \cdot x}.$$

Proof This is a straightforward computation, using

$$Lx \cdot D_x = \frac{1}{i}\sum_{1 \le j,k \le n} L_{jk}\frac{x_j D_k + D_k x_j}{2}$$

and, for $\omega \in \mathbb{R}^n$,

$$(\omega.D_x)\mathrm{e}^{\frac{i}{2}\Gamma x\cdot x} = (\Gamma x\cdot\omega)\mathrm{e}^{\frac{i}{2}\Gamma x\cdot x}.$$

□

Lemma 19

$$(GD_x\cdot D_x)\mathrm{e}^{\frac{i}{2}\Gamma x\cdot x} = (G\Gamma x\cdot\Gamma x - i\mathrm{Tr}(G\Gamma))\,\mathrm{e}^{\frac{i}{2}\Gamma x\cdot x}.$$

Proof As above, we get

$$\widehat{H}\mathrm{e}^{\frac{i}{2}\Gamma x\cdot x} = \left(\frac{1}{2}Kx\cdot x + x\cdot L\Gamma x + \frac{1}{2}G\Gamma x\cdot\Gamma x - \frac{i}{2}\mathrm{Tr}(L+G\Gamma)\right)\mathrm{e}^{\frac{i}{2}\Gamma x\cdot x} \quad (3.22)$$

□

We are now ready to solve the equation

$$i\frac{\partial}{\partial t}\psi = \hat{H}\psi, \quad (3.23)$$

with

$$\psi|_{t=0}(x) = g(x) := \pi^{-n/4}e^{-x^2/2}.$$

For simplicity we assume here that $\Gamma = i\mathbb{1}$, the proof can be easily generalized to $\Gamma \in \mathfrak{S}_n^+$.

We try the ansatz

$$\psi(t,x) = a(t)e^{\frac{i}{2}\Gamma_t x\cdot x}, \quad (3.24)$$

which gives the equations

$$\dot{\Gamma}_t = -K - 2\Gamma_t^T L - \Gamma_t G\Gamma_t, \quad (3.25)$$

$$\dot{a}(t) = -\frac{1}{2}\left(Tr(L+G\Gamma_t)\right)a(t) \quad (3.26)$$

with the initial conditions

$$\Gamma_0 = i\mathbb{1},\quad a(0) = (\pi)^{-n/4}.$$

We note that $\Gamma^T L$ et $L\Gamma$ determine the same quadratic forms. So the first equation is a Ricatti equation and can be written as

$$\dot{\Gamma}_t = -K - \Gamma_t L^T - L\Gamma_t - \Gamma_t G\Gamma_t, \quad (3.27)$$

where L^T denotes the transposed matrix for L. We shall now see that Eq. (3.27) can be solved using Hamilton equation

$$\dot{F}_t = J \begin{pmatrix} K & L \\ L^T & G \end{pmatrix} F_t \tag{3.28}$$

$$F_0 = \mathbb{1} \tag{3.29}$$

We know that

$$F_t = \begin{pmatrix} A_t & B_t \\ C_t & D_t . \end{pmatrix}$$

is a symplectic matrix $\forall t$. So using the next lemma, we have $\det(A_t + iB_t) \neq 0 \quad \forall t$. Let us denote

$$M_t = A_t + iB_t, \quad N_t = C_t + iD_t. \tag{3.30}$$

We shall prove that $\Gamma_t = N_t M_t^{-1}$. By an easy computation, we get

$$\begin{aligned} \dot{M}_t &= L^T M_t + G N_t \\ \dot{N}_t &= -K M_t - L N_t. \end{aligned} \tag{3.31}$$

Now, compute

$$\begin{aligned} \frac{d}{dt}(N_t M_t^{-1}) &= \dot{N} M^{-1} - N M^{-1} \dot{M} M^{-1} \\ &= -K - LNM^{-1} - NM^{-1}(L^T M + GN)M^{-1} \\ &= -K - LNM^{-1} - NM^{-1}L^T - NM^{-1}GNM^{-1}, \end{aligned} \tag{3.32}$$

which is exactly Eq. (3.27).

Now we compute $a(t)$, using the following equality,

$$\mathrm{Tr}\left(L^T + G(C + iD)(A + iB)^{-1}\right) = \mathrm{Tr}(\dot{M})M^{-1} = \mathrm{Tr}\,(L + G\Gamma_t)$$

using $\mathrm{Tr}L = \mathrm{Tr}L^T$. Let us recall the Liouville formula

$$\frac{d}{dt} \log(\det M_t) = \mathrm{Tr}(\dot{M}_t M_t^{-1}) \tag{3.33}$$

which give directly

$$a(t) = (\pi)^{-n/4} \left(\det(A_t + iB_t)\right)^{-1/2}. \tag{3.34}$$

To complete the proof, we need to prove the following

Lemma 20 *Let F be a symplectic matrix.*

$$F = \begin{pmatrix} A & B \\ C & D \end{pmatrix}$$

Then $\det(A + iB) \neq 0$ *and* $\Im(C + iD)(A + iB)^{-1}$ *is positive definite.*

We shall prove a more general result concerning the Siegel space $\mathfrak{S}^+(n)$.

Lemma 21 *If*

$$F = \begin{pmatrix} A & B \\ C & D \end{pmatrix}.$$

is a symplectic matrix and $Z \in \mathfrak{S}_n^+$ *then* $A+BZ$ *et* $C+DZ$ *are non singular and* $(C + DZ)(A + BZ)^{-1} \in \mathfrak{S}^+(n)$.

Proof Let us denote $M := A + BZ, \quad N := C + DZ$. F is symplectic, so we have $F^T JF = J$. Using

$$\begin{pmatrix} M \\ N \end{pmatrix} = F \begin{pmatrix} I \\ Z \end{pmatrix}.$$

we get

$$(M^T, N^T)J \begin{pmatrix} M \\ N \end{pmatrix} = (I, Z)J \begin{pmatrix} I \\ Z \end{pmatrix} = 0. \tag{3.35}$$

Which gives

$$M^T N = N^T M.$$

In the same way, we have

$$\frac{1}{2i}(M^T, N^T)J \begin{pmatrix} \bar{M} \\ \bar{N} \end{pmatrix} = \frac{1}{2i}(I, Z)F^T JF \begin{pmatrix} I \\ \bar{Z} \end{pmatrix} \tag{3.36}$$

$$= \frac{1}{2i}(I, Z)J \begin{pmatrix} I \\ \bar{Z} \end{pmatrix} = \frac{1}{2i}(\bar{Z} - Z) = -\Im Z$$

We get the following equation

$$N^T \bar{M} - M^T \bar{N} = 2i\Im Z \tag{3.37}$$

Because $\Im Z$ is non degenerate, from (3.37), we get that M and N are injective. If $x \in \mathbb{C}^n, \ Ex = 0$, we have

$$\bar{M}\bar{x} = x^T M^T = 0$$

hence

$$x^T \Im Z \bar{x} = 0$$

then $x = 0$.

So, we can define,

$$\alpha(F)(Z) = (C + DZ)(A + BZ)^{-1} \tag{3.38}$$

Let us prove that $\alpha(F)Z \in \mathfrak{S}^+(n)$. We have:

$$\alpha(F)Z = NM^{-1} \Rightarrow (\alpha(F)Z)^T = (M^{-1})^T N^T = (M^{-1})^T M^T N M^{-1}$$
$$= NM^{-1} = \alpha(F)Z.$$

We have also:

$$M^T \frac{NM^{-1} - \bar{N}\bar{M}^{-1}}{2i} \bar{M} = \frac{N^T \bar{M} - M^T \bar{N}}{2i} = \Im Z.$$

and this proves that $\Im(\alpha(F)(Z))$ is positive and non degenerate.

This finishes the proof of the Theorem for $z = 0$. □

The map $F \mapsto \alpha(F)$ defines a representation of the symplectic group $\mathrm{Sp}(n)$ in the Siegel space $\mathfrak{S}^+(n)$. For later use it is useful to introduce the determinant: $\delta(F, Z) = \det(A + BZ), F \in \mathrm{Sp}(n), Z \in \mathfrak{S}^+(n)$. The following results are easy algebraic computations.

Proposition 30 *We have, for every $F_1, F_2 \in \mathrm{Sp}(n)$,*
(i) $\alpha(F_1F_2) = \alpha(F_1)\alpha(F_2)$.
(ii) $\delta(F_1F_2, Z) = \delta(F_1, \alpha(F_2)(Z))\delta(F_2, Z)$.
(iii) For every $Z_1, Z_2 \in \mathfrak{S}^+(n)$ there exists $F \in \mathrm{Sp}(n)$ such that $\alpha(F)(Z_1) = Z_2$. In other words the representation α is transitive in $\mathfrak{S}^+(n)$.

Many other properties of the representation α are studied in [104, 183].

For completeness, we state the following

Corollary 13 *The propagator $\mathcal{U}_t$ is well defined and it is a unitary operator in $L^2(\mathbb{R}^n)$.*

Proof For every coherent state φ_z, $\mathcal{U}_t\varphi_z$ is solution of the Schrödinger equation. As we have seen in Chap. 1, the family $\{\varphi_z\}_{z\in\mathbb{R}^{2n}}$ is overcomplete in $L^2(\mathbb{R}^n)$. So formula (3.20) wholly determines the unitary group $\mathcal{U}_t$. In a preliminary step we can see that $\mathcal{U}_t\psi$ is well defined for $\psi \in \mathcal{S}(\mathbb{R}^n)$ using inverse Fourier–Bargmann transform, that $\mathcal{U}_t\psi \in \mathcal{S}(\mathbb{R}^n)$ and $\|\mathcal{U}_t\psi\| = \|\psi\|$. So we can extend $\mathcal{U}_t$ in $L^2(\mathbb{R}^n)$.

In particular it results that $\widehat{U}_t$ is a unitary operator and that $\hat{H}_t$ has a unique self-adjoint extension in $L^2(\mathbb{R}^n)$. □

It will be useful to compute the Fourier–Bargmann transform of $\mathcal{U}_t\varphi_z$.

Recall that $\mathcal{U}(t) = \hat{R}(F_t)$ where $\hat{R}(F)$ is the metaplectic operator corresponding to the symplectic $2n \times 2n$ matrix F and that F_t has a four blocks decomposition

$$F_t = \begin{pmatrix} A_t & B_t \\ C_t & D_t \end{pmatrix}$$

Now we define $n \times n$ complex matrices Y_t, Z_t as follows:

$$Y_t = A_t + iB_t - i(C_t + iD_t), \quad Z_t = A_t + iB_t + i(C_t + iD_t)$$

One has the following property, using the symplectic relations satisfied by the blocks A_t, B_t, C_t, D_t of F_t:

Lemma 22 *We have*

$$Z^*Z = Y^*Y - 4\mathbb{1}.$$

Moreover Y_t is invertible.

One can define the matrix W_t as follows:

$$W_t = Z_t Y_t^{-1}.$$

which satisfies the following property:

Lemma 23 *(i)*

$$W_0 = 0.$$

(ii)

$$W_t^* W_t < \mathbb{1}.$$

(iii)

$$(\Gamma + i\mathbb{1})^{-1} = \frac{1}{2i}(\mathbb{1} + W).$$

In particular W is a symmetric matrix.

Proof (i) is an easy consequence of the fact that $F_0 = \mathbb{1}$, hence $Y_0 = 2\mathbb{1}, Z_0 = 0$.
(ii) We have

$$W^*W = (Y^*)^{-1}Z^*ZY^{-1} = (Y^*)^{-1}(Y^*Y - 4\mathbb{1})Y^{-1} = \mathbb{1} - 4(YY^*)^{-1}.$$

(iii) Is a simple algebraic computation. □

Theorem 17 *The matrix elements $\langle \varphi_z, \mathcal{U}_t \varphi_{z'} \rangle$ are given by the following formula*

$$\begin{aligned}\langle \varphi_z, \mathcal{U}_t \varphi_{z'} \rangle &= 2^{n/2} \det{}^{-1/2}(A_t + D_t + i(B_t - C_t)) \mathrm{e}^{-i/2\sigma(F_t z', z)} \\ &\times \mathrm{e}^{-(x^2+\xi^2)/4} \mathrm{e}^{-W(\xi+ix)\cdot(\xi+ix)/4}\end{aligned} \tag{3.39}$$

where $z - F_t z' = (x, \xi)$ and

$$W_t = (A_t - D_t + i(B_t + C_t))(A_t + D_t + i(B_t - C_t))^{-1}.$$

Proof For simplicity we forget the time index t everywhere.
It is enough to assume that $z' = 0$. From the metaplectic invariance, we get

$$\begin{aligned}\langle \varphi_z, \mathcal{U}_t \varphi_{z'} \rangle &= \langle \varphi_z, \hat{T}(Fz')\varphi^{(\Gamma)} \rangle \\ &= \mathrm{e}^{-i\sigma(Fz',z)/2} \langle \varphi_{z-Fz'}, \varphi^{(\Gamma)} \rangle.\end{aligned} \tag{3.40}$$

So we have to compute $\langle \varphi_X, \varphi^{(\Gamma)} \rangle$.

We have

$$\langle \varphi_X, \varphi^{(\Gamma)} \rangle = \pi^{-n/2} \det{}^{-1/2}(A + iB) e^{\frac{1}{2}(ip\cdot q - q^2)} \int_{\mathbb{R}^n} e^{\frac{i}{2}(\Gamma + i)x\cdot x} e^{-ix\cdot(p+iq)} dx. \tag{3.41}$$

So the result follows from computation of the Fourier transform of a generalized Gaussian (or squeezed state). □

3.3 The Metaplectic Transformations

Recall that a metaplectic transformation associated with a linear symplectic transformation $F \in \mathrm{Sp}(n)$ in $\mathbb{R}^{2n}$, is a unitary operator $\widehat{R}(F)$ in $L^2(\mathbb{R}^n)$ satisfying one of the following equivalent conditions

$$\widehat{R}(F)^* \widehat{A} \widehat{R}(F) = \widehat{A \circ F}, \quad \forall A \in \mathscr{S}(\mathbb{R}^{2n}). \tag{3.42}$$

$$\widehat{R}(F)^* \widehat{T}(X) \widehat{R}(F) = \widehat{T}[F^{-1}(X)], \quad \forall X \in \mathbb{R}^{2n}. \tag{3.43}$$

$$\widehat{R}(F)^* \widehat{A} \widehat{R}(F) = \widehat{A \circ F},$$
$$\text{for } A(q,p) = q_j,\ 1 \le j \le n \text{ and} A(q,p) = p_k,\ 1 \le k \le n. \tag{3.44}$$

$\widehat{A}$ is the Weyl quantization of the classical symbol $A(q,p)$ and we recall that the operator $\widehat{T}(X)$ is defined by

$$\widehat{T}(X) = \exp\left(\frac{i}{\hbar}(\xi \cdot \hat{Q} - x \cdot \hat{P})\right) \tag{3.45}$$

when $X = (x, \xi) \in \mathbb{R}^{2n}$.

We shall prove below that for every $F \in \mathrm{Sp}(2n)$ there exists a metaplectic transformation $\widehat{R}(F)$. This transformation is unique up to a multiplication by a complex number of modulus 1.

Lemma 24 *If $\widehat{R_1}(F)$ and $\widehat{R_2}(F)$ are two metaplectic operators associated to the same symplectic map F then there exists $\lambda \in \mathbb{C}$, $|\lambda| = 1$, such that $\widehat{R_1}(F) = \lambda \widehat{R_2}(F)$.*

Proof Denote $\hat{R} = \widehat{R_1}(F)\widehat{R_2}(F)^{-1}$. Then we have $\hat{R}^* \hat{T}(X) \hat{R} = \hat{T}(X)$ for all $X \in \mathbb{R}^{2n}$. Applying Schur lemma (see Chap. 1) we get that $\hat{R} = \lambda \mathbb{1}$, $\lambda \in \mathbb{C}$. But $\hat{R}$ is unitary so that $|\lambda| = 1$. □

We shall prove here that $F \to \widehat{R}(F)$ defines a projective representation of the real symplectic group $\mathrm{Sp}(n)$ with sign indetermination only. More precisely, let us denote by $\mathrm{Mp}(n)$ the group of metaplectic transformations and π_p the natural projection: $\mathrm{M_p} \to \mathrm{Sp}(2n)$ then the metaplectic representation is a group homomorphism $F \mapsto$

$\hat{R}(F)$, from Sp(n) onto Mp(n)/$\{\mathbb{1}, -\mathbb{1}\}$, such that $\pi_p[\widehat{R}(F)] = F$, $\forall F \in \mathrm{Sp}(2n)$ For more details about the metaplectic transformations see [176].

Proposition 31 *For every $F \in \mathrm{Sp}(n)$ we can find a C^1—smooth curve F_t, $t \in [0, 1]$, in* Sp(n)*, such that $F_0 = \mathbb{1}$ and $F_1 = F$.*

Proof An explicit way to do that is to use the polar decomposition of F, $F = V|F|$ where V is a symplectic orthogonal matrix and $|F| = \sqrt{F^T F}$ is positive symplectic matrix. Each of these matrices have a logarithm, so $F = \mathrm{e}^K \mathrm{e}^L$ with K, L Hamiltonian matrices, and we can choose $F_t = \mathrm{e}^{tK}\mathrm{e}^{tL}$. □

Let F_t be as in Proposition 31. F_t is the linear flow defined by the quadratic Hamiltonian $H_t(z) = \frac{1}{2} S_t z \cdot z$ where $S_t = -J\dot{F}_t F_t^{-1}$. So using above results, **we define** $\widehat{R}(F) = \widehat{U_1}$. Here $\mathscr{U}_t$ is the solution of the Schrödinger equation

$$i\hbar \frac{d}{dt}\mathscr{U}_t = \widehat{H}(t)\mathscr{U}_t \tag{3.46}$$

that obeys $\mathscr{U}_0 = \mathbb{1}$. Namely it is the quantum propagator of the quadratic Hamiltonian $\widehat{H}(t)$. That the metaplectic operator so defined satisfies the required properties follows from Theorem 15.

Proposition 32 *Let us consider two symplectic paths F_t and F'_t joining $\mathbb{1}$ $(t = 0)$ to F $(t = 1)$. Then we have $\widehat{U_1} = \pm\widehat{U'_1}$ (with obvious notations).*

Moreover, if $F^1, F^2 \in \mathrm{Sp}(2n)$ then we have

$$\boxed{\hat{R}(F^1)\hat{R}(F^2) = \pm\widehat{R}(F^1F^2)}. \tag{3.47}$$

Proof We first remark that the propagator of a quadratic Hamiltonian is determined by its action on squeezed states φ^Γ and its classical flow. So using (3.20) we see that the phase shift between the two paths comes from variation of argument between 0 and 1 of the complex numbers $b(t) = \det(A_t + iB_t)$ and $b'(t) = \det(A'_t + iB'_t)$.

We have $\arg[b(t)] = \Im\left(\int_0^t \frac{\dot{b}(s)}{b(s)} ds\right)$ and by a complex analysis argument, we have

$$\Im\left(\int_0^1 \frac{\dot{b}(s)}{b(s)} ds\right) = \Im\left(\int_0^1 \frac{\dot{b}'(s)}{b'(s)} ds\right) + 2\pi N$$

with $N \in \mathbb{Z}$. So we get

$$b(1)^{-1/2} = \mathrm{e}^{iN\pi} b'(1)^{-1/2}.$$

The second part of the proposition is an easy consequence of Theorem 16 concerning propagation of squeezed coherent states and Proposition 30. More precisely, the sign indetermination in (3.47) is a consequence of variations for the phase of $\det(A + iB)$ concerning $F = F^1$ and $F = F^2$. To compare with F^1F^2 we apply Proposition 30. □

Remark 13 A geometrical consequence of Proposition 30 is the following. The map $F \mapsto \hat{R}(F)$ induces a group isomorphism between the symplectic group $\mathrm{Sp}(n)$ and the quotient of the metaplectic group $\mathrm{Mp}(n)/\{-\mathbb{1}, 1\}$. In other words the group $\mathrm{Mp}(n)$ is a two-cover of $\mathrm{Sp}(n)$.

An interesting property of the metaplectic representation is the following.

Proposition 33 *The metaplectic representation $\hat{R}$ has two irreductible non equivalent components in $L^2(\mathbb{R}^n)$. These components are the subspaces $L^2_{\mathrm{od}}(\mathbb{R}^n)$ of odd states and $L^2_{\mathrm{ev}}(\mathbb{R}^n)$ of even states.*

Proof Let us first remark that the subspaces $L^2_{\mathrm{od,\,ev}}(\mathbb{R}^n)$ are invariant for $\hat{R}(F)$ because quadratic Hamiltonians commute with the parity operator $\Pi\psi(x) = \psi(-x)$.

Now we have to prove that $L^2_{\mathrm{od,\,ev}}(\mathbb{R}^n)$ are irreducible for $\hat{R}$.

Let us begin by considering the subspace $L^2_{\mathrm{ev}}(\mathbb{R}^n)$. Let $\hat{B}$ be a bounded operator in $L^2_{\mathrm{ev}}(\mathbb{R}^n)$ such that $\hat{R}(F)\hat{B} = \hat{B}\hat{R}(F)$ for every $F \in \mathrm{Sp}(n)$. According to Schur lemma, we have to prove that $\hat{B} = \lambda\mathbb{1}$, $\lambda \in \mathbb{C}$.

In particular $\hat{B}$ commutes with the propagator of the harmonic oscillator $U_t = \mathrm{e}^{it\hat{H}_{\mathrm{os}}}$. We can suppose that $\hat{B}$ is Hermitian. So $\hat{B}$ is diagonal in the Hermite basis ϕ_α (see Chap. 1). We have $\hat{B}\phi_\alpha = \lambda_\alpha\phi_\alpha$ for every $\alpha \in \mathbb{N}^n$.

To conclude we have to prove that $\lambda_\alpha = \lambda_\beta$ if $|\alpha|$ and $|\beta|$ are even.

Assume for simplicity that $n = 1$ (the proof is also valid for $n \geq 2$).

Let us consider the metaplectic transformation $\hat{R}_t = \mathrm{e}^{itx^2}$ associated to the symplectic transform $F_t = \exp\begin{pmatrix} 0 & t \\ 0 & 0 \end{pmatrix}$.

We have

$$\hat{R}_t\phi_k = \sum_{j\geq 0} c(t,k,j)\phi_j.$$

Using that $\hat{R}_t\hat{B} = \hat{B}\hat{R}_t$ we get

$$c(t,k,j)\lambda_j = c(t,k,j)\lambda_k. \tag{3.48}$$

Now we shall prove that if $k - j$ is even then $c(t,k,j) \neq 0$ for some t hence $\lambda_j = \lambda_k$.

This is a consequence of the following

$$c(t,k,j) = \int \mathrm{e}^{itx^2}\phi_k(x)\phi_j(x)dx.$$

If $c(t,k,j) = 0$ for every t then $\int_{\mathbb{R}} x^{2m}\phi_k(x)\phi_j(x)dx = 0$ for every $m \in \mathbb{N}$. But this is not possible if $k - j$ is even (see properties of Hermite functions).

So we have proved that $L^2_{\mathrm{ev}}(\mathbb{R})$ is irreducible. With the same proof we also get that $L^2_{\mathrm{od}}(\mathbb{R})$ is also irreducible.

Assume now that $\hat{B}\hat{R}(F) = \hat{R}(F)\hat{B}$ and that $\hat{B}$ is a linear transformation from $L^2_{\mathrm{ev}}(\mathbb{R})$ in $L^2_{\mathrm{odd}}(\mathbb{R})$. We still have $\hat{B}\phi_k = \lambda_k\phi_k$. But an Hermite function is odd or even so $\lambda_k = 0$ for all k and $\hat{B} = 0$. So these representations are non equivalent. □

3.4 Representation of the Quantum Propagator in Terms of the Generator of Squeezed States

We have seen in the proof of Proposition 31 that any symplectic matrix has a symplectic polar decomposition. In this section our aim is to analyse more deeply this polar decomposition and its consequence on the metaplectic representation. More precisely let $\mathcal{U}_t := \mathcal{U}(t,0)$ be the quantum propagator associated with a time dependent quadratic Hamiltonian $H(t,X)$ (continuous in time t), we want to analyze the squeezed state $\mathcal{U}_t\varphi_z$ ($z \in \mathbb{R}^{2n}$). Recall that $\mathcal{U}_t = \hat{R}(F_t)$ where $F_t = F(t,0)$ is the classical, linear flow.

These results were obtained in [61, 65]. For simplicity we forget the time in our notations.

Let us start with classical Hamiltonian mechanics in the complex model $\mathbb{C}^n$, $\zeta = \frac{q-ip}{\sqrt{2}}$. As above, F is a classical flow for a quadratic, time dependent Hamiltonian. Let us denote by F^c the same flow in $\mathbb{C}^n$. We easily get

$$F^c\zeta = \frac{1}{2}(Y\zeta + \bar{Z}\bar{\zeta}), \tag{3.49}$$

where

$$Y = A + D + i(B - C), \quad Z = A - D + i(B + C). \tag{3.50}$$

Recall that all these matrices are time dependent and Y is invertible. So we have $Y = |Y|V$ (polar decomposition), where $|Y|^2 = YY^*$ and V is a unitary transformation of $\mathbb{C}^n$ ($V \in \mathrm{SU}(n)$).

We already introduced $W = ZY^{-1}$ and we know that $0 \le W^*W < \mathbb{1}$. So we can factorize F^c in the following way.

$$F^c = D^c \cdot S^c, \tag{3.51}$$

$$S^c\zeta = V\zeta, \tag{3.52}$$

$$D^c\zeta = (\mathbb{1} - W^*W)^{-1/2}\zeta + (\mathbb{1} - W^*W)^{-1/2}W^*\bar{\zeta}. \tag{3.53}$$

Coming back to the real representation in $\mathbb{R}^{2n}$, S is an orthogonal symplectic transformation (a rotation in the phase space). Let us compute $\hat{R}(S)$. To do that, we write $V = \mathrm{e}^{iL}$ (this is possible locally in time). L is an Hermitian complex matrix. V is the flow at time 1[2] of the quadratic Hamiltonian

$$H_S^c(\zeta,\bar{\zeta}) = \frac{1}{2}\left(\zeta \cdot L^T\bar{\zeta} + \bar{\zeta} \cdot L\zeta\right).$$

Let H be the real representation of H^c, then we have $\hat{R}(S) = \mathrm{e}^{-i\hat{H}_S}$. So $\hat{R}(S)$ is a quantum rotation because we have, for every observable O,

[2] This time is a new time, which has nothing to do with t.

$$e^{i\hat{H}_S}\hat{O}e^{-i\hat{H}_S} = \widehat{O \cdot S}.$$

It is more difficult to compute $\hat{R}(D)$ (the dilation or squeezing part).

We write down the polar decomposition of W, $W = U|W|$, $|W|^2 = W^*W$, U unitary transformation in $\mathbb{C}^n$. We are looking for a generator at (new) time 1 for the transformation D. Let us introduce a complex transformation B in $\mathbb{C}^n$ with the polar decomposition $B = U|B|$ ($|B|^2 = B^*B$). After standard computations, we get

$$\exp\begin{pmatrix} 0 & B^* \\ B & 0 \end{pmatrix} = \begin{pmatrix} \cosh|B| & -\sinh|B|U^* \\ U\sinh|B| & -U\cosh|B|U^* \end{pmatrix}. \tag{3.54}$$

Comparing with previous computation of D^c, we get

$$\cosh|B| = (\mathbb{1} - W^*W)^{-1/2} \tag{3.55}$$

$$\sinh|B|U^* = \cosh|B|W^* \tag{3.56}$$

$$\sinh|B| = \cosh|B||W|. \tag{3.57}$$

We can solve the last equation:

$$|B| = \arg\tanh|W|. \tag{3.58}$$

More explicitely we have

$$B = W\sum_{n\geq 0}\frac{(W^*W)^n}{2n+1}. \tag{3.59}$$

In particular B is a symmetric matrix ($B^T = B$) because W is a symmetric matrix.

As for the rotation part, we can get now a dilation generator,

$$H_D^c = \frac{i}{2}(\zeta \cdot B\zeta - \bar{\zeta} \cdot B^*\bar{\zeta})$$

such that the (complex) equation of motion is

$$\dot{\zeta} = B^*\bar{\zeta}.$$

Finally we restore the time t. We have a decomposition $F_t = D(B_t)S(L_t)$ such that $\mathcal{U}_t = \lambda_t\hat{D}(B_t)\hat{S}(L_t)$, where λ_t is a complex number, $|\lambda_t| = 1$ and

$$\hat{D}(B_t) = e^{\frac{1}{2}(\mathbf{a}^\dagger \cdot B_t\mathbf{a}^\dagger - \mathbf{a}\cdot B_t{}^*\mathbf{a})} \tag{3.60}$$

is the generator of squeezed states. More properties of $D(B)$ will will given at the end of this section. We can get now

Proposition 34 *For every time t we have*

$$\mathcal{U}_t\varphi_0 = (\det{}^{1/2}V_t)\phi_{B_t}, \tag{3.61}$$

where ϕ_{B_t} is the squeezed state defined by

$$\phi_{B_t} = \hat{D}(B_t)\varphi_0 \tag{3.62}$$

and $\mathrm{det V_t}^{1/2}$ *is defined by continuity, starting from $t = 0$ ($V_0 = \mathbb{1}$).*

Proof It is enough to show that φ_0 is an eigenstate of $\hat{S}(L)$ with eigenvalue $\gamma = \frac{1}{2}\mathrm{Tr}(L)$. Clearly

$$(\mathbf{a}^\dagger \cdot L^t\mathbf{a} + \mathbf{a} \cdot L\mathbf{a}^\dagger)\varphi_0 = \sum_{i,j}(L_{j,i}\mathbf{a}_j^\dagger \mathbf{a}_i + L_{i,j}\mathbf{a}_i\mathbf{a}_j^\dagger)\varphi_0 = \sum_i L_{i,i}\varphi_0,$$

since $\mathbf{a}_i\varphi_0 = 0, \quad \forall i = 1, \ldots n$. We get the result by exponentiating. Let us remark here that even if L is defined in a small time interval we can conclude because the prefactor of ϕ_{B_t} must be continuous in time t. □

We can also demonstrate that the quantum evolution of a coherent state φ_z is simply a displaced along the classical motion of a squeezed state:

Proposition 35 *Let $z_t = (q_t, p_t)$ be the phase space point of the classical flow at time t starting with initial conditions $z = (q, p)$. Thus*

$$\begin{pmatrix} q_t \\ p_t \end{pmatrix} = F_t \begin{pmatrix} q \\ p \end{pmatrix}.$$

One has:

$$\mathcal{U}(t)\varphi_z = \mathrm{det V_t}^{1/2}\hat{T}(z_t)\Phi_{B_t}.$$

Proof One uses that due to the fact that $\mathcal{U}(t) = \hat{R}(F_t)$

$$\mathcal{U}(t)\varphi_z = \mathcal{U}(t)\hat{T}(z)\varphi_0 = \hat{T}(z_t)\mathcal{U}(t)\varphi_0.$$

□

More on n-dimensional squeezed states

Consider now any complex symmetric $n \times n$ matrix W such that $W^*W < \mathbb{1}$. Take as before the polar decomposition of W to be:

$$W = U|W|, \quad |W| = (W^*W)^{1/2},$$

U being unitary. We define the $n \times n$ complex symmetric matrix B to be:

$$B = U \arg\tanh |W|.$$

More explicitly writing the Taylor expansion of arg tanh u at 0 we find

$$B = W \sum_{n=0}^{\infty} \frac{(W^* W)^n}{2n+1} \text{ and we have } |W| = \tanh |B|. \tag{3.63}$$

Now we construct the unitary operator $D(B)$ in $L^2(\mathbb{R}^n)$ as

$$\hat{D}(B) = \exp\left(\frac{1}{2}(\mathbf{a}^\dagger \cdot B\mathbf{a}^\dagger - \mathbf{a} \cdot B^*\mathbf{a})\right).$$

We have

Lemma 25 *(i) $D(B)$ is unitary with inverse $D(-B)$.*
(ii)

$$\hat{D}(B)\begin{pmatrix}\mathbf{a}\\ \mathbf{a}^\dagger\end{pmatrix}\hat{D}(-B) = \begin{pmatrix}(\mathbb{1} - WW^*)^{-1/2} & -W(\mathbb{1} - W^*W))^{-1/2}\\ -(\mathbb{1} - W^*W)^{-1/2}W^* & (\mathbb{1} - W^*W)^{-1/2}\end{pmatrix}\begin{pmatrix}\mathbf{a}\\ \mathbf{a}^\dagger\end{pmatrix}.$$

Proof Let us denote

$$\mathbf{a}_B(t) = \hat{D}(tB)\mathbf{a}\hat{D}(-tB), \quad \mathbf{a}_B^\dagger(t) = \hat{D}(tB)\mathbf{a}^\dagger\hat{D}(-tB).$$

Computing

$$\frac{d}{dt}\mathbf{a}_B(t) = \hat{D}(tB)\left[\frac{1}{2}\left(\mathbf{a}^\dagger \cdot B\mathbf{a}^\dagger - \mathbf{a} \cdot B^*\mathbf{a}\right), \mathbf{a}\right]\hat{D}(-tB)$$

we get, using the commuting relations:

$$\frac{d}{dt}\mathbf{a}_B(t) = -B\mathbf{a}_B(t).$$

and the same for $\mathbf{a}^\dagger$:

$$\frac{d}{dt}\mathbf{a}_B^\dagger(t) = B^*\mathbf{a}_B^*(t).$$

We get the result by computing $\exp\left(\begin{pmatrix}0 & B^*\\ B & 0\end{pmatrix}\right)$ as in (3.54). □

We define the n-dimensional squeezed state as

$$\psi^{(B)} = \hat{D}(B)\varphi_0$$

where φ_0 is the standard Gaussian (ground state of the Harmonic oscillator).

We shall first compute $\psi^{(B)}$ in the Fock–Bargmann representation. We recall that in the Fock–Bargmann representation one has

$$\mathscr{B}[\varphi_0](\zeta) = (2\pi\hbar)^{-n/2}.$$

(independent of $\zeta \in \mathbb{C}^n$).

Then we try the following ansatz

$$\mathscr{B}[\psi^{(B)}](\zeta) = a \exp\left(\frac{1}{2\hbar}\zeta \cdot M\zeta\right). \tag{3.64}$$

where M is a complex symmetric matrix that we want to compute. From now on we assume $\hbar = 1$.

Let us consider the antihermitian Hamiltonian H_B to be

$$H_B = \frac{1}{2}(\mathbf{a}^\dagger \cdot B\mathbf{a}^\dagger - \mathbf{a} \cdot B^*\mathbf{a}).$$

Take a fictitious time t to vary between 0 and 1 and consider the evolution operator $U_B(t) = e^{tH_B}$. Then $\psi^{(B)}(t) = U_B(t)\varphi_0$ satisfies the differential equation

$$\frac{d}{dt}\psi^{(B)}(t) = H_B\psi^{(B)}(t)$$

with the following limiting values

$$\psi^{(B)}(0) = \varphi_0, \quad \psi^{(B)}(1) = \psi^{(B)}.$$

In the Fock–Bargmann representation H_B has the following form:

$$H_B(\zeta, \partial_\zeta) = \frac{1}{2}(\zeta \cdot B\zeta - \partial_\zeta \cdot B^*\partial_\zeta).$$

Using the ansatz (3.64) for $\psi^{(B)}(t)$ we see that one has

$$\left(\dot{a} + a\frac{1}{2}\zeta \cdot \dot{M}\zeta\right)\exp\left(\frac{1}{2}\zeta \cdot M\zeta\right) = a(t)H_B(\zeta, \partial_\zeta)\exp\left(\frac{1}{2}\zeta \cdot M\zeta\right).$$

Now we compute the right hand side using the convention of summation over repeated indices; we get:

$$\frac{1}{2}a(t)\left(\frac{1}{2}\zeta \cdot B\zeta - \frac{1}{2}\partial_{\zeta_j}B^*{}_{ij}M_{ik}\zeta_k - \frac{1}{2}B^*_{ij}\partial_{\zeta_j}\zeta_k M_{ki}\right)\exp\left(\frac{1}{2}\zeta \cdot M\zeta\right),$$

$$= \frac{1}{2}a\left(\zeta \cdot B\zeta - \mathrm{Tr}(B^*M) - \zeta \cdot MB^*M\zeta\right)\exp\left(\frac{1}{2}\zeta \cdot M\zeta\right)$$

Identifying we get:

$$\dot{M} = B - MB^*M, \quad \dot{a} = -\frac{1}{2}a\mathrm{Tr}(B^*M).$$

Let us solve the differential equation

$$\partial_t M = -MB^*M + B \tag{3.65}$$

with $M_{t=0} = 0$.

We consider N such as

$$M = UN$$

with U independent of t given by the polar decomposition of B.

Then Eq. (3.65) becomes

$$U\partial_t N = -UN|B|N + U|B|.$$

Thus it reduces to

$$\partial_t N = -N|B|N + |B|.$$

At time $t = 0$ $N = 0$ thus the solution at time $t = 1$ is a function of $|B|$ (thus commuting with $|B|$) given by

$$N = \tanh |B|.$$

This implies that

$$M = U \tanh |B| = U|W| = W.$$

Now consider the differential equation satisfied by $a(t)$. We get:

$$2\dot{a} = -a\mathrm{Tr}(B^*M) = -a\mathrm{Tr}(|B|U^*UN) = -a\mathrm{Tr}(|B| \tanh |B|).$$

Since $a(0) = (2\pi)^{-n/2}$ we get

$$\begin{aligned} a(t) &= (2\pi)^{-n/2} \exp\left(-\frac{1}{2}\int_0^t ds\mathrm{Tr}(|B| \tanh s|B|)\right) \\ &= (2\pi)^{-n/2} \exp\left(-\frac{1}{2}\mathrm{Tr} \log \cosh t|B|\right) \end{aligned}$$

$$a(1) = (2\pi)^{-n/2}(\det \cosh |B|)^{-1/2} = (2\pi)^{-n/2}(\det(\mathbb{1} - |W|^2))^{1/4},$$

where we have used that $e^{\mathrm{Tr} A} = \det e^A$.

Thus the squeezed state $\psi^{(B)} = D(B)\varphi_0$ has the following Fock–Bargmann representation

$$\boxed{\psi^{(B)}(\zeta) = (2\pi)^{-n/2}(\det\cosh|B|)^{-1/2}\exp\left(\frac{1}{2}\zeta\cdot W\zeta\right)} \tag{3.66}$$

Now using the inverse Fock–Bargmann transform we go back to the coordinate representation of $\psi^{(B)}$. One has to compute the following integral:

$$\int dqdp\exp\left(-\left(\frac{x^2}{2} - x\sqrt{2}\bar{\zeta} + \frac{\bar{\zeta}^2}{2}\right) - \zeta\cdot\bar{\zeta} + \frac{1}{2}\zeta\cdot W\zeta\right) \tag{3.67}$$

The argument of the exponential can be rewritten as:

$$\begin{aligned} &-\frac{x^2}{2} - \frac{1}{4}q^2 - \frac{ip\cdot q}{2} + \frac{p^2}{4} + x\cdot(q+ip) - \frac{1}{2}(q^2+p^2)\\ &\quad + \frac{1}{4}(q\cdot Wq - ip\cdot Wq - iq\cdot Wp - p\cdot Wp)\\ &= -\frac{x^2}{2} + x(q+ip) - \frac{3}{4}q^2 - \frac{1}{4}p^2 + \frac{1}{4}(q\cdot Wq - ip\cdot Wq - iq\cdot Wp - p\cdot Wp). \end{aligned} \tag{3.68}$$

Thus it appears a quadratic form in $q,\ p$ which can be written in matricial form as:

$$-\frac{1}{2}(q,p)\mathcal{M}\begin{pmatrix} q\\ p\end{pmatrix}$$

with

$$\mathcal{M} = \frac{1}{2}\begin{pmatrix} 3\mathbb{1} - W & -i(W+\mathbb{1})\\ -i(W+\mathbb{1}) & W+\mathbb{1}\end{pmatrix}.$$

One has

$$\mathcal{M}^{-1} = \frac{1}{2}\begin{pmatrix} \mathbb{1} & i\\ i & (3\mathbb{1}-W)(\mathbb{1}+W)^{-1}\end{pmatrix}.$$

Thus the integral over q, p in (3.67) will be the exponential of the following quadratic form

$$-\frac{1}{2}(x, ix)\mathcal{M}^{-1}\begin{pmatrix} x\\ ix\end{pmatrix} - \frac{x^2}{2} \tag{3.69}$$

with a phase which is exactly

$$(\det(\mathcal{M}/2\pi))^{-1/2} = (2\pi)^{n/2}(\det(\mathbb{1}+W))^{-1/2}.$$

It is easy to compute the expression in (3.69). It gives

$$-\frac{1}{2}x\cdot(\mathbb{1}-W)(\mathbb{1}+W)^{-1}x$$

Thus restoring the $\hbar$ dependence and the factors π we get again a formula proved differently (by solving a Ricatti equation) at the beginning of this chapter.

Proposition 36 *The squeezed state $\psi^{(B)}$ is actually a gaussian in the position representation given by*

$$\boxed{\psi^{(B)}(x) = a_\Gamma \exp\left(\frac{i}{2\hbar}x \cdot \Gamma x\right)} \tag{3.70}$$

with

$$\Gamma = i(\mathbb{1} - W)(\mathbb{1} + W)^{-1}$$

and

$$a_\Gamma = (\pi\hbar)^{-n/2}(\det(\mathbb{1} - |W|^2))^{-1/2}(\mathbb{1} + W)^{-1/2}.$$

3.5 Representation of the Weyl Symbol of the Metaplectic Operators

See Chap. 2 for the definitions of covariant and contravariant Weyl symbols. We have shown that $\hat{R}(F) = \mathcal{U}_1$ where $\mathcal{U}_t$ is the quantum propagator of the quadratic Hamiltonian

$$H(t) = \frac{1}{2}(\hat{Q}, \hat{P})M_t \begin{pmatrix} \hat{Q} \\ \hat{P} \end{pmatrix} \tag{3.71}$$

with $M_t = -J\dot{F}_t F_t^{-1}$, F_t being a continuous path in the space $\mathrm{Sp}(n)$ joining $\mathbb{1}$ at $t = 0$ to F at $t = 1$. In [66] the authors show the following result

Theorem 18 *(i) If* $\det(\mathbb{1} + F) \neq 0$ *the contravariant Weyl symbol of* $\widehat{R}(F)$ *has the following form:*

$$\boxed{R(F, X) = e^{i\pi\nu}|\det(\mathbb{1} + F)|^{-1/2}\exp(-iJ(\mathbb{1} - F)(\mathbb{1} + F)^{-1}X \cdot X)} \tag{3.72}$$

where $\nu \in \mathbb{Z}$ *if* $\det(\mathbb{1} + F) > 0$ *and* $\nu \in \mathbb{Z} + 1/2$ *if* $\det(\mathbb{1} + F) < 0$.

(ii) If $\det(\mathbb{1} - F) \neq 0$ *the covariant Weyl symbol of* $\widehat{R}(F)$ *has the following form:*

$$\boxed{R^\#(F, X) = e^{i\pi\mu}|\det(\mathbb{1} - F)|^{-1/2}\exp\left(-\frac{i}{4}J(\mathbb{1} + F)(\mathbb{1} - F)^{-1}X \cdot X\right)} \tag{3.73}$$

where $\mu = \bar{\nu} + \frac{n}{2}$ *and* $\bar{\nu} \in \mathbb{Z}$.

This formula has been heuristically proposed by Mehlig and Wilkinson [193] without the computation of the phase. See also [75].

We can restore the $\hbar$ dependence of $R(F, X)$ and $R^\#(F, X)$ by putting a factor $\hbar^{-1}$ in the argument of the exponentials.

Proof Let us state the following proposition which is a direct consequence of (3.39) after algebraic computations.

Proposition 37 *The matrix elements of $\widehat{R}(F)$ on coherent states φ_z, are given by the following formula:*

$$\langle\varphi_{z+X}|\hat{R}(F)\varphi_z\rangle = 2^n\,(\det(\mathbb{1}+F+iJ(\mathbb{1}-F)))^{-1/2}$$
$$\times\exp\left(-\left|z+\frac{X}{2}\right|^2+\frac{1}{2}i\sigma(X,z)+K_F\left(z+\frac{X-iJX}{2}\right)\cdot\left(z+\frac{X-iJX}{2}\right)\right) \tag{3.74}$$

where

$$K_F = (\mathbb{1}+F)(\mathbb{1}+F+iJ(\mathbb{1}-F))^{-1}. \tag{3.75}$$

Now we can compute the distribution covariant symbol of $\hat{R}(F)$ by plugging formula (3.74) in formula (2.29).

Let us begin with the regular case $\det(\mathbb{1}-F)\neq 0$ and compute the covariant symbol.

Using Proposition (37) and formula (3.39), we have to compute a Gaussian integral with a complex, quadratic, non degenerate covariance matrix (see [153]).

This covariance matrix is $K_F-\mathbb{1}$ and we have clearly

$$K_F-\mathbb{1} = -iJ(\mathbb{1}-F)(\mathbb{1}+F+iJ(\mathbb{1}-F))^{-1} = -(\mathbb{1}-i\Lambda)^{-1}$$

where $\Lambda = (\mathbb{1}+F)(\mathbb{1}-F)^{-1}J$ is a real symmetric matrix. So we have

$$\Re(K_F-\mathbb{1}) = -(\mathbb{1}+\Lambda^2)^{-1},\quad \Im(K_F-\mathbb{1}) = -\Lambda(\mathbb{1}+\Lambda^2)^{-1}. \tag{3.76}$$

So that $\mathbb{1}-K_F$ is in the Siegel space $\mathfrak{S}^+(n)$ and Theorem (7.6.1) of [153] can be applied. The only serious problem is to compute the index μ.

Let us define a path of $2n\times 2n$ symplectic matrices as follows: $G_t = \mathrm{e}^{t\pi J_n}$ if $\det(\mathbb{1}-F)>0$ and:
$G_t = G_t^2\otimes \mathrm{e}^{t\pi J_{2n-2}}$ if $\det(\mathbb{1}-F)<0$, where

$$G^2 = \begin{pmatrix}\eta(t) & 0\\ 0 & \frac{1}{\eta(t)}\end{pmatrix}$$

where η is a smooth function on $[0,1]$ such that $\eta(0)=1$, $\eta(t)>1$ on $]0,1]$ and where J_n is the $2n\times 2n$ matrix defining the symplectic matrix on the Euclidean space $\mathbb{R}^{2n}$.

G_1 and F are in the same connected component of $\mathrm{Sp}_\star(n)$ where $\mathrm{Sp}_\star(2n) = \{F\in \mathrm{Sp(n)},\ \det(\mathbb{1}-F)\neq 0\}$. So we can consider a path $s\mapsto F'_s$ in $\mathrm{Sp}_\star\mathrm{(n)}$ such that $F'_0=G_1$ and $F'_1=F$.

Let us consider the following "argument of determinant" functions for families of complex matrices.

$$\theta[F_t] = \arg_c[\det(\mathbb{1}+F_t+iJ(\mathbb{1}-F_t)] \tag{3.77}$$
$$\beta[F] = \arg_+[\det(\mathbb{1}-K_F)^{-1}] \tag{3.78}$$

where $\arg_c$ means that $t \mapsto \theta[F_t]$ is continuous in t and $\theta[\mathbb{1}] = 0$ ($F_0 = \mathbb{1}$), and $S \mapsto \arg_+[\det(S)]$ is the analytic determination defined on the Siegel space $\mathfrak{S}^+(n)$ such that $\arg_+[\det(S)] = 0$ if S is real (see [153, vol.1, Sect. (3.4)]).

With these notations we have

$$\mu = \frac{\beta[F] - \theta[F]}{2\pi}. \tag{3.79}$$

Let us consider first the case $\det(\mathbb{1} - F) > 0$.

Using that J has the spectrum $\pm i$, we get: $\det(\mathbb{1} + G_t + iJ(\mathbb{1} - G_t)) = 4^n e^{nt\pi i}$ and $\mathbb{1} - K_{G_1} = \mathbb{1}$.

Let us remark that $\det(\mathbb{1} - K_F)^{-1} = \det(\mathbb{1} - F)^{-1}\det(\mathbb{1} - F + iJ(\mathbb{1} + F))$. Let us introduce $\Delta(E, \mathscr{M}) = \det(\mathbb{1} - E + \mathscr{M}(\mathbb{1} + E))$ for $E \in \mathrm{Sp}(2n)$ and $\mathscr{M} \in \mathrm{sp}_+(2n, \mathbb{C})$. Let consider the closed path $\mathscr{C}$ in $\mathrm{Sp}(2n)$ defined by adding $\{G_t\}_{0\le t\le 1}$ and $\{F'_s\}_{0\le s\le 1}$. We denote by $2\pi\bar{\nu}$ the variation of the argument for $\Delta(\bullet, \mathscr{M})$ along $\mathscr{C}$. Then we get easily

$$\beta(F) = \theta[F] + 2\pi\bar{\nu} + n\pi, \; n \in \mathbb{Z}. \tag{3.80}$$

When $\det(\mathbb{1} - F) < 0$, by an explicit computation, we find $\arg_+[\det(\mathbb{1} - K_{G_1})] = 0$. So we can conclude as above.

The formula for the contravariant symbol can be easily deduced from the covariant formula using a symplectic Fourier transform. □

Remark 14 When the quadratic Hamiltonian H is time independent then $F_t = e^{tJS}$, S is a symmetric matrix. So if $\det(e^{tJS} + \mathbb{1}) \neq 0$, then we get applying (3.72),

$$\boxed{R(e^{tJS}, X) = e^{i\pi\nu}\left|\det\left[\cosh\left(\frac{t}{2}JS\right)\right]\right|^{-1/2}\exp\left(i\sigma\left(\tanh\left(\frac{t}{2}JS\right)X\cdot X\right)\right).} \tag{3.81}$$

This formula was obtained in [154]. In particular the Melher formula for the Harmonic oscillator is obtained with $S = \mathbb{1}$.

In [75] the author discusses the Maslov index related with the metaplectic representation.

Remark 15 In the paper [66] the authors give a different method to compute the contravariant Weyl symbol $R(F, X)$ inspired by [96]. They consider a smooth family F_t of linear symplectic transformations associated with a family of time dependent quadratic Hamiltonians H_t. After quantization we have a quantum propagator $\mathscr{U}_t$ with its contravariant Weyl symbol $U_t^w(X)$. We make the ansatz $U_t^w(X) = \alpha_t e^{X\cdot M_t X}$ where α_t is a complex number, M_t is a symmetric matrix. Using the Schrödinger equation $\partial_t U_t^w = H_t \circledast U_t^w$, and the Moyal product, we find for M_t a Riccati equation which is solved with the classical motion. Afterwards, α_t is found by solving a Liouville equation hence we recover the previous results (see [66] for details).

This approach will be adapted later in this book in the fermionic setting.

3.6 Traps

We now give an application in physics of our computations concerning quadratic Hamiltonians.

The quantum motion of an ion in a quadrupolar radiofrequency trap is solved exactly in terms of the classical trajectories. It is proven that the quantum stability regions coincide with the stability regions of the associated Mathieu equation. By quantum stability we mean that the quantum evolution over one period (the so-called Floquet operator) has only pure-point spectrum. Thus the quantum motion is "trapped" in a suitable sense. We exhibit the set of eigenstates of the Floquet operator.

3.6.1 *The Classical Motion*

Let us consider a three-dimensional Hamiltonian of the following form:

$$H(t) = \frac{\vec{p}^2}{2m} + \frac{e}{r_0^2}\left(z^2 - \frac{1}{2}(x^2 + y^2)\right)(V_1 - V_0\cos(\omega t)). \tag{3.82}$$

Here $\vec{p} = (p_x, p_y, p_z)$ is the three-dimensional momentum, m the mass of the ion, and $\vec{r} = (x, y, z)$ is the three-dimensional position. r_0 is the size of the trap and e the charge of the electron. V_1 (resp. V_0) is the constant (resp. alternating) voltage. Such time-periodic hamiltonians are realized by using traps that are hyperboloids of revolution along the z-axis that are submitted to a direct current plus an alternating current voltage. Also known as Paul traps they allow to confine isolated ions like Cesium for rather long times, and also a few ions together in the same trap.

It is obvious to see that this Hamiltonian is purely quadratic and that it decouples into three one-dimensional Hamiltonians:

$$H(t) = h_x(t) \otimes \mathbb{1} \otimes \mathbb{1} + \mathbb{1} \otimes h_y(t) \otimes \mathbb{1} + \mathbb{1} \otimes \mathbb{1} \otimes h_z(t)$$

where $h_x(t) = h_y(t) = \frac{p_x^2}{2m} - \frac{x^2}{2}(\alpha - \beta\cos(\omega t)),\ h_z(t) = \frac{p_z^2}{2m} + z^2(\alpha - \beta\cos(\omega t))$ with $\alpha = \frac{e}{r_0^2}V_1,\ \beta = \frac{e}{r_0^2}V_0$.

Thus $x(t), y(t), z(t)$ evolve according to the Mathieu equations:

$$\ddot{x}(t) - x(t)(\alpha/m - \cos(\omega t)\beta/m) = 0, \qquad \ddot{z}(t) + z(t)(2\alpha/m - \cos(\omega t)2\beta/m) = 0. \tag{3.83}$$

It is known that each of these equations have stability regions parametrized by (α, β, ω), in which the motion remains bounded. Furthermore it can be shown that it is quasiperiodic with Floquet exponent ρ. See [192].

Theorem 19 *There exist $\rho, \rho' \in \mathbb{R}$ and rapidly converging sequences $c_n, c_n' \in \mathbb{R}$ depending on $a = 4\alpha/m\omega^2$, $b = 2\beta/m\omega^2$ such that the solution of equations (3.83)* *with* $x(0) = u, \dot{x}(t) = v/m, z(0) = u', \dot{z}(0) = v'/m$ *are given by*

$$x(t,u,v) = \frac{u}{c}\sum_{-\infty}^{+\infty} c_n \cos\left((2n+\rho)\frac{\omega t}{2}\right) + \frac{2v}{m\omega d}\sum_{-\infty}^{+\infty} c_n \sin\left((2n+\rho)\frac{\omega t}{2}\right)$$

$$z(t,u',v') = \frac{u'}{c'}\sum_{-\infty}^{+\infty} c'_n \cos\left((2n+\rho')\frac{\omega t}{2}\right) + \frac{2v'}{m\omega d'}\sum_{-\infty}^{+\infty} c'_n \sin\left((2n+\rho')\frac{\omega t}{2}\right)$$

where $c = \sum_{\mathbb{Z}} c_n$, $d = \sum_{\mathbb{Z}}(2n+\rho)c_n$, *and similarly for* c', d'. *Note that the Floquet exponent* ρ *is the same for* x, y *but* $\rho' \neq \rho$.

The proof depends heavily on the linearity of Mathieu's equations. Note that the stability regions are delimited by the curves C_j for which (3.83) has periodic solutions, ie for which $\rho = j \in \mathbb{N}$. For given α, β, there exists $\omega_1, \omega_2 \in \mathbb{R}^+$ such that for any $\omega \in]\omega_1, \omega_2[$ the classical equations of motion (3.83) have stable solutions. ω_1 has to be large enough hence the name "radio-frequency traps".

3.6.2 *The Quantum Evolution*

Since ions are actually quantum objects it is relevant to consider now the quantum problem. As known from the general considerations of this Chapter on quadratic Hamiltonians the quantum evolution for Hamiltonians of the form (3.82) is completely determined by the classical motion. The Hilbert space of quantum states is $\mathcal{H} = L^2(\mathbb{R}^3)$. One thus has that the time-periodic Hamiltonian

$$\hat{H}(t) := -\frac{\hbar^2}{2m}\Delta + (\alpha - \beta\cos(\omega t))\left(z^2 - \frac{x^2+y^2}{2}\right),$$

where Δ is the 3-dimensional Laplacian and generates an unitary operator $\mathcal{U}(t,s)$ that evolves a quantum state from time s to time t. The Floquet operator is the operator on time evolution over one period $T = 2\pi/\omega$:

$$\mathcal{U}(T,0) = \mathcal{U}_x(T,0)\mathcal{U}_y(T,0)\mathcal{U}_z(T,0)$$

We shall denote by $\mathcal{U}_F$ (resp $\mathcal{U}'_F$) the operator $\mathcal{U}_x(T,0)$ (resp. $\mathcal{U}_z(T,0)$). Then one has the following results:

Theorem 20 *(i) Given* $\alpha, \beta \in \mathbb{R}$ *there exists* $\omega_1, \omega_2 \in \mathbb{R}^+$ *as in the previous section such that for any* $\psi \in \mathcal{H}$ *and any* $\varepsilon > 0$ *there exists* $R \in \mathbb{R}^+$ *such that*

$$\sup_t \|F(|\vec{r}| > R)\mathcal{U}(t,s)\psi\| < \varepsilon,$$

$F(|\vec{r}| > R)$ being the characteristic function of the exterior of the ball $|\vec{r}| \leq R$.
(ii) For α, β, ω as above, $\mathcal{U}_F$ has pure point spectrum of the form $\{\exp(-i\rho\pi(k+1/2))\}_{k\in\mathbb{N}}$ where ρ is as in the preceding section the classical Floquet exponent. Similarly for $\mathcal{U}'_F$.

Remark 16 A complete proof can be seen in [59].
(i) says that the time evolution of any quantum state remains essentially localized along the quantum evolution.

Proof (*A sketch*) To prove (i) it is enough to state the result in one dimension, and to assume $m = \hbar = 1$. We establish the result for $\psi \in \mathcal{C}_0^\infty$ which is a dense set. Let $W(-2x, \xi, t)$ be the Wigner function of the state $\psi(t) := \mathcal{U}(t, 0)\psi$. Then $W(x, ., t) \in L^1(\mathbb{R}) \quad \forall t \in \mathbb{R}, \quad \forall x \in \mathbb{R}$. Furthermore we have

$$\left|\psi\left(t, -\frac{x}{2}\right)\right|^2 = \int d\xi\, W(x, \xi, t).$$

It follows that

$$\|F(|x| > R)\psi(t)\|^2 = \int_{|x|>2R} dxd\xi\, W(x, \xi, t) = \int_{|x|>2R} dxd\xi\, W(x(t), \dot{x}(t), 0),$$

where $x(t)$ is a solution of (3.83) with $x(0) = x$, $\dot{x}(0) = \xi$. Then we perform a change of variables with uniform (in t) jacobian, the linearity of Mathieu's equation and the boundedness in time of its solution in the stable region, together with the fact that $W(x, \xi, t) \in L^1(\mathbb{R}^2)$ to conclude for (i). □

An extension of Theorem 20 to Hermite-like wavefunctions is the following: define

$$\Phi_k(t, x) = (L_t)^{-1/2}\left(\frac{L_t^*}{L_t}\right)^{k/2} H_k(x/\sqrt{\hbar}|L_t|)\exp\left(\frac{iN_t x^2}{2\hbar L_t}\right),$$

where H_k are the normalized Hermite polynomials and the determination of the square-root is followed by continuity from $t = 0$. We have

$$\mathcal{U}(t, 0)\Phi_k(0, .) = \Phi_k(t, .), \qquad \forall t \in \mathbb{R}$$

This can be proven by using Trotter product formula, a steplike approximation $f_N(t)$ of the function $f(t) = \alpha - \beta\cos(\omega t)$ and the continuity of solutions of (3.83) when $f_N \to f$. See [59, 130, 131] for details.

We start now a sketch proof of (ii) of Theorem 20.

The normalized eigenstates of $\mathcal{U}_F$ are generalized squeezed states. Let us assume $m = 1$ for simplicity. Let $F(t)$ be the symplectic 2×2 matrix solution of

$$\dot{F} = JMF$$

where

$$M(t) = \begin{pmatrix} 1 & 0 \\ 0 & f(t) \end{pmatrix}$$

$f(t) = \alpha - \beta \cos(\omega t)$

$$F(t) = \begin{pmatrix} A_t & B_t \\ C_t & D_t \end{pmatrix}.$$

We choose suitable initial data $F(0) = \begin{pmatrix} g & 0 \\ 0 & g^{-1} \end{pmatrix}$ with

$$g = \left(\frac{\omega d}{2c}\right)^{-1/2}$$

We define $L_t = A_t + iB_t, N_t = C_t + iD_t$. For these initial data we have

$$L_t = \frac{g}{c} e^{i\rho\omega t/2} \sum_{-\infty}^{+\infty} c_n e^{in\omega t},\ N_t = \frac{1}{gd} e^{i\rho\omega t/2} \sum_{-\infty}^{+\infty} c_n e^{in\omega t}.$$

It is clear that $|L_t|$, $\frac{N_t}{L_t}$ are T-periodic where $T = 2\pi/\omega$, and furthermore

$$e^{i(k+1/2)\frac{\rho\omega t}{2}} (L_t)^{-1/2} \left(\frac{L_t^*}{L_t}\right)^{k/2}$$

is also T-periodic. Then we introduce the self-adjoint quasi-energy operator

$$K = -i\hbar \frac{\partial}{\partial t} + h_x(t)$$

acting in the Hilbert space

$$\mathscr{K} = L^2(\mathbb{R}) \otimes L^2(\mathbb{T}_\omega)$$

of functions depending on both x and t, T-periodic in t. This formalism has been introduced by Howland [157] and Yajima [266]. K is closely related to the quantum evolution operator. We have the following result:

Lemma 26 *(i) Assume $\Psi \in \mathscr{K}$ is an eigenstate of K with eigenvalue λ. Then for any $t \in \mathbb{T}_\omega$, $\Psi(t, .) \in L^2(\mathbb{R})$ and satisfies*

$$\mathscr{U}(t + T, t)\Psi(t) = e^{-i\lambda T/\hbar}\Psi(t).$$

(ii) Conversely let $\psi \in L^2(\mathbb{R})$ satisfy

$$\mathscr{U}(T, 0)\psi = e^{-iT\lambda/\hbar}\psi.$$

Then $\Psi = e^{i\lambda t/\hbar}\mathscr{U}(t,0)\psi \in \mathscr{K}$ *and satisfies*

$$K\Psi = \lambda\Psi.$$

Proof Define

$$\Psi_k(t,x) = e^{(k+1/2)\rho\omega t/2}\Phi_k(x).$$

Then using the periodicity properties we have that $\Psi_k \in \mathscr{K}$ and

$$\mathscr{U}(T,0)\Phi_k(0,.) = \exp\left(-i\pi\rho\left(k+\frac{1}{2}\right)\right)\Phi_k(0,.)$$

and thus

$$K\Psi_k = \left(k+\frac{1}{2}\right)\frac{\hbar\rho\omega}{2}\Psi_k.$$

□

Chapter 4
The Semiclassical Evolution of Gaussian Coherent States

Abstract In this chapter we consider semiclassical asymptotics of the quantum evolution of coherent states at any order in the Planck constant $\hbar$. We consider a control in time of the remainder term depending explicitly on $\hbar$ and on the stability matrix of the classical system. We find that the quantum evolved coherent state is in L^2-norm well approximated by a squeezed state located around the phase-space point z_t of the classical flow reached at time t, with a dispersion controlled by the stability matrix at point z_t. The idea goes back to Hepp [148] and was further developed by Hagedorn [130, 131]. The method that we develop here follows the paper [65] where we consider general time-dependent Hamiltonians and use the squeezed states formalism and the metaplectic transformation (see Chap. 3). The difference between the exact and the semiclassical evolution is estimated in time t and in the semiclassical parameter $\hbar$ giving in particular the well known Ehrenfest time of order $\log(\hbar^{-1})$. We then provide two applications of the semiclassical estimates: the first one concerns the semiclassical estimate of the spreading of quantum wave packets, which are coherent states, in terms of the Lyapunov exponents of the classical flow. The second application is to the scattering theory for general short range interactions: then the large time asymptotics can be controlled and the quantum scattering operator acts on coherent states following the classical scattering theory with good estimates in $\hbar$. More accurate estimates can be obtained using the Fourier–Bargmann transform [218]. We consider Gevrey type estimates for the semiclassical coefficients and $\hbar$-exponentially small remainder estimates in Sobolev norms for solutions of time dependent Schrödinger equations.

4.1 General Results and Assumptions

We shall consider the quantum Hamiltonian $\hat{H}(t)$ of a possibly time-dependent problem. We assume that the corresponding time-dependent Schrödinger equation defines a unique quantum unitary propagator $\mathcal{U}(t, s)$[1]. Then we consider the canonical gaussian coherent states φ_z where $z = (q, p) \in \mathbb{R}^{2n}$ is a phase-space point. Then we let

[1] We shall see later that this condition is fulfilled if $H(t, X)$ is subquadratic in X.

D. Robert and M. Combescure, *Coherent States and Applications in Mathematical Physics*, Theoretical and Mathematical Physics,
https://doi.org/10.1007/978-3-030-70845-0_4

it evolve with the quantum propagator that is we consider the quantum state at time t defined by

$$\psi_z(t) := \mathcal{U}(t,0)\varphi_z.$$

Semiclassically (when $\hbar$ is small) $\psi_z(t)$ is well approximated in L^2-norm by a superposition of "squeezed states" centered around the phase-space point $z_t = \Phi^{t,0}z$ of the classical flow. This study goes back to the pioneering paper by Hepp [148] and was later developed by G. Hagedorn in a series of papers [130, 131]. In [61] an approach is developed connecting the semiclassical propagation of coherent states to the so-called squeezed states. We develop here a generalization of these results, following the paper [65], allowing general time-dependent Hamiltonians and estimating the error term with respect to time t, to z and to $\hbar$.

The knowledge of the time evolution for any Gaussian φ_z is a way to get many properties for the full propagator $\mathcal{U}(t,0)$ (we can always assume that the initial time is 0). This is easy to understand using that the family $\{\varphi_z,\ z \in \mathbb{R}^{2n}\}$ is overcomplete (Chap. 1). Let us denote by $K_t(x,y)$ the Schwartz-distribution kernel of $\mathcal{U}(t,0)$. From overcompleteness we get the following formula:

$$K_t(x,y) = (2\pi\hbar)^{-n}\int_{\mathbb{R}^{2n}} dz[\mathcal{U}(t,0)\varphi_z](x)\overline{\varphi_z(y)}. \tag{4.1}$$

This equality holds as Schwartz distributions on $\mathbb{R}^{2n}$ and explains why it is very useful to solve the Schrödinger equation with coherent state φ_z as initial state:

$$i\hbar\partial_t\psi_z(t) = \hat{H}(t)\psi_z(t), \quad \psi_z(0) = \varphi_z. \tag{4.2}$$

Several applications will be given later as well for time dependent and time independent Schrödinger equations type.

4.1.1 Assumptions and Notations

Let $\hat{H}(t)$ be a selfadjoint Schrödinger Hamiltonian in $L^2(\mathbb{R}^n)$ obtained by quantizing a general time-dependent symbol $H(t,X)$ which is a classical Hamiltonian. We use the $\hbar$- Weyl quantization (see Chap. 2). H is assumed to be a $\mathscr{C}^\infty$-smooth function for $X = (x,\xi) \in \mathbb{R}^n \times \mathbb{R}^n$, continuous in $t \in]-T,T[\ \ 0 \le T \le +\infty$ and satisfying a global estimate:

(A.0) There exist some nonnegative constants $m,\ M,\ K_{H,T}$ such that

$$\langle X\rangle^{-M/2}\left|\partial_X^\gamma H(t,X)\right| \le K_{H,T},$$

uniformly in $X \in \mathbb{R}^{2n}$, $t \in]-T, T[$, for $|\gamma| \geq m$.[2] So H may be a very general Hamiltonian including time-dependent magnetic fields or non Euclidean metrics. We furthermore assume $H(t, x, \xi)$ to be such that the classical and quantum evolutions exist from time 0 to time t for t in some interval $]-T, T[$ where $T < +\infty$ or $T = +\infty$. More precisely:

(A.1) Given some $z = (q_0, p_0) \in \mathbb{R}^{2n}$ there exists a positive T such that the Hamilton's equations

$$\dot{q}_t = \frac{\partial H}{\partial p}(q_t, p_t, t), \qquad \dot{p}_t = -\frac{\partial H}{\partial q}(q_t, p_t, t)$$

have a unique solution for any $t \in]-T, T[$ starting from initial data $z := (q_0, p_0)$. We denote $z_t = (q_t, p_t) := \Phi_H^t(z)$ the phase space point reached at time t starting by z at time 0.

(A.2) There exists a unique quantum propagator $\mathcal{U}(t, s)$, $(t, s) \in \mathbb{R}^2$ with the following properties:

(i) $\mathcal{U}(t, s)$ is unitary in $L^2(\mathbb{R}^n)$ with

$$\mathcal{U}(t, s) = \mathcal{U}(t, r)\mathcal{U}(r, s), \quad \forall (r, s, t) \in \mathbb{R}^3.$$

(ii) $\mathcal{U}(.,.)$ is a strongly continuous operator-valued function for the operator norm topology in $L^2(\mathbb{R}^n)$. The usual norm in $L^2(\mathbb{R}^n)$ is denoted by $\|.\|$.

Remark 17 For general Hamiltonian $H(t)$ the existence time T in (A.1) may depend on z.

In Chap. 1 we have described the construction of standard coherent states by applying the Weyl-Heisenberg operator to the ground state φ_0 the n-dimensional harmonic oscillator with Hamiltonian

$$K_0 := \frac{1}{2}(\hat{P}^2 + \hat{Q}^2) \tag{4.3}$$

$$\varphi_z = \hat{T}(z)\varphi_0$$

In Chap. 3 we have computed an explicit formula for the time evolution of coherent states driven by any quadratic Hamiltonian in (q, p).

We shall now define generalized coherent states by applying $\hat{T}(z)$ to the excited states ϕ_ν of the n-dimensional harmonic oscillator; given a multi-index $\nu = (\nu_1, \dots \nu_n)$, ϕ_ν is the normalized eigenstate of (4.3) with eigenvalue $|\nu| + n/2$. We recall the notation

$$|\nu| = \sum_{j=1}^{n} \nu_j$$

[2] $M = 0$ and $m = 2$ mean that $H(t)$ is subquadratic, (see footnote. 1).

We thus define

$$\phi_z^{(\nu)} = \hat{T}(z)\phi_\nu$$

and φ_z is simply $\phi_z^{(0)}$.

Similarly we define generalized squeezed states $\phi_{z,B}^{(\nu)}$ centered around the phase-space point z. Let W be a symmetric $n \times n$ matrix with polar decomposition $W = U|W|$ where $|W| = (W^*W)^{1/2}$ and U a unitary $n \times n$ matrix. We assume that

$$W^*W \le \mathbb{1}$$

and define

$$B := U \mathrm{Argtanh}|W|. \tag{4.4}$$

Now as in Chap. 3 we construct the unitary operator $\hat{D}(B)$ in $L^2(\mathbb{R}^n)$ as

$$\hat{D}(B) = \exp\left(\frac{1}{2}(\mathbf{a}^* \cdot B\mathbf{a}^* - \mathbf{a} \cdot B^*\mathbf{a})\right),$$

$\mathbf{a}^*$ and $\mathbf{a}$ being the creation and annihilation operators. We define

$$\phi_{z,B}^{(\nu)} := \hat{T}(z)\hat{D}(B)\phi_\nu.$$

Of course we have:

$$\phi_{0,B}^{(\nu)} = \varphi^{(B)}$$

where $\varphi^{(B)}$ is the standard squeezed state, using the notations of Chap. 2 and

$$\phi_{z,0}^{(0)} = \varphi_z$$

using the notation of Chap. 1.

4.1.2 The Semiclassical Evolution of Generalized Coherent States

In this section we consider the quantum evolution of superpositions of generalized coherent states of the form $\phi_z^{(\nu)}$ and prove under the above assumptions that, up to an error term which can be controlled in $t, z, \hbar$, it is close in L^2-norm to a superposition of squeezed states of the form $\phi_{z_t,B_t}^{(\nu)}$ where $z_t := \Phi^{t,0}z$ is the phase-space point reached at time t by the classical flow $\Phi^{t,0}$ of $H(t)$, and B_t is well-defined through the linear stability problem at point z_t. We follow the approach developed in [65] where we use:

- the algebra of the generators of coherent and squeezed states

– the so-called Duhamel's principle which is nothing but the following identity:

$$\mathcal{U}_1(t,s) - \mathcal{U}_2(t,s) = \frac{1}{i\hbar}\int_s^t d\tau\, \mathcal{U}_1(t,\tau)(\hat{H}_1(\tau) - \hat{H}_2(\tau))\mathcal{U}_2(\tau,s), \tag{4.5}$$

where $\mathcal{U}_i(t,s)$ is the quantum propagator generated by the time-dependent Hamiltonian $\hat{H}_i(t)$, $i = 1, 2$. We take $\hat{H}_1(t) = \hat{H}(t)$ and $\hat{H}_2(t)$ to be the "Taylor expansion up to order 2" of $\hat{H}(t)$ around the classical path z_t. More precisely:

$$\begin{aligned}\hat{H}_2(t) = H(q_t, p_t, t) + (\hat{Q} - q_t)\cdot\frac{\partial H}{\partial q}(q_t, p_t, t) + (\hat{P} - p_t)\cdot\frac{\partial H}{\partial p}(q_t, p_t, t)\\ +\frac{1}{2}(\hat{Q} - q_t, \hat{P} - p_t)M_t\begin{pmatrix}\hat{Q} - q_t\\ \hat{P} - p_t\end{pmatrix},\end{aligned} \tag{4.6}$$

M_t being the Hessian of $H(t)$ computed at point $z_t = (q_t, p_t)$:

$$M_t = \left(\frac{\partial^2 H}{\partial z^2}\right)|_{z=z_t}. \tag{4.7}$$

The interesting point is that since $\hat{H}_2(t)$ is at most quadratic, its quantum propagator is written uniquely through the generators of coherent and squeezed states. In Chap. 3 we have shown the link between the quantum propagator of purely quadratic Hamiltonians and the metaplectic transformations. It is shown that the quantum propagator of purely quadratic Hamiltonians can be decomposed into a quantum rotation times a squeezing generator. Let us be more explicit: Let $\hat{H}_Q(t)$ be a purely quadratic quantum Hamiltonian of the form

$$\hat{H}_Q(t) = (\hat{Q}, \hat{P})S(t)\begin{pmatrix}\hat{Q}\\ \hat{P}\end{pmatrix}$$

with $S(t)$ a $2n \times 2n$ symmetric matrix of the form

$$S(t) = \begin{pmatrix}G_t & \tilde{L}_t\\ L_t & K_t\end{pmatrix}$$

where G_t, K_t are symmetric and $\tilde{L}$ denotes the transpose of L. In what follows $S(t)$ will be simply M_t. Let $F(t)$ be the symplectic matrix solution of

$$\dot{F}(t) = JM_tF(t) \tag{4.8}$$

with initial data $F_0 = \mathbb{1}$, where

$$J = \begin{pmatrix} 0 & \mathbb{1} \\ -\mathbb{1} & 0 \end{pmatrix}.$$

It has the four block decomposition

$$F(t) = \begin{pmatrix} A(t) & B(t) \\ C(t) & D(t) \end{pmatrix}.$$

In Chap. 3 it has been established that the quantum propagator $\mathscr{U}_q(t, 0)$ of $\hat{H}_Q(t)$ is nothing but the metaplectic operator $\hat{R}(F(t))$ associated to the symplectic matrix $F(t)$. It implies that the Heisenberg observables $\hat{Q}(t),\ \hat{P}(t)$ obey the classical Newton's equations for the Hamiltonian $\hat{H}_Q(t)$ as expected. Therefore we have:

Lemma 27

$$\mathscr{U}_q(0, t) \begin{pmatrix} \hat{Q} \\ \hat{P} \end{pmatrix} \mathscr{U}_q(t, 0) = F(t) \begin{pmatrix} \hat{Q} \\ \hat{P} \end{pmatrix}.$$

Passing from the $(\hat{Q}, \hat{P})$ representation to the $(\mathbf{a}^*, \mathbf{a})$ representation we easily get

Lemma 28

$$\mathscr{U}_q(0, t) \begin{pmatrix} \mathbf{a}^* \\ \mathbf{a} \end{pmatrix} \mathscr{U}_q(t, 0) = \frac{1}{2} \begin{pmatrix} Y_t & \bar{Z}_t \\ Z_t & \bar{Y}_t \end{pmatrix} \begin{pmatrix} \mathbf{a}^* \\ \mathbf{a} \end{pmatrix}.$$

Furthermore one can show that $\hat{R}(F(t))$ decomposes into a product of a rotation part and a "squeezing" part. Defining the complex matrices:

$$Y_t = A(t) + iB(t) - i(C(t) + iD(t)), \quad Z_t = A(t) + iB(t) + i(C(t) + iD(t)) \tag{4.9}$$

we have the following identity:

$$Z^*Z = Y^*Y - 4\mathbb{1}$$

and Y_t is invertible. The matrix $W_t = Z_t Y_t^{-1}$ is such that

$$W_t^* W_t < \mathbb{1}, \quad W_0 = 0.$$

Furthermore it has been shown in Chap. 3 (Lemmas 21 and 23) that it is a symmetric matrix. Thus one can define the matrix B_t according to (4.4); note that B_t is not to be confused with the matrix $B(t)$ of the four-block decomposition of $F(t)$. Introducing the polar decomposition of Y_t:

$$Y_t = |Y_t| V_t^*$$

where

$$|Y|^2 = YY^*$$

we see that V_t is a smooth function of t and we can define (at least locally in time) a smooth self-adjoint matrix Γ_t by

$$V_t = \exp(i\Gamma_t).$$

Let $\hat{R}(t)$ be the following unitary operator in $L^2(\mathbb{R}^n)$ (metaplectic transformation):

$$\hat{R}(t) = \exp\left\{\frac{i}{2}(\mathbf{a}^*, \mathbf{a})\begin{pmatrix} 0 & \bar{\Gamma}_t \\ \Gamma_t & 0 \end{pmatrix}\begin{pmatrix} \mathbf{a}^* \\ \mathbf{a} \end{pmatrix}\right\}.$$

It has the following property:

Lemma 29

$$\hat{R}(t)\begin{pmatrix} \mathbf{a}^* \\ \mathbf{a} \end{pmatrix}\hat{R}(t)^{-1} = \begin{pmatrix} V_t\mathbf{a}^* \\ (\bar{V}_t)^*\mathbf{a} \end{pmatrix}.$$

Remark 18 $\hat{R}(t)$ will be the rotation part of the metaplectic transformation $\mathscr{U}_q(t, 0)$ while $\hat{D}(B_t)$ will be the squeezing part.

One has the following property:

Proposition 38 *The quantum propagator solving the time-dependent Schrödinger equation*

$$i\hbar\frac{d}{dt}\mathscr{U}_q(t, s) = \hat{H}_Q(t)\mathscr{U}_q(t, s), \quad \mathscr{U}_q(s, s) = \mathbb{1},$$

is given by

$$\mathscr{U}_q(t, s) = \hat{D}(B_t)\hat{R}(t)\hat{R}(s)^{-1}\hat{D}(-B_s).$$

Due to the chain rule it is enough to show that

$$\mathscr{U}_q(t, 0) = \hat{D}(B_t)\hat{R}(t).$$

The detailed proof can be found in the paper [65].

Now we can derive an explicit formula for the quantum propagator of $\hat{H}_2(t)$. Let $S_t(z)$ be the classical action along the trajectory for the classical Hamiltonian $H(x, \xi, t)$ starting at phase-space point $z = (q, p)$ at time 0 and reaching $z_t = (q_t, p_t)$ at time t:

$$S_t(z) = \int_0^t ds[\dot{x}_s \cdot \xi_s - H(x_s, \xi_s, s)]$$

and define

$$\delta_t = S_t(z) - \frac{q_t \cdot p_t - q \cdot p}{2}. \tag{4.10}$$

The following result holds true:

Proposition 39 *Let $\mathscr{U}_2(t,s)$ be the quantum propagator for the hamiltonian $\hat{H}_2(t)$ given by (4.6). Then we have*

$$\mathscr{U}_2(t,s) = \exp[i(\delta_t - \delta_s)/\hbar]\hat{T}(z_t)\hat{D}(B_t)\hat{R}(t)\hat{R}(s)^{-1}\hat{D}(-B_s)\hat{T}(-z_s). \quad (4.11)$$

Proof Here we use the notation $z := \frac{q+ip}{\sqrt{2\hbar}}$.

Using the Baker–Campbell–Hausdorff formula (and omitting the index t in the following formulas) we get:

$$\frac{d}{dt}\hat{T}(z) = \hat{T}(z)\left[\dot{\bar{z}}\cdot\mathbf{a}^* - \dot{z}\cdot\mathbf{a} + \frac{1}{2}\left(z\cdot\dot{\bar{z}} - \dot{z}\cdot\bar{z}\right).\right]$$

But

$$\frac{1}{2}\left(z\cdot\dot{\bar{z}} - \dot{z}\cdot\bar{z}\right) = \frac{i}{2\hbar}\left(\dot{p}\cdot q - \dot{q}\cdot p\right), \text{ and } i\dot{\delta} = i\left[\dot{S} - \frac{1}{2}\left(p\cdot q + \dot{p}\cdot q\right)\right],$$

so that taking the derivative of (4.11) with respect to time t we get

$$\begin{aligned} i\hbar\frac{\partial\mathscr{U}_2}{\partial t} &= \left[-\dot{S} + \frac{1}{2}(p\cdot\dot{q} + \dot{p}\cdot q) - \frac{1}{2}(\dot{p}\cdot q - \dot{q}\cdot p) + \dot{q}\cdot\left(\hat{P} - p\right) - \dot{p}\cdot\left(\hat{Q} - q\right)\right]\mathscr{U}_2 \\ &\quad + \hat{T}(z_t)H_q(t)\hat{T}(-z_t)\mathscr{U}_2 \\ &= \left\{\dot{q}\cdot p - \dot{S} + \dot{q}\cdot\left(\hat{P} - p\right) - \dot{p}\cdot\left(\hat{Q} - q\right) + \frac{1}{2}\left(\hat{Q} - q, \hat{P} - p\right)M_t\begin{pmatrix}\hat{Q} - q \\ \hat{P} - p\end{pmatrix}\right\}\mathscr{U}_2 \\ &= \hat{H}_2(t)\mathscr{U}_2, \end{aligned} \quad (4.12)$$

where we have used that $\dot{q}\cdot p - \dot{S} = H(q,p,t)$. □

The important fact here will be that $\mathscr{U}_2$ propagates coherent states into Weyl translated squeezed states so that using (4.5) we get a comparison between the exact quantum evolution of coherent states and the Weyl-displaced squeezed state centered around the phase-space point z_t.

Recall the notations of Chap. 1. Let $\phi_{\mathbf{k}}$ be the multidimensional Hermite functions, $\mathbf{k} \in \mathbb{N}^n$. Let us denote $\phi_{\mathbf{k},z} := \hat{T}(z)\phi_{\mathbf{k}}$. Consider as an initial state a coherent state $\varphi_z = \hat{T}(z)\varphi_0$. We get

$$\mathscr{U}_2(t,0)\varphi_z = e^{i\delta_t/\hbar}\hat{T}(z_t)\hat{D}(B_t)\hat{R}(t)\Psi_0 = e^{i(\delta_t/\hbar+\gamma_t)}\phi_{z_t}^{(B_t}$$

using the fact that

$$\hat{R}(t)\phi_0 = \exp(i\gamma_t)\phi_0, \text{ where } \gamma_t = \frac{1}{2}\mathrm{tr}(\Gamma_t).$$

Therefore applying Duhamel's formula (4.5) to φ_z we get

$$\mathscr{U}(t,0)\varphi_z - e^{i(\delta_t/\hbar+\gamma_t)}\phi_{z_t}^{(E_t)} = \frac{1}{i\hbar}\int_0^t ds\,\mathscr{U}(t,s)[\hat{H}(s)-\hat{H}_2(s)]e^{i(\delta_s/\hbar+\gamma_s)}\,\phi_{z_s}^{(B_s)}. \tag{4.13}$$

This will be the starting point of our semiclassical estimate. Taking an arbitrary multiindex $\mathbf{k} = (k_1, \dots k_n)$ we denote

$$\phi_{\mathbf{k},z_t}^{(B_t)}(t) := \hat{T}(z_t)\hat{D}(B_t)\hat{R}(t)\phi_{\mathbf{k}}.$$

Starting from (4.13) we deduce that for any integer $l \geq 1$ there exist indexed functions $c_\nu(t,\hbar)$ such that

$$\left\| \mathscr{U}(t,0)\varphi_z - \sum_{|\mathbf{k}|=0}^{3(l-1)} c_{\mathbf{k}}(t,\hbar)\phi_{\mathbf{k},z_t}^{(B_t)}(t)e^{i\delta_t/\hbar} \right\| \leq C_t\hbar^{l/2}. \tag{4.14}$$

Furhermore the constant C_t can be controlled in the time t and in the center z of the initial state. Let us now explicit the Taylor expansion of the Hamiltonian around the classical phase-space point z_t at time t: we denote

$$\frac{f^{(\nu)}(\zeta)}{\nu!}\cdot(\zeta')^\nu = \prod_1^{2n}\frac{1}{\nu_j!}\frac{\partial^{\nu_j}f(\zeta)}{\partial\zeta_j^{\nu_j}}(\zeta_j')^{\nu_j}, \quad \nu = (\nu_1, \dots \nu_{2n})$$

for f being a real function of $\zeta \in \mathbb{R}^{2n}$.

Taking for ζ the phase-space point z we can write the Taylor expansion of $H(t,q,p)$ around the phase-space point $z_t = (q_t, p_t)$ at time t as:

$$\begin{aligned} H(\zeta,t) &= \sum_{|\nu|=0}^{l+1}\frac{H^{(\nu)}(z_t,t)}{\nu!}\cdot(\zeta-z_t)^\nu \\ &+ \sum_{|\nu|=l+2}\int_0^1\frac{H^{(\nu)}(t,z_t+\theta(\zeta-z_t))}{(\nu-1)!}\cdot(\zeta-z_t)^\nu(1-\theta)^{l+1}\,d\theta \\ &= \sum_{|\nu|=0}^{l+1}\frac{H^{(\nu)}(t,z_t)}{\nu!}\cdot(\zeta-z_t)^\nu + \sum_{|\nu|=l+2} r_{\nu,t}(\zeta-z_t,t)\cdot(\zeta-z_t)^\nu. \end{aligned} \tag{4.15}$$

Now we perform the $\hbar$-quantization of (4.6) denoting $\Omega := (\hat{Q}, \hat{P})$; we get:

$$\hat{H}(t) - \hat{H}_2(t) = \sum_{|\nu|=3}^{l+1}\frac{H^\nu(z_t,t)}{\nu!}\cdot(\Omega-z_t)^\nu + \sum_{|\nu|=l+2}R_\nu(t), \tag{4.16}$$

where

$$R_\nu(t) = \hat{T}(-z_t)\mathrm{Op}_\hbar^w[\zeta^\nu\cdot r_{\nu,t}(\zeta)]\hat{T}(z_t).$$

We now insert (4.16) into (4.13) and obtain the semiclassical estimate of order $\sqrt{\hbar}$ for the propagation of generalized coherent states, using in particular the Calderon-Vaillancourt estimate (Chap. 2). Let us introduce some notation to control the estimates for large times: we consider only nonnegative time and define

$$\sigma(z,t) := \sup_{0\le s\le t}(1+|z_s|)$$

$$\theta(z,t) = \sup_{0\le s\le t}[\mathrm{tr}(F^*(s)F(s))]^{1/2}$$

where $F(t)$ is the symplectic matrix solution of (4.8). Let M_1 be any fixed integer not smaller than $(M+(m-2)_+)/2$. Let us define

$$\rho_l(z,t,\hbar) = \sigma(z,t)^{lM_1}\sum_{1\le j\le l}\left(\frac{|t|}{\hbar}\right)^j(\sqrt{\hbar}\theta(z,t))^{2j+l}.$$

Let us recall that the constants $M, m, K_{H,T}$ were defined in Assumption (A.0). Note that if (A.0) is satisfied with $m=2$ and $M=0$, ($H(t)$ is said subquadratic) then we have $M_1=0$. The result will be a generalization of (4.14) using as initial state a superposition of generalized coherent states (defined with higher order Hermite functions).

Theorem 21 *Assume $H(t,x,\xi)$ and z are such that (A.0)–(A.2) are satisfied. Then for any integers $l\ge 1$, $J\ge 1$ and any real number $\kappa>0$ there exist a universal constant $\Gamma>0$ such that for every family of complex numbers $\{c_\mu,\ \mu\in\mathbb{N}^n,\ |\mu|\le J\}$ there exist $c_\nu(t,\hbar)$ for $\nu\in\mathbb{N}^n$, $|\nu|\le 3(l-1)+J$ such that for $0<\hbar+\sqrt{\hbar}\theta(t)<\kappa$ the following L^2-estimate holds:*

$$\left\|\mathscr{U}(t,0)\left(\sum_{|\mathbf{j}|=0}^{J}c_{\mathbf{j}}\phi_{\mathbf{j},z}\right)-e^{i\delta_t/\hbar}\sum_{|\mathbf{k}|=0}^{J+3(l-1)}c_{\mathbf{k}}(t,\hbar)\phi^{(B_t)}_{\mathbf{k},z_t}(t)\right\|\le$$
$$\Gamma K_{H,T}\rho_l(z,\hbar,t)\left(\sum_{0\le|\mu|\le J}|c_\mu|^2\right)^{1/2}. \qquad (4.17)$$

Moreover the coefficients $c_\mu(t,\hbar)$ can be computed by the following formula:

$$c_\mu(t,\hbar)-c_\mu=$$
$$\sum_{\substack{|\nu|\le J\\|\mu-\nu|\le 3l-3}}\ \sum_{1\le p\le l-1}\left(\sum_{\substack{k_1+\cdots+k_p\le 2p+l-1\\k_j\ge 3}}\hbar^{(k_1+\cdots+k_p)/2-p}a_{p,\mu,\nu}(t)\right)c_\nu, \qquad (4.18)$$

where the entries $a_{p,\mu,\nu}(t)$ are given by the evolution of the classical system and are universal polynomials in $\partial_z^\gamma H(t, z_t)$ for $|\gamma| \le l + 2$ satisfying $a_{p,\mu,\nu}(0) = 0$.

Remark 19 Comments on the error estimate and the Ehrenfest time
The error term seems accurate but not very explicit in our general setting. Let us assume for simplicity that $T = +\infty$ and that the classical trajectory z_t is bounded and unstable with a Lyapounov exponent $\lambda > 0$. So there exists some constant $C > 0$ such that

$$\theta(z, t) \le Ce^{\lambda t}, \quad \forall t \ge 0.$$

Then for every $\varepsilon > 0$ there exists C_ε and $h_\varepsilon > 0$ such that

$$0 < t \le \frac{1 - 3\varepsilon}{6\lambda} \log(\frac{1}{\hbar}) \Rightarrow \rho_\ell(z, \hbar, t) \le C_\varepsilon \hbar^{\varepsilon\ell} \tag{4.19}$$

for $0 < \hbar < h_\varepsilon$. So roughly speaking we can say that the semi-classical expansion (4.17) is still valid for times smaller than the Ehrenfest time $T_E := \frac{1}{6\lambda}\log(\frac{1}{\hbar})$.

When the classical trajectory is stable ($\theta(z, t) \le C(1 + |t|), \forall t \in \mathbb{R}$) then (4.17) is valid for much longer time interval: $0 \le t \le C_\varepsilon \hbar^{-\frac{1}{2}-\varepsilon}$.

Remark 20 Let us denote $\psi^{(l)}(t, x) = e^{i\delta_t/\hbar}\sum_{|\mathbf{k}|=0}^{J+3(l-1)} c_{\mathbf{k}}(t, \hbar)\phi_{\mathbf{k},z_t}^{(B_t)}(t, x)$ and assume for simplicity that $\psi(0) = \varphi_z$ and denote $\psi_z(t) = \mathscr{U}(t, 0)\varphi_z$. $\psi^{(l)}(t, x)$ is clearly localized very close to the point z_t of the classical path. Moreover the shape of $\psi^{(l)}(t, x)$ is close to a Gaussian shape with center at z_t and with a complex covariance matrix Γ_t depending on the stability matrix of the classical system. This shape evolves with time (see the next figure). It is exponentially small outside any ball $\{|x - q_t| \le \hbar^{\frac{1}{2}-\varepsilon}\}$, for any ε and its $\hbar$-Fourier transform $\tilde{\psi}_z(t, \xi)$ exponentially small outside any ball $\{|\xi - p_t| \le \hbar^{\frac{1}{2}-\varepsilon}\}$. From (4.17) we get easily for the exact evolved state $\psi_z(t)$ the following probability estimates to be true outside of a narrow tube around the classical path:

$$\int_{|x-q_t|\ge\hbar^{\frac{1}{2}-\varepsilon}} dx|\psi_z(t, x)|^2 + \int_{|\xi-p_t|\ge\hbar^{\frac{1}{2}-\varepsilon}} d\xi|\tilde{\psi}_z(t, \xi)|^2 = O(\hbar^{+\infty}). \tag{4.20}$$

Proof We only give here the strategy of the proof. For the technical details we refer the reader to [65] or to [218] for a different approach. The idea of the proof is to use repeatidly the Duhamel Formula to yield a Dyson series expansion of $\mathscr{U}(t, 0) - \mathscr{U}_2(t, 0)$ and to use (4.16) to calculate $\hat{H}(t) - \hat{H}_2(t)$. One thus has for every integer $p \ge 1$:

$$\mathcal{U}(t,0) - \mathcal{U}_2(t,0) =$$
$$\sum_{1\le j\le p} (i\hbar)^{-j} \int_0^t dt_1 \int_{t_1}^t dt_2 \cdots \int_{t_{j-1}}^t dt_{j-1} \mathcal{U}_2(t,t_j)(\hat{H}(t_j) - \hat{H}_2(t_j))\mathcal{U}_2(t_j,t_{j-1})$$
$$\times(\hat{H}(t_{j-1}) - \hat{H}_2(t_{j-1}))\mathcal{U}_2(t_{j-1},t_{j-2})\cdots(\hat{H}(t_1) - \hat{H}_2(t_1))\mathcal{U}_2(t_1,0)$$
$$+(i\hbar)^{-p-1} \int_0^t dt_1 \int_{t_1}^t dt_2 \cdots \int_{t_p}^t dt_{p+1} \mathcal{U}(t,t_{p+1})(\hat{H}(t_{p+1}) - \hat{H}_2(t_{p+1}))\mathcal{U}_2(t_{p+1},t_p)$$
$$\times(\hat{H}(t_p) - \hat{H}_2(t_p))\mathcal{U}_2(t_p,t_{p-1})\cdots(\hat{H}(t_1) - \hat{H}_2(t_1))\mathcal{U}_2(t_1,0). \tag{4.21}$$

We shall apply this formula with $p = l - 1$. It is convenient to introduce the following notation:

$$\Pi(t_1,\ldots t_j; B_1,\ldots B_j) = \mathcal{U}_2(0,t_j)B_j\mathcal{U}_2(t_j,t_{j-1})\ldots\mathcal{U}_2(t_2,t_1)B_1\mathcal{U}_2(t_1,0)$$

where $B_k = \mathrm{Op}_\hbar^w b_k$ are quantum observables. By the chain rule we have:

$$\Pi(t_1,\ldots t_j; B_1,\ldots B_j) = \prod_{1\le k\le j} \mathcal{U}_2(0,t_k)B_k\mathcal{U}_2(t_k,0)$$

where the product is ordered from the right to the left. Denoting

$$\hat{H}_k(t) := \sum_{|\nu|=k} \frac{H^{(\nu)}(z_t,t)}{\nu!} \cdot (\Omega - z_t)^\nu$$

we have that $\Pi(t_1,\ldots t_p; k_1,\ldots k_p) := \Pi(t_1,\ldots t_p; \hat{H}_{k_1},\ldots \hat{H}_{k_p})$ is a homogeneous non-commutative polynomial in $\hat{Q}$, $\hat{P}$ of degree $k_1 + \cdots k_p$ using Lemma (27). Therefore the last integral in (4.21) will yield the error term in the theorem. For the detailed proof we refer the reader to [65]. □

A more general result is obtained by taking as initial state a coherent state with arbitrary profile (see Chap. 1, Sect. 1.1.2): let $f \in \mathcal{S}(\mathbb{R}^n)$ to be the profile of a state. Define

$$(\Lambda_\hbar f)(x) = \hbar^{-n/4} f\left(\frac{x}{\sqrt{\hbar}}\right). \tag{4.22}$$

Then for any $z \in \mathbb{R}^{2n}$ we construct a coherent state centered at z by

$$f_z := \hat{T}(z)\Lambda_\hbar f$$

One has the following result:

Theorem 22 *There exists a family of differential operators with time-dependent coefficients,*

$$p_{kj}(x, D_x, t), \quad j \ge 1,\ k \ge 3,$$

depending only on the Hamiltonian along the classical path z_s, $0 \le s \le t$, $p_{jk}(x, \xi, t)$ *being a polynomial in* (x, ξ) *of degree* $\le k$ *such that for any real number* $\kappa > 0$ *and any integer* $l \ge 1$ *and any* $f \in \mathcal{S}(\mathbb{R}^n)$ *there exists* $\Gamma > 0$ *such that the following* L^2*-norm estimate holds:*

$$\left\| \mathcal{U}(t,0)\hat{T}(z)\Lambda_\hbar f - \hat{T}(z_t)\mathcal{U}_2(t,0)\Lambda_\hbar P_l(f,t,\hbar) \right\| \le \Gamma K_{H,t}\rho_l(z,t,\hbar), \quad (4.23)$$

where $P_l(.,t,\hbar)$ *is the* $(\hbar, t)$*-dependent differential operator defined by*

$$P_l(f,t,\hbar) = f + \sum_{(k,j)\in I_l} \hbar^{k/2-j} p_{jk}(t,x,D_x)f,$$

with $I_l = \{(k,j) \in \mathbb{N}\times\mathbb{N} : \ 1 \le j \le l-1, \ k \ge 3j, \ 1 \le k-2j \le l\}$.

Moreover the polynomials $p_{kj}(t,x,\xi)$ *can be computed explicitly in terms of the Weyl symbol of the following differential operators defined above*

$$\int_0^t dt_l \int_0^{t_l} dt_{l-1} \ldots \int_0^{t_2} dt_1 \Pi(t_1,\ldots t_l; k_1,\ldots k_l).$$

Remark 21 If we consider an $\hbar$-metaplectic transformation V in $L^2(\mathbb{R}^n)$ we consider the following rescaling $\tilde{V} = \Lambda_\hbar^{-1} V \Lambda_\hbar$ (with no confusion with the notation for the transpose of a matrix). By definition since V has a quadratic generator $\tilde{V}$ is $\hbar$-independent. So we have that $\mathcal{U}_2(t,0)\Lambda_\hbar P_l(f,t,\hbar)$ is actually a coherent state centered at z_t with profile $\tilde{D}(B_t)\tilde{R}(t)f$. More explicitly

$$\mathcal{U}_2(t,0)\Lambda_\hbar P_l(f,t,\hbar) = e^{i\delta_t/\hbar}\hat{T}(z_t)\Lambda_\hbar \left(\sum_{0\le j\le l-1} \hbar^{j/2} p_j(t,x,D_x)\tilde{D}(B_t)\tilde{R}(t)f \right)$$

where $p_j(x, D_x, t)$ are differential operators with polynomial coefficients depending smoothly on t as long as the classical flow $z \to z_t$ exists.

4.1.3 Related Works and Other Results

In the physics literature the quantum propagation of coherent states has been considered by many authors, in particular by Heller [144, 145] and Littlejohn [182]. In the mathematical literature gaussian wave packets have been introduced and studied in many respects, particularly under the name "gaussian beams" (see [10, 212, 213]). Somehow related to the subject of this Chapter is the study by Paul and Uribe [205, 206] of the $\hbar$-asymptotics of the inner products of the eigenfunctions of a Schrödinger type Hamiltonian with a coherent state and of "semiclassical trace formulas" (see Chap. 5). However their approach differs from the one presented here by the use of Fourier-integral operators, which were introduced in connection with wave packets propagation in the classical paper by Cordoba and Fefferman [68].

4.2 Application to the Spreading of Quantum Wave Packets

In this section we give an application of the estimate of the preceding section to the spreading (in phase space) of a quantum wave packet which is, at time 0 localized in the neighborhood of a fixed point of the corresponding classical motion. Let $z = (q, p)$ be such a fixed point and take as an initial quantum state the coherent state φ_z. The quantum state at time t is

$$\psi_z(t) = \mathscr{U}(t, 0)\varphi_z.$$

We have seen that we can approximate $\Psi(t)$ by a Gaussian wavepacket again localized around $z_t = z$ but with a spreading governed by the stability matrix M_0 of the corresponding classical motion (given by (4.7)). A way of measuring the spreading of wave packets around the point z in phase space is to compute

$$\begin{aligned} S(t) &= \left\langle \hat{T}(-z)\psi_z(t), \sum_{j=1}^{n}(a_j^* a_j + a_j a_j^*)\hat{T}(-z)\psi_z(t)\right\rangle \\ &= \|\mathbf{a}\hat{T}(-z)\psi_z(t)\|^2 + \|\mathbf{a}^*\hat{T}(-z)\psi_z(t)\|^2. \end{aligned} \tag{4.24}$$

The intuition behind this definition is the following. Let $\mathscr{W}_{z,t}$ be the Wigner function of the states $\psi_z(t)$. According to the properties seen in Chap. 2, $(2\pi\hbar)^{-n}\mathscr{W}_{z,t}(X)$ is a quasi-probability on the phase space $\mathbb{R}^{2n}_X$ and we have

$$(2\pi\hbar)^{-n}\int dX \mathscr{W}_{z,t}(X)A(X) = \langle\Psi(t), \hat{A}\psi_z(t)\rangle.$$

Applying this relation to $\hat{A} = \hat{T}(z)(\mathbf{a}^* \cdot \mathbf{a} + \mathbf{a} \cdot \mathbf{a}^*)\hat{\mathbf{T}}(-\mathbf{z})$ which has the Weyl-symbol $A(q, p) = |q - x|^2 + |p - \xi|^2$, we see that $S(t)$ is the variance of the quasi-probability $(2\pi\hbar)^{-n}\mathscr{W}_{z,t}(X)$.

Let us notice that $S(t)$ is well defined if the estimate (4.14) holds in the Sobolev space $\Sigma_2^\hbar(\mathbb{R}^n)$ (defined in Chap. 1) and with some more assumption on the quantum evolution $\mathscr{U}(t, 0)$ one can get the estimate (4.14) in $\Sigma_2^\hbar$-norm as we shall see now. More refined estimates in other Sobolev norms will be given in the last section of this chapter.

Let us consider some symbol g satisfying assumption (A.0) with $m = 0$, such that $\mathrm{Op}_\hbar^w(g)$ is invertible in the Schwartz space $\mathscr{S}(\mathbb{R}^n)$. Let us assume that the following L^2-operator norm estimate holds:

$$\|(\mathrm{Op}_\hbar^w g(\mathcal{U}(t,0)(\mathrm{Op}_\hbar^w g)^{-1}\| \leq C_{t,g}.$$

Then an estimate analogous to (4.14) holds true:

$$\|\mathrm{Op}_\hbar{}^w g\mathcal{U}(t,0)\hat{T}(z)\Lambda_\hbar f - \mathcal{U}_2(t,0)\Lambda_\hbar P_l(f,t,\hbar)\| \leq C_{t,g}\Gamma K_{H,tp_l(z,t,\hbar)} \quad (4.25)$$

Obviously $S(0) = n$ and we are interested in the difference

$$\Delta S(t) := S(t) - S(0)$$

Let us first calculate

$$T(t) := \left\langle \hat{T}(-z)\phi_z(t), \sum_{j=1}^{n}(a_j^* a_j + a_j a_j^*)\hat{T}(-z)\phi_z(t) \right\rangle,$$

where $\phi_z(t)$ is the approximant of $\psi_z(t)$ given by

$$\phi_z(t) = e^{i\delta_t/\hbar}\hat{T}(z)\mathcal{U}_0(t)\psi_0.$$

Then

$$T(t) = n + 2\|\mathbf{a}\mathcal{U}_0(t)\psi_0\|^2 = n + \frac{1}{2}\mathrm{tr}(Z_t^* Z(t),$$

where Z_t is defined bt (4.9) and using Lemma (28). Therefore

$$T(t) - T(0) = \frac{1}{2}\mathrm{tr}(Z_t^* Z_t).$$

We show now that this is the dominant behavior of $\Delta S(t)$ up to small correction terms that we can estimate.

Let us assume that the classical flow at phase-space point z has finite Lyapunov exponents, with a greatest Lyapunov exponent $\lambda \in \mathbb{R}$. (for notions concerning the stability and Lyapunov exponents for ordinary differential equations we refer to [49]). Then by definition there exists some constant $C > 0$ such that $\|F(t)\| \leq Ce^{\lambda t}$, $\forall t \geq 0$ where C is independent of t. In what follows we denote by C a generic constant independent of t, $\hbar$. Then under the above assumptions we get

$$\boxed{\|\psi_z(t) - \varphi_z(t)\|_{\Sigma_\hbar^2(\mathbb{R}^n)} \leq C\sqrt{\hbar}te^{3\lambda t}, \ \forall t \geq 0}.$$

We deduce the following result:

Theorem 23 *Under the above assumptions we have the long time asymptotics* $\Delta S(t) = \Delta T(t) + O(\hbar^\varepsilon)$ *if one of the two following conditions is fulfilled:*

(i) $\lambda \leq 0$ *("stable case") and* $0 \leq t \leq \hbar^{\varepsilon-1/2}$

(ii) $\lambda > 0$ *("unstable case") and* $\exists \varepsilon' > \varepsilon$ *such that*

$$0 \le t \le ((1 - 2\varepsilon')/6\lambda)\log(1/\hbar).$$

In particular we have

Corollary 14 *Let us assume that the Hamiltonian* H *is time independent and that the greatest Lyapunov exponent is* $\lambda > 0$*. Then* $S(t) - S(0)$ *behaves like* $e^{2\lambda t}$ *as* $t \to +\infty$ *and* $\hbar \to 0$ *as long as* $t[\log(1/\hbar)]^{-1}$ *stays small enough.*

(i) More precisely there exists $C > 0$ *such that*

$$\frac{e^{2\lambda t}}{C} \le \Delta T(t) \le Ce^{2\lambda t}, \quad \forall t \ge 0.$$

$$\Delta S(t) = \Delta T(t) + O(\hbar^{\varepsilon}), \text{ for } 0 \le t \le \frac{1 - 2\varepsilon'}{6\lambda} \log\frac{1}{\hbar} \text{ for some } \varepsilon' > \varepsilon.$$

(ii) In particular for $n = 1$ *we have a more explicit result:*

$$\Delta S(t) = \frac{4b^2 + (a - c)^2}{2(b^2 - ac)} \sinh^2(\lambda t) + O(\hbar^{\varepsilon}), \tag{4.26}$$

under the above condition for t *where* $\begin{pmatrix} a & b \\ b & c \end{pmatrix} = M_0$ *is the Hessian matrix of* H *at* z*.*

Proof We get (i) using that the matrix JM_0 has at least one eigenvalue with real part λ. To prove (4.26) we compute explicitly the exponential of the matrix tJM_0 which gives $F(t)$. Its eigenvalues are $\lambda = \sqrt{b^2 - ac}$ and $1/\lambda$. So we get the formula

$$\exp(tJM_0) = \begin{pmatrix} \cosh(\lambda t) + \frac{b}{\lambda}\sinh(\lambda t) & \frac{c}{\lambda}\sinh(\lambda t) \\ -\frac{a}{\lambda}\sinh(\lambda t) & \cosh(\lambda t) - \frac{b}{\lambda}\sinh(\lambda t) \end{pmatrix}$$

hence we have

$$\Delta T(t) = \operatorname{tr}(Z_t^* Z_t) = \frac{4b^2 + (a - c)^2}{2(b^2 - ac)} \sinh^2(\lambda t).$$

□

4.3 Evolution of Coherent States and Bargmann Transform

In Sect. 4.1 we have studied the evolution of coherent states using the generators of coherent states and the Duhamel formula. Here we present a different approach following [218], working essentially on the Fourier–Bargmann side (see Chap. 1). This approach is useful to get estimates in several norms of Banach spaces of functions and also to get analytic type estimates.

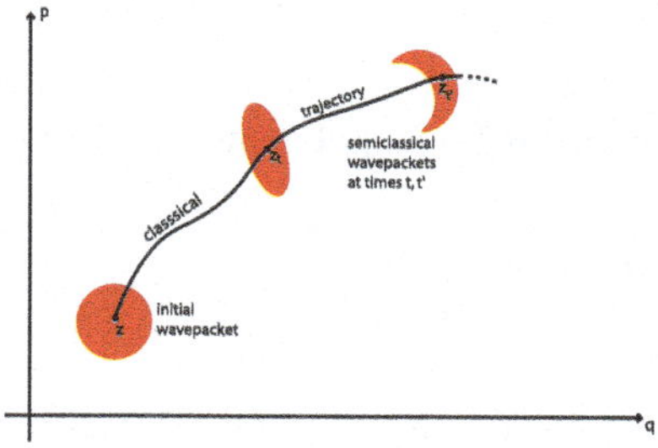

Time evolution of acoherent state

We keep the notations of Sect. 4.1. We revisit now the algebraic computations of this section in a different presentation.

Recall that we want to solve the Cauchy problem

$$i\hbar\frac{\partial\psi_z(t)}{\partial t} = \widehat{H}(t)\psi_z(t), \quad \psi(0) = \varphi_z, \tag{4.27}$$

where φ_z is a coherent state localized at a point $z \in \mathbb{R}^{2n}$. Our first step is to transform this problem with suitable unitary transformations such that the singular perturbation problem in $\hbar$ becomes a regular perturbation problem. A mathematical statement is in Theorem 24.

4.3.1 *Formal Computations*

We rescale the evolved state $\psi_z(t)$ by defining f_t such that $\psi_z(t) = \hat{T}(z_t)\Lambda_\hbar f_t$. Then f_t satisfies the following equation.

$$i\hbar\partial_t f_t = \Lambda_\hbar^{-1}\hat{T}(z_t)^{-1}\left(\widehat{H}(t)\hat{T}(z_t) - i\hbar\partial_t\hat{T}(z_t)\right)\Lambda_\hbar f_t \tag{4.28}$$

with the initial condition $f_{t=0} = g$ where $g(x) = \pi^{-n/4}e^{-\frac{1}{2}|x|^2}$. We get easily the formula

$$\Lambda_\hbar^{-1}\hat{T}(z_t)^{-1}\widehat{H}(t)\hat{T}(z_t)\Lambda_\hbar = Op_1^w H(t, \sqrt{\hbar}x + q_t, \sqrt{\hbar}\xi + p_t). \tag{4.29}$$

Using the Taylor formula we get the formal expansion

$$H(t, \sqrt{\hbar}x + q_t, \sqrt{\hbar}\xi + p_t) = H(t, z_t) + \sqrt{\hbar}\partial_q H(t, z_t)x + \sqrt{\hbar}\partial_p H(t, z_t)\xi \\ + \hbar K_2(t; x, \xi) + \hbar \sum_{j\geq 3} \hbar^{j/2-1} K_j(t; x, \xi), \quad (4.30)$$

where $K_j(t)$ is the homogeneous Taylor polynomial of degree j in $X = (x, \xi) \in \mathbb{R}^{2n}$.

$$K_j(t; X) = \sum_{|\gamma|=j} \frac{1}{\gamma!} \partial_X^\gamma H(t; z_t) X^\gamma.$$

We shall use the following notation for the remainder term of order $k \geq 1$,

$$R_k(t; X) = \left(H(t, z_t + \sqrt{\hbar}X) - \sum_{j<k} \hbar^{j/2} K_j(t; X) \right). \quad (4.31)$$

It is clearly a term of order $\hbar^{k/2}$ from the Taylor formula. By a straightforward computation, the new function $f_t^\# = \exp\left(-i\frac{\delta_t}{\hbar}\right) f_t$ satisfies the following equation

$$i\partial_t f_t^\# = \mathrm{Op}_1^w[K_2(t)]f^\# + Op_1^w[R_z^{(3)}(t)]f^\#, \quad f_{t=t_0}^\# = g. \quad (4.32)$$

Recall the definition of δ_t in (4.10) and Op_1^w in (2.2).

In the r.h.s of Eq. (4.32) the second term is a (formal) perturbation series in $\sqrt{\hbar}$. We change again the unknown function $f_t^\#$ by $b(t)g$ such that $f_t^\# = \hat{R}[F_t]b(t)g$. Let us recall that the metaplectic transformation $\hat{R}[F_t]$ is the quantum propagator associated with the Hamiltonian $K_2(t)$ (see Sect. 4.1). The new unknown function $b(t, x)$ satisfies the following differential equation which is now a regular perturbation differential equation in the the small parameter $\hbar$.

$$\begin{aligned} i\partial_t b(t, x)g(x) &= Op_1^w[R_z^{(3)}(t, F_t(x, \xi)](b(t)g)(x) \\ b(0, x) &= 1. \end{aligned} \quad (4.33)$$

Now we can solve Eq. (4.33) semiclassically by the ansatz

$$b(t, x) = \sum_{j\geq 0} \hbar^{j/2} b_j(t, x).$$

Let us identify powers of $\hbar^{1/2}$, denoting

$$K_j^\#(t, X) = K_j(t, F_t(X)), \quad X \in \mathbb{R}^{2n},$$

we thus get that the $b_j(t, x)$ are uniquely defined by the following induction formula for $j \geq 1$, starting with $b_0(t, x) \equiv 1$,

$$\partial_t b_j(t,x)g(x) = \sum_{k+\ell=j+2,\ \ell\geq 3} Op_1^w[K_\ell^\#(t)](b_k(t,\cdot)g)(x) \tag{4.34}$$

$$b_j(0,x) = 0. \tag{4.35}$$

Let us remark that $Op_1^w[K_\ell^\#(t)]$ is a differential operator with polynomial symbols of degree ℓ in (x,ξ). So it is not difficult to see, by induction on j, that $b_j(t)$ is a polynomial of degree $\leq 3j$ in variable $x \in \mathbb{R}^n$ with complex time dependent coefficient depending on the center z of the Gaussian in the phase space. Moreover, coming back to the Schrödinger equation, using our construction of the $b_j(t,x)$, we easily get for every $N \geq 0$,

$$i\hbar\partial_t \psi_z^{(N)}(t) = \widehat{H}(t)\psi_z^{(N)}(t) + R_z^{(N)}(t), \tag{4.36}$$

where

$$\psi_z^{(N)}(t) = \mathrm{e}^{i\delta_t/\hbar}\hat{T}(z_t)\Lambda_\hbar\hat{R}[F_t]\left(\sum_{0\leq j\leq N} \hbar^{j/2}b_j(t)g\right) \tag{4.37}$$

and

$$R_z^N(t,x) = \mathrm{e}^{i\delta_t/\hbar}\left(\hbar^{(N+3)/2}\sum_{\substack{j+k=N+3\\ k\geq 3}} \hat{T}(z_t)\Lambda_\hbar\hat{R}[F_t]Op_1^w[R_k(t)\circ F_t](b_j(t)g)\right). \tag{4.38}$$

Thus, we have an algorithm to build approximate solutions $\psi_z^{(N)}(t,x)$ of the Schrödinger equation (4.27) modulo the error term $R_z^{(N)}(t,x)$. Of course the hard mathematical work is to estimate accurately this error term.

Remark that all these computations use only existence of the classical trajectory.

We need some technical estimates concerning the Fourier–Bargmann transform to have a bridge between the Bargmann side to the usual configuration space. We refer to [218] for the proofs of the following subsection.

4.3.2 Weighted Estimates and Fourier–Bargmann Transform

We restrict here our study to properties we need later. For other interesting properties of the Fourier–Bargmann transform the reader may consult the book [188].

Recall that in Chap. 1, the Fourier–Bargmann transform $\mathcal{F}^{\mathcal{B}}$ (here $\hbar = 1$) was defined as follows

$$\mathcal{F}^{\mathcal{B}}v(X) =: v^\sharp(z) = (2\pi\hbar)^{-n/2}\langle\varphi_z, v\rangle,\ X = (q,p) \in \mathbb{R}^{2n}.$$

Let us begin with the following formulae, easy to prove by integration by parts. With the notations $X = (q, p) \in \mathbb{R}^{2n}$, $x \in \mathbb{R}^n$ and $u \in \mathscr{S}(\mathbb{R}^n)$, we have

$$\mathscr{F}^{\mathscr{B}}(xu)(X) = i(\partial_p - \frac{i}{2}q)\mathscr{F}^{\mathscr{B}}(u)(X). \tag{4.39}$$

$$\mathscr{F}^{\mathscr{B}}(\partial_x u)(X) = (\frac{3}{2i}p - \partial_q)\mathscr{F}^{\mathscr{B}}(u)(X) \tag{4.40}$$

$$= (i(p - \partial_p) + \frac{q}{2})\mathscr{F}^{\mathscr{B}}(u)(X). \tag{4.41}$$

Recall that $\mathrm{e}^{(p^2+q^2)/4}\mathscr{F}^{\mathscr{B}}(u)(q, p)$ is holomorphic in the complex variable $q - ip$ (see Chap. 1).

So, let us introduce the weighted Sobolev spaces, denoted here $\mathscr{K}_m(n)$, $m \in \mathbb{N}$. $u \in \mathscr{K}_m(n)$ means that $u \in L^2(\mathbb{R}^n)$ and $x^\alpha \partial_x^\beta u \in L^2(\mathbb{R}^n)$ for every multiindex α, β such that $|\alpha + \beta| \le m$, with its natural norm. Then we have easily

Proposition 40 *The Fourier–Bargmann transform is a linear continuous function from $\mathscr{K}_m(n)$ into $\mathscr{K}_m(2n)$ for every $m \in \mathbb{N}$.*

Now we shall give an estimate in exponential weighted Lebesgue spaces.

Proposition 41 *For every $p \in [1, +\infty]$, for every $a \ge 0$ and every $b > a\sqrt{2}$ there exists $C > 0$ such that for all $u \in \mathscr{S}(\mathbb{R}^n)$ we have,*

$$\|\mathrm{e}^{a|x|}u(x)\|_{L^p(\mathbb{R}^n_x)} \le C\|\mathrm{e}^{b|X|}\mathscr{F}^{\mathscr{B}}u(X)\|_{L^2(\mathbb{R}^{2n}_X)}. \tag{4.42}$$

More generally, for every $a \ge 0$ and every $b > a\frac{\sqrt{2}}{|S|}$ there exists $C > 0$ such that for all $u \in \mathscr{S}(\mathbb{R}^n)$ and all $S \in \mathrm{Sp}(n)$ we have

$$\|\mathrm{e}^{a|x|}\left[\hat{R}(S)u\right](x)\|_{L^p(\mathbb{R}^n_x)} \le C\left\|\mathrm{e}^{b|X|}\mathscr{F}^{\mathscr{B}}u(X)\right\|_{L^2(\mathbb{R}^{2n}_X)}. \tag{4.43}$$

We need to control the norms of Hermite functions (see Chap. 1) in some weighted Lebesgue spaces. Let μ be a C^∞-smooth and positive function on $\mathbb{R}^m$ such that

$$\lim_{|x|\to+\infty} \mu(x) = +\infty \tag{4.44}$$

$$|\partial^\gamma \mu(x)| \le \theta|x|^2,\ \forall x \in \mathbb{R}^m, |x| \ge R_\gamma. \tag{4.45}$$

for some $R_\gamma > 0$ and $\theta < 1$.

Lemma 30 *For every real $p \in [1, +\infty]$, for every $\ell \in \mathbb{N}$, there exists $C > 0$ such that for every $\alpha, \beta \in \mathbb{N}^m$ we have:*

$$\|\mathrm{e}^{\mu(x)}x^\alpha\partial_x^\beta(\mathrm{e}^{-|x|^2})\|_{\ell,p} \le C^{|\alpha+\beta|+1}\Gamma\left(\frac{|\alpha+\beta|}{2}\right), \tag{4.46}$$

where $\| \bullet \|_{\ell,p}$ *is the norm on the Sobolev space*[3] $W^{\ell,p}$, Γ *is the Euler Gamma function.*[4]

More generally, for every real $p \in [1, +\infty]$, *for every* $\ell \in \mathbb{N}$, *there exists* $C > 0$ *such that*

$$\|e^{\mu(\Im(\Gamma)^{-1/2}x)}x^\alpha\partial^\beta(e^{-|x|^2})\|_{\ell,p} \le C^{|\alpha+\beta|+1}\left(|\Im(\Gamma)^{1/2}| + |\Im(\Gamma)^{-1/2}|\Gamma\left(\frac{|\alpha+\beta|}{2}\right)\right). \tag{4.47}$$

4.3.3 Large Time Estimates and Fourier–Bargmann Analysis

In this section we try to control the semi-classical error term $R_z^{(N)}(t, x)$, for large time, in the Fourier–Bargmann representation. This is also a preparation to control the remainder of order N in $\hbar$, N for analytic Hamiltonians considered in the following sub-section.

Let us introduce the Fourier–Bargmann transform of $b_j(t)g$,

$$B_j(t, X) = \mathcal{F}^{\mathcal{B}}[b_j(t)g](X) = \langle g_X, b_j(t)g\rangle, \text{ for } X \in \mathbb{R}^{2n}.$$

The induction Eq. (4.34) becomes for $j \ge 1$,

$$\partial_t B_j(t, X) = \int_{\mathbb{R}^{2n}}\left(\sum_{\substack{k+\ell=j+2\\ \ell\ge 3}} < g_X, Op_1^w[K_\ell^\#(t)]g_{X'} >\right) B_k(t, X')dX', \tag{4.48}$$

with initial condition $B_j(0, X) = 0$ for $j \ge 1$ and with $B_0(t, X) = \exp\left(-\frac{|X|^2}{4}\right)$.

We have seen in the Sect. 4.1 that we have

$$\langle Op_1^w[K_\ell^\sharp(t)]g_{X'}, g_X\rangle = (2\pi)^{-n}\int_{\mathbb{R}^{2n}} K_\ell^\sharp(t, Y)\mathcal{W}_{X,X'}(Y)dY, \tag{4.49}$$

where $\mathcal{W}_{X,X'}$ is the Wigner function of the pair $(g_{X'}, g_X)$. Let us now compute the remainder term in the Fourier–Bargmann representation. Using that $\mathcal{F}^{\mathcal{B}}$ is an isometry we get

$$\mathcal{F}^{\mathcal{B}}\left[Op_1^w[R_\ell(t)\circ F_{t,t_0}\right](b_j(t)g)](X) =$$
$$\int_{\mathbb{R}^{2n}} B_j(t, X')\langle g_X, Op_1^w[R_\ell(t)\circ F_{t,t_0}]g_{X'}\rangle dX', \tag{4.50}$$

[3] Recall that $u \in W^{\ell,p}$ means that $\partial_x^\alpha u \in L^p$ for every $|\alpha| \le \ell$.

[4] The Euler classical Gamma function Γ must not be mixed up with the covariance matrix Γ_t.

where $R_\ell(t)$ is given by the Taylor integral formula (4.51):

$$R_\ell(t,X) = \frac{\hbar^{\ell/2-1}}{k!}\sum_{|\gamma|=\ell}\int_0^1 \partial_X^\gamma H(t, z_t+\theta\sqrt{\hbar}X)X^\gamma(1-\theta)^{\ell-1}d\theta. \tag{4.51}$$

We shall use (4.50) to estimate the remainder term $R_z^{(N)}$, using estimates (4.42) and (4.43).

Now we shall consider long time estimates for $B_j(t,X)$.

Lemma 31 *For every $j \geq 0$, every $s \in \mathbb{N}$, $r \geq 1$, there exists $C(j,\alpha,\beta)$ such that for $|t| \leq T$, we have*

$$\left\| e^{\mu(X/4)}X^\alpha\partial_X^\beta B_j(t,X)\right\|_{s,r} \leq C(j,\alpha,\beta)\sigma(t,z)^{NM_1}|F|_T^{3j}(1+T)^j, \tag{4.52}$$

where $|F|_T = \sup_{|t|\leq T}|F_t|$. M_1 and $\sigma(t,z)$ were defined in section1, M_1 depends on assumption (A.0) on $H(X,t)$. $\|\bullet\|_{s,r}$ is the norm in the Sobolev space $W^{s,r}(\mathbb{R}^{2n})$[5]

Proof The main idea of the proof is as follows (see [218] for details). We proceed by induction on j. For $j=0$ (4.52) results from (4.46).

Let us assume inequality proved up to $j-1$. We have the induction formula ($j \geq 1$)

$$\partial_t B_j(t,X) = \sum_{\substack{k+\ell=j+2\\ \ell\geq 3}}\int_{\mathbb{R}^{2n}} K_\ell(t,X,X')B_k(t,X')dX', \tag{4.53}$$

where

$$K_\ell(t,X,X') = \sum_{|\gamma|=\ell}\frac{1}{\gamma!}\partial_X^\gamma H(t,z_t)\langle Op_1^w(F_tY)^\gamma g_{X'}, g_X\rangle, \text{ and} \tag{4.54}$$

$$\langle Op_1^w(F_tY)^\gamma g_{X'}, g_X\rangle = 2^{2n}\int_{\mathbb{R}^{2n}}(F_tY)^\gamma \mathscr{W}_{X,X'}(Y)dY. \tag{4.55}$$

By a Fourier transform computation on Gaussian functions, we get the following more explicit expression

$$\langle Op_1^w(F_tY)^\gamma g_{X'}, g_X\rangle = \sum_{\beta\leq\gamma} C_\beta^\gamma 2^{-|\beta|}\left(F_t\left(\frac{X+X'}{2}\right)\right)^{\gamma-\beta}\cdot$$
$$\cdot H_\beta\left(F_t\left(\frac{J(X-X')}{2}\right)\right)e^{-|X-X'|^2/4}e^{-(i/2)\sigma(X',X)}. \tag{4.56}$$

Estimate (4.52) follows easily. □

[5]The Sobolev norm is defined here as $\|f\|_{s,r} = \left(\sum_{|\alpha|\leq s}\int dx|f(x)|^r\right)^{1/r}$ for $s\in\mathbb{N}$, $r \geq 1$.

Now we have to estimate the remainder term. Let us compute the Fourier–Bargmann transform of the error term:

$$\tilde{R}_z^{(N+1)}(t,X) = \mathscr{F}^{\mathscr{B}}\left[\hbar^{j/2}\sum_{\substack{j+k=N+3\\k\geq 3}} Op_1^w[R_k](t)\circ F_t](b_j(t)g)\right](X)$$

$$= \sum_{\substack{j+k=N+3\\k\geq 3}} \int_{\mathbb{R}^{2n}} B_j(t,X')\langle Op_1^w[R_k(t)\circ F_t]g_{X'}, g_X\rangle dX'$$

Using estimates on the B_j we get the following estimate for the error term:

Lemma 32 *For every $\kappa > 0$, for every $\ell \in \mathbb{N}$, $s \geq 0$, $r \geq 1$, there exists $C_{N,\ell}$ such that for all T and t, $|t| \leq T$, we have*

$$\left\| X^\alpha \partial_X^\beta \widetilde{R}_z^{(N+1)}(t,X)\right\|_{s,r} \leq C_{N,\ell} M_{N,\ell}(T,z)|F|_T^{3N+3}(1+T)^{N+1}\hbar^{\frac{N+3}{2}} \tag{4.57}$$

for $\sqrt{\hbar}|F|_T \leq \kappa$, $|\alpha|+|\beta| \leq \ell$, where $M_{N,\ell}(T,z)$ is a continuous function of $\sup_{\substack{|t-t_0|\leq T\\3\leq|\gamma|\leq N_\ell}} |\partial_X^\gamma H(t,z_t)|$.

Proof As above for estimation of the $B_j(t,X)$, let us consider the integral kernels

$$N_k(t,X,X') = \langle Op_1^w[R_k(t)\circ F_t]g_{X'}, g_X\rangle. \tag{4.58}$$

We have

$$N_k(t,X,X') = \hbar^{(k+1)/2}\sum_{|\gamma|=k+1}\frac{1}{k!}\int_0^1 (1-\theta)^k$$

$$\times\left(\int_{\mathbb{R}^{2n}} \partial_Y^\gamma H(t, z_t+\theta\sqrt{\hbar}F_tY)(F_tY)^\gamma.\mathscr{W}_{X',X}(Y)dY\right)d\theta. \tag{4.59}$$

Let us denote by $N_{k,t}$ the operator with the kernel $N_k(t,X,X')$. Using the change of variable $Z = Y - \frac{X+X'}{2}$ and integrations by parts in X' as above, we can estimate $N_{k,t}[B_j(t,\bullet)](X)$. Then using the estimates on the $B_j(t,X)$ we get estimate (4.57). □

Now, it is not difficult to convert these results in the configuration space, using (4.42). Let us define $\lambda_{\hbar,t}(x) = \left(\frac{|x-q_t|^2+1}{\hbar|F_t|^2}\right)^{1/2}$.

Theorem 24 *Let us assume that (A.0) is satisfied. Then we have for the remainder term,*

$$R_z^{(N)}(t,x) = i\hbar\frac{\partial}{\partial t}\psi_z^{(N)}(t,x) - \widehat{H}(t)\psi_z^{(N)}(t,x),$$

the following estimate. For every $\kappa > 0$, for every ℓ, $M \in \mathbb{N}$, $r \geq 1$ there exist $C_{N,M,\ell}$ and N_ℓ such that for all T and t, $|t| \leq T$, we have:

$$\left\| \lambda_{\hbar,t}^{M} R_z^{(N)}(t) \right\|_{\ell,r} \leq C_{N,\ell} \hbar^{(N+3-\ell)/2} \sigma(z,t)^{NM_1} |F|_T^{3N+3} (1+T)^{N+1}, \tag{4.60}$$

for every $\hbar \in]0,1]$, $\sqrt{\hbar}|F_t| \leq \kappa$.

Moreover, if $\hat{H}(t)$ admits a unitary propagator (see condition (A.2)), then under the same conditions as above, we have

$$\|\mathscr{U}_t \varphi_z - \psi_z^{(N)}(t)\|_2 \leq C_{N,\ell} \sigma(z,t)^{NM_1} |F|_T^{3N+3} (1+T)^{N+2} \hbar^{(N+1)/2}. \tag{4.61}$$

Proof Using the inverse Fourier–Bargmann transform, we have

$$R_z^{(N)}(t,x) = \hat{T}(z_t) \Lambda_\hbar \left(\int_{\mathbb{R}^{2n}} (\hat{R}[F_t]\varphi_X)(x) \widetilde{R}_z^{(N)}(t,X) dX \right).$$

Let us remark that using estimates on the $b_j(t,x)$, we can assume that N is arbitrary large. We can apply previous results on the Fourier–Bargmann estimates to get (4.60). The second part is a consequence of the first part and of the Duhamel principle. □

We see that the estimate (4.60) is much more accurate in norm than estimate (4.61), we have lost much information applying the propagator $\mathscr{U}_t$. The reason is that in general we only know that the propagator is bounded on L^2 and no more. Sometimes it is possible to improve (4.61) if we know that $\mathscr{U}_t$ is bounded on some weighted Sobolev spaces. Let us give here the following example.

Let $\hat{H} = -\hbar^2 \Delta + V(x)$. Assume that V satisfies:

$$V \in C^\infty(\mathbb{R}^n), \; |\partial^\alpha V(x)| \leq C_\alpha V(x), \quad V \geq 1, |V(x) - V(y)| \leq C(1 + |x-y|)^M,$$

for some $M \in \mathbb{R}$. So the time dependent Schrödinger equation for $\hat{H}$ has a unitary propagator $\mathscr{U}_t = \mathrm{e}^{it\hbar^{-1}\hat{H}}$. The domain of $\hat{H}^m$ can be determined for every $m \in \mathbb{N}$ (see for example [217]).

$$D(\hat{H}^m) = \{u \in W^{2m,2}(\mathbb{R}^n), \; V^m u \in L^2(\mathbb{R}^n)\}.$$

It is an Hilbert space with the norm defined by

$$\|u\|_{2m,V}^2 = \sum_{|\alpha| \leq 2m} \|\hbar^{|\alpha|} \partial^\alpha u\|_{L^2(\mathbb{R}^n)}^2 + \|Vu\|_{L^2(\mathbb{R}^n)}^2$$

Using the Sobolev theorem we get the supremum norm estimate for the error:

$$\sup_{x \in \mathbb{R}^n} |(\mathscr{U}_t \varphi_z)(x) - \psi_z^{(N)}(t,x)| \leq C_{N,\ell} \sigma(z,t)^{NM_1} |F|_T^{3N+3} (1+T)^{N+2} \hbar^{(N-n/2)/2} \tag{4.62}$$

4.3.4 Exponentially Small Estimates

Up to now the order N of the semi-classical approximations was fixed, even arbitrary large, but the error term was not controlled for N large. Here we shall give estimates with a control for large N. The method is the same as on the previous section, using systematically the Fourier–Bargmann transform. The proof is not given here, we refer to [218] for more details. For a different approach see [133, 134]. To get exponentially small estimates for asymptotic expansions in small $\hbar$ it is quite natural to assume that the classical Hamiltonian $H(t, X)$ is analytic in X, where $X = (x, \xi) \in \mathbb{R}^{2n}$. This problem was studied in a different context in [115] concerning Borel summability for semi-classical expansions for systems of bosons.

So, in what follows we introduce suitable assumptions on $H(t, X)$. As before we assume that $H(t, X)$ is continuous in time t and C^∞ in X and that the quantum and classical dynamics are well defined.

Let us define a complex neighborhood of $\mathbb{R}^{2n}$ in $\mathbb{C}^{2n}$,

$$\Omega_\rho = \{X \in \mathbb{C}^{2n}, |\Im X| < \rho\} \tag{4.63}$$

where $\Im X = (\Im X_1, \ldots, \Im X_{2n})$ and $|\cdot|$ is the Euclidean norm in $\mathbb{R}^{2n}$ or the Hermitian norm in $\mathbb{C}^{2n}$. Our main assumptions are the following.

($\mathbf{A}_\omega$) (Analyticity assumption) There exists $\rho > 0$, $T \in]0, +\infty]$, $C > 0$, $\nu \geq 0$, such that $H(t)$ is holomorphic in Ω_ρ and for $t \in I_T$, $X \in \Omega_\rho$, we have

$$\begin{aligned} &|H(t, X)| \leq C\mathrm{e}^{\nu|X|}, \text{ and} \\ &|\partial_X^\gamma H(t, z_t + Y)| \leq R^\gamma \gamma! \mathrm{e}^{\nu|Y|}, \quad \forall t \in \mathbb{R},\ Y \in \mathbb{R}^{2n} \end{aligned} \tag{4.64}$$

for some $R > 0$ and all γ, $|\gamma| \geq 3$.

We begin by giving the results on the Fourier–Bargmann side. It is the main step and gives accurate estimates for the propagation of Gaussian coherent states in the phase space. We have seen that it is not difficult to transfer these estimates in the configuration space to get approximations of the solution of the Schrödinger equation, by applying the inverse Fourier–Bargmann transform as we did in the C^∞ case.

The main results are stated in the following theorem.

Theorem 25 *Let us assume that conditions* (A_0) *and* (A_ω) *are satisfied. Then the following uniform estimates hold.*

$$\begin{aligned} &\left\| X^\alpha \partial_X^\beta B_j(t, X) \right\|_{L^2(\mathbb{R}^{2d}, \mathrm{e}^{\lambda|X|}dX)} \leq \\ &C_{\lambda,T}^{3j+1+|\alpha|+|\beta|} |F|_T^{3j} (1 + |t - t_0|)^j j^{-j} (3j + |\alpha| + |\beta|)^{\frac{3j+|\alpha|+|\beta|}{2}}. \end{aligned} \tag{4.65}$$

where $C_\lambda > 0$ depends only on $\lambda \geq 0$ and is independent on $j \in \mathbb{N}$, $\alpha, \beta \in \mathbb{N}^{2n}$ and $|t| \leq T$.

Concerning the remainder term estimate we have:

$$\left\| X^{\alpha}\partial_X^{\beta}\tilde{R}_z^{(N)}(t,X)\right\|_{L^2(\mathbb{R}^{2n},e^{\lambda|X|}dX)} \le \hbar^{(N+3)/2}(1+|t|)^{N+1} \tag{4.66}$$
$$|F|_T^{3N+3}(C'_\lambda)^{3N+3+|\alpha|+|\beta|}(N+1)^{-N-1}(3N+3+|\alpha|+|\beta|)^{\frac{3N+3+|\alpha|+|\beta|}{2}}.$$

where $\lambda < \rho$, $\alpha, \beta \in \mathbb{N}^{2n}$, $N \ge 1$, $\nu\sqrt{\hbar}|F|_T \le 2(\rho-\lambda)$, C'_λ depends on λ and is independent on the other parameters ($\hbar$, T, N, α, β).

From Theorem 25 we easily get weighted estimates for approximate solutions and remainder term for the time dependent Schrödinger equation. Let us recall the Sobolev norms in the Sobolev space $W^{m,r}(\mathbb{R}^n)$.

$$\|u\|_{r,m,\hbar} = \left(\sum_{|\alpha|\le m}\hbar^{|\alpha|/2}\int_{\mathbb{R}^n}|\partial_x^{\alpha}u(x)|^r dx\right)^{1/r}$$

and a function $\mu \in C^{\infty}(\mathbb{R}^d)$ such that $\mu(x) = |x|$ for $|x| \ge 1$.

Proposition 42 *For every $m \in \mathbb{N}$, $r \in [1, +\infty]$, $\lambda > 0$ and $\varepsilon \le \min\{1, \frac{\lambda}{|F|_T}\}$, there exists $C_{r,m,\lambda,\varepsilon} > 0$ such that for every $j \ge 0$ and every $t \in I_T$ we have*

$$\|\hat{R}[F_t]b_j(t)ge^{\varepsilon\mu}\|_{r,m,1} \le (C_{r,m,\lambda,\varepsilon})^{j+1}(1+|F|_T)^{3j+2d}j^{j/2}(1+|t|)^j \tag{4.67}$$

Theorem 26 *With the above notations and under the assumptions of Theorem 25, $\psi_z^{(N)}(t,x)$ satisfies the Schrödinger equation*

$$i\hbar\partial_t\psi_z^{(N)}(t,x) = \widehat{H}(t)\psi_z^{(N)}(t,x) + R_z^{(N)}(t,x), \tag{4.68}$$

$$\text{where } \psi_z^{(N)}(t,x) = e^{i\delta_t/\hbar}\hat{T}(z_t)\Lambda_\hbar\hat{R}[F_t]\left(\sum_{0\le j\le N}\hbar^{j/2}b_j(t)g\right) \tag{4.69}$$

is estimated in Proposition 42 and the remainder term is controlled with the following weighted estimates:

$$\|R_z^{(N)}(t)e^{\varepsilon\mu_{\hbar,t}}\|_{r,m,\hbar} \le$$
$$C^{N+1}(N+1)^{(N+1)/2}\left(\sqrt{\hbar}|F|_T^3\right)^{N+3}\hbar^{-m'}(1+|t|)^{N+1}, \tag{4.70}$$

where C depends only on m, r, ε and not on $N \ge 0$, $|t| \le T$ and $\hbar > 0$, with the condition $\sqrt{\hbar}|F|_T \le \kappa$. The exponential weight is defined by $\mu_{\hbar,t}(x) = \mu\left(\frac{x-q_t}{\sqrt{\hbar}}\right)$. $m' \ge 0$ and $h_\varepsilon < \min\{1, \frac{\rho}{|F|_T}\}$.

Remark 22 We see that the order in j of the coefficient $b_j(t)g$ in the asymptotic expansion in $\hbar^{j/2}$ is $C^j j^{j/2}$ or using Stirling formula $C'^j\Gamma(j)^{1/2}$ for some constant $C > 0$, $C' > 0$. So we have obtained that the renormalized evolved state $b(t,x)g(x)$ obtained from $\psi_z(t,x)$ has a Gevrey-2 asymptotic expansion in $\hbar^{1/2}$. Recall that a formal complex series $\sum_{j\geq 0} c_j\kappa^j$ is a Gevrey series of index $\mu > 0$ if there exists constants $C_0 > 0$, $C > 0$ such that

$$|c_j| \leq C_0 C^j \Gamma(j)^{1/\mu}, \forall j \geq 1.$$

Any holomorphic function $f(\kappa)$ in a complex neighborhood of 0 has a convergent Gevrey-1 Taylor series. But in many physical examples we have a non-convergent Gevrey asymptotic series $f[\kappa]$ for a function f holomorphic in some sector with apex 0. Under some technical conditions on f it is possible to define the Borel sum $B_f(\tau)$ for the formal power series $f[\kappa]$ and to recover $f(\kappa)$ from its Borel sum performing a Laplace transform on $B_f(\tau)$ (see [239] for details and bibliography).

When it is not possible to apply Borel summability, there exists a well known method to minimize the error between $\sum_{1\leq j\leq N} c_j\kappa^j$ and $f(\kappa)$. It is called the astronomers method and consists in stopping the expansion after the smallest term of the series (it is also called "the least term truncation method" for a series). Concerning the semiclassical expansion found for $b(t,x)g(x)$ it is not clear that it is Borel summable or summable in some weaker sense. A sufficient condition would be that the propagator $\mathscr{U}_t$ can be extended holomorphically in $\kappa := \hbar^{1/2}$ in a (small) sector $\{re^{i\theta},\ 0 < r < r_0,\ |\theta| < \varepsilon\}$. In a different context (quantum field theory for bosons), Borel summability was proved in [115].

Using the astronomers method Theorem 26 we easily get the following consequences.

Corollary 15 (Finite Time, Large N) *Let us assume here that $T < +\infty$. There exist $c > 0$, $\hbar_0 > 0$, $a > 0$, $\varepsilon > 0$, such that if we choose $N_\hbar = [\frac{a}{\hbar}] - 1$ we have*

$$\|R_z^{(N_\hbar)}(t)e^{\varepsilon\mu_{\hbar,t}}\|_{L^2} \leq \exp\left(-\frac{c}{\hbar}\right), \tag{4.71}$$

for every $|t| \leq T$, $\hbar \in]0, \hbar_0]$. Moreover, we have

$$\|\psi_z^{(N_\hbar)}(t) - \mathscr{U}(t,t_0)\varphi_z\|_{L^2} \leq \exp\left(-\frac{c}{\hbar}\right). \tag{4.72}$$

And

Corollary 16 (Large Time, Large N) *Let us assume that $T = +\infty$ and there exist $\gamma \geq 0$, $\delta \geq 0$, $C_1 \geq 0$, such that $|F_{t,t_0}| \leq \exp(\gamma|t|)$, $|z_t| \leq \exp(\delta|t|)$ for every $\theta \in]0, 1[$ there exists $a_\theta > 0$ such that if we choose $N_{\hbar,\theta} = [\frac{a_\theta}{\hbar^\theta}] - 1$ there exist $c_\theta > 0$, $\eta_\theta > 0$ such that*

$$\|\hbar^{(N_{\hbar,\theta}+2)/2} R_z^{(N_{\hbar,\theta}+1)}(t) \mathrm{e}^{\varepsilon \mu_{\hbar,t}}\|_{L^2} \leq \exp\left(-\frac{c_\theta}{\hbar^\theta}\right) \tag{4.73}$$

for every $|t| \leq \frac{1-\theta}{6\gamma} \log(\hbar^{-1})$, $\forall \hbar \in \hbar \in]0, \eta_\theta]$. *Moreover we have:*

$$\|\psi_z^{(N_{\hbar,\theta})}(t) - \mathscr{U}(t, t_0)\varphi_z\|_{L^2} \leq \exp\left(-\frac{c_\theta}{\hbar^\theta}\right), \tag{4.74}$$

under the conditions of (4.73).

Remark 23 We have considered here standard Gaussian. All the results are true and proved in the same way for Gaussian coherent states defined by g^Γ, for any $\Gamma \in \Sigma_n^+$. These results have been proved in [218] and in [133, 134] using different methods.

All the results in this subsection can be easily deduced from Theorem 25. Proposition 42 and Theorem 26 are easily proved using the estimates of Sect. 2.2. The proof of the corollaries are consequences of Theorem 26 and Stirling formula for the Euler Gamma function.

4.4 Application to the Scattering Theory

In this section we assume that the interaction satisfies a short range assumption and we shall prove results for the action of the scattering operator acting on the squeezed states. One gets a semiclassical asymptotics for the action of the scattering operator on a squeezed state located at point z_- in terms of a squeezed state located at point z_+ where $z_+ = S^{cl}(z_-)$, S^{cl} being the classical scattering matrix. For the basic classical and quantum scattering theories we refer the reader to [81, 215]. Let us first recall some basic facts on classical and quantum scattering theory. We consider a classical Hamiltonian H for a particle moving in a curved space and in an electromagnetic field;

$$H(q, p) = \frac{1}{2} g(q) p \cdot p + a(q) \cdot p + V(q), \; q \in \mathbb{R}^n, \; p \in \mathbb{R}^n, \tag{4.75}$$

$g(q)$ is a smooth positive definite matrix and there exists $c > 0, \; C > 0$ such that

$$c|p|^2 \leq g(q) p \cdot p \leq C|p|^2, \; \forall (q, p) \in \mathbb{R}^{2n},$$

$a(q)$ is a smooth linear form on $\mathbb{R}^n$ and $V(q)$ a smooth scalar potential. In what follows it will be assumed that $H(q, p)$ is a short range perturbation of $H^{(0)} = \frac{p^2}{2}$ in the following sense: there exist $\rho > 1$, and C_α, for $\alpha \in \mathbb{N}^n$ such that

$$|\partial_q^\alpha(\mathbb{1} - g(q))| + |\partial_q^\alpha a(q)| + |\partial_q^\alpha V(q)| \leq C_\alpha \langle q \rangle^{-\rho-|\alpha|}, \; \forall q \in \mathbb{R}^n. \tag{4.76}$$

H and $H^{(0)}$ define two Hamiltonian flows Φ^t, Φ_0^t on the phase space $\mathbb{R}^{2n}$ for all $t \in \mathbb{R}$. The classical scattering theory establishes a comparison of the two dynamics Φ^t, Φ_0^t in the large time limit. Note that the free dynamics is explicit:

$$\Phi_0^t(q, p) = (q + tp, p).$$

The methods of [81, 215] can be used to prove the existence of the classical wave operators defined by

$$\Omega_\pm^{cl} X = \lim_{t \to \pm\infty} \Phi^{-t}(\Phi_0^t X). \tag{4.77}$$

This limit exists for every $X \in \mathscr{Z}_0$ where $\mathscr{Z}_0 = \{(q, p) \in \mathbb{R}^{2n},\ p \neq 0\}$ and is uniform on every compact of $\mathscr{Z}_0$. We also have for all $X \in \mathscr{Z}_0$

$$\lim_{t \to \pm\infty} \left(\Phi^t \Omega_\pm^{cl}(X) - \Phi_0^t(X)\right) = 0.$$

Moreover $\Omega_\pm^{cl}$ are $\mathscr{C}^\infty$-smooth symplectic transformations. They intertwine the free and the interacting dynamics:

$$H \circ \Omega_\pm^{cl} X = \Omega_\pm^{cl} \circ H^{(0)}(X),\ \forall X \in \mathscr{Z}_0, \text{ and } \Phi^t \circ \Omega_\pm^{cl} = \Omega_\pm^{cl} \circ \Phi_0^t.$$

Then the classical scattering matrix S^{cl} is defined by

$$S^{cl} = (\Omega_+^{cl})^{-1} \Omega_-^{cl}.$$

This definition makes sense since one can prove (see [215]) that modulo a closed set $\mathscr{N}_0$ of Lebesgue measure zero in $\mathscr{Z}$ ($\mathscr{Z} \setminus \mathscr{Z}_0 \subseteq \mathscr{N}_0$) one has

$$\Omega_+^{cl}(\mathscr{Z}_0) = \Omega_-^{cl}(\mathscr{Z}_0)$$

Moreover S^{cl} is smooth in $\mathscr{Z} \setminus \mathscr{N}_0$ and commutes with the free evolution:

$$S^{cl} \Phi_0^t = \Phi_0^t S^{cl}.$$

The scattering operator has the following kinematic interpretation: let us consider $X_- \in \mathscr{Z}_0$ and its free evolution $\Phi_0^t X_-$. There exists a unique interacting evolution $\Phi^t(X)$ which is close to $\Phi_0^t(X_-)$ for $t \searrow -\infty$. Moreover there exists a unique point $X_+ \in \mathscr{Z}_0$ such that $\Phi^t(X)$ is close to $\Phi_0^t(X_+)$ for $t \nearrow +\infty$. X, X_+ are given by

$$X = \Omega_-^{cl} X_- \text{ and } X_+ = S^{cl} X_-.$$

Using [81] we can get a more precise result. Let I be an open interval of $\mathbb{R}$ and assume that I is "non trapping" for H which means that for every X such that $H(X) \in I$ we have $\lim_{t \to \pm\infty} |\Phi^t(X)| = +\infty$. Then we have:

Proposition 43 *If I is a non trapping interval for H then S^{cl} is defined everywhere in $H^{-1}(I)$ and is a $\mathscr{C}^\infty$ smooth symplectic map.*

On the quantum side one can define the wave operators and the scattering operator in a similar way. Let us note that the quantization $\hat{H}$ of H is essentially self-adjoint so that the unitary group $\mathscr{U}(t) = \exp\left(-\frac{it}{\hbar}\hat{H}\right)$ is well defined in $L^2(\mathbb{R}^n)$. The free evolution $\mathscr{U}_0(t) := \exp\left(-\frac{it}{\hbar}\hat{H}^{(0)}\right)$ is explicit:

$$(\mathscr{U}_0(t)\psi)(x) = (2\pi\hbar)^{-n}\int\int_{\mathbb{R}^{2n}} \exp\left(\frac{i}{\hbar}(-t\frac{\xi^2}{2} + (x-y)\cdot\xi)\right)\psi(y)\,dy\,d\xi. \tag{4.78}$$

The assumption (4.76) implies that we can define the wave operators $\Omega_\pm$ and the scattering operator $S^{(\hbar)} = (\Omega_+)^*\Omega_-$ (see [81, 215]). Recall that

$$\Omega_\pm = \lim_{t\to\pm\infty} \mathscr{U}(-t)\mathscr{U}_0(t).$$

The ranges of $\Omega_\pm$ are equal to the absolutely continuous subspace of $\hat{H}$ and we have:

$$\Omega_\pm\mathscr{U}_0(t) = \mathscr{U}(t)\Omega_\pm, \quad S^{(\hbar)}\mathscr{U}_0(t) = \mathscr{U}_0(t)S^{(\hbar)}. \tag{4.79}$$

One wants to obtain a correspondence between $\lim_{\hbar\to 0} S^{(\hbar)}$ and S^{cl}. There are many works on the subject (see [130, 134, 221, 265]). Here we want to check this classical limit using the coherent states approach like in [130, 134]. We present here a different technical approach extending these results to more general perturbations of the Laplace operator.

We recall some notations of Chap. 3: $\mathfrak{S}^+(n)$ is the Siegel space namely the space of complex symmetric $n \times n$ matrices Γ such that $\Im\Gamma$ is positive and non degenerate. Given F any $2n \times 2n$ symplectic matrix the unitary operator $\hat{R}(F)$ is the metaplectic transformation associated to F. g^Γ is the gaussian function of L^2 norm 1 defined by

$$g^\Gamma(x) = a_\Gamma \exp\left(\frac{i}{2\hbar}\Gamma x\cdot x\right) \tag{4.80}$$

and we denote

$$\varphi_z^\Gamma = \hat{T}(z)g^\Gamma$$

Finally $\Lambda_\hbar$ is the unitary operator defined in (4.22).

The main result of this section states a relationship between the quantum scattering and the classical scattering.

Theorem 27 *For every $N \geq 1$, every $z_- \in \mathscr{Z}\setminus\mathscr{N}_0$ and every $\Gamma_- \in \Sigma_n$ we have the following semiclassical approximation for the scattering operator $S^{(\hbar)}$ acting on the Gaussian coherent state $\varphi_{z_-}^{\Gamma_-}$:*

$$S^{(\hbar)}\varphi_{z_-}^{\Gamma_-} = e^{i\delta_+/\hbar}\hat{T}(z_+)\Lambda_\hbar\hat{R}(G_+)\left(\sum_{0\le j\le N}\hbar^{j/2}b_j g^{\Gamma_-}\right) + \mathscr{O}(\hbar^{(N+1)/2}), \qquad (4.81)$$

where we define

$$z_+ = S^{cl}z_-, \quad z_\pm = (q_\pm, p_\pm).$$

$z_t = (q_t, p_t)$ is the interacting scattering trajectory $z_t = \Phi^t(\Omega_-^{cl}z_-)$.
$\delta_+ = \int_{-\infty}^{+\infty}(p_t q_t - H(z_t))\,dt - \frac{q_+p_+ - q_-p_-}{2}$.
$G_+ = \frac{\partial z_+}{\partial z_-}$.
b_j is a polynomial of degree $\le 3j$, $b_0 = 1$.
The error term $\mathscr{O}(\hbar^{(N+1)/2})$ is estimated in the L^2-norm.

Let us denote

$$\psi_- = \varphi_{z_-}^{\Gamma_-}, \text{ and } \psi_+ := S^{(\hbar)}\psi_-.$$

Using the definition of $S^{(\hbar)}$ we have

$$\psi_+ = \lim_{t\to+\infty}\left(\lim_{s\to-\infty}\mathscr{U}_0(t)\mathscr{U}(t-s)\mathscr{U}_0(s)\right)\psi_-. \qquad (4.82)$$

The strategy of the proof consists in applying the estimate (4.14) at fixed time t to $U(t-s)$ in (4.82) and then to see what happens in the limits $s\to-\infty$, $t\to+\infty$.

Let us denote by F_t^0 the Jacobi stability matrix for the free evolution and by $F_t(z)$ the Jacobi stability matrix along the trajectory $\Phi^t(z)$. We have

$$F_t^0 = \begin{pmatrix} \mathbb{1}_n & t\,\mathbb{1}_n \\ 0 & \mathbb{1}_n \end{pmatrix}.$$

We need large time estimates concerning classical scattering trajectories and their Jacobi stability matrices.

Proposition 44 *Under the assumptions of Theorem 27 there exists a unique scattering solution of the Hamilton's equation $\dot{z}_t = J\nabla H(z_t)$ such that*

$$\dot{z}_t - \partial_t\Phi_0^t z_+ = \mathscr{O}(\langle t\rangle^{-\rho}), \text{ for } t\to+\infty.$$

$$\dot{z}_t - \partial_t\Phi_0^t z_- = \mathscr{O}(\langle t\rangle^{-\rho}), \text{ for } t\to-\infty.$$

Proposition 45 *Let us denote*

$$G_{t,s} := F_{t-s}(\Phi_0^s z_-)F_s^0.$$

Then we have:

(i) $\lim_{s\to-\infty} G_{t,s} = G_t$ *exists* $\forall t\ge 0$.

(ii) $\lim_{t\to+\infty} F_0^{-t} G_t = G_+$ *exists.*

(iii) $G_t = \frac{\partial z_t}{\partial z_-}$, and $G_+ = \frac{\partial z_+}{\partial z_-}$.

These two propositions will be proven later together with the following one. The main step in the proof of Theorem 27 will be to solve the following asymptotic Cauchy problem for the Schrödinger equation with data given at time $t = +\infty$:

$$i\hbar\partial_s\psi_{z_-}^{N)} = \hat{H}\psi_{z_-}^{(N)}(s) + \mathcal{O}(\hbar^{(N+3)/2} f_N(s)) \tag{4.83}$$

$$\lim_{s\to-\infty} \mathcal{U}_0(-s)\psi_{z_-}^{(N)}(s) = \varphi_{z_-}^{\Gamma_-}$$

where $f_N \in L^1(\mathbb{R}) \cap L^\infty(\mathbb{R})$ is independent of $\hbar$. The following result is an extension for infinite times of results proven in Sect. 1 for finite times.

Proposition 46 *The problem (4.83) has a solution which can be computed in the following way:*

$$\boxed{\psi_{z_-}^{(N)}(t,x) = e^{i\delta(z_t)/\hbar}\hat{T}(z_-)\Lambda_\hbar\hat{R}(G_t)\left(\sum_{0\le j\le N}\hbar^{j/2}b_j(t,z_-)g^{\Gamma_-}\right)}.$$

The $b_j(t, z_-, x)$ are uniquely defined by the following induction formula for $j \ge 1$ starting with $b_0(t, x) \equiv 1$:

$$\partial_t b_j(t,z_-,x)g(x) = \sum_{k+l=j+2,\ l\ge 3} \mathrm{Op}_1^w[K_l^\sharp(t)](b_k(t,.)g)(x), \tag{4.84}$$

$$\lim_{t\to-\infty} b_j(t,z_-,x) = 0\ , \tag{4.85}$$

with

$$K_j^\sharp(t,X) = K_j(t,G_t(X)) = \sum_{|\gamma|=j}\frac{1}{\gamma!}\partial_X^\gamma H(z_t)(G_tX)^\gamma,\ X \in \mathbb{R}^{2n},$$

$b_j(t, z_-, x)$ is a polynomial of degree $\le 3j$ in variable $x \in \mathbb{R}^n$ with complex time dependent coefficients depending on the scattering trajectory z_t starting from z_- at time $t = --\infty$. Moreover we have the remainder uniform estimate

$$i\hbar\partial_t\psi_{z_-}^{(N)}(t) = \hat{H}\psi_{z_-}^{(N)}(t) + \mathcal{O}(\hbar^{(N+3)/2}\langle t\rangle^{-\rho}), \tag{4.86}$$

uniformly in $\hbar \in]0,1]$, and $t \ge 0$.

Proof (Theorem 27). Without going into the details which are similar to the finite time case, we remark that in the induction formula (4.84) we can use the following estimates to get uniform decrease in time estimates for $b_j(t, z_-, x)$. First there exist

$c > 0$ and $T_0 > 0$ such that for $t \geq T_0$ we have $|q_t| \geq ct$. Using the short range assumption and conservation of the classical energy we have that for $|\gamma| \geq 3$ there exists $C_\gamma > 0$ such that

$$|\partial_X^\gamma H(z_t)| \leq C_\gamma \langle t \rangle^{-\rho-1}. \tag{4.87}$$

Therefore we deduce (4.86) from (4.84) and (4.87).

Using Proposition 46 and Duhamel's formula we get

$$\mathscr{U}(t)\psi_{z_-}^{(N)}(s) = \psi_{z_-}^{(N)}(t+s) + \mathscr{O}(\hbar^{(N+1)/2}),$$

uniformly in $t,\ s \in \mathbb{R}$. But we have:

$$\|\psi_{z_-}^{(N)}(t) - \mathscr{U}(t-s)\mathscr{U}_0(s)\psi_-\| \leq \tag{4.88}$$

$$\|\psi_{z_-}^{(N)}(t) - \mathscr{U}(t-s)\psi_{z_-}^{(N)}(s)\| + \|\mathscr{U}_0(s)\psi_- - \psi_{z_-}^{(N)}(s)\|. \tag{4.89}$$

We know that

$$\lim_{s\to-\infty} \|\mathscr{U}_0(s)\psi_- - \psi_{z_-}^{(N)}(s)\| = 0.$$

Then going to the limit $s \to -\infty$ we get uniformly in $t \geq 0$

$$\|\psi_{z_-}^{(N)}(t) - \mathscr{U}(t)\Omega_-\psi_-\| = \mathscr{O}(\hbar^{(N+1)/2}).$$

Then we can compute $\mathscr{U}_0(-t)\psi_{z_-}^{(N)}(t)$ in the limit $t \to +\infty$ and we find out that $S^{(\hbar)}\psi_- = \psi_+^{(N)} + \mathscr{O}(\hbar^{(N+1)/2})$ where

$$\psi_+^{(N)} = \lim_{t\to+\infty} \mathscr{U}_0(-t)\psi_{z_-}^{(N)}(t).$$

So we have proved Theorem 27. □

Let us prove now Proposition 44 following the book [215]:

Proof Let us denote $u(t) := z_t - \Phi_t^0 z_-$. We have to solve the integral equation

$$u(t) = \Phi_t^0(z_-) + \int_{-\infty}^t (J\nabla H(u(s)) + \Phi_s^0(z_-))\, ds$$

We can choose $T_1 < 0$ such that the map K defined by

$$Ku(t) = \int_{-\infty}^t (J\nabla H(u(s)) + \Phi_s^0(z_-))\, ds.$$

is a contraction in the complete metric space $\mathscr{C}_{T_1}$ of continuous functions u from $]-\infty, T_1]$ into $\mathbb{R}^{2n}$ such that $\sup_{t\leq T_1}|u(t)| \leq 1$, with the natural distance. So we can apply the fixed point theorem to prove Proposition 44 using standard technics. □

Proposition 45 can be proved by the same method.
Let us now prove Proposition 46.

Proof Let us denote $z_s^0 := \Phi_s^0(z_-)$. Furthermore if S is a symplectic matrix

$$S = \begin{pmatrix} A & B \\ C & D \end{pmatrix}$$

and $\Gamma \in \mathfrak{S}^+(n)$ ($\mathfrak{S}^+(n)$ is the Siegel space) we define

$$\Sigma_S(\Gamma) = (C + D\Gamma)(A + B\Gamma)^{-1} \in \mathfrak{S}^+(n).$$

Then let

$$\Gamma_s^0 = \Sigma_{F_s^0}(\Gamma_-).$$

One has for every $N \geq 0$:

$$i\hbar\partial_t\psi_z^{(N)}(t, s, x) = \hat{H}(t)\psi_z^{(N)}(t, s, x) + R_z^{(N)}(t, s, x),$$

where

$$\psi_z^{(N)}(t, s, x) = e^{i\delta_s s,t/\hbar}\hat{T}(z_t)\Lambda_\hbar\hat{R}(F_{t,s}F_s^0)\left(\sum_{0\leq j\leq N}\hbar^{j/2}b_j(t, s)g^{\Gamma_-}\right) \tag{4.90}$$

and

$$R_{z_-}^{(N)}(t, s, x) = e^{i\delta_{s,t}/\hbar}\hbar^{(N+3)/2}\times$$

$$\left(\sum_{j+k=N+2,\ k\geq 3}\hat{T}(z_-)\Lambda_\hbar\hat{R}(F_{t,s}F_s^0)\mathrm{Op}_1^w(R_k(t, s)\circ[F_{t,s}F_s^0])(b_j(t, s)g^{\Gamma_-})\right). \tag{4.91}$$

One denotes $F_{t,s} = F_{t-s}(\Phi_0^s z_-)$ the stability matrix at $\Phi_{t-s}(\Phi_s^0(z_-))$. Moreover the polynomials $b_j(t, s, x)$ are uniquely defined by the following induction formula for $j \geq 1$ starting with $b_0(s, s, x) \equiv 1$:

$$\partial_t b_j(t, s, x) = \sum_{k+l=j+2,\ l\geq 3}\mathrm{Op}_1^w[K_l^\sharp(t, s)](b_k(t, .)g^{\Gamma_-})(x),$$

$$b_j(s, s, x) = 0$$

where

$$K_l^\sharp(t,s,X) = \sum_{|\gamma|=l} \frac{1}{\gamma!}\partial_X^\gamma H(\Phi^{t-s}(\Phi_s^0 z_-))(F_{t-s}F_s^0 X)^\gamma, \quad X \in \mathbb{R}^{2n}.$$

So using Propositions 44 and 45 we can control the limit $s \to -\infty$ in Eqs. (4.90) and (4.91) and we get the proof of the Proposition 46. □

The following corollary is an immediate consequence of Theorem 27 and of the properties of the metaplectic transformation:

Corollary 17 *For every $N \in \mathbb{N}$ we have*

$$S^{(\hbar)}\varphi_{z_-}^{\Gamma_-} = e^{i\delta_+/\hbar} \sum_{0\le j\le N} \hbar^{j/2}\pi_j\left(\frac{x-q_+}{\sqrt{\hbar}}\right)\varphi_{z_+}^{\Gamma_+}(x) + \mathcal{O}(\hbar^\infty),$$

where $z_+ = S^{cl}(z_-)$, $\Gamma_+ = \Sigma_{G_+}(\Gamma_-)$, $\pi_j(y)$ are polynomials of degree $\le 3j$ in $y \in \mathbb{R}^n$. In particular $\pi_0 = 1$.

Recall that S_0^m is the space of smooth classical observables L such that for every $\gamma \in \mathbb{R}^{2n}$ there exists $C_\gamma \ge 0$ such that

$$|\partial_X^\gamma L(X)| \le C_\gamma \langle X\rangle^m, \ \forall X \in \mathbb{R}^{2n}.$$

The Weyl quantization $\hat{L}$ of L is well defined (see Chap. 2). One has the following result:

Corollary 18 *For any symbol $L \in S_0^m$, $m \in \mathbb{R}$, we have*

$$\boxed{\langle S^{(\hbar)}\varphi_{z_-}, \hat{L}S^{(\hbar)}\varphi_{z_-}\rangle = L(S^{cl}(z_-)) + \mathcal{O}(\sqrt{\hbar})}.$$

In particular one recovers the classical scattering operator from the quantum scattering operator in the semiclassical limit.

Proof Using Corollary 17 one gets

$$\langle S^{(\hbar)}\varphi_{z_-}, \hat{L}S^{(\hbar)}\varphi_{z_-}\rangle = \langle \varphi_{z_-}^{\Gamma_+}, \hat{L}\varphi_{z_-}^{\Gamma_+}\rangle + \mathcal{O}(\sqrt{\hbar})$$

and the result follows using results proved in Chap. 2. □

Remark 24 A similar result was proven for the time-delay operator in [253]. The proof given here is different and doesn't use a global non-trapping assumption.

Further study concerns the scattering evolution of Lagrangian states (also called WKB states). It was considered by Yajima [265] in the momentum representation and by Robinson [224] for the position representation. The approach developed here provides a more direct and general proof that is detailed in [218].

Note also that under the analytic and Gevrey assumption one can recover the result of [134] for the semiclassical propagation of coherent states with exponentially small estimate.

Chapter 5
Trace Formulas and Coherent States

Abstract The best known trace formula in mathematical physics is certainly the Gutzwiller trace formula linking the eigenvalues of the Schrödinger operator $\widehat{H}$ as Planck's constant goes to zero (the semiclassical régime) with the closed orbits of the corresponding classical mechanical system. Gutzwiller gave a heuristic proof of this trace formula, using the Feynman integral representation for the propagator of $\widehat{H}$. In mathematics this kind of trace formula was first known as Poisson formula which gives a relationship between the eigenvalues of the operator $\frac{d}{id\theta}$ on the circle and the periods of the rotation group. The generalization to the Laplace operator on a compact manifold, then for more general elliptic operators, was performed using microlocal analysis and the theory of Fourier integral operators. Our goal here is to show how the use of coherent states allows us to give a rather simple and direct rigorous proof of the Gutzwiller trace formula in a rather general setting.

5.1 Introduction

A quantum system is described by its Hamiltonian $\hat{H}$ and its admissible energies are the eigenvalues $E_j(\hbar)$ (we suppose that the spectrum of the self-adjoint operator $\hat{H}$ in the Hilbert space $\mathscr{H} = L^2(\mathbb{R}^n)$ is discrete). The frequency transition between energies $E_j(\hbar)$ and $E_k(\hbar)$ is $\omega_{j,k} = \frac{E_k(\hbar)-E_j(\hbar)}{\hbar}$.

If $n = 1$, or if the system is integrable, it is possible to prove semi-classical expansion for individual eigenvalues $E_j(\hbar)$ when $\hbar \searrow 0$. For more general systems it is very difficult and almost impossible to analyze individual eigenvalues. But it is possible to give a statistical description of the energy spectrum in the semi-classical régime by considering mean values

$$\mathrm{Tr}\left(f(\hat{H})\right) = \sum f(E_j(\hbar)). \tag{5.1}$$

A first result can be obtained if we suppose that the $\hbar$-Weyl symbol H of $\hat{H}$ is smooth and satisfies the assumption of the functional calculus in Chap. 2 (Theorem 10).

D. Robert and M. Combescure, *Coherent States and Applications in Mathematical Physics*, Theoretical and Mathematical Physics,
https://doi.org/10.1007/978-3-030-70845-0_5

Consider an interval $I_\varepsilon = [\lambda_1 - \varepsilon, \lambda_2 + \varepsilon]$ such that $H^{-1}(I)$ is compact for ε small enough. The following result is proved in [141]

Proposition 47 *(i) For every smooth function f supported in I_ε we have the asymptotic expansion at any order in $\hbar$*

$$\mathrm{Tr} f(\hat{H}) \asymp (2\pi\hbar)^{-n} \int_{\mathbb{R}^{2n}} dX f(H(X)) + \sum_{j\geq 1} \hbar^{j-n} C_j(f), \tag{5.2}$$

where $C_j(f)$ are computable distributions in the test function f.

(ii) If λ_1 and λ_2 are noncritical values for H[1] *and if N_I denotes the number of eigenvalues of $\hat{H}$ in $I := [\lambda_1, \lambda_2]$ then we have the Weyl asymptotic formula*

$$\boxed{N_I = (2\pi\hbar)^{-n}\mathrm{Vol}(H^{-1}(I)) + O(\hbar^{1-n})}. \tag{5.3}$$

Remark 25 (i) The first part of the Proposition is an easy application of the functional calculus.

Using (i) it is possible to prove a Weyl formula with an error term $O(\hbar^{\theta-n})$ with $\theta < \frac{2}{3}$. The error term with $\theta = 1$ is optimal (in general) and can be obtained using a method initiated by Hörmander [52, 141, 152]. Furthermore using a trick initiated by Duistermaat–Guillemin [89] the remainder term can be improved in $o(\hbar^{1-n})$ if the measure of closed classical path on $H^{-1}(\lambda_i)$ is zero, for $i = 1, 2$ (see [211]).

The density of states of a quantum system $\hat{H}$ is the sum of delta distributions $D(E) = \sum \delta(E - E_j(\hbar))$. The integrated density of states is the spectral repartition function $N(E) = \sharp\{j,\ E_j(\hbar) < E\}$ where $\sharp\mathscr{E}$ denotes the number of terms in a series $\mathscr{E}$. So that we have $D(E) = \frac{d}{dE}N(E)$.

The Gutzwiller trace formula is a semi-classical formula which expresses the density of states of a quantum system in terms of the characteristics of the corresponding classical system (invariant tori for the completely integrable systems, periodic orbits otherwise). Remark that properties of the classical system may have consequence on the error term in the Weyl formula (see the Remark above). The prototype of the trace formula is the Poisson summation formula:

$$\sum_{n=-\infty}^{+\infty} f(n) = \sum_{k=-\infty}^{+\infty} \tilde{f}(2\pi k) \tag{5.4}$$

for any $f \in \mathscr{S}(\mathbb{R})$. Recall that $\tilde{f}$ is the Fourier transform $\tilde{f}(\tau) = \int_{\mathbb{R}} dt e^{-it\tau} f(t)$.

We show in which respect it is a Trace Formula. Consider the quantum momentum operator $\hat{P}$ in dimension one $\hat{P} = -i\frac{d}{dx}$ acting in $L^2([0, 2\pi])$ with periodic boundary conditions $u(0) = u(2\pi)$. $\hat{P}$ is an unbounded operator with purely discrete spectrum and one has

$$\mathrm{sp}(\hat{P}) = \mathbb{Z}.$$

[1] See definition in Sect. 5.2.

Therefore one has

$$\mathrm{Tr} f(\hat{P}) = \sum_{n=-\infty}^{+\infty} f(n).$$

The classical Hamiltonian is $H(q, p) = p$. Therefore the solutions of the classical equations of motion are

$$q = t \mod(2\pi).$$

So, the classical trajectories are periodic in phase-space $\mathbb{T}^1 \times \mathbb{R}$ and are k-repetitions of the primitive orbit of period 2π, $\forall k \in \mathbb{Z}$. Thus the periods of the classical flow are equal to $2k\pi$, $k \in \mathbb{Z}$. Then the Poisson Summation Formula expresses that the trace of a function of a quantum Hamiltonian equals the sum over the periodic orbits of the corresponding classical flow of the Fourier Transform of this function taken at the periods of the classical flow.

From the Poisson Summation formula one deduces a Trace Formula for the one-dimensional Harmonic Oscillator: take

$$\hat{H}_{\mathrm{os}} = \frac{1}{2}(\hat{P}^2 + \hat{Q}^2)$$

the Hamiltonian of the Harmonic Oscillator. We assume $\hbar = 1$ for simplicity. The spectrum of $\hat{H}_0$ is $\{n + \frac{1}{2}, \quad n \in \mathbb{N}\}$. Therefore for any $f \in \mathscr{S}([0, +\infty[)$ one has

$$\mathrm{Tr}\left(f(\hat{H}_{\mathrm{os}})\right) = \sum_{n=0}^{\infty} f(n + \frac{1}{2}).$$

Replace f by $\hat{T}(q, 0)f$ in Eq. (5.4). One gets

$$\sum_{n\in\mathbb{Z}} f(n+q) = \sum_{k\in\mathbb{Z}} e^{2i\pi kq} \tilde{f}(2k\pi).$$

Therefore for $q = \frac{1}{2}$ one gets

$$\mathrm{Tr}\left(f(\hat{H}_{\mathrm{os}})\right) = \sum_{k=-\infty}^{+\infty} (-1)^k \tilde{f}(2k\pi).$$

But $2k\pi$ are the periods of the orbits for the classical Harmonic oscillator of frequency 1, which are all repetitions of the primitive orbit of period 2π. One notes the apparition of a factor $(-1)^k$ in the trace formula which is the manifestation of the so-called Maslov index of the periodic orbit of period $2k\pi$.

The paper of Gutzwiller appeared in 1971. Between 1973 and 1975 several authors gave rigorous derivations of trace formulas, generalizing the classical Poisson summation formula from $d^2/d\theta^2$ on the circle to elliptic operators on compact manifolds:

Colin de Verdière [56], Chazarain [51], Duistermaat-Guillemin [89]. The first article is based on a parametrix construction for the associated heat equation, while the two others replace this with a parametrix, constructed as a Fourier integral operator, for the associated wave equation.

The pioneering works in quantum physics of Gutzwiller [127, 128] and Balian–Bloch [14, 15] (1972–74) showed that the trace of a quantum observable $\hat{A}$, localized in a spectral neighborhood of size $O(\hbar)$ of an energy E for the quantum Hamiltonian $\hat{H}$, can be expressed in terms of averages of the classical observable A associated with $\widehat{A}$ over invariant sets for the flow of the classical Hamiltonian H associated with $\widehat{H}$. This is related to the spectral asymptotics for $\hat{H}$ in the semi-classical limit, and it can be understood as a "correspondence principle" between classical and quantum mechanics as Planck's constant $\hbar$ goes to zero.

More recently, papers by Guillemin–Uribe [126](1989), Paul–Uribe [203, 204] (1991, 1995), Meinrenken [194] (1992) and Dozias [87](1994) have developed the necessary tools from microlocal analysis [153] in a nonhomogeneous (semiclassical) setting to deal with Schrödinger-type Hamiltonians. Extensions and simplifications of these methods have been given by Petkov–Popov [210] and Charbonnel–Popov [50].

In this Chapter we show how to recover the semiclassical Gutzwiller Trace Formula from the coherent states method.

The coherent states approach presented here seems particularly suitable when one wishes to compare the phase space quantum picture with the phase space classical flow. Furthermore, it avoids problems with caustics, and the Maslov indices appear naturally. In short, it implies the Gutzwiller trace formula in a very simple and transparent way, without any use of the global theory of Fourier integral operators. In their place we use the coherent states approximation (gaussian beams) and the stationary phase theorem.

The use of gaussian wave packets is such a useful idea that one can trace it back to the very beginning of quantum mechanics, for instance, Schrödinger [233] (1926). However, the realization that these approximations are universally applicable, and that they are valid for arbitrarily long times, has developed gradually. In the mathematical literature these approximations have never become textbook material, and this has lead to their repeated rediscovery with a variety of different names, e.g. coherent states and gaussian beams. The first place that we have found where they are used in some generality is Babich [10] (1968) (see also [11]). Since then they have appeared, often as independent discoveries, in the work of Arnaud [7] (1973), Keller and Streifer [165] (1971), Heller [144, 145] (1975, 1987), Ralston [212, 213] (1976,1982), Hagedorn [130, 131] (1980-85), and Littlejohn [182] (1986)—and probably many more that we have not found. Their use in trace formulas was proposed (heuristically) by Wilkinson [262] (1987). The propagation formulas of [130, 131] were extended in Combescure–Robert [66], with a detailed estimate on the error both in time and in Planck's constant $\hbar$. This propagation formula of coherent states which is described in Chap. 4 allows us to avoïd the whole machinery of Fourier Integral operator theory. The early application of these methods in [10] was for the construction of quasi-modes, and this has been pursued further in [212] and [203]. They have also been applied to the pointwise behaviour of semiclassical measures [206].

The proof we shall present here for the Gutzwiller trace formula is inspired from the paper by Combescure–Ralston–Robert [63].

5.2 The Semiclassical Gutzwiller Trace Formula

We consider a quantum system in $L^2(\mathbb{R}^n)$ with the Schrödinger Hamiltonian

$$\widehat{H} = -\hbar^2 \Delta + V(x), \tag{5.5}$$

where Δ is the Laplacian in $L^2(\mathbb{R}^n)$ and $V(x)$ a real, $C^\infty(\mathbb{R}^n)$ potential.

The corresponding Hamiltonian for the classical motion is

$$H(q, p) = p^2 + V(q),$$

and for a given energy $E \in \mathbb{R}$ we denote by Σ_E the "energy shell"

$$\Sigma_E := \{(q, p) \in \mathbb{R}^{2n} \ : \ H(q, p) = E\}. \tag{5.6}$$

More generally we shall consider Hamiltonians $\widehat{H}$ obtained by the $\hbar$-Weyl quantization of the classical Hamiltonian H, so that $\widehat{H} = \mathrm{Op}_\hbar^w(H)$, where

$$\mathrm{Op}_\hbar^w(H)\psi(x) = (2\pi\hbar)^{-n} \iint_{\mathbb{R}^{2n}} H\left(\frac{x+y}{2}, \xi\right) \psi(y) \mathrm{e}^{\frac{i(x-y)\cdot\xi}{\hbar}} dy d\xi \tag{5.7}$$

The Hamiltonian H is assumed to be a smooth, real valued function of $z = (x, \xi) \in \mathbb{R}^{2n}$, and to satisfy the following global estimates

- (H.0) there exist non-negative constants C, m, C_γ such that

$$|\partial_z^\gamma H(z)| \leq C_\gamma < H(z) >, \quad \forall z \in \mathbb{R}^{2n}, \ \forall \gamma \in \mathbb{N}^{2n} \tag{5.8}$$

$$< H(z) > \ \leq \ C < H(z') > \cdot < z - z' >^m, \quad \forall z, \ z' \in \mathbb{R}^{2n} \tag{5.9}$$

where we have used the notation $< u >= (1 + |u|^2)^{1/2}$ for $u \in \mathbb{R}^m$.

Remark 26 (i) $H(q, p) = p^2 + V(q)$ satisfies (H.0), if $V(q)$ is bounded below by some constant $a > 0$ and satisfies the property (H.0) in the variable q.

(ii) The technical condition (H.0) implies in particular that $\widehat{H}$ is essentially self-adjoint on $L^2(\mathbb{R}^n)$ for $\hbar$ small enough and that $\chi(\widehat{H})$ is a $\hbar$-pseudodifferential operator if $\chi \in C_0^\infty(\mathbb{R})$ (see Chap. 2 and [141]).

Let us denote by ϕ^t the classical flow induced by Hamilton's equations with Hamiltonian H, and by $S(q, p; t)$ the classical action along the trajectory starting at

(q, p) at time $t = 0$, and evolving during time t

$$S(q, p; t) = \int_0^t (p_s \cdot \dot{q}_s - H(q, p))\, ds, \tag{5.10}$$

where $(q_t, p_t) = \phi_t(q, p)$, and dot denotes the derivative with respect to time. We shall also use the notation: $\alpha_t = \phi^t(\alpha)$ where $\alpha = (q, p) \in \mathbb{R}^{2n}$, is a phase space point. Recall that the Hamiltonian H is constant along the flow ϕ^t.

If γ is a periodic trajectory parametrized as $t \mapsto \alpha_t$, $\alpha_{T_\gamma^*} = \alpha_0$ where T_γ^* is the primitive period (the smallest positive period), the classical action along γ is

$$S_\gamma = \int_0^{T_\gamma^*} dt p_t \dot{q}_t := \oint_\gamma pdq.$$

An important role in what follows is played by the "linearized flow" around the classical trajectory, which is defined as follows. Let

$$H''(\alpha_t) = \left.\frac{\partial^2 H}{\partial \alpha^2}\right|_{\alpha=\alpha_t} \tag{5.11}$$

be the Hessian of H at point $\alpha_t = \phi^t(\alpha)$ of the classical trajectory. Let J be the symplectic matrix

$$J = \begin{pmatrix} 0 & \mathbb{1} & - \mathbb{1} & 0 \end{pmatrix} \tag{5.12}$$

where 0 and $\mathbb{1}$ are respectively the null and identity $n \times n$ matrices. Let F_t be the $2n \times 2n$ real symplectic matrix solution of the linear differential equation

$$\dot{F}_t = J\, H''(\alpha_t)\, F_t, \tag{5.13}$$

$$F_0 = \begin{pmatrix} \mathbb{1} & 0 \\ 0 & \mathbb{1} \end{pmatrix} = \mathbb{1}.$$

F_t depends on $\alpha = (q, p)$, the initial point for the classical trajectory, α_t.

Let γ be a *closed* orbit on $\sum_E$ with period T_γ, and let us denote simply by F_γ the matrix $F_\gamma = F(T_\gamma)$. F_γ is usually called the "monodromy matrix" of the closed orbit γ. Of course, F_γ does depend on α, but its eigenvalues do not, since the monodromy matrix with a different initial point on γ is conjugate to F_γ. F_γ has 1 as eigenvalue of algebraic multiplicity at least equal to 2. In all that follows, we shall use the following definition

Definition 13 We say that γ is a nondegenerate orbit if the eigenvalue 1 of F_γ has algebraic multiplicity 2.

Let σ denote the usual symplectic form on $\mathbb{R}^{2n}$

$$\sigma(\alpha, \alpha') = p \cdot q' - p' \cdot q \qquad \alpha = (q, p);\ \alpha' = (q', p') \tag{5.14}$$

($\cdot$ is the usual scalar product in $\mathbb{R}^n$). We denote by $\{\alpha_1, \alpha_1'\}$ a basis for the eigenspace of F_γ belonging to the eigenvalue 1, and by V its orthogonal complement in the sense of the symplectic form σ

$$V = \left\{\alpha \in \mathbb{R}^{2n} \ :\ \sigma(\alpha, \alpha_1) = \sigma(\alpha, \alpha_1') = 0\right\} \ . \tag{5.15}$$

Then, the restriction P_γ of F_γ to V is called the (linearized) "Poincaré map" for γ.

In more general cases the Hamiltonian flow will contain manifolds of periodic orbits with the same energy. When this happens, the periodic orbits will necessarily be degenerate, but the techniques we use here can still apply. The precise hypothesis for this ("Hypothesis C") will be given later. Following Duistermaat and Guillemin we call this a "clean intersection hypothesis", it is more explicit than other versions of this assumption. Since the statement of the trace formula is simpler and more informative when one does assume that the periodic orbits are nondegenerate, we will give only that formula in this case.

We shall now assume the following. Let $(\Gamma_E)_T$ be the set of all periodic orbits on $\sum_E$ with periods T_γ, $0 < |T_\gamma| \le T$ (including repetitions of primitive orbits and assigning negative periods to primitive orbits traced in the opposite sense).

- (H.1) There exists $\delta E > 0$ such that $H^{-1}([E - \delta E, E + \delta E])$ is a compact set of $\mathbb{R}^{2n}$ and E is a noncritical value of H (i.e. $H(z) = E \Rightarrow \nabla H(z) \neq 0$).
- (H.2) All γ in $(\Gamma_E)_T$ are nondegenerate, i.e. 1 is not an eigenvalue for the corresponding "Poincaré map". P_γ. In particular, this implies that for any $T > 0$, $(\Gamma_E)_T$ is a discrete set, with periods $-T \le T_{\gamma_1} < \cdots < T_{\gamma_N} \le T$.
  We can now state the Gutzwiller trace formula. Let $\widehat{A} = \mathrm{Op}_\hbar^w(A)$ be a quantum observable, such that A satisfies the following
- (H.3) there exists $\delta \in \mathbb{R}$, $C_\gamma > 0$ ($\gamma \in \mathbb{N}^{2n}$), such that

$$|\partial_z^\gamma A(z)| \le C_\gamma < H(z) >^\delta \quad \forall z \in \mathbb{R}^{2n},$$

- (H.4) g is a $\mathscr{C}^\infty$ function whose Fourier transform $\tilde{g}$ is of compact support with Supp $\tilde{g} \subset [-T, T]$ and let χ be a smooth function with a compact support contained in $]E - \delta E, E + \delta E[$, equal to 1 in a neighborhood of E. Then the following "regularized density of states" $\rho_A(E)$ is well defined

$$\rho_A(E) = \mathrm{Tr}\left(\chi(\widehat{H})\widehat{A}\chi(\widehat{H})g\left(\frac{E - \widehat{H}}{\hbar}\right)\right). \tag{5.16}$$

Note that (H.1) implies that the spectrum of $\widehat{H}$ is purely discrete in a neighborhood of E so that $\rho_A(E)$ is well defined. We have also, more explicitly,

$$\rho_A(E) = \sum_{1 \le j \le N} g\left(\frac{E - E_j}{\hbar}\right)\chi^2(E_j) < \hat{A}\varphi_j, \varphi_j >, \tag{5.17}$$

where $E_1 \leq \cdots \leq E_N$ are the eigenvalues of $\hat{H}$ in $]E - \delta E, E + \delta E[$ (with multiplicities) and φ_j is the corresponding eigenfunction ($\hat{H}\varphi_j = E_j\varphi_j$). Let us remark here that the $\hbar$ scaling: $\frac{E-E_j}{\hbar}$ is the right one to have a nice semiclassical limit. The first argument is that if $n = 1$ (and for integrable systems), in regular case, eigenvalues are given by Bohr–Sommerfeld formula [142] and their mutual distance is of order $\hbar$. The second argument is included in the following result [140, 211].

Under assumption (H.1) the Liouville measure dL_E is well defined on the energy surface Σ_E :

$$dL_E = \frac{d\Sigma_E}{|\nabla H|} \qquad (d\Sigma_E \text{ is the Euclidean measure on } \Sigma_E).$$

Now we can state the Gutzwiller Trace Formula.

Theorem 28 (Gutzwiller Trace Formula) *Assume* (H.0)–(H.2) *are satisfied for* H, (H.3) *for* A *and* (H.4) *for* g. *Then the following asymptotic expansion holds true, modulo* $O(\hbar^\infty)$,

$$\rho_A(E) \equiv (\pi)^{-n/2}\widehat{g}(0)\hbar^{-(n-1)} \int_{\Sigma_E} A(\alpha)d\sigma_E(\alpha) + \sum_{k\geq -n+2} c_k(\widehat{g})\hbar^k +$$
$$\sum_{\gamma\in(\Gamma_E)_T} (2\pi)^{n/2-1} \mathrm{e}^{i(S_\gamma/\hbar+\sigma_\gamma\pi/2)} \left\{\widehat{g}(T_\gamma)|\det(\mathbb{1} - P_\gamma)|^{-1/2} \int_0^{T_\gamma^*} A(\alpha_s)ds + \right.$$
$$\left. \sum_{j\geq 1} d_j^\gamma(\widehat{g})\hbar^j \right\} \tag{5.18}$$

where T_γ^* *is the primitive period of* γ, σ_γ *is the Maslov index of* γ ($\sigma_\gamma \in \mathbb{Z}$ *and is computed in the proof),* $c_k(\tilde{g})$ *are distributions in* $\tilde{g}$ *with support in* $\{0\}$, $d_j^\gamma(\tilde{g})$ *are distributions in* $\tilde{g}$ *with support* $\{T_\gamma\}$.

Remark 27 We can include more general Hamiltonians depending explicitly in $\hbar$, $H = \sum_{j=1}^{K} \hbar^j H^{(j)}$ such that $H^{(0)}$ satisfies (H.0) and for $j \geq 1$,

$$|\partial^\gamma H^{(j)}(z)| \leq C_{\gamma,j} < H^{(0)}(z) > . \tag{5.19}$$

It is useful for applications to consider Hamiltonians like $H^{(0)} + \hbar H^{(1)}$ where $H^{(1)}$ may be, for example, a spin term. In that case the formula (5.18) is true with different coefficients. In particular the first term in the contribution of T_γ is multiplied by $\exp\left(-i \int_0^{T_\gamma^*} H^{(1)}(\alpha_s)ds\right)$.

Remark 28 For Schrödinger operators we only need smoothness of the potential V. In this case the trace formula (5.18) is still valid without any assumptions at infinity for V when we restrict ourselves to a compact energy surface, assuming $E < \liminf_{|x|\to\infty} V(x)$. Using exponential decrease of the eigenfunctions [143] we can prove that, modulo an error term of order $O(\hbar^{+\infty})$, the potential V can be replaced by a potential $\tilde{V}$ satisfying the assumptions of the Remark (2.1).

It is also possible to get a trace formula for Hamiltonians with symmetries [48].

5.3 Preparations for the Proof

We shall make use of the standard coherent states introduced in Chap. 1 and their propagation by the time dependent Schrödiger equation established in Chap. 4. We denote by

$$\varphi_\alpha = \hat{T}(\alpha)\psi_0 \tag{5.20}$$

the usual coherent states centered at the point α of the phase space $\mathbb{R}^{2n}$. Then it is known that any operator $\hat{B}$ with a symbol decreasing sufficiently rapidly is in trace class [217], and its trace can be computed by (see Chap. 1)

$$\mathrm{Tr}\hat{B} = (2\pi\hbar)^{-n}\int_{\mathbb{R}^{2n}} < \varphi_\alpha, \hat{B}\varphi_\alpha > d\alpha. \tag{5.21}$$

The regularized density of states $\rho_A(E)$ can now be rewritten as

$$\rho_A(E) = (2\pi)^{-n-1}\hbar^{-n}\int_{\mathbb{R}\times\mathbb{R}^{2n}} \check{g}(t)\, e^{iEt/\hbar} < \varphi_\alpha, \widehat{A}_\chi\, \mathscr{U}(t)\, \varphi_\alpha > dtd\alpha, \tag{5.22}$$

where $\mathscr{U}(t)$ is the quantum unitary group :

$$\mathscr{U}(t) = e^{-it\widehat{H}/\hbar}, \text{ and } \widehat{A}_\chi = \chi(\widehat{H})\widehat{A}\chi(\widehat{H}). \tag{5.23}$$

Our strategy for computing the behavior of $\rho_A(E)$ as $\hbar$ goes to zero is first to compute the bracket

$$m(\alpha, t) =< \varphi_\alpha, \widehat{A_\chi \mathscr{U}}(t)\varphi_\alpha >, \tag{5.24}$$

where we drop the subscript χ in A_χ for simplicity.

First of all we shall use Lemma 14 of Chap. 2, giving the action of an $\hbar$-pseudodifferential operator on a Gaussian.

Thus $m(t, \alpha)$ is a linear combination of terms like

$$m_\gamma(\alpha, t) =< \Psi_{\gamma,\alpha}, \mathscr{U}(t)\varphi_\alpha > . \tag{5.25}$$

Now we compute $\mathscr{U}(t)\varphi_\alpha$, using the semiclassical propagation of coherent states result of Chap. 4. We recall that F_t is a time dependent symplectic matrix (Jacobi matrix) defined by the linear equation (5.13). $\hat{R}(F)$ denotes the metaplectic representation of the linearized flow F (see Chap. 3), and the $\hbar$-dependent metaplectic representation is defined by

$$\hat{R}_\hbar(F) = \Lambda_\hbar \hat{R}(F)\Lambda_\hbar^{-1}. \tag{5.26}$$

We will also use the notation

$$\delta(\alpha, t) = \int_0^t p_s \cdot \dot{q}_s ds - tH(\alpha) - \frac{p_t \cdot q_t - p \cdot q}{2}. \tag{5.27}$$

From Chap. 4 we have the following propagation estimates in the L^2-norm: for every $N \in \mathbb{N}$ and every $T > 0$ it exists $C_{N,T}$ such that

$$\left\| \mathcal{U}(t)\varphi_\alpha - \exp\left(\frac{i\delta(\alpha, t)}{\hbar}\right) \hat{T}(\alpha_t)\hat{R}_\hbar(F_t)\Lambda_\hbar P_N(x, D_x, t, \hbar)\widetilde{\psi_0}\right) \| \leq C_{N,T}\hbar^N, \tag{5.28}$$

where $\widetilde{\psi_0}(x) = \pi^{-n/4}\exp(-|x|^2/2)$, and $P_N(t, \hbar)$ is the $(\hbar, t)$-dependent differential operator defined by

$$P_N(x, D_x, t, \hbar) = \mathbb{1} + \sum_{(k,j)\in I_N} \hbar^{k/2-j} p^w_{kj}(x, D, t)$$

$$\text{with } I_N = \{(k, j) \in \mathbb{N} \times \mathbb{N},\ 1 \leq j \leq 2N - 1,\ k \geq 3j,\ 1 \leq k - 2j < 2N\}, \tag{5.29}$$

where the differential operators $p_{kj}(x, D_x, t)$ are products of j Weyl quantization of homogeous polynomials of degree k_s with $\sum_{1\leq s\leq j} k_s = k$ (see [66] Theorem (3.5) and its proof). So that we get

$$p^w_{kj}(x, D_x, t)\widetilde{\psi_0} = Q_{kj}(x)\widetilde{\psi_0}(x), \tag{5.30}$$

where $Q_{kj}(x)$ is a polynomial (with coefficients depending on (α, t)) of degree k having the same parity as k. This is clear from the following facts: homogeneous polynomials have a definite parity, and Weyl quantization behaves well with respect to symmetries: $Op^w(A)$ commutes to the parity operator $\Sigma f(x) = f(-x)$ if and only if A is an even symbol and anticommutes with Σ if and only if A is an odd symbol) and $\widetilde{\psi_0}(x)$ is an even function.

So we get

$$m(\alpha, t) = \sum_{(j,k)\in I_N; |\gamma|\leq 2N} c_{k,j,\gamma}\hbar^{\frac{k+|\gamma|}{2}-j} \exp\left(\frac{i\delta(t, \alpha)}{\hbar}\right) \cdot$$

$$\cdot\left\langle \hat{T}(\alpha)\Lambda_\hbar Q_\gamma\widetilde{\psi_0}, \hat{T}(\alpha_t)\Lambda_\hbar Q_{k,j}\hat{R}(F_t)\widetilde{\psi_0}\right\rangle + O(\hbar^N), \tag{5.31}$$

where $Q_{k,j}$ respectively Q_γ are polynomials in the x variable with the same parity as k respectively $|\gamma|$. This remark will be useful in proving that we have only entire powers in $\hbar$ in (5.18), even though half integer powers appear naturally in the asymptotic propagation of coherent states.

By an easy computation we have

$$\left\langle \hat{T}(\alpha)\Lambda_\hbar Q_\gamma \widetilde{\psi_0}, \hat{T}(\alpha_t)\Lambda_\hbar Q_{k,j}\hat{R}(F_t)\widetilde{\psi_0}\right\rangle =$$
$$\exp\left(-i\frac{1}{2\hbar}\sigma(\alpha,\alpha_t)\right)\left\langle \hat{T}_1\left(\frac{\alpha-\alpha_t}{\sqrt{\hbar}}\right) Q_\gamma\widetilde{\psi_0}, Q_{k,j}\hat{R}(F_t)\widetilde{\psi_0}\right\rangle \tag{5.32}$$

where $\hat{T}_1(\cdot)$ is the Weyl translation operator with $\hbar = 1$.

We set

$$m_{k,j,\gamma}(\alpha,t) = \left\langle \hat{T}_1\left(\frac{\alpha-\alpha_t}{\sqrt{\hbar}}\right) Q_\gamma\widetilde{\psi_0}, Q_{k,j}\hat{R}(F_t)\widetilde{\psi_0}\right\rangle \tag{5.33}$$
$$m_0(\alpha,t) = \left\langle \hat{T}_1\left(\frac{\alpha-\alpha_t}{\sqrt{\hbar}}\right) \widetilde{\psi_0}, \hat{R}(F_t)\widetilde{\psi_0}\right\rangle. \tag{5.34}$$

We compute $m_0(\alpha,t)$ first. We shall use the fact that the metaplectic group transforms Gaussian wave packets into Gaussian wave packets in a very explicit way. If we denote by A_t, B_t, C_t, D_t the four $n\times n$ matrices of the block form of F_t

$$F_t = \begin{pmatrix} A_t & B_t \\ C_t & D_t \end{pmatrix} \tag{5.35}$$

We have already seen in Chaps. 3 and 4, since F_t is symplectic, that $U_t = A_t + iB_t$ is invertible. So we have defined

$$\Gamma_t = V_t U_t^{-1}, \quad \text{where} \quad V_t = (C_t + iD_t). \tag{5.36}$$

We have from our Chap. 3 (see also [104], Chap. 4)

$$m_0(\alpha,t) = [\det U_t]_c^{-1/2}\pi^{-n/2}\,.$$
$$\cdot\int_{\mathbb{R}^n}\exp\left\{\frac{i}{2}(\Gamma_t + i\mathbb{1})x\cdot x - \frac{i}{\sqrt{\hbar}}(x - \frac{q-q_t}{2})\right.$$
$$\left.\times(p - p_t + i(q-q_t))\right\}dx \tag{5.37}$$

But the integration in x is a Fourier transform of a Gaussian and can be performed (see in Appendix Sect. A.1). The complex matrix $\Gamma_t + i\mathbb{1}$ is invertible and we have

$$(\Gamma_t + i\mathbb{1})^{-1} = \frac{\mathbb{1} + W_t}{2i} \tag{5.38}$$

where use the following notations

$$W_t = Z_t Y_t^{-1}, \quad Z_t = U_t + iV_t, \; Y_t = U_t - iV_t. \tag{5.39}$$

It is clear that Y is invertible (see Chap. 3). So we get

$$m_0(\alpha, t) = 2^n \pi^{n/2} [\det Y_t U_t^{-1}]_*^{-1/2} [\det U_t]_c^{-1/2} e^{\frac{i}{\hbar}\Psi_E(t,\alpha)} \tag{5.40}$$

where $[\det Y_t U_t^{-1}]_*^{-1/2}$, $[\det U_t]_c^{-1/2}$ are defined below, the phase $\Psi_E(t,\alpha)$ is given by

$$\Psi_E(t,\alpha) = t(E - H(\alpha)) + \frac{1}{2}\int_0^t \sigma(\alpha_s - \alpha, \dot\alpha_s)ds + \frac{i}{4}(\mathbb{1} + W_t)(\breve\alpha - \breve\alpha_t)\cdot\overline{(\breve\alpha - \breve\alpha_t)}, \tag{5.41}$$

with $\breve\alpha = q + ip$ if $\alpha = (q, p)$.

In (5.40) we have a product of square root of determinant. $[\det U_t]_c^{-1/2}$ is a branch for $[\det U_t]^{-1/2}$ with the phase (or argument) obtained by continuity in time from $U_{t=0} = \mathbb{1}$. For a complex symmetric matrix M with definite-positive real part, $[\det M]_*^{-1/2}$ is a branch for $(\det M)^{-1/2}$ with the phase obtained by continuity along a path joining $\Re M$ to M, the eigenvalues of $M^{1/2}$ having positive real part. Following carefully these phases will give the Maslov correction index.

Remark 29 There is here a difference with the paper [63] were the phase was obtained before integration in $y \in \mathbb{R}^n$, so computations here will be a little bit more natural and easier.

The same phase $\Psi_E(t,\alpha)$ appears when computing $m_{k,j,\gamma}(\alpha, t)$ with non trivial amplitudes. Then the formula for the regularized density of states in (5.22) takes the form

$$\rho_A(E) = \int_{\mathbb{R}} dt \int_{\mathbb{R}^{2n}} d\alpha a(t,\alpha,\hbar) e^{\frac{i}{\hbar}\Psi_E(t,\alpha)} \tag{5.42}$$

where Ψ_E is given in (5.41) and $a(t,\alpha,\hbar) \asymp \sum_{j\in\mathbb{N}} a_j(t,\alpha)\hbar^j$

Our plan is to prove Theorem 28 by expanding (5.42) by the method of stationary phase. The necessary stationary phase lemma for complex phase functions can easily be derived from Theorem 7.7.5 in [153]. There is also an extended discussion of complex phase functions depending on parameters in [153] leading to Theorem 7.7.12, but the form of the stationary manifold here permits us to use the following result proved in Appendix A.2.

Theorem 29 (stationary phase expansion) *Let $\mathcal{O} \subset \mathbb{R}^d$ be an open set, and let $a, f \in \mathcal{C}^\infty(\mathcal{O})$ with $\Im f \geq 0$ in $\mathcal{O}$ and supp $a \subset \mathcal{O}$. We define*

$$M = \{x \in \mathcal{O}, \Im f(x) = 0, f'(x) = 0\} \quad ,$$

and assume that M is a smooth, compact and connected submanifold of $\mathbb{R}^d$ of dimension k such that for all $x \in M$ the Hessian, $f''(x)$, of f is nondegenerate on the normal space N_x to M at x.

Under the conditions above, the integral $J(\omega) = \int_{R^d} e^{i\omega f(x)} a(x)dx$ has the following asymptotic expansion as $\omega \to +\infty$, modulo $O(\omega^{-\infty})$,

$$J(\omega) \equiv \left(\frac{2\pi}{\omega}\right)^{\frac{d-k}{2}} \sum_{j \geq 0} c_j \omega^{-j} \quad .$$

The coefficient c_0 is given by

$$c_0 = e^{i\omega f(m_0)} \int_M \left[\det\left(\frac{f''(m)|N_m}{i}\right)\right]_*^{-1/2} a(m)dV_M(m), \tag{5.43}$$

where $dV_M(m)$ is the canonical Euclidean volume in M, $m_0 \in M$ is arbitrary, and $[\det P]_*^{-1/2}$ denotes the product of the reciprocals of square roots of the eigenvalues of P chosen with positive real parts. Note that, since $\Im f \geq 0$, the eigenvalues of $\frac{f''(m)|N_m}{i}$ lie in the closed right half plane.

5.4 The Stationary Phase Computation

In this section we compute the stationary phase expansion of (5.22) with phase Ψ_E given by (5.41). Note that $a(t, \alpha, \hbar)$ is actually, according to (5.31), a polynomial in $\hbar^{1/2}$ and $\hbar^{-1/2}$. Hence the stationary phase theorem (with $\hbar$ independent symbol a) applies to each coefficient of this polynomial.

We need to compute the first and second order derivatives of $\Psi_E(t, \alpha)$. Let us introduce the $2n \times 2n$ complex symmetric matrix

$$W_t^\sharp = \begin{pmatrix} W_t & -iW_t \\ -iW_t & -W_t \end{pmatrix}.$$

It is enough to compute first derivatives for Ψ_E up to $O(|\alpha_t - \alpha|^2)$.

$$\partial_t \Psi_E(t, \alpha) \equiv E - H(\alpha) - \frac{1}{2}(\alpha_t - \alpha) \cdot J\dot{\alpha}_t + \frac{1}{2i}(W^\sharp - \mathbb{1})\dot{\alpha}_t \cdot (\alpha_t - \alpha). \tag{5.44}$$

$$\partial_\alpha \Psi_E(t, \alpha) \equiv \frac{1}{2}(\mathbb{1} + F^T)J(\alpha_t - \alpha) + \frac{1}{2i}(F^T - \mathbb{1})(W^\sharp - \mathbb{1})(\alpha_t - \alpha). \tag{5.45}$$

The critical set for the stationary phase theorem is defined as

$$\mathscr{C}_E = \{(\alpha, t) \in \mathbb{R}^{2n} \times \mathbb{R},\ \Im(\Psi_E(\alpha, t)) = 0,\ \partial_t \Psi_E(t, \alpha) = 0,\ \partial_\alpha \Psi_E(t, \alpha) = 0\}.$$

We have seen in Chap. 3 that since F is symplectic, one has $W^*W < \mathbb{1}$, so if $\Im(\Psi_E(\alpha, t)) = 0$ then $\alpha_t = \alpha$. Using (5.44) we get

$$\mathscr{C}_E = \{(\alpha, t) \in \mathbb{R}^{2n} \times \mathbb{R}, \quad H(\alpha) = E; \ \alpha_t = \alpha\}.$$

Hence (t, α) is a critical point means that α is on a periodic path of energy E, for the Hamiltonian H, and period t.

The second derivatives of Ψ_E restricted on $\mathscr{C}_E$ can be computed as follows

$$\partial^2_{t,t}\Psi_E(t, \alpha) = \frac{1}{2i}(W_t^\sharp - \mathbb{1})\dot{\alpha} \cdot \dot{\alpha}. \tag{5.46}$$

$$\partial^2_{t,\alpha}\Psi_E(t, \alpha) = -\partial_\alpha H(\alpha) + \frac{1}{2i}(F_t^T - \mathbb{1})(W_t^\sharp - \mathbb{1})(\dot{\alpha}). \tag{5.47}$$

$$\partial^2_{\alpha,\alpha}\Psi_E(t, \alpha) = \frac{1}{2}(JF_t - F_t^T J) + \frac{1}{2i}(F_t^T - \mathbb{1})(W_t^\sharp - \mathbb{1})(F_t - \mathbb{1}). \tag{5.48}$$

Let $\Psi''_E(t_0, \alpha_0)$ be the Hessian matrix of Ψ_E at point (t_0, α_0) of $\mathscr{C}_E$. We have to compute the kernel of $\Psi''_E(\alpha_0, t_0)$.

Lemma 33 *For every $(t_0, \alpha_0) \in \mathscr{C}_E$ we have*

$$\ker(\Psi''_E(t_0, \alpha_0)) = \{(\tau, v) \in \mathbb{R} \times \mathbb{R}^{2n}, \quad v \cdot \partial_\alpha H = 0, \quad (F_{t_0} - \mathbb{1})v + \tau\dot{\alpha} = 0\}. \tag{5.49}$$

Proof Using the Taylor formula we have

$$\Im(\psi_E(t, \alpha)) = \frac{1}{2}\Im\Psi''_E(t_0, \alpha_0)(t - t_0, \alpha - \alpha_0) \cdot (t - t_0, \alpha - \alpha_0) + O(|t - t_0|^3 + |\alpha - \alpha_0|^3). \tag{5.50}$$

From $W^*W < 1$ we get that, for some $c > 0$,

$$\Im(\psi_E(t, \alpha)) \geq c|\alpha - \alpha_t|^2. \tag{5.51}$$

Using that $\alpha - \alpha_t = (\alpha - \alpha_0) + (\alpha_0 - \alpha_{0,t_0}) + (\alpha_{0,t_0} - \alpha_{t_0}) + (\alpha_{t_0} - \alpha_t)$ we get

$$\alpha - \alpha_t = (\mathbb{1} - F_{t_0})(\alpha - \alpha_0) + (t_0 - t)\dot{\alpha}_t + O(|t - t_0|^2 + |\alpha - \alpha_0|^2)$$

Then from (5.50) and (5.51) we get, for some $c > 0$,

$$|\Im\Psi''_E(t_0, \alpha_0)(t - t_0, \alpha - \alpha_0) \cdot (t - t_0, \alpha - \alpha_0)| \geq c\|(F_{t_0} - \mathbb{1})(\alpha - \alpha_0) + (t - t_0)\dot{\alpha}\|^2. \tag{5.52}$$

So we have proved the part "$\subseteq$" in (5.49). The part "$\supseteq$" is obvious. □

The first thing to check, in order to apply the stationary phase theorem is that the support of α in (5.42) can be taken as compact, up to an error $O(\hbar^\infty)$. We do this in the following way : let us recall some properties of $\hbar$-pseudodifferential calculus proved in [83, 141]. The function $m_H(z) := < H(z) >$ is a weight function. In [83] it is proved that $\chi(\hat{H}) = \hat{H}_\chi$ where $H_\chi \in \mathrm{S}(m_H^{-k})$, for every k (χ is like in (H.4)). More precisely, we have in the $\hbar$ asymptotic sense in $\mathrm{S}(m_H^{-k})$,

$$H_\chi = \sum_{j\geq 0} H_{\chi j}\hbar^j$$

and support $[H_{\chi,j}]$ is in a fixed compact set for every j (see (H.4) and [141] for the computations of $H_{\chi,j}$). Let us recall that the symbol space $\mathrm{S}(m_H)$ is equipped with the family of semi-norms

$$\sup_{z\in\mathbb{R}^{2n}} m_H^{-1}(z)\left|\frac{\partial^\gamma}{\partial z^\gamma}u(z)\right|.$$

Now we can prove the following lemma

Lemma 34 *There is a compact set K in $\mathbb{R}^{2n}$ such that for*

$$m(\alpha,t) :=< \varphi_\alpha, \hat{A}_\chi \mathscr{U}(t)\varphi_\alpha >$$

we have

$$\int_{\mathbb{R}^{2n}/K} |m(\alpha,t)|d\alpha = O(\hbar^{+\infty})$$

uniformly in every bounded interval in t.

Proof Let $\tilde{\chi} \in C_0^\infty(]E-\delta E, E+\delta E[)$ such that $\tilde{\chi}\chi = \chi$. Using (H.3) and the composition rule for $\hbar$-pseudodifferential operators we can see that $\hat{A}_\chi(\hat{H})$ is bounded on $L^2(\mathbb{R}^n)$. So there exists a $C > 0$ such that

$$|m(\alpha,t)| \leq C \parallel \tilde{\chi}(\hat{H})\varphi_\alpha \parallel^2 \quad .$$

But we can write

$$\parallel \tilde{\chi}(\hat{H})\varphi_\alpha \parallel^2 =< \tilde{\chi}(\hat{H})^2\varphi_\alpha, \varphi_\alpha > \quad .$$

Let us introduce the Wigner function, w_α, for φ_α (i.e. the Weyl symbol of the orthogonal projection on φ_α). We have

$$< \tilde{\chi}(\hat{H})^2\varphi_\alpha, \varphi_\alpha >= (\pi\hbar)^{-n}\int H_{\chi^2}(z)w_\alpha(z)dz$$

where

$$w_\alpha(z) = (\pi\hbar)^{-n}\, e^{-\frac{|z-\alpha|^2}{\hbar}} \quad .$$

Using remainder estimates from [141] we have, for every N large enough,

$$\hat{H}_{\chi^2} = \sum_{0\leq j\leq N} H_{\chi^2,j}\hbar^j + \hbar^{N+1}R_N(\hbar)$$

where the following estimate in Hilbert-Schmidt norm holds

$$\sup_{0<\hbar\leq 1} \| R_N(\hbar) \|_{HS} < +\infty \quad .$$

Now there is an $R > 0$ such that for every j, we have $\mathrm{Supp}[H_{\chi^2,j}] \subseteq \{z, |z| < R\}$. So the proof of the lemma follows from

$$\| R_N(\hbar) \|_{HS}^2 = (2\pi\hbar)^{-n} \int \| R_N(\hbar)\varphi_\alpha \|^2 \, d\alpha$$

and from the elementary estimate, which holds for some $C, c > 0$,

$$\iint_{\{|z|\leq R, |\alpha|\geq R+1\}} e^{-\frac{|\alpha-r|^2}{\hbar}} dz d\alpha \leq Ce^{-\frac{c}{\hbar}}.$$

□

From (5.49) we see that a sufficient condition to apply the stationary phase theorem is the following "clean intersection condition" for the Hamiltonian flow ϕ^t. **Clean Intersection Condition (CI)**: We assume that $\mathscr{C}_E$ is a union of smooth compact connected components and on each component, the tangent space $T_{(t_0,\alpha_0)}\mathscr{C}_E$ to $\mathscr{C}_E$ at (t_0, α_0) coïncides with the linear space $\{(\tau, v) \in \mathbb{R} \times \mathbb{R}^{2n}, \quad v \cdot \partial_\alpha H = 0, \quad (F_{t_0}\mathbb{1})v + \tau\dot{\alpha} = 0\}$.

So under condition (CI) the kernel of $\Psi_E''(t_0, \alpha_0)$ coïncides with the tangent space $T_{(t_0,\alpha_0)}(\mathscr{C}_E)$. Hence $\Psi_E''(t_0, \alpha_0)$ is non-degenerate on the normal space at $\mathscr{C}_E$ on (t_0, α_0) as it is required to apply the stationary phase theorem. At this point we have already proved that there exists an asymptotic expansion for the regularized density of states $\rho_A(E)$. A more difficult problem is to compute this asymptotics in general.

The simpler case is the period 0 of the flow: $\mathscr{C}_E = \{0, \alpha), \ H(\alpha) = E\}$. Then the property (CI) is satisfied if E is non critical for H. Remark that 0 is not an accumulation point in the periods of classical paths.

The Hessian matrix on $\mathscr{C}_E$ is

$$\Psi_E''(0, \alpha) = \begin{pmatrix} -\frac{1}{2}\nabla H & i\nabla H \\ -\nabla H & 0 \end{pmatrix},$$

where $\nabla H = \partial_\alpha H$. The normal space $\mathscr{N}_\alpha$ to $\mathscr{C}_E$ has the basis $\{(1, 0), \ (0, \nabla H)\}$. So the determinant of $\Psi_E''(0, \alpha)$ restricted to $\mathscr{N}_{(0,\alpha)}$ is $\|\nabla H\|^4$. The stationary phase theorem gives us

Proposition 48 *Let g be such that $\tilde{g}$ is supported in $]-T_0, T_0[$ where $T_0 > 0$ and ϕ^t has no periodic trajectory on Σ_E with a period in $]-T_0, T_0[\setminus\{0\}$. Then we have*

$$\rho_A(E) = \tilde{g}(0)(2\pi)^{-n} \left(\int_{\Sigma_E} \frac{\chi(H\alpha)}{\|\nabla H(\alpha)\|} A(\alpha) d\Sigma_E(\alpha) \right) \hbar^{1-n} + O(\hbar^{2-n}). \quad (5.53)$$

Moreover the asymptotics can be extended as a full asymptotics in $\hbar$.

As an application of (5.53) we have the following

Theorem 30 (Weyl asymptotic formula) *Assume that H satisfies condition (H.0) (5.8). Consider $\lambda_1 < \lambda_2$ such that $H^{-1}[\lambda_1 - \varepsilon, \lambda_2 + \varepsilon]$ is compact and are non critical values for H. Let N_I be the number of eigenvalues $E_j(\hbar)$ of $\hat{H}$ in $I = [\lambda_1, \lambda_2]$.*
Then we have

$$N_I = (2\pi\hbar)^{-n}\mathrm{Vol}\{\alpha \in \mathbb{R}^{2n},\ H(\alpha) \in I\} + O(\hbar^{1-n}). \quad (5.54)$$

Proof Using a partition of unity it is enough to consider

$$\sigma(\lambda) = \sum_{E_j(\hbar) \le \lambda} \chi(E_j(\hbar))$$

where χ is supported in a small neighborhood of λ_1 or λ_2 (between λ_1 and λ_2 we can apply the functional calculus to get an asymptotic expansion (see [141]).

To prove an asymptotic expansion for $\sigma(\lambda)$ we use (5.53) choosing $A = \chi(H)$, $\tilde{g}$ even, $\tilde{g}(0) = 1$, $g \ge 0$ and $g(\lambda) \ge \delta_0$ for $|\lambda| \le \varepsilon_0$ for some $\delta_0 > 0, \varepsilon_0 > 0$. We have

$$\sigma(\lambda) - \frac{1}{\hbar} \int g\left(\frac{\mu - \lambda}{\hbar}\right) \sigma(\mu) d\mu = \int (\sigma(\lambda) - \sigma(\lambda + \tau\hbar))\, g(\tau) d\tau. \quad (5.55)$$

From (5.53) we have, after integration,

$$\hbar^{-1} \int g\left(\frac{\mu - \lambda}{\hbar}\right) \sigma(\mu) d\mu = (2\pi\hbar)^{-n} \int_{H(\alpha)\le\lambda} d\alpha \chi(H(\alpha)) + O(\hbar^{-n}).$$

Using the following estimate, for some $C > 0$,

$$|\sigma(\lambda + \tau\hbar) - \sigma(\lambda)| \le C(1 + |\tau|)\hbar^{1-n}, \quad (5.56)$$

we get

$$\sigma(\lambda) = (2\pi\hbar)^{-n} \int_{H(\alpha)\le\lambda} d\alpha \chi(H(\alpha)) + O(\hbar^{-n}),$$

then (5.54) follows.

Now we prove (5.56). It is enough to consider the case $\tau \ge 0$.
Suppose $\tau \le \varepsilon_0$. Then

$$\delta_0 \int_\lambda^{\lambda+\tau\hbar} \sigma(\mu) \leq \int d\mu g\left(\frac{\mu-\lambda}{\hbar}\right) = O(\hbar^{1-n}).$$

For $\tau = \ell\varepsilon_0$ with $\ell \in \mathbb{N}$, using the triangle inequality, we get

$$|\sigma(\lambda + \ell\varepsilon_0\hbar) - \sigma(\lambda)| \leq C\ell\hbar^{1-n}.$$

Finally for $\ell\varepsilon_0 < \tau < (\ell+1)\varepsilon_0$ using again the triangle inequality we get (5.56). □

Remark 30 Assuming that the set of all periodical trajectories of H in Σ_E is of Liouville-measure 0, it is possible to prove by the same method the following result. For every $C > 0$ we have

$$\lim_{\hbar \searrow 0} \hbar^{n-1} \sharp\{j,\ E - C\hbar \leq E_j(\hbar) \leq E + C\hbar\} = \int_{\Sigma_E} dL_E =: L_E(\Sigma_E). \qquad (5.57)$$

This result was already proved in [140, 211] using Fourier Integral operators.

Now we come to the proof of the Gutzwiller Trace Formula (5.18).

Note that for isolated periodic orbits on Σ_E the non-degenerate assumption is equivalent to the condition (CI). So it results from our discussion that in this case the Hessian matrix Ψ_E'' at (t_0, α_0), where γ is a periodic path with period $t_0 = kT_\gamma^*$ and $\alpha_0 \in \gamma$ is non-degenerate on the normal space $\mathscr{N}_{t_0,\alpha_0}$ at $\mathscr{C}_E$. Here $\mathscr{N}_{t_0,\alpha_0}$ is the linear space $\{\mathbb{R}(1,0) + \{(0,v),\ v \in \mathbb{R}^{2n}, \sigma(v, \nabla H) = 0\}$. Our main problem is to compute the determinant of the restriction $\Psi_{E,\perp}''(t_0, \alpha_0)$ of $\Psi_E''(t_0, \alpha_0)$ to $\mathscr{N}_{t_0,\alpha_0}$. We shall denote $\Pi_{\dot\alpha}$ the orthogonal projection in $\mathbb{R}^{2n}$ on $J\nabla H(\alpha_0) := \dot\alpha$ (tangent vector to γ).

It is convenient to introduce the notations

$$G = \frac{1}{2}(W_{t_0}^\sharp - \mathbb{1}), \quad K = (F_{t_0}^T - \mathbb{1})(G + \frac{i}{2}J) + iJ.$$

Using that F_{t_0} is symplectic we have

$$\partial_{\alpha,\alpha}^2 \Psi_E(t_0, \alpha_0) = K(F_{t_0} - \mathbb{1}).$$

So we have

$$\Psi_E''(t_0, \alpha_0) = i^{-1}\begin{pmatrix} G\dot\alpha \cdot \dot\alpha & K\dot\alpha \\ K\dot\alpha & K(F_{t_0} - \mathbb{1}) \end{pmatrix}. \qquad (5.58)$$

This formula is general. Furthermore we have the very useful result

Lemma 35 *K is a $2n \times 2n$ invertible matrix and we have*

$$K^{-1} = -\frac{1}{2}\begin{pmatrix} U - \mathbb{1} & -i(\mathbb{1} + U) \\ i\mathbb{1} + V & -(\mathbb{1} + iV) \end{pmatrix}. \qquad (5.59)$$

In particular we have

$$\det K = (-1)^n \det \left(\frac{Y}{2}\right)^{-1}, \tag{5.60}$$

where $U = U_{t_0}$, $V = V_{t_0}$, $Y = Y_{t_0}$.

Proof We have, using definition of W,

$$W^\sharp - \mathbb{1} + iJ = \begin{pmatrix} W - \mathbb{1} & -i(W - \mathbb{1}) \\ -i(W + \mathbb{1}) & -(W + \mathbb{1} \end{pmatrix} \tag{5.61}$$

$$= \begin{pmatrix} 2iV & 2V \\ -2iU & -2U \end{pmatrix} \begin{pmatrix} (U - iV)^{-1} & 0 \\ 0 & (U - iV)^{-1} \end{pmatrix}. \tag{5.62}$$

After some algebraic computations, using in particular the symplectic relations, we find

$$K = \begin{pmatrix} -\mathbb{1} - iV & i(\mathbb{1} + U) \\ -i\mathbb{1} - V & -\mathbb{1} + U \end{pmatrix} \begin{pmatrix} (U - iV)^{-1} & 0 \\ 0 & (U - iV)^{-1} \end{pmatrix}$$

So we get the Lemma. □

Now we begin to use the non-degeneracy condition to compute the determinant of $\Psi''_{E,\perp}(t_0, \alpha_0)$. We have

$$\det\left(i^{-1}\Psi''_{E,\perp}(t_0, \alpha_0)\right) = i^{-1} \det \begin{pmatrix} d & K\dot\alpha \\ K\dot\alpha & K(F_{t_0} - \mathbb{1}) + i\Pi_{\dot\alpha} \end{pmatrix}. \tag{5.63}$$

where $d = \frac{i}{2}(W^\sharp - \mathbb{1})$.

Let us introduce now convenient coordinates. We define a Poincaré section $\mathscr{S}$ by the equation $\mathscr{T}(\alpha) = 0$ where $\mathscr{T}$ is a classical observable such that $\{\mathscr{T}, H\}(\alpha) = 1$, $\mathscr{T}(\alpha_0) = 0$, $\mathscr{T}$ is defined in an open neighborhood $\mathscr{V}_0$ of α_0.

The first return Poincaré map $\mathscr{P}(\alpha) = \phi^{T(\alpha)}(\alpha)$ is defined in $\mathscr{V}_0 \cap S$ such that $\mathscr{T}(\phi^{T(\alpha)}(\alpha)) = 0$ with $T(\alpha_0) = t_0$, $T(\alpha)$ is the first return time.

In $\mathscr{V}_0$ we can define new symplectic coordinates: (e, τ, ϖ), where $e = H(\alpha)$, $\tau = \mathscr{T}(\alpha)$, $\varpi(\alpha) \in \mathbb{R}^{2(n-1)}$. The differential $\mathscr{P}'(\alpha_0)$ of $\mathscr{P}$ at α_0 is related with the stability matrix $F = (\partial_\alpha \phi^{t_0})(\alpha_0)$:

$$\mathscr{P}'(\alpha_0)v = Fv - (F^T \nabla \mathscr{T} \cdot v)\dot\alpha. \tag{5.64}$$

For e near E, the Poincaré map $\mathscr{P}_e$ is defined in $\mathscr{V}_0 \cap \mathscr{S} \cap \Sigma_e$ into $\mathscr{V}_1 \cap \mathscr{S} \cap \Sigma_e$, where $\mathscr{V}_1$ is a neighborhood of α_0, by $\mathscr{P}(\alpha) = \phi^{T(\alpha)}(\alpha)$, $T(\alpha_0) = t_0$. It is a symplectic map and for $e = E$ its differential P_γ is the restriction of $\mathscr{P}'(\alpha_0)$ to $N_\gamma := T_{\alpha_0}(\mathscr{S} \cap \Sigma_E)$ (for more details we refer to [135]).

Note that $N_\gamma = \{v \in \mathbb{R}^{2n},\ v \cdot \nabla H = v \cdot \nabla \mathscr{T} = 0\}$.

When the energy e is varying around E we have a smooth family of closed trajectories of period $T(e)$ parametrized by $\alpha(e) \in \mathscr{V}_0 \cap \mathscr{S}$ such that $\alpha(E) = \alpha_0$ and $T(E) = t_0$. T and α are smooth in e. This result is known as the cylinder Theorem

[135]. It is a consequence of the implicit function theorem applied to the equation $\phi^T(e)(\alpha(e)) = \alpha(e)$. In particular we have

$$(F - \mathbb{1})\alpha'(E) = T'(E)\dot{\alpha}. \tag{5.65}$$

Note that $\alpha'(E) \cdot \nabla H = 1$ so $\alpha'(E) \neq 0$ and the non-degeneracy assumption implies that $\{\dot{\alpha}, \alpha'(E)\}$ is a basis for the generalized eigenspace E_1 for the eigenvalue 1 of F (we have a non trivial Jordan block if $T'(E) \neq 0$).

Let V be the symplectic orthogonal of E_1: $V = \{v|\sigma(v, \dot{\alpha}) = \sigma(v, \alpha'(E)) = 0\}$. The restriction of F to V is the algebraic linear Poincaré map $P_\gamma{}^{(al)}$. Using (5.64) we can easily proved that P_γ and $P_\gamma{}^{(al)}$ are conjugate: $MP_\gamma = P_\gamma{}^{(al)}$ where M is an invertible linear map from N_γ onto V. So we have

$$\det(P_\gamma - \mathbb{1}) = \det(P_\gamma{}^{(al)} - \mathbb{1}).$$

In particular if γ is non-degenerate then $P_\gamma - \mathbb{1}$ is invertible. The strategy is to simplify as far as possible the r.h.s in (5.63).

To simplify our discussion we shall assume that $T'(E) \neq 0$. It is not a restriction because if $T' = 0$ we can perturb a little F by F^ε, $\varepsilon > 0$, such that

$$F^\varepsilon \dot{\alpha} = F \quad \text{on } V, \quad F^\varepsilon \dot{\alpha} = \alpha, \quad F^\varepsilon \alpha'(E) = \alpha'(E) + \varepsilon\dot{\alpha}$$

The determinant we have to compute depends only on the symplectic map F, so we can compute with $F^\varepsilon \dot{\alpha}$ and take the limit as $\varepsilon \to 0$.

The first step is to find $v \in \mathbb{C}^{2n}$ such that

$$(F - \mathbb{1})v + v \cdot \dot{\alpha} K^{-1}\dot{\alpha} = \dot{\alpha}. \tag{5.66}$$

With this $v := v_0$ we get

$$\det\left(i^{-1}\Psi''_{E,\perp}(t_0, \alpha_0)\right) = i^{-1} \det\begin{pmatrix} d - v_0 \cdot K\dot{\alpha} & K\dot{\alpha} \\ 0 & K(F - \mathbb{1}) + i\Pi_{\dot{\alpha}} \end{pmatrix} \tag{5.67}$$

where $\Pi_{\dot{\alpha}} = \frac{(v \cdot \dot{\alpha})}{|\dot{\alpha}|^2}\dot{\alpha}$.

A direct computation gives

$$K^{-1}\dot{\alpha} = -\frac{i}{2}(F + \mathbb{1})\nabla H,$$

so Eq. (5.66) is transformed in

$$(F - \mathbb{1})v = \frac{i}{2} v \cdot \dot{\alpha}(F + \mathbb{1})\nabla H.$$

Using $(F - \mathbb{1})^T \nabla H = 0$ we have $v \cdot \dot{\alpha} = 0$, so we have to solve

$$(F - \mathbb{1})v_0 = \dot{\alpha}. \tag{5.68}$$

We are looking for $v_0 = \lambda\dot{\alpha} + \mu\alpha'(E)$ and we find

$$v_0 = \frac{1}{T'(E)}\left(\frac{\alpha'(E)\cdot\dot{\alpha}}{|\dot{\alpha}|^2}\dot{\alpha} - \alpha'(E)\right). \tag{5.69}$$

So our first simplification gives the expression

$$\det\left(i^{-1}\Psi''_{E,\perp}(t_0,\alpha_0)\right) = i^{-1}(d - v_0\dot{K}\dot{\alpha})\det K \det(F - \mathbb{1} + iK^{-1}\Pi_{\dot{\alpha}})$$

For the first term we get

$$d - v_0\dot{K}\dot{\alpha} = -iv_0\cdot J\dot{\alpha} = i\frac{\sigma(\alpha'(E),\dot{\alpha})}{T'(E)}$$

$\det K$ is already computed. We shall compute $\det(F - \mathbb{1} + iK^{-1}\Pi_{\dot{\alpha}})$ in a symplectic basis $\{v_1, v_2, v_3, \ldots, v_{2n}\}$ where $v_1 = \dot{\alpha}$, $v_2 = \frac{1}{\nabla H\cdot\alpha'(E)}\alpha'(E)$ where $\sigma(v_{2j-1}, v_{2j}) = 1$ for $1 \le j \le n$ and $\sigma(v_j, v_k) = 0$ if $|j-k| \ne 1$. In this basis we have

$$K^{-1}\Pi_{\dot{\alpha}}v_j = \frac{v_j\cdot\dot{\alpha}}{|\dot{\alpha}|^2}K^{-1}\Pi_{\dot{\alpha}}\dot{\alpha}$$

So combining with the first column we can eliminate the terms $K^{-1}\Pi_{\dot{\alpha}}v_j$ for $j \ge 2$ and using that $F - \mathbb{1}$ is invertible on V we can assume that in the first column only the two first terms are not zero. Finally we have obtained

$$\det\left(F - \mathbb{1} + iK^{-1}\Pi_{\dot{\alpha}}\right) = \det\begin{pmatrix}\begin{pmatrix}x_1 & \delta\\ x_2 & 0\end{pmatrix} & \mathbf{0}\\ \mathbf{0} & [P_\gamma - \mathbb{1}]\end{pmatrix} = -\delta x_2 \det(P_\gamma - \mathbb{1}). \tag{5.70}$$

We are left to compute x_2 and δ. We have

$$\delta = \sigma((F - \mathbb{1})v_2, v_2) = -\frac{T'(E)}{(\nabla H\cdot\alpha'(E))^2}\sigma(\dot{\alpha},\alpha'(E)) \tag{5.71}$$

$$x_2 = -\frac{1}{2}\sigma((F + \mathbb{1})\nabla H, v_1) = |\nabla H|^2. \tag{5.72}$$

So we get

$$\det(i^{-1}\Psi''_{E,\perp}(t_0,\alpha_0)) = 2^n\|\nabla H\|^2\det(Y)^{-1}\det(P_\gamma - \mathbb{1}). \tag{5.73}$$

Using the expression (5.40) we find the leading term $\rho_{1,\gamma}(E)$ for the contribution of the periodic path γ in formula (5.18), assuming for simplicity that $A = 1$,

$$\rho_{1,\gamma}(E) = (2\pi)^{n/2-1}[\det(YU^{-1})]_*^{-1/2}[\det U]_c^{-1/2}[\det Y]^{1/2}$$
$$\widehat{\mathrm{g}}(T_\gamma)[\det(P_\gamma - \mathbb{1})]^{-1/2}\mathrm{e}^{i\psi_E}\|\nabla H\|^{-1} \tag{5.74}$$

where $[u]^{1/2}$ denotes a suitable branch for the square root. So we get

$$\rho_{1,\gamma}(E) = (2\pi)^{n/2-1}\mathrm{e}^{i(S_\gamma/\hbar+\sigma_\gamma\pi/2)}\widehat{\mathrm{g}}(T_\gamma)T_\gamma^*|\det(\mathbb{1} - P_\gamma)|^{-1/2} \tag{5.75}$$

with $\sigma_\gamma \in \mathbb{Z}$ and $S_\gamma = \oint_\gamma pdq$.

Let us remark that, because P_γ is symplectic and 1 is not eigenvalue of P_γ then we have $\det(P_\gamma - \mathbb{1}) = (-1)^{\sigma'}|\det(P_\gamma - \mathbb{1})|$ where σ' is the number of eigenvalues of P_γ smaller than 1. So we see that

$$\mathrm{e}^{i\sigma_\gamma\pi/2} = \pm\mathrm{e}^{i\sigma'\pi/2}. \tag{5.76}$$

Thus we get that the contribution of the Maslov index in Theorem 28 is to determine the sign in (5.76).

We have given here an analytical method for its computation. We do not consider its geometrical interpretation ("Maslov cycle") for which we refer to the literature on this subject [122, 177, 223] and references in these works.

The other coefficients, d_j^γ are spectral invariants which have been studied by Guillemin and Zelditch. In principle we can compute them using this explicit approach. This completes the proof of Theorem 28.

5.5 A Pointwise Trace Formula and Quasi-Modes

From the well known Bohr–Sommerfeld quantization rules it is believed that there exists strong connections between periodic trajectories of a classical system H and bound states of its quantization $\hat{H}$. In this section we discuss some properties of localization for bound states or approximate bound states (quasimodes) near periodic trajectories in the simplest cases. More general results are proved in [206].

5.5.1 A Pointwise Trace Formula

The idea of this formula have appeared in [206]. We give here a proof of the main result of [206] for Gaussian coherent states. We assume that properties (H.0) and (H.1) are satisfied. Consider the local density of states defined for every $\alpha \in \mathbb{R}^{2n}$ by

$$\rho_E(\alpha) \equiv \sum_j \mathrm{g}\left(\frac{E - E_j(\hbar)}{\hbar}\right)\chi(E_j(\hbar))\left|\langle\psi_\alpha, \psi_j^\hbar\rangle\right|^2,$$

where $\psi_j^\hbar$ are the normalized eigenfunctions for $\hat{H}$, $\hat{H}\psi_j^\hbar = E_j(\hbar)\psi_j^\hbar$.

Theorem 31 *The local density of states $\rho_E(\alpha)$ has the following asymptotic behavior as $\hbar \to 0$,*

$$\rho_E(\alpha) \equiv \sum_{k\in\mathbb{N}} \ell_k(g,\alpha)\hbar^{\frac{1}{2}+k}. \tag{5.77}$$

The coefficients $\ell_k(g,\alpha)$ are smooth in α and are distributions in $\tilde{g}$. Their expressions depend on the behavior of the path $t \mapsto \phi^t\alpha$.
(i) If the path $t \mapsto \phi^t\alpha$ has no periodic point with period in $\mathrm{supp}\tilde{g}$ then $\ell_k(g,\alpha)$ are distributions in $\tilde{g}$ supported in $\{0\}$. In particular the leading term is

$$\ell_0(g,\alpha) = \frac{1}{\sqrt{2}}\pi^{-\frac{n+1}{2}}\frac{1}{\|\nabla H(\alpha)\|}\tilde{g}(0) \tag{5.78}$$

(ii) If $t \mapsto \phi^t\alpha$ has a primitive period T^*, $\ell_k(g,\alpha)$ are distributions in $\tilde{g}$ supported in $\{mT^*,\ m\in\mathbb{Z}\}$. In particular the leading term is

$$\ell_0(g,\alpha) = \frac{1}{\sqrt{2}}\pi^{-\frac{n+1}{2}}\sum_{m\in\mathbb{Z}}\tilde{g}(mT^*)C(m) \tag{5.79}$$

where

$$C(m) = \left(\mathbb{1} - W^\sharp_{mT^*}\dot{\alpha}\cdot\dot{\alpha}\right)^{-\frac{1}{2}}.$$

Recall that $W_t^\sharp$ depends on the monodromy matrix F_t.

Proof As for the trace formula (5.18), we first give a time dependent formula for $\rho_E(\alpha)$ with the propagator $\mathscr{U}(t) = \mathrm{e}^{-i\frac{t}{\hbar}\hat{H}}$.

If Π_z is the orthogonal projection on the coherent state φ_z we have

$$\rho_E(\alpha) = \mathrm{Tr}\left(g\left(\frac{E-\hat{H}}{\hbar}\right)\chi(\hat{H})\Pi_\alpha\right),$$

by computing the trace on the basis $\psi_j^\hbar$. So we get

$$\rho_E(\alpha) = \frac{1}{2\pi}\int dt\tilde{g}(t)\mathrm{e}^{i\frac{t}{\hbar}E}\langle\varphi_\alpha, \mathscr{U}(t)\chi(\hat{H})\varphi_\alpha\rangle. \tag{5.80}$$

In (5.80) the integrand is the same as in the proof of the trace formula. The difference is that here we have only a time integration. So the stationary phase theorem is much simpler to apply: α is fixed such that $H(\alpha) = E$, the critical set of the phase ψ_E is defined by the equation $\phi^t\alpha = \alpha$ so we have $t = mT^*$ where T^* is the primitive period ($T^* = 0$ if α is not a periodic point of the flow).

$\ddot{\psi}_E(t,\alpha)$ is here the second derivative in time of ψ_E. So we have

$$\ddot{\psi}_E(t, \alpha) = \frac{i}{2}(\mathbb{1} - W_t^\sharp)\dot{\alpha} \cdot \dot{\alpha}.$$

For $t = 0$ we have $\ddot{\psi}_E(\alpha) = \frac{i}{2}\|\nabla H\|^2$. Using that $\nabla H \neq 0$, the stationary phase theorem gives the part (i) of the Theorem.

For the periodic case we have to recall that W_t is a complex symmetric $n \times n$ matrix and that $W^*W < 1$. With this properties we have easily that for every $T > 0$ there exists $c_T > 0$ such that

$$\Re(\mathbb{1} - W_t^\sharp)\dot{\alpha} \cdot \dot{\alpha} \geq c_t\|\dot{\alpha}\|^2 \ \text{ for } \ t \in [-T, T].$$

So the critical points $t = mT^*$, $m \in \mathbb{Z}$, are non degenerate and the stationary phase theorem gives the part (ii) of the Theorem. □

5.5.2 *Quasi-Modes and Bohr–Sommerfeld Quantization Rules*

Quasi-modes (or approximated eigenfunctions) can be considered in more general and more interesting cases (see [166, 206, 212, 213]) but for simplicity we shall consider here mainly the fully periodic case. In applications we can have to consider semi-classical $H(z) = H_0(z) + \hbar H_1(z) + \hbar^2 R_2(\hbar, z)$.

We always assume that (H.0) 131 and (H.1) 133 are satisfied for $H_0, H_1 \in S(m_{H_0})$ and $R(\hbar, \cdot) \in S_{sc}(m_{H_0})$. This class of symbols is defined in Chap. 3. All our results are still valid for these $\hbar$-dependent Hamiltonians.

We introduce the condition (H.P) For every $E \in [E_-, E_+]$, $\Sigma_E := H_0^{-1}(E)$ is connected and the Hamiltonian flow $\Phi^t_{H_0}$ is periodic on Σ_E with a period T_E. Moreover the average of H_1 along any closed trajectory is constant (depends only on the energy E).

Remark 31 For $n = 1$ the periodicity condition is always satisfied. For $d > 1$ this condition is rather strong. Nevertheless it is satisfied for integrable systems and for systems with a large group of symmetries.

Let us first recall a result in classical mechanics (Guillemin–Sternberg, [125]):

Proposition 49 *Let us assume that above conditions are satisfied. Let γ be a closed path of energy E and period T_E. Then the action integral $\mathcal{J}(E) = \int_\gamma pdq$ defines a function of E, C^∞ in $]E_-, E_+[$ and such that $\mathcal{J}'(E) = T_E$. In particular for one degree of freedom systems we have*

$$\mathcal{J}(E) = \int_{H_0(z) \leq E} dz.$$

Now we can extend $\mathcal{J}$ to an increasing function on $\mathbb{R}$, linear outside a neighborhood of I. Let us introduce the rescaled Hamiltonian $\hat{K} = (2\pi)^{-1}\mathcal{J}(\hat{H})$. Using properties concerning the functional calculus [141], we can see that $\hat{K}$ has all the properties of $\hat{H}$ and furthermore the Hamiltonian flow of its principal symbol K_0 has a constant period 2π in $\Sigma_\lambda^{K_0} = K_0^{-1}(\lambda)$ for $\lambda \in [\lambda_-, \lambda_+]$ where $\lambda_\pm = \frac{1}{2\pi}\mathcal{J}(E_\pm)$. So in what follows we replace $\hat{H}$ by $\hat{K}$, its "energy renormalization". Indeed, the mapping $\frac{1}{2\pi}\mathcal{J}$ is a bijective correspondance between the spectrum of $\hat{H}$ in $[E_-, E_+]$ and the spectrum of $\hat{K}$ in $[\lambda_-, \lambda_+]$, including mutiplicities, such that $\lambda_j = \frac{1}{2\pi}\mathcal{J}(E_j)$.

Let us denote by a the average of the action of a periodic path on $\Sigma_\lambda{}^{K_0}$ and by $\mu \in \mathbb{Z}$ its Maslov index. ($a = \frac{1}{2\pi}\int_\gamma pdx - 2\pi F$). Under the above assumptions the following results were proved in [141], using semi-classical Fourier-integral operators and ideas introduced before by Colin de Verdière [55] and Weinstein [257].

5.5.2.1 Statements of Results Concerning Spectral Asymptotics

Theorem 32 [55, 141, 257] *Here we assume that $H = H_0$ (without subprincipal terms).There exist $C_0 > 0$ and $\hbar_1 > 0$ such that*

$$\mathrm{spect}(\hat{K})\bigcap[\lambda_-, \lambda_+] \subseteq \bigcup_{k\in\mathbb{Z}} I_k(\hbar), \tag{5.81}$$

with

$$I_k(\hbar) = [-a + (k - \frac{\mu}{4})\hbar - C_0\hbar^2, -a + (k - \frac{\mu}{4})\hbar + C_0\hbar^2].$$

for $\hbar \in]0, \hbar_1]$.

Let us remark for $\hbar$ small enough, the intervals $I_k(\hbar)$ do not intersect and this theorem gives the usual Bohr–Sommerfeld quantization conditions for the energy spectrum, more explicitly,

$$\lambda_k = \frac{1}{2\pi}\mathcal{J}(E_k) = (k - \frac{\mu}{4})\hbar - a + O(\hbar^2).$$

Under a stronger assumption on the flow, it is possible to estimate the number of states in each cluster $I_k(\hbar)$. (H.F) $\Phi_{K_0}^t$ has no fixed point in $\Sigma_F^{K_0}$, $\forall\lambda \in [\lambda_- - \varepsilon, \lambda_+ + \varepsilon]$ and $\forall t \in]0, 2\pi[$.

Let us denote by $d_k(\hbar)$ the number of eigenvalues of $\hat{K}$ in the interval $I_k(\hbar)$.

Theorem 33 [52, 55, 142] *Under the above assumptions, for $\hbar$ small enough and $-m + (k - \frac{\mu}{4})\hbar \in [\lambda_-, \lambda_+]$, we have :*

$$d_k(\hbar) \equiv \sum_{j\geq 1} \Gamma_j(-a + (k - \frac{\mu}{4})\hbar)\hbar^{j-d}, \tag{5.82}$$

with $\Gamma_j \in C^\infty([\lambda_-, \lambda_+])$. In particular

$$\Gamma_1(\lambda) = (2\pi)^{-d} \int_{\Sigma_\lambda} d\nu_\lambda.$$

In the particular case $n = 1$ we have $\mu = 2$ and $a = \min(H_0)$ hence $d_k(\hbar) = 1$. Furthermore the Bohr–Sommerfeld conditions take the following more accurate form

Theorem 34 [141] *Let us assume $n = 1$ and $m = 0$. Then there exists a sequence $f_k \in C^\infty([F_-, F_+])$, for $k \geq 2$, such that*

$$\lambda_\ell + \sum_{k\geq 2} h^k f_k(\lambda_\ell) = (\ell + \frac{1}{2})\hbar + O(\hbar^\infty), \tag{5.83}$$

for $\ell \in \mathbb{Z}$ such that $(\ell + \frac{1}{2})\hbar \in [\lambda_-, \lambda_+]$.

In particular there exists $g_k \in C^\infty([\lambda_-, \lambda_+])$ such that

$$\lambda_\ell = (\ell + \frac{1}{2})\hbar + \sum_{k\geq 2} h^k g_k((\ell + \frac{1}{2})\hbar) + O(\hbar^\infty), \tag{5.84}$$

where $\ell \in Z$ such that $(\ell + \frac{1}{2})\hbar \in [F_-, F_+]$.

We can deduce from the above theorem and Taylor formula the Bohr–Sommerfeld quantization rules for the eigenvalues E_n at all order in $\hbar$.

Corollary 19 *There exist $\lambda \mapsto b(\lambda, \hbar)$ and C^∞ functions b_j defined on $[\lambda^-, \lambda^+]$ such that $b(\lambda, \hbar) = \sum_{j\in\mathbb{N}} b_j(\lambda)\hbar^j + O(\hbar^\infty)$ and the spectrum E_n of $\hat{H}$ is given by*

$$\boxed{E_n = b\left((n + \frac{1}{2})\hbar, \hbar\right) + O(\hbar^\infty)}, \tag{5.85}$$

for n such that $(n + \frac{1}{2})\hbar \in [\lambda^-, \lambda^+]$. In particular we have $b_0(\lambda) = \mathcal{J}^{-1}(2\pi\lambda)$ and $\boxed{b_1 = 0}$.

When $H^{-1}(I)$ is not connected but such that the M connected components are mutually symmetric, under linear symplectic maps, then the above results still hold [141].

Remark 32 For $n = 1$, the methods usually used to prove existence of a complete asymptotic expansion for the eigenvalues of $\hat{H}$ are not suitable to compute the coefficients $b_j(\lambda)$ for $j \geq 2$. This was done recently in [57] using the coefficients d_{jk} appearing in the functional calculus.

5.5.2.2 A Proof of the Quantization Rules and Quasi-Modes

We shall give here a direct proof for the Bohr–Sommerfeld quantization rules by using coherent states, following [32]. A similar approach, with more restrictive assumptions, was considered before in [204].

The starting point is the following remark. Let $r > 0$ and suppose that there exists C_r such that for every $\hbar \in]0, 1]$, there exist $E \in \mathbb{R}$ and $\psi \in L^2(\mathbb{R}^d)$, such that

$$\|(\hat{H} - E)\psi\| \le C_r\hbar^r, \quad \text{and} \quad \liminf_{\hbar\to 0} \|\psi\| := c > 0. \tag{5.86}$$

If these conditions are satisfied, we shall say that $\hat{H}$ has a quasi-mode of energy E with an error $O(\hbar^r)$. With quasi-modes we can find some points in the spectrum of $\hat{H}$ close to the energy E. More precisely, if $\delta > \frac{C_r}{c}$, the interval $[E - \delta\hbar^r, E + \delta\hbar^r]$ meets the point spectrum of $\hat{H}$. This is easily proved by contradiction, using that $\hat{H}$ is self-adjoint. So if the spectrum of $\hat{H}$ is discrete in a neighborhood of E, then we know that $\hat{H}$ has at least one eigenvalue in $[E - \delta\hbar^r, E + \delta\hbar^r]$.

Let us assume that the Hamiltonian $\hat{H}$ satisfies conditions (H.0), (H.1), (H.P) 150.

Using Proposition 49, we can assume that the Hamiltonian flow Φ_H^t has a constant period 2π in $H^{-1}]E_- - \varepsilon, E_+ + \varepsilon[$, for some $\varepsilon > 0$.

Following an old idea in quantum mechanics (A. Einstein), let us try to construct a quasimode for $\hat{H}$ with energies $E^{(\hbar)}$ close to $E \in [E_-, E_+]$, related with a 2π periodic trajectory $\gamma_E \subset \Sigma_E^{H_0}$, by the Ansatz

$$\psi_{\gamma_E} = \int_0^{2\pi} e^{\frac{itE^{(\hbar)}}{\hbar}} \mathcal{U}(t)\varphi_z dt, \tag{5.87}$$

where $z \in \gamma_E$ (ψ_{γ_E} is a state living on γ_E). Let us introduce the real numbers

$$\sigma(\hbar) = \frac{1}{2\pi\hbar}\int_0^{2\pi} [\dot{q}(t)p(t) - H_0(q(t), p(t))]dt + \frac{\mu}{4},$$

where $t \mapsto (q(t), p(t))$ is a 2π-periodic trajectory γ_E in $H_0{}^{-1}(E)$, $E \in [E_-, E_+]$, μ is the Maslov index of γ. In order that the Ansatz (5.87) provides a good quasimode, we must first check that its mass is not too small.

Proposition 50 *Assume that 2π is the primitive period of γ_E. Then there exists a real number $m_E > 0$ such that*

$$\|\psi_{\gamma_E}\| = m_E\hbar^{1/4} + O(\hbar^{1/2}) \tag{5.88}$$

Proof Using the propagation of coherent states and the formula giving the action of metaplectic transformations on Gaussians, up to an error term $O(\sqrt{\hbar})$, we have

$$\|\psi_{\gamma_E}\|^2 = (\pi\hbar)^{-d/2}\int_0^{2\pi}\int_0^{2\pi}\int_{\mathbb{R}^d} e^{\frac{i}{\hbar}\Phi(t,s,x)}\,(\det(A_t+iB_t))^{-1/2}\cdot \cdot\left(\overline{\det(A_s+iB_s)}\right)^{-1/2} dtdsdx,$$

where the phase Φ is

$$\Phi(t,s,x) = (t-s)E + (\delta_t-\delta_s) + \frac{1}{2}(q_s\cdot p_s - q_t\cdot p_t) + x\cdot(p_t-p_s) + \frac{1}{2}\left(\Gamma_t(x-q_t)\cdot(x-q_t) - \overline{\Gamma_s}(x-q_s)\cdot(x-q_s)\right)). \quad (5.89)$$

Γ_t is the complex matrix defined in (5.36). Let us show that we can compute an asymptotics for $\|\psi_{\gamma_E}\|^2$ with the stationary phase Theorem. Using that $\Im(\Gamma_t)$ is positive non-degenerate, we find that

$$\Im(\Phi(t,s,x)) \geq 0, \quad \text{and } \{\Im(\Phi(t,s,x)) = 0\} \Leftrightarrow \{x = q_t = q_s\}. \quad (5.90)$$

On the set $\{x = q_t = q_s\}$ we have $\partial_x\Phi(t,s,x) = p_t - p_s$. So if $\{x = q_t = q_s\}$ then we have $t = s$ (2π is the primitive period of γ_E) and we get easily that $\partial_s\Phi(t,s,x) = 0$. In the variables (s,x) we have found that $\Phi(t,s,x)$ has one critical point: $(s,x) = (t,q_t)$. Let us compute the hessian matrix $\partial^{(2)}_{s,x}\Phi$ at (t,t,q_t).

$$\partial^{(2)}_{s,x}\Phi(t,t,q_t) = \begin{pmatrix} -(\overline{\Gamma_t}\dot{q}_t - \dot{p}_t)\cdot\dot{q}_t & [\overline{\Gamma_t}(\dot{q}_t - \dot{p}_t)]^T \\ \overline{\Gamma_t}(\dot{q}_t - \dot{p}_t) & 2i\Im\Gamma_t. \end{pmatrix}. \quad (5.91)$$

To compute the determinant, we use the identity, for $r \in \mathbb{C}$, $u \in \mathbb{C}^d$, $R \in GL(\mathbb{C}^d)$

$$\begin{pmatrix} r & u^T \\ u & R \end{pmatrix}\begin{pmatrix} 1 & 0 \\ -R^{-1}u & \mathbb{1} \end{pmatrix} = \begin{pmatrix} r - u^T\cdot R^{-1}u & u^T \\ 0 & R. \end{pmatrix}. \quad (5.92)$$

Then we get

$$2\det[-i\partial^2_{s,x}\Phi(t,t,q_t)] = \Gamma_t\dot{q}_t\cdot\dot{q}_t + (\Im(\Gamma_t))^{-1}(\Re\Gamma_t\dot{q}_t - \dot{p}_t)\cdot(\Re\Gamma_t\dot{q}_t - \dot{p}_t)\det[2\Im\Gamma_t]. \quad (5.93)$$

But E is not critical, so $(\dot{q}_t, \dot{p}_t) \neq (0,0)$ and we find that $\det[-i\partial^2_{s,x}\Phi(t,t,q_t)] \neq 0$. The stationary phase Theorem (see Appendix), gives

$$\|\psi_{\gamma_E}\|^2 = m_E{}^2\sqrt{\hbar} + O(\hbar) \quad (5.94)$$

with

$$m_E^2 = 2^{(d+1)/2}\sqrt{\pi}\int_0^{2\pi} |\det(A_t+iB_t)|^{-1/2}|\det[-i\partial^2_{s,x}\Phi(t,t,q_t)]|^{-1/2}dt. \quad (5.95)$$

□

We now give one formulation of the Bohr–Sommerfeld quantization rule. This is an alternative proof of Theorem 32 with quasi-modes but the localization of eigenvalues is a little less accurate.

Theorem 35 *Let us assume that the Hamiltonian $\hat{H}$ satisfies conditions (H.0) 131, (H.1) 133, (H.P) 150 with period $T_E = 2\pi$ and that 2π is a primitive period for a periodic trajectory $\gamma_E \subseteq \Sigma_E$.*

Then $\hbar^{-1/4}\psi_{\gamma_E}$ is a quasi-mode for $\hat{H}$, with an error term $O(\hbar^{7/4})$, if E satisfies the quantization condition:

$$\sigma(\hbar) := \frac{\mu}{4} + b + \frac{1}{2\pi\hbar}\int_{\gamma_E} pdq \in \mathbb{Z}. \tag{5.96}$$

Moreover, the number $a := \frac{1}{2\pi}\int_{\gamma_E} pdq - E$ is constant on $[E_-, E_+]$. Chosen $C > 0$ large enough and consider the intervals

$$I(k,\hbar) = [(-\frac{\mu}{4} - b + k)\hbar + a - C\hbar^{7/4}, (-\frac{\mu}{4} - b + k)\hbar + a + C\hbar^{7/4}],$$

where b is the average of the subprincipal term H_1 along the flow. Then if $I(k,\hbar) \cap [E_-, E_+] \neq \emptyset$ then $\hat{H}$ has at least one eigenvalue in $I(k,\hbar)$.

Proof We are looking for an energy $E^\hbar$ close to E such that

$$\hat{H}\psi_{\gamma_E} = E^\hbar\psi_{\gamma_E} + O(\hbar^2).$$

By integration by parts, we get

$$\begin{aligned}\hat{H}\psi_{\gamma_E} &= i\hbar\int_0^{2\pi} e^{\frac{itE}{\hbar}}\partial_t\mathcal{U}(t)\varphi_z dt \\ &= i\hbar\left(e^{\frac{2i\pi E}{\hbar}}\mathcal{U}(2\pi)\varphi_z - \varphi_z\right) + E\psi_{\gamma_E}.\end{aligned} \tag{5.97}$$

Using the propagation of coherent states result obtained in Chap. 4 and the 2π periodicity of the flow we get that

$$\mathcal{U}(2\pi)\varphi_z = e^{i(\mu/4+b)}e^{\frac{i}{\hbar}(\int_0^{2\pi} p\dot{q}dt - 2\pi E)}\varphi_z + \mathcal{O}(\hbar). \tag{5.98}$$

Notice here that the contribution in $\sqrt{\hbar}$ vanishes. This can be checked.
Using (5.97), (5.98) and (5.88) we get finally a quasi-mode with an error $O(\hbar^{7/4})$ and quantization of a series of eigenvalues. □

Remark 33 Later, as an application of the Herman–Kluk representation formula (Chap. 11), we shall prove that under the assumptions of Theorem 35 we have

$$\mathrm{Spect}[\hat{H}] \cap [E_-, E_+] \subseteq \cup_{k\in\mathbb{Z}} I(k,\hbar).$$

This is the clustering phenomenon for the eigenvalues when the Hamiltonian flow is periodic. On the contrary if there exits one non periodical trajectory of energy $E \in [E_-, E_+]$ then the clusters disappear. This will be proved also later. More accurate results on quasi-modes using coherent states are proved in particular in [212], [166] and [218].

Part II
Coherent States in Non Euclidian Geometries

Chapter 6
Quantization and Coherent States on the 2-Torus

Abstract The two dimensional torus $\mathbb{T}^2$ is a very simple symplectic space. Nevertheless it gives non trivial examples of chaotic dynamical systems. These systems can be quantized in a natural way. We shall study some dynamical and spectral properties for them.

6.1 Introduction

The 2-torus $\mathbb{T}^2$, with its canonical symplectic form, is seen here as a phase space. It is useful to consider classical systems and quantum systems built on $\mathbb{T}^2$ for at least two purposes.

Dynamical properties of classical non-integrable Hamiltonian systems in the phase space $\mathbb{R}^d \times \mathbb{R}^d$ ($d \geq 2$), are quite difficult to study. In particular there are not so many explicit models of chaotic systems. But on the 2-torus it is very easy to get a discrete chaotic system by considering a 2×2 matrix F with entries in $\mathbb{Z}$ and such that $|\mathrm{Tr} F| > 2$. So we get a discrete flow $t \mapsto F^t z$ for $t \in \mathbb{Z}$, $z \in \mathbb{T}^2$,[1] where $\mathbb{T}^2 = \mathbb{R}^2/\mathbb{Z}^2$ is the 2-dimensional torus.

In 1980 Hannay–Berry succeeded to construct a "good quantization" $\hat{R}(F)$ corresponding to the "classical system" $(\mathbb{T}^2, F)$. From this starting point many results were obtained concerning consequences of classical chaos on the behavior of the eigenstates of the unitary family of operators $\hat{R}(F)$ as well by physicists and mathematicians. In this section we shall explain some of these results and their relationship with periodic coherent states.

[1] For $t \in \mathbb{N}$, $F^t z$ means that we apply F t-times starting from z and if $t > 0$ then $F^t = (F^{-t})^{-1}$.

D. Robert and M. Combescure, *Coherent States and Applications in Mathematical Physics*, Theoretical and Mathematical Physics,
https://doi.org/10.1007/978-3-030-70845-0_6

6.2 The Automorphisms of the 2-Torus

We have already seen in Chap. 1 that $\mathbb{R}^2$ is a symplectic linear space with the canonical symplectic bilinear form $\sigma = dq \wedge dp$. $\mathbb{T}^2$ is also a symplectic (compact) manifold with the symplectic two-form $\sigma = dq \wedge dp$ identified with the plane Lebesgue measure.

Here we call automorphism of the 2-torus $\mathbb{T}^2$ any map F induced by a symplectic matrix $F \in SL(2, \mathbb{Z})$.

Let F be of the form

$$F = \begin{pmatrix} a & b \\ c & d \end{pmatrix}, \tag{6.1}$$

with entries in $\mathbb{Z}$ safisfying $\det(F) = ad - bc = 1$; the corresponding map of the 2-torus is given by

$$(q, p) \in \mathbb{T}^2 \longmapsto (q', p') \in \mathbb{T}^2, \text{ with } q' = aq + bp \ (\text{mod } 1), \quad p' = cq + dp \ (\text{mod } 1)$$

So F is a symplectic diffeomorphism of $\mathbb{T}^2$. In particular it preserves the Lebesgue measure m_L on $\mathbb{T}^2$.

We shall consider now the discrete dynamical system in $\mathbb{T}^2$ generated by F.

Let us first recall the definitions and properties concerning classical chaos (ergodicity, mixing). For more details we refer to the books [69, 163, 187].

A (discrete) dynamical system is a triplet (X, Φ, m) where X is a measurable space, m a probability measure on X and Φ a measurable map on X preserving the measure m:

For any mesurable set $E \subset M$ one has $m(\Phi^{-1}E) = m(E)$.

The orbit (or trajectory) of a point $x \in X$ is $O(x) := \{\Phi^k(x),\ k \in \mathbb{Z}\}$. The orbit is periodic if $\Phi^T(x) = x$ for some $T \in \mathbb{Z}, T \neq 0$.

Definition 14 For a dynamical system $D = (X, \Phi, m)$ let us consider the time average or Birkhoff average $E_T(f, x) = \frac{1}{T}\sum_{t=0}^{t=T} f(\Phi^t(x))$, where f is measurable.

Φ is ergodic if for any function $f \in L^1(X, m)$ one has

$$\lim_{T\to\infty} E_T(f, x) = m(f) \quad m - \text{everywhere} \tag{6.2}$$

where $m(f) := \int_X f dm$ is the spatial average.

Remark 34 If a dynamical system is ergodic its time average (in the sense of the left hand side of (6.2) equals the "space average", and doesn't depend on the initial point $x \in X$ almost surely.

Without dynamical assumptions we have the following Birkhoff Ergodic Theorem (for a proof see [252]).

Theorem 36 *For any $f \in L^1(X, m)$, for m-almost $x \in X$, the following time average exists:*

$$\lim_{T\to+\infty} E_T(f,x) = \mu_x^B(f). \tag{6.3}$$

Moreover $f \mapsto \mu_x^B(f)$ *defines a (Birkhoff)* Φ*-invariant measure on* X.

Proposition 51 *A dynamical system* $D = (X, \Phi, m)$ *is ergodic if and only if one of the following statement is satisfied:*
(i) Any measurable set $E \subseteq X$ *which is* Φ*-invariant is such that* $m(E) = 0$ *or* $m(X\backslash E) = 0$.
(ii) If $f \in L^\infty(X, m)$ *is* Φ*-invariant* ($f \circ \Phi = f$) *then it is constant* m*-everywhere.*

See [252] for proofs.

This means in particular that the periodic orbits of an ergodic dynamical system are rather "rare" from a measurable point of view:

Proposition 52 *Let* Φ *be a continuous map on a compact topological space* X *endowed with a probability measure* m *which is* Φ*-invariant and such that* $m(U) > 0$ *for any open set* U. *If* (X, Φ, m) *is ergodic, then the set of periodic orbits in* X *is of measure* 0.

Although relatively rare, the periodic orbit have a strong importance in the framework of ergodic theory since they allow the construction of invariant measures in the following way. Let $x \in X$ and let $O(x)$ be a periodic orbit of period $p(x) \in \mathbb{R}$. The following probability measure is clearly invariant:

$$m_x = \frac{1}{p(x)} \sum_{k=1}^{p(x)} \delta_{\Phi^k(x)}$$

where δ_a is the Dirac distribution at point $a \in \mathbb{R}$.

Given a map Φ in X and m an invariant measure, it is not always true that m is the unique invariant measure. If it is the case the map Φ is said "uniquely ergodic" and the unique invariant measure is ergodic (see [163]).

Well known examples are irrational rotations on the circle (or translations on the torus $\mathbb{T}^1$). If α is an irrational number, $\Phi(x) = x + \alpha$, mod.1 defines a unique ergodic transformation in $\mathbb{T}^1$, the unique invariant measure is the Lebesgue measure (see [163]). In the topological framework one has a characterization of such maps [69]:

Proposition 53 *Let* $D = (X, \Phi, m)$ *be a dynamical system with* X *a compact metric space, and* Φ *a continuous map.* D *is uniquely ergodic if and only if* $\forall f \in \mathscr{C}(X)$:

$$\lim_{k\to\infty} \|\frac{1}{T}\sum_{l=0}^{T-1} f \circ \Phi^l - m(f)\|_\infty = 0,$$

where $\|\cdot\|_\infty$ *is the norm of the uniform convergence.*

There is a stronger property of dynamical systems which is the "mixing" property:

Definition 15 A dynamical system $D = (X, \Phi, m)$ is said to be mixing if $\forall f, g \in L^2(X, m)$ one has

$$\lim_{k\to\infty} \int_X f(\Phi^k(x))g(x)dm(x) = m(f)m(g).$$

The following result is useful and easy to prove.

Proposition 54 *$D = (X, \Phi, m)$ is mixing if and only if there exists a total set $\mathcal{T}$ in $L^2(X, m)$ such that for every $f, g \in \mathcal{T}$ we have*

$$\lim_{k\to\infty} \int_X f(\Phi^k(x))g(x)dm(x) = m(f)m(g).$$

In the framework where X is a differentiable manifold, one can define the notion of Anosov system (see [163, 183]):

Definition 16 A diffeomorphism Φ of a differentiable manifold M is Anosov if $\forall x \in M$, there exists a decomposition of the tangent space at x in direct sum of two subspaces E_x^u and E_x^s and constants $K > 0$, $0 < \lambda < 1$ satisfying

$$(D_x\Phi)E_x^s = E_{\Phi(x)}^s, \qquad (D_x\Phi)E_x^u = E_{\Phi(x)}^u$$

and

$$\|(D_x\Phi^n)E_x^s\| \le K\lambda^n, \qquad \|(D_x\Phi^{-n})E_x^u\| \le K\lambda^n$$

$\forall x \in M, n \in \mathbb{N}$.

We have the following useful stability result (see[183]).

Theorem 37 *Let M be a compact manifold and Φ an Anosov diffeomorphism on M. There exists $\varepsilon > 0$ small enough such that if $\|\Phi - \Psi\|_{C^1(M)} < \varepsilon$ then Ψ is an Anosov diffeomorphism on M, where*

$$\|\Psi\|_{C^1(M)} = \sup_{x\in M} (|\Phi(x)| + \|D_x\Phi(x)\|).$$

Theorem 38 *If Φ is a diffeomorphism Anosov on $\mathbb{T}^2$ then the dynamical system $D = (\mathbb{T}^2, \Phi, \mu)$ is mixing, μ being the normalized Lebesgue measure on $\mathbb{T}^2$.*

Let $F \in SL(2, \mathbb{Z})$. The hyperbolic automorphism of the 2-torus defined by F represents the simplest examples of hyperbolic dynamical systems when $|\mathrm{Tr} F| > 2$. Namely if this is satisfied then F has two eigenvalues $\lambda_+ = \lambda > \lambda_- = \lambda^{-1}$ with $\lambda > 1$. Denote $T_x(\mathbb{T}^2)$ the tangent space at point $x \in \mathbb{T}^2$, E_x^+ (resp. E_x^-) the eigenspace associated to the eigenvalue λ (resp. λ^{-1}) and $D_xF : T_x(\mathbb{T}^2) \longmapsto T_{Fx}(\mathbb{T}^2)$ the differential of F. One has

$$\|D_x F(v)\| = |\lambda| \|v\| \text{ if } v \in E_x^+,$$

$$\|D_x F(v)\| = |\lambda^{-1}| \|v\| \text{ if } v \in E_x^-,$$

where $\| \cdot \|$ is the norm associated to the Riemannian metric $ds^2 = dq^2 + dp^2$ on $\mathbb{T}^2$. This proves that F is an Anosov diffeomorphism, and is therefore ergodic and mixing.

We can also give a more direct proof that F is mixing using Proposition 54. Let us consider the total family in $L^2(\mathbb{T}^2)$, $e_k(x) = \mathrm{e}^{2i\pi k \cdot x}$, where $k \in \mathbb{Z}^2$. We have

$$\int_{\mathbb{T}^2} \overline{e_k(x)} e_\ell(\Phi^n(x)) dm(x) = \int_{\mathbb{T}^2} \overline{e_k(x)} e_{(\Phi^T)^n \ell}(x) dm(x).$$

If $\ell \neq 0$, using that A has eigenvalues λ and λ^{-1} with $\lambda > 1$, we get that $(\Phi^T)^n \ell$ is large for $\pm n$ large hence we get $\int_{\mathbb{T}^2} \overline{e_k(x)} e_\ell(\Phi^n(x)) dm(x) = 0$. We can conclude using Proposition 54.

One can easily identify the periodic points of F:

Proposition 55 *The periodic points $(q, p) \in \mathbb{T}^2$ of an hyperbolic automorphism of $\mathbb{T}^2$ are exactly points (q, p) such that $(q, p) \in \mathbb{Q}^2/\mathbb{Z}^2$.*

Proof Let A be an hyperbolic automorphism of $\mathbb{T}^2$, and $n \in \mathbb{N}^*$. Then the finite set $L_n = \left\{(\frac{r}{n}, \frac{s}{n}),\ r, s = 1, \dots, n\right\}$ is invariant under A and so all elements of L_n are periodic for A. Let $m \neq n \in \mathbb{N}^*$. Since one has

$$L_{m,n} = \left\{\left(\frac{r}{m}, \frac{s}{n}\right),\ r = 1, \dots, m,\ s = 1, \dots, n\right\} \subset L_{mn}$$

all points of $L_{m,n}$ are also periodic. Thus all points in $\mathbb{Q}^2/\mathbb{Z}^2$ are periodic for A. No other point can be periodic. Namely a point $(q, p) \in \mathbb{T}^2$ is periodic of period $k \in \mathbb{N}^*$ if and only if there exists $(m, n) \in \mathbb{Z}^2$ such that

$$(A^k - \mathbb{1}) \begin{pmatrix} q \\ p \end{pmatrix} = \begin{pmatrix} m \\ n \end{pmatrix}.$$

But the matrix $A^k - \mathbb{1}$ is invertible and has only rational entries. Thus

$$\begin{pmatrix} q \\ p \end{pmatrix} = (A^k - \mathbb{1})^{-1} \begin{pmatrix} m \\ n \end{pmatrix} = \begin{pmatrix} r_k \\ s_k \end{pmatrix},$$

with $(r_k, s_k) \in \mathbb{Q}^2$. This completes the proof. □

Remark 35 An hyperbolic automorphism of $\mathbb{T}^2$ is always mixing (so ergodic) but never uniquely ergodic since every periodic point x gives an invariant probability measure m_x.

6.3 The Kinematics Framework and Quantization

We follow closely the approaches of [12, 34, 38, 39, 73, 137].

Let us recall that we consider as phase space the 2-torus $\mathbb{T}^2 = \mathbb{R}^2/\mathbb{Z}^2$ with its canonical symplectic two form.

Using the correspondence principle between classical and quantum mechanics, it seems natural to look for the quantum states ψ having the same periodicity in position and momentum (q, p) as the underlying classical system.

The Weyl–Heisenberg translation operators $\hat{T}(q, p)$ "translate" the quantum state by a vector $z = (q, p) \in \mathbb{R}^2$. So we are looking for some Hilbert space $\mathscr{H}$, included in the Schwartz temperate distribution space $\mathscr{S}'(\mathbb{R})$, such that for every $\psi \in \mathscr{H}$ we have

$$\hat{T}(1, 0)\psi = e^{-i\theta_1}\psi, \tag{6.4}$$

$$\hat{T}(0, 1)\psi = e^{i\theta_2}\psi, \tag{6.5}$$

where we allow a phase $\theta = (\theta_1, \theta_2)$ since two wavefunctions ψ_1, ψ_2 satisfying $\psi_2 = e^{i\alpha}\psi_1$ define the same quantum state and more importantly, we shall recover the plane model as θ runs over the square $[0, 2\pi[\times[0, 2\pi[$.

Equation (6.5) means that the $\hbar$-Fourier transform $\mathscr{F}_\hbar\psi$ satisfies

$$\mathscr{F}_\hbar\psi(p+1) = e^{-i\theta_2}\mathscr{F}_\hbar\psi(p). \tag{6.6}$$

Recall that $\mathscr{F}_\hbar\psi(p) = (2\pi\hbar)^{-1/2}\int_{\mathbb{R}} \mathrm{e}^{-iqp/\hbar}\psi(q)dq$.

From Eqs. (6.4), (6.5) we see that ψ must be a joint eigenvector for the Weyl–Heisenberg operators $\hat{T}(q, p)$ and we get

$$\hat{T}(0, 1)\hat{T}(1, 0)\psi = \hat{T}(1, 0)\hat{T}(0, 1)\psi$$

Since we have that

$$\hat{T}(0, 1)\hat{T}(1, 0) = e^{i/\hbar}\hat{T}(1, 0)\hat{T}(0, 1)$$

conditions (6.4), (6.5) entail the following quantification condition $\frac{1}{2\pi\hbar} = N$ where $N \in \mathbb{N}$ and $\hbar$ is the Planck constant. Moreover, the quantum states ψ live in a N-dimensional complex vector space.

This result can be obtained using the powerful methods of the geometric quantization [73]. Here we follow a more elementary approach as in [38, 39].

Let us denote by $\mathscr{H}_N(\theta)$ the linear space of temperate distributions ψ satisfying periodicity conditions (6.4), (6.5) with $\hbar = \frac{1}{2\pi N}$ (remark that if $\hbar \neq \frac{1}{2\pi N}$ and if ψ satisfies (6.4), (6.5) then $\psi = 0$). So in all this chapter it is assumed that $\hbar = \frac{1}{2\pi N}$ for some $N \in \mathbb{N}$.

Proposition 56 *$\mathscr{H}_N(\theta)$ is an N dimensional complex linear sub-space of the temperate distribution space $\mathscr{S}'(\mathbb{R})$.*

Proof Let $\psi \in \mathscr{H}_N(\theta)$. From condition (6.5) we get that the support of ψ is in the discrete set $\{q_j = \frac{2\pi j+\theta_2}{2\pi N},\ j \in \mathbb{Z}\}$. So ψ is a sum of derivatives of Dirac distributions $\psi = \sum c_j^\alpha \delta_{q_j}^{(\alpha)}$. Using uniqueness of this decomposition we can prove that $c_j^{(\alpha)} = 0$ for $\alpha \neq 0$, so we have $\psi = \sum c_j \delta_{q_j}$ where $c_j = c_j^{(0)}$. Now, using (6.4) we get a periodicity condition on the coefficient c_j. So we have

$$c_{j+N} = \mathrm{e}^{i\theta_1} c_j, \quad \forall j \in \mathbb{Z} \text{ and } \psi = \sum_{0 \le j \le N-1} c_j \left(\sum_{k \in \mathbb{Z}} \mathrm{e}^{ik\theta_1} \delta_{q_j+k} \right). \tag{6.7}$$

Conversely it is easy to see that if ψ satisfies (6.7) then $\psi \in \mathscr{H}_N(\theta)$. So the proposition is proven. □

From the proof of the proposition, we get a basis of $\mathscr{H}_N(\theta)$:

$$e_j^{(\theta)} = N^{-1/2} \sum_{k \in \mathbb{Z}} \mathrm{e}^{ik\theta_1} \delta_{q_j+k}, \quad 0 \le j \le N-1.$$

$e_j^{(\theta)}$ obviously satisfies (6.4). Let us check that it satisfies (6.6) by computing its Fourier transform. As a consequence of the usual Poisson formula:

$$\sum_{k \in \mathbb{Z}} \mathrm{e}^{2i\pi kx} = \sum_{\ell \in \mathbb{Z}} \delta_\ell(x),$$

we get after some easy computations

$$\mathscr{F}_\hbar e_j^{(\theta)}(p) = N^{-1} \mathrm{e}^{-2i\pi p(j+\theta_2/2\pi)} \sum_{\ell \in \mathbb{Z}} \delta_{\frac{\ell}{N}+\frac{\theta_1}{2\pi}}. \tag{6.8}$$

Let us introduce $p_\ell = \frac{\ell}{N} + \frac{\theta_1}{2\pi}$ and $\varepsilon_\ell^{(\theta)} = N^{-1/2} \sum_{k \in \mathbb{Z}} \mathrm{e}^{-ik\theta_2} \delta_{\frac{\ell}{N}+\frac{\theta_1}{2\pi}+k}$, for $\ell = 0, \ldots,$ $N-1$. We have now

$$\mathscr{F}_\hbar e_j^{(\theta)} = \sum_{0 \le \ell \le N-1} F_{j,\ell} \varepsilon_\ell^{(\theta)} \tag{6.9}$$

where the matrix element $F_{j,\ell}$ is given by

$$F_{j,\ell} = N^{-1/2} \exp\left(-\frac{i}{N}\left(2\pi j\ell + \theta_2\ell + \theta_1 j + \frac{\theta_1\theta_2}{2\pi}\right)\right).$$

We put on $\mathscr{H}_N(\theta)$ the unique Hilbert space structure such that $\{e_j^{(\theta)}\}_{0\leq j\leq N-1}$ is an orthonormal basis. So we see that $\mathscr{F}_\hbar$ is a unitary transformation from $\mathscr{H}_N(\theta_1,\theta_2)$ onto $\mathscr{H}_N(-\theta_2,\theta_1)$. In particular if $\theta = (0,0)$, the matrix $\{F_{j,\ell}\}$ is the matrix of the discrete Fourier transform.

For all $\psi \in \mathscr{H}_{(\theta_1,\theta_2)}$ we have

$$\psi = \sum_{j=0}^{N-1} c_j(\psi) e_j^{(\theta)}.$$

Then the vector

$$\left(c_j(\psi)\right)_{j=0}^{N-1}$$

is interpreted physically as the quantum state of the particle in the position representation. Similarly in the momentum representation one has that $\tilde{\psi}^\theta \in \mathscr{H}_N(-\theta_2,\theta_1)$ is decomposed as

$$\tilde{\psi}^\theta = \sum_{j=0}^{N-1} d_j(\psi)\varepsilon_j^{(\theta)}.$$

One goes from the position to the momentum representation via a generalized Discrete Fourier Transform:

$$d_k(\hat{\psi}) = \exp\left(-i\frac{\theta_2}{N}\left(\frac{\theta_1}{2\pi} + k\right)\right)\frac{1}{\sqrt{N}}\sum_{j=0}^{N-1} c_j(\psi)\exp\left(-ij(\frac{2\pi k}{N} + \frac{\theta_1}{N})\right).$$

For $\theta = (0,0)$ we recognize the discrete Fourier operator that we have introduced above.

A convenient representation formula for elements of $\mathscr{H}_N(\theta)$ can be obtained using the following symmetrization operator

$$\Sigma_N^{(\theta)} = \sum_{z\in\mathbb{Z}^2} (-1)^{Nz_1z_2} \mathrm{e}^{i(\theta_1 z_1 - \theta_2 z_2)} \hat{T}(z). \tag{6.10}$$

Let us remark that $\psi \in \mathscr{H}_N(\theta)$ if and only if $\psi \in \mathscr{S}'(\mathbb{R})$ satisfies

$$\hat{T}(z)\psi = (-1)^{Nz_1z_2} \mathrm{e}^{i\sigma((\theta_2,\theta_1),(z_1,z_2))}\psi. \tag{6.11}$$

Proposition 57 $\Sigma_N^{(\theta)}$ *defines a linear continuous map from* $\mathscr{S}(\mathbb{R})$ *in* $\mathscr{S}'(\mathbb{R})$. *Its range is* $\mathscr{H}_N(\theta)$. *Moreover for every* $\psi \in \mathscr{S}(\mathbb{R})$ *we have*

$$\langle e_j^{(\theta)}, \Sigma_N^{(\theta)}\psi\rangle = N^{-1/2}\sum_{\ell\in\mathbb{Z}} \mathrm{e}^{i\ell\theta_1}\psi(q_j-\ell) = e_j^{(\theta)}(\psi) \tag{6.12}$$

and

$$\int_{\mathbb{R}} |\psi(x)|^2 dx = \frac{1}{4\pi^2}\iint_{[0,2\pi[^2} |e_j^{(\theta)}(\psi)|^2 d\theta. \tag{6.13}$$

The map $\psi \mapsto \{e_j^{(\theta)}(\psi)\}_{0\leq j\leq N-1}$ *can be extended as an isometry from* $L^2(\mathbb{R})$ *onto the Hilbert space* $L^2\left([0,2\pi[^2, \mathbb{C}^N, \frac{d\theta}{4\pi^2}\right)$.

Proof Recall that we have

$$\hat{T}(z)\psi(x) = \mathrm{e}^{-iz_1z_2/2\hbar}\mathrm{e}^{ixz_2/\hbar}\psi(x-z_1).$$

So we have

$$\Sigma_N^{(\theta)}\psi = \sum_{z_1,z_2\in\mathbb{Z}} \mathrm{e}^{i(\theta_1z_1-\theta_2z_2)}\mathrm{e}^{ixz_2/\hbar}\psi(x-z_1).$$

We first compute the z_2-sum using the Poisson formula:

$$\sum_{z_2\in\mathbb{Z}} \mathrm{e}^{i(xz_2/\hbar-\theta_2z_2)} = \frac{1}{N}\sum_{k\in\mathbb{Z}} \delta_{\frac{k}{N}+\frac{\theta_2}{2\pi N}}$$

we get

$$\Sigma_N^{(\theta)}\psi = \frac{1}{N}\sum_{k\in\mathbb{Z}} \mathrm{e}^{ik\theta_1}\sum_{0\leq j\leq N-1}\left(\sum_{\ell\in\mathbb{Z}} \mathrm{e}^{i\ell\theta_1}\psi(q_j-\ell)\right)\delta_{q_j+k}. \tag{6.14}$$

The equalities (6.12) and (6.13) follow easily from (6.14).

In particular we see that for every $j=0,\ldots,N-1$ we have $e_j^{(\theta)} = \Sigma_N^{(\theta)}\psi_j$, where $\psi_j(x) = \psi_0(q_j-x)$, ψ_0 is C^∞, with support in $[-\frac{1}{4N}, \frac{1}{4N}]$ and $\psi_0(0)=1$. This proved that $\Sigma_N^{(\theta)}(\mathscr{S}(\mathbb{R})) = \mathscr{H}_N(\theta)$.

Let us define the map $\mathscr{I}(\psi) = \{e_j^{(\theta)}(\psi)\}_{0\leq j\leq N-1}$. We know that $\mathscr{I}$ defines an isometry from $L^2(\mathbb{R})$ into $L^2([0,2\pi[^2, \mathbb{C}^N, \frac{d\theta}{4\pi^2})$. We have to prove now that $\mathscr{I}$ is onto.

It is enough to prove that the conjugate operator $\mathscr{I}^*$ is injective on $L^2\left([0,2\pi[^2, \mathbb{C}^N, \frac{d\theta}{4\pi^2}\right)$. To do that we have to compute $\langle \mathscr{I}^* f, \psi\rangle$ where $f = (f_0,\ldots,f_{N-1})$, f_j are periodical functions on the lattice $2\pi\mathbb{Z}\times 2\pi\mathbb{Z}$ and $\psi \in \mathscr{S}(\mathbb{R})$. This is an exercise left to the reader. □

This leads to a direct integral decomposition of $L^2(\mathbb{R})$:

$$L^2(\mathbb{R}) \cong \left(\frac{1}{2\pi}\right)^2\int_0^{2\pi}\int_0^{2\pi} d\theta\,\mathscr{H}_N(\theta),$$

$$\psi \cong \left(\frac{1}{2\pi}\right)^2 \int_0^{2\pi}\int_0^{2\pi} d\theta\psi(\theta), \text{ where } \psi(\theta) = \Sigma_N^{(\theta)}\psi.$$

This is a Bloch decomposition of $L^2(\mathbb{R})$ analogous to the description of electrons in a periodic structure.

It appears that $\mathscr{H}_N(\theta)$ is equipped with the natural inner product and the spaces $\mathscr{H}_N(\theta)$ are the natural quantum Hilbert spaces of states having the torus as phase space.

Let us explain now in more details the identification

$$L^2(\mathbb{R}) \cong \left(\frac{1}{2\pi}\right)^2 \int_0^{2\pi}\int_0^{2\pi} d\theta \mathscr{H}_N(\theta).$$

For every $\psi \in \mathscr{S}(\mathbb{R})$ we define $\tilde{\psi}(\theta, j) = e_j^{(\theta)}(\psi)$ where $\theta \in [0, 2\pi[^2$ and $j \in \mathbb{Z}$. We have seen that $\psi \mapsto \tilde{\psi}$ is an isometry from $L^2(\mathbb{R})$ onto $L^2([0, 2\pi[^2 \times (\mathbb{Z}/N\mathbb{Z}), \frac{d\theta}{4\pi^2} \otimes d\mu_N)$ where μ_N is the uniform probability on $\mathbb{Z}/N\mathbb{Z}$.

Let $\hat{A}$ be some bounded operator in $L^2(\mathbb{R})$. Assume that $\hat{A}$ is a linear continuous operator from $\mathscr{S}(\mathbb{R})$ to $\mathscr{S}(\mathbb{R})$ and from $\mathscr{S}'(\mathbb{R})$ to $\mathscr{S}'(\mathbb{R})$ and that $\hat{A}$ commutes with $\Sigma_N^{(\theta)}(\hat{A}\Sigma_N^{(\theta)} = \Sigma_N^{(\theta)}\hat{A})$, for every $\theta \in [0, 2\pi[^2$. Then $\hat{A}$ is a decomposable operator (see Reed-Simon [215], t.1, p. 281). More precisely we have the following useful result.

Proposition 58 *Let us denote by $\hat{A}_{N,\theta}$ the restriction of $\hat{A}$ to $\mathscr{H}_N(\theta)$. Then we have, for every $\psi_1, \psi_2 \in L^2(\mathbb{R})$,*

$$\langle \psi_2, \hat{A}\psi_1\rangle_{L^2(\mathbb{R})} = \iint_{[0,2\pi[^2} \langle \psi_2(\theta), \hat{A}_{N,\theta}\psi_1(\theta)\rangle_{\mathscr{H}_N(\theta)} \frac{d\theta}{4\pi^2}, \tag{6.15}$$

where $\psi_1(\theta) = \Sigma_N^{(\theta)}\psi_1$.

Proof This is easily proved using that $\psi(\theta) = \sum_{0\le j\le N-1} \tilde{\psi}(\theta, j)e_j^{(\theta)}$ and (6.13). □

We shall apply the following results proved using Reed-Simon, [215].

Corollary 20 *Let $\hat{A}$ be a decomposable operator like above. Then we have*

$$\|\hat{A}\|_{L^2(\mathbb{R})} = \sup_{\theta\in[0,2\pi[^2} \|\hat{A}_{N,\theta}\|_{\mathscr{H}_N^{(\theta)}} \tag{6.16}$$

and $\hat{A}$ is an isometry in $L^2(\mathbb{R})$ if and only if $\hat{A}_{N,\theta}$ is an isometry in $\mathscr{H}_N(\theta)$ for every $\theta \in [0, 2\pi[^2$.

First examples are the Weyl–Heisenberg translations.

Lemma 36 *Let $z = (z_1, z_2) \in \mathbb{R}^2$. Then $\hat{T}(z)\Sigma_N^{(\theta)} = \Sigma_N^{(\theta)}\hat{T}(z)$ if and only if $Nz \in \mathbb{Z}^2$. Moreover if $z_1 = \frac{n_1}{N}$ and $z_2 = \frac{n_2}{N}$ we have*

$$\hat{T}_{N,\theta}\left(\frac{n_1}{N}, \frac{n_2}{N}\right) e_j^{(\theta)} = \exp\left(i\pi\frac{n_1 n_2}{N}\right)\exp\left(i(\theta_2 + 2\pi j)\frac{n_2}{N}\right) e_{j+n_1}^{(\theta)}. \qquad (6.17)$$

Proof Exercise.

Corollary 21 *The unitary (projective) representation $(n_1, n_2) \mapsto \hat{T}_{N,\theta}(\frac{n_1}{N}, \frac{n_2}{N})$ of the group $\mathbb{Z}^2$ in $\mathscr{H}_N^{(\theta)}$ is irreducible.*

Proof Let V be an invariant subspace of $\mathscr{H}_N^{(\theta)}$ and $v = \sum_{0 \le j \le N-1} a_j e_j^{(\theta)}$, $v \neq 0$. If $a_j = 0$ for $j \neq j_0$, then using the translation $\hat{T}_{N,\theta}(\frac{n_1}{N}, 0)$ we get that $e_j^{(\theta)} \in V$ $\forall j$. But playing with $\hat{T}_{N,\theta}(0, \frac{n_2}{N})$, if m coefficients a_j are not 0 there exists a non zero vector of V with $m-1$ non null coefficients. So we can conclude that $V = \mathscr{H}_N(\theta)$. □

Remark 36 It has been proved that all irreducible unitary representations of the discrete Heisenberg group are equivalent to $(\hat{T}_{N,\theta}(\frac{n_1}{N}, \frac{n_2}{N}), \mathscr{H}_N(\theta))$, for some $(N, \theta) \in \mathbb{N}^* \times [0, 2\pi[^2$ [77].

For the particular case $\theta = (0, 0)$, the states e_j^0 can be identified with the natural basis in $\mathbb{C}^N$. Then the translation operators $\hat{T}(1/N, 0)$, $\hat{T}(0, 1/N)$ are simply $N \times N$ matrices of the following form:

$$\hat{T}(0, 1/N) := Z = \mathrm{diag}(1, \omega, \omega^2, \ldots, \omega^{N-1}) \qquad (6.18)$$

where $\omega = e^{2i\pi/N}$ is the primitive Nth root of unity.

$$\hat{T}(1/N, 0) := X = \begin{pmatrix} 0\ 0 \ldots 0\ 1 \\ 1\ 0 \ldots 0\ 0 \\ .\ .\ \ldots\ .\ . \\ 0\ 0 \ldots 1\ 0 \end{pmatrix}. \qquad (6.19)$$

These operators (matrices) have been introduced by Schwinger [233] as "generalized Pauli matrices" and are intensively used in quantum information theory for the mutually unbiased bases problem in $\mathbb{C}^N$. See [62, 233]. They have the following properties:

Proposition 59 *(i) X and Z are unitary.*
(ii) They are idempotent, namely

$$X^N = Z^N = \mathbb{1}$$

(the identity matrix in $\mathbb{C}^N$).
(iii) They ω-commute:

$$XZ = \omega ZX$$

(iv) X is diagonalized by the discrete Fourier transform $\mathscr{F}$:

$$\mathscr{F}^* X \mathscr{F} = Z$$

where $\mathscr{F}_{j,k} = \frac{1}{\sqrt{N}}\omega^{jk},\ \forall j,k = 1,\dots,N$

Remark 37 A complex $N \times N$ matrix is an Hadamard matrix if all its entries have equal modulus. Note that $\mathscr{F}$ is an unitary Hadamard matrix of the Vandermonde form, and that X and its powers generate the commutative algebra of the "circulant" matrices. An $N \times N$ matrix C is said to be circulant if all its rows and columns are successive circular permutations of the first:

$$C = \mathrm{circ}(c_1, c_2, \dots, c_N) = \begin{pmatrix} c_1 & c_2 & \dots & c_N \\ c_N & c_1 & \dots & c_{N-1} \\ . & . & \dots & . \\ c_2 & c_3 & \dots & c_1 \end{pmatrix}$$

$$C = c_1 \mathbb{1} + c_N X + \cdots + c_2 X^{N-1}. \tag{6.20}$$

(see [72]) The Discrete Fourier transform in $\mathbb{C}^N$ is very natural in this context since it transforms any basis vector in the position representation into any basis vector in the momentum representation.

Lemma 37 *For any circulant matrix C there exists a diagonal matrix D such that*

$$\mathscr{F}^* C \mathscr{F} = D.$$

Furthermore

$$D_{j,j} = \sqrt{N}\hat{c}_j = \sum_{0}^{N-1} c_{k+1}\omega^{-jk}.$$

Proof Use Proposition (59) (iv) and (6.20). □

These properties are very useful to construct the $N+1$ mutually unbiased bases in Quantum Information Theory for N a prime number. See [62].

6.4 The Coherent States of the Torus

Already used in the physical literature in [173] we introduce now the coherent states adapted to the torus structure of the phase-space. They will be the image by the periodisation operator $\Sigma_N^{(\theta)}$ of the usual Gaussian coherent states studied in Chap. 1.

In dimension 1 one has, for $z = (q, p) \in \mathbb{R}^2$. and $\gamma \in \mathbb{C}$, $\Im\gamma > 0$,

$$\varphi_{\gamma,z}(x) = \left(\frac{\Im\gamma}{\pi\hbar}\right)^{1/4} \exp\left(-\frac{iqp}{2\hbar} + \frac{ixp}{\hbar} + i\gamma\frac{(x-q)^2}{2\hbar}\right), \tag{6.21}$$

$$\varphi_{\gamma,z}^{(\theta)} = \Sigma_N^{(\theta)}\varphi_{\gamma,z}. \tag{6.22}$$

It is easily seen from the definition properties of $\Sigma_N^{(\theta)}$ and the product rules for $\hat{T}(z)$ that

$$\varphi_{\gamma,z}^{(\theta)} = \sum_{n_1,n_2\in\mathbb{Z}} (-1)^{Nn_1n_2} e^{i(\theta_1 n_1 - \theta_2 n_2) + \frac{i}{2\hbar}\sigma(n,z)} \hat{T}(n+z)\varphi_{\gamma,0}. \tag{6.23}$$

For every $z' = (z_1', z_2') \in \mathbb{Z}^2$ we get

$$\varphi_{\gamma,z+z'}^{(\theta)} = (-1)^{Nz_1'z_2'} e^{i(z_2'\theta_2 - z_1'\theta_1)} e^{i\pi N\sigma(z,z')} \varphi_{\gamma,z}^{(\theta)}. \tag{6.24}$$

Thus the states $\varphi_{\gamma,z+z'}^{(\theta)}$, $\varphi_{\gamma,z}^{(\theta)}$ are equal modulo phase factor, so they describe the same physical system and we can identify them. Recall that σ is the symplectic form:

$$\sigma((a,b),(c,d)) = ad - bc.$$

The set $\left\{\varphi_{\gamma,z}^{(\theta)}\right\}_{z\in\mathbb{T}^2}$ therefore constitutes a coherent states system adapted to the torus.

In the basis $e_j^{(\theta)}$ of the position representation we have

$$c_j(q,p) := \langle e_j^{(\theta)}, \varphi_{q,p}^{(\theta)}\rangle = \left(\frac{\Im\gamma}{\pi\hbar}\right)^{1/4} \frac{1}{\sqrt{N}} e^{-i\frac{qp}{2\hbar}} \sum_{m\in\mathbb{Z}} e^{i\theta_1 m} e^{\frac{i}{\hbar}(x_j^m p)} \exp\left(i\gamma\frac{1}{2\hbar}(x_j^m - q)^2\right), \tag{6.25}$$

where $x_j^m = \frac{j}{N} + \frac{\theta_2}{2\pi N} - m$.

Similarly we have in the momentum representation (for $\gamma = i$):

$$d_k(q,p) = \left(\frac{1}{\pi\hbar}\right)^{1/4} \frac{1}{\sqrt{N}} e^{-\frac{i}{2\hbar}qp} \sum_{m\in\mathbb{Z}} e^{-im\theta_2 + \frac{i}{\hbar}q\xi_m^k} \exp\left(\frac{-1}{2\hbar}(\xi_m^k - p)^2\right),$$

with $\xi_m^k = \frac{k}{N} + \frac{\theta_1}{2\pi N} - m$.

An important property which is inherited from the overcompleteness character of the set of coherent states in $L^2(\mathbb{R}^n)$ is that the $\left\{\varphi_{\gamma,z}^{(\theta)}\right\}_{z\in\mathbb{T}^2}$ form an overcomplete system of $\mathcal{H}_N(\theta)$ with a resolution of the identity operator $\mathbb{1}_{\mathcal{H}_N(\theta)}$:

Proposition 60 *$\forall\theta \in [0, 2\pi)^2$ and $\forall\hbar = 1/2\pi N$ we have*

$$\mathbb{1}_{\mathscr{H}_N(\theta)} = \int_{\mathbb{T}^2} \frac{dqdp}{2\pi\hbar} |\varphi^{(\theta)}_{\gamma,q,p}\rangle\langle\varphi^{(\theta)}_{\gamma,q,p}|$$

where we use the bra-ket notation for the projector on the coherent state $\varphi^{(\theta)}_{q,p}$.

Proof For simplicity we assume $\gamma = i$. Since $\mathscr{H}_N(\theta)$ is finite dimensional it is enough to prove that $\forall (j,k) \in [0, N-1]^2$ we have

$$\int_{\mathbb{T}^2} \frac{dqdp}{2\pi\hbar} \langle\varphi^{\theta}_{q,p}, e^{\theta}_k\rangle\langle e^{\theta}_j, \varphi^{\theta}_{q,p}\rangle = \delta_{j,k}.$$

Now using (6.25) together with Fubini's Theorem we get

$$\begin{aligned}\int_{\mathbb{T}^2} \frac{dqdp}{2\pi\hbar} \bar{c}_k(q,p)c_j(q,p) = \textstyle\sum_{m,n} e^{i\theta_1(m-n)} \int_0^1 dp \exp\left(2\pi i(j-k-N(m-n))p\right) \\ \times \textstyle\int_0^1 dq \bar{\varphi}^{\theta}_{0,0}(x^m_j - q)\varphi^{\theta}_{0,0}(x^m_j - q). \qquad (6.26)\end{aligned}$$

If $j = k$ then the first integral in the right hand side of (6.26) is zero except for $m = n$ in which case we get 1. Thus we get

$$\int_{(q,p)\in\mathbb{T}^2} \frac{dqdp}{2\pi\hbar} |c_j(q,p)|^2 = \sum_{m\in\mathbb{Z}} \int_0^1 dq \|\varphi^{\theta}_{0,0}(x^j_m - q)\|^2 = \|\varphi^{\theta}_{0,0}\|^2 = 1.$$

If $j \neq k$ the same integral is zero since $j - k \in \mathbb{N}^*$. □

We also get the Fourier-Bargmann transform $\psi^{\sharp}$ of any state $\psi \in \mathscr{H}_N(\theta)$ as

$$\psi^{\sharp} = \sqrt{N}\langle\varphi^{\theta}_{q,p}, \psi\rangle.$$

The map $W^{\theta} : \psi \in \mathscr{H}_N(\theta) \longmapsto W^{\theta}\psi = \sqrt{N}\psi^{\sharp} \in L^2(\mathbb{T}^2, \frac{dqdp}{2\pi\hbar})$ is obviously isometric. The quantity

$$H^{\theta}(q,p) = |\langle\varphi^{\theta}_{q,p}, \psi\rangle|^2$$

is called the Husimi function of $\psi \in \mathscr{H}_N(\theta)$.

We have as a corollary an analogous result as Proposition 6:

Corollary 22 *Let $\hat{A}_{\theta} \in \mathscr{L}(\mathscr{H}_{\theta})$. Then*

$$\mathrm{Tr}(\hat{A}_{\theta}) = \int_{\mathbb{T}^2} \langle\varphi^{\theta}_{q,p}, \hat{A}_{\theta}\varphi^{\theta}_{q,p}\rangle \frac{dqdp}{2\pi\hbar}.$$

We have the following very useful semiclassical result.

Proposition 61 *For every complex numbers γ, γ' with positive imaginary part we have:*
(i) There exist constants $C > 0, c > 0$ such that for any $z, z' \in \mathbb{T}^2$, $N \geq 1$,

$$|\langle \varphi^{\theta}_{\gamma',z'}, \varphi^{\theta}_{\gamma,z} \rangle| \leq C\sqrt{N} e^{-d(z,z')^2 cN}, \tag{6.27}$$

where $d(z, z')$ is the distance between z and z' on the torus $\mathscr{T}^2$. In particular for $\gamma = \gamma' = i$ we can choose $c = \pi$.
(ii) There exists $c > 0$ such that $\forall \theta \in [0, 2\pi)^2$ we have $\forall z = (q, p) \in \mathbb{T}^2$

$$\|\varphi^{\theta}_{\gamma,z}\|^2 = 1 + O(\mathrm{e}^{-cN}).$$

Proof For simplicity, let assume that $\gamma = \gamma' = i$. The proof is the same for arbitrary γ, γ'.

We recall that in the continuous case one has

$$|\langle \varphi_{z'}, \varphi_z \rangle|^2 = (\pi\hbar)^{-1} \exp\left(-\frac{|z' - z|^2}{2\hbar}\right)$$

so that

$$\|\varphi_z\| = 1.$$

Thus we shall prove that the analogous properties (i) and (ii) hold for the coherent states of the 2-torus but only in the semiclassical limit $N \to \infty$. A weaker result is given in [38], here we shall give a different proof.

We rewrite (6.24): for every $z \in \mathscr{T}^2$, $m = (m_1, m_2) \in \mathbb{Z}^2$,

$$\varphi^{(\theta)}_{z+m} = \mathrm{e}^{i\pi N(\sigma(z,m)+m_1 m_2)} \mathrm{e}^{i(m_2\theta_2 - m_1\theta_1)} \mathrm{e}^{i\pi N\sigma(z,z')} \varphi^{(\theta)}_z. \tag{6.28}$$

Let us denote $f_{z,z'}(\theta) = \langle \varphi^{\theta}_z, \varphi^{\theta}_{z'} \rangle$ and consider $f_{z,z'}$ as a periodic function in θ for the lattice $(2\pi\mathbb{Z})^2$. Its Fourier coefficient $c_m(z, z')$ can be computed using (6.28),

$$c_m(z, z') = \int_{[0,2\pi[^2} \mathrm{e}^{-im\cdot\theta} \langle \varphi^{\theta}_z, \varphi^{\theta}_{z'} \rangle \frac{d\theta}{4\pi^2} \tag{6.29}$$

$$= \mathrm{e}^{i\pi N(\sigma(z,m)+m_1 m_2)} \int_{[0,2\pi[^2} \langle \varphi^{\theta}_{z'}, \varphi^{\theta}_{z+\check{m}} \rangle \frac{d\theta}{4\pi^2} \tag{6.30}$$

$$= \mathrm{e}^{i\pi N(\sigma(z,m)+m_1 m_2)} \langle \varphi_{z'}, \varphi_{z+\check{m}} \rangle. \tag{6.31}$$

where $\check{m} = (m_1, -m_2)$. $f_{z,z'}$ being a smooth function in θ, we get

$$|f_{z,z'}(\theta)| \leq \sum_{m\in\mathbb{Z}^2} |c_m(z, z')|. \tag{6.32}$$

But $|c_m(z, z')| = (\pi\hbar)^{-1} \exp\left(-\frac{|z'-z-\check{m}|^2}{2\hbar}\right)$. So

$$\sum_{m\in\mathbb{Z}^2} |c_m(z, z')| = (\pi\hbar)^{-1} \sum_{m\in\mathbb{Z}^2} \exp\left(-\frac{|z'-z-\check{m}|^2}{2\hbar}\right).$$

Now we have

$$|z'-z-\check{m}|^2 \geq |z'-z|^2 + |m|^2 - 2|m||z-z'|.$$

So we get

$$\sum_{m\in\mathbb{Z}^2, |m|\geq 4\sqrt{2}} \exp\left(-\frac{|z'-z-\check{m}|^2}{2\hbar}\right) \leq \mathrm{e}^{-|z-z'|^2\pi N} \sum_{m\in\mathbb{Z}^2} \mathrm{e}^{-|m|^2\pi N/2}.$$

So for every $N \geq 1$ we get

$$\sum_{m\in\mathbb{Z}^2, |m|\geq 4\sqrt{2}} \exp\left(-\frac{|z'-z-\check{m}|^2}{2\hbar}\right) \leq C\mathrm{e}^{-|z-z'|^2\pi N}.$$

For the finite sum we have easily

$$\sum_{m\in\mathbb{Z}^2, |m|\leq 4\sqrt{2}} \exp\left(-\frac{|z'-z-\check{m}|^2}{2\hbar}\right) \leq Ce^{-d(z,z')^2\pi N}.$$

so we get (i).

Concerning the proof with any γ, γ', we have to use the inequality

$$|\langle\varphi_{\gamma',z'}, \varphi_{\gamma,z}\rangle| \leq C\hbar^{-1/2}\mathrm{e}^{-c\frac{|z-z'|^2}{\hbar}}.$$

where $C > 0, c > 0$ depend on γ, γ', but not in z, z'.

The proof of (ii) uses the same method with $z' = z$. So we get

$$|f_{z,z}(\theta) - 1| \leq \sum_{m\neq(0,0)} |\langle\varphi_z, \varphi_{z+m}\rangle|$$

and

$$|f_{z,z}(\theta) - 1| \leq \sqrt{N} \sum_{m\neq(0,0)} \mathrm{e}^{-|m|^2\pi N}.$$

So we have proved (ii). □

6.5 The Weyl and Anti-Wick Quantizations on the 2-Torus

We will show how a phase space function (classical Hamiltonian) $H \in \mathscr{C}^\infty(\mathbb{T}^2)$ can be quantized as a selfadjoint operator in the Hilbert space $\mathscr{H}_N(\theta)$. These functions have to be real.

6.5.1 *The Weyl Quantization on the 2-Torus*

We identify the functions H with the functions $\mathscr{C}^\infty$ on $\mathbb{R}^2$ of period $(1, 1) \in \mathbb{R}^2$. Then we have

$$H(q, p) = \sum_{(m,n)\in\mathbb{Z}^2} H_{m,n} e^{2i\pi\sigma((q,p),(m,n))}.$$

So we define, following [78, 137]:

Definition 17

$$\mathrm{Op}_\hbar^W(H) = \sum_{m,n} H_{m,n} \hat{T}\left(\frac{m}{N}, \frac{n}{N}\right). \tag{6.33}$$

Recall that $\hbar = \frac{1}{2\pi N}$, $N \in \mathbb{N}^*$.

One has the following property:

Proposition 62 *Let $\theta \in [0, 2\pi)^2$ and $\hbar > 0$. Then for any function $H \in \mathscr{C}^\infty(\mathbb{T}^2)$ one has*

$$\mathrm{Op}_\hbar^W(H)\mathscr{H}_N(\theta) \subseteq \mathscr{H}_N(\theta).$$

Proof This follows directly from the definition of $\mathscr{H}_N(\theta)$ and from

$$\hat{T}(m, n)\mathrm{Op}_\hbar^W(H)\hat{T}(m, n)^* = \mathrm{Op}_\hbar^W(H), \ if\ m, n \in \mathbb{Z}.$$

In other words $\mathrm{Op}_\hbar^W(H)$ commutes with $\Sigma_N^{(\theta)}$. □

Thus we define the operator $\mathrm{Op}_{\hbar,\theta}^W(H) \in \mathscr{L}(\mathscr{H}_\theta)$ as the restriction of (6.33) to $\mathscr{H}_N(\theta)$.

In the decomposition of $L^2(\mathbb{R})$ as a direct integral, $\mathrm{Op}_{\hbar,\theta}^W(H)$ is the fiber at θ of $\mathrm{Op}_\hbar^W(H)$.

$$\mathrm{Op}_\hbar^W(H) = \iint_{[0,2\pi[^2} \mathrm{Op}_{\hbar,\theta}^W(H)\frac{d\theta}{4\pi^2}. \tag{6.34}$$

We have also the following formula by restriction to $\mathscr{H}_N(\theta)$,

$$\mathrm{Op}_{\hbar,\theta}^W(H) = \sum_{n,m\in\mathbb{Z}} H_{n,m}\hat{T}\left(\frac{n}{N}, \frac{m}{N}\right). \tag{6.35}$$

In particular we see that the map $H \mapsto \mathrm{Op}_{\hbar,\theta}^W(H)$ cannot be injective, so the Weyl symbol H of $\mathrm{Op}_{\hbar,\theta}^W(H)$ is not unique. It becomes unique by restricting to trigonometric polynomials symbols. Let us denote by $\mathscr{T}_N$ the linear space spanned by $\pi_{n,m}(q, p) = \mathrm{e}^{2i\pi(nq-mp)}$, for $n, m = 0, \ldots, N-1$. Then we have

Proposition 63 $\mathrm{Op}_{\hbar,\theta}^W$ *is a unitary map from* $\mathscr{T}_N$ *(with the norm of* $L^2([0,1]^2)$*) onto* $\mathscr{L}(\mathscr{H}_N(\theta))$*, equipped with its Hilbert-Schmidt norm. In particular we have, for every* $H, K \in \mathscr{T}_N$*,*

$$\mathrm{Tr}(\mathrm{Op}_{\hbar,\theta}^W(H)\mathrm{Op}_{\hbar,\theta}^W(K)^*) = N \iint_{\mathscr{T}^2} H(z)\overline{K(z)}dz. \tag{6.36}$$

Proof Let us recall the formula

$$\hat{T}\left(\frac{k}{N}, \frac{\ell}{N}\right) e_j^{(\theta)} = \mathrm{e}^{i\pi k\ell/N}\mathrm{e}^{i(\theta_2+2\pi j)\ell/N} e_{j+k}^{(\theta)}. \tag{6.37}$$

Using that the discrete Fourier transform is unitary, we get

$$\mathrm{Tr}\left(\hat{T}\left(\frac{k}{N}, \frac{\ell}{N}\right)\hat{T}\left(\frac{k'}{N}, \frac{\ell'}{N}\right)^*\right) = N\delta_{k,k'}\delta_{\ell,\ell'}. \tag{6.38}$$

So the system $\{N^{-1/2}\hat{T}(\frac{k}{N}, \frac{\ell}{N})\}_{0\le k,\ell\le N-1}$ is an orthonormal basis in $\mathscr{L}(\mathscr{H}_N(\theta))$, equipped with its Hilbert-Schmidt norm and we get the proposition. □

Corollary 23 *Every linear operator* $\hat{H}$ *in* $\mathscr{H}_N(\theta)$ *has a unique Weyl symbol* $H \in \mathscr{T}_N$*,*

$$H(z) = \sum_{0\le m,n\le N-1} H_{m,n} e^{2i\pi\sigma(z,(m,n))},$$

where

$$H_{m,n} = N^{-1/2}\mathrm{Tr}\left(\hat{H}\hat{T}\left(\frac{m}{N}, \frac{n}{N}\right)^*\right). \tag{6.39}$$

A semiclassical result for the Weyl quantization is the following:

Proposition 64 *For all* $H \in \mathscr{C}^\infty(\mathbb{T}^2)$ *one has*

$$\lim_{N\to\infty} \frac{1}{N}\mathrm{Tr}\left(\mathrm{Op}_{\hbar,\theta}^W(H)\right) = \int_{\mathbb{T}^2} dz H(z).$$

Proof Using the orthonormal position basis $e_j^{(\theta)}$ one has

$$\mathrm{Tr}\left(\mathrm{Op}_{\hbar,\theta}^W(H)\right) = \sum_{j=0}^{N-1}\sum_{m,n} H_{n,m}\langle e_j^\theta, \hat{T}\left(\frac{m}{N}, \frac{n}{N}\right) e_j^\theta\rangle.$$

Now we use the property (6.37):

$$\frac{1}{N}\mathrm{Tr}(\mathrm{Op}^W_{\hbar,\theta}(H)) = \frac{1}{N}\sum_{j,k=0}^{N-1}\sum_{\ell,n\in\mathbb{Z}} H_{n,\ell N+k}$$
$$\times \exp\left(i\frac{\pi n}{N}(\ell N+k) - i\theta_1\ell + i\frac{2\pi n}{N}(\frac{\theta_2}{2\pi}+j+k)\right)\langle e^\theta_j, e^\theta_{j+k}\rangle$$
$$= \sum_{\ell,n} H_{n,\ell N}(-1)^{\ell n} e^{-i\ell\theta_1 + i\frac{n\theta_2}{N}}\frac{1}{N}\sum_{j=0}^{N-1} e^{i\frac{2\pi nj}{N}}. \tag{6.40}$$

Thus we conclude

$$\frac{1}{N}\mathrm{Tr}(\mathrm{Op}^W_{\hbar,\theta}(H)) = H_{0,0} + \sum_{\ell,n\in\mathbb{Z}^*} H_{\ell N,nN} e^{i\sigma((\ell,n),(\theta_1,\theta_2))}.$$

The last term tends to 0 because of the regularity of H, and the first one is $\int_{\mathbb{T}^2} dz H(z)$, which completes the proof. □

6.5.2 The Anti-Wick Quantization on the 2-Torus

As in the continuous case (see Chap. 2) the Anti-Wick quantization is associated to the system of coherent states defined above on the torus $\mathbb{T}^2$.

Definition 18 Let $H \in L^\infty(\mathbb{T}^2)$. Then $\varphi^\theta_{q,p}$ being the system of coherent states defined in the previous section, we define

$$\mathrm{Op}^{AW}_{\hbar,\theta}(H) := \int_{\mathbb{T}^2} H(z)|\varphi^\theta_z\rangle\langle\varphi^\theta_z|\frac{dz}{2\pi\hbar}.$$

Remark 38 Note that $\mathrm{Op}^{AW}_{\hbar,\theta}(H)$ for $H \in \mathscr{C}^\infty(\mathbb{T}^2)$ is simply the restriction of $\mathrm{Op}^{AW}_\hbar(H)$ (considered as an operator on $\mathscr{S}'(\mathbb{R})$) to $\mathscr{H}_N(\theta)$.

Let us recall that we always assume $2\pi\hbar N = 1$.

As for Weyl quantization, anti-Wick quantization on $\mathbb{R}^2$ and on $\mathbb{T}^2$ are related with a direct integral decomposition

Proposition 65 *Let $H \in C^\infty(\mathbb{T}^2)$. Then we have the direct integral decomposition*

$$\mathrm{Op}^{AW}_\hbar(H) = \iint_{[0,2\pi[^2} \mathrm{Op}^{AW}_{\hbar,\theta}(H)\frac{d\theta}{(2\pi)^2} \tag{6.41}$$

In particular we have the uniform norm estimate

$$\|\mathrm{Op}_{\hbar,\theta}^{AW}(H)\| \leq \|H\|_{\infty}. \tag{6.42}$$

Proof Using periodicity of H and direct integral decomposition of $\psi \in \mathscr{S}(\mathbb{R})$ $(\psi(\theta) = \Sigma_N^{(\theta)}\psi)$, we get

$$\mathrm{Op}_{\hbar}^{AW}(H)\psi = \sum_{n=(n_1,n_2)\in\mathbb{Z}^2} \iint_{[0,2\pi[^2} \iint_{\mathscr{T}^2} H(z)\langle\varphi_{z+n}^{\theta}, \psi(\theta)\rangle\varphi_{z+n} dz \frac{d\theta}{4\pi^2}. \tag{6.43}$$

Using periodicity in z of φ_z and $\varphi_{z+n} = \mathrm{e}^{i\sigma(n,z)/2\hbar}\hat{T}(n)\varphi_z$, we get

$$\mathrm{Op}_{\hbar}^{AW}(H)\psi = \iint_{[0,2\pi[^2} \iint_{\mathscr{T}^2} H(z)\langle\varphi_{z}^{\theta}, \psi(\theta)\rangle\varphi_{z+n} dz \frac{d\theta}{4\pi^2}. \tag{6.44}$$

So we have proved $\left(\mathrm{Op}_{\hbar}^{AW}(H)\right)_{\theta} = \mathrm{Op}_{\hbar,\theta}^{AW}(H)$. □

Now we show a link between Anti-Wick quantization and the Husimi function:

Proposition 66 *One has for any $H \in \mathscr{C}^{\infty}(\mathbb{T}^2)$ and for any $\psi \in \mathscr{H}_{\theta}$ that:*

$$\langle\psi, \mathrm{Op}_{\hbar,\theta}^{AW} H\psi\rangle = N\int_{\mathbb{T}^2} dz H(z) H_{\psi}(z).$$

And we have the following semiclassical limit:

Proposition 67 *For any $z \in \mathbb{T}^2$, any $\theta \in [0; 2\pi[^2$, and any $H \in \mathscr{C}^{\infty}(\mathbb{T}^2)$ we have:*

$$\lim_{N\to\infty} \langle\varphi_z^{\theta}, \mathrm{Op}_{\hbar,\theta}^{AW} H\varphi_z^{\theta}\rangle = H(z)$$

Proof We denote $z = (q, p) \in \mathbb{T}^2$ and $B_{\varepsilon}(z)$ the ball of center z and radius ε and by $B_{\varepsilon}^c(z)$ its complementary set. Take $\varepsilon \ll 0$. We have:

$$\langle\varphi_z^{\theta}, \mathrm{Op}_{\hbar,\theta}^{AW}(H)\varphi_z^{\theta}\rangle = \int_{B_{\varepsilon}^c(z)} \frac{dz'}{2\pi\hbar} H(z')|\langle\varphi_z^{\theta}, \varphi_{z'}^{\theta}\rangle|^2 + \int_{B_{\varepsilon}(z)} \frac{dz'}{2\pi\hbar} H(z')|\langle\varphi_z^{\theta}, \varphi_{z'}^{\theta}\rangle|^2$$

It is clear that the first term in the right hand side tends to 0 as $\hbar \to 0$ because of Proposition (61) (i). For the second term we denote $g(z, z') = N|\langle\varphi_z^{\theta}, \varphi_{z'}^{\theta}\rangle|^2$. We have:

$$\left|\int_{B_\varepsilon(z)} dz' H(z')g(z,z') - H(z)\right| \leq$$

$$\int_{B_\varepsilon(z)} dz'|H(z') - H(z)|g(z,z') + |H(z)|\left|\int_{B_\varepsilon(z)} g(z,z')dz' - 1\right| \leq$$

$$\varepsilon\|\nabla H\|_\infty \int_{B_\varepsilon(z)} g(z,z')dz' + |H(z)|\left|\int_{B_\varepsilon(z)} dz' g(z,z') - 1\right|. \quad (6.45)$$

Using the resolution of identity we have that

$$\int_{B_\varepsilon(z)} dz' g(z,z') = \|\varphi_z^{(\theta)}\|^2 - \int_{B_\varepsilon^c(z)} dz' g(z,z').$$

Using Proposition (61) (ii) the first term in the right hand side tends to 1 as $N \to \infty$, and it is clear that the second is small as $N \to \infty$. This completes the proof. □

As in the continuous case the Weyl and Anti-Wick quantizations are equivalent in the semiclassical regime:

Proposition 68 *For any $H \in \mathscr{C}^\infty(\mathbb{T}^2)$ and any $\theta \in [0, 2\pi)^2$ we have*

$$\|\mathrm{Op}_{\hbar,\theta}^W(H) - \mathrm{Op}_{\hbar,\theta}^{AW}(H)\|_{\mathscr{L}(\mathscr{H}_N(\theta))} = \mathscr{O}(N^{-1}), \text{ as } N \to \infty. \quad (6.46)$$

Proof This result follows from the similar one in the continuous case (see Chap. 2, Proposition 27) using the estimate

$$\|\mathrm{Op}_{\hbar,\theta}^W(H) - \mathrm{Op}_{\hbar,\theta}^{AW}(H)\|_{\mathscr{L}(\mathscr{H}_N(\theta))} \leq \|\mathrm{Op}_\hbar^W(H) - \mathrm{Op}_\hbar^{AW}(H)\|_{\mathscr{L}(L^2(\mathbb{R}))}. \quad (6.47)$$

□

6.6 Quantum Dynamics and Exact Egorov's Theorem

6.6.1 Quantization of SL(2, $\mathbb{Z}$)

We now consider a dynamics in phase space induced by symplectic transformations $F \in SL(2,\mathbb{Z})$. It creates a discrete time evolution in $\mathbb{T}^2$ and the n-step evolution is provided by F^n.

One wants here to quantize F as a natural operator in $\mathscr{H}_N(\theta)$.

We have seen in Chap. 2 (2.41) that F is quantized in $\mathscr{L}(L^2(\mathbb{R}))$ by the metaplectic transformation $\hat{R}(F)$. Let us recall the following property

$$\hat{R}(F)^*\hat{T}(z)\hat{R}(F) = \hat{T}(F^{-1}z). \quad (6.48)$$

We shall see now how to associate to F an unitary operator in $\mathscr{H}_N(\theta)$.

Proposition 69 *Let $F \in SL(2,\mathbb{Z})$. Then for any $\theta \in [0, 2\pi)^2$ there exists $\theta' \in [0, 2\pi)^2$ such that*

$$\hat{R}(F)\mathcal{H}_\theta \subseteq \mathcal{H}_{\theta'}$$

Furthermore θ' is defined as follows:

$$\begin{pmatrix}\theta_2'\\ \theta_1'\end{pmatrix} = F\begin{pmatrix}\theta_2\\ \theta_1\end{pmatrix} + \pi N\begin{pmatrix}ab\\ cd\end{pmatrix}, \quad \begin{pmatrix}\text{mod } 2\pi\\ \text{mod } 2\pi\end{pmatrix} \tag{6.49}$$

Moreover we have

$$\hat{R}(F)\Sigma_N^{(\theta)} = \Sigma_N^{(\theta')}\hat{R}(F). \tag{6.50}$$

Proof We use here Proposition 57 and formula (6.10). From (6.48) we get, if $\psi \in \mathcal{H}_N(\theta)$ and $z = (z_1, z_2) \in \mathbb{Z}^2$,

$$\hat{T}(z)\hat{R}(F)\psi = \hat{R}(F)\hat{T}(F^{-1}(z))\psi = \mathrm{e}^{-i(\sigma(z',(\theta_2,\theta_1))+\pi N z_1'z_2')}\hat{R}(F)\psi, \tag{6.51}$$

where

$$z' = F^{-1}z = \begin{pmatrix}z_1'\\ z_2'\end{pmatrix} = \begin{pmatrix}d & -b\\ -c & a\end{pmatrix}\begin{pmatrix}z_1\\ z_2\end{pmatrix}$$

We have $\sigma(z', (\theta_2, \theta_1)) = \sigma(z, F(\theta_2, \theta_1))$ and

$$z_1'z_2' = -cdz_1^2 + (ad + bc)z_1z_2 - abz_2^2$$

But modulo 2 we have $z_1^2 \equiv -z_1$, $z_2^2 \equiv -z_2$, $ad + bc \equiv 1$. So we get, modulo 2π,

$$\sigma(z', (\theta_2, \theta_1)) + \pi N z_1'z_2' \equiv z_1(d\theta_1 + c\theta_2 + \pi Ncd) - z_2(b\theta_1 + a\theta_2 - \pi Nab) + \pi N z_1z_2.$$

So we have $\hat{R}(F)\mathcal{H}_\theta \subset \mathcal{H}_{\theta'}$ with θ' given by (6.49). Moreover it is easy to check formula (6.50). □

Let us denote $\theta' := \pi_F(\theta)$. So π_F is a smooth map from the torus $\mathbb{R}^2/(2\pi\mathbb{Z})^2$ into itself.

Remark 39 We can easily see that $\hat{R}_\theta(J) = \mathcal{F}^*$ for $\theta = (0, 0)$ in the basis of $\left\{e_j^{(\theta)}\right\}_{j=1}^N$, up to a phase.

Definition 19 For every $F \in SL(2,\mathbb{Z})$, we shall denote $\hat{R}_{N,\theta}(F)$ the restriction of $\hat{R}(F)$ to $\mathcal{H}_N(\theta)$. It is the quantization of F in $\mathcal{H}_N(\theta)$. $\hat{R}_{N,\theta}(F)$ is a linear operator from $\mathcal{H}_N(\theta)$ in $\mathcal{H}_N(\theta')$.

Proposition 70 *$\hat{R}_{N,\theta}(F)$ is a one-to-one linear map from $\mathcal{H}_N(\theta)$ in $\mathcal{H}_N(\theta')$.*

Furthermore we have the following relationship between $\hat{R}_{N,\theta}(F)$ and $\hat{R}(F)$.

$$\hat{R}(F) = \int_{[0,2\pi[^2}^{\oplus} V_{N,\theta}\hat{R}_{N,\theta}(F)\frac{d\theta}{(2\pi)^2} \tag{6.52}$$

where $V_{N,\theta}$ *is the canonical isometry from* $\mathscr{H}_N(\theta')$ *onto* $\mathscr{H}_N(\theta)$ *defined by* $V_{N,\theta}e_j^{(\theta')} = e_j^{(\theta)}$.

In particular for every $\theta \in [0, 2\pi[^2$, $\hat{R}_{N,\theta}(F)$ *is a unitary transformation from* $\mathscr{H}_N(\theta)$ *onto* $\mathscr{H}_N(\theta')$

Proof We know that $\hat{R}(F)$ is an isomorphism from $\mathscr{S}(\mathbb{R})$ onto $\mathscr{S}(\mathbb{R})$ and from $\mathscr{S}'(\mathbb{R})$ onto $\mathscr{S}'(\mathbb{R})$. Using that $\mathscr{H}_N(\theta)$ is finite dimensionnal we get that $\hat{R}_{N,\theta}(F)$ is a one-to-one linear map from $\mathscr{H}_N(\theta)$ in $\mathscr{H}_N(\theta')$.

We can easily check that $V_{N,\theta}\Sigma_N^{(\pi_F(\theta))} = \Sigma_N^{(\theta)}$ for every $\theta \in [0, 2\pi[^2$. So using that π_F is an aera preverving transformation we get, for every $\psi, \eta \in \mathscr{S}(\mathbb{R})$,

$$\langle \eta, \hat{R}(F)\psi \rangle_{L^2(\mathbb{R})} = \frac{1}{4\pi^2} \int_{[0,2\pi[^2} \langle \eta(\theta), V_{N,\theta}\hat{R}_{N,\theta}\psi(\theta)\rangle_{\mathscr{H}_N(\theta)} d\theta.$$

So the proposition is proved using standard properties of direct integral decompositions for operators. □

One has the following results concerning the interesting case $\theta' = \theta$. The proofs are left to the reader or see [38, 39, 137].

Proposition 71 *Consider* $F \in SL(2, \mathbb{Z})$ *with* $|\mathrm{Tr} F| > 2$. *Then* $\forall N \in \mathbb{N}^*$ *there exists* $\theta \in [0, 2\pi)^2$ *so that*

$$\hat{R}(F)\mathscr{H}_N(\theta) \subseteq \mathscr{H}_N(\theta)$$

where θ *can be chosen independent of N if and only if F is of the form*

$$\begin{pmatrix} \text{even} & \text{odd} \\ \text{odd} & \text{even} \end{pmatrix}, \quad or \quad \begin{pmatrix} \text{odd} & \text{even} \\ \text{even} & \text{odd} \end{pmatrix}$$

The case $|\mathrm{Tr} F| = 3$ *is the only case where the choice of* θ *is unique with* $\theta = (\pi, \pi)$ *for N odd and* $\theta = (0, 0)$ *for N even.*

Moreover in the case

$$F = \begin{pmatrix} \text{even} & \text{odd} \\ \text{odd} & \text{even} \end{pmatrix}$$

the value $\theta = (0, 0)$ *is a solution of the fixed point equation* $\theta = \pi_F(\theta)$.

Remark 40 It has been shown in [78] that for $F \in SL(2, \mathbb{Z})$ of the following form

$$F = \begin{pmatrix} 2g & 1 \\ 2g^2 - 1 & 2g \end{pmatrix}$$

the operator $\hat{R}(F)$ has matrix elements in the basis $\left\{e_j^{(0)}\right\}_{j=0}^{N-1}$ of the form

$$\hat{R}(F)_{j,k} = \frac{C_N}{\sqrt{N}} \exp\left(\frac{2i\pi}{N}(gj^2 - jk + gk^2)\right)$$

where $|C_N| = 1$, so that it is represented as a unitary Hadamard matrix. We don't know at present whether this property is shared by more general maps F.

6.6.2 The Egorov Theorem is Exact

As in the continuous case the Egorov theorem is exact since $\hat{R}(F)$ is the metaplectic representation of the linear symplectic map F:

Theorem 39 *For any $H \in \mathscr{C}^\infty(\mathbb{T}^2)$ one has*

$$\hat{R}_{N,\theta}(F)^* \mathrm{Op}^W_{\hbar,\theta}(H) \hat{R}_{N,\theta}(F) = \mathrm{Op}^W_{\hbar,\theta}(H \circ F) \tag{6.53}$$

Proof By denoting $\hat{T}_\theta(z)$ the restriction of $\hat{T}(z)$ to $\mathscr{H}_N(\theta)$ one has

$$\mathrm{Op}^W_{\hbar,\theta}(H) = \sum_{m,n\in\mathbb{Z}} H_{m,n} \hat{T}_\theta\left(\frac{m}{N}, \frac{n}{N}\right)$$

So

$$\hat{R}_{N,\theta}(F)^* \mathrm{Op}^W_{\hbar,\theta}(H) \hat{R}_{N,\theta}(F) = \sum_{m,n} H_{m,n} \hat{R}_{N,\theta}(F)^* \hat{T}_\theta\left(\frac{m}{N}, \frac{n}{N}\right) \hat{R}_{N,\theta}(F).$$

But we know that

$$\hat{R}_{N,\theta}(F)^* \hat{T}_\theta(z) \hat{R}_{N,\theta}(F) = \hat{T}_\theta(F^{-1}z), \ \forall z = (m/N, n/N)$$

We do the change of variables

$$\begin{pmatrix} m' \\ n' \end{pmatrix} = F^{-1} \begin{pmatrix} m \\ n \end{pmatrix}$$

Then we get

$$\hat{R}_{N,\theta}(F)^* \mathrm{Op}^W_{\hbar,\theta}(H) \hat{R}_{N,\theta}(F) = \sum_{m',n'\in\mathbb{Z}} (H \circ F)_{m',n'} \hat{T}_\theta\left(\frac{m'}{N}, \frac{n'}{N}\right) = \mathrm{Op}^W_{\hbar,\theta}(H \circ F)$$

This completes the proof. □

6.6.3 Propagation of Coherent States

As in the continuous case the quantum propagation of coherent states is explicit and "imitates" the classical evolution of phase space points. Here the phase space is the 2-torus and $\forall z \in \mathbb{T}^2$ the time evolution of the point $z = (q, p)$ is given by

$$z' = \begin{pmatrix} q' \\ p' \end{pmatrix} = F \begin{pmatrix} q \\ p \end{pmatrix}, \quad \text{mod} \begin{pmatrix} 1 \\ 1 \end{pmatrix}.$$

We shall use "generalized coherent states" which are actually "squeezed states". Take $\gamma \in \mathbb{C}$ with $\Im\gamma > 0$. The normalized Gaussian $\varphi_\gamma \in L^2(\mathbb{R})$ were defined in 6.21.

$$\varphi_\gamma(x) := \left(\frac{\Im\gamma}{\pi\hbar}\right)^{1/4} \exp\left(\frac{i\gamma x^2}{2\hbar}\right).$$

Then the generalized coherent states in $L^2(\mathbb{R})$ are

$$\varphi_{\gamma,z} := \hat{T}(z)\varphi^\gamma. \tag{6.54}$$

The generalized coherent states on the 2-torus are as above

$$\varphi_{\gamma,z}^{(\theta)} = \Sigma_N^{(\theta)} \varphi_{\gamma,z} \in \mathcal{H}_N(\theta).$$

We take such coherent state as initial state and apply to it the quantum evolution operator $\hat{R}_\theta(F)$. One has the following result:

Proposition 72 *Let $F \in SL(2, \mathbb{Z})$ be given by*

$$F = \begin{pmatrix} a & b \\ c & d \end{pmatrix}.$$

If $\pi_F(\theta) = \theta$, then

$$\hat{R}_{N,\theta}(F)\varphi_{\gamma,z}^\theta = \left(\frac{|b\gamma + a|}{b\gamma + a}\right)^{1/2} \varphi_{F\cdot\gamma,Fz}^\theta$$

where $F \cdot \gamma = \frac{d\gamma + c}{b\gamma + a}$.

Proof We know that $\Sigma_N^{(\theta)}\hat{R}(F) = \hat{R}(F)\Sigma_N^{(\theta)}$. Let $z = (q, p) \in \mathbb{T}^2$. Then we have

$$\hat{R}_{N,\theta}(F)\varphi_{\gamma,z}^\theta = \hat{R}_{N,\theta}(F)\Sigma_N^{(\theta)}\varphi_{\gamma,z} = \Sigma_N^{(\theta)}\hat{R}(F)\varphi_{\gamma,z}.$$

The result follows from the propagation of coherent states by metaplectic transformations in the plane (see Chap. 3). □

6.7 Equipartition of the Eigenfunctions of Quantized Ergodic Maps on the 2-Torus

One of the simplest trace of the ergodicity of a map F on $\mathbb{T}^2$ in the quantum world is the equipartition of the eigenfunctions of $\hat{R}_{N,\theta}(F)$ in the classical limit $N \to \infty$. It has been established in the literature in different contexts: for the geodesic flow on a compact Riemannian manifold it was proven by [56, 232, 267]. For Hamiltonian flows in $\mathbb{R}^n$ it was established in [140], and for smooth convex ergodic billiards in [111]. For the case of the d-torus this problem has been investigated in [37]. Here we restrict ourselves on the case of the 2-torus.

Theorem 40 *(Quantum ergodicity) Let F be an ergodic area preserving map on $\mathbb{T}^2$, and $\hat{R}_{N,\theta}(F) \in \mathscr{U}(\mathscr{H}_\theta)$ its quantization where $\theta = \pi_F(\theta)$. Denote by $\left\{\phi_j^N\right\}_{j=1,\dots,N}$ the eigenfunctions of $\hat{R}_{N,\theta}(F)$. Then there exists $E(N) \subset \{1, \dots, N\}$ satisfying*

$$\lim_{N\to\infty} \frac{\#E(N)}{N} = 1$$

such that $\forall A \in \mathscr{C}^\infty(\mathbb{T}^2)$ and all maps $j : N \in \mathbb{N} \longmapsto j(N) \in E(N)$ we have:

$$\boxed{\lim_{N\to\infty} \langle \phi_{j(N)}^N, \mathrm{Op}_{\hbar,\theta}^W(A)\phi_{j(N)}^N\rangle = \int_{\mathbb{T}^2} A(z)dz}, \tag{6.55}$$

$$\boxed{\lim_{N\to\infty} \langle \phi_{j(N)}^N, \mathrm{Op}_{\hbar,\theta}^{AW}(A)\phi_{j(N)}^N\rangle = \int_{\mathbb{T}^2} A(z)dz} \tag{6.56}$$

uniformly with respect to the map $j(N)$.

Remark 41 This Theorem says that the Wigner distribution and Husimi distribution (when divided by N) converge in the sense of distributions to the Liouville distribution along subsequences of density one.

We begin with a lemma:

Lemma 38 *Let us introduce the Radon probability measures μ_j^N, $\bar{\mu}_N$ as follows:*

$$\mu_j^N(A) = \langle \phi_j^N, \mathrm{Op}_{\hbar,\theta}^{AW}(A)\phi_j^N\rangle, \quad \bar{\mu}_N(A) = \frac{1}{N}\sum_{j=1}^N \mu_j^N(A)$$

These measures are F-invariant. Because ϕ_j^N are eigenstates for $\hat{R}_{N,\theta}(F)$.

One has $\forall A \in \mathscr{C}^\infty(\mathbb{T}^2)$:

$$\lim_{N\to\infty} \bar{\mu}_N(A) = \mu(A),$$

where μ is the Liouville measure on the 2-torus $\mathcal{T}^2$.

Proof We get that the measures μ_j^N are F-invariant, modulo $O(N^{-1})$, using Egorov theorem and that ϕ_j^N are eigenstates for $\hat{R}_{N,\theta}(F)$.

Clearly we have

$$\bar{\mu}_N(A) = \frac{1}{N}\mathrm{Tr}(\mathrm{Op}_{\hbar,\theta}^{AW}(A))$$

So we have

$$|\bar{\mu}_N(A) - \mu(A)| = \left|\frac{1}{N}\mathrm{Tr}(\mathrm{Op}_{\hbar,\theta}^{AW}(A)) - \mu(A)\right|$$

$$\leq \|\mathrm{Op}_{\hbar,\theta}^{AW}(A) - \mathrm{Op}_{\hbar,\theta}^{W}(A)\|_{\mathcal{L}(\mathcal{H}_\theta)} + \left|\frac{1}{N}\mathrm{Tr}(\mathrm{Op}_{\hbar,\theta}^{W}(A)) - \mu(A)\right|$$

We deduce the result using Propositions 64 and 68. □

Remark 42 The Lemma is still true for the Schwartz distributions $\nu_j^N(A) = \langle \phi_j^N, \mathrm{Op}_{\hbar,\theta}^W(A)\phi_j^N\rangle$ and $\bar{\nu}_N(A) = \frac{1}{N}\sum_{j=1}^N \nu_j^N(A)$. The ν_j^N are exactly F-invariant.

Let us prove now

Proposition 73 *For every $A \in \mathscr{C}^\infty(\mathbb{T}^2)$ we have*

$$\lim_{N\to+\infty} \frac{1}{N}\sum_{0\leq j\leq N} |\mu_j^N(A) - \mu(A)|^2 = 0 \tag{6.57}$$

Proof We can replace A by $A - \mu(A)$ and assume that $\mu(A) = 0$.

Define for $n \in \mathbb{N}^*$ the "time-average" of A:

$$A_n = \frac{1}{n}\sum_{k=1}^{k=n} A \circ F^k.$$

Using the Remark after Lemma 38, we can replace μ_j^N by ν_j^N.

We have $\nu_j(A) = \nu_j(A_n)$ for every $n \geq 1$. So we get using the Cauchy-Schwarz inequality and F invariance of ν_j^N,

$$\begin{aligned}|\nu_j^N(A)|^2 = |\langle \mathrm{Op}_\hbar^W(A_n)\phi_j^N, \phi_j^N\rangle|^2 &\leq \|\mathrm{Op}_\hbar^W(A_n)\phi_j^N\|^2 \\ &\leq \langle \mathrm{Op}_\hbar^W(A_n)^*\mathrm{Op}_\hbar^W(A_n)\phi_j^N, \phi_j^N\rangle \end{aligned} \tag{6.58}$$

But from the composition rule for $\hbar$ Weyl quantization (Chap. 2) we have

$$\mathrm{Op}_\hbar^W(A_n)^*\mathrm{Op}_\hbar^W(A_n) = \mathrm{Op}_\hbar^W(|A_n|^2) + \mathscr{O}\left(\frac{1}{N}\right) \tag{6.59}$$

For every n, consider the limit $N \to +\infty$. Using Lemma 38 we get

$$\limsup_{N\to+\infty} |\mu_j^N(A)|^2 = \limsup_{N\to+\infty} |\nu_j^N(A)|^2 \le \int_{\mathscr{T}^2} |A_n|^2 d\mu$$

Using the ergodicity assumption and the Lebesgue dominated convergence theorem we have

$$\lim_{n\to+\infty} \int_{\mathbb{T}^2} |A_n|^2 d\mu = 0.$$

The limit (6.57) follows if $\mu(A) = 0$. □

Now, the Bienaymé-Tchebichev inequality gives the following result according which "almost-all" eigenstates are equidistributed on the torus.

Proposition 74 *For any $H \in \mathscr{C}^\infty(\mathbb{T}^2)$ and $\forall \varepsilon > 0$*

$$\lim_{N\to\infty} \frac{\#\left\{j :\, |\mu_j^N(H) - \mu(H)| < \varepsilon\right\}}{N} = 1.$$

Along the same lines as in [140] one can conclude for the existence of a H-independent set $E(N)$ such that the theorem holds true.

Remark 43 A natural question is "is the quantum ergodic theorem true with $E(N) = N$" (unique quantum ergodicity)? The answer is negative. In [74] the following result is proved. Let $\mathscr{C} = \{\tau_1, \dots, \tau_K\}$, K periodic orbits for F. Consider the probability measure $\mu_{\mathscr{C},\alpha} = \sum_{1\le j\le K} \alpha_j \mu_{\tau_j}$, where $\alpha_j \in [0, 1]$ and $\sum_{1\le j\le K} \alpha_j = 1$. Then there exists a sequence $N_k \to +\infty$ such that

$$\lim_{k\to\infty} \langle \phi_{N_k}^N, \mathrm{Op}_{\hbar,\theta}^W(A)\phi_{N_k}^N\rangle = \frac{1}{2}\int_{\mathbb{T}^2} A(z)dz + \frac{1}{2}\mu_{\mathscr{C},\alpha}(A). \tag{6.60}$$

This result shows that some eigenstates can concentrate along periodic orbits, this phenomenon is named scarring.

6.8 Spectral Analysis of Hamiltonian Perturbations

The previous results can be extended to some perturbations of automorphisms of the torus $\mathbb{T}^2$.

Let H be a real periodic Hamiltonian, $H \in C^\infty(\mathbb{T}^2)$ and $F \in SL(2, \mathbb{Z})$. Let $\theta \in [0, 2\pi[^2$ be such that $\theta = \pi_F(\theta)$. We consider here the following unitary operator in $\mathscr{H}_N(\theta)$, where $2\pi\hbar N = 1$,

$$U_\varepsilon = \exp\left(-i\frac{\varepsilon}{\hbar}\mathrm{Op}^w_{\hbar,\theta}(H)\right)\hat{R}_{N,\theta}(F).$$

We shall see that if F is hyperbolic and ε small enough the quantum ergodic theorem is still true.

To prepare the proof we begin by some useful properties concerning the propagators $V(t) = \mathrm{e}^{-i\frac{t}{\hbar}\hat{H}}$, $V_{N,\theta}(t) = \mathrm{e}^{-i\frac{t}{\hbar}\hat{H}_{N,\theta}}$.

Let us introduce the following Hilbert spaces $\mathscr{K}_s = \mathrm{Dom}(\hat{H}_{osc}+1)^{s/2}$, where $s \geq 0$, with the norm $\|\psi\|_s^2 = \|(\hat{H}_{osc}+1)^{s/2}\psi\|^2$ ($\hat{H}_{osc} = -\hbar^2\frac{d^2}{dx^2}+x^2$).

Lemma 39 *For every $s \geq 0$ and every $T > 0$, there exists $C_{s,T} > 0$ such that*

$$\|V(t)\psi\|_s \leq C_{s,T}\|\psi\|_s,\ \forall\psi \in \mathscr{K}_s, \forall t \in [-T,T], \forall\hbar \in]0,2\pi].$$

Proof It is sufficient to assume that $s \in \mathbb{N}$ (using complex interpolation). For $s = 0$ we know that $V(t)$ is unitary. Let us denote $\Lambda = (\hat{H}_{osc}+1)^{1/2}$. Let us assume $s = 1$. It is enough to prove that $\Lambda V(t)\Lambda^{-1}$ is bounded from $L^2(\mathbb{R})$ into $L^2(\mathbb{R})$.

We have

$$\frac{\hbar}{i}\frac{d}{dt}V(-t)\Lambda V(t) = V(-t)[\hat{H},\Lambda]V(t).$$

Using the semi-classical calculus (Chap. 2), and that H is periodic, we know that $\frac{i}{\hbar}[\hat{H},\Lambda]$ is bounded on $L^2(\mathbb{R})$. So the lemma is proved for $s = 1$.

Now we will prove the result for every $s \in \mathbb{N}$ by induction. Assume the lemma is proved for $k \leq s-1$. Compute

$$\frac{\hbar}{i}\frac{d}{dt}V(-t)\Lambda^s V(t) = V(t)[\hat{H},\Lambda^s]V(t).$$

But $\frac{i}{\hbar}[\hat{H},\Lambda^s]$ is an $\hbar$ pseudodifferential operator of order $s-1$ for the weight $\mu(x,\xi) = (1+x^2+\xi^2)^{1/2}$. In particular the operator $\frac{i}{\hbar}[\hat{H},\Lambda^s]\Lambda^{1-s}$ is bounded on $L^2(\mathbb{R})$. So we get the result for s using the induction assumption. □

The following result will be useful to transform properties from the space $L^2(\mathbb{R})$ to the spaces $\mathscr{H}_N(\theta)$.

Let $\psi \in \mathscr{S}(\mathbb{R})$ and for $\theta \in [0,2\pi[^2$ be such that $\tilde{\psi}(\theta) = (\tilde{\psi}(\theta,0),\ldots,\tilde{\psi}(\theta,N-1)) \in \mathbb{C}^N$ (coefficient of $\psi(\theta)$ in the canonical basis of $\mathscr{H}_N(\theta)$). Let us denote $H^s_N([0,2\pi[^2)$ the periodic Sobolev space of order $s \geq 0$ of functions from $[0,2\pi[^2$ into $\mathbb{C}^N$ Its norm is denoted $\|\cdot\|_{N,s}$.

Lemma 40 *For every $s \geq 0$ there exists C_s such that*

$$\|\tilde{\psi}\|_{N,s} \leq C_s\|\psi\|_s,\ \forall\psi \in \mathscr{S}(\mathbb{R}). \tag{6.61}$$

In particular we have the following pointwise estimate: for every $s > 1$ there exists C_s such that

$$|\tilde{\psi}(\theta)| \leq C_s\|\psi\|_s,\ \forall\psi \in \mathscr{K}_s. \tag{6.62}$$

Proof Let us recall that, for every $\psi \in \mathscr{S}(\mathbb{R})$, $\theta = (\theta_1, \theta_2) \in]0, 2\pi]^2$,

$$\tilde{\psi}(\theta, j) = N^{-1/2} \sum_{\ell \in \mathbb{Z}} e^{i\ell\theta_1} \psi(q_j - \ell),\ q_j = \frac{j}{N} + \frac{\theta_2}{2\pi N}.$$

So we have

$$\partial_{\theta_1} \tilde{\psi}(\theta, j) = N^{-1/2} \sum_{\ell \in \mathbb{Z}} e^{i\ell\theta_1} i(\ell - q_j)\psi(q_j - \ell) + iN^{-1/2} q_j \tilde{\psi}(\theta, j), \quad (6.63)$$

$$\partial_{\theta_2} \tilde{\psi}(\theta, j) = N^{-1/2} \sum_{\ell \in \mathbb{Z}} e^{i\ell\theta_1} i\hbar \frac{d}{dx} \psi(q_j - \ell). \quad (6.64)$$

Reasonning by induction on $|m| = m_1 + m_2$, $m = (m_1, m_2)$, we get easily

$$\int_{[0,2\pi[^2} |\partial_\theta^m \tilde{\psi}(\theta, j)|^2 d\theta \le C_m \sum_{k+\ell \le |m|} \int_{\mathbb{R}} (|(\hbar\partial_x)^k \psi(x)|^2 + |x^\ell \psi(x)|^2) dx. \quad (6.65)$$

So estimate (6.61) follows. Estimate (6.62) is a consequence of Sobolev estimate in dimension 2. □

Let us now consider the propagation of coherent states $\varphi_{\gamma,z}^{(\theta)}$ under the dynamics $V_{N,\theta}(t)$ in $\mathscr{H}_N(\theta)$. We shall prove that estimates can be obtained from the corresponding evolution in $L^2(\mathbb{R})$ (see Chap. 4), using the two previous lemmas.

Recall these results. We have checked approximate solutions for the Schrödinger equation:

$$i\hbar\partial_t \psi_t = \hat{H}\psi_t,\ \psi_0 = \varphi_{\gamma,z},\ \Im\gamma > 0.$$

We have found $\psi_{z,t}^{(M)}$ such that

$$i\hbar\partial_t \psi_{z,t}^{(M)} = \hat{H}\psi_{z,t}^{(M)} + \hbar^{(N+3)/2} R_{z,t}^{(M)},\ \psi_{z,0}^{(M)} = \varphi_{\gamma,z} \quad (6.66)$$

where, for every $s \ge 0$, $\|R_{z,t}^{(M)}\|_{\mathscr{K}_s} = O(1)$ for $\hbar \to 0$.

$\psi_{z,t}^{(M)}$ has the following expression

$$\psi_{z,t}^M(x) = e^{i\frac{\delta_t}{\hbar}} \sum_{0 \le j \le M} \hbar^{j/2} \pi_j(t, \frac{x - q_t}{\sqrt{\hbar}}) \varphi_{z_t}^{\Gamma_t}(x), \quad (6.67)$$

where $z_t = (q_t, p_t)$ is the classical path in the phase space $\mathbb{R}^2$ such that $z_0 = z$ satisfying

$$\begin{cases} \dot{q}_t = \frac{\partial H}{\partial p}(t, q_t, p_t) \\ \dot{p}_t = -\frac{\partial H}{\partial q}(t, q_t, p_t),\ q_0 = q,\ p_0 = p \end{cases} \quad (6.68)$$

and

$$\varphi_{z_t}^{\Gamma_t} = \hat{T}(z_t)\varphi^{\Gamma_t}. \tag{6.69}$$

φ^{Γ_t} is the Gaussian state:

$$\varphi^{\Gamma_t}(x) = (\pi\hbar)^{-d/4}a(t)\exp\left(\frac{i}{2\hbar}\Gamma_t x.x\right), \tag{6.70}$$

Γ_t is a complex number with positive non degenerate imaginary part, δ_t is a real function, $a(t)$ is a complex function, $\pi_j(t,x)$ is a polynomial in x (of degree $\leq 3j$) with time dependent coefficients.

More precisely Γ_t is given by the Jacobi stability matrix of the Hamiltonian flow $z \mapsto \Phi_H^t z := z_t$. If we denote

$$A_t = \frac{\partial q_t}{\partial q}, \; B_t = \frac{\partial p_t}{\partial q}, \; C_t = \frac{\partial q_t}{\partial p}, \; D_t = \frac{\partial p_t}{\partial p}, \tag{6.71}$$

then we have

$$\Gamma_t = (C_t + \gamma D_t)(A_t + \gamma B_t)^{-1}, \;\; \Gamma_0 = \gamma, \tag{6.72}$$

$$\delta_t(z) = \int_0^t (p_s q_s - H(z_s))ds - \frac{q_t p_t - q_0 p_0}{2}, \tag{6.73}$$

$$a(t) = [\det(A_t + \gamma B_t)]^{-1/2}, \tag{6.74}$$

where the complex square root is computed by continuity from $t = 0$.

Using the two lemmas and the Duhamel formula, we get, (using the notation $\psi^{(\theta)} = \Sigma_N^{(\theta)}\psi$),

Proposition 75 *For every $m \geq 0$ and every $\theta \in [0, 2\pi[^2$ we have*

$$\|V_{N,\theta}\varphi_{\gamma,z}^{(\theta)} - \psi_{z,t}^{(m,\theta)}\|_{\mathscr{H}_N(\theta)} = O(N^{-(m+1)/2}). \tag{6.75}$$

In particular we have $\psi_{z,t}^{(0,\theta)} = \varphi_{\Gamma_t,z_t}^{(\theta)}$ with the notation of Sect. 6.4.

Let us come back to Hamiltonian perturbations of hyperbolic automorphism F. Let us denote $F_H^\varepsilon = \Phi_H^\varepsilon \circ F$. F_H^ε is symplectic on $\mathscr{T}^2$ (it preserves the area). By the C^1 stability of Anosov dynamical systems, for ε small enough, F_H^ε is Anosov. The quantum analogue of F_H^ε is the unitary operator:

$$\hat{R}_{N,\theta,\varepsilon}(F) = \exp\left(-i\frac{\varepsilon}{\hbar}\mathrm{Op}_{\hbar,\theta}^w(H)\right)\hat{R}_{N,\theta}(F)$$

We have the following semi-classical correspondence

Proposition 76 *The following estimates hold true uniformly in $z \in \mathscr{T}^2$ and $\varepsilon \in [0,1]$*

$$\hat{R}_{N,\theta,\varepsilon}(F)\varphi_{\gamma,z}^{(\theta)} = C_\varepsilon \varphi_{\gamma_\varepsilon, F_H^\varepsilon(z)}^{\theta} \tag{6.76}$$

where $\gamma_\varepsilon = \frac{C_\varepsilon + F\cdot\gamma D_\varepsilon}{A_\varepsilon + F\cdot\gamma B_\varepsilon}$ *and*

$$C_\varepsilon = \mathrm{e}^{i\delta_\varepsilon(F(z))/\hbar}\left(\frac{|\gamma b + a|}{\gamma b + a}\right)^{1/2}\left(\frac{|A_\varepsilon + F\cdot\gamma B_\varepsilon|}{A_\varepsilon + F\cdot\gamma B_\varepsilon}\right)^{1/2}.$$

In particular C_ε *is a complex number of modulus one.*

Proof This is a direct consequence of propagation of coherent states (Sect. 4.3). □

Now we shall prove some spectral properties for $\hat{R}_{N,\theta,\varepsilon}(F)$ for ε small enough. Let us denote $\eta_j^N, 0 \le j \le N-1$ the eigenvalues of $\hat{R}_{N,\theta,\varepsilon}(F)$, so that $\hat{R}_{N,\theta,\varepsilon}(F)\psi_j^N = \eta_j^N\psi_j^N$ where $\{\psi_j^N\}_{0\le j\le N-1}$ is an orthonormal basis of $\mathscr{H}_N(\theta)$ and $\eta_j^N \in \mathbb{S}^1$, the unit circle of the complex plane.

Theorem 41 *For* $\varepsilon > 0$ *small enough, when* $N \to +\infty$, *the eigenvalues* $\{\eta_j^N\}_{0\le j\le N-1}$ *are uniformly distributed on* $\mathbb{S}^1$ *i.e. for every interval* I *on* $\mathbb{S}^1$) *we have*

$$\lim_{N\to+\infty}\frac{\sharp\{j;\, \eta_j^N \in I\}}{N} = \mu(I) \tag{6.77}$$

were μ *is the Lebesgue probability measure on* $\mathbb{S}^1$.

Proof In a first step we will prove that for every $f \in C^1(\mathbb{S}^1)$ we have

$$\lim_{N\to+\infty}\frac{1}{N}\mathrm{Tr}\left[f\left(\hat{R}_{N,\theta,\varepsilon}(F)\right)\right] = \int_{\mathbb{S}^1} f(x)d\mu(x). \tag{6.78}$$

Using Fourier decomposition of f it is enough to prove (6.78) for $f(z) = z^k, k \in \mathbb{Z}$. Hence we have to prove that for every $k \neq 0$,

$$\lim_{N\to+\infty}\frac{1}{N}\mathrm{Tr}(\left(\hat{R}_{N,\theta,\varepsilon}(F)\right)^k) = 0. \tag{6.79}$$

We assume $k \ge 1$ (for $k \le -1$ there are obvious modifications).

Using that the coherent states are an overcomplete system in $\mathscr{H}_N(\theta)$, we have

$$\mathrm{Tr}(\left(\hat{R}_{N,\theta,\varepsilon}(F)\right)^k) = \int_{\mathcal{T}^2}\langle\varphi_z^\theta, \left(\hat{R}_{N,\theta,\varepsilon}(F)\right)^k\varphi_z^\theta\rangle.$$

Using the propagation of coherent states we get

$$|\langle\varphi_z^\theta, \left(\hat{R}_{N,\theta,\varepsilon}(F)\right)^k\varphi_z^\theta\rangle| = |\langle\varphi_z^\theta, \varphi_{\gamma(\varepsilon,k),(F_H^\varepsilon)^k}\rangle| + O(\varepsilon N^{-1/2})$$

and using Proposition 61 we have

$$|\langle\varphi_z^\theta, \left(\hat{R}_{N,\theta,\varepsilon}(F)\right)^k\varphi_z^\theta\rangle| \le C(\varepsilon,k)\sqrt{N}\mathrm{e}^{-c(\varepsilon,k)d((F_H^\varepsilon)^k z,z)^2 N} + O(\varepsilon N^{-1/2}).$$

But we know that for ε small enough F_H^ε is Anosov so it is ergodic and its periodic set has zero measure. So for every $\delta > 0$ we have $\mu\{z \in \mathscr{T}^2,\ d((F_H^\varepsilon)^k z, z) \geq \delta\} = 0$. Using that $O(\varepsilon N^{-1/2})$ is uniform in $z \in \mathscr{T}^2$ we get (6.79) hence (6.78).

Now we get easily the Theorem considering $f_\pm \in C^1(\mathbb{S}^1)$ such that $f_- \leq 1\!\!1_I \leq f_+$ and $\int_{\mathbb{S}^1}(f_+ - f_-)d\mu < \delta$ with $\delta \to 0$. □

Chapter 7
Spin Coherent States

Abstract In this chapter we consider the unit sphere $\mathbb{S}^2$ of the Euclidean space $\mathbb{R}^3$ with its canonical symplectic structure is a phase space. Then coherent states are labelled by points on $\mathbb{S}^2$ and allow us to build a quantization of the two-sphere $\mathbb{S}^2$. They are defined in each finite dimensional space of an irreducible unitary representation of the symmetry rotation group $SO(3)$ (or its covering $SU(2)$) of $\mathbb{S}^2$ and give a semiclassical interpretation for the spin. As an application we state the Berezin-Lieb inequalities and compute the thermodynamic limit for large spin systems.

7.1 Introduction

Up to now we have considered Gaussian coherent states and their relationship with the Heisenberg group, the symplectic group and the harmonic oscillator. These states are used to describe field coherent states (Glauber [118]). For the description of assembly of two-levels atom, physicists have introduced what they have called "atomic coherent states" [6]. These states are defined in Hilbert space irreducible representations of some symmetry Lie group.

As we shall see later it is possible to associate coherent states to any irreducible representation of a Lie group. This general construction is due to Perelomov [209]. In this chapter we consider the rotation group $SO(3)$ of the Euclidean space $\mathbb{R}^3$ and its companion $SU(2)$. Irreducible representations of these groups are related with the spin of particles as it was discovered by Pauli [207].

In this chapter (and in the rest of the book) we shall use freely some basic notions concerning Lie groups, Lie algebra and their representations. We have recalled most of them in an appendix.

D. Robert and M. Combescure, *Coherent States and Applications in Mathematical Physics*, Theoretical and Mathematical Physics,
https://doi.org/10.1007/978-3-030-70845-0_7

7.2 The Groups $SO(3)$ and $SU(2)$

Let us consider the Euclidean space $\mathbb{R}^3$ equipped with the usual scalar product, $x = (x_1, x_2, x_3)$, $y = (y_1, y_2, y_3)$, $x \cdot y = x_1 y_1 + x_2 y_2 + x_3 y_3$ and the Euclidean norm $\|x\| = (x_1^2 + x_2^2 + x_3^2)^{1/2}$. The isometry group of $\mathbb{R}^3$ is denoted $O(3)$.[1] $A \in O(3)$ means that $\|Ax\| = \|x\|$ for every $x \in \mathbb{R}^3$. $SO(3)$ is the subgroup of direct isometries i.e. $A \in SO(3)$ means that $A \in O(3)$ and $\det A = 1$. It is well known that $A \in SO(3)$ is a rotation characterized by a unitary vector $v \in \mathbb{R}^3$ (rotation axis) and an angle $\theta \in [0, 2\pi[$. More precisely, v is an 1-eigenvector for A, $Av = v$ and A is a rotation of angle θ in the plane orthogonal to v. So we have the following formula, for every $x \in \mathbb{R}^3$,

$$Ax := R(\theta, v)x = (1 - \cos\theta)(v \cdot x)v + (\cos\theta)x + \sin\theta(v \wedge x). \tag{7.1}$$

Recall that the wedge product $v \wedge x$ is the unique vector in $\mathbb{R}^3$ such that

$$\det[v, x, w] = (v \wedge x) \cdot w, \quad \forall w \in \mathbb{R}^3.$$

It is easy to compute the Lie algebra $\mathfrak{so}(3)$ of $SO(3)$,

$$\mathfrak{so}(3) = \{A \in \mathrm{Mat}(3, \mathbb{R}),\ A^T + A = 0\}$$

where A^T is the transposed matrix of A.

Considering rotations around vectors of the canonical basis $\{e_1, e_2, e_3\}$ of $\mathbb{R}^3$ we get a basis $\{E_1, E_2, E_3\}$ of $\mathfrak{so}(3)$ where

$$E_k = \frac{d}{d\theta} R(\theta, e_k)|_{\theta=0}$$

It satisfies the commutation relation

$$[E_k, E_\ell] = E_m \tag{7.2}$$

for every circular permutation (k, ℓ, m) of $(1, 2, 3)$.

It is well known that any rotation matrix is an exponential.

Proposition 77 *For every $v \in \mathbb{R}^3$, $\|v\| = 1$ and $\theta \in [0, 2\pi[$ we have*

$$R(\theta, v) = \mathrm{e}^{\theta M(v)} \tag{7.3}$$

where $M(v) = \sum_{1 \le k \le 3} v_j E_j$.

[1] An isometry in an Euclidean space is automatically linear so that $O(3)$ is a subgroup of $GL(\mathbb{R}^3)$.

Proof $\theta \mapsto R(\theta, v)$ and $\theta \mapsto e^{\theta M(v)}$ are one parameter groups, so it is enough to see that their derivatives at $\theta = 0$ are the same. This is true because we have $v \wedge x = M(v)x$. □

In the same way as $SO(2)$ (identified to the circle $\mathbb{S}^1$) $SO(3)$ is connected but not simply connected. To compute irreducible representations of $SO(3)$ it is convenient to consider a simply connected cover of $SO(3)$ which can be realized as the complex Lie group $SU(2)$. $SU(2)$ is the group of unitary 2×2 matrices A with complex coefficients, $A = \begin{pmatrix} a & b \\ -\bar{b} & \bar{a} \end{pmatrix}$, such that $|a|^2 + |b|^2 = 1$.

The Lie algebra $\mathfrak{su}(2)$ is the real vector space of dimension 3 of 2×2 complex antihermitian matrices of zero trace:

$$\mathfrak{su}(2) = \left\{X \in \mathfrak{gl}(2, \mathbb{C}) \mid X^* + X = 0,\ \mathrm{Tr}X = 0\right\}.$$

The 3 linearly independent matrices:

$$A_1 = \frac{1}{2}\begin{pmatrix} 0 & 1 \\ -1 & 0 \end{pmatrix}, \quad A_2 = \frac{1}{2}\begin{pmatrix} 0 & i \\ i & 0 \end{pmatrix}, \quad A_3 = \frac{1}{2}\begin{pmatrix} i & 0 \\ 0 & -i \end{pmatrix}$$

form a basis of $\mathfrak{su}(2)$ on $\mathbb{R}$ and satisfy the commutation relations

$$[A_k, A_\ell] = A_m, \tag{7.4}$$

provided k, l, m is a circular permutation of $1, 2, 3$.

Let us consider the adjoint representation of $SU(2)$. This representation is defined in the real vector space $\mathfrak{su}(2)$ by the formula

$$\rho_U(A) = UAU^{-1}, \quad U \in SU(2),\ A \in \mathfrak{su}(2). \tag{7.5}$$

In physics the spin is defined by considering the Pauli matrices, which are hermitian 2×2 matrices given by

$$\sigma_1 = \begin{pmatrix} 0 & 1 \\ 1 & 0 \end{pmatrix}, \quad \sigma_2 = \begin{pmatrix} 0 & -i \\ i & 0 \end{pmatrix}, \quad \sigma_3 = \begin{pmatrix} 1 & 0 \\ 0 & -1 \end{pmatrix} \tag{7.6}$$

which satisfy the commutation relations

$$[\sigma_k, \sigma_l] = 2i\varepsilon_{k,\ell,m}\sigma_m \tag{7.7}$$

which is equivalent to (7.4) with the correspondance $A_1 = \frac{i}{2}\sigma_2$, $A_2 = \frac{i}{2}\sigma_1$, $A_3 = \frac{i}{2}\sigma_3$. Noticing that $\{\sigma_1, \sigma_2, \sigma_3\}$ is an orthonormal basis for the 3 dimensional real linear space $H_{2,0}$ of Hermitian 2×2 matrices with trace 0, which will be identified with $\mathbb{R}^3$. The scalar product in $H_{2,0}$ is

$$\langle A, B\rangle = \frac{1}{2}\mathrm{Tr}(A^* B).$$

Let us denote R_U the 3×3 matrix of ρ_U in this basis. The following proposition gives the basic relationship between the groups $SO(3)$ and $SU(2)$.

Proposition 78 *For every $U \in SU(2)$ we have:*
(i) R_U has real coefficients.
(ii) R_U is an isometry in $H_{2,0}$ (the 3-dimensional linear space defined above).
(iii) $\det R_U = 1$.
(iv) The map $U \mapsto R_U$ is a surjective group morphism from $SU(2)$ onto $SO(3)$.
(v) The kernel of $U \mapsto R_U$ is $\ker R = \{\mathbb{1}, -\mathbb{1}\}$.

Proof (i) From $\langle R_U\sigma_k, \sigma_\ell\rangle = \frac{1}{2}\mathrm{Tr}(U\sigma_k U^{-1}\sigma_\ell)$ we get $\overline{\langle R_U\sigma_k, \sigma_\ell\rangle} = \langle R_U\sigma_k, \sigma_\ell\rangle$.
(ii) Using commutativity of trace we have $\langle R_U A, R_U B\rangle = \frac{1}{2}\mathrm{Tr}AA^* = \langle A, A\rangle$.
(iii) We have $\det R_U = \pm 1$ because R_U is an isometry. But $\det R_{\mathbb{1}} = 1$ and $SU(2)$ is connected so $\det R_U = 1$.
(iv) It is easy to see that R is a group morphism.
Let us consider the following generators of $SU(2)$.

$$U_1(\varphi) = \begin{pmatrix} \mathrm{e}^{-i\varphi/2} & 0 \\ 0 & \mathrm{e}^{i\varphi/2} \end{pmatrix}, \quad U_2(\theta) = \begin{pmatrix} \cos(\theta/2) & -\sin(\theta/2) \\ \sin(\theta/2) & \cos(\theta/2) \end{pmatrix}.$$

Let us remark that we have, for every $\varphi, \varphi', \theta$,

$$U_1(\varphi)U_2(\theta)U_1(\varphi') = \begin{pmatrix} \cos(\theta/2)\mathrm{e}^{-i/2(\varphi+\varphi')} & -\sin(\theta/2)\mathrm{e}^{i/2(\varphi'-\varphi)} \\ \sin(\theta/2)\mathrm{e}^{-i/2(\varphi'-\varphi)} & \cos(\theta/2)\mathrm{e}^{i/2(\varphi+\varphi')} \end{pmatrix}. \tag{7.8}$$

Then compute the image:

$$R_{U_1(\varphi)} = \begin{pmatrix} \cos\varphi & \sin\varphi & 0 \\ -\sin\varphi & \cos\varphi & 0 \\ 0 & 0 & 1 \end{pmatrix}, \quad R_{U_2(\theta)} = \begin{pmatrix} \cos\theta & 0 & -\sin\theta \\ 0 & 1 & 0 \\ \sin\theta & 0 & \cos\theta \end{pmatrix}$$

$R_{U_1(\varphi)}$ is the rotation of angle φ with axis e_3, $R_{U_2(\theta)}$ is the rotation of angle θ with axis e_2.

So, if $\varphi, \theta, \eta \in [0, 2\pi[$ we have

$$R_{U_1(\varphi)U_2(\theta)U_1(\eta)} = R_{U_1(\varphi)}R_{U_2(\theta)}R_{U_1(\eta)}$$

where (φ, θ, η) are the Euler angles of the rotation $R(\varphi, \theta, \eta) := R_{U_1(\varphi)}R_{U_2(\theta)}R_{U_1(\eta)}$. But any rotation can be defined with its Euler angles, so R is surjective.
(v) Let $U \in SU(2)$ be such that $UAU^{-1} = A$ for every $A \in H_{2,0}$. Then we easily get that $UAU^{-1} = A$ for every $A \in \mathrm{Mat}(2, \mathbb{C})$ hence $U = \lambda\mathbb{1}$ with $\lambda = \pm 1$. □

We shall see now that every irreducible representation of $SO(3)$ comes from an irreducible representation of its companion $SU(2)$.

Corollary 24 *ρ is an irreducible representation of $SO(3)$ if and only if ρ is an irreducible representation of $SU(2)$ such that*

$$\rho R_U = \rho R_{-U}, \quad \forall U \in SU(2).$$

Proof Let ρ be a represention of $SU(2)$ in a finite dimensional linear space E such that $\rho R_U = \rho R_{-U}$. Then we define a representation $\tilde{\rho}$ of $SO(3)$ in E by the equality $\tilde{\rho}(R_U) = \rho(U)$. Conversely every representation of $SO(3)$ comes from a representation of $SU(2)$ like above. In other words the following diagram is commutative

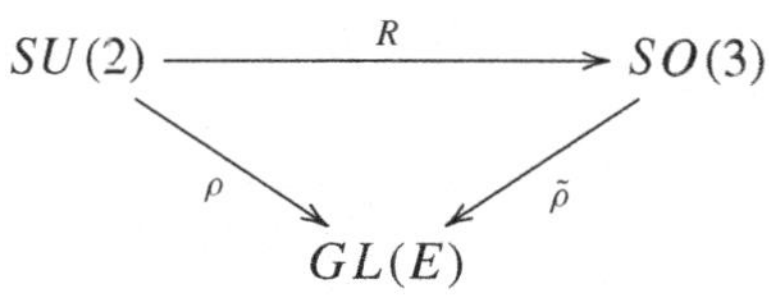

where $GL(E)$ is the group of invertible linear maps in E. □

Corollary 25 *The Lie algebras $\mathfrak{so}(3)$ and $\mathfrak{su}(2)$ are isomorph though the isomorphism $DR(\mathbb{1})$ (differential of R at the unit of the Lie group $SU(2)$). In particular we have $DR(\mathbb{1})A_k = E_k$, $k = 1, 2, 3$.*

Remark 44 The generators $\{L_k\}_{1 \le k \le 3}$ of the rotations with axis e_k give a basis of the Lie algebra $\mathfrak{so}(3)$ as a real linear space. Recall that $\mathbf{L} = (L_1, L_2, L_3)$ is the angular momentum.[2] L_k belongs to the complex Lie algebra $\mathfrak{so}(3) \oplus i\mathfrak{so}(3)$ and is sometimes denoted J_k, $J = ix \wedge \nabla_x$. For example $L_3 = i(x_2\partial_{x_1} - x_1\partial_{x_2})$. We have the commutation relations, for every circular permutation (k, ℓ, m) of $(1, 2, 3)$,

$$[L_k, L_\ell] = iL_m. \tag{7.9}$$

This basis can be identified with the matrix basis (iE_1, iE_2, iE_3) considered before.

7.3 The Irreducible Representations of $SU(2)$

The group $SU(2)$ is simply connected (it has the topology of the sphere $\mathbb{S}^3$), so we know that all its representations are determined by the representations of its Lie algebra $\mathfrak{so}(2)$ (see Appendix). Moreover, $SU(2)$ is a compact Lie group so all its irreducible representations are finite dimensional.

[2] Multiplication by i gives self-adjoint generators instead of anti-self-adjoint operators.

7.3.1 The Irreducible Representations of $\mathfrak{su}(2)$

We shall first consider the representation of the Lie algebra $\mathfrak{su}(2)$ and determine all its irreducible representations.

Recall that $\mathfrak{su}(2)$ is a real Lie algebra and it is more convenient to consider its complexification $\mathfrak{su}(2) + i\mathfrak{su}(2)$. But any matrix can be decomposed as a sum of an Hermitian and anti-Hermitian part, so we have

$$\mathfrak{sl}(2, \mathbb{C}) = \mathfrak{su}(2) + i\mathfrak{su}(2) \tag{7.10}$$

$\mathfrak{sl}(2, \mathbb{C})$ is the space of matrices $A = \begin{pmatrix} a & b \\ c & -a \end{pmatrix}$, $a, b, c \in \mathbb{C}$. It is the Lie algebra of the group of 2×2 complex matrices g such that $\det g = 1$. It results from (7.10) that irreducible representations of the real Lie algebra $\mathfrak{su}(2)$ are determined by irreducible representations of the complex Lie algebra $\mathfrak{sl}(2, \mathbb{C})$.

One considers $\mathfrak{sl}(2, \mathbb{C})$ endowed with the basis $\{H, K_+, K_-\}$ in which the commutation relations are

$$[H, K_\pm] = \pm K_\pm, \qquad [K_+, K_-] = 2H, \tag{7.11}$$

where

$$H = \frac{1}{2}\begin{pmatrix} 1 & 0 \\ 0 & -1 \end{pmatrix}, \quad K_+ = \begin{pmatrix} 0 & 1 \\ 0 & 0 \end{pmatrix}, \quad K_- = \begin{pmatrix} 0 & 0 \\ 1 & 0 \end{pmatrix}.$$

It is also convenient to introduce Hermitian generators (see footnote 2).

$$K_3 = H = \frac{\sigma_3}{2}, \; K_1 = \frac{K_+ + K_-}{2} = \frac{\sigma_1}{2}, \; K_2 = \frac{K_+ - K_-}{2i} = \frac{\sigma_2}{2}. \tag{7.12}$$

Let (E, R) be a (finite dimensional) irreducible representation of $\mathfrak{sl}(2, \mathbb{C})$. For convenience let us denote $\hat{H}$ the operator $R(H)$. $\hat{H}$ admits at least an eigenvalue λ and an eigenvector $v \neq 0$:

$$\hat{H}v = \lambda v.$$

From the commutation relations (7.11) we have

$$\hat{H}\hat{K}_+ v = (\hat{K}_+\hat{H} + \hat{K}_+)v = (\lambda + 1)\hat{K}_+ v.$$

$$\hat{H}\hat{K}_- v = (\hat{K}_-\hat{H} - \hat{K}_-)v = (\lambda - 1)\hat{K}_- v.$$

Since there must be only a finite number of distinct eigenvalues of $\hat{H}$, there exists an eigenvalue λ_0 of $\hat{H}$ and an eigenvector v_0 such that

$$\hat{H}v_0 = \lambda_0 v_0, \qquad \hat{K}_- v_0 = 0,$$

λ_0 is the smallest eigenvalue of $\hat{H}$. One defines then

$$v_k = (\hat{K}_+)^k v_0.$$

It must obey

$$\hat{H} v_k = (\lambda_0 + k) v_k.$$

One can show by induction on k that

$$\hat{K}_- v_k = c_k v_{k-1}, \text{ where } c_{k+1} = c_k - 2(\lambda_0 + k), \quad \forall k \in \mathbb{N}.$$

So we get

$$c_k = -k(2\lambda_0 + k - 1), \ k \in \mathbb{N}.$$

Since the vectors $v_k \neq 0$ are linearly independent and the vector space E is finite dimensional there exists an integer n such that

$$v_0 \neq 0, v_1 \neq 0, \ldots, v_n \neq 0, v_{n+1} = 0$$

Thus from $\hat{K}_- v_{n+1} = 0$ one deduces that $2\lambda_0 + n = 0$.

Then $\forall k \in \mathbb{N}$ one has

$$[\hat{K}_+, \hat{K}_-] v_k = 2\hat{H} v_k, \quad [\hat{H}, \hat{K}_\pm] v_k = \pm \hat{K}_\pm v_k. \tag{7.13}$$

One deduces that the vectors $\{v_k\}_{k=0}^n$ generate a subspace of E invariant by the representation R and since the representation we look for is irreducible, they generate the complex linear space E which is therefore of finite dimension $n+1$. The elements of the basis $\{v_k\}_{k=0}^n$ are called Dicke states in [6].

In conclusion we have found necessary conditions to get an irreducible representation $(E^{(n)}, R^{(n)})$ of dimension $n+1$ of $\mathfrak{sl}(2, \mathbb{C})$ with a basis $\{v_k\}_{k=0}^n$ of $E^{(n)}$ such that

$$\begin{aligned} R^{(n)}(H) v_k &= (k - \frac{n}{2}) v_k \\ R^{(n)}(K_+) v_k &= v_{k+1} \\ R^{(n)}(K_-) v_k &= k(n - k + 1) v_{k-1} \end{aligned} \tag{7.14}$$

for $0 \le k \le n$ and $v_{-1} = v_{n+1} = 0$.

We have to check that these conditions can be realized in some concrete linear space. Let $E^{(n)}$ the complex linear space generated by homogeneous polynomials of degree n in $(z_1, z_2) \in \mathbb{C}$ with the basis $v_k = \frac{z_1^k z_2^{n-k}}{(n-k)!}$ and

$$R^{(n)}(K_-) = z_2 \frac{\partial}{\partial z_1}, \ R^{(n)}(K_+) = z_1 \frac{\partial}{\partial z_2}, \ R^{(n)}(H) = \frac{1}{2}\left(z_1 \frac{\partial}{\partial z_1} - z_2 \frac{\partial}{\partial z_2}\right)$$

So we have proved:

Proposition 79 *Every irreducible representation of* $\mathfrak{sl}(2,\mathbb{C})$ *of finite dimension is equivalent to* $(E^{(n)}, R^{(n)})$ *for some* $n \in \mathbb{N}$.

In the physics literature one considers j such that $n = 2j$; j is thus either integer or half-integer and represents the angular momentum of the particles. We shall see later that representations of $SO(3)$ correspond to $j \in \mathbb{N}$ so n is even. j is the greatest eigenvalue of $R^{(n)}(H)$.

In quantum mechanics the representation $(E^{(2j)}, R^{(2j)})$ is denoted $(V^{(j)}, D^{(j)})$ and one considers the basis of $V^{(j)}$ indexed by the number m, $-j \le m \le j$, where m is integer if j is, and half-integer if j is.

States in $V^{(j)}$ represent spin states and the operators in $V^{(j)}$ are spin observables. So we introduce the notation $\hat{S}_\ell = R^{(2j)}(K_\ell)$ (it is the spin observable along the axis $0x_\ell$, $1 \le \ell \le 3$) and $\hat{S}_\pm = R^{(2j)}(K_\pm)$. This basis is usually written in the "ket" notation of Dirac as $|j, m\rangle$. The correspondence is the following:

$$|j, m\rangle = (-1)^{j+m}\sqrt{\frac{(j-m)!}{(j+m)!}}v_{j+m}.$$

In this basis the representation D^j of the elements K_3, K_+, K_- (basis of the complex Lie algebra $\mathfrak{sl}(2,\mathbb{C})$ defined at the begining) act as follows

$$\begin{aligned}\hat{S}_3|j, m\rangle &= m|j, m\rangle\\ \hat{S}_+|j, m\rangle &= \sqrt{(j-m)(j+m+1)}|j, m+1\rangle\\ \hat{S}_-|j, m\rangle &= \sqrt{(j+m)(j-m+1)}|j, m-1\rangle.\end{aligned}$$

Hence

$$\hat{S}_-|j, -j\rangle = 0, \quad |j, m\rangle = \sqrt{\frac{(j-m)!}{(j+m)!(2j)!}}(\hat{S}_+)^{j+m}|j, -j\rangle.$$

We recall that the two components L_1, L_2 of the angular momentum are related to the operators $L_\pm$ as follows:

$$L_+ = L_1 + iL_2, \quad L_- = L_1 - iL_2$$

In the representation space $(V^{(j)}, D^j)$ of $\mathfrak{sl}(2,\mathbb{C})$ one can consider the spin operator $\mathbf{S} = (\hat{S}_1, \hat{S}_2, \hat{S}_3)$ and

$$\hat{S}_1^2 + \hat{S}_2^2 + \hat{S}_3^2 = \mathbf{S}^2$$

$\mathbf{S}^2$ can be rewritten as

$$\mathbf{S}^2 = \hat{S}_-\hat{S}_+ + \hat{S}_3(\hat{S}_3 + \mathbb{1}).$$

It is clear that for the representation D^j, the vector $|j, m\rangle$ is eigenstate of $\mathbf{S}^2$:

$$\mathbf{S}^2|j, m\rangle = j(j+1)|j, m\rangle$$

Thus $\mathbf{S}^2$ acts as a multiple of the identity and is called the Casimir operator of the representation D^j.

One defines a scalar product on $E^{(2j)}$ by imposing that the basis $|j, m\rangle$ is an orthonormal basis. The representation of the Dicke states $|j, m\rangle$ in the polynomials space $V^{(j)}$ is given by

$$\mathscr{D}_m^{(j)}(z_1, z_2) := |j, m\rangle(z_1, z_2) = \frac{z_1^{j+m} z_2^{j-m}}{\sqrt{(j+m)!(j-m)!}}, \quad -j \leq m \leq j.$$

7.3.2 *The Irreducible Representations of SU(2)*

We shall see now that for every $j \in \frac{\mathbb{N}}{2}$ we can get a representation $T^{(j)}$ of $SU(2)$ such that its differential $dT^{(j)}$ coincides with the representations $D^{(j)}$ of $\mathfrak{su}(2)$ that we have studied in the previous section. Furthermore they are the only irreducible representations of $SU(2)$.

Since every unitary matrix is diagonalizable with unitary passage matrices we have $\forall A \in SU(2)$:

$$A = g \begin{pmatrix} e^{it} & 0 \\ 0 & e^{-it} \end{pmatrix} g^{-1}$$

for some $t \in \mathbb{R}$. Then we show that the exponential map from $\mathfrak{su}(2)$ to $SU(2)$ is surjective:

From the relation

$$\begin{pmatrix} e^{it} & 0 \\ 0 & e^{-it} \end{pmatrix} = \exp(it\sigma_3)$$

we deduce

$$A = \exp(itg\sigma_3 g^{-1})$$

with $ig\sigma_3 g^{-1} \in \mathfrak{su}(2)$, hence the result.

Taking the Pauli matrices as a basis we have that every $A \in \mathfrak{su}(2)$ can be written as

$$A = i\mathbf{a} \cdot \sigma, \quad \mathbf{a} = (a_1, a_2, a_3) \in \mathbb{R}^3.$$

We deduce easily that:

$$\det A = \|\mathbf{a}\|^2$$

and

$$A^2 = -(\det A)\,\mathbb{1}.$$

Therefore we have proven the following lemma

Lemma 41 *For any* $A \in \mathfrak{su}(2)$ *one has*

$$A^2 = -(\det A)\mathbb{1}.$$

Furthermore one has the following result:

Proposition 80 *For any* $A \in \mathfrak{su}(2)$ *such that* $\det A = 1$, *one has* $\forall t \in \mathbb{R}$*:*

$$\exp(tA) = \cos t\,\mathbb{1} + (\sin t)A \tag{7.15}$$

Proof Both members of (7.15) have A as derivative at $t = 0$. It is therefore enough to show that the map $t \in \mathbb{R} \to \mathbb{1}\cos t + A\sin t$ is a one parameter subgroup of $GL(2, \mathbb{C})$. Take $s \in \mathbb{R}$. One has

$$\begin{aligned}(\mathbb{1}\cos t + A\sin t)(\mathbb{1}\cos s + A\sin s) &= \mathbb{1}\cos t\cos s + A^2\sin s\sin t \;+ \\ &\quad(\sin s\cos t + \cos s\sin t)A \\ &= \mathbb{1}\cos(s+t) + A\sin(s+t).\end{aligned} \tag{7.16}$$

□

As a consequence we see that every element $g \in SU(2)$ can be written as

$$g = \alpha_1\mathbb{1} + \alpha_2\mathscr{I} + \alpha_3\mathscr{J} + \alpha_4\mathscr{K}$$

where

$$\mathscr{I} = \begin{pmatrix} 0 & i \\ i & 0 \end{pmatrix}, \quad \mathscr{J} = \begin{pmatrix} 0 & -1 \\ 1 & 0 \end{pmatrix}, \quad \mathscr{K} = \begin{pmatrix} i & 0 \\ 0 & -i \end{pmatrix}$$

the vector $(\alpha_1, \alpha_2, \alpha_3, \alpha_4)$ of $\mathbb{R}^4$ being of norm 1. Thus $SU(2)$ can be identified with the group of quaternions of norm 1.

The group $SU(2)$ acts on $\mathbb{C}^2$ by the usual matrix action.

$$g = \begin{pmatrix} a & b \\ -\bar{b} & \bar{a} \end{pmatrix} \in SU(2) \tag{7.17}$$

with $|a|^2 + |b|^2 = 1$. Then

$$g\begin{pmatrix} z_1 \\ z_2 \end{pmatrix} = \begin{pmatrix} az_1 + bz_2 \\ -\bar{b}z_1 + \bar{a}z_2 \end{pmatrix}$$

g induces a natural action on functions $f : \mathbb{C}^2 \to \mathbb{C}$:

$$g \cdot f := f \circ g^{-1}.$$

Since g has determinant one its inverse g^{-1} equals

$$g^{-1} = \begin{pmatrix} \bar{a} & -b \\ \bar{b} & a \end{pmatrix}$$

thus

$$g \cdot f(z_1, z_2) = f(\bar{a}z_1 - bz_2, \bar{b}z_1 + az_2).$$

The linear $V^{(j)}$ space of homogeneous polynomials in z_1, z_2 of degree $2j$ (we recall that $j \in \frac{1}{2}\mathbb{N}$) is stable by the action of $SU(2)$. Moreover in $V^{(j)}$, the map $p \mapsto g \cdot p$ is unitary for the scalar product defined above. The action of $g \in SU(2)$ on an homogeneous polynomial p of degree $2j$ will be also denoted:

$$T^j(g)p(z_1, z_2) = g \cdot p(z_1, z_2) = p \circ g^{-1}(z_1, z_2).$$

Let us prove the following Lemma.

Lemma 42 *Consider the homogeneous polynomial p:*

$$p(z_1, z_2) = \sum_{l=0}^{2j} c_l z_1^l z_2^{2j-l}.$$

Then the map

$$p \mapsto \|p\|^2 = \sum_{l=0}^{2j} l!(2j-l)!|c_l|^2$$

defines an Hilbertian norm of $V^{(j)}$ that is invariant by the action of $SU(2)$.

Proof It is enough to check that $\|p \circ g^*\|^2 = \|p\|^2$ which implies that the representation $(V^{(j)}, T^j)$ is unitary. Take

$$p(z_1, z_2) = p_{\alpha,\beta}(z_1, z_2) = (\alpha z_1 + \beta z_2)^{2j} \quad \alpha, \ \beta \in \mathbb{C}$$

which generate $V^{(j)}$. Then if

$$p(z_1, z_2) = \sum_{l=0}^{2j} c_l z_1^l z_2^{2j-l}, \quad p'(z_1, z_2) = \sum_{l=0}^{2j} c'_l z_1^l z_2^{2j-l}$$

$$p \circ g^{-1}(z_1, z_2) = \sum_{l=0}^{2j} d_l z_1^l z_2^{2j-l}, \quad p' \circ g^{-1}(z_1, z_2) = \sum_{l=0}^{2j} d'_l z_1^l z_2^{2j-l}$$

one has

$$\langle p', p \rangle = \langle p' \circ g^{-1}, p \circ g^{-1} \rangle = \sum_{l=0}^{2j} c_l \bar{c}'_l l!(2j-l)!$$
$$= \sum_{l=0}^{2j} d_l \bar{d}'_l l!(2j-l)! = (2j)!(\alpha\bar{\alpha}' + \beta\bar{\beta}')^{2j}. \tag{7.18}$$

Namely the Hermitian scalar product in $\mathbb{C}^2$ of (α, β) with (α', β') is invariant under $SU(2)$. □

We shall study now the representation $T^{(j)}$ of $SU(2)$ in $V^{(j)}$. One defines

$$f^j_m(z_1, z_2) = z_1^{j+m} z_2^{j-m}.$$

We first consider diagonal matrices in $SU(2)$. They are of the form

$$g_t = \exp(-2itK_3) = \begin{pmatrix} e^{it} & 0 \\ 0 & e^{-it} \end{pmatrix}$$

Then

$$(T^{(j)}(g_t)f^j_m)(z_1, z_2) = f^j_m(z_1 e^{-it}, z_2 e^{it}) = e^{-2imt} f^j_m(z_1, z_2) \tag{7.19}$$

Thus every f^j_m is eigenstate of $T^{(j)}(g_t)$ with eigenvalue e^{-2imt}.

We shall now consider the differential $dT^{(j)}(g)$ for the basis elements $K_3, K_\pm$ of $\mathfrak{sl}(2, \mathbb{C})$: one considers $X \in \mathfrak{sl}(2, \mathbb{C})$,

$$X = \begin{pmatrix} \alpha & \beta \\ \gamma & -\alpha \end{pmatrix}$$

and

$$g_t = \exp(tX) = \begin{pmatrix} a(t) & b(t) \\ c(t) & d(t) \end{pmatrix}.$$

Then $g(0) = \mathbb{1}$ and $g'(0) = X$ so that

$$a(0) = d(0) = 1,\ b(0) = c(0) = 0, \qquad a'(0) = \alpha, b'(0) = \beta, c'(0) = \gamma, d'(0) = -\alpha.$$

For any polynomial in two variables $f(z_1, z_2)$ one has:

$$((dT^{(j)})(-X)f)(z_1, z_2) = \frac{d}{dt}(\rho(g_t)^{-1} f)(z_1, z_2)|_{t=0} = \frac{d}{dt}(f \circ g_t)(z_1, z_2)|_{t=0}$$
$$= (\alpha z_1 + \beta z_2)\partial_1 f(z_1, z_2) + (\gamma z_1 + \delta z_2)\partial_2 f(z_1, z_2).$$

Therefore

$$(dT^{(j)})(K_3) = \frac{1}{2}(z_1\partial_1 - z_2\partial_2)$$

$$(dT^{(j)})(K_+) = z_1\partial_2, \qquad (d\rho)(K_-) = z_2\partial_1.$$

We shall determine the action of $d\rho(K_3)$, $dT^{(j)}(K_\pm)$ on the basis vectors f_m^j of $V^{(j)}$:

$$dT^{(j)}(K_3)f_m^j = mf_m^j$$

Furthermore using

$$\partial_1 f_m^j = (j+m)z_1^{j+m-1}z_2^{j-m}, \quad \partial_2 f_m^j = (j-m)z_1^{j+m}z_2^{j-m-1}$$

we get

$$dT^{(j)}(K_+)f_m^j = (j-m)f_{m+1}^j, \quad dT^{(j)}(K_-)f_m^j = (j+m)f_{m-1}^j.$$

Denoting

$$|j,m\rangle = \frac{1}{\sqrt{(j-m)!(j+m)!}}f_m^j$$

we get

$$dT^{(j)}(K_3)|j,m\rangle = m|j,m\rangle \tag{7.20}$$
$$dT^{(j)}(K_+)|j,m\rangle = \sqrt{j(j+1)-m(m+1)}|j,m+1\rangle \tag{7.21}$$
$$dT^{(j)}(K_-)|j,m\rangle = \sqrt{j(j+1)-m(m-1)}|j,m-1\rangle. \tag{7.22}$$

So we deduce the following results:

Proposition 81 *The differential of the representation $T^{(j)}$ of $SU(2)$ coincides with the representation $D^{(j)}$ of $\mathfrak{su}(2)$.*

Proposition 82 *For any $j \in \frac{1}{2}\mathbb{N}$, $(V^{(j)}, T^{(j)})$ is an irreducible representation of $SU(2)$.*

It defines an irreducible representation of $SO(3)$ if and only if $j \in \mathbb{N}$.
If n is odd ($j \in \frac{\mathbb{N}}{2}$, $j \notin \mathbb{N}$) then $T^{(j)}$ is a projective representation of $SO(3)$.

Proof It follows from general results about the differential of the representations of Lie groups that the differential of $T^{(j)}$ is an irreducible representation.

The second part comes from the following fact: $T^{(j)}(-g) = T^{(j)}(g)$, $\forall g \in SU(2)$ if and only if n is even.

The proof of the last part is left to the reader. □

Corollary 26 *Every irreducible representation of $SU(2)$ is equivalent to one of the representations $(V^{(j)}, T^{(j)})$, $j \in \frac{1}{2}\mathbb{N}$.*

Proof Since $SU(2)$ is compact we know that every irreducible representation of $SU(2)$ is finite-dimensional. We have seen that every irreducible representation of finite dimension of $\mathfrak{su}(2)$ is one of the D^j, $j \in \frac{1}{2}\mathbb{N}$. This implies the result since $SU(2)$ is connected and simply connected. □

7.3.3 Irreducible Representations of $SO(3)$ and Spherical Harmonics

We have seen above that irreducible representations of $SO(3)$ are described by $(T^{(j)}, V^{(j)})$ for $j \in \mathbb{N}$. A more concrete equivalent representation can be obtained with a spectral decomposition of the Laplace operator on $\mathbb{S}^2$.

Recall that in spherical coordinates (r, θ, φ) we have

$$\Delta = \frac{\partial^2}{\partial x_1^2} + \frac{\partial^2}{\partial x_2^2} + \frac{\partial^2}{\partial x_3^2} = \frac{\partial^2}{\partial r^2} + \frac{2}{r}\frac{\partial}{\partial r} + \frac{1}{r^2}\Delta_{\mathbb{S}^2}$$

where $\Delta_{\mathbb{S}^2}$ is the spherical Laplace operator on $\mathbb{S}^2$,

$$\Delta_{\mathbb{S}^2} := \frac{\partial^2}{\partial \theta^2} + \frac{1}{\tan\theta}\frac{\partial}{\partial \theta} + \frac{1}{\sin^2\theta}\frac{\partial^2}{\partial \varphi^2}. \tag{7.23}$$

$SO(3)$ has a natural representation Σ in the function space $L^2(\mathbb{R}^3)$ (and in $L^2(\mathbb{S}^2)$) defined as follows. $\Sigma(g)f(x) = f(g^{-1}x)$ where $g \in SO(3)$, $f \in L^2(\mathbb{R}^3)$ and the Laplace operator commutes with Σ.

Let us introduce the linear space $\mathscr{H}_3^{(j)}$ of homogeneous polynomials f in (x_1, x_2, x_3) of total degree j and satisfying $\Delta f = 0$ and restricted to the sphere $\mathbb{S}^2$; $\mathscr{H}_3^{(j)}$ is the space of spherical harmonics.

Recall that the Euclidean measure on $\mathbb{S}^2$ is $d\mu_2(\theta, \varphi) = \sin\theta d\theta d\varphi$.

Theorem 42 *$\mathscr{H}_3^{(j)}$ is a subspace of $C^\infty(\mathbb{S}^2)$ of dimension 2j+1, invariant for the action Σ. The representation $(\Sigma, \mathscr{H}_3^{(j)})$ is irreducible and is unitary equivalent to the representation $(T^{(j)}, V^{(j)})$.*

Recall the following expression for the measure $d\mu_2$ on the sphere: $d\mu_2(\theta, \varphi) = \sin\theta d\theta d\varphi$.

Proof We shall prove some properties of spherical harmonics which are proved in more details for example in [171].

The space $\mathscr{H}_3^{(j)}$ is invariant by the generators L_1, L_2, L_3 of rotations. In spherical coordinates we have

$$L_3 = \frac{1}{i}\frac{\partial}{\partial \varphi} \tag{7.24}$$

$$L_2 = \frac{1}{i}\left(\cos\varphi\frac{\partial}{\partial\theta} - \frac{\sin\varphi}{\tan\theta}\frac{\partial}{\partial\varphi}\right) \tag{7.25}$$

$$L_1 = i\left(\sin\varphi\frac{\partial}{\partial\theta} + \frac{\cos\varphi}{\tan\theta}\frac{\partial}{\partial\varphi}\right). \tag{7.26}$$

We can compute the Casimir operator: $\mathbf{L}^2 := L_1^2 + L_2^2 + L_3^2 = -\Delta_{\mathbb{S}^2}$. In particular we have $[L_3, \Delta_{\mathbb{S}^2}] = 0$.

If $L_\pm := L_1 \pm iL_2$ then we have

$$[L_3, L_\pm] = \pm L_\pm, \quad [L_+, L_-] = 2L_3 \tag{7.27}$$

$$L_+L_- = \mathbf{L}^2 - L_3(L_3 - \mathbb{1}), \quad L_-L_+ = \mathbf{L}^2 - L_3(L_3 + \mathbb{1}). \tag{7.28}$$

Let $f \in \mathscr{H}_3^{(j)}$. In polar coordinates we have $f(r,\theta,\varphi) = r^j Y(\theta,\varphi)$. So we get that

$$\Delta f = 0 \iff -\Delta_{\mathbb{S}^2} Y = j(j+1)Y.$$

L_3 can be diagonalized in $\mathscr{H}_3^{(j)}$

$$L_3 Y = \lambda Y \iff Y(\varphi,\theta) = \mathrm{e}^{im\varphi} f(\theta), m \in \mathbb{Z}, \ -j \le m \le j.$$

So admitting that $\mathscr{H}_3^{(j)}$ has dimension $2j+1$ we see that the representation $(\Sigma, \mathscr{H}_3^{(j)})$ is unitary equivalent to the representation $(T^{j)}, V^{(j)})$. □

Remark 45 Using the same method as in Sect. 7.3, for every $j \in \mathbb{N}$ we have an orthonormal basis $\{Y_j^k\}_{-j\le k\le j}$ of $\mathscr{H}_3^{(j)}$ where Y_j^k are eigenfunctions of L_3: $L_3 Y_j^k = kY_j^k, -j \le k \le j$. In other words Y_j^k are Dicke states. Moreover they have the following expression:

$$Y_j^k(\theta,\varphi) = \sqrt{\frac{2j+1}{4\pi}}\mathrm{e}^{ik\varphi} P_j(\cos\theta) \tag{7.29}$$

where P_j are the Legendre polynomials

$$P_j(u) = \frac{1}{2^j j!}\frac{d^j}{du^j}(u^2-1)^j.$$

For $k \neq 0$ we can use the following formula

$$\begin{aligned} L_+ Y_j^k &= \sqrt{j(j+1) - k(k+1)}\, Y_j^{k+1}, \\ L_- Y_j^k &= \sqrt{j(j+1) - k(k-1)}\, Y_j^{k-1}. \end{aligned} \tag{7.30}$$

Let us prove now two useful properties of the spherical harmonics

Proposition 83 *(i) For every $j \in \mathbb{N}$, $\mathscr{H}_3^{(j)}$ has dimension $2j + 1$.*
(ii) $\{Y_j^k,\ -j \le k \le j,\ j \in \mathbb{N}\}$ is an orthonormal basis of $L^2(\mathbb{S}^2)$ or, equivalently,

$$\oplus_{j\in\mathbb{N}}\mathscr{H}_3^{(j)} = L^2(\mathbb{S}^2).$$

Proof Let us introduce the space $\mathscr{P}_3^{(j)}$ of homogeneous polynomials in (x_1, x_2, x_3) of total degree j. The dimension of $\mathscr{P}_{3,j}$ is $\frac{(j+1)(j+2)}{2}$. It easy to prove that Δ is surjective so we get (i):

$$\dim(\ker\Delta) = \frac{(j+1)(j+2)}{2} - \frac{(j-1)j}{2} = 2j+1$$

To prove (ii) let us introduce on $\mathscr{P}_3^{(j)}$ a scalar product such that we have an orthonormal basis $\{\frac{x_1^{k_1}x_2^{k_2}x_3^{k_3}}{\sqrt{k_1!k_2!k_3!}}\}_{k_1+k_2+k_3=j}$. Let us introduce the Hilbert space $\mathscr{H} := \oplus\mathscr{P}_3^{(j)}$. So the linear operators $\frac{\partial}{\partial x_k}$ and x_k are hermitian conjugate. Hence Δ is conjugate to $r^2 = x_1^2 + x_2^2 + x_3^2$.

We have $r^2\mathscr{P}_3^{(j)} \subseteq \mathscr{P}_3^{(j+2)}$ and $\Delta\mathscr{P}_3^{(j)} = \mathscr{P}_3^{(j-2)}$. Using the formula (Fredholm property) $\ker A = (\mathrm{Im}A^*)^\perp$, we get $\mathscr{P}_3^{(j)} = H_3^j \oplus r^2\mathscr{P}_3^{(j-2)}$. Step by step we get

$$\mathscr{P}_3^{(j)} = \mathscr{H}_3^{(j)} \oplus r^2\mathscr{H}_3^{(j-2)} \oplus \cdots \oplus r^{2\ell}\mathscr{H}_3^{(j-2\ell)}, \tag{7.31}$$

where $j - 1 \le 2\ell \le j$.

Now we can prove that the spherical harmonics is a total system in $L^2(\mathbb{S}^2)$. Let us remark that the algebra $\cup_{j\in\mathbb{N}}\mathscr{P}_{3,j}$ is dense in $C(\mathbb{S}^2)$ for the sup-norm (consequence of Stone-Weierstrass Theorem). So using (7.31) we get that $\cup_{j\in\mathbb{N}}\mathscr{H}_3^{(j)}$ is dense in $C(\mathbb{S}^2)$ for the sup-norm, so $\cup_{j\in\mathbb{N}}\mathscr{H}_3^{(j)}$ is dense in $L^2(\mathbb{S}^2)$. □

7.4 The Coherent States of $SU(2)$

7.4.1 Definition and First Properties

Let us start with a reference (non zero) vector $\psi_0^{(j)} \in V^{(j)}$ and consider elements of the orbit of $\psi_0^{(j)}$ in $V^{(j)}$ by the action of the representation $T^{(j)}$. We get a family of states of the form

$$\psi_g^{(j)} := |g\rangle := T^{(j)}(g)\psi_0^{(j)}.$$

In a more explicit form we have

$$\psi_g^{(j)}(z_1, z_2) = \psi_0^{(j)}(g^{-1}(z_1, z_2)),\ \ (z_1, z_2) \in \mathbb{C}^2.$$

In principle any vector $\psi_0 \in V^{(j)}$ can be taken as reference state to built coherent states. However the states $\psi_0 = |j, \pm j\rangle$ are better because the total spin operator $\mathbf{S} = (\hat{S}_1, \hat{S}_2, \hat{S}_3)$, is minimal on these states which are in some sense the closest to the classical states. In practice we shall choose here $\psi_0^{(j)} = |j, j\rangle$ but the choice $\psi_0^{(j)} = |j, -j\rangle$ is also possible with some modifications in the computations.

Let us recall the definition of the dispersion for an observable $\hat{A}$ for a state ψ, where ψ is a normalized state in an Hilbert space $\mathscr{H}$ and $\hat{A}$ a self-adjoint operator in $\mathscr{H}$. For ψ in the domain of $\hat{A}$ the average is $\langle \hat{A} \rangle_\psi := \langle \psi, \hat{A}\psi \rangle$ and the dispersion is defined like the variance for a random variable in probability:

$$\Delta_\psi \hat{A} := \langle (\hat{A} - \langle \hat{A} \rangle_\psi \mathbb{1})^2 \rangle_\psi = \langle \hat{A}^2 \rangle_\psi - \langle \hat{A} \rangle_\psi^2.$$

Let us recall here the Heisenberg uncertainty principle: if $\hat{A}$, $\hat{B}$ are self-adjoint operators in $\mathscr{H}$ and $\psi \in \mathscr{H}$, $\|\psi\| = 1$ then we have

$$(\Delta_\psi \hat{A})(\Delta_\psi \hat{B}) \geq \frac{1}{4}\left(\langle i[\hat{A}, \hat{B}] \rangle_\psi\right)^2. \tag{7.32}$$

Application to the total spin observable gives

$$\Delta_\psi \mathbf{S} = \sum_{1\leq k\leq 3} \Delta_\psi \hat{S}_k = \sum_{1\leq k\leq 3} \|\hat{S}_k \psi\|^2 - \langle \hat{S}_k \psi, \psi \rangle^2.$$

For $\psi_0 = |j, m\rangle$ we get

$$\Delta_{\psi_0} \mathbf{S} = j(j+1) - m^2.$$

So the dispersion is minimal for $m = \pm j$.

Heisenberg inequality for spin operators reads (see (7.32))

$$\langle \hat{S}_1^2 \rangle \langle \hat{S}_2^2 \rangle \geq \frac{1}{4} \langle \hat{S}_3 \rangle^2. \tag{7.33}$$

It is easy to see that this inequality is an equality for $\psi_0 = |j, \pm j\rangle$.

Now our goal is to study the main properties of the coherent states $|g\rangle$. Let us first remark that the full group $G = SU(2)$ is not a good set to parametrize these coherent states because the map $g \mapsto |g\rangle$ is not injective. So we introduce the so-called isotropy group H defined as follows

$$H = \{g \in G, \exists \delta \in \mathbb{R},\ T(g)\psi_0 = \mathrm{e}^{i\delta}\psi_0\}$$

We find that H is the subgroup of diagonal matrices

$$H = \left\{ \begin{pmatrix} \alpha & 0 \\ 0 & \bar{\alpha} \end{pmatrix} \right\}, \qquad \alpha = \exp(i\psi)$$

Now the map $\dot{g} \mapsto |g\rangle$ is a bijection from the quotient space $X := G/H$ onto the orbit of ψ_0, where G/H is the set of left coset gH of H in G and $g \mapsto \dot{g}$ is the canonical map: $G \to G/H$.

Let us denote $X_0 = X\backslash\{\begin{pmatrix} 0 & 1 \\ -1 & 0 \end{pmatrix}\}$.

Lemma 43 *X_0 is isomorphic to the set of elements of the form*

$$\left\{\begin{pmatrix} \alpha & \beta \\ -\bar{\beta} & \alpha \end{pmatrix}, \quad \alpha \in \mathbb{R},\ \alpha \neq 0,\ \beta = \beta_1 + i\beta_2 \in \mathbb{C}, \quad \alpha^2 + \beta_1^2 + \beta_2^2 = 1\right\}.$$

Proof Let us denote $g(a, b)$ a generic element of $SU(2)$, $g = \begin{pmatrix} a & b \\ -\bar{b} & \bar{a} \end{pmatrix}$, $a, b \in \mathbb{C}$, $|a|^2 + |b|^2 = 1$.

We first remark that for every $|b| = 1$, $g(0, b)$ is in the coset of $g(0, 1)$.

Now if $|a|^2 + |b|^2 = 1$ and $a \neq 0$ then we have a unique decomposition

$$g(a, b) = g(a', b')g(\alpha, 0)$$

with $a' > 0$, $\alpha = \frac{a}{|a|}$, $(a')^2 + |b'|^2 = 1$.

So we get the lemma. □

Remark 46 Concerning the orbit with our choice of ψ_0 the image of X_0 does not contain the monomial z_1^{2j}, which are obtained with $g(0, 1)$.

Choosing the parametrization

$$\alpha = \cos\frac{\theta}{2}, \quad \beta = -\sin\frac{\theta}{2}e^{-i\varphi}, \quad 0 \leq \theta < 2\pi, \quad 0 \leq \varphi < 2\pi$$

we see that the space X_0 is just a representation of the two-dimensional sphere $\mathbb{S}^2$ minus the north pole, namely the set of unit three-dimensional vectors

$$\mathbf{n} = (\sin\theta\cos\varphi, \sin\theta\sin\varphi, \cos\theta),\ 0 \leq \theta < \pi,\ 0 \leq \varphi < 2\pi;$$

and any element $g \in X$ can be written as

$$g = g_{\mathbf{n}} = \exp\left[i\frac{\theta}{2}(\sin\varphi\sigma_1 - \cos\varphi\sigma_2)\right] \tag{7.34}$$

where σ_1, σ_2 are the Pauli matrices.

Thus $g_{\mathbf{n}}$ describes a rotation by the angle θ around the vector $\mathbf{m} = (\sin\varphi, -\cos\varphi, 0)$ belonging to the equatorial plane of the sphere and perpendicular to $\mathbf{n}$ (it is well defined because $\theta \in [0, \pi[$).

Definition 20 The coherent states of $SU(2)$ are the following states, defined in the representation space $V^{(j)}$,

$$|\mathbf{n}\rangle = T(g_{\mathbf{n}})\psi_0 := G(\mathbf{n})\psi_0 \tag{7.35}$$

In the physics literature they are called the spin coherent states because the spin is classified with the irreducible representations of $SU(2)$ (see [207]). These coherent states have several other names: atomic coherent states, Bloch coherent states.

Choosing $g = g_{\mathbf{n}}$ the coherent state of the $SU(2)$ group can now be written as

$$|\mathbf{n}\rangle = T(g_{\mathbf{n}})\psi_0 = \exp(i\theta\mathbf{m}\cdot\mathbf{K})\psi_0$$

where

$$\mathbf{m} = (\sin\varphi, -\cos\varphi, 0), \quad \mathbf{K} = (K_1, K_2, K_3)$$

$\mathbf{m}$ is the unit vector orthogonal to both $\mathbf{n}$ and $\mathbf{n}_0 = (0, 0, 1)$. Note that this definition excludes the south pole $\mathbf{n}_S = (0, 0, -1)$.

Thus a coherent state of $SU(2)$ corresponds to a point of the two-dimensional sphere $\mathbb{S}^2$ which may be considered as the phase-space of a classical dynamical system, the "classical spin". The coherent states associated with the south pole will be the image of $(g(0, \beta))$, $|\beta| = 1$ giving monomials z_1^{2j}.

So we have parametrized the spin coherent state by the sphere $\mathbb{S}^2$.

Another useful parametrization can be obtained with the complex plane, using the stereographic projection from the south pole of the sphere $\mathbb{S}^2$ onto the complex plane $\mathbb{C}$.

If $\mathbf{n} = (n_1, n_2, n_3) \in \mathbb{S}^2$ then the stereographic projection of $\mathbf{n}$ from the south pole is the complex number $\zeta(\mathbf{n}) = \frac{n_1+in_2}{1+n_3}$. So in polar coordinates we have $\zeta(\mathbf{n}) = \tan(\theta/2)e^{i\varphi}$.

As we have already remarked, the group $SU(2)$ naturally embeds into the complex group $SL(2, \mathbb{C})$ which is the group of complex matrices having determinant one. The following gaussian decomposition in $SL(2, \mathbb{C})$ will be useful. The proof is an easy exercise.

Lemma 44 *For any $g \in SL(2, \mathbb{C})$ of the form*

$$g = \begin{pmatrix} \alpha & \beta \\ \gamma & \delta \end{pmatrix}, \quad \text{with } \delta \neq 0$$

one has a unique (Gaussian) decomposition

$$g = t_+ \cdot d \cdot t_-$$

where d is diagonal and $t_\pm$ are triangular matrices of the form

$$t_+ = \begin{pmatrix} 1 & \zeta \\ 0 & 1 \end{pmatrix}, \quad t_- = \begin{pmatrix} 1 & 0 \\ z & 1 \end{pmatrix} \tag{7.36}$$

and

$$d = \begin{pmatrix} \varepsilon^{-1} & 0 \\ 0 & \varepsilon \end{pmatrix}.$$

We have the formulas

$$\varepsilon = \delta, \quad z = \frac{\gamma}{\delta}, \quad \zeta = \frac{\beta}{\delta} \tag{7.37}$$

Moreover if $g \in SU(2)$ *then we have*

$$|\varepsilon|^2 = (1 + |\zeta|^2)^{-1} = (1 + |z|^2)^{-1} \tag{7.38}$$

This allows to write as consequence of Gauss decomposition,

$$T^{(j)}(g) = T^{(j)}(t_+)T^{(j)}(d)T^{(j)}(t_-)$$

Let us write $\varepsilon = re^{is}$ with $r > 0$ and $s \in \mathbb{R}$.

Taking $\psi_0 = |j, j\rangle$ as the reference state one gets in the representation $(T^{(j)}, V^{(j)})$

$$T^{(j)}(t_-)\psi_0 = e^{z_+\hat{S}_-}\psi_0 = \psi_0 \tag{7.39}$$

$$T^{(j)}(d)\psi_0 = \mathrm{e}^{-2j(\log r + is)}\psi_0 \tag{7.40}$$

So we have

$$T^{(j)}(g)\psi_0 = e^{i\varphi}NT(t_+)\psi_0. \tag{7.41}$$

From (7.38) we get $N = (1 + |\zeta|^2)^{-j}$. φ is a real number (argument of δ). If $g = g_{\mathbf{n}}$ then δ is real so we get $s = \varphi = 0$.

In conclusion we have obtained an identification of the coherent states $|\mathbf{n}\rangle$ with the state $|\zeta\rangle$ defined as follows:

$$|\zeta\rangle = (1 + |\zeta|^2)^{-j} \exp(\zeta \hat{S}_+)|j, j\rangle.$$

More precisely we denote $|\zeta\rangle = |g_{\mathbf{n}}\rangle$ with the following correspondence:

$$\mathbf{n} = (\sin\theta\cos\varphi, \sin\theta\sin\varphi, \cos\theta), \quad \zeta = -\tan\frac{\theta}{2}e^{-i\varphi}.$$

The geometrical interpretation is that $-\bar{\zeta}$ is the stereographic projection of $\mathbf{n}$.

Recall the following expression of $g_{\mathbf{n}}$:

$$g_{\mathbf{n}} = \begin{pmatrix} \cos\frac{\theta}{2} & -\sin\frac{\theta}{2}\mathrm{e}^{-i\varphi} \\ \sin\frac{\theta}{2}\mathrm{e}^{i\varphi} & \cos\frac{\theta}{2} \end{pmatrix}. \tag{7.42}$$

Another form for $|\mathbf{n}\rangle$ is given by the following equivalent definition using that $g_{\mathbf{n}} = e^{i\theta(\sin\varphi\hat{S}_1 - \cos\varphi\hat{S}_2)}$,

$$|\mathbf{n}\rangle = G(\xi)\psi_0$$

where

$$G(\xi) = \exp(\xi\hat{S}_+ - \bar{\xi}\hat{S}_-)$$

and $\xi = \tan\left(\frac{\theta}{2}\right)e^{i\varphi}$.

The gaussian decomposition also provides a "normal form" of $D(\xi)$:

$$G(\xi) = \exp(\zeta\hat{S}_+)\exp(\eta\hat{S}_3)\exp(\zeta'\hat{S}_-)$$

with

$$\eta = -2\log|\xi|,\ \zeta' = -\bar{\zeta}.$$

Since $\zeta,\ \zeta',\ \eta$ do not depend on j it is enough to check this formula in the representation where $j = \frac{1}{2}$, $\mathbf{S} = \frac{1}{2}R(\sigma)$, where σ is the three component Pauli matrix.

For each $\mathbf{n} \in \mathbb{S}^2$ the coherent state $\psi_{\mathbf{n}}$ minimizes Heisenberg inequality obtained by translation of (7.33) by $G(\mathbf{n})$ i.e. putting $\tilde{S}_k = G(\mathbf{n})\hat{S}_k G(\mathbf{n})^{-1}$ instead of $\hat{S}_k$, $1 \le k \le 3$.

7.4.2 Some Explicit Formulas

Many explicit formulas can be proved for the spin coherent states. These formulas have many similarities with formulas already proved for the Heisenberg coherent states and can be written as well with the coordinates $\mathbf{n}$ on the sphere $\mathbb{S}^2$ or in the coordinates ζ in the complex plane $\mathbb{C}$.

Let us first remark that $\mathbb{S}^2$ and $\mathbb{C}$ can be identified with a classical phase space. $\mathbb{S}^2$ is equipped with the symplectic two form $\sigma = \sin\theta d\theta \wedge d\varphi$. In the stereographic projection it is transformed in $\sigma = 2i\frac{d\zeta\wedge d\bar{\zeta}}{(1+|\zeta|^2)^2}$. This is an easy computation using $\zeta = -\tan\frac{\theta}{2}e^{-i\varphi}$.

Let us consider first some properties of operators $G(\mathbf{n})$. We shall also use the notations $D(\xi)$ or $D(\zeta)$ where ξ and ζ are given by $\xi = \tan\left(\frac{\theta}{2}\right)e^{i\varphi}$ and $\zeta = -\tan\frac{\theta}{2}e^{-i\varphi}$ using polar coordinates for $\mathbf{n}$. The multiplication law for the operators $G(\mathbf{n})$ is given by the following formula

Proposition 84 *(i) For every $\mathbf{n}_1, \mathbf{n}_2$ outside the south pole of $\mathbb{S}^2$ we have*

$$G(\mathbf{n}_1)G(\mathbf{n}_2) = G(\mathbf{n}_3)\exp(-i\Phi(\mathbf{n}_1, \mathbf{n}_2)J_3) \tag{7.43}$$

where $\Phi(\mathbf{n}_1, \mathbf{n}_2)$ is the oriented area of the geodesic triangle on the sphere with vertices at the points $[\mathbf{n}_0, \mathbf{n}_1, \mathbf{n}_2]$.

$\mathbf{n}_3$ is determined by

$$\mathbf{n}_3 = R_{g_{\mathbf{n}_1}} \mathbf{n}_2 \tag{7.44}$$

where R_g is the rotation associated to $g \in SU(2)$ as in Proposition 78 and

$$g_{\mathbf{n}} = \exp\left(i\frac{\theta}{2}(\sigma_1 \sin\varphi - \sigma_2 \cos\varphi)\right). \tag{7.45}$$

(ii) More generally for every $g \in SU(2)$ and every $\mathbf{n} \in \mathbb{S}^2$ such that $\mathbf{n}$ and $g \cdot \mathbf{n}$ are outside the south pole we have

$$G^{(j)}(g)\psi_{\mathbf{n}} = \exp(-ij\mathscr{A}(g, \mathbf{n}))\psi_{g\cdot\mathbf{n}}, \tag{7.46}$$

where $g \cdot \mathbf{n} = R_g(\mathbf{n})$ and $\mathscr{A}(g, \mathbf{n})$ is the aera of the spherical triangle $[\mathbf{n}_0, \mathbf{n}, g \cdot \mathbf{n}]$.

Proof We prove the result in the two-dimensional representation of $SU(2)$. The vector $\mathbf{n}_3$ is determined only by geometrical rule and is thus independent of the representation. We choose the representation in $V^{1/2}$.

Define $R(g)$ to be the rotation in $SO(3)$ induced by any $g \in SU(2)$. By the definition (7.45) of $g_{\mathbf{n}}$ we have that

$$R(g_{\mathbf{n}})\mathbf{n}_0 = \mathbf{n}, \quad \forall \mathbf{n} = (\sin\theta\cos\varphi, \sin\theta\sin\varphi, \cos\theta).$$

We need to compute $g = g_{\mathbf{n}_1} g_{\mathbf{n}_2}$. We use the following lemma:

Lemma 45 $\forall g \in SU(2)$ $\exists \mathbf{m} \in \mathbb{S}^2$ *and* $\delta \in \mathbb{R}$ *such that*

$$g = g_{\mathbf{m}} r_3(\delta)$$

where $r_3(\delta) = \exp\left(i\frac{\delta}{2}\sigma_3\right)$.

Applying the lemma to $g = g_{\mathbf{n}_1} g_{\mathbf{n}_2}$ we get

$$g = g_{\mathbf{m}} r_3(\delta)$$

and we need to identify $\mathbf{m}$ with $\mathbf{n}_3$ given by (7.44). We have that

$$R(r_3(\delta))\mathbf{n}_0 = \mathbf{n}_0.$$

Then

$$R(g) = R(g_{\mathbf{n}_1})R(g_{\mathbf{n}_2}) = R(g_{\mathbf{m}})R(r_3(\delta))$$

Applying this identity to the vector $\mathbf{n}_0$ we get

$$R(g_{\mathbf{n}_1})R(g_{\mathbf{n}_2})\mathbf{n}_0 = R(g_{\mathbf{n}_1})\mathbf{n}_2 = R(g_{\mathbf{m}})n_0 = \mathbf{m}.$$

Thus we have proven that $\mathbf{m} = \mathbf{n}_3 = R(g_{\mathbf{n}_1})\mathbf{n}_2$.

So we have seen that the displacement operator $G(\mathbf{n})$ transforms any spin coherent state $|\mathbf{n}_1\rangle$ into another coherent state of the system up to a phase:

$$G(\mathbf{n})|\mathbf{n}_1\rangle = G(\mathbf{n})D(\mathbf{n}_1)\psi_0 = G(\mathbf{n}_2)\exp(i\hat{S}_3\Phi(\mathbf{n},\mathbf{n}_1))\psi_0 = \exp(-ij\Phi(\mathbf{n},\mathbf{n}_1))|\mathbf{n}_2\rangle$$

where $\mathbf{n}_2 = R(g_\mathbf{n})\mathbf{n}_1$. The second factor in the right hand side does depend on the representation. The computation of $\Phi(\mathbf{n}_1, \mathbf{n}_2)$ will be done later.

In the proof of (7.46) the nontrivial part is to compute the phase $\mathscr{A}(g, \mathbf{n})$ which also will be done later. □

The following lemma shows that the spin is independent of the direction.

Lemma 46 *One has:*

$$G(\mathbf{n})\hat{S}_3 G(\mathbf{n})^{-1} = \mathbf{n}\cdot\mathbf{S}.$$

We shall prove the lemma in the representation of the Pauli matrices. First note that

$$G(\mathbf{n}) = \cos\frac{\theta}{2} + i\sin\frac{\theta}{2}(\sin\varphi\sigma_1 - \cos\varphi\sigma_2)$$

Then

$$\begin{aligned} G(\mathbf{n})\sigma_3 G(\mathbf{n})^{-1} = &\left(\cos\tfrac{\theta}{2} + i\sin\tfrac{\theta}{2}(\sin\varphi\sigma_1 - \cos\varphi\sigma_2)\right)\sigma_3 \\ &\left(\cos\tfrac{\theta}{2} - i\sin\tfrac{\theta}{2}(\sin\varphi\sigma_1 - \cos\varphi\sigma_2)\right) \end{aligned} \tag{7.47}$$

We use the properties of the Pauli matrices:

$$\sigma_1\sigma_3 = -i\sigma_2,\ \sigma_2\sigma_3 = i\sigma_1,\ \sigma_2\sigma_1 = -i\sigma_3$$

to compute the right hand side. One gets

$$G(\mathbf{n})\sigma_3 G(\mathbf{n})^{-1} = \cos\theta\sigma_3 + \sin\theta(\sin\varphi\sigma_2 + \cos\varphi\sigma_1) = \mathbf{n}\cdot\sigma$$

The following consequence is that $|\mathbf{n}\rangle$ is always an eigenvector of the operator $\mathbf{n}\cdot\mathbf{S}$:

Proposition 85 *One has:*

$$\boxed{\mathbf{n}\cdot\mathbf{S}|\mathbf{n}\rangle = j|\mathbf{n}\rangle}. \tag{7.48}$$

Proof Denote by $|\mathbf{n}_0\rangle$ the vector

$$e^{iS_3\theta}\psi_0$$

Then

$$S_3|\mathbf{n}_0\rangle = je^{i\theta\hat{S}_3}\psi_0 = j|\mathbf{n}_0\rangle$$

since we take $\psi_0 = |j, j\rangle$.

Now we shall use the above lemma:

$$\mathbf{n}\cdot\mathbf{S}|\mathbf{n}\rangle = G(\mathbf{n})J_3\psi_0 = j|\mathbf{n}\rangle$$

This completes the proof of the proposition. □

Let us give now a result which is more general than (7.48).

Proposition 86 *With the above notations, in particular $\psi_{\mathbf{n}}^{(j)}$ being the coherent state located at $\mathbf{n}\in\mathbb{S}^2$, for any $j\in\mathbb{N}/2$ we have*

$$\mathbf{S}\psi_{\mathbf{n}}^{(j)} = j\mathbf{n}\psi_{\mathbf{n}}^{(j)} + \sqrt{2j}\mathbf{v}(\mathbf{n})g_{\mathbf{n}}\cdot\mathcal{D}_{j-1}^{(j)}, \tag{7.49}$$

where $\mathbf{v}(\mathbf{n}) = (v_1(\mathbf{n}), v_2(\mathbf{n}), v_3(\mathbf{n}))\in\mathbb{C}^3$ is given by

$$\begin{aligned}
v_1(\mathbf{n}) &= \cos^2\left(\frac{\theta}{2}\right) - \sin^2\left(\frac{\theta}{2}\right)\mathrm{e}^{-2i\varphi}\\
v_2(\mathbf{n}) &= \frac{i}{2}\left(\cos^2\left(\frac{\theta}{2}\right) + \sin^2\left(\frac{\theta}{2}\right)\mathrm{e}^{-2i\varphi}\right)\\
v_3(\mathbf{n}) &= -\frac{\sin\theta}{2}\mathrm{e}^{-i\varphi}.
\end{aligned}$$

In particular we have $\mathbf{n}\cdot\mathbf{v}(\mathbf{n}) = 0$.

Proof It is enough to compute $\hat{S}_k\psi_g^{(j)}$ for $k = 1, 2, 3$ for

$$g = \begin{pmatrix} a & b\\ -\bar{b} & a\end{pmatrix}\in SU(2),\ a\in\mathbb{R},\ b\in\mathbb{C}.$$

It is more convenient to compute $g^{-1}\cdot\hat{S}_k g\cdot\psi_0^{(j)}$. Elementary computations give the following formulas

$$\begin{aligned}
g^{-1}\cdot\hat{S}_3 g\cdot\psi_0^{(j)} &= j(a^2-|b|^2)\psi_0^{(j)} + \sqrt{2j}ab\mathcal{D}_{j-1}^{(j)},\\
g^{-1}\cdot\hat{S}_+ g\cdot\psi_0^{(j)} &= -2jab\psi_0^{(j)} - \sqrt{2j}b^2\mathcal{D}_{j-1}^{(j)},\\
g^{-1}\cdot\hat{S}_- g\cdot\psi_0^{(j)} &= -2ja\bar{b}\psi_0^{(j)} - \sqrt{2j}a^2\mathcal{D}_{j-1}^{(j)}.
\end{aligned}$$

So (7.49) follows directly. Moreover with a little more computation we can check that $\mathbf{n}\cdot\mathbf{v}(\mathbf{n}) = 0$ which proves again (7.48). □

As in the Heisenberg setting, the spin-coherent states family $|\mathbf{n}\rangle$ is not an orthogonal system. One can compute the scalar product of two coherent states $|\mathbf{n}\rangle,\ |\mathbf{n}'\rangle$:

Proposition 87 *One has*

$$\boxed{\langle \mathbf{n}'|\mathbf{n}\rangle = e^{ij\Phi(\mathbf{n},\mathbf{n}')}\left(\frac{1+\mathbf{n}\cdot\mathbf{n}'}{2}\right)^j}. \tag{7.50}$$

where $\Phi(\mathbf{n},\mathbf{n}')$ is a real number. If the spherical triangle with vertices $\{\mathbf{n}_0,\mathbf{n},\mathbf{n}'\}$ is an Euler triangle then $\Phi(\mathbf{n},\mathbf{n}')$ is the oriented area of this triangle.

Proof To each point $\mathbf{n}$ on the sphere $\mathbb{S}^2$ we associate its spherical coordinates $\theta \in [0,\pi)$, $\varphi \in [0,2\pi)$ as usual:

$$x = \sin\theta\cos\varphi,\ \ y = \sin\theta\sin\varphi,\ \ z = \cos\theta.$$

The corresponding element $g_{\mathbf{n}} \in SU(2)$ is defined as

$$g_{\mathbf{n}} = \exp\left(i\frac{\theta}{2}(\sin\varphi\sigma_1 - \cos\varphi\sigma_2)\right).$$

The matrix $i(\sin\varphi\sigma_1 - \cos\varphi\sigma_2)$ can be viewed as a pure quaternion that we denote q. Using (7.15) we have that

$$g_{\mathbf{n}} = \cos\frac{\theta}{2} + q\sin\frac{\theta}{2}.$$

Taking $\mathbf{n}' \in \mathbb{S}^2$ with spherical coordinates θ', φ' we get (using quaternion calculus)

$$\begin{aligned} g_{\mathbf{n}}g_{\mathbf{n}'} &= \cos\frac{\theta}{2}\cos\frac{\theta'}{2} - \sin\frac{\theta}{2}\sin\frac{\theta'}{2}\cos(\varphi-\varphi') \\ +\sigma_3 i\sin\frac{\theta}{2}\sin\frac{\theta'}{2}\sin(\varphi'-\varphi) &+ q'\cos\frac{\theta}{2}\sin\frac{\theta'}{2} + q\sin\frac{\theta}{2}\cos\frac{\theta'}{2}. \end{aligned} \tag{7.51}$$

Therefore $g_{\mathbf{n}}g_{\mathbf{n}'}$ is of the form (7.17) with

$$a = \cos\frac{\theta}{2}\cos\frac{\theta'}{2} - \sin\frac{\theta}{2}\sin\frac{\theta'}{2}e^{i(\varphi-\varphi')}.$$

The element $g_{\mathbf{n}}$ of $SU(2)$ can be written as

$$g_{\mathbf{n}} = \begin{pmatrix} \cos\frac{\theta}{2} & -\sin\frac{\theta}{2}e^{-i\varphi} \\ \sin\frac{\theta}{2}e^{i\varphi} & \cos\frac{\theta}{2} \end{pmatrix}.$$

Now we turn to the representation $T^{(j)}(g)$ in the space $V^{(j)}$ of homogeneous polynomials of degree $2j$ in z_1, z_2. The coherent state $|\mathbf{n}\rangle$ is of the form $T(g_{\mathbf{n}})\psi_0$ for some $\mathbf{n} \in \mathbb{S}^2$ and ψ_0 being a reference state. We choose $\psi_0 = \frac{z_1^{2j}}{\sqrt{(2j)!}}$ in the homogeneous polynomial representation. Note that this is coherent with the choice $|j,-j\rangle$. The overlap between two coherent states is given by the scalar product:

$$\langle T^{(j)}(g_{\mathbf{n}})\psi_0, T^{(j)}(g_{\mathbf{n}'})\psi_0\rangle = \frac{1}{(2j)!}\langle (z_2\cos\frac{\theta}{2} + z_1\sin\frac{\theta}{2}e^{i\varphi})^{2j}, (z_2\cos\frac{\theta'}{2} + z_1\sin\frac{\theta'}{2}e^{i\varphi'})^{2j}\rangle. \tag{7.52}$$

We make use of the following result:

Lemma 47 *Let* $\Pi_{\alpha,\beta}(z_1, z_2) = (\alpha z_1 + \beta z_2)^{2j}$ *Then the scalar product in* $V^{(j)}$ *of two such polynomials equals:*

$$\langle \Pi_{\alpha',\beta'}, \Pi_{\alpha,\beta}\rangle = (2j)!(\alpha\bar{\alpha}' + \beta\bar{\beta}')^{2j}.$$

Then we get

$$\langle \mathbf{n}'|\mathbf{n}\rangle = \left(\cos\frac{\theta}{2}\cos\frac{\theta'}{2} + \sin\frac{\theta}{2}\sin\frac{\theta'}{2}e^{i(\varphi-\varphi')}\right)^{2j}.$$

By an easy calculus we obtain

$$\left|\cos\frac{\theta}{2}\cos\frac{\theta'}{2} + \sin\frac{\theta}{2}\sin\frac{\theta'}{2}e^{i(\varphi-\varphi')}\right|^2 = \frac{1+\mathbf{n}\cdot\mathbf{n}'}{2}.$$

Let us now compute the phase of the overlap $\langle \mathbf{n}_{'}|\mathbf{n}\rangle$. It is a non trivial and interesting computation related with Berry phase as we shall see. It can be extended to a more general setting for coherent states on Kähler manifolds [41]. We follow here the elementary proof of the paper [5].

Let us denote

$$\eta = \arg\left(\cos\frac{\theta}{2}\cos\frac{\theta'}{2} + \sin\frac{\theta}{2}\sin\frac{\theta'}{2}e^{i(\varphi-\varphi')}\right).$$

Using classical trigonometric formula we get

$$\tan\eta = \frac{\sin\theta\sin\theta'\sin(\varphi-\varphi')}{(1+\cos\theta)(1+\cos\theta') + \sin\theta\sin\theta'\cos(\varphi-\varphi')}. \tag{7.53}$$

We now compare this formula with the following spherical geometric formula already known by Euler and Lagrange (see [95] for a detailed proof).

Let 3 points $\mathbf{n}_1, \mathbf{n}_2, \mathbf{n}_3$ be on the unit sphere $\mathbb{S}^2$, not all on the same great circle and such that the spherical triangle with vertices $\mathbf{n}_1, \mathbf{n}_2, \mathbf{n}_3$ is an Euler triangle i.e. the angles and the sides are all smaller than π. Let ω be the aera of this triangle. Then we have

$$\tan\frac{\omega}{2} = \frac{|\det[\mathbf{n}_1, \mathbf{n}_2, \mathbf{n}_3]|}{1+\mathbf{n}_1\cdot\mathbf{n}_2 + \mathbf{n}_2\cdot\mathbf{n}_3 + \mathbf{n}_3\cdot\mathbf{n}_1}. \tag{7.54}$$

Notice that this formula takes account of the orientation of the piecewise geodesic curve with vertices $\mathbf{n}_1, \mathbf{n}_2, \mathbf{n}_3$. The orientation is positive if the frame $\{O\mathbf{n}_1, O\mathbf{n}_2, O\mathbf{n}_3\}$ is direct.

From (7.53) and (7.54) we get directly that $\tan\frac{\eta}{2} = \tan\frac{\omega}{2}$. But $\omega, \eta \in]-\pi, \pi[$ so we can conclude that $\eta = \omega$. □

As it is expected, the spin coherent state system provides a "resolution of the identity" in the Hilbert space $V^{(j)}$:

Proposition 88 *We have the formula*

$$\boxed{\frac{2j+1}{4}\int_{\mathbb{S}^2} d\mathbf{n}|\mathbf{n}\rangle\langle\mathbf{n}| = \mathbb{1}}. \tag{7.55}$$

Or using complex coordinates $|\zeta\rangle$,

$$\boxed{\int_{\mathbb{C}} d\mu_j(\zeta)|\zeta\rangle\langle\zeta| = \mathbb{1}} \tag{7.56}$$

where the measure $d\mu_j$ *is*

$$d\mu_j(\zeta) = \frac{2j+1}{\pi}\frac{d^2\zeta}{(1+|\zeta|^2)^2}$$

with $d^2\zeta = \frac{|d\zeta\wedge d\bar{\zeta}|}{2}$.

Proof The two formulas are equivalent by the change of variables $\zeta = -\tan\frac{\theta}{2}\mathrm{e}^{-i\varphi}$. So it is sufficient to prove the complex version.

Let us recall the analytic expression for $|\mathbf{n}\rangle$ and $|\zeta\rangle$.

$$|\mathbf{n}\rangle = \psi_{\mathbf{n}}(z_1, z_2) = \frac{1}{\sqrt{(2j)!}}(\cos\frac{\theta}{2}z_1 + \sin\frac{\theta}{2}\mathrm{e}^{-i\varphi}z_2)^{2j}. \tag{7.57}$$

$$|\zeta\rangle = \psi_\zeta(z_1, z_2) = \frac{1}{\sqrt{(2j)!}}\frac{1}{(1+|\zeta|^2)^j}(z_1 - \zeta z_2)^{2j}. \tag{7.58}$$

Recall that we have an orthonormal basis of Dicke states $\{\mathscr{D}_k^{(j)}\}_{-j\le k\le j}$ in $V^{(j)}$ where

$$\mathscr{D}_k^{(j)}(z_1, z_2) = \frac{z_1^{j+k}z_2^{j-k}}{\sqrt{(j+k)!(j-k)!}}.$$

So we get

$$\langle\mathscr{D}_k^{(j)}, \psi_\zeta\rangle = \left(\frac{(2j)!}{(j+k)!(j-k)!}\right)^{1/2}(1+|\zeta|^2)^{-j}\zeta^{j-k}. \tag{7.59}$$

And using Parseval formula we get

$$\langle \eta | \zeta \rangle = (1 + |\zeta|^2)^{-j}(1 + |\eta|^2)^{-j}(1 + \bar{\eta}\zeta)^{2j}. \tag{7.60}$$

(7.60) is the complex version for the overlap formula of two coherent states (7.52).

It is convenient to introduce now the spin Bargmann transform (see Chap. 1 for the Bargmann transform in the Heisenberg setting).

For every $v \in V^{(j)}$ we define the following polynomial in the complex variable ζ

$$v^{j,\sharp}(\zeta) = \langle \psi_\zeta | v \rangle (1 + |\zeta|^2)^j.$$

If $v = \sum_{-j \le k \le j} c_k e_k$ a direct computation gives

$$\int_{\mathbb{C}} |v^{j,\sharp}(\zeta)|^2 \frac{d^2\zeta}{(1 + |\zeta|^2)^{2j+2}} = \frac{\pi}{2j+1} \sum_{-j \le k \le j} |c_k|^2$$
$$= \frac{\pi}{2j+1} \|v\|^2. \tag{7.61}$$

Or equivalently

$$\int_{\mathbb{C}} |\langle \psi_\zeta | v \rangle|^2 \frac{d^2\zeta}{(1 + |\zeta|^2)^2} = \frac{\pi}{2j+1} \|v\|^2. \tag{7.62}$$

This formula is equivalent to the overcompleteness formula by polarisation. □

Remark 47 From the proof we have obtained that the spin-Bargmann transform: $\mathscr{B}^j v(\zeta) := v^{j,\sharp}(\zeta)$ is an isometry from $V^{(j)}$ onto the space $\mathscr{P}_{2j}$ of polynomials of degree at most $2j$ equipped with the scalar product

$$\langle P, Q \rangle = \frac{2j+1}{\pi} \int_{\mathbb{C}} \bar{P}(\zeta) Q(\zeta) \frac{d^2\zeta}{(1 + |\zeta|^2)^{2j+2}}$$

In particular we have

$$\psi_\eta^{j,\sharp}(\zeta) = (1 + |\eta|^2)^{-j}(1 + \eta\bar{\zeta})^{2j}.$$

We now extend the computation of the phase $\Phi(\mathbf{n}, \mathbf{n}')$ in the general case by giving for it a different expression related with the well known geometric phase. A similar computation was done in [196].

Proposition 89 *For any* $\mathbf{n}, \mathbf{n}' \in \mathbb{S}^2$ *we have*

$$\Phi(\mathbf{n}, \mathbf{n}') = -\frac{i}{j} \oint_{[\mathbf{n}, \mathbf{n}']} \langle \psi_n, d\psi_n \rangle \tag{7.63}$$

where the integral is computed on the shortest geodesic arc joining $\mathbf{n}$ *to* $\mathbf{n}'$ *of the one differential form* $\langle \psi_n, d\psi_n \rangle$

To explain the formula we recall here the main idea behind the geometric phase discovered by Berry [28] and Pancharatnam [202] (see also [2]).

Let us consider a closed loop $\mathbf{n}(t) : [0, 1] \to \mathbb{S}^2$ which is continuous by part, $\mathbf{n}(0) = \mathbf{n}(1)$. We define the time-dependent Hamiltonian as

$$\hat{H}(t) = \mathbf{n}(t) \cdot \mathbf{S}.$$

The solution of the time-dependent Schrödinger equation with this Hamiltonian is denoted $\psi(t)$. Let us consider $\eta(t)$ in the Hilbert space and $\alpha(t) \in \mathbb{R}$ such that

$$\psi(t) = e^{i\alpha(t)}\eta(t).$$

We choose $\alpha(t)$ such that $\langle \eta(t), \dot{\eta}(t) \rangle = 0$. In other terms $\eta(t)$ describes a parallel transport along the curve. Then the geometrical phase $\alpha(t)$ obeys

$$i\dot{\alpha}(t) + \langle \psi(t), \dot{\psi}(t) \rangle = 0.$$

We thus have

$$\alpha(1) := \alpha(\gamma) = i \oint_\gamma \langle \psi, d\psi \rangle.$$

If γ delimitates a portion Γ of $\mathbb{S}^2$ we have by Stokes theorem

$$\alpha(\gamma) = \int_\Gamma \langle d\psi, d\psi \rangle$$

where the product of the differentials is the external product.

Let $\psi_\mathbf{n}$ be the coherent state $|\mathbf{n}\rangle$ obtained at $t = 1$ from $\psi(0) = \psi_{\mathbf{n}_0}$. In the homogeneous polynomial representation we get:

$$\psi_\mathbf{n} = \frac{1}{\sqrt{(2j)!}} \left(\cos\frac{\theta}{2} z_1 + \sin\frac{\theta}{2} \mathrm{e}^{-i\varphi} z_2 \right)^{2j}.$$

Thus

$$\frac{\partial \psi_\mathbf{n}}{\partial \theta} = \frac{j}{\sqrt{(2j)!}} \left(z_1 \cos\frac{\theta}{2} e^{i\varphi} - z_2 \sin\frac{\theta}{2} \right) \left(\cos\frac{\theta}{2} z_1 + \sin\frac{\theta}{2} \mathrm{e}^{-i\varphi} z_2)^{2j} \right)^{2j-1}.$$

$$\frac{\partial \psi_\mathbf{n}}{\partial \varphi} = \frac{2j}{\sqrt{(2j)!}} i \sin\frac{\theta}{2} e^{i\varphi} z_1 \left(z_1 \sin\frac{\theta}{2} e^{i\varphi} + z_2 \cos\frac{\theta}{2} \right)^{2j-1}$$

Using the invariance of the scalar product in $V^{(j)}$ under $SU(2)$ transformations we perform the change of variables

$$Z_1 = z_1 \cos\frac{\theta}{2} - z_2 e^{-i\varphi}\sin\frac{\theta}{2},$$
$$Z_2 = z_1 e^{i\varphi}\sin\frac{\theta}{2} + z_2\cos\frac{\theta}{2}. \tag{7.64}$$

One thus have using the orthogonality relations in $V^{(j)}$:

$$\langle\psi_{\mathbf{n}}, \frac{\partial\psi_n}{\partial\theta}\rangle = \frac{j}{(2j)!}e^{i\varphi}\langle Z_1 Z_2^{2j-1}, Z_2^{2j}\rangle = 0 \tag{7.65}$$

One can calculate the scalar product of $\frac{\partial\psi_{\mathbf{n}}}{\partial\theta}$ and $\frac{\partial\psi_{\mathbf{n}}}{\partial\varphi}$:

$$\begin{aligned}\langle\frac{\partial\psi_{\mathbf{n}}}{\partial\theta}, \frac{\partial\psi_{\mathbf{n}}}{\partial\varphi}\rangle &= i\frac{2j^2}{(2j)!}e^{-i\varphi}\sin\frac{\theta}{2}\langle Z_1 Z_2^{2j-1}, \left(Z_1\cos\frac{\theta}{2}e^{i\varphi} + Z_2\sin\frac{\theta}{2}\right)Z_2^{2j-1}\rangle\\ &= \frac{1}{2}ij\sin\theta,\end{aligned} \tag{7.66}$$

since

$$\langle Z_1 Z_2^{2j-1}, Z_1 Z_2^{2j-1}\rangle = (2j-1).$$

We shall now calculate the phase of the scalar product $\langle\psi_{\mathbf{n}}, \psi_{\mathbf{n}'}\rangle$ by calculating the geometric phase along the geodesic triangle $\mathscr{T} = [\mathbf{n}_0, \mathbf{n}, \mathbf{n}', \mathbf{n}_0]$. Denote by Ω the domain on $\mathbb{S}^2$ delimited by $\mathscr{T}$. We have

$$\alpha(\mathscr{T}) = i\oint_{\mathscr{T}}\langle\psi_{\mathbf{n}}, d\psi_{\mathbf{n}}\rangle = i\int_{\Omega}\langle d\psi_{\mathbf{n}}, d\psi_{\mathbf{n}}\rangle. \tag{7.67}$$

This yields

$$\alpha(\mathscr{T}) = -j\int_{\Omega}\sin\theta d\theta d\varphi = -j\mathrm{Area}(\Omega).$$

We denote by $[\mathbf{n}_1, \mathbf{n}_2]$ the portion of great circle on $\mathbb{S}^2$ between $\mathbf{n}_1$ and $\mathbf{n}_2$. We now integrate (7.67) successively along $[\mathbf{n}_0, \mathbf{n}]$, $[\mathbf{n}, \mathbf{n}']$ and $[\mathbf{n}', \mathbf{n}_0]$. From the fact that $[\mathbf{n}_0, \mathbf{n}]$, $[\mathbf{n}', \mathbf{n}_0]$ lie in verticle planes we have

$$\int_{[\mathbf{n}_0,\mathbf{n}]}\langle\psi_{\mathbf{n}}, d\psi_{\mathbf{n}}\rangle = \int_{[\mathbf{n}',\mathbf{n}_0]}\langle\psi_{\mathbf{n}}, d\psi_{\mathbf{n}}\rangle = 0.$$

Thus for an Euler triangle we get:

$$-j\mathrm{Area}(\Omega) = -j\Phi(\mathbf{n}, \mathbf{n}') = i\int_{[\mathbf{n},\mathbf{n}']}\langle\psi_{\mathbf{n}}, d\psi_{\mathbf{n}}\rangle$$

In the general case we can subdivise the triangle in several Euler triangles with vertex at $\mathbf{n}_0$ by adding vertices between $\mathbf{n}$ and $\mathbf{n}'$. Then we get that

$$\Phi(\mathbf{n}, \mathbf{n}') = -\frac{i}{j} \int_{[\mathbf{n},\mathbf{n}']} \langle \psi_{\mathbf{n}}, d\psi_{\mathbf{n}} \rangle.$$

□

As a consequence of our study of the geometric phase, let us now compute the phase in formula (7.43).

We already know that

$$G^{(j)}(\mathbf{n}_1) G^{(j)}(\mathbf{n}_2) \psi_0 = G^{(j)}(\mathbf{n}_3) e^{i\alpha}$$

and we have to compute α.

We have

$$\begin{aligned} \langle \psi_0, G^{(j)}(\mathbf{n}_1) D(\mathbf{n}_2) \psi_0 \rangle &= e^{i\alpha} \langle \psi_0, G^{(j)}(\mathbf{n}_3) \psi_0 \rangle \\ &= \langle G^{(j)}(\mathbf{n}_1)^* \psi_0, G^{(j)}(\mathbf{n}_2) \psi_0 \rangle \end{aligned} \tag{7.68}$$

From Lemma 48 we know that $\langle \psi_0, G^{(j)}(\mathbf{n}_3) \psi_0 \rangle \geq 0$. But $G^{(j)}(\mathbf{n}_1)^* = G^{(j)}(\mathbf{n}_1^*)$ where $\mathbf{n}_1^*$ is the symmetric of $\mathbf{n}_1$ on the great circle determined by $\mathbf{n}_0$ and $\mathbf{n}_1$. Applying the computation of the phase in (7.52) we get

$$\alpha = \arg(\langle \mathbf{n}_1^*, \mathbf{n}_2 \rangle).$$

With an elementary geometric argument we get the phase in formula (7.43).

As a consequence, we can get the phase $\mathscr{A}(g, \mathbf{n})$ in formula (7.46). If $g = g_{\mathbf{m}}$ then $\mathscr{A}(g_m, \mathbf{n}) = \Phi(\mathbf{m}, \mathbf{n})$. For a generic $g \in SU(2)$ we have $g = g_{\mathbf{m}} r_3(\delta)$ for some $\delta \in \mathbb{R}$.

So we have $g g_{\mathbf{n}} = g_{\mathbf{m}} r_3(\delta) g_{\mathbf{n}}$. Let $\mathbf{n} = (\theta, \varphi)$(polar coordinates). We have $r_3(\delta) g_{\mathbf{n}} = g_{\mathbf{n}'} r_3(\delta)$ where $\mathbf{n}' = (\theta, \varphi + \delta)$. So using computation of Φ in (7.43) and elementary geometry we get that $\mathscr{A}(g, \mathbf{n})$ is equal to the area of the spherical triangle $[n_0, n, g \cdot \mathbf{n}]$. □.

Let us close our discussion concerning the geometric phase for coherent states by the following result:

Lemma 48 *Let ψ_1 and ψ_2 be two different states on a great circle of $\mathbb{S}_j$ (unit sphere on $V^{(j)}$). One parametrizes this by the angle θ in the following way:*

$$\psi(\theta) = x_1(\theta) \psi_1 + x_2(\theta) \psi_2$$

where $x_i(\theta) \in \mathbb{R}$ and $\psi(0) = \psi_1$, $\psi(\theta_0) = \psi_2$.

One assumes that

$$\langle \psi(\theta), \dot{\psi}(\theta) \rangle = 0.$$

Then ψ_1 and ψ_2 are in phase, namely

$$\langle \psi_1, \psi_2 \rangle > 0$$

Proof One can assume that $a = \Re\langle \psi_1, \psi_2 \rangle > 0$. Then by an easy calculus we get

$$x_1(\theta) = \cos\theta - \frac{a}{\sqrt{1-a^2}} \sin\theta, \quad x_2(\theta) = \frac{1}{\sqrt{1-a^2}} \sin\theta, \quad a = \cos\theta_0$$

$$\Im\langle \psi(\theta), \dot{\psi}(\theta) \rangle = \frac{1}{2\sqrt{1-a^2}} \Im\langle \psi_1, \psi_2 \rangle.$$

One deduces that $\langle \psi_1, \psi_2 \rangle$ is real therefore positive. □

Remark 48 For two coherent states $|\mathbf{n}\rangle$ and $|\mathbf{n}'\rangle$ there are two natural geodesics joining them: the geodesic on the two sphere $\mathbb{S}^2$ and the geodesic on the sphere $\mathbb{S}_j$ (sphere $(2j+1)$-dimensional). It is a consequence of results proved above that the geometric phases for these two curves in $V^{(j)}$ are the same. This was not obvious before computations.

7.5 Coherent States on the Riemann Sphere

We have seen that it is convenient to compute on the sphere $\mathbb{S}^2$ using complex coordinates given by the stereographic projection. We shall give here more details about this. In particular this gives a semi-classical interpretation for the spin coherent states and a quantization of the sphere $\mathbb{S}^2$. The picture is analogous to the harmonic oscillator coherent states and the associated Wick quantization of the phase space $\mathbb{R}^{2d}$.

The stereographic projection of the sphere $\mathbb{S}^2$ from its south pole is the transformation $\pi_s(\mathbf{n}) = \zeta$, defined by

$$\zeta = \frac{n_1 + in_2}{1 + n_3}$$

where $\mathbf{n} = (n_1, n_2, n_3)$. π_s is an homeomorphism from $\mathbb{S}^2_\star := \mathbb{S}^2 \backslash \{(0,0,1)\}$ on the complex plane $\mathbb{C}$. Moreover π_s can be extended in an homeomorphism from $\mathbb{S}^2$ on $\tilde{\mathbb{C}} := \mathbb{C} \cup \{\infty\}$ such that $\pi_s(0,0,1) = \infty$. $\tilde{\mathbb{C}}$ is a one dimensional complex and compact manifold called the Riemann sphere.

π_s^{-1} is determined by the formula $\pi_s^{-1}\zeta = (n_1, n_2, n_3)$ where

$$n_1 = \frac{\zeta + \bar{\zeta}}{1 + |\zeta|^2}, \quad n_2 = \frac{\zeta - \bar{\zeta}}{i(1 + |\zeta|^2)}, \quad n_3 = \frac{1 - |\zeta|^2}{1 + |\zeta|^2}.$$

It is known that the group of automorphisms of $\tilde{\mathbb{C}}$ (bijective and biholomorphic transformations) is the Möbius group, the group of homographic transformations $h(z) = \frac{az+b}{cz+d}$ where $a, b, c, d \in \mathbb{C}$ such that $ad - bc = 1$. The conventions are: if $c = 0$ then $h(\infty) = \infty$; if $c \neq 0$ then $h(\infty) = \frac{a}{c}$, $h(-\frac{d}{c}) = \infty$.

Let us denote $g = \begin{pmatrix} a & b \\ c & d \end{pmatrix}$, $g \in SL(2, \mathbb{C})$ and $f = h_g$. Then is not difficult to see that $h_g = \mathbb{1}$ if and only if $g = \pm\mathbb{1}_2$.

We have seen that if $g \in SU(2)$ then R_g defines a rotation in $\mathbb{S}^2$. In $\tilde{\mathbb{C}}$ we get that R_g becomes a Möbius transformation:

Lemma 49 *Let* $\tilde{R}_g = \pi_s R_g \pi_s^{-1}$ *and* $g \in SU(2)$, $g = \begin{pmatrix} a & b \\ -\bar{b} & \bar{a} \end{pmatrix}$. *Then we have*

$$\tilde{R}_g \zeta = \frac{a\zeta + b}{-\bar{b}\zeta + \bar{a}}, \quad \forall \zeta \in \tilde{\mathbb{C}}. \tag{7.69}$$

Proof We only give a sketch. First it is enough to consider $g = g_2(\theta)$. After some computations we get the result using that Möbius transformations preserve the cross ratio $\frac{\zeta_1 - \zeta_3}{\zeta_2 - \zeta_3}$. □

For simplicity we denote $\tilde{R}_g \zeta := g \cdot \zeta$.

Now our aim is to realize the representation $(T^{(j)}, V^j)$ in a space of holomorphic functions on the Riemann sphere $\tilde{\mathbb{C}}$. This is achieved easily with the spin-Bargmann transform $\mathscr{B}^j$ introduced above. Recall that $\mathscr{B}^j v(\zeta) = \langle \psi_\zeta, v \rangle (1 + |\zeta|^2)^j$.

We get the images of the Dicke basis and of the coherent states:

$$\tilde{d}_\ell(\zeta) := \mathscr{B}^j(d_k)(\zeta) = \left(\frac{(2j)!}{\ell!(2j - \ell)!} \right)^{1/2} \zeta^\ell, \; j + k = \ell. \tag{7.70}$$

$$\tilde{\psi}_\zeta(z) = \langle \psi_z, \psi_\zeta \rangle (1 + |z|^2)^j = (1 + |\zeta|^2)^{-j} (1 + \bar{\zeta} z)^{2j}, \; z, \zeta \in \mathbb{C}. \tag{7.71}$$

Let us remark that the Hilbert space V^j is transformed in the Hilbert space $\mathscr{P}_{2j}$ (polynomials of degree $\leq 2j + 1$) and that $\mathscr{P}_{2j}$ coïncides with the space of holomorphic functions P on $\mathbb{C}$ such that

$$\int_{\mathbb{C}} |P(\zeta)|^2 (1 + |\zeta|^2)^{-2j-2} d^2\zeta < +\infty.$$

So the exponential weight of the usual Bargmann space is replaced here by a polynomial weight.

We see now that the action of $SU(2)$ in the Bargmann space $\mathscr{P}_{2j}$ is simple and has a nice semi-classical interpretation.

Let us denote $\tilde{T}^j(g) := \mathscr{B}^j T^j(g) \mathscr{B}^{j\star}$.

On the Riemann sphere the representation $\tilde{T}^j$ has the following expression.

Proposition 90 *For every $\psi \in \mathscr{P}_{2j}$ and $g \in SU(2)$ we have*

$$\tilde{T}^j(g)\psi(\zeta) = \mu_j(g, \zeta)\psi(\tilde{R}_{g^{-1}}(\zeta)) \tag{7.72}$$

where $g = \begin{pmatrix} a & b \\ -\bar{b} & \bar{a} \end{pmatrix}$, $\mu_j(g, \zeta) = (a + \bar{b}\zeta)^{2j}$ *and* $\tilde{R}_{g^{-1}}(\zeta) = \frac{\bar{a}\zeta - b}{\bar{b}\zeta + a}$.

Proof It is enough to prove the proposition for $\psi(\zeta) = \zeta^\ell$. From definition of $\mathscr{B}^j$ we get

$$(\mathscr{B}^j T^j(g) d_k(\zeta) = (2j)! \frac{(j+k)!}{\sqrt{(j-k)!}} \langle (\bar{\zeta} z_1 + z_2)^{2j}, (\bar{a} z_1 - b z_2)^{j+k} (\bar{b} z_1 + a z_2)^{j-k} \rangle_{V^j}.$$

Using that the scalar product in V^j is invariant for the $SU(2)$ action we obtain

$$(\mathscr{B}^j T^j(g) d_\ell)(\zeta) = \left(\frac{(2j)!}{\ell!(2j-\ell)!} \right)^{1/2} (a + \bar{b}\zeta)^{2j} \left(\frac{\bar{a}\zeta - b}{\bar{b}\zeta + a} \right)^\ell \tag{7.73}$$

Hence we get the proposition □

Now, following Onofri [201] we shall give a classical mechanical interpretation of the term $\mu(g, \zeta)$. This interpretation can be extended to any semi-simple Lie group, as we shall see later.

Let us introduce $K(\zeta, \bar{\zeta}) = 2\log(1 + \zeta\bar{\zeta})$ (Kähler potential), d the exterior differential, ∂ the exterior differential in ζ, $\bar{\partial}$ the exterior differential in $\bar{\zeta}$.

We have $d = \partial + \bar{\partial}$ and $d\partial = \bar{\partial}\partial = -\partial\bar{\partial}$.

We introduce the one form $\theta = -i\partial K$ and the two form

$$\omega = d\theta = 2i \frac{d\zeta \wedge d\bar{\zeta}}{(1 + |\zeta|^2)^2}.$$

ω is clearly a non-degenerate antisymmetric two form. So $(\tilde{\mathbb{C}}, \omega)$ is a symplectic manifold. Moreover it is a Kähler one dimensional complex manifold for the Hermitian metric

$$ds^2 = 4 \frac{d\zeta d\bar{\zeta}}{(1 + |\zeta|^2)^2}.$$

It is not difficult to see that ω is invariant by the action of $SU(2)$: $g_\star\omega = \omega$. In other words $SU(2)$ acts in $\tilde{\mathbb{C}}$ by canonical transformations.

$\tilde{\mathbb{C}}$ is connected and simply connected, so there exists a smooth function $S(g, \zeta)$ such that $dS(g, \zeta) = \theta - g_\star\theta$.

Now let us compute $d\mu$ as follows.

From (7.72) with $\psi = \tilde{\psi}_0 = 1$ and using that $\langle \psi_\zeta, \psi_0 \rangle = (1 + |\zeta|^2)^j$ we get:

$$\mu_j(g, \zeta) = \frac{\langle \psi_0, T^j(g^{-1} g_\zeta)^\star \psi_0 \rangle}{\langle \psi_0, \psi_\zeta \rangle}.$$

Then we compute $d\mu = (ij(\theta - g_\star\theta))\mu = (ijdS)\mu$. Hence we get the classically mechanical interpretation for $\mu(g, \zeta)$

$$\mu(g, \zeta) = \mu(g, 0)e^{ij\int_0^\zeta(\theta - g_\star\theta)}. \tag{7.74}$$

□

7.6 Application to High Spin Inequalities

One of the first successful use of spin coherent states was the thermodynamic limit of spin systems as an application of Berezin-Lieb inequalities. Berezin-Lieb inequality holds true for general coherent states.

7.6.1 *Berezin-Lieb Inequalities*

We shall follow here the notations of Sect. 2.6 concerning Wick quantization. We assume here that the Hilbert space $\mathscr{H}$ is finite dimensional (this is enough for our application). Let $\hat{A} \in \mathscr{L}(\mathscr{H})$ with a covariant symbol A_c and contravariant symbol A^c defined on some metric space M with a probability Radon measure $d\mu(m)$. It is not difficult to see that these two symbols satisfy the following duality formulas

$$\mathrm{Tr}(\hat{A}\hat{B}) = \int_M A_c(m)B^c(m)d\mu(m), \tag{7.75}$$

$$A_c(m') = \int_M |\langle e_m|e_{m'}\rangle|^2 A^c(m)d\mu(m). \tag{7.76}$$

In particular we have

$$\mathrm{Tr}(\hat{A}) = \int_M A_c(m)d\mu(m). \tag{7.77}$$

Let us remark while A_c is well defined, A^c is not uniquely defined in general.

Theorem 43 *Let $\hat{H}$ be a self-adjoint operator in $\mathscr{H}$ and χ a convex function on $\mathbb{R}$. Then we have the inequalities*

$$\int_M \chi(H_c(m))d\mu(m) \le \mathrm{Tr}(\chi(\hat{H})) \le \int_M \chi(H^c(m))d\mu(m). \tag{7.78}$$

Proof Let us recall the Jensen inequality [228], which is the main tool for proving the Berezin-Lieb inequalities. For any probability measure ν on M, any convex function χ on $\mathbb{R}$ and any $f \in L^1(M, d\nu)$ we have

$$\chi\left(\int_M f d\nu(m)\right) \leq \int_M \chi(f(m))d\nu(m). \tag{7.79}$$

We start with the formula

$$\mathrm{Tr}\left(\chi(\hat{H})\right) = \int_M \langle e_m, \chi(\hat{H})e_m\rangle d\mu(m).$$

Using the spectral decomposition for self-adjoint operators, we have

$$\langle e_m, \chi(\hat{H})e_m\rangle = \int_{\mathbb{R}} \chi(\lambda)d\nu_m(\lambda)$$

where ν_m is the spectral measure of the state e_m. It is a probability (discrete) measure. So Jensen inequality gives

$$\langle e_m, \chi(\hat{H})e_m\rangle \geq \chi(A_c(m)).$$

Integrating in m we get the first Berezin-Lieb inequality

$$\int_M \chi(H_c(m))d\mu(m) \leq \mathrm{Tr}(\chi(\hat{H})).$$

For the second inequality we introduce an orthonormal basis of $\mathscr{H}$: $\{v_n\}_{1\leq n\leq N}$ of eigenfunctions of $\hat{A}$. So we have

$$\langle v_n|\chi(\hat{H})v_n\rangle = \chi(\langle v_n|\hat{H}v_n\rangle) = \chi\left(\int_M H^c(m)|\langle v_n|e_m\rangle|^2 d\mu(m)\right).$$

From Jensen inequality applied with the probability measure $|\langle v_n|e_m\rangle|^2 d\mu(m)$ we get

$$\langle v_n|\chi(\hat{H})v_n\rangle \leq \int_M \chi(H^c(m))|\langle v_n|e_m\rangle|^2 d\mu(m).$$

Summing over n we get the second Berezin-Lieb inequality. □

7.6.2 High Spin Estimates

We consider a one dimensional Heisenberg chain of N spin $\mathbf{S}^n = (S_1^n, S_2^n, S_3^n)$, $1 \leq n \leq N$. The Hamiltonian of this system is

$$\hat{H} = -\sum_{1\leq n\leq N-1} \mathbf{S}^n\mathbf{S}^{n+1}.$$

$\hat{H}$ is an Hermitian operator in the finite dimensional Hilbert space $\mathcal{H}_N = \otimes^N \mathcal{H}^n$ where $\mathcal{H}^n = V^{(j)}$ for every n.

From the coherent states systems in $V^{(j)}$ we get in a standard way a coherent system in $\mathcal{H}_N$ parametrized on $(\mathbb{S}^2)^N$ (or $\mathbb{C}^N$ using the complex parametrisation). In the sphere representation, if $\Omega = (\mathbf{n}^1, \ldots, \mathbf{n}^N) \in \mathbb{S}^N$ we define the coherent state $\psi_\Omega = \psi_{\mathbf{n}^1} \otimes \psi_{\mathbf{n}^2} \ldots \psi_{\mathbf{n}^N}$. We get, as for $N = 1$, an overcomplete system with a resolution of identity. We denote $d\Omega_N$ the probability measure $(4\pi)^{-N} d\mathbf{n}^1 \otimes \cdots \otimes d\mathbf{n}^N$. Wick and anti-Wick quantization are also well defined as explained in Chap. 1. In particular the Berezin-Lieb inequalities are true in this setting.

As usual the following convention is used: if $\hat{A} \in \mathcal{L}(\mathcal{H}^m)$ and $\hat{B} \in \mathcal{L}(\mathcal{H}^{m'})$ then $\hat{A}\hat{B} = \mathbb{1} \otimes \mathbb{1} \cdots \otimes \hat{A} \otimes \cdots \otimes \hat{B} \otimes \mathbb{1} \cdots \otimes \mathbb{1}$ with $\hat{A}$ at position m, $\hat{B}$ at position m'.

In particular for the covariant and contravariant symbols of $\hat{A}\hat{B}$ we have the following obvious properties

$$(AB)_c(\Omega) = A_c(\mathbf{n}^m) B_c(\mathbf{n}^{m'}), \tag{7.80}$$
$$(AB)^c(\Omega) = A^c(\mathbf{n}^m) B^c(\mathbf{n}^{m'}). \tag{7.81}$$

We shall prove the following results concerning the symbols of one spin operators (S_1, S_2, S_3)

Proposition 91 *The covariant symbols are*

$$S_{1,c}(\mathbf{n}) = j \sin\theta \cos\varphi, \tag{7.82}$$
$$S_{2,c}(\mathbf{n}) = j \sin\theta \sin\varphi, \tag{7.83}$$
$$S_{3,c}(\mathbf{n}) = j \cos\theta. \tag{7.84}$$

The contravariant symbols are

$$S_1^c(\mathbf{n}) = (j+1) \sin\theta \cos\varphi, \tag{7.85}$$
$$S_2^c(\mathbf{n}) = (j+1) \sin\theta \sin\varphi, \tag{7.86}$$
$$S_3^c(\mathbf{n}) = (j+1) \cos\theta. \tag{7.87}$$

Proof It seems more convenient to compute in the complex representation ψ_ζ for the coherent states. In the space $V^{(j)}$ the spin operators are represented by differential operators

$$S_1 = \frac{1}{2}(z_1 \partial_{z_2} + z_2 \partial_{z_1}), \tag{7.88}$$
$$S_2 = \frac{1}{2i}(z_1 \partial_{z_2} - z_2 \partial_{z_1}), \tag{7.89}$$
$$S_3 = \frac{1}{2}(z_1 \partial_{z_1} - z_2 \partial_{z_2}). \tag{7.90}$$

The coherent state ψ_ζ^j has the expression

$$\psi_\zeta(z_1, z_2) = \frac{1}{\sqrt{(2j)!}}(1+|\zeta|^2)^{-j}(z_1 - \zeta z_2)^{2j}.$$

So a direct computation gives

$$\langle \zeta | S_1 | \zeta \rangle = j \sin\theta \cos\varphi. \tag{7.91}$$

In the same way we can compute $S_{2,c}(\mathbf{n})$ and $S_{3,c}(\mathbf{n})$.

Computing contravariant symbols is more difficult because we have no direct formula.

The trick is to start with a large enough set of functions in complex variables $(\zeta, \bar\zeta)$ and to compute their anti-Wick quantizations.

Let us denote $A_{\alpha,\beta,\tau}(\zeta,\bar\zeta) = (1+|\zeta|^2)^{-\tau}\zeta^\alpha\bar\zeta^\beta$ and $\hat A_{\alpha,\beta,\tau}(k,\ell)$ the matrix elements of the operator $\hat A_{\alpha,\beta,\tau}$, $-j \le k, \ell \le j$, in the canonical basis of $V^{(j)}$. We shall use the following formula

$$\begin{aligned}\langle v_\ell, \hat A v_k\rangle &= \quad \textstyle\int_{\mathbb{C}} A^c(\zeta)\langle v_\ell, \Pi_\zeta v_k\rangle d\mu_j(\zeta)\\ \langle v_\ell, \Pi_\zeta v_k\rangle &= \langle v_\ell, \psi_\zeta\rangle\langle \psi_\zeta, v_k\rangle = c_k c_\ell \bar\zeta^{j+\ell}\zeta^{j+k}\end{aligned} \tag{7.92}$$

where $c_k = \left(\frac{(2j)!}{(j+k)!(j-k)!}\right)^{1/2}$.
So we get

$$\hat A_{\alpha,\beta,\tau}(k,\ell) = c_k c_\ell \int_{\mathbb{C}} (d^2\zeta)\zeta^{j+k+\alpha}\bar\zeta^{j+\ell+\beta}(1+|\zeta|^2)^{-2j-\tau-2}.$$

Using polar coordinates $\zeta = re^{i\gamma}$ we get

$$\hat A_{\alpha,\beta,\tau}(k,\ell) = c_k c_\ell \delta_{\alpha+k,\beta+\ell} 2(2j+1)\int_0^\infty dr \frac{r^{2j+2\alpha+2k+1}}{(1+r^2)^{2j+\tau+2}}.$$

We compute the one variable integral using beta and gamma functions

$$\int_0^\infty dr \frac{r^s}{(1+r^2)^t} = \frac{1}{2}B\left(\frac{s+1}{2}, t - \frac{s+1}{2}\right) = \frac{1}{2}\frac{\Gamma(\frac{s+1}{2})\Gamma(t-\frac{s+1}{2})}{\Gamma(t)}.$$

Finally we have the formula

$$\hat A_{\alpha,\beta,\tau}(k,\ell) = (2j+1)c_k c_\ell \delta_{\alpha+k,\beta+\ell}\frac{\Gamma(j+k+\alpha+1)\Gamma(j-\alpha-k+\tau+1)}{\Gamma(2j+\tau+2)}. \tag{7.93}$$

For example if $\tau = 1, \alpha = 0, 1, \beta = 0, 1, 2$ we find

$$\hat{A}_{1,0,1}(k, k+1) = \frac{1}{2j+2}\sqrt{(j+k+1)(j-k)}, \tag{7.94}$$

$$\hat{A}_{0,1,1}(k, k+1) = \frac{1}{2j+2}\sqrt{(j+k)(j-k+1)}, \tag{7.95}$$

$$\hat{A}_{1,1,1}(k, k+1) = \frac{1}{2j+2}(j+k+1). \tag{7.96}$$

Using these results we easily get a contravariant symbol for **S** as in (7.85). □

Now we can compute the covariant and contravariant symbols of the Hamiltonian $\hat{H}$. Applying the above results, translated in the sphere variables, we get

$$H_c(\Omega) = j^2 \sum_{1\le m\le N-1} \mathbf{n}^m \mathbf{n}^{m+1}, \tag{7.97}$$

$$H^c(\Omega) = (j+1)^2 \sum_{1\le m\le N-1} \mathbf{n}^m \mathbf{n}^{m+1}. \tag{7.98}$$

We are ready now to prove the main result which is a particular case of much more general results proved in [180, 236]. Let us introduce the quantum partition function

$$Z^q(\beta, j) = \frac{1}{(2j+1)^N}\mathrm{Tr}e^{-\beta\hat{H}}, \ \beta \in \mathbb{R}.$$

The corresponding classical partition function is obtained taking the average of spin operators on coherent states. So we define

$$Z^{c\ell}(\beta, j) = \int_{(\mathbb{S}^2)^N} \mathrm{e}^{-\beta H_c(\Omega)} d\mu(\Omega).$$

Putting together all the necessary results we have proved the following Lieb's inequalities:

Proposition 92 *We have the following inequalities,*

$$\boxed{Z^{c\ell}(\beta, j) \le Z^q(\beta, j) \le Z^{c\ell}(\beta, j+1), \quad \forall j, \ \text{integer or half}-\text{integer}}. \tag{7.99}$$

This result can be used to study the thermodynamic limit of large spin systems (see [180]).

Corollary 27 *With the notations of the previous proposition we have*

$$\lim_{j\to+\infty} \frac{Z^q(\beta, j)}{Z^{c\ell}(\beta, j)} = 1. \tag{7.100}$$

7.7 More on High Spin Limit: From Spin Coherent States to Harmonic-Oscillator Coherent States

We want to give here more details concerning the transition between spin coherent states and Heisenberg coherent states.

Let us begin by the following easy connection between spin coherent states and harmonic oscillator coherent state. We compute on the Bargmann side. We start from

$$\psi_{\eta}^{j,\sharp}(\zeta) = (1+|\eta|^2)^{-j}(1+\bar{\eta}\zeta)^{2j}.$$

If we replace η and ζ by $\frac{\eta}{\sqrt{2j}}$ and $\frac{\zeta}{\sqrt{2j}}$ and let $j \to +\infty$ then we get

$$\lim_{j\to+\infty} \psi_{\frac{\eta}{\sqrt{2j}}}^{j,\sharp}\left(\frac{\zeta}{\sqrt{2j}}\right) = \mathrm{e}^{\bar{\eta}\zeta-|\eta|^2/2} = \sqrt{2\pi}\varphi_\eta(\zeta).$$

In the right side $\varphi_\eta(\zeta)$ is the Bargmann transform of the Gaussian coherent state located in η. In this sense the spin coherent states converge, in the high spin limit, to the "classical" coherent states.

In the same way we shall prove that the Dicke states converge to the Hermite basis of the harmonic oscillator.

Let us recall that in the Dicke basis of $V^{(j)}$ the Hamiltonian $\hat{H} = S_3$ is diagonal: $\hat{H}d_{j,k} = kd_{j,k}$, $-j \leq k \leq j$. The spin-Bargmann transform of $\mathscr{D}_k^{(j)}$ is easily computed:

$$\mathscr{D}_k^{(j),\sharp}(\zeta) = \frac{(2j)!}{(j+k)!(j-k)!}\zeta^{j+k}.$$

Recall that (see Chap. 1) the Bargmann transform of the Hermite functions ψ_ℓ ($\ell \in \mathbb{N}$) is

$$\psi_\ell^\sharp(\zeta) = \frac{\zeta^\ell}{\sqrt{2\pi\ell!}}.$$

Then we have the following asymptotic result.

Proposition 93 *For every $\ell \in \mathbb{N}$ and every $r > 0$ we have*

$$\boxed{\lim_{\substack{j\to+\infty, k\to-\infty\\ j-k\to\ell}} \mathscr{D}_k^{(j),\sharp}(\zeta)\left(\frac{\zeta}{\sqrt{2j}}\right) = \sqrt{2\pi}\psi_\ell^\sharp(\zeta)}. \tag{7.101}$$

uniformly in $|\zeta| \leq r$.

Proof The result follows easily using that for j, k large enough such that $j - k \approx \ell$ we have

$$\frac{(2j)!}{(j+k)!(j-k)!} \approx \frac{(2j)!}{\ell!(2j-\ell)!} \approx \frac{(2j)^\ell}{\ell!}.$$

□

A different approach concerning the high spin limit of spin coherent states is to consider contractions of the Lie group $SU(2)$ in the Heisenberg Lie group $\mathbf{H}_1$ (see [160]). We are not going to consider here the theory of Lie group contractions in general but to compute on the example of $SU(2)$ and its representations.

Let us consider the irreducible representation of index j of $SU(2)$ in the Bargmann space $\mathscr{P}_{2j}$. $\mathscr{P}_{2j}$ is clearly a finite dimensional subspace of the Bargmann-Fock space $\mathscr{F}(\mathbb{C})$.

Let us compute in this representation the images $L_{\pm}^{\sharp}$, $L_3^{\sharp}$ of the generators $K_{\pm}$, K_3 of the Lie algebra $\mathfrak{sl}(\mathbb{C})$. We get easily

$$L_3^{\sharp} = \zeta\frac{\partial}{\partial\zeta} - j, \tag{7.102}$$

$$L_+^{\sharp} = 2j\zeta - \zeta^2\frac{\partial}{\partial\zeta}, \tag{7.103}$$

$$L_-^{\sharp} = \frac{\partial}{\partial\zeta}. \tag{7.104}$$

Let us introduce a small parameter $\varepsilon > 0$ and denote

$$L_+^{\varepsilon} = \varepsilon L_+^{\sharp}, \tag{7.105}$$

$$L_-^{\varepsilon} = \varepsilon L_-^{\sharp}, \tag{7.106}$$

$$L_3^{\varepsilon} = L_3^{\sharp} + \frac{1}{2\varepsilon^2}\mathbb{1}. \tag{7.107}$$

We have the following commutation relations

$$[L_+^{\varepsilon}, L_-^{\varepsilon}] = 2\varepsilon^2 L_3^{\varepsilon} - \mathbb{1}, \quad [L_3^{\varepsilon}, L_{\pm}^{\varepsilon}] = \pm L_{\pm}^{\varepsilon}. \tag{7.108}$$

As $\varepsilon \to 0$ equations (7.105) define a family of singular transformations of the Lie algebra $\mathfrak{su}(2)$ and for $\varepsilon = 0$ we get (formally)

$$[L_+^0, L_-^0] = -\mathbb{1}, \ [L_3^0, L_{\pm}^0] = \pm L_{\pm}^0 \tag{7.109}$$

These commutation relations are those satisfied by the harmonic oscillator Lie algebra: $L_+^0 \equiv a^{\dagger}$, $L_-^0 \equiv a$, $L_3^0 \equiv N := a^{\dagger}a$.

We can give a mathematical proof of this analogy by computing covariant symbols.

Proposition 94 *Assume that $\varepsilon \to 0$ and $j \to +\infty$ such that* $\lim(2j\varepsilon^2) = 1$. *Then we have*

$$\lim\langle\psi^j_{\zeta/\sqrt{2j}}|L_3^{\varepsilon}|\psi^j_{\zeta/\sqrt{2j}}\rangle = |\zeta|^2 = \langle\varphi_\zeta|a^{\dagger}a|\varphi_\zeta\rangle. \tag{7.110}$$

$$\lim\langle\psi^j_{\zeta/\sqrt{2j}}|L_+^{\varepsilon}|\psi^j_{\zeta/\sqrt{2j}}\rangle = \bar{\zeta} = \langle\varphi_\zeta|a^{\dagger}|\varphi_\zeta\rangle. \tag{7.111}$$

$$\lim \langle \psi^j_{\zeta/\sqrt{2j}} | L^{\varepsilon}_{-} | \psi^j_{\zeta/\sqrt{2j}} \rangle = \zeta = \langle \varphi_\zeta | a | \varphi_\zeta \rangle. \tag{7.112}$$

Proof From computations of Sect. 7.5 we have

$$\langle \psi^j_\zeta | L^\sharp_3 | \psi^j_\zeta \rangle = j \frac{1 - |\zeta|^2}{1 + |\zeta|^2}.$$

So we have

$$\langle \psi^j_{\zeta/\sqrt{2j}} | L^{\varepsilon}_3 | \psi^j_{\zeta/\sqrt{2j}} \rangle = |\zeta^2| + O(\frac{1}{j}).$$

and we get (7.110).

The other formulas are proved in the same way using the following relations

$$\langle \psi^j_\zeta | L^\sharp_+ | \psi^j_\zeta \rangle = \frac{2j\bar{\zeta}}{1 + |\zeta|^2}, \quad \langle \psi^j_\zeta | L^\sharp_- | \psi^j_\zeta \rangle = \frac{2j\zeta}{1 + |\zeta|^2}.$$

□

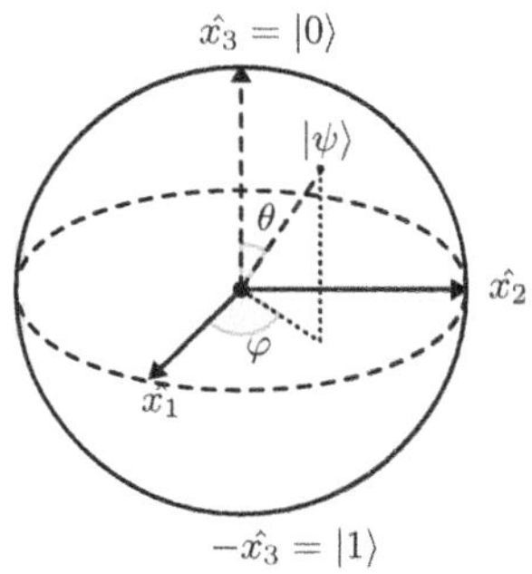

The sphere $\mathbb{S}^2$ as a phase space for the spin is often named the Bloch sphere in quantum optics.

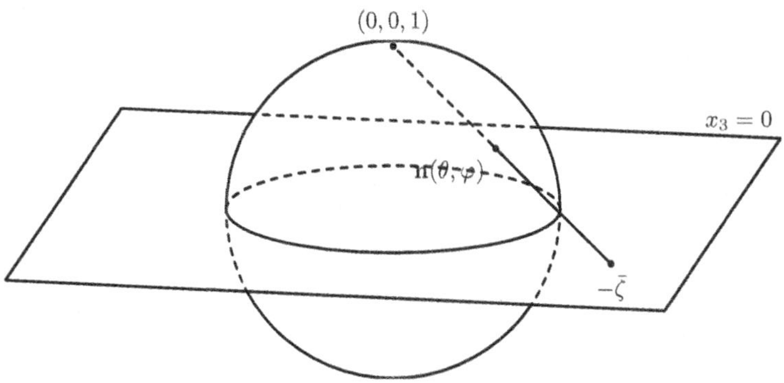

The stereographic projection.

Chapter 8
Pseudo-Spin Coherent States

Abstract We have seen before that spin coherent states are strongly linked with the algebraic and geometric properties of the Euclidean 2-sphere $\mathbb{S}^2$. We shall now consider analogue setting when the sphere is replaced by the 2-pseudosphere i.e. the Euclidean metric in $\mathbb{R}^3$ is replaced by the Minkowski metric and the group $SO(3)$ by the symmetry group $SO(2, 1)$. The main big difference with the Euclidean case is that in the Minkowski case the pseudosphere and its symmetry groupe $SO(2, 1)$ are not compact and that all irreducible unitary representations of $SO(2, 1)$ are infinite dimensional.

8.1 Introduction to the Geometry of the Pseudosphere, $SO(2, 1)$ and $SU(1, 1)$

In this section we shall give a brief introduction to pseudo-euclidean geometry (often named hyperbolic geometry). There exists a huge literature on this subject. For more details we refer to in [13] or to any standard text book on hyperbolic geometry like [46].

8.1.1 Minkowski Model

On the linear space $\mathbb{R}^3$ we consider the Minkowski metric defined by the symmetric bilinear form $\langle x, y\rangle_{\mathrm{M}} := x \boxdot y := x_1y_1 + x_2y_2 - x_0y_0$ where $x, y \in \mathbb{R}^3$, $x = (x_0, x_1, x_2)$, $y = (y_0, y_1, y_2)$.

So we get the three dimensional Minkowski space $(\mathbb{R}^3, \langle\cdot, \cdot\rangle_{\mathrm{M}})$ with its canonical orthonormal basis $\{e_0, e_1, e_2\}$.

Let us remark that the Minkowski metric is the restriction of the Lorentz relativistic metric, defined by the quadratic form $x_1^2 + x_2^2 + x_3^2 - x_0^2$, to the 3 dimensional subspace of $\mathbb{R}^4$ defined by $x_3 = 0$.

The surface in $\mathbb{R}^3$ defined by the equation $= \{x \in \mathbb{R}^3,\ \langle x, x\rangle_{\mathrm{M}} = -1\}$ is a hyperboloid with two symmetric sheets. The pseudosphere $P\mathbb{S}^2$ is one of this sheet. So we

D. Robert and M. Combescure, *Coherent States and Applications in Mathematical Physics*, Theoretical and Mathematical Physics,
https://doi.org/10.1007/978-3-030-70845-0_8

can choose the upper sheet:

$$P\mathbb{S}^2 = \{x = (x_0, x_1, x_2),\ x_1^2 + x_2^2 - x_0^2 = -1,\ x_0 > 0\}$$

$P\mathbb{S}^2$ is a surface which can be parametrized with the pseudopolar coordinates (τ, φ):

$$x_0 = \cosh\tau,\ x_1 = \sinh\tau\cos\varphi,\ x_2 = \sinh\tau\sin\varphi,\quad \tau \in [0, +\infty[,\ \varphi \in [0, 2\pi[.$$

$P\mathbb{S}^2$ is a Riemann surface for the metric induced on $P\mathbb{S}^2$ by the Minkowski metric. In coordinates (x_0, x_1, x_2) $P\mathbb{S}^2$ is defined by the equation $x_0 = \sqrt{1 + x_1^2 + x_2^2}$. So, in coordinates $\{x_1, x_2\}$ on $P\mathbb{S}^2$, the metric $ds^2 = -dx_0^2 + dx_1^2 + dx_2^2$ is given by the following symmetric matrix

$$G = \begin{pmatrix} 1 - \frac{x_1^2}{1+x_1^2+x_2^2} & \frac{x_1x_2}{1+x_1^2+x_2^2} \\ \frac{x_1x_2}{1+x_1^2+x_2^2} & 1 - \frac{x_2^2}{1+x_1^2+x_2^2} \end{pmatrix}.$$

Hence we see that ds^2 is positive definite on $P\mathbb{S}^2$. In polar coordinates we have a simpler expression: $ds^2 = d\tau^2 + \sinh\tau^2 d\varphi^2$. The curvature of $P\mathbb{S}^2$ is -1 everywhere (compare with the sphere $\mathbb{S}^2$ with curvature is +1 everywhere). By analogy with the Euclidean sphere we shall denote $\mathbf{n}$ the generic point on $P\mathbb{S}^2$.

The Riemannian surface measure in pseudopolar coordinates is given by computing the density $\sqrt{\det G}$, where G is the matrix of the metric in coordinates (τ, φ). So the surface measure on $P\mathbb{S}^2$ is

$$d^2\mathbf{n} = \sinh\tau d\tau d\varphi.$$

We can see that the geodesics on the pseudo-sphere $P\mathbb{S}^2$ are determined by their intersection with planes through the origin 0.

We consider now the symmetries of $P\mathbb{S}^2$. Let us denote $L = \begin{pmatrix} -1 & 0 & 0 \\ 0 & 1 & 0 \\ 0 & 0 & 1 \end{pmatrix}$ the matrix of the Minkowski quadratic form: $\langle x, y\rangle_{\mathrm{M}} = Lx \cdot y$ (recall that the $\cdot$ denotes the usual scalar product in $\mathbb{R}^3$).

Its invariance group is denoted $O(2, 1)$. So $A \in O(2, 1)$ means $Ax \boxdot Ax = x \boxdot x$ for every $x \in \mathbb{R}^3$ or equivalently, $A^T LA = L$ (A^T is the transposed matrix of A). In particular if $A \in O(2, 1)$ then $\det A = \pm 1$. The direct invariance group is the subgroup $SO(2, 1)$ defined by $A \in O(2, 1)$ and $\det A = 1$. This group is not connected so we introduce $SO_0(2, 1)$ the component of $\mathbb{1}$ in $SO(2, 1)$, it is a closed subgroup of $SO(2, 1)$.

The pseudosphere $P\mathbb{S}^2$ is clearly invariant by $SO_0(2, 1)$.

It is not difficult to see that 1 is always an eigenvalue for every $A \in SO_0(2, 1)$. Let $v \in \mathbb{R}^2$ be such that $\|v\| = 1$, $Av = v$ and $\langle v, v\rangle_{\mathrm{M}} \neq 0$. The orthogonal complement of $\mathbb{R}v$ for the Minkowski form $\langle ., . \rangle_M$ is a two-dimensional plane invariant by A. So A looks like a rotation in Euclidean geometry.

Let us give the following examples of transformations in $SO_0(2, 1)$.

$$R_\varphi = \begin{pmatrix} 1 & 0 & 0 \\ 0 & \cos\varphi & -\sin\varphi \\ 0 & \sin\varphi & \cos\varphi \end{pmatrix} : \text{ rotation in the plane } \{e_1, e_2\} \tag{8.1}$$

$$B_{1,\tau} = \begin{pmatrix} \cosh\tau & \sinh\tau & 0 \\ \sinh\tau & \cosh\tau & 0 \\ 0 & 0 & 1 \end{pmatrix} : \text{ boost in the direction } e_1 \tag{8.2}$$

$$B_{2,\tau} = \begin{pmatrix} \cosh\tau & 0 & \sinh\tau \\ 0 & 1 & 0 \\ \sinh\tau & 0 & \cosh\tau \end{pmatrix} : \text{ boost in the direction } e_2. \tag{8.3}$$

These three transformations generate all the group $SO_0(2, 1)$. This can be easily proved using the following remark: if $Av = v$ and if U is a transformation then $A_U(Uv) = Uv$ where $A_U = UAU^{-1}$.

$SO_0(2, 1)$ is a Lie group. In particular it is a three dimensional manifold.

8.1.2 *Lie Algebra*

We can get a basis for the Lie algebra $\mathfrak{so}(2, 1)$ of $SO(2, 1)$ by computing the generators of the three one parameter subgroups defined in (8.1). We get

$$E_0 := \frac{d}{d\varphi} R_\varphi|_{\varphi=0} = \begin{pmatrix} 0 & 0 & 0 \\ 0 & 0 & -1 \\ 0 & 1 & 0 \end{pmatrix}. \tag{8.4}$$

$$E_1 := \frac{d}{d\tau} B_{1,\tau}|_{\tau=0} = \begin{pmatrix} 0 & 1 & 0 \\ 1 & 0 & 0 \\ 0 & 0 & 0 \end{pmatrix}. \tag{8.5}$$

$$E_2 := \frac{d}{d\tau} B_{2,\tau}|_{\tau=0} = \begin{pmatrix} 0 & 0 & 1 \\ 0 & 0 & 0 \\ 1 & 0 & 0 \end{pmatrix}. \tag{8.6}$$

The commutation relations of the Lie algebra are the following

$$[E_0, E_1] = E_2, \;\; [E_2, E_0] = E_1, \;\; [E_1, E_2] = -E_0. \tag{8.7}$$

If we compare with generators of $\mathfrak{so}(3)$ we remark the minus sign in the last relation.

Let us consider the exponential map $\exp : \mathfrak{so}(2, 1) \to SO(2, 1)$. If $A = x_0E_0 + x_1E_1 + x_2E_2$ where $x_0^2 + x_1^2 + x_2^2 = 1$, we have the one parameter subgroup of

$SO(2,1)$, $R(\theta) = e^{\theta A}$. $R(\theta)$ is a pseudo-rotation with axis $v = (x_0, -x_2, x_1)$. Its geometrical properties are classified by the sign of $\langle v, v\rangle_M$.

- $\langle v, v\rangle_M > 0$ ("time-like" axis): $R(1)$ has a unique fixed point on $P\mathbb{S}^2$, each orbit is bounded. $R(1)$ is said to be elliptic
- $\langle v, v\rangle_M < 0$ ("space-like" axis) : there exists a unique geodesic on $P\mathbb{S}^2$ invariant by $U(1)$, $R(1)$ is said hyperbolic
- $\langle v, v\rangle_M = 0$ ("light-like" axis) : the geodesics asymptotically going to $\mathbb{R}v$ are invariant by $R(1)$. $R(1)$ is said parabolic

These classification will be more explicit on other models of $P\mathbb{S}^2$ as we shall see. As in the Euclidean case we can realize the Lie algebra relations (8.7) in a Lie algebra of complex 2×2 matrices.

We replace the group $SU(2)$ by the group $SU(1,1)$ of pseudo-unitary unimodular matrices of the form:

$$g = \begin{pmatrix} \alpha & \beta \\ \bar{\beta} & \bar{\alpha} \end{pmatrix} \qquad |\alpha|^2 - |\beta|^2 = 1$$

$SU(1,1)$ is a Lie group of real dimension 3. Let us introduce now a convenient parametrization of $SU(1,1)$

$$\alpha = \cosh\frac{t}{2} e^{i(\varphi+\psi)/2},\ \beta = \sinh\frac{t}{2} e^{i(\varphi-\psi)/2}$$

where the triple (φ, t, ψ) runs through the domain $0 \le \varphi < 2\pi,\ 0 < t < \infty,\ -2\pi \le \psi < 2\pi$.

In this way one gets the factorization:

$$g(\varphi, t, \psi) = \begin{pmatrix} \cosh\frac{t}{2} e^{i(\varphi+\psi)/2} & \sinh\frac{t}{2} e^{i(\varphi-\psi)/2} \\ \sinh\frac{t}{2} e^{i(\psi-\varphi)/2} & \cosh\frac{t}{2} e^{-i(\varphi+\psi)/2} \end{pmatrix} = g(\varphi, 0, 0) g(0, t, 0) g(0, 0, \psi).$$

In the group $SU(1,1)$ we choose three one-parameter subgroups consisting of the following matrices

$$\omega_1(t) = g(0, t, 0), \quad \omega_2(t) = \begin{pmatrix} \cosh\frac{t}{2} & i\sinh\frac{t}{2} \\ -i\sinh\frac{t}{2} & \cosh\frac{t}{2} \end{pmatrix}, \quad \omega_0(t) = g(t, 0, 0)$$

The generators of these one parameter subgroups are

$$b_1 = \frac{1}{2}\begin{pmatrix} 0 & 1 \\ 1 & 0 \end{pmatrix}, \quad b_2 = \frac{i}{2}\begin{pmatrix} 0 & 1 \\ -1 & 0 \end{pmatrix}, \quad b_0 = \frac{i}{2}\begin{pmatrix} 1 & 0 \\ 0 & -1 \end{pmatrix}.$$

These three matrices form a basis of the Lie algebra $\mathfrak{su}(1,1)$ with the commutation relations:

$$[b_1, b_2] = -b_0, \quad [b_2, b_0] = b_1, \quad [b_0, b_1] = b_2 \tag{8.8}$$

We see that the Lie algebras $\mathfrak{so}(2, 1)$ and $\mathfrak{su}(1, 1)$ are isomorph and we can identify the basis $\{E_0, E_1, E_2\}$ and $\{b_0, b_1, b_2\}$. Now we shall see that $SU(1, 1)$ is for $SO_0(2, 1)$ what is $SU(2)$ for $SO(3)$. Let us consider the adjoint representation of $SU(1, 1)$. For every $g \in SU(1, 1)$ and $x = (x_0, x_1, x_2)$ we have

$$g\left(\sum_{0\le k\le 2} x_k b_k\right)g^{-1} = \sum_{0\le k\le 2} y_k b_k.$$

Let us denote $y = \rho(g)x$, where $y = (y_0, y_1, y_2)$. As in the Euclidean case we have the following result.

Proposition 95 *For every $g \in SU(1, 1)$ we have:*
(i) $g \mapsto \rho(g)$ is a group morphism from $SU(1, 1)$ onto $SO(2, 1)_0$
(ii) $\rho(g) = \mathbb{1}$ if and only if $g = \pm\mathbb{1}$.
In particular the groups $SO(2, 1)_0$ and $SU(1, 1)/\{\mathbb{1}, -\mathbb{1}\}$ are isomorph.
(iii) If $A = v_0 b_0 + v_1 b_1 + v_2 b_2$ then for every $\tau \in \mathbb{R}$, $\rho\left(e^{\tau A}\right)$ is a pseudorotation of axis $(v_0, -v_2, v_1)$.

Proof It is easy to see that $\rho(g) \in SO(2, 1)$ and ρ is a group morphism. Let us remark that we have $\omega_k(t) = e^{tb_k}$ for $k = 0, 1, 2$. Then we get that $\rho(\omega_0(t))$ is the rotation R_t of axis e_0 in the Minkowski space, $\rho(\omega_1(t))$ is the boost $B_{1,t}$ and $\rho(\omega_2(t))$ is the boost $B_{2,t}$. Then it results that the range of ρ is the full group $SO_0(2, 1)$.
(ii) is easy to prove, as in the Euclidean case.
We assume $v = (v_0, v_1, v_2) \neq (0, 0, 0)$. The kernel of $v_0E_0 + v_1E_1 + v_2E_2$ is clearly generated by the vector $(v_0, -v_2, v_1)$so we get (iii). □

It will be useful to remark that the group $SU(1, 1)$ is isomorph to the $SL(2, \mathbb{R})$ group of real 2×2 matrices A such that $\det A = 1$. The isomorphism is simply the conjugation $A \mapsto \Lambda^{-1}A\Lambda$ where $A \in SL(2, \mathbb{R})$, $\Lambda = \frac{1}{\sqrt{2}}\begin{pmatrix} 1 & i \\ i & 1 \end{pmatrix}$.

Let us remark that we have $\mathfrak{su}(1, 1) + i\mathfrak{su}(1, 1) = \mathfrak{sl}(2, \mathbb{C})$ so $\mathfrak{su}(1, 1)$ and $\mathfrak{su}(2)$ have the same complexification as Lie algebras.

We shall see now that this isomorphism has a simple geometry interpretation and this is very useful to get a better insight of the pseudosphere geometry.

8.1.3 The Disc and the Half-Plane Poincaré Representations of the Pseudosphere

Considering the stereographic projection from the apex $(0, 0, -1)$ onto the plane $\{e_1, e_2\}$, the upper hyperboloïd is transformed into the disc $\mathbb{D}$ of radius 1. By this transformation the metric on the pseudosphere becomes a metric on $\mathbb{D}$. Choosing in $P\mathbb{S}^2$ polar coordinates (τ, φ) and in $\mathbb{D}$ polar coordinates $(r, -\varphi)$, we have

$$x_0 = \cosh\tau,\ x_1 = \sinh\tau\cos\varphi,\ x_2 = \sinh\tau\sin\varphi,\quad r = \tanh\frac{\tau}{2}.$$

In $\mathbb{D}$ the metric and the surface element of the pseudosphere have the following expression

$$ds^2 = \frac{4(dr^2 + r^2 d\varphi^2)}{(1-r^2)^2}, \qquad d^2\zeta = \frac{4rdrd\varphi}{(1-r^2)^2}$$

So we get another model of the pseudosphere named the Poincaré disc model.

It is also convenient to introduce a complex representation of $\mathbb{D}$, $\zeta = re^{-i\varphi}$. In $\mathbb{D}$ geodesics are the straight lines through the origin and arcs circle orthogonal to the boundary (see textbooks on the subject).

Let us remark that the metric is conformal so the angles for the metric coincïde with the Euclidean angles. So an isometry of $\mathbb{D}$ is a conformal transformation of $\mathbb{D}$. We know that such direct transformations f are homographic:

$$f(\zeta) = \frac{\alpha\zeta + \beta}{\bar{\beta}\zeta + \bar{\alpha}}, \quad \text{such that } |\alpha|^2 - |\beta|^2 = 1$$

In other words the direct symmetry group of $\mathbb{D}$ is given by a representation of $SU(1,1)$, $g \mapsto H_g$, where $H_g(\zeta) = f(\zeta)$, $g = \begin{pmatrix}\alpha & \beta \\ \bar{\beta} & \bar{\alpha}\end{pmatrix}$. Moreover we see easily that $H_g = \mathbb{1}_{\mathbb{D}}$ if and only if $g = \pm\mathbb{1}$. So we recover geometrically the fact that $SO(2,1) \approx SU(1,1)/\{\mathbb{1}, \pm\mathbb{1}\}$.

It is well known that the unit disc can be conformally transformed into the half complex plane $\mathbb{H} = \{z = u + iv,\ v > 0\}$ by the homography $H_0\zeta = \frac{-i\zeta + i}{\zeta + 1}$ or $H_0^{-1}z = \frac{-z+i}{z+i}$, where $\zeta \in \mathbb{D}$, $z \in \mathbb{H}$. Explicitly, if $\zeta = re^{i\varphi}$ and $z = u + iv$, we have

$$u = \frac{2r\sin\varphi}{1 + r^2 + 2r\cos\varphi}, \quad v = \frac{1 - r^2}{1 + r^2 + 2r\cos\varphi}.$$

In $\mathbb{H}$ the metric and the surface element of the pseudosphere have the following expression

$$ds^2 = \frac{du^2 + dv^2}{v^2}, \qquad d^2z = \frac{dudv}{v}.$$

So we get another model of the pseudosphere named the Poincaré half-plane model.

In this model the boundary of the disc is transformed into the real axis. In $\mathbb{H}$ the geodesics are vertical lines and half circles orthogonal to the real axis.

As for the disc we can see that the direct symmetry group of $\mathbb{H}$ is the group $SL(2,\mathbb{R})$ with the homographic action $H_g(z) = \frac{\alpha z + \beta}{\gamma z + \delta}$, $\alpha, \beta, \gamma, \delta \in \mathbb{R}$, $\alpha\delta - \gamma\beta = 1$ and $H_g = \mathbb{1}_{\mathbb{H}}$ if and only if $g = \pm\mathbb{1}$. We recover the fact that $SU(1,1)$ and $SL(2,\mathbb{R})$ are isomorph groups.

Finally there is another realization of the pseudosphere which is important to define coherent states: $P\mathbb{S}^2$ can be seen as a quotient of $SU(1,1)$ by its compact maximal subgroup $U(1)$.

Lemma 50 *For every* $\mathbf{n} = (\cosh\tau, \sinh\tau\sin\varphi, \sinh\tau\cos\varphi)$ *define*

$$g_{\mathbf{n}} = \begin{pmatrix} \cosh(\tau/2) & \sinh(\tau/2)\mathrm{e}^{-i\varphi} \\ \sinh(\tau/2)\mathrm{e}^{i\varphi} & \cosh(\tau/2) \end{pmatrix}.$$

Then the map $\mathbf{n} \mapsto \{g_{\mathbf{n}}\omega_0(t),\ t \in [-2\pi, 2\pi[\}$ *is a bijection from* $\mathbb{S}^2$ *in the right cosets of* $SU(1,1)$ *modulo* $U(1)$ *where* $U(1)$ *is identified with the diagonal matrices* $\begin{pmatrix} \mathrm{e}^{it/2} & 0 \\ 0 & \mathrm{e}^{-it/2} \end{pmatrix}$, $t \in [0, 4\pi[$.

Proof Let us denote $g(\alpha,\beta) = \begin{pmatrix} \alpha & \beta \\ \bar{\beta} & \bar{\alpha} \end{pmatrix}$, $\alpha, \beta \in \mathbb{C}$, $|\alpha|^2 - |\beta|^2 = 1$. The Lemma is a direct consequence of the following decomposition : there exist $\alpha' > 0$, $\beta' \in \mathbb{C}$, $t \in [0, 4\pi[$, unique such that

$$g(\alpha,\beta) = g(\alpha',\beta')\omega_0(t)$$

More precisely we have $\alpha' = |\alpha|$, $t = 2\arg\alpha$, $\beta' = \mathrm{e}^{it/2}\beta$. This proves the lemma. □

So modulo composition by a rotation of axis e_0 on the right, every $g \in SU(1,1)$ is equivalent to a unique $g_{\mathbf{n}}$, $\mathbf{n} \in P\mathbb{S}^2$. We can see that $g_{\mathbf{n}}$ is a pseudorotation with axis $v = (0, \sin\varphi, \cos\varphi)$ (remark a change of sign).

To get that, remark

$$g_{\mathbf{n}}^{-1}\frac{d}{d\tau}g_{\mathbf{n}} = \frac{1}{2}(\cos\varphi\sigma_1 + \sin\varphi\sigma_2) = (\cos\varphi b_1 - \sin\varphi b_2)$$

so $g_{\mathbf{n}} = \mathrm{e}^{\tau(\cos\varphi b_1 - \sin\varphi b_2)}$, this is indeed a pseudorotation of axis direction $(0, \sin\varphi, \cos\varphi)$.

8.2 Unitary Representations of $SU(1,1)$

Let us begin by introducing a useful and important invariant operator for a Lie group representation : the Casimir operator. We first define the Killing form on a Lie algebra $\mathfrak{g}$,

$$\langle X, Y\rangle_0 = \frac{1}{2}\mathrm{Tr}(ad(X)ad(Y)) \tag{8.9}$$

where ad is the Lie adjoint-representation: $ad(X)Y = [X, Y]$, $\forall Y \in \mathfrak{g}$.

$ad(X)$ is the infinitesimal generator of the one parameter group transformation in $\mathfrak{g}$:

$$G_\theta(Y) = e^{\theta X} Y e^{-\theta X}.$$

We have the antisymmetric property:

$$\langle ad(X)Y, Z\rangle_0 = -\langle Y, ad(X)Z\rangle_0, \quad \forall X, Y, Z \in \mathfrak{g}.$$

Consider a basis $\{X_j\}$ of $\mathfrak{g}$. The Killing form in this basis has the matrix $g_{j,k} = \mathrm{tr}(ad(X_j)ad(X_k))$. We denote $g^{j,k}$ the inverse of the matrix $g_{j,k}$.

Let us now consider a representation R of a Lie group G in the linear space V and $\rho = dR$ the corresponding representation of its Lie algebra $\mathfrak{g}$ in $\mathscr{L}(V)$. The Casimir operator C^ρ_{as} is defined as follows

$$C^\rho_{as} = \sum g^{j,k} \rho(X_j)\rho(X_k).$$

Let us remark that if V is infinite dimensional some care is necessary to check the domain of C^ρ_{as}. Nevertheless a standard computation gives the following important property.

Lemma 51 *The Casimir operator C^ρ_{as} commutes with the representation ρ, $\rho(X)C^\rho_{as} = C^\rho_{as}\rho(X)$, $\forall X \in \mathfrak{g}$.*
In particular if the representation is irreducible then $C^\rho_{as} = c^\rho_{as}\mathbb{1}$, $c_{as} \in \mathbb{R}$.

Let us now remark that every irreducible unitary representation of $SU(1,1)$ is infinite dimensional, this is a big difference compare to $SU(2)$.

Proposition 96 *Let ρ be a unitary representation of $SU(1,1)$ in a finite dimensional Hilbert space $\mathscr{H}$. Then ρ is trivial, i.e. $\rho(g) = \mathbb{1}_{\mathscr{H}}$, $\forall g \in SU(1,1)$*

Proof Using the isomorphism from $SL(2,\mathbb{R})$ onto $SU(1,1)$, it is enough to prove the proposition for a unitary representation of $SL(2,\mathbb{R})$.

For every $x \in \mathbb{R}$, $a \in \mathbb{R}$, $a \neq 0$, we have

$$\begin{pmatrix} a & 0 \\ 0 & a^{-1} \end{pmatrix} \begin{pmatrix} 1 & x \\ 0 & 1 \end{pmatrix} \begin{pmatrix} a & 0 \\ 0 & a^{-1} \end{pmatrix}^{-1} = \begin{pmatrix} 1 & a^2x \\ 0 & 1 \end{pmatrix}.$$

Then we see that $\rho\begin{pmatrix} 1 & x \\ 0 & 1 \end{pmatrix}$ are all conjugate for $x > 0$. By continuity and compactness they are conjugate to $\mathbb{1}_{\mathscr{H}} = \rho\begin{pmatrix} 1 & 0 \\ 0 & 1 \end{pmatrix}$ because the representation is finite dimensional. The same property holds true for $x < 0$ and for matrices $\begin{pmatrix} 1 & 0 \\ y & 1 \end{pmatrix}$. But the group $SL(2,\mathbb{R})$ is generated by the two matrices $\begin{pmatrix} 1 & x \\ 0 & 1 \end{pmatrix}$ and $\begin{pmatrix} 1 & 0 \\ y & 1 \end{pmatrix}$. So we can conclude that $\rho(A) = \mathbb{1}_{\mathscr{H}}$ for every $A \in SL(2,\mathbb{R})$. □

Unitary irreducible representations of $SU(1, 1)$ have been computed independently by Gelfand-Neumark [109] and by Bargmann [18]. We shall follow here the presentation by Bargmann with some minor modifications.

8.2.1 *Classification of the Possible Representations of $SU(1, 1)$*

Let ρ be an irreducible unitary representation of $SU(1, 1)$ in some Hilbert space $\mathscr{H}$. From Proposition 96 we know that $\mathscr{H}$ is infinite dimensional.

In the representation space the generators b_j define the operators

$$B_j := i\frac{d}{dt}\rho(\omega_j(t))|_{t=0},\ j = 0, 1, 2,$$

with the commutation relations

$$[B_1, B_2] = -iB_0,\ [B_2, B_0] = iB_1,\quad [B_0, B_1] = iB_2. \tag{8.10}$$

Or with the complex notation $B_\pm = B_2 \pm iB_1$, we have

$$[B_-, B_+] = 2B_0,\quad [B_0, B_\pm] = \pm B_\pm. \tag{8.11}$$

We have $\rho(\omega_0(t)) = \mathrm{e}^{-itB_0}$ hence $\mathrm{e}^{-i4\pi B_0} = 1$. So the spectrum of B_0 is a subset of $\{\frac{k}{2},\ k \in \mathbb{Z}\}$. There exists $\psi_0 \in \mathscr{H}$, $\|\psi_0\| = 1$ and $B_0\psi_0 = \lambda\psi_0$, $\lambda = \frac{k_0}{2}$, $k_0 \in \mathbb{Z}$.

Using the commutation relation we have

$$B_0B_+\psi_0 = (\lambda + 1)B_+\psi_0, \tag{8.12}$$

$$B_0B_-\psi_0 = (\lambda - 1)B_-\psi_0. \tag{8.13}$$

Reasoning by induction, we get for every $k \in \mathbb{N}$,

$$B_0(B_+^k)\psi_0 = (\lambda + k)(B_+)^k\psi_0 \tag{8.14}$$

$$B_0(B_-)^k\psi_0 = (\lambda - k)(B_-)^k\psi_0 \tag{8.15}$$

Introduce now the Casimir operator C_{as} of the representation which is supposed irreducible, so $C_{as} = c_{as}\mathbb{1}$ where

$$C_{as} = B_0^2 - \frac{1}{2}(B_-B_+ + B_+B_-).$$

Hence we have the equations

$$(B_-B_+ + B_+B_-)\psi_0 = 2(\lambda^2 - c_{as})\psi_0 \tag{8.16}$$
$$(B_-B_+ - B_+B_-)\psi_0 = 2\lambda\psi_0. \tag{8.17}$$

So we get

$$B_-B_+\psi_0 = (\lambda(\lambda + 1) - c_{as})\psi_0 \tag{8.18}$$
$$B_+B_-\psi_0 = (\lambda(\lambda - 1) - c_{as})\psi_0. \tag{8.19}$$

Using now $B_\pm^k\psi_0$ instead of ψ_0 we have proved for every $k \in \mathbb{N}$,

$$B_-B_+^k\psi_0 = ((\lambda + k - 1)(\lambda + k) - c_{as})\psi_0 \tag{8.20}$$
$$B_+B_-^k\psi_0 = ((\lambda - k + 1)(\lambda - k) - c_{as})\psi_0 \tag{8.21}$$

Let us denote $\nu_k^+ = ((\lambda + k - 1)(\lambda + k) - c_{as})$ and $\nu_k^- = ((\lambda - k + 1)(\lambda - k) - c_{as})$.

Using that $B_+ = B_-{}^*$ and $B_- = B_+{}^*$ we get from (8.20),

$$\|B_+^{k+1}\psi_0\|^2 = \nu_{k+1}^+\|B_+^k\psi_0\|^2 \tag{8.22}$$
$$\|B_-^{k+1}\psi_0\|^2 = \nu_{k+1}^-\|B_-^k\psi_0\|^2. \tag{8.23}$$

From Eq. (8.22) we can start the discussion.

(I) Suppose that for all $k \in \mathbb{N}$, $B_+^k\psi_0 \neq 0$ and $B_-^k\psi_0 \neq 0$. Then for every $k \in \mathbb{N}$, $\lambda \pm k$ is an eigenvalue for B_0.
(I-1) If 0 is in this family (i.e. $\lambda \in \mathbb{Z}$) then we can suppose that $B_0\psi_0 = 0$ so we can choose $\lambda = 0$. From (8.22) we find the necessary condition $c_{as} < 0$.
(I-2) If 0 is not in the family $\lambda \pm k$, $\lambda = \frac{1}{2} + k_0$, $k_0 \in \mathbb{Z}$ and using $B_\pm^k$ we can assume that $\lambda = \frac{1}{2}$. From (8.22) we get that $c_{as} < 1/4$.
In these two cases we get an orthonormal basis $\{\varphi_m\}_{m\in\mathbb{Z}}$ for $\mathscr{H}$, such that $B_0\varphi_m = (m + \frac{\varepsilon}{2})\varphi_m$. where $\varepsilon = 0$ in the first case and $\varepsilon = 1$ in the second case. This is really a basis because the linear space span by the φ_m is invariant by the representation which is irreducible.

(II) Suppose now that there exists $k_0 \in \mathbb{N}$ such that $B_+^{k_0+1}\psi_0 = 0$ and $B_+^{k_0}\psi_0 \neq 0$. Using (8.20) we see that for every $\ell \in \mathbb{N}$, $B_-^\ell\psi_0$ is proportional to $B_-^{\ell'}B_+^{k_0}\psi_0$, so we can replace ψ_0 by $B_+^{k_0}\psi_0$. So we have $B_0\psi_0 = \lambda\psi_0$, $B_+\psi_0 = 0$. Hence this gives $\nu_1^+ = 0$ and $c_{as} = \lambda(\lambda + 1)$. If $\lambda = 0$ we get $B_-\psi_0 = 0$ and the Hilbert space is unidimensional. So we have $\lambda > 0$. As above we get an orthonormal basis of $\mathscr{H}$ $\{\varphi_m\}_{m\in\mathbb{N}}$ such that $B_0\varphi_m = (\lambda - m)\varphi_m$.
If there exists $k_0 \in \mathbb{N}$ such that $B_-^{k_0+1}\psi_0 = 0$ and $B_-^{k_0}\psi_0 \neq 0$ we have a similar result with eigenvalues $\lambda + m$ for B_0.

In case (I) the Casimir parameter c_{as} varies in an interval and we say that the representation belongs to the continuous series; in case (II) the Casimir parameter varies in a discrete set and we say that the representation belongs to the discrete series.

8.2.2 Discrete Series Representations of $SU(1,1)$

In the last section we have found necessary conditions satisfied by any irreducible representation of $SU(1,1)$. Now we have to prove that this conditions can be realized in some concrete Hilbert spaces.

8.2.2.1 The Hilbert Spaces $\mathscr{H}_n(\mathbb{D})$

Let n be a real number, $n \geq 2$, $\mathscr{H}_n(\mathbb{D})$ is the Hilbert space of holomorphic functions f on the unit disc $\mathbb{D}$ of the complex plane $\mathbb{C}$ satisfying

$$\|f\|^2_{\mathscr{H}_n(\mathbb{D})} := \frac{n-1}{\pi}\int_{\mathbb{D}} |f(z)|^2(1-|z|^2)^{n-2}dxdy < +\infty, \quad z = x+iy. \tag{8.24}$$

The measure $d\nu_n(z) := \frac{n-1}{\pi}(1-|z|^2)^{n-2}dxdy$ is a probability measure on $\mathbb{D}$ and $\mathscr{H}_n(\mathbb{D})$ is a complete linear space, with the Hilbert norm defined by (8.24), is a consequence of standard properties of holomorphic functions.

It is useful to produce the following characterization of $\mathscr{H}_n(\mathbb{D})$ using the series expansion of f

$$f(z) = \sum_{k\geq 0} c_k(f)z^k$$

which is absolutely convergent inside the disc $\mathbb{D}$.

In (8.24) let us compute the integral in polar coordinates $z = re^{i\theta}$. From the Parseval formula for Fourier series we get

$$\|f|_{\mathscr{H}_n^2(\mathbb{D})} = 2(n-1)\pi \sum_{k\geq 0} |c_k(f)|^2 \int_0^1 r^{2k+1}(1-r^2)^{n-2}dr. \tag{8.25}$$

So we have

$$\|f\|^2_{\mathscr{H}_n(\mathbb{D})} = \sum_{k\geq 0} |c_k(f)|^2 \frac{\Gamma(n)\Gamma(k+1)}{\Gamma(n+k)} \tag{8.26}$$

This gives a unitary equivalent definition of $\mathscr{H}_n(\mathbb{D})$ as an Hilbert space of functions on the unit circle. In particular the scalar product in $\mathscr{H}_n(\mathbb{D})$ of f_1 and f_2 has the following expression

$$\langle f_1, f_2\rangle_{\mathscr{H}_n(\mathbb{D})} = \sum_{k\geq 0} \overline{c_k(f_1)}c_k(f_2)\gamma_{n,k} \tag{8.27}$$

where $\gamma_{n,\ell} = \frac{\Gamma(n)\Gamma(\ell+1)}{\Gamma(n+\ell)}$.

From formula (8.26) and (8.27) we get easily that $e_\ell(z) := \frac{z^\ell}{\sqrt{\gamma_{n,\ell}}}$, for $\ell \geq 0$, is an orthonormal basis of $\mathscr{H}_n(\mathbb{D})$.

8.2.2.2 Discrete Series Realization of $SU(1, 1)$ in $\mathscr{H}_n(\mathbb{D})$

These representations can be introduced using Gauss decomposition in the complex Lie group $SL(2, \mathbb{C})$:

$$\begin{pmatrix} \alpha & \beta \\ \gamma & \delta \end{pmatrix} = \begin{pmatrix} 1 & 0 \\ \frac{\gamma}{\alpha} & 1 \end{pmatrix} \begin{pmatrix} \alpha & 0 \\ 0 & \frac{1}{\alpha} \end{pmatrix} \begin{pmatrix} 1 & \frac{\beta}{\alpha} \\ 0 & 1 \end{pmatrix} \tag{8.28}$$

where $\alpha, \beta, \gamma, \delta$ are complex numbers such that $\alpha\delta - \beta\delta = 1, \alpha \neq 0$.

Moreover this decomposition as product like

$$\begin{pmatrix} 1 & 0 \\ z & 1 \end{pmatrix} \begin{pmatrix} u & 0 \\ 0 & \frac{1}{u} \end{pmatrix} \begin{pmatrix} 1 & w \\ 0 & 1 \end{pmatrix}$$

is unique. So this allows to define natural actions of $SU(1, 1)$ in the disc $\mathbb{D}$.

Let us denote by $t_-(z)$ the matrices $\begin{pmatrix} 1 & 0 \\ z & 1 \end{pmatrix}$, $t_+(z)$ the matrices $\begin{pmatrix} 1 & z \\ 0 & 1 \end{pmatrix}$ and $d(u) = \begin{pmatrix} u & 0 \\ 0 & u^{-1} \end{pmatrix}$.

Let us denote g a generic element of $SU(1, 1)$,

$$g = \begin{pmatrix} \alpha & \beta \\ \bar{\beta} & \bar{\alpha} \end{pmatrix}$$

Consider the t_- matrix in the Gaussian decomposition of $gt_-(z)$. We have $t_- = t_-(\tilde{z})$ where the complex number $\tilde{z}$ is $\tilde{z} = M_-(g)(z) = \frac{\bar{\beta}+\bar{\alpha}z}{\alpha+\beta z}$. We get easily that $|M_-(g)(z)| = 1$ if $|z| = 1$ and from the maximum principle, $|M_-(g)(z)| < 1$ if $|z| < 1$. So we have defined a right action of $SU(1, 1)$ in $\mathbb{D}$. In the same way we get a left action considering the t_+ matrix in the decomposition of $\begin{pmatrix} 1 & z \\ 0 & 1 \end{pmatrix} g^{-1}$. So we get the action $M_+(g)(z) = \frac{-\beta+\alpha z}{\bar{\alpha}-\bar{\beta}z}$.

Now we want to define unitary actions of $SU(1, 1)$ in the space $\mathscr{H}_n(\mathbb{D})$ as follows

$$\mathscr{D}_n^-(g)f(z) = m_g^-(z)f(M_-(g^{-1})z) \tag{8.29}$$

where the multiplier $m_g^-(z)$ is chosen such that $\mathscr{D}_n^-$ defines an unitary representation of $SU(1, 1)$. We prove now that it is true with the choice $m_g^-(z) = (\bar{\alpha} - \beta z)^{-n}$.

Theorem 44 *For every integer $n \geq 2$ we have the following unitary representation of $SU(1, 1)$ in the Hilbert space $\mathscr{H}(\mathbb{D})$:*

$$\mathscr{D}_n^-(g)f(z) = (\bar{\alpha} - \beta z)^{-n} f\left(\frac{-\bar{\beta} + \alpha z}{\bar{\alpha} - \beta z}\right). \tag{8.30}$$

$$\mathscr{D}_n^+(g)f(z) = (\alpha + \bar{\beta} z)^{-n} f\left(\frac{\beta + \bar{\alpha} z}{\alpha + \bar{\beta} z}\right). \tag{8.31}$$

Proof It is not difficult to see that $\mathscr{D}_n^{\pm}$ are $SU(1,1)$ actions in the linear space $\mathscr{H}_n(\mathbb{D})$. Let us prove that $\mathscr{D}_n^-$ is unitary. This follows with the holomorphic change of variables $Z = \frac{-\bar{\beta}+\alpha z}{\bar{\alpha}-\beta z}$, $Z = X + iY$, $z = x + iy$. We get $z = \frac{\bar{\beta}+\bar{\alpha} Z}{\alpha+\beta Z}$, $\frac{dz}{dZ} = (\beta Z + \alpha)^{-2}$. We conclude using that for an holomorphic change of variable in the plane, we have from Cauchy conditions,

$$\left|\det \frac{\partial(x,y)}{\partial(X,Y)}\right| = \left|\frac{dz}{dZ}\right|^2.$$

So we get for any $f \in \mathscr{H}_n(\mathbb{D})$,

$$\int_{\mathbb{D}} |\bar{\alpha} - \beta z|^{-2n} \left| f\left(\frac{-\bar{\beta} + \alpha z}{\bar{\alpha} - \beta z}\right)\right|^2 (1 - |z|^2)^{n-2} dx dy = \int_{\mathbb{D}} |f(Z)|^2 (1 - |Z|^2)^{n-2} dX dY$$

which says that $\mathscr{D}_n^-$ is unitary. Let us remark here that the above computations show that the multiplier m_g^- is necessary to prove unitarity. □

The next step is to prove that these representations are irreducible. To do that we first compute the corresponding Lie algebra representation.

Let us compute the image of this basis by the representation $\mathscr{D}_n^-$. Straightforward computations give

$$\frac{d}{dt}\mathscr{D}_n^-(\omega_0(t))f(z)|_{t=0} = \frac{1}{2i}(n + 2z\frac{d}{dz})f(z), \tag{8.32}$$

$$\frac{d}{dt}\mathscr{D}_n^-(\omega_1(t))f(z)|_{t=0} = \frac{1}{2}(nz + (z^2 - 1)\frac{d}{dz})f(z), \tag{8.33}$$

$$\frac{d}{dt}\mathscr{D}_n^-(\omega_2(t))f(z)|_{t=0} = \frac{1}{2i}(nz + (1 + z^2)\frac{d}{dz})f(z). \tag{8.34}$$

So we get the three self-adjoint generators B_0, B_1, B_2, $B_j = id\mathscr{D}_n^-(b_j)$, where d denotes the differential on the group at $\mathbb{1}$,

$$B_0 = \frac{n}{2} + z\frac{d}{dz}, \tag{8.35}$$

$$B_1 = \frac{i}{2}\left(nz + (z^2 - 1)\frac{d}{dz}\right), \tag{8.36}$$

$$B_2 = \frac{1}{2}\left(nz + (z^2 + 1)\frac{d}{dz}\right), \tag{8.37}$$

with the commutation relations

$$[B_1, B_2] = -iB_0, \ [B_2, B_0] = -iB_1, \ [B_0, B_1] = iB_2. \tag{8.38}$$

Using the notation $B_\pm = B_2 \mp iB_1$, we have

$$B_- = \frac{d}{dz}, \tag{8.39}$$

$$B_+ = nz + z^2\frac{d}{dz}, \tag{8.40}$$

and

$$[B_-, B_+] = 2B_0, \ [B_0, B_\pm] = \pm B_\pm. \tag{8.41}$$

Remark 49 The operators B_a for $a = 0, 1, 2, \pm$ are non-bounded operators in the Hilbert space $\mathscr{H}_n(\mathbb{D})$ so we have to define their domains. Here we know that the representation $\mathscr{D}_n^-$ is unitary. So Stone's theorem gives that B_0 is essentially self-adjoint and the linear space $\mathscr{P}^\infty$ of all polynomials in z is a core for B_0. Moreover the spectrum of B_0 is discrete, with simple eigenvalues $\{n/2 + k, \ k \in \mathbb{N}\}$.

B_1 and B_2 also have a unique closed extension. Moreover it could be possible to characterize their domains (left to the reader!). Operators $B_\pm$ are closable in $\mathscr{H}_n(\mathbb{D})$. We keep the same notation $B_\pm$ for their closures.

We have the useful property

Lemma 52 *$B_\pm$ are adjoint of each other : $B_\pm^* = B_\mp$. In particular for every $\zeta \in \mathbb{C}$ the operator $i(\zeta B_- - \bar{\zeta} B_+)$ is self-adjoint.*

Proof It is enough to prove $B_\pm^* = B_\mp$. This is formally obvious because we know that B_1, B_2 are self-adjoint. We left to the reader to check that the domains are the same. □

Let us compute the Casimir operator C_{as}^n for the representation $\mathscr{D}_n^-$.

A direct computation of the coefficients $g_{j,k}$ of the Killing form shows that $g_{j,k} = 0$ if $j \neq k$ and $g_{0,0} = -1$, $g_{1,1} = g_{2,2} = 1$. So we get

$$C_{as}^n = B_0^2 - B_1^2 - B_2^2 = B_0^2 - \frac{1}{2}(B_+B_- + B_-B_+). \tag{8.42}$$

Let us assume for the moment that $\mathscr{D}_n^-$ is irreducible. Then by Schur's lemma we know that C_{as}^n is a number; this number can be computed using the monomial z^0. We find easily

$$C_{as}^n = \frac{n}{2}\left(\frac{n}{2} - 1\right) = k(k-1), \ k =: \frac{n}{2}. \tag{8.43}$$

Let us remark that $k := \frac{n}{2}$ is the lowest eigenvalue of B_0 and is called the Bargmann index.

8.2.3 Irreducibility of Discrete Series

Here we prove that for every integer $n \geq 2$, the representation $\mathscr{D}_n^-$ is irreducible.

Let E be a closed invariant subspace in $\mathscr{H}_n(\mathbb{D})$. The restriction of $\mathscr{D}_n^-$ to the compact commutative subgroup $g(\theta, 0, 0)$ is a sum of one dimensionnal unitary representations. So there exists $u \in E$, $\nu \in \mathbb{R}$ such that

$$\mathscr{D}_n^-(g(\theta, 0, 0))u = e^{i\nu\theta}u, \ \forall\theta \in \mathbb{R}.$$

But u has a series expansion $u(z) = \sum a_j z^j$. So by identification we have $e^{i\nu\theta}a_j = e^{-i(n+2j)\theta}a_j$. From this we get that there exists j_0 such that $a_{j_0} \neq 0$ so we find $\nu = n + 2j_0$. But this entails $a_j = 0$ if $j \neq j_0$. Hence we conclude that the monomial z^{j_0} belongs to E.

Now playing with $B_\pm$ we conclude easily that E contains all the monomials z^j, $j \in \mathbb{N}$. We have proved above that the monomials is a total system in $\mathscr{H}_n(\mathbb{D})$. So E=$\mathscr{H}_n(\mathbb{D})$.

The discrete series representations have an important property: they are square integrable (see Appendix).

On the group $SU(1, 1)$ we have a left and right invariant Haar measure μ. μ is positive on each non empty open set and unique up to a positive constant (see [169]).

The following result is proved in [170].

Proposition 97 *For every $f \in \mathscr{H}_n(\mathbb{D})$ we have*

$$\int_{SU(1,1)} |\langle \mathscr{D}_n^-(g)f, f\rangle_{\mathscr{H}_n(\mathbb{D})}|^2 d\mu(g) < +\infty$$

where dg is the Haar measure on $SU(1, 1)$.

There exists other unitary irreducible representations for $SU(1, 1)$: the principal series and the complementary series (see also the book of Knapp [168] for more details). These representations are *not square integrable*.

Up to equivalence, discrete series, principal series and complementary series are the only irreducible unitary representations. Let us first explain now principal series.

8.2.4 Principal Series

These representations can also be realized in Hilbert spaces of functions on the unit circle.

They are defined in the following way: take a nonnegative number λ and a point z on the unit circle. Then the homographic transformation

$$z \mapsto \frac{\alpha z + \bar{\beta}}{\beta z + \bar{\alpha}}$$

obviously maps the unit circle into itself. One defines

$$\mathscr{P}_{i\lambda}(g)f(z) = |\beta z + \bar{\alpha}|^{-1+2i\lambda} f\left(\frac{\alpha z + \bar{\beta}}{\beta z + \bar{\alpha}}\right).$$

One considers the Hilbert space $L^2(\mathbb{S}^1)$ with the scalar product

$$\langle f_1, f_2 \rangle = \frac{1}{2\pi} \int_0^{2\pi} d\theta \bar{f}_1(\theta) f_2(\theta).$$

To prove the unitarity of the representation $\mathscr{P}_{i\lambda}(g)$ in the Hilbert space we perform the change of variable $\theta \mapsto \theta'$ where

$$e^{i\theta'} = \frac{\alpha e^{i\theta} + \bar{\beta}}{\beta e^{i\theta} + \bar{\alpha}}.$$

The jacobian satisfies

$$\left|\frac{d\theta'}{d\theta}\right| = |\beta e^{i\theta} + \bar{\alpha}|^{-2}$$

Thus we have for any real λ:

$$\int_0^{2\pi} d\theta |(\mathscr{P}_{i\lambda}(g)f)(e^{i\theta})|^2 = \int_0^{2\pi} d\theta' |f(e^{i\theta'})|^2.$$

Now we can prove:

Proposition 98 *For any $\lambda \in \mathbb{R}$, $\mathscr{P}_{i\lambda}$ is a unitary irreducible representation of $SU(1,1)$ in the Hilbert space $L^2(\mathbb{S}^1)$. Moreover its Casimir operator is: $C_{i\lambda} = -\left(\frac{1}{4} + \lambda^2\right)$.*

For the generator $\omega_0(t)$ of the Lie group $\mathfrak{su}(1,1)$ one gets

$$(\mathscr{P}_{i\lambda}(\omega_0(t))f)(e^{i\theta}) = f(e^{i(\theta+t)}).$$

Thus the corresponding generator of $SU(1,1)$ is simply

$$L_0 = \frac{d}{d\theta}.$$

To find the generator associated to ω_1 we need to calculate

$$\left|\sinh\frac{t}{2}e^{i\theta} + \cosh\frac{t}{2}\right|^2 = \cosh t + \cos\theta \sinh t.$$

Thus

$$(\mathscr{P}_{i\lambda}(\omega_{-}(t)f)(e^{i\theta}) = (\cosh t + \sinh t \cos\theta)^{-1/2+i\lambda} f\left(\frac{\cosh\frac{t}{2}e^{i\theta} + \sinh\frac{t}{2}}{\sinh\frac{t}{2}e^{i\theta} + \cosh\frac{t}{2}}\right).$$

Thus the generator L_1 of $SU(1, 1)$ is given by

$$L_1 = \left(-\frac{1}{2} + i\lambda\right)\cos\theta - \sin\theta\frac{d}{d\theta}.$$

Similarly using the third generator ω_2 one finds

$$L_2 = -\left(-\frac{1}{2} + i\lambda\right)\sin\theta - \cos\theta\frac{d}{d\theta}.$$

Therefore the usual linear combinations $B_\pm = \pm L_1 + iL_2$ satisfy

$$B_+ = \left(-\frac{1}{2} + i\lambda\right)e^{-i\theta} - ie^{-i\theta}\frac{d}{d\theta}.$$

$$B_- = -\left(-\frac{1}{2} + i\lambda\right)e^{i\theta} - ie^{i\theta}\frac{d}{d\theta}.$$

We define

$$B_0 = iL_0 = i\frac{d}{d\theta}.$$

Using the same method as for discrete series, we can prove that these representations are irreducible (start with B_0 and use $B\pm$).

To calculate the Casimir operator $B_0^2 - \frac{1}{2}(B_-B_+ + B_+B_-)$ it is enough to apply it to the constant function. Thus $B_0 1 = 0,\ B_\pm 1 = \pm(-\frac{1}{2} + i\lambda)e^{\mp i\theta}$. One finds

$$C := B_0^2 - \frac{1}{2}(B_-B_+ + B_+B_-) = -\left(\frac{1}{4} - \lambda^2\right)\mathbb{1}.$$

8.2.5 Complementary Series

When the parameter λ of the principal series is imaginary the representation is not unitary in the space $L^2(\mathbb{S}^1)$. So, following Bargmann [20] we introduce a different Hilbert space. Let us introduce the sesquilinear form depending of the real parameter $\sigma \in]0, \frac{1}{2}[$,

$$\langle f_1, f_2\rangle_\sigma = c\iint_{[0,2\pi]^2}(1-\cos(\theta_1-\theta_2))^{\sigma-1/2}\overline{f_1(\theta_1)}f_2(\theta_2)d\theta_1 d\theta_2.$$

$\langle f_1, f_2\rangle_\sigma$ is well defined if f_1, f_2 are continuous on $\mathbb{S}^1$. The constant c is computed such that $\langle 1, 1\rangle_\sigma = 1$,

$$c = 2^{1/2-\sigma}\pi B(\sigma, 1/2)^{-1}.$$

The integral $\int_0^{2\pi}(1-\cos\theta)^{\sigma-1/2}d\theta$ is computed using the change of variable $x = \cos(\theta)$ so we get c. The following properties are useful to build the Hilbert space $\mathscr{H}_\sigma$.

Proposition 99 *(i) For every $f_1, f_2 \in C(\mathbb{S}^1)$ we have*

$$\iint_{[0,2\pi]^2}(1-\cos(\theta_1-\theta_2))^{\sigma-1/2}|f_1(\theta_1)||f_2(\theta_2)|d\theta_1 d\theta_2 \le \|f_1\|\|f_2\|,$$

(ii)

$$\langle e_k, e_\ell\rangle_\sigma = 0, \quad \text{if } k \neq \ell,$$

where $e_k(\theta) = \mathrm{e}^{ik\theta}$.
(iii)

$$\langle e_k, e_k\rangle_\sigma = \lambda_k(\sigma), s$$

where

$$\lambda_k(\sigma) = \frac{\Gamma(1/2+\sigma)\Gamma(|k|+1/2-\sigma)}{\Gamma(1/2-\sigma)\Gamma(|k|+1/2+\sigma)}. \tag{8.44}$$

In particular $\lambda_0 = 1$ and $\lambda_k(\sigma) > 0$ for every $k \in \mathbb{Z}$ and $\sigma \in]0, 1/2[$.

Proof (i) is proved using the change of variable $u = \theta_1 - \theta_2$ and Cauchy-Schwarz inequality.
(ii) It is a consequence of the following equality, for $k \neq \ell$,

$$\langle e_k, e_\ell\rangle_\sigma = \iint_{[0,2\pi]^2}(1-\cos(u))^{\sigma-1/2}\mathrm{e}^{-ik\theta}\ e^{i\ell(\theta+u)}d\theta du = 0.$$

(iii) We have $\lambda_{-k} = \lambda_k$, so it is enough to consider the case $k \geq 0$.
Hence we have

$$\langle e_k, e_k\rangle_\sigma = 2c\int_0^\pi (1-\cos(u))^{\sigma-1/2}\cos(ku)du.$$

We compute the integral using the change of variable $x = \cos u$, so $\cos(ku) = T_k(x)$, where T_k is the Tchebichev polynomial of order k.

$$\langle e_k, e_k \rangle_\sigma = 2c \int_{-1}^{1} (1-x)^{\sigma-1/2}(1-x^2)^{-1/2} T_k(x) dx.$$

But we have the following expression known as the Rodrigues formula [70]

$$T_k(x) = (-1)^k 2^{k-1} \frac{(k-1)!}{(2k)!} (1-x^2)^{1/2} \frac{d^k}{dx^k}((1-x^2)^{k-1/2}).$$

Hence we get the result by integrations by parts and well known formulas for gamma and beta special functions. □

So if $f_1, f_2 \in C(\mathbb{S}^1)$, $f_j = \sum_{k\in\mathbb{Z}} c_k^j e_k$ is the Fourier decomposition of f_j. Then it results that

$$\langle f_1, f_2 \rangle_\sigma = \sum_{k\in\mathbb{Z}} \lambda_k(\sigma) \overline{c_k^1} c_k^2.$$

This shows that $(f_1, f_2) \mapsto \langle f_1, f_2 \rangle_\sigma$ is a positive-definite sesquilinear form on $C(\mathbb{S}^1)$.

Now we can define the complementary series $\mathscr{C}_\sigma$, $0 < \sigma < 1/2$, as follows. It is realized in the Hilbert space $\mathscr{H}_\sigma$ of functions f on $\mathbb{S}^1$ such that $\sum_{k\in\mathbb{Z}} \lambda_k(\sigma)|c_k|^2 < +\infty$ equipped with the scalar product $\langle f_1, f_2 \rangle_\sigma$ (see proposition (iii)).

So we can define, for $f \in \mathscr{H}_\sigma$,

$$\mathscr{C}_\sigma(g)f(z) = |\beta z + \bar{\alpha}|^{-1+2\sigma} f\left(\frac{\alpha z + \bar{\beta}}{\beta z + \bar{\alpha}}\right).$$

Proposition 100 *For all $0 < \sigma < 1/2$, $\mathscr{C}_\sigma$ is a unitary irreducible representation of $SU(1, 1)$ in $\mathscr{H}_\sigma$. Moreover its Casimir operator is $C_\sigma = \sigma^2 - \frac{1}{4}$.*

Proof We use the same methods as for the discrete and continuous series. In particular the computations are the same as for the continuous series with σ in place of $i\lambda$. The main difference here is in the definition of the Hilbert space which is necessary to get a unitary representation. □

8.2.6 Realizations for Bosonic Systems

Let us start with a one boson system. We consider the usual annihilation and creation operators a, $a^\dagger$ in $L^2(\mathbb{R})$ (see Chap. 1). The following operators satisfy the commutation relations (8.41) of the Lie algebra $\mathfrak{su}(1, 1)$

$$B_+ = \frac{1}{2}(a^\dagger)^2, \quad B_- = \frac{1}{2}a^2, \quad B_0 = \frac{1}{4}(aa^\dagger + a^\dagger a). \tag{8.45}$$

We have seen in Chap. 3 that the metaplectic representation is a projective representation of the group $\mathrm{Sp}(1) = SL(2,\mathbb{R})$ and it is decomposed into two irreducible representations in the Hilbert subspaces of $L^2(\mathbb{R})$, $L^2_{ev}(\mathbb{R})$ of even states and $L^2_{od}(\mathbb{R})$ of odd states. But the group $SL(2,\mathbb{R})$ is isomorphic to the group $SU(1,1)$ by the explicit map $g \mapsto F_g := M_0 g M_0^{-1}, g \in SU(1,1), F_g \in SL(2,\mathbb{R})$ with $M_0 = \begin{pmatrix} 1 & i \\ i & 1 \end{pmatrix}$.

So the metaplectic representation defines a representation of the group $SU(1,1)$ in the space $L^2(\mathbb{R})$ with two irreducible components $\hat{R}_{ev,od}$ in the space $L^2_{ev,od}(\mathbb{R})$.

In quantum mechanics it is natural to consider ray-representations (or projective representations) instead of genuine representations. For example the metaplectic representation is a ray-representation.

Let us compute the Casimir operators $C_{ev,od}$ for each components.

We compute C_{ev} using the bound state ψ_0 for the harmonic oscillator ($\psi_0 \in L^2_{ev}(\mathbb{R})$).

Using that $\hat{H}_{osc} = \frac{1}{2}(aa^\dagger + a^\dagger a) = aa^\dagger - \frac{1}{2}$ we get $C_{ev}\psi_0 = -\frac{3}{16}\psi_0$ and $C_{od}\psi_1 = -\frac{3}{16}\psi_1$ ($B_0\psi_0 = \frac{1}{4}\psi_0$ and $B_0\psi_1 = \frac{3}{4}\psi_1$).

We see that the commutation relations (8.45) define two irreducible "representations" of $SU(1,1)$ which are neither in the discrete series neither in the continuous series. The reason is that they are ray representations corresponding with the even part and odd part of the metaplectic representation.

More details concerning ray representations can be found in [19, 260]. In particular these ray-representations are genuine representations of the covering group $\widetilde{SU}(1,1)$ (which is simply connected but $SU(1,1)$ is not). A "double valued representation" ρ in a linear space satisfies

$$\rho(gh) = C(g,h)\rho(g)\rho(h), \text{ with } C(g,h) = \pm 1.$$

Let us now consider the two bosons system. We consider two annihilation and creation operators $a_\pm$, $a_\pm^\dagger$ in $L^2(\mathbb{R}^2)$ (see Chap. 1). The following operators satisfies the commutation relations (8.41) of the Lie algebra $\mathfrak{su}(1,1)$

$$B_+ = a_1^\dagger a_2^\dagger, \quad B_- = a_1 a_2, \quad B_0 = \frac{1}{2}(a_1^\dagger a_1 + a_2^\dagger a_2 + 1). \tag{8.46}$$

The Casimir operator is

$$C_{as} = -\frac{1}{4} + \frac{1}{4}(a_2^\dagger a_2 - a_1^\dagger a_1)^2.$$

We know from Chap. 1 that we have an orthonormal basis of $L^2(\mathbb{R}^2)$, $\{\phi_{m_1,m_2}\}_{(m_1,m_2)\in\mathbb{N}^2}$, where $\phi_{m_1,m_2} = (a_1^\dagger)^{m_1}(a_2^\dagger)^{m_2}\phi_{0,0}$ such that

$$B_0\phi_{m_1,m_2} = \frac{m_1 + m_2 + 1}{2}\phi_{m_1,m_2} \tag{8.47}$$
$$B_+\phi_{m_1,m_2} = \phi_{m_1+1,m_2+1} \tag{8.48}$$
$$B_-\phi_{m_1,m_2} = \phi_{m_1-1,m_2-1}. \tag{8.49}$$

A direct computation gives

$$C_{as}\phi_{m_1,m_2} = \left(-\frac{1}{4} + (m_1 - m_2)^2\right)\phi_{m_1,m_2}. \tag{8.50}$$

So if we introduce $k = \frac{1}{2}(1 + |n_0|)$, we get easily (assuming $n_0 \geq 0$) the following lemma.

Lemma 53 *For every positive half integer k, the Hilbert space spanned by $\{\phi_{m_2+2k-1,m_2},\ m_2 \in \mathbb{N}\}$, is an irreducible space for the representation of the Lie algebra with generators (8.45).*

We know now that this Lie algebra representation defines a unitary representation of $\widetilde{SU}(1, 1)$ but only a projective representation of $SU(1, 1)$.

8.3 Pseudo-Spin-Coherent States for Discrete Series

We can now proceed to the construction of coherent states by analogy with the harmonic oscillator case (Glauber states) and the spin coherent states.

We consider here the discrete series representation.

8.3.1 Definition of Coherent States for Discrete Series

Let us consider the representation $(\mathcal{D}_n^-, \mathcal{H}(\mathbb{D}))$. It could be possible to work with $\mathcal{D}_n^+$ as well. Every $g \in SU(1, 1)$ can be decomposed as $g = g_{\mathbf{n}}h$ where $h \in U(1)$ and $\mathbf{n} \in P\mathbb{S}^2$. It is convenient to start with $\psi_0 \in \mathcal{H}(\mathbb{D})$ such that $h\psi_0 = \psi_0$ so we take as a fiducial state $\psi_0(\zeta) = \gamma_{n,0}^{-1/2}\zeta^0$ and we define $\psi_{\mathbf{n}} = \mathcal{D}_n^-(g_{\mathbf{n}})\psi_0$. Most of properties of $\psi_{\mathbf{n}}$ will follow from suitable formula for the operator family $D(\mathbf{n}) = \mathcal{D}_n^-(g_{\mathbf{n}})$. There are many similarities with the spin setting. We shall explain now these similarities in more details.

Using polar coordinates for $\mathbf{n}$ we have

$$g_{\mathbf{n}} = \begin{pmatrix} \cosh(\tau/2) & \sinh(\tau/2)\mathrm{e}^{-i\varphi} \\ \sinh(\tau/2)\mathrm{e}^{i\varphi} & \cosh(\tau/2) \end{pmatrix}.$$

So using the definition of the representation $\mathscr{D}_n^-$ we have the straightforward formula for the pseudospin coherent states.

$$\psi_\zeta(z) = (1-|\zeta|^2)^{n/2}(1-\bar{\zeta}z)^{-n}, \text{ where } \zeta = \sinh(\tau/2)\mathrm{e}^{i\varphi}. \tag{8.51}$$

Now we shall give a Lie group interpretation of the coherent states. Let us recall that $B_m = id\mathscr{D}_n^-(\mathbb{1})b_m$, $m = 0, 1, 2$, and $B_+ = B_2 + iB_1$, $B_- = B_2 - iB_-$. Then we have

$$D(\mathbf{n}) = \exp\left(-i\tau(\cos\varphi B_1 - \sin\varphi B_2)\right) \tag{8.52}$$

$$= \exp\left(\tau/2(B_-\mathrm{e}^{i\varphi} - B_+\mathrm{e}^{-i\varphi})\right). \tag{8.53}$$

The second formula reads

$$D(\mathbf{n}) = D(\xi) = \exp\left(\bar{\xi}B_- \xi B_+\right), \quad \text{with } \xi = \frac{\tau}{2}\mathrm{e}^{-i\varphi}. \tag{8.54}$$

We can get a simpler formula using an heuristic following from Gauss decomposition

$$g_\mathbf{n} = \begin{pmatrix} \cosh(\tau/2) & \sinh(\tau/2)\mathrm{e}^{-i\varphi} \\ \sinh(\tau/2)\mathrm{e}^{i\varphi} & \cosh(\tau/2) \end{pmatrix} =$$

$$\begin{pmatrix} 1 & 0 \\ \tanh(\tau/2)\mathrm{e}^{i\varphi} & 1 \end{pmatrix} \begin{pmatrix} \cosh(\tau/2) & 0 \\ 0 & \frac{1}{\cosh(\tau/2)} \end{pmatrix} \begin{pmatrix} 1 & \tanh(\tau/2)\mathrm{e}^{-i\varphi} \\ 0 & 1 \end{pmatrix}. \tag{8.55}$$

Recall that $B_0\psi_0 = \frac{n}{2}\psi_0$ and $|\zeta| = \tanh(\tau/2)$. Moreover if $b \in \mathfrak{su}(1,1)$ then we have

$$\mathscr{D}_n^-\left(\mathrm{e}^{tb}\right) = \mathrm{e}^{-itB}, \quad \text{with } B = id\mathscr{D}_n^-(\mathbb{1})b. \tag{8.56}$$

Suppose that (8.56) can be used for $b_1 \pm b_2$ (which are not in the Lie algebra $\mathfrak{su}(1,1)$. Then we get the following representation of pseudocoherent states in the Poincaré disc $\mathbb{D}$, where ζ and $\mathbf{n}$ represents the same point on the pseudosphere $P\mathbb{S}^2$,

$$\psi_\mathbf{n} = \psi_\zeta = (1-|\zeta|^2)^{n/2}\mathrm{e}^{\bar{\zeta}B_+}\psi_0. \tag{8.57}$$

Let us remark that in the spin case this heuristic is rigorous because the representation $\mathscr{D}^j$ is well defined on $SL(2,\mathbb{C})$ which is not true for $\mathscr{D}_n^-$. Nevertheless it is possible to give a rigorous meaning to formula (8.57) as we shall see in the next section.

8.3.2 Some Explicit Formulae

We follow more or less the computations done in the spin case. We shall give details only when the proofs are really different.

It is convenient to compute in the canonical basis $\{e_\ell\}_{\ell\in\mathbb{N}}$ of the representation space $\mathscr{H}_n(\mathbb{D})$ (analogue of Dicke states or Hermite basis). We get easily the formula

$$B_+e_\ell = \sqrt{(n+\ell)(\ell+1)}e_{\ell+1} \tag{8.58}$$

$$B_-e_\ell = \sqrt{\ell(n+\ell-1)}e_{\ell-1}, \quad B_-e_0 = 0 \tag{8.59}$$

$$B_0e_\ell = (\frac{n}{2}+\ell)e_\ell. \tag{8.60}$$

Let us remark that the linear space $\mathscr{P}_j$ of polynomials of degree $\leq j$ is stable for B_0 and B_- but not for B_+. For every $\ell \in \mathbb{N}$ we have

$$B_+^\ell e_0 = (n(n+1)\cdots(n+\ell-1)\ell!)^{1/2}\, e_\ell.$$

Following our heuristic argument we expand the exponent $e^{\bar{\zeta}B_+}$ as a Taylor series (which is not allowed because B_+ is unbounded) and we recover the formula:

$$\psi_\zeta(z) = (1-|\zeta|^2)^k(1-\bar{\zeta}z)^{-2k}, \quad k = n/2. \tag{8.61}$$

Let us give now a rigorous proof for this. It is enough to explain what is $e^{tB_+}e_\ell$ for every $t \in \mathbb{D}$ and and every $\ell \in \mathbb{N}$. For simplicity we assume $t \in]-1, 1[$.

Proposition 101 *For every $m \in \mathbb{N}$, the differential equation*

$$\dot{\phi}_t = B_+\phi_t, \quad \phi_0(z) = z^m,$$

has a unique solution holomorphic in $(t, z) \in \mathbb{D}\times\mathbb{D}$ given by the following formulas. For $m = 0$

$$\phi_t(z) = (1-tz)^{-n}, \tag{8.62}$$

for $m \geq 1$,

$$\phi_t(z) = (1-tz)^{-n} - 1 + z^m(1-tz)-n-m. \tag{8.63}$$

Proof We check $\phi_t(z) = \sum_{\ell\in\mathbb{N}} x_\ell(t)z^\ell$. So we can compute $x_\ell(t)$ using the induction formula

$$x_{\ell+1}(t) = x_{\ell+1}(0) + (n+\ell)\int_0^t x_\ell(s)ds.$$

The result follows easily. □

From our computations we get the expansion of ψ_ζ in the canonical basis

$$\psi_\zeta = (1-|\zeta|^2)^k \sum_{\ell\in\mathbb{N}} \left(\frac{\Gamma(2k+\ell)}{\Gamma(\ell+1)\Gamma(2k)}\right)^{1/2} \bar{\zeta}e_\ell. \tag{8.64}$$

Proposition 102 *For every* $\mathbf{n}_1, \mathbf{n}_2 \in P\mathbb{S}^2$ *we have*

$$D(\mathbf{n}_1)D(\mathbf{n}_2) = D(\mathbf{n}_3)\exp(-i\Phi(\mathbf{n}_1, \mathbf{n}_2)B_0), \tag{8.65}$$

where $\Phi(\mathbf{n}_1, \mathbf{n}_2)$ *is the oriented area of the geodesic triangle on the pseudosphere with vertices at the points* $[\mathbf{n}_0, \mathbf{n}_1, \mathbf{n}_2]$.
$\mathbf{n}_3$ *is determined by*

$$\mathbf{n}_3 = R(g_{\mathbf{n}_1})\mathbf{n}_2, \tag{8.66}$$

where $R(g)$ *is the rotation associated to* $g \in SU(1,1)$ *and*

$$g_{\mathbf{n}} = \exp\left(-i\frac{\tau}{2}(\sigma_1 \sin\varphi + \sigma_2 \cos\varphi)\right). \tag{8.67}$$

Proof Computation of $\mathbf{n}_3$ is easy using the following lemma. The phase will be detailed later. □

Lemma 54 *For all* $g \in SU(1,1)$ *there exists* $\mathbf{m} \in \mathbb{PS}^2$ *and* $\delta \in \mathbb{R}$ *such that*

$$g = g_{\mathbf{m}} r_3(\delta),$$

where $r_3(\delta) = \exp\left(i\frac{\delta}{2}B_0\right)$.

The following lemma shows that the pseudospin is also independent of the direction.

Lemma 55 *One has:*

$$D(\mathbf{n})\hat{B}_0 D(\mathbf{n})^{-1} = -\mathbf{n} \boxdot \mathbf{B}, \tag{8.68}$$

where $\boxdot$ *is the notation for the Minkowski scalar product 235*

Proof Let (τ, φ) be the pseudopolar coordinates of $\mathbf{n}$, $\mathbf{n} = \mathbf{n}(\tau, \varphi)$. We have $D(\mathbf{n}(\tau)) = \exp(-i\tau(\cos\varphi B_1 - \sin\varphi B_2))$.
Let us use the notation $A(\tau) := D(\mathbf{n}(\tau))AD(\mathbf{n}(\tau))^{-1}$ where A is any operator in $\mathscr{H}(\mathbb{D})$. Then we have the equalities

$$\frac{d}{d\tau}B_0(\tau) = -\cos\varphi B_2(\tau) - \sin\varphi B_1(\tau), \tag{8.69}$$

$$\frac{d}{d\tau}B_1(\tau) = -\sin\varphi B_0(\tau), \tag{8.70}$$

$$\frac{d}{d\tau}B_2(\tau) = -\cos\varphi B_0(\tau). \tag{8.71}$$

We have the following consequences

$$\frac{d^2}{d\tau^2}B_0(\tau) = B_0(\tau), \quad \frac{d}{d\tau}B_0(0) = -\cos\varphi B_2 - \sin\varphi B_1, \tag{8.72}$$

hence we get that $B_0(\tau) = -\mathbf{n}(\tau) \boxdot \mathbf{B}$. □

The following consequence is that $|\mathbf{n}\rangle$ is an eigenvector of the operator $\mathbf{n} \boxdot \mathbf{B}$:

Proposition 103 *One has:*

$$\mathbf{n} \boxdot \mathbf{B}|\mathbf{n}\rangle = -k|\mathbf{n}\rangle, \tag{8.73}$$

where $k = \frac{n}{2}$.

As in the Heisenberg and spin settings, the pseudospincoherent states family $|\mathbf{n}\rangle$ is not an orthogonal system in $\mathscr{H}(\mathbb{D})$. One can compute the scalar product of two coherent states $|\mathbf{n}\rangle$, $|\mathbf{n}'\rangle$:

Proposition 104 *One has*

$$\langle \mathbf{n}'|\mathbf{n}\rangle = e^{-ik\Phi(\mathbf{n},\mathbf{n}')} \left(\frac{1 - \mathbf{n} \boxdot \mathbf{n}'}{2}\right)^{-k}, \tag{8.74}$$

where $\Phi(\mathbf{n}, \mathbf{n}')$ *is the oriented area of the hyperbolic triangle* $\{\mathbf{n}_0, \mathbf{n}, \mathbf{n}'\}$.

Proof We use the complex representation of coherent states, starting from the definition, we get

$$\psi_\zeta(z) = (1 - |\zeta|^2)^k (1 - \bar{\zeta} z)^{-2k}. \tag{8.75}$$

Next we use the series expansion

$$(1 - \bar{\zeta} z)^{-2k} = \sum_{\ell \geq 0} \frac{(2k + \ell - 1)!}{(2k - 1)!\ell!} (\bar{\zeta} z)^\ell$$

to compute the Fourier coefficient of the coherent state $|\zeta\rangle$ in the basis e_ℓ:

$$\langle e_\ell|\zeta\rangle = \left(\frac{\Gamma(2k + \ell)}{\Gamma(\ell + 1)\Gamma(2k)}\right)^{1/2} (1 - |\zeta|^2)^k \zeta^\ell. \tag{8.76}$$

The Parseval identity gives

$$\langle \mathbf{n}'|\mathbf{n}\rangle = (1 - |\zeta|^2)^k ((1 - |\zeta'|^2)^k (1 - \bar{\zeta}'\zeta)^{-2k}. \tag{8.77}$$

We can translate this equality in pseudopolar cordinates ($\zeta = \tanh(\tau/2)\mathrm{e}^{-i\varphi}$) and we get

$$\langle \mathbf{n}'|\mathbf{n}\rangle = \left(\cosh(\tau/2)\cosh(\tau'/2) - \sinh(\tau/2)\sinh(\tau'/2)\mathrm{e}^{i(\varphi'-\varphi)}\right)^{-2k}. \tag{8.78}$$

□

An easy computation now gives the following lemma

Lemma 56

$$|\langle \mathbf{n}'|\mathbf{n}\rangle|^2 = \left(\frac{1 - \mathbf{n} \boxdot \mathbf{n}'}{2}\right)^{-2k}. \tag{8.79}$$

The computation of the phase Φ in formula (8.74) can be done as for the spin case using the geometric phase method.

As it is expected, the pseudospin coherent state system provides a "resolution of the identity" in the Hilbert space $\mathscr{H}(\mathbb{D})$:

Proposition 105 *We have the formula*

$$\boxed{\frac{2k-1}{4}\int_{P\mathbb{S}^2} d\mathbf{n}|\mathbf{n}\rangle\langle\mathbf{n}| = \mathbb{1}}. \tag{8.80}$$

Or using complex coordinates $|\zeta\rangle$,

$$\boxed{\int_{\mathbb{D}} d\nu_{2k}(\zeta)|\zeta\rangle\langle\zeta| = \mathbb{1}}. \tag{8.81}$$

where the measure $d\nu_n$ *is*

$$d\nu_n(\zeta) = \frac{n-1}{\pi}\frac{d^2\zeta}{(1-|\zeta|^2)^2},$$

with $d^2\zeta = \frac{|d\zeta\wedge d\bar{\zeta}|}{2}$.

Proof The two formulas are equivalent by the change of variables $\zeta = \tanh\frac{\tau}{2}\mathrm{e}^{-i\varphi}$. So it is sufficient to prove the complex version.

We introduce

$$\tilde{f}(\zeta) = \langle\zeta|f\rangle_{\mathscr{H}_n(\mathbb{D})}.$$

Decompose f in the basis of $\mathscr{H}_n(\mathbb{D})$, $f = \sum_{\ell\geq 0} c_\ell e_\ell$, we have

$$|f^\sharp(\zeta)|^2 = \sum_{\ell\geq 0}\frac{\Gamma(2k+\ell)}{\Gamma(\ell+1)\Gamma(2k)}(1-|\zeta|^2)^{2k}|\zeta|^{2\ell}|c_\ell|^2.$$

After integration in ζ we have

$$\frac{2k-1}{\pi}\int_{\mathbb{D}}|f^\sharp(\zeta)|^2\frac{d^2\zeta}{(1-|\zeta|^2)^2} = \int_{\mathbb{D}}|f(z)|^2 d^2z.$$

So we get the resolution of identity by a polarisation argument. □

8.3.3 *Bargmann Transform and Large k Limit*

Here we introduce the (pseudospin) Bargmann transform and prove that as $k \to +\infty$ the representation $\mathcal{D}^-_{2k}$ contracts to the Harmonic oscillator representation or Heisenberg-Schrödinger-Weyl representation.

Let us denote

$$\varphi^{k,\sharp}(\zeta) = \langle \zeta | \varphi \rangle_{\mathscr{H}_n(\mathbb{D})}(1 - |\zeta|^2)^{-k}, \ \varphi \in \mathscr{H}_{2k(\mathbb{D})}, \ \zeta \in \mathbb{D}.$$

In fact this transformation is trivial, here it is identity! But it is convenient to see this as a Bargmann transform. Using Parseval formula we get easily

$$\varphi^{k,\sharp}(\zeta) = \sum_{\ell \in \mathbb{N}} \zeta^\ell \gamma_{2k,\ell}^{-1/2} \langle e_\ell, \varphi \rangle_{\mathscr{H}_n(\mathbb{D})} = \sum_{\ell \in \mathbb{N}} e_\ell(\zeta) \langle e_\ell, \varphi \rangle_{\mathscr{H}_n(\mathbb{D})}.$$

Here we shall note the dependence in the Bargmann index k, so we denote the pseudospin coherent state $\psi^k_\zeta(z)$.

Proposition 106 *The pseudocoherent states ψ^k_ζ converge to the Glauber coherent state φ_ζ (see Chap. 1) as $k \to +\infty$ in the following Bargmann sense and the Dicke states e_k converge to the the Hermite function ψ_ℓ, for every $\ell \in \mathbb{N}$.*

$$\lim_{k \to +\infty} \psi^{k,\sharp}_{\zeta'/\sqrt{2k}}(\zeta/\sqrt{2k}) = \varphi^\sharp_{\zeta'}(\zeta), \ \ \forall \zeta, \zeta' \in \mathbb{C}. \tag{8.82}$$

Proof It is an easy exercise, knowing that

$$\varphi^\sharp_\zeta(\zeta') = \exp\left(\bar{\zeta}\zeta' - \frac{|\zeta|^2}{2}\right)$$

and that

$$\psi^\sharp_\ell(\zeta) = \frac{\zeta^\ell}{\sqrt{2\pi\ell!}}.$$

□

As for the spin case we have analogous results for the generators of the Lie algebras.

Let us introduce a small parameter $\varepsilon > 0$ and denote

$$B^\varepsilon_\pm = \varepsilon B_\pm, \tag{8.83}$$

$$B^\varepsilon_0 = B_0 - \frac{1}{2\varepsilon^2}\mathbb{1}. \tag{8.84}$$

We have the following commutation relations

$$[B_+^\varepsilon, B_-^\varepsilon] = 2\varepsilon^2 B_3^\varepsilon - \mathbb{1}, \quad [B_3^\varepsilon, B_\pm^\varepsilon] = \pm B_\pm^\varepsilon. \tag{8.85}$$

As $\varepsilon \to 0$ Eq. (8.83) define a family of singular transformations of the Lie algebra $\mathfrak{su}(1,1)$ and for $\varepsilon = 0$ we get (formally)

$$[B_+^0, B_-^0] = -\mathbb{1}, \ [B_3^0, B_\pm^0] = \pm B_\pm^0. \tag{8.86}$$

These commutation relations are those satisfied by the harmonic oscillator Lie algebra : $B_+^0 \equiv a^\dagger$, $B_-^0 \equiv a$, $B_0^0 \equiv N := a^\dagger a$.

We can give a mathematical proof of this analogy by computing the averages.

Proposition 107 *Assume that* $\varepsilon \to 0$ *and* $k \to +\infty$ *such that* $\lim 2k\varepsilon^2 = 1$. *Then we have*

$$\lim \langle \psi^k_{\zeta/\sqrt{2k}} | B_0^\varepsilon | \psi^k_{\zeta/\sqrt{2k}} \rangle = |\zeta|^2 = \langle \varphi_\zeta | a^\dagger a | \varphi_\zeta \rangle. \tag{8.87}$$

$$\lim \langle \psi^k_{\zeta/\sqrt{2k}} | B_+^\varepsilon | \psi^k_{\zeta/\sqrt{2k}} \rangle = \bar{\zeta} = \langle \varphi_\zeta | a^\dagger | \varphi_\zeta \rangle. \tag{8.88}$$

$$\lim \langle \psi^k_{\zeta/\sqrt{2k}} | B_-^\varepsilon | \psi^k_{\zeta/\sqrt{2k}} \rangle = \zeta = \langle \varphi_\zeta | a | \varphi_\zeta \rangle. \tag{8.89}$$

Proof From the proof of Lemma 55 we can compute the following averages.

$$\langle \psi_{\mathbf{n}}^k, B \psi_{\mathbf{n}}^k \rangle = k\mathbf{n}.$$

Using the ζ parametrization we get the result as in the spin case. □

8.4 Coherent States for the Principal Series

As for discrete series we can consider coherent states for the principal and complementary series. The principal series is realized in the Hilbert space $L^2(\mathbb{S}^1)$ with the Haar probability measure on the circle $\mathbb{S}^1$ and with its orthonormal basis $e_\ell(\theta) = \mathrm{e}^{i\ell\theta}$, $\ell \in \mathbb{Z}$.

e_0 being invariant by the rotations subgroup of $S(1,1)$ we define the coherent states $\psi^\lambda_{\mathbf{n},\zeta}(z)$, $z \in \mathbb{S}^1$, as

$$\psi_{\mathbf{n}}^{i\lambda}(z) = |\cosh(\tau/2) + \sinh(\tau/2)\mathrm{e}^{-i\varphi} z|^{2i\lambda - 1}, \tag{8.90}$$

$$\psi_\zeta^{i\lambda}(\theta) = |1 - |\zeta|^2|^{\frac{1}{2} - i\lambda} |1 - \bar{\zeta} z|^{2i\lambda - 1}. \tag{8.91}$$

In the first formula coherent states are parametrized by the pseudo-sphere and in the second formula they are parametrized by the complex plane.

Properties of these coherent states are analyzed in the book [208] (pp. 77–83).

8.5 Generator of Squeezed States. Application

We shall prove here that the $SU(1,1)$ generalized coherent states considered above (introduced by Perelomov [208]) are nothing but the one dimensional squeezed states introduced in Chap. 3.

We consider the realization of the Lie algebra $\mathfrak{su}(1,1)$ defined by the generators

$$B_0 = \frac{1}{4}(\mathbf{a}\mathbf{a}^\dagger + \mathbf{a}^\dagger\mathbf{a}), \tag{8.92}$$

$$B_+ = \frac{1}{2}(\mathbf{a}^\dagger)^2, \tag{8.93}$$

$$B_- = \frac{1}{2}\mathbf{a}^2. \tag{8.94}$$

These generators are defined as closed operators in the Hilbert space $L^2(\mathbb{R})$. They obey the commutation rules (8.41). Furthermore for the Casimir operator we have $C_{as} = -\frac{3}{16}\mathbb{1}$.

We have already remarked in Sect. 8.2.6 that this representation of the Lie algebra $\mathfrak{su}(1,1)$ gives a projective representation of $SU(1,1)$ (not a genuine group representation).

8.5.1 *The Generator of Squeezed States*

Consider now one dimensional squeezed states. Recall the following definition. Take a complex number ω such that $|\omega| < 1$. We define

$$\beta(\omega) = \frac{\omega}{|\omega|}\arg\tanh(|\omega|),$$

$$D(\beta) = \exp(\beta\hat{B}_+ - \bar{\beta}\hat{B}_-).$$

$D(\beta)$ is also known as the "Bogoliubov transformation" and generates squeezing.

For $|0\rangle = \varphi_0$ being the ground state of B_0, let the generalized coherent state ψ_β be defined as:

$$\psi_\beta = D(\beta)|0\rangle$$

Remark 50 It is not difficult to show that in terms of the operators $\hat{Q}$, $\hat{P}$ of quantum mechanics one has:

$$D(\beta) = \exp\left(\frac{i}{2}\Im\beta(\hat{Q}^2 - \hat{P}^2) - \frac{i}{2}\Re\beta(\hat{Q}\hat{P} + \hat{P}\hat{Q})\right)$$

We first recall the fundamental property of $D(\beta)$ proved in Chap. 3 in any dimension.

Lemma 57 *On $\mathscr{D}(\hat{Q}) \cap \mathscr{D}(\hat{P})$ the following identities holds true:*
(i) $D(\beta)$ is unitary and satisfies:

$$D(\beta)^{-1} = D(-\beta),$$

(ii)

$$D(\beta)\mathbf{a}D(-\beta) = (1 - |\omega|^2)^{-1/2}(\mathbf{a} - \omega\mathbf{a}^\dagger),$$

(iii)

$$D(\beta)B_0D(-\beta) = \cosh(2r)B_0 - \frac{\sinh(2r)}{2}(B_+e^{i\theta} + B_-e^{-i\theta}),$$

with $\beta = re^{i\theta}$ being the polar decomposition of β into modulus and phase.

The following results are direct consequences of Chap. 3.

Proposition 108 *(i) Define $\delta = \frac{1-\omega}{1+\omega}$. One has that*

$$\Re\delta > 0$$

and

$$\psi_\beta(x) = \left(\frac{\Re\delta}{\pi}\right)^{1/4} \left(\frac{1+\omega}{|1+\omega|}\right)^{1/2} \exp\left(-\delta\frac{x^2}{2}\right).$$

(ii) More generally if ϕ_k is the k-th normalized eigenstate of B_0 (Hermite function) one has

$$D(\beta)\phi_k = \frac{2^{-k/2}}{\sqrt{n!}}\left(\frac{\Re\delta}{\pi}\right)^{1/4} \left(\frac{1+\omega}{|1+\omega|}\right)^{k+1/2} H_k(x\sqrt{\Re\delta}) \exp\left(-\frac{\delta x^2}{2}\right).$$

where H_k is the normalized k-th Hermite polynomial.

Now we address the following question: what is the Wigner function of a squeezed state? It will appear that it is a Gaussian in q, p but with squeezing in some direction and dilatation in the other direction. One has the following result:

Proposition 109 *The Wigner function $W_{\psi_\beta}(q, p)$ is given by*

$$W_{\psi_\beta}(q, p) = 2\exp\left(-\frac{q^2\Re\delta}{\hbar} - \frac{1}{\hbar\Re\delta}(p + q\Im\delta)^2\right).$$

Remark 51 (i) For $\beta = 0$, $\delta = 1$, and thus we recover the Wigner function of φ_0.
(ii) It is clear that

$$\frac{1}{2\pi\hbar}\int dqdp W_{\psi_\beta}(q, p) = 1.$$

The proof is an easy computation of Gaussian integrals.

8.5.2 Application to Quantum Dynamics

Consider the time dependent quadratic Hamiltonian

$$\hat{H}_2(t) = \lambda(t)B_+ + \bar{\lambda}(t)B_- + \mu(t)B_0, \tag{8.95}$$

where λ and μ are C^1 functions of t, λ is complex and μ is real. Its propagator is denoted $U_2(t, s)$. This is a particular case of general quadratic Hamiltonian studied in Chap. 1 and in Chap. 3.

We revisit here the computation of $U_2(t, s)$ using the $\mathfrak{su}(1, 1)$ Lie algebra relations satisfied by $\{B_0, B_+, B_-\}$.

It is convenient to formulate the result in an abstract setting.

Proposition 110 *Assume that $B_0, B_\pm$ are closed operators with a dense domain in an Hilbert space $\mathscr{H}$ such that $B_0^* = B_0$, $B_+^* = B_-$ and satisfying the commutation relations :*

$$[B_-, B_+] = 2B_0, \quad [B_0, B_\pm] = \pm B_\pm.$$

Then $\hat{H}_2(t)$ defined by (8.95) has a propagator given by

$$U_2(t, s) = D(\beta_t)\exp\left(i(\gamma_t - \gamma_s)B_0\right)D(-\beta_t), \tag{8.96}$$

where the complex function β_t and the real function γ_t satisfy the differential equations

$$i\dot{\omega}_t = \bar{\lambda}\omega_t^2 + \mu\omega_t + \lambda, \quad \omega_0 = 0. \tag{8.97}$$

$$\dot{\gamma} = -\lambda\bar{\omega} - \bar{\lambda}\omega - \mu, \quad \gamma_0 = 0. \tag{8.98}$$

Proof The first step is to compute the following derivatives

$$i\frac{d}{dt}D(\beta_t) = (\alpha_t B_+ + \bar{\alpha}_t B_- + \rho_t B_0)D(\beta_t), \tag{8.99}$$

where

$$\alpha_t = i\frac{\dot{\omega}_t}{1 - |\omega|^2}, \tag{8.100}$$

$$\rho_t = i\frac{\omega_t\dot{\bar{\omega}}_t - \dot{\omega}_t\bar{\omega}_t}{1 - |\omega_t|^2}. \tag{8.101}$$

Using (8.99) we can compute

$$i\frac{d}{dt}U_2(t,s) = (\alpha_t B_+ + \bar{\alpha}_t B_- + \rho_t B_0 - \dot{\gamma} D(\beta_t) B_0)\, D(-\beta_t)$$

and we get directly equations (8.97).

Let us prove now (8.99).

The method is the following. Denote $L(t) = \beta_t B_+ - \bar{\beta}_t B_-$. We have

$$L(t+\delta t) = L(t) + \delta L(t) \approx L(t) + \delta \frac{d}{dt} L(t).$$

Applying Duhamel formula we get

$$\mathrm{e}^{L(t+\delta t)} - \mathrm{e}^{L(t)} = \int_0^1 ds \mathrm{e}^{sL(t+\delta t)} \delta L(t) \mathrm{e}^{(1-s)L(t)}.$$

Then as $\delta t \to 0$ we have

$$\frac{d}{dt}\mathrm{e}^{L(t)} = \int_0^1 ds \mathrm{e}^{sL(t)} \dot{L}(t) \mathrm{e}^{(1-s)L(t)}.$$

Now we have $\dot{L}(t) = \dot{\beta}_t B_+ - \dot{\bar{\beta}}_t B_-$ and

$$\frac{d}{ds}\mathrm{e}^{sL(t)} B_+ \mathrm{e}^{-sL(t)} = -2\bar{\beta} \mathrm{e}^{sL(t)} B_0 \mathrm{e}^{-sL(t)}$$

and

$$\frac{d}{ds}\mathrm{e}^{sL(t)} B_- \mathrm{e}^{-sL(t)} = -2\beta \mathrm{e}^{sL(t)} B_0 \mathrm{e}^{-sL(t)}.$$

Using Lemma 57 we get formula (8.99). □

Remark 52 The differential equation satisfied by ω_t is a Ricatti equation. We have seen in Chap. 4 that this equation comes from a classical flow. In particular ω_t is defined for every time t and satisfied $|\omega_t| < 1$ ($\omega_0 = 0$).

Let us now consider the time dependent Hamiltonian

$$\hat{H}_g(t) = \frac{1}{2}\left(\hat{P}^2 + f(t)\hat{Q}^2\right) + \frac{g^2}{2\hat{Q}^2},$$

where g is a coupling constant and f a function of time t. Properties of this Hamiltonian have been considered by [208] and used in [60] to study the quantum dynamics for ions in a Paul trap.

The $\mathfrak{su}(1,1)$ Lie algebra relations are satisfied by

$$B_0 = \frac{\hat{P}^2 + \hat{Q}^2}{4} + \frac{g^2}{4\hat{Q}^2},\quad B_\pm = \frac{\hat{Q}^2 - \hat{P}^2}{4} - \frac{g^2}{4\hat{Q}^2} \mp \frac{QP + PQ}{4}. \tag{8.102}$$

So we get

$$\hat{H}_g(t) = \frac{1}{2}(f(t) - 1)B_+ + \frac{1}{2}(f(t) - 1)B_- + (1 + f(t))B_0.$$

The algebra is the same as above but here the potential $\frac{g^2}{2\hat{Q}^2}$ has a non integrable singularity and we have to take care of the domain of definition for operators $B_0, B_\pm$. Let us consider the Hilbert space $L^2(\mathbb{R}_+)$. Recall the Hardy inequality

$$\int_0^{+\infty} dx \frac{|u(x)|^2}{|x|^2} \leq 4 \int_0^{+\infty} dx |u'(x)|^2, \ \forall u \in H_0^1(\mathbb{R}_+).$$

Recall that $H_0^1(\mathbb{R}_+)$ is the Sobolev space $H^1(\mathbb{R}_+)$ with the condition $u(0) = 0$. Denote by $L^2{}_1(\mathbb{R}_+)$ the space $\{u \in L^2(\mathbb{R}_+), xu \in L^2(\mathbb{R}_+)\}$. The sesquilinear form $(u, v) \mapsto \langle u, B_0 v\rangle$ is well defined on $V := H_0^1(\mathbb{R}_+) \cap L_1^2(\mathbb{R}_+)$ and is Hermitian, non negative. So B_0 has a self-adjoint extension as a unbounded operator in $L^2(\mathbb{R}_+)$. Furthermore we see that $B_\pm$ are also defined as forms on V and have closed extensions in $L^2(\mathbb{R}_+)$ such that $B_-^* = B_-$. These extensions also satisfy the $\mathfrak{su}(1, 1)$ Lie algebra relations. Hence the unitary operators $D(\beta)$, $e^{i\gamma B_0}$ are well defined in $L_1^2(\mathbb{R}_+)$ with β complex and γ real.

So we can apply Proposition 110 to the propagator $U_g(t, s)$ of $H_g(t)$.

Corollary 28 *The Hamiltonian $H_g(t)$ has a time dependent propagator $U_g(t, s)$ given by (8.96)*

$$U_g(t, s) = D(\beta_t) \exp\left(i(\gamma_t - \gamma_s)B_0\right) D(-\beta_t), \tag{8.103}$$

where $B_0, B_\pm$ are defined by Eq. (8.102) and β_t, γ_t are determined by (8.97). Moreover they are related to complex solutions of the Newton equation

$$\ddot{\xi}_t = f(t)\xi_t, \quad \ddot{\xi}_0 = ig,$$

$$\omega_t = \frac{\xi + i\dot{\xi}}{\xi - i\dot{\xi}}, \quad \gamma_t = -\frac{1}{2}\arg(\xi - i\dot{\xi}). \tag{8.104}$$

Remark 53 Note that the solution of the classical equation of motion for $H_0(t)$ solves the quantum evolution problem for $\hat{H}_g$ for every $g \in \mathbb{R}$.

The $\mathfrak{su}(1, 1)$ Lie algebra can also be used to solve the stationary Schrödinger equations for the hydrogen atom. It is nothing else than a group theoretic approach of a method already used by Schrödinger himself [234].

Let us consider the radial Hamiltonian for the hydrogen atom with mass 1, $\hbar = 1$, charge e, energy $E > 0$.

$$\left(\frac{d^2}{dr^2} + \frac{2}{r}\frac{d^2}{dr^2} + 2\frac{e^2}{r} - \frac{\ell(\ell+1)}{r^2} + 2E\right) R(r) = 0. \tag{8.105}$$

We transform this equation by the change of variable $r = x^2$ and of function $R(r) = x^{-3/2}f(x)$. Then we get

$$\left(\frac{d^2}{dx^2} - 8Ex^2 - \frac{4\ell(\ell+1)+3/4}{x^2} + 8e^2\right)f = 0.$$

Another change of variable $x = \lambda u$ with $\lambda = (-\frac{1}{8E})^{1/4}$ gives

$$\left(\frac{d^2}{du^2} - u^2 - \frac{4\ell(\ell+1)+3/4}{u^2} + 8\lambda^2 e^2)\right)f = 0. \tag{8.106}$$

This equation is the eigenvalue equation for the generator B_0 with $g^2 = 4\ell(\ell+1)+3/4$. So the negative energies of the radial Schrödinger equation (8.105) are determined by the eigenvalues of B_0.

Lemma 58 *The self adjoint operator B_0 has a compact resolvent. Its spectrum is a discrete set of simple eigenvalues given by*

$$\lambda_k = \frac{2s+1}{4} + k, \quad k \in \mathbb{N}, \text{ where } s = \frac{1}{2} + \left(\frac{1}{4} + g^2\right)^{1/2}. \tag{8.107}$$

Proof The domain of B_0 is included in $H_0^1(\mathbb{R}_+) \cap L_1^2(\mathbb{R}_+)$ so we deduce that its resolvent is compact.

The computation of the spectrum is standard, using B_- and B_+ as annihilation and creation operators on the ground state ψ_0. using the results of Sect. 8.2.1.

We compute the ground state by solving equation $B_-\psi_0 = 0$. This a singular differential equation. We put $\psi_0(x) = x^s\varphi(x)$ We can eliminate the singularity by choosing $s = \frac{1}{2} + \left(\frac{1}{4} + g\right)^{1/2}$. Then the equation is satisfied if $\varphi(x) = \exp\left(\frac{-x^2}{2}\right)$. So we have $\psi_0(x) = C_0 x^s \exp\left(\frac{-x^2}{2}\right)$ where s is like in (8.107) and $B_0\psi_0 = \frac{2s+1}{4}\psi_0$. Then we get all the spectrum of B_0 and all the bounded states $\psi_k = C_k B_+^k \psi_0$ where the constants C_K are chosen to have an orthonormal basis in $L^2(\mathbb{R}_+)$. □

Applying this lemma and formula (8.106) we get that Eq. (8.105) has non trivial solutions for $\boxed{E = E_n = -\frac{e^4}{2n^2}}$, $n \geq 1$, the well known energy levels of the hydrogen atom. More properties will be given in the next chapter.

Remark 54 The Casimir operator is here $C_{as} = c_{as}\mathbb{1}$. c_{as} is computed by

$$C_{as}\psi_0 = (B_0^2 - \frac{1}{2}(B_+B_- + B_+B_-))\psi_0 = k(k-1)\psi_0,$$

with $k = \frac{2s+1}{4}$. This is not compatible with a discrete representation of $SU(1,1)$ except if s is half an integer. What we have considered here is an irreducible representation of the universal cover $\widetilde{SU}(1,1)$.

It could be possible to study coherent states $\varphi_\beta = D(\beta)\psi_0$ in this representation too as we have done for the discrete representations.

8.6 Wavelets and Pseudo-Spin Coherent States

As it is well known wavelets are associated with the affine group of transformations of the real axis $\mathbb{R}$: $t \mapsto at + b$ where $a > 0$ and $b \in \mathbb{R}$. We denote $g(a, b)$ this affine transformation and $\mathbf{A}_F$ the group of all affine transformations.

Wavelets are real functions defined by the action of the group $\mathbf{A}_F$ on a given function ψ so we have

$$\psi_{a,b}(t) = \frac{1}{\sqrt{a}}\psi\left(\frac{t-b}{a}\right).$$

If $\varphi_{a,b}$ is the Fourier transform of $\psi_{a,b}$ and φ the Fourier transform of ψ, we have

$$\varphi_{a,b}(s) = \sqrt{a}\mathrm{e}^{-ibs}\varphi(as).$$

Now we shall see, following ideas taken from the paper [29], that wavelets and pseudo-spin coherent states are closely related.

This is not surprising using the following facts: the affine group is isomorph to a subgroup of the group $SL(2, \mathbb{R})$[1] which is isomorph to $SL(2, \mathbb{R})/SO(2)$ and this one is isomorph to $SU(1, 1)/U(1)$.[2] Recall that the groups $SU(1, 1)$ and $SL(2, \mathbb{R})$ are conjugate

$$CSU(1, 1)C^{-1} = SL(2, \mathbb{R}), \quad \text{where } C = \frac{1}{\sqrt{2}}\begin{pmatrix} 1 & i \\ i & 1 \end{pmatrix}.$$

Considering consequences of these facts, we shall get that the discrete series representations of $SU(1, 1)$ have realizations with strong connections with the affine group hence relationship between pseudo-coherent states and generalized wavelets will follow.

Remark that the affine group can be identify to $\mathbb{R}_+^* \times \mathbb{R}$ with the group law: $(a, b) \times (a', b') = (aa', ab' + b)$. Let us consider the mapping

$$\mathscr{M} : (a, b) \mapsto \begin{pmatrix} \sqrt{a} & \frac{b}{\sqrt{a}} \\ 0 & \frac{1}{\sqrt{a}} \end{pmatrix}.$$

It is easy to see that $\mathscr{M}$ is a group isomorphism from $\mathbf{A}_F$ into $SL(2, \mathbb{R})$. Its image is denoted $\mathbf{A}_{sl}$.

---

[1] Recall that $SL(2, \mathbb{R})$ is the group of 2×2 real matrices of determinant one.

[2] Recall that $U(1)$ is identified here with the group of matrices $\begin{pmatrix} \mathrm{e}^{-i\theta} & 0 \\ 0 & \mathrm{e}^{i\theta} \end{pmatrix}$, $\theta \in \mathbb{R}$.

$SO(2)$ is a compact subgroup of $SL(2,\mathbb{R})$ and it is not difficult to prove that the quotient space $SL(2,\mathbb{R})/SO(2)$ can be identified to $\mathbf{A}_{sl}$:

Lemma 59 *For every $A \in SL(2,\mathbb{R})$ there exits a rotation $R(\theta) = \begin{pmatrix} \cos\theta & \sin\theta \\ -\sin\theta & \cos\theta \end{pmatrix}$ and a unique affine transformation $(a, b) \in \mathbb{R}_+^* \times \mathbb{R}$ such that*

$$A = \mathscr{M}(a, b)R(\theta).$$

In particular the left cosets set $SL(2,\mathbb{R})/SO(2)$ is isomorph to $\mathbf{A}_{sl}$.

Let us consider the discrete series $\mathscr{D}_n^+$ which will be now denoted $\mathscr{D}_n$ ($n \geq 2$ is an integer, $n = 2k$ where k is the Bargmann index). Using the isomorphism $\zeta \mapsto z(\zeta) = \frac{\zeta+i}{1+i\zeta}$ from the unit disc $\mathbb{D}$ onto the Poincaré half plane $\mathbb{H}$, $\mathscr{D}_n$ can be realized in the Hilbert space $\mathscr{H}_n(\mathbb{H})$. $\mathscr{H}_n(\mathbb{H})$ is the space of holomorphic functions f in $\mathbb{H}$ such that

$$\|f\|^2_{\mathscr{H}_n(\mathbb{H})} = \frac{n-1}{\pi}\int_{\mathbb{H}} |f(X+iY)|^2 Y^{n-2} dXdY < +\infty,$$

with the natural norm. So we have a unitary map $f \mapsto F$ from $\mathscr{H}_n(\mathbb{H})$ onto $\mathscr{H}_n(\mathbb{D})$ where $F(\zeta) = 2(1+i\zeta)^{-n} f(z(\zeta))$. In $\mathscr{H}_n(\mathbb{H})$ the discrete series $\mathscr{D}_n$ gives naturally a unitary representation of $SL(2,\mathbb{R})$

$$\left(\mathscr{D}_n\begin{pmatrix} a & b \\ c & d\end{pmatrix} f\right)(z) = (d - bz)^{-n} f\left(\frac{az - c}{-bz + d}\right). \tag{8.108}$$

To establish a connection with the affine group it is convenient to realize the representation $\mathscr{D}_n$ in the space $\check{\mathscr{H}}_n(\mathbb{H})$ of anti-holomorphic functions on $\mathbb{H}$, so that $f \in \check{\mathscr{H}}_n(\mathbb{H})$ means that $\check{f}(z) = f(\bar{z})$ with $f \in \mathscr{H}_n(\mathbb{H})$. $f \mapsto \check{f}$ is a unitary map. We denote $\check{\mathscr{I}}_n F = \check{f}$. Then $\mathscr{D}_n$ is unitary equivalent to the representation $\check{\mathscr{D}}_n$

$$\left(\check{\mathscr{D}}_n\begin{pmatrix} a & b \\ c & d\end{pmatrix} f\right)(z) = (a - cz)^{-n} f\left(\frac{dz - b}{-cz + a}\right). \tag{8.109}$$

Note that the unitary equivalence between $\check{\mathscr{D}}_n$ and $\mathscr{D}_n$ is implemented by the group, isomorphism in $SL(2,\mathbb{R})$,

$$\begin{pmatrix} a & b \\ c & d\end{pmatrix} \mapsto \begin{pmatrix} d & c \\ b & a\end{pmatrix}.$$

The restriction of $\check{\mathscr{D}}_n$ to the affine group has the following expression

$$\left(\check{\mathscr{D}}_n(\mathscr{M}(a, b))f\right)(z) = a^{-n/2} f\left(\frac{z - b}{a}\right). \tag{8.110}$$

Wavelets are functions of a real variable, so the last step is to find a realization of $\mathscr{D}_n$ in the space

$$L_n^2(\mathbb{R}_+) = \{\varphi | \int_0^{+\infty} t^{1-n}|\varphi(t)|^2 dt < +\infty\}.$$

with the natural norm.

Let us introduce the (anti-holomorphic) Fourier-Laplace transform:

$$(\mathscr{L}_n\varphi)(z) = c_n \int_0^{+\infty} \varphi(t)\mathrm{e}^{-it\bar{z}}dt,\ z \in \mathbb{C},\ \Im(z) > 0,$$

c_n is a normalization constant.

Using the Fourier inverse formula and the Plancherel formula we have

$$\varphi(t) = \frac{1}{2\pi c_n}\mathrm{e}^{ty} \int_{-\infty}^{+\infty} \mathscr{L}_n\varphi(x - iy)\mathrm{e}^{itx}dx,\ y > 0,\ t > 0. \tag{8.111}$$

$$\frac{1}{2\pi c_n}\int_{\mathbb{H}} |\mathscr{L}_n\varphi(x - iy)|^2 y^{n-2}dxdy = \Gamma(n-1)2^{1-n}\int_0^{+\infty} t^{1-n}|\varphi(t)|^{2-n}dt. \tag{8.112}$$

So we get isometries between the spaces $L_n^2(\mathbb{R}_+)$, $\mathscr{H}(\mathbb{H})$ and $\check{\mathscr{H}}(\mathbb{H})$ choosing $c_n = \frac{2^{n-2}}{\pi(n-2)!}$. We have obtained the following irreducible unitary representation of the affine group in the space $L_n^2(\mathbb{R}_+)$.

$$\mathscr{W}_n(a, b) := \mathscr{L}_n^{-1}\check{\mathscr{D}}_n(\mathscr{M}(a, b))\mathscr{L}_n.$$

$$\varphi_{a,b}(t) := (\mathscr{W}_n(a, b)\varphi)\,(t) = a^{1-n/2}\mathrm{e}^{-ibt}\varphi(at),$$

which represent wavelets on the Fourier side. Coherent states for $SU(1, 1)$ where defined in the Hilbert space $\mathscr{H}_n(\mathbb{D})$ by an action of $SU(1, 1)$, starting from a fiducial states ψ_0 invariant by the action of the unit circle $U(1)$ (isomorph to $SO(2)$). Then $SU(1, 1)$ coherent states are parametrized by the quotient $SU(1, 1)/U(1)$.

But from Lemma 59 and the isomorphism between $SU(1, 1)$ and $SL(2, \mathbb{R})$ we get that $SU(1, 1)/U(1)$ can also be parametrized by the affine group: $(a, b) \mapsto g_{a,b}$ where we choose one element $g_{a,b}$ in each left coset. We have seen above in the construction of coherent states that $SU(1, 1)/U(1)$ can be parametized by $\mathbb{C}$ (or by by pseudosphere) :$\xi \mapsto g_\xi$. If $g_{a,b}$ and g_ξ are in the same coset then we have $g_{a,b} = g_\xi h$ where $h \in U(1)$. We have choosen ψ_0 rotation invariant so the actions of $g_{a,b}$ and g_ξ define the same coherent state.

Let us move this construction in $\mathscr{H}_n(\mathbb{H})$ and in $L_n^2(\mathbb{R}_+)$. We get respectively fiducial states $f_0(z) = d_n(1 - iz)^n$ and $\varphi_0(t) = e_n t^{n-1}\mathrm{e}^{-t}$ where d_n and e_n are suitable constants.

Then using properties of the representation $\mathscr{D}_n$ we get a bijective correspondence between $SU(1, 1)$ coherent states defined in $\mathscr{H}_n(\mathbb{D})$ for $\mathscr{D}_n^+$ and wavelets in $L_n^2(\mathbb{R}_+)$. More precisely we have obtained

$$\varphi_{a,b}(t) := (\mathscr{W}_n(a,b)\varphi_0)(t) = a^{1-n/2}e^{-ibt}\varphi(at),$$

which represent wavelets on the Fourier side. Their relationship with the $SU(1,1)$ coherent states is given by

$$\mathscr{L}_n^{-1}\check{\mathscr{I}}_n\psi_\xi = \varphi_{a(\xi),b(\xi)}, \quad \forall\xi \in \mathbb{C}, \tag{8.113}$$

$$\check{\mathscr{I}}_n^{-1}\mathscr{L}_n\varphi_{a,b} = \psi_{\xi(a,b)}, \quad \forall(a,b) \in \mathbb{R}_+^* \times \mathbb{R}, \tag{8.114}$$

where $\xi \mapsto (a(\xi), b(\xi))$ is a bijection from $\mathbb{C}$ onto $\mathbb{R}_+^* \times \mathbb{R}$ and $(a,b) \mapsto \xi(a,b)$ is a bijection from $\mathbb{R}_+^* \times \mathbb{R}$ onto $\mathbb{C}$.

In particular we also have a resolution of identity for wavelets which can be obtained from 8.81 or by a direct computation.

$$f = \frac{n-1}{4\pi}\iint\limits_{\mathbb{R}_+^*\times\mathbb{R}} \frac{dadb}{a^2}\langle\varphi_{ab}, f\rangle\varphi_{a,b}, \quad \forall f \in L_n^2(\mathbb{R}_+). \tag{8.115}$$

$\mathscr{H}_n(\mathbb{D})$, $\mathscr{H}_n(\mathbb{H})$ and $L_n^2(\mathbb{R}_+)$.

Finally remark that $\mathscr{W}_n$ is a representation in $L_n^2(\mathbb{R}_+)$ of a subgroup of $SU(1,1)$ conjugated to the restriction of $\mathscr{D}_n^+$. But all the representations $\mathscr{W}_n$ are equivalent contrary to the representations $\mathscr{D}_n^+$ which are non-equivalent.

If M_n is the unitary map $M_n\varphi(t) = t^{1-n/2}\varphi(t)$ from $L_n^2(\mathbb{R}_+)$ onto $L^2(\mathbb{R}_+)$ then we have clearly $M_n\mathscr{W}_n = \mathscr{W}_2 M_n$ so $\mathscr{W}_n$ and $\mathscr{W}_2$ are conjugate for every $n \geq 2$.

Hyperbolic triangles in the Poincaré disc

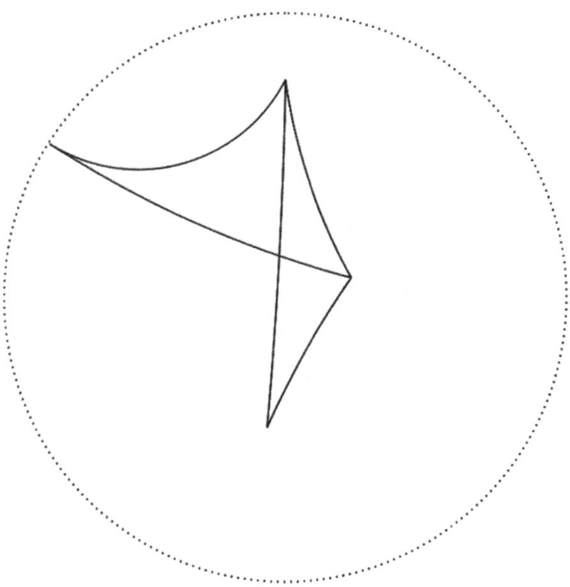

Chapter 9
The Coherent States of the Hydrogen Atom

Abstract The aim of this chapter is to present a construction of a set of coherent states for the hydrogen atom proposed by Villegas-Blas (The Laplacian on the n-sphere, the hydrogen atom and the Bargmann space representation. Ph.D. thesis, Mathematics Department, University of Virginia, 1996, [251]), Thomas and Villegas-Blas (Commun Math Phys 187:623–645, 1997, [244]). We show that in a semiclassical sense they concentrate essentially around the Kepler orbits (in configuration space) of the classical motion. A suitable unitary transformation (the Fock operator) maps the pure point subspace of the hydrogen atom Hamitonian onto the Hilbert space $L^2(\mathbb{S}^3)$ for the $\mathbb{S}^3$ sphere. We study the coherent states for the $\mathbb{S}^3$ sphere (as introduced by Uribe (J Funct Anal 59:535–556, 1984, [247]) and show that the action of the group $SO(4)$ is irreducible in the space generated by the spherical harmonics of a given degree. Note that coherent states for the hydrogen atom have been extensively studied by J. Klauder and his school. We have chosen not to present them here and refer the interested reader to Klauder and Skagerstam (Coherent states, Worlds Scientific Publishing, Singapore, 1985, [167]).

9.1 The $\mathbb{S}^3$ Sphere and the Group $SO(4)$

9.1.1 Introduction

It is well known that the non-relativistic quantum model for the hydrogen atom is the quantization $\hat{H}$ of the Kepler Hamiltonian $H(x, p) = \frac{|p|^2}{2} - \frac{1}{|x|}$, $p, x \in \mathbb{R}^3$.

The natural symmetry group for H seems to be the rotation group $SO(3)$. We shall see in Sect. 9.2 that the hydrogen atom has "hidden symmetries" and its symmetry group is the larger group $SO(4)$ which explain the large degeneracies of the energy levels of $\hat{H}$. This is why we start by studying the group $SO(4)$, its irreducible representations and hyperspherical harmonics.

Recall that $SO(4)$ is the group of direct isometries of the Euclidean space $\mathbb{R}^4$ or its unit sphere $\mathbb{S}^3$,

$$\mathbb{S}^3 = \left\{(x = (x_1, \ldots, x_4) \,|x_1^2 + x_2^2 + x_3^2 + x_4^2 = 1\right\}.$$

D. Robert and M. Combescure, *Coherent States and Applications in Mathematical Physics*, Theoretical and Mathematical Physics,
https://doi.org/10.1007/978-3-030-70845-0_9

Let us introduce the Laplacian $\Delta_{\mathbb{S}^3}$ for the sphere $\mathbb{S}^3$, as the restriction to the unit sphere $\mathbb{S}^3$ of the Laplace operator $\Delta_{\mathbb{R}^4} := \sum_{1\leq j\leq 4} \frac{\partial^2}{\partial x_j^2}$. More explicitly computing $\Delta_{\mathbb{R}^4}$ in hyperspherical coordinates:

$$\begin{aligned} x_1 &= r \sin\chi \sin\theta \cos\varphi \\ x_2 &= r \sin\chi \sin\theta \sin\varphi \\ x_3 &= r \sin\chi \cos\theta \\ x_4 &= r \cos\chi, \quad \text{where } \chi, \theta \in [0, \pi[, \varphi \in [0, 2\pi[, \end{aligned} \tag{9.1}$$

we get

$$\Delta_{\mathbb{R}^4} = \frac{1}{r^3}\frac{\partial}{\partial r}\left(r^3 \frac{\partial}{\partial r}\right) + \frac{1}{r^2}\Delta_{\mathbb{S}^3} \tag{9.2}$$

where

$$\Delta_{\mathbb{S}^3} = \frac{\partial^2}{\partial \chi^2} + \frac{2}{\tan\chi}\frac{\partial}{\partial\chi} + \frac{1}{\sin^2\chi}\Delta_{\mathbb{S}^2}, \text{ where} \tag{9.3}$$

$$\Delta_{\mathbb{S}^2} = \frac{1}{\tan\theta}\frac{\partial}{\partial\theta} + \frac{\partial^2}{\partial\theta^2} + \frac{1}{\sin^2\theta}\frac{\partial^2}{\partial\varphi^2}. \tag{9.4}$$

Equation (9.2) defines the operator $\Delta_{\mathbb{S}^3}$. It is a self adjoint operator in the Hilbert space $L^2(\mathbb{S}^3)$ for the Euclidean measure $d\mu_3(\theta, \varphi, \chi) = \sin^2\chi \sin\theta d\theta d\varphi d\chi$. The measure $d\mu_3$ and the operator $\Delta_{\mathbb{S}^3}$ are invariant by the group $SO(4)$. The spectrum of $\Delta_{\mathbb{S}^3}$ can be described has follows: there exists an explicit constant $c \in \mathbb{R}$ such that if $\Delta_3 := -\Delta_{\mathbb{S}^3} + c$ then Δ_3 has the discrete spectrum

$$\left\{\lambda_k = (k+1)^2 \, |k \in \mathbb{N}\right\}.$$

Each λ_k is known to have multiplicity $(k+1)^2$ (see [197] or what follows), which coïncides with multiplicities of bound states of hydrogen atom as we shall see later.

9.1.2 Irreducible Representations of $SO(4)$

$SO(4)$ is a compact Lie group so we know that all its irreducible representations are finite dimensional.

The Lie algebra $\mathfrak{so}(4)$ of $SO(4)$ is the algebra of antisymmetric 4×4 real matrices; $\mathfrak{so}(4)$ has dimension 6 so the Lie group $SO(4)$ has dimension 6.

We shall see that its irreducible representations can be deduced from irreducible representations of $SU(2)$ (computed in the chapter "Spin Coherent States" which will be denoted (SCS)).

It is convenient to use the quaternion field **H** and its generators $\{\mathbf{1}, \mathscr{I}, \mathscr{J}, \mathscr{K}\}$. **H** is a 4-dimensional real linear space which can be represented as the space of 2×2 matrices $q = \begin{pmatrix} a & b \\ -\bar{b} & \bar{a} \end{pmatrix}$ where $a, b \in \mathbb{C}$. The basis is related with Pauli matrices: $\mathbf{1} = \sigma_0, \mathscr{I} = i\sigma_3, \mathscr{J} = i\sigma_2, \mathscr{K} = i\sigma_1$.

The following properties are easy to prove

1. $\{\mathbb{1}, \mathscr{I}, \mathscr{J}, \mathscr{K}\}$ is an orthonormal basis for the scalar product $\langle q, q' \rangle := \frac{1}{2}\mathrm{tr}(q^\star \cdot q')$. So $\underline{\mathbf{H}}$ can be identified with the Euclidean space $\mathbb{R}^4$.
2. $q^\star \cdot q = q \cdot q^\star = (|a|^2 + |b|^2)\mathbb{1}$. In particular if $q \neq 0$, q is invertible and $q^{-1} = \frac{q^\star}{|q|^2}$ where $|q|^2 = |a|^2 + |b|^2$.
3. $SU(2) = \{q \in \mathbf{H},\ |q| = 1\}$
4. $\mathfrak{su}(2) = \{q \in \mathbf{H},\ \ q^\star + q = 0\}$.
   A quaternion q is said pure (or imaginary) if $q^\star + q = 0$ and real if $q^\star = q$.
5.

$$g \in SU(2) \iff g = \mathrm{e}^{\theta A} = \cos\theta + (\sin\theta)A, \quad \theta \in \mathbb{R}, \ A \text{ a pure quaternion}.$$

Now we can identify $SO(4)$ with the direct isometries group of the Euclidean space **H**. In particular if for any $g_1, g_2 \in SU(2)$ we define $\tau(g_1, g_2)q = g_1 \cdot q \cdot g_2^{-1}$ then $\tau(g_1, g_2) \in SO(4)$. Furthermore we have

Proposition 111 *τ is a group morphism from $SU(2) \times SU(2)$ in $SO(4)$.*
The kernel of τ is $\ker\tau = \{(\mathbb{1}, \mathbb{1}), (-\mathbb{1}, -\mathbb{1})\}$ *and τ is surjective.*
In particular the group $SO(4)$ is isomorphic to the quotient group
$SU(2) \times SU(2)/\{(\mathbb{1}, \mathbb{1}), (-\mathbb{1}, -\mathbb{1})\}$ and its Lie algebra $\mathfrak{so}(4)$ is isomorphic to the Lie algebra $\mathfrak{su}(2) \oplus \mathfrak{su}(2)$.

Proof It is clear that τ is a group morphism. $(g_1, g_2) \in \ker\tau$ means that $g_1 \cdot q = q \cdot g_2, \forall q \in \mathbf{H}$. So we get successively :$g_1 = g_2$, $g_1 = \lambda\mathbf{1}$, $\lambda \in \mathbb{C}$, $\lambda = \pm 1$ because $g_1 \in SU(2)$.

To prove that τ is surjective, we use that $\tau(SU(2) \times SU(2))$ is a subgroup of $SO(4)$, its action on **H** is transitive and that we have a surjective group homomorphism $g \mapsto R_g$ from $SU(2)$ in $SO(3)$ (acting in pure quaternions). Let $A \in SO(4)$. If $A\mathbf{1} = \mathbf{1}$ then there exists $g \in SU(2)$ such that $A = \tau(g, g)$. If $A\mathbf{1} = q$ then there exist $g_1, g_2 \in SU(2)$ such that $q = \tau(g_1, g_2)\mathbf{1}$ so we can write $A = \tau(g_1 \cdot g, g_2 \cdot g)$. □

Now we use the following classical result to deduce irreducible representations of $SO(4)$ (for a proof see [44]).

Proposition 112 *Let G_1, G_2 be two compact Lie groups, (ρ_1, V_1) and (ρ_2, V_2) two irreducible representations of G_1 and G_2 respectively. Then $(\rho_1 \otimes \rho_2, V_1 \otimes V_2)$ is an irreducible representation of $G_1 \times G_2$. Conversely every irreducible representation of $G_1 \times G_2$ is like this.*

Corollary 29 *Let (ρ, V) be an irreducible unitary representation of $SO(4)$. Then there exists $j_1, j_2 \in \frac{\mathbb{N}}{2}$ such that $j_1 + j_2 \in \mathbb{N}$ and such that (ρ, V) is unitarily equivalent to $(T^{(j_1)} \otimes T^{(j_2)}, V^{(j_1)} \otimes V^{(j_2)})$.*

Proof Using Proposition 111 we can asume that (ρ, V) is an irreducible representation of $SU(2) \times SU(2)/\{(\mathbb{1}, \mathbb{1}), (-\mathbb{1}, -\mathbb{1})\}$. So (ρ, V) is an irreducible representation of $SU(2) \times SU(2)$ and from Proposition 112 we have $(\rho, V) \equiv (T^{(j_1)} \otimes T^{(j_2)}, V^{(j_1)} \otimes V^{(j_2)})$. To get a representation of $SU(2) \times SU(2)/\{(\mathbb{1}, \mathbb{1}), (-\mathbb{1}, -\mathbb{1})\}$ it is necessary and sufficient that

$$T^{(j_1)}(-\mathbb{1}) \otimes T^{(j_2)}(-\mathbb{1}) = \mathbb{1}_{V^{(1)} \otimes V^{(2)}}.$$

So we find the condition $j_1 + j_2 \in \mathbb{N}$. □

9.1.3 Hyperspherical Harmonics and Spectral Decomposition of $\Delta_{\mathbb{S}^3}$

As we have already seen in Chap. 7 for $SO(3)$ we shall see now that irreducible representations of $SO(4)$ are closely related with the hyperspherical harmonics and spectral decomposition of $\Delta_{\mathbb{S}^3}$.

Let us introduce the space $\mathscr{P}_4^{(k)}$ of homogeneous polynomials $f(x_1, x_2, x_3, x_4)$ in the four variables (x_1, x_2, x_3, x_4) of total degree $k \in \mathbb{N}$ and $\tilde{\mathscr{H}}_4^{(k)}$ the space of $f \in \mathscr{P}_4^{(k)}$ such that $\Delta_{\mathbb{R}^4} f = 0$.

$\tilde{\mathscr{H}}_4^{(k)}$ is determined by its restriction $\mathscr{H}_4^{(k)}$ to the sphere $\mathbb{S}^3$ which is by definition the space of hyperspherical harmonics of degree k.

Let $f \in \mathscr{P}_4^{(k)}$. For $x \in \mathbb{R}^4$ we have $x = r\omega, r > 0, \omega \in \mathbb{S}^3$ and $f(x) = r^k \psi(\omega)$. So, using (9.2) we have

$$\Delta_{\mathbb{S}^3} \psi = -k(k+2)\psi, \quad \forall \psi \in \mathscr{H}_4^{(k)}. \tag{9.5}$$

Let us introduce the modified Laplacian $\Delta_3 = -\Delta_{\mathbb{S}^3} + 1$. Then $\mathscr{H}_4^{(k)}$ is an eigenspace for Δ_3 with eigenvalue $\lambda_k = (k+1)^2$.

The group $SO(4)$ has a natural unitary representation in $L^2(\mathbb{S}^3)$ defined by the formula

$$\rho_g \psi(\omega) = \psi(g^{-1}\omega), \quad g \in SO(4), \quad \omega \in \mathbb{S}^3.$$

ρ commutes with $\Delta_{\mathbb{S}^3}$ and each $\mathscr{H}_4^{(k)}$ is invariant by ρ. c The main goal of this sub-section is to prove the following results

Theorem 45 *(i) The Hilbert space $L^2(\mathbb{S}^3)$ is the direct Hilbertian sum of hyperspherical subspaces:*

$$L^2(\mathbb{S}^3) = \bigoplus_{k\in\mathbb{N}} \mathscr{H}_4^{(k)}, \quad (\mathscr{H}_4^{(0)} = \mathbb{C}). \tag{9.6}$$

Moreover we have

$$\dim(\mathscr{H}_4^{(k)}) = (k+1)^2. \tag{9.7}$$

(ii) For every $k \in \mathbb{N}$ the representation $(\rho, \mathscr{H}_4^{(k)})$ is irreducible and equivalent to the representation $(T^{(k/2)} \otimes T^{(k/2)}, V^{(k/2)} \otimes V^{(k/2)})$ where $(T^{(j)}, V^{(j)})$ is the irreducible representation of $SU(2)$ defined in Chap. 7.

Proof We shall follow more or less the proof of the similar Proposition in Chap. 7. We first get the following decomposition for homogeneous polynomial spaces in $\mathbb{R}^4$,

$$\mathscr{P}_4^{(k)} = \tilde{\mathscr{H}}_4^{(k)} \oplus r^2 \mathscr{P}_4^{(k-2)} \tag{9.8}$$

$$\mathscr{P}_4^{(k)} = \tilde{\mathscr{H}}_4^{(k)} \oplus r^2 \tilde{\mathscr{H}}_4^{(k-2)} \oplus \cdots \oplus r^{2\ell} \tilde{\mathscr{H}}_4^{(k-2\ell)} \tag{9.9}$$

where $k-1 \le 2\ell \le k$ and r^2 is multiplication by $r^2 := x_1^2 + x_2^2 + x_3^2 + x_4^2$.

Using the Stone-Weierstrass theorem it follows that $\bigcup_{k\in\mathbb{N}} \mathscr{H}_4^{(k)}$ is dense in $L^2(\mathbb{S}^3)$ hence we get the decomposition formula (9.6).

Furthermore we have

$$\dim(\mathscr{P}_4^{(k)}) = \dim(\mathscr{H}_4^{(k)}) + \dim(\mathscr{P}_4^{(k-2)}), \quad \text{and} \quad \dim(\mathscr{P}_4^{(k)}) = \binom{k+3}{k}$$

so we get formula (9.7) by the binomial Newton formula.

Let us prove now that $\mathscr{H}_4^{(k)}$ is irreducible for the representation ρ.

Denote by $\Sigma_4^{(k)}$ the restriction to $\mathbb{S}^3$ of $\mathscr{P}_4^{(k)}$. Let us consider a fixed point on $\mathbb{S}^3$, for example $\mathbf{n}_0 = (0, 0, 0, 1)$. We can identify with $SO(3)$ the subgroup of $SO(4)$ fixing $\mathbf{n}_0$. Let $\Sigma_{4,3}^{(k)}$ be the subspace of $\psi \in \Sigma_4^{(k)}$ such that $\rho_g \psi = \psi, \forall g \in SO(3)$. Irreducibility of $\mathscr{H}_4^{(k)}$ will follow from the following lemma.

Lemma 60 *(i) The dimension of $\Sigma_{4,3}^{(k)}$ is $[\frac{k}{2}] + 1$ ($[\lambda]$ is the greatest integer n such that $n \le \lambda$).*
(ii) Let $E \neq \{0\}$ be a finite dimensional $SO(4)$-invariant subspace of the space $C(\mathbb{S}^3)$. Then there exists $\psi_0 \neq 0$, $\psi_0 \in E$ such that $\rho_g \psi_0 = \psi_0$ $\forall g \in SO(3)$.

Admitting this lemma for a moment let us finish the proof of the theorem.

For every k we have the following decomposition of $\Sigma_4^{(k)}$ into irreducible factors:

$$\Sigma_4^{(k)} = \bigoplus_{1\le\ell\le L} E_\ell$$

where each E_ℓ is equivalent to some $V^{(j)} \otimes V^{(j')}$. Let us apply the lemma (ii) to each E_ℓ. We get

$$L \leq \dim(\Sigma_{4,3}^{(k)}) = \left[\frac{k}{2}\right] + 1 \tag{9.10}$$

But from (9.8) we get

$$\Sigma_4^{(k)} = \bigoplus_{2j \leq k} \mathscr{H}_4^{(k-2j)}.$$

So if one of the $\mathscr{H}_4^{(k-2j)}$ is reducible then $\Sigma_4^{(k)}$ would have at least $[\frac{k}{2}] + 2$ irreducible factors which is not possible because of (9.10). So all the spaces $\mathscr{H}_4^{(k)}$ are irreducible for ρ. □

Let us prove now that $(\rho, \mathscr{H}_4^{(k)})$ is equivalent to the representation $(T^{(k/2)} \otimes T^{(k/2)}, V^{(k/2)} \otimes V^{(k/2)})$. This is a consequence of the Peter-Weyl theorem whose statement is given now.

Theorem 46 ([44]) *Let G be a compact Lie group with Haar measure $d\mu$ and let $(\rho_\lambda, V_{\lambda\in\Lambda}^{(\lambda)})$ be the set of all its irreducible unitary representations (up to unitary equivalence). For each $\lambda \in \Lambda$ consider an orthonormal basis of $V^{(\lambda)}$: $\{e_k^{(\lambda)}\}_{1\leq k\leq k_\lambda}$ and the matrix elements $a_{k,\ell}^{(\lambda)}(g) = \sqrt{k_\lambda}\langle e_k^{(\lambda)}, \rho_\lambda(g)(e_\ell^{(\lambda)})\rangle$, $g \in G$.*

Then $\{a_{k,\ell}^{(\lambda)},\ 1 \leq k, \ell \leq k_\lambda,\ \lambda \in \Lambda\}$ is an orthonormal basis of $L^2(G)$.

In the quaternion model we can see that the manifolde $\mathbb{S}^3$ is isomorphic to the Lie group $SU(2)$ and up to a normalization constant the Hilbert spaces $L^2(\mathbb{S}^3)$ and $L^2(SU(2))$ coincide.

In $V^{(j)}$ consider the orthonormal basis v_k, $0 \leq k \leq 2j$ and define the linear map by $v_k \otimes v_\ell \mapsto a_{k,\ell}^{(j)}$ from $V^{(j)} \otimes V^{(j)}$ in $L^2(SU(2))$. We get a unitary map $U^{(j)}$ from $V^{(j)} \otimes V^{(j)}$ on a subspace $E^{(j)}$ of $L^2(SU(2))$. Now applying the Peter-Weyl theorem we have the following unitary equivalences

$$L^2(\mathbb{S}^3) \sim L^2(SU(2)) \sim \bigoplus_{j\in\mathbb{N}/2} E^{(j)} \sim \bigoplus_{j\in\mathbb{N}/2} V^{(j)} \otimes V^{(j)}.$$

Using uniqueness for the decomposition into irreducible representations we can conclude that $\mathscr{H}_4^{(k)} \sim V^{(k/2)} \otimes V^{(k/2)}$. □

9.1.4 *The Coherent States for $\mathbb{S}^3$*

We want to introduce coherent states defined on $\mathbb{S}^3$. Following Uribe [247] we consider a pair of unit vectors $\mathbf{a}$, $\mathbf{b}$ in $\mathbb{S}^3$ such that

$$\mathbf{a} \cdot \mathbf{b} = 0$$

($\cdot$ denotes the usual bilinear form in $\mathbb{C}^4$)

Let

$$\alpha = \mathbf{a} + i\mathbf{b} \in \mathbb{C}^4$$

The geometrical interpretation is that α represents a tangent vector $\mathbf{b}$ at the point $\mathbf{a} \in \mathbb{S}^3$. So the set $\mathscr{A} = \{\alpha = \mathbf{a} + i\mathbf{b} \mid \mathbf{a} \cdot \mathbf{b} = 0, |\mathbf{a}| = |\mathbf{b}| = 1\}$ can be identified with the unit tangent bundle $S(\mathbb{S}^3)$ over $\mathbb{S}^3$. Moreover the intersection of $\mathbb{S}^3$ and the real plane containing the vectors $\mathbf{a}, \mathbf{b}$ is a geodesic on $\mathbb{S}^3$ which will be denoted $\mathring{\alpha}$. We have clearly $\mathring{\alpha} = \{\omega \in \mathbb{S}^3, \ \ |\omega \cdot \alpha| = 1\}$.

For every $\alpha \in \mathscr{A}$, $\omega \in \mathbb{S}^3$ and $k \in \mathbb{N}$ we define

$$\Psi_{\alpha,k}(\omega) = (\alpha \cdot \omega)^k = (\mathbf{a} \cdot \omega + i\mathbf{b} \cdot \omega)^k.$$

Uribe ([247]) has defined these coherent states for sphere $\mathbb{S}^n$ in any dimension n. We shall call them spherical coherent states.

Using definition of $\mathscr{A}$ we see that $\Psi_{\alpha,k}$ is a spherical harmonics of degree k namely it is an eigenfunction of Δ_3 with eigenvalue $(k+1)^2$.

Let us remark that if ω is not on the geodesic $\mathring{\alpha}$ then we have $|\alpha \cdot \omega| < 1$ hence $\Psi_{\alpha,k}(\omega)$ is exponentially small as $k \to +\infty$, so $\Psi_{\alpha,k}(\omega)$ lives close to the geodesic $\mathring{\alpha}$ when k is large.

The L^2-norm of $\Psi_{\alpha,k}$ in $L^2(\mathbb{S}^3)$ is given by

$$\|\Psi_{\alpha,k}\|^2 = \int_{\mathbb{S}^3} |\alpha \cdot \omega|^{2k} d\mu_3(\omega)$$

Using hyperspherical coordinates we get

$$\|\Psi_{\alpha,k}\|^2 = 2\pi \int_0^\pi \sin^{2k+2} \chi d\chi \int_0^\pi \sin^{2k+1} \theta d\theta$$

Using well known expression for the Wallis integrals $w_n = \int_0^{\pi/2} \sin^n(\theta) d\theta$, we get

$$\boxed{\|\Psi_{\alpha,k}\|^2 = \frac{2\pi^2}{k+1}} \tag{9.11}$$

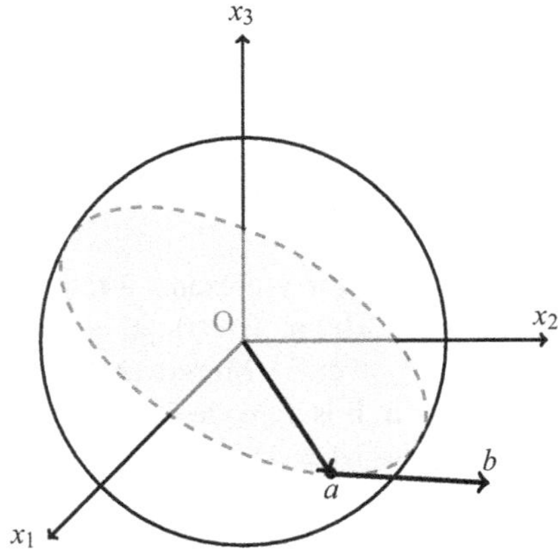

Coherent states on the sphere $\mathbb{S}^3$

Let us check now the completeness of the coherent states $\Psi_{\alpha,k}$.

Proposition 113 *The coherent states $\Psi_{\alpha,k}$ form a complete set on the irreducible eigenspace $\mathcal{H}_4^{(k)}$ of the modified Laplacian Δ_3 associated with the eigenvalue λ_k:*

$$P_k = C(k) \int_{\mathring{\alpha} \in \Gamma} |\Psi_{\alpha,k}\rangle\langle\Psi_{\alpha,k}| d\mu(\mathring{\alpha}) \tag{9.12}$$

where P_k is the projector onto $\mathcal{H}_4^{(k)}$ of Δ_3 belonging to the eigenvalue λ_k, $C(k)$ is a constant of normalization, Γ is the space of geodesics $\mathring{\alpha}$, and $d\mu(\mathring{\alpha})$ is the $SO(4)$ invariant probability measure on Γ. Moreover we can compute $C(k)$:

$$\boxed{C(k) = \frac{(k+1)^3}{2\pi^2}}$$

Proof Let $g \in SO(4)$. One defines

$$g\alpha = g\mathbf{a} + ig\mathbf{b}.$$

It is not difficult to see that $SO(4)$ acts on $\mathcal{A}$ transitively so that

$$\mathcal{A} = \{g\alpha_0, \ |g \in SO(4)\},$$

where $\alpha_0 \in \mathcal{A}$ is fixed. Then the Haar probability measure $d\mu_H$ on $SO(4)$ induces a pushed forward measure (or image measure) $d\mu(\alpha)$ on $\mathcal{A}$.

The integral operator in the right hand side of (9.12) commutes with any rotation operator ρ_g given by a rotation g in $SO(4)$. Therefore by Schur's lemma this integral operator must be a multiple of the identity on each irreducible subspace $\mathcal{H}_4^{(k)}$.

The computation of $C(k)$ is straightforward taking the trace in (9.12) and using (9.11). □

As we have already remarked for spin coherent states in Chap. 7, the $\Psi_{\alpha,k}$ are not mutually orthogonal. Here it is more difficult to compute $\langle \Psi_{\alpha,k}, \Psi_{\alpha',k}\rangle$ for $\alpha \neq \alpha'$ but it is possible to compute this overlap asymptotically for $k \to +\infty$ as it is shown in [244, 251]. Here we only state the result.

Proposition 114 *For any $\delta > 0$ we have for $k \to +\infty$*

$$\langle \Psi_{\alpha,k}, \Psi_{\alpha',k}\rangle = \frac{2\pi^2}{k}\left(\frac{\alpha^\star \cdot \alpha'}{2}\right)^k \left(1 + O(k^{\delta-\frac{1}{2}})\right) + O(k^{-\infty}). \tag{9.13}$$

Remark 55 α and α' define the same geodesic if and only if $|\alpha^\star \cdot \alpha'| = 1$. So (9.13) shows that the overlap is exponentially small if and only if $\mathring{\alpha} \neq \mathring{\alpha}'$.

Remark that on $\mathscr{A}$ the geodesic flow is multiplication by e^{it}, $t \in \mathbb{R}$.

9.2 The Hydrogen Atom

9.2.1 Generalities

We consider the Hydrogen atom Hamiltonian

$$\hat{H} := \frac{\hat{P}^2}{2} - \frac{1}{|\hat{Q}|} = -\frac{1}{2}\Delta_{\mathbb{R}^3} - \frac{1}{|x|} \tag{9.14}$$

$\hat{H}$ is the quantization of the Kepler Hamiltonian $H(\mathbf{q}, \mathbf{p}) = \frac{\mathbf{p}^2}{2} - \frac{1}{|\mathbf{q}|}$, $(\mathbf{q}, \mathbf{p}) \in \mathbb{R}^3\setminus\{0\} \times \mathbb{R}^3$.

The notations are the same as in Chap. 1. The first expression is more often used in physics the second in mathematics. $\hat{H}$ is a selfadjoint operator in $L^2(\mathbb{R}^3)$ with domain

$$\mathscr{D}(\hat{H}) = \mathscr{D}(\hat{P}^2) \cap \mathscr{D}(\frac{1}{|\hat{Q}|}) = H^2(\mathbb{R}^3) \;\; \text{(Sobolev space, Kato's result)}$$

For simplicity we have taken $m = e = \hbar = 1$.

It is easy to prove that $\hat{H}$ commutes with the angular momentum operator $\hat{L} = \hat{Q} \wedge \hat{P}$ which is a consequence of its $SO(3)$ symmetry property: $\hat{H}$ commutes with the three generators $\hat{L} = (\hat{L}_1, \hat{L}_2, \hat{L}_3)$ of $SO(3)$.

Other symmetries were discovered a long time ago for the Kepler problem (Laplace, Runge, Lenz) and give, after quantization, symmetries for the hydrogen atom. Let us consider first the classical Hamiltonian setting and introduce the Laplace–Runge–Lenz vector:

$$\mathbf{M} := \mathbf{p} \wedge \mathbf{L} - \frac{\mathbf{q}}{|\mathbf{q}|} = (M_1, M_2, M_3) \tag{9.15}$$

where the classical angular momentum is $\mathbf{L} = \mathbf{q} \wedge \mathbf{p} = (L_1, L_2, L_3)$. We have the following properties

1. $\{M_k, H\} = 0, \; k = 1, 2, 3.$
2. $\mathbf{L} \cdot \mathbf{M} = 0$
3. $\{M_j, L_k\} = \epsilon_{j,k,\ell} M_\ell$
4. $\{M_j, M_k\} = -2H \epsilon_{j,k,\ell} M_\ell$

where $\epsilon_{j,k,\ell}$ is the usual antisymmetric tensor.

In particular we can deduce form these properties that if the energy of H is fixed and negative ($H = E < 0$) then the six integrals $\mathbf{L}, \mathbf{M}$ span a Lie algebra (for the Poisson bracket) isomorphic to the Lie algebra $\mathfrak{so}(4)$ (see Sect. 9.1.2 of this chapter). In these sense the Kepler problem has "hidden symmetries" contained in the Laplace–Runge–Lenz vector $\mathbf{M}$. For an historical point of view about $\mathbf{M}$ we refer to [124].

After quantization we get a Laplace–Runge–Lenz operator:

$$\hat{M} = \frac{1}{2}(P \wedge \hat{L} - \hat{L} \wedge P) - \frac{X}{|X|}$$

One has the following commutation rules, corresponding to the above classical one (see [241] for detailed computations).

1. $[\hat{L}_j, \hat{L}_k] = i\epsilon_{jkl}\hat{L}_l$
2. $[\hat{L}_j, \hat{M}_k] = i\epsilon_{jkl}\hat{M}_l$
3. $[\hat{M}_j, \hat{M}_k] = -2i\epsilon_{jkl}\hat{M}_l\hat{H}$
4. $\hat{L} \cdot \hat{M} = \hat{M} \cdot \hat{L} = 0$
5. $\hat{M}^2 - 1 = 2\hat{H}(\hat{L}^2 + \mathbb{1})$

It is known that $\hat{H}$ has a purely absolutely spectrum on $[0, +\infty)$ and a negative point spectrum of the form

$$E_n = -\frac{1}{2n^2}, \quad n \in \mathbb{N}^*.$$

The degeneracy of the eigenvalue E_n is n^2. We shall see that this is due to the "hidden symmetries" contained in the Laplace–Runge–Lenz operator $\mathbf{M}$ commuting with $\hat{H}$ ($\hat{L}$ is also commuting with $\hat{H}$ but it generates apparent spherical symmetries of $\hat{H}$). Let us now recall the usual proof for the following result.

Lemma 61 *The eigenvalue* $E_n = -\frac{1}{2n^2}$ *of* $\hat{H}$ *has degeneracy* n^2.

Proof We give here a sketch of proof, for the details we refer to any text book in quantum mechanics.

In spherical coordinates we have

$$\hat{H} = -r\frac{\partial^2}{\partial r^2}r - \frac{1}{r^2}\Delta_{\mathbb{S}^2} - \frac{1}{r}.$$

Recall that the spherical harmonics Y_ℓ^m satisfy: $\hat{L}^2 Y_\ell^m = \ell(\ell+1)Y_\ell^m$ (see Chap. 7).

Eigenvalues are obtained by solving the radial equation

$$-r\frac{\partial^2}{\partial r^2}r + \frac{\ell(\ell+1)}{r^2} - \frac{1}{r}f(r) = Ef(r).$$

So we get the eigenvalues E_n for $n \geq 1$ and a basis of the eigenspace:

$$\mathscr{E}_n := \{\psi_{n,\ell,m},\ -\ell \leq m \leq \ell,\ 0 \leq \ell \leq n\}.$$

The degeneracy of E_n equals to the dimension of $\mathscr{E}_n$:

$$\dim(\mathscr{H}_n) = \sum_{p=0}^{n-1}(2p+1).$$

To this sum of odd numbers up to $2n-1$ we add and subtracts the sum of even numbers up to $2n-2$. This yields

$$\dim(\mathscr{H}_n) = n(2n-1) - n(n-1) = n^2.$$

□

In the following section we shall recover this result using the $SO(4)$ symmetry in a transparent way.

9.2.2 *The Fock Transformation: A Map from $L^2(\mathbb{S}^3)$ Towards the Pure-Point Subspace of $\hat{H}$*

We follow a presentation of Bander–Itzykson [17]. Consider the eigenvalue problem of the Hydrogen atom:

$$\hat{H}\psi = \left(\frac{\hat{P}^2}{2} - \frac{1}{|\hat{Q}|}\right)\psi = E\psi$$

In Fourier variable $\mathbf{p}$ one gets

$$\left(\frac{\mathbf{p}^2}{2} - E\right)\tilde{\psi}(\mathbf{p}) = \frac{1}{2\pi^2}\int_{\mathbb{R}^3}\frac{\tilde{\psi}(\mathbf{q})}{|\mathbf{p}-\mathbf{q}|^2}d\mathbf{q} \tag{9.16}$$

where we have set $\hbar = 1$ for simplicity and $\tilde{\psi}(\mathbf{p}) = (2\pi)^{-3/2}\int_{\mathbb{R}^3} d\mathbf{q}e^{-i\mathbf{q}\cdot\mathbf{p}}\psi(\mathbf{q})$.

Since the bound states of the hydrogen atom have negative eigenenergies E we define $p_0 > 0$ such that

$$2E = -p_0^2$$

Then we define the stereographic projection from the momentum space $\mathbb{R}^3$ onto $\mathbb{S}_0^3$ (the sphere $\mathbb{S}^3$ with the north pole $(0, 0, 0, 1)$ removed): consider $\mathbb{S}^3$ divided into two hemispheres by the momentum space $\mathbb{R}^3$. Given a vector $\mathbf{p}/p_0 \in \mathbb{R}^3$ (homogeneous coordinates), take the line from the north pole to this point. It will intersect the sphere $\mathbb{S}^3$ at a point $\mathbf{w} \in \mathbb{S}_0^3$. We have:

$$\mathbf{w} = (w_1, w_2, w_3, w_4) \tag{9.17}$$

$$w_i = \frac{2p_0}{\mathbf{p}^2 + p_0^2} p_i, \quad i = 1, 2, 3, \tag{9.18}$$

$$w_4 = \frac{\mathbf{p}^2 - p_0^2}{\mathbf{p}^2 + p_0^2}. \tag{9.19}$$

$\mathbf{w} := F(\mathbf{p})$ defines a new parametrization of the sphere $\mathbb{S}_0^3$. The inverse transformation is simply

$$F^{-1}(\mathbf{w}) = p_i(\mathbf{w}) = \frac{p_0 w_i}{1 - w_4}, \quad i = 1, 2, 3.$$

In this parametrization the Euclidean measure $d\mu_3$ of $\mathbb{S}^3$ can be computed using the formula

$$d\mu_3(\mathbf{w}) = \det\left(\langle \frac{\partial F}{\partial p_k}, \frac{\partial F}{\partial p_\ell} \rangle\right) d^3\mathbf{p}.$$

So we get

$$d\mu_3(\mathbf{w}) = 2\delta(\mathbf{w}^2 - 1)d^4\mathbf{w} = \left(\frac{2p_0^2}{\mathbf{p}^2 + p_0^2}\right)^3 d^3\mathbf{p}.$$

When $p_0 = k$ we denote by $\mathbf{w}_k(\mathbf{p})$ the corresponding stereographic transformation (9.17).

To the change of variables F is associated the following unitary transform U_F from $L^2(\mathbb{R}^3)$ into $L^2(\mathbb{S}^3)$:

$$U_F(\tilde{\psi})(\mathbf{w}) := \Phi(\mathbf{w}) = \frac{1}{\sqrt{p_0}} \left(\frac{\mathbf{p}(\mathbf{w})^2 + p_0^2}{2p_0}\right)^2 \tilde{\psi}(\mathbf{p}(\mathbf{w})). \tag{9.20}$$

Now we can show that the L^2 norms of $\hat{\psi}$ and Φ are the same:

$$\|\Phi(\mathbf{w})\|_{L^2(\mathbb{S}^3)}^2 = \int_{\mathbb{S}^3} |\Phi(\mathbf{w})|^2 d\mu_3(\mathbf{w}) = \int_{\mathbb{R}^3} \frac{\mathbf{p}^2 + p_0^2}{2p_0^2} |\tilde{\psi}(\mathbf{p})|^2 d\mathbf{p}.$$

Due to the virial theorem[1] one has

$$E\int_{\mathbb{R}^3}|\tilde{\psi}(\mathbf{p})|^2 d\mathbf{p} = -\int \frac{\mathbf{p}^2}{2}|\tilde{\psi}(\mathbf{p}|^2 d\mathbf{p} \tag{9.21}$$

Thus

$$\|\Phi(\mathbf{w})\|_{L^2(\mathbb{S}^3)} = \|\hat{\psi}(\mathbf{p})\|_{L^2(\mathbb{R}^3)}.$$

One has the following remarkable property (just compute):

Lemma 62 *Given* $\mathbf{q} \in \mathbb{R}^3$ *we define the point* $\mathbf{v} \in \mathbb{S}_0^3$ *by the stereographic equations (9.17) with* $\mathbf{q}$ *instead of* $\mathbf{p}$ *and the same* p_0. *One has:*

$$|\mathbf{p}-\mathbf{q}|^2 = \frac{(\mathbf{p}^2+p_0^2)(\mathbf{q}^2+p_0^2)}{(2p_0)^2}|\mathbf{w}-\mathbf{v}|^2.$$

Therefore from Eq. (9.16) one gets that the equation obeyed by Φ is simply

$$\Phi(\mathbf{w}) = \frac{1}{2\pi^2 p_0}\int_{\mathbb{S}^3}\frac{\Phi(\mathbf{v})}{|\mathbf{v}-\mathbf{w}|^2}d\mu_3(\mathbf{v}). \tag{9.22}$$

Consider the operator T in $L^2(\mathbb{S}^3)$ defined by

$$(T\Phi)(\mathbf{w}) = \int_{\mathbb{S}^3}\frac{\Phi(\mathbf{v})}{|\mathbf{w}-\mathbf{v}|^2}d\mu_3(\mathbf{v}).$$

Note that T commutes with rotations $\rho_{\mathscr{R}}$ defined by

$$(\rho_{\mathscr{R}}\Phi)(\mathbf{v}) = \Phi(\mathscr{R}^{-1}\mathbf{v}),$$

with $\mathscr{R} \in SO(4)$. For $n \in \mathbb{N}^*$, $\mathscr{H}_4^{(n-1)}$ is the finite-dimensional space generated by the harmonic polynomials of degree $n-1$ in the variables (v_1, v_2, v_3, v_4) restricted to $\mathbb{S}^3$. We have shown that the operators $\rho_{\mathscr{R}}$ restricted to the space $\mathscr{H}_4^{(n-1)}$ give an irreducible representation of $SO(4)$. Thus due to Schur's lemma, the operator T acts as a multiple of the identity in $\mathscr{H}_4^{(n-1)}$

$$T|_{\mathscr{H}_4^{(n-1)}} = \lambda_n \mathbb{1}.$$

We shall now calculate λ_n.

Proposition 115 *One has:*

$$\lambda_n = \frac{2\pi^2}{n}$$

---

[1]Recall that the virial theorem says: if ψ is a bound state of $\hat{H}$ and $\hat{A} = \frac{x\cdot\nabla_x+\nabla_x\cdot x}{2i}$ then $\langle[\hat{H},\hat{A}]\psi,\psi\rangle = 0$. For the hydrogen atom we have $i^{-1}[\hat{H},\hat{A}] = -\Delta - \frac{1}{|x|}$. So we have (9.21).

It is enough to take a particular function in $\mathscr{H}_4^{(n-1)}$ say:

$$(TG)(\mathbf{v}) = (v_3 + i v_4)^{n-1}$$

which satisfies the equation:

$$G(\mathbf{w}) = \int_{\mathbb{S}^3} \frac{G(\mathbf{v})}{|\mathbf{w}-\mathbf{v}|^2} d\mu_3(\mathbf{v}) = \lambda_n G(\mathbf{w}).$$

We introduce the spherical coordinates (χ, θ, ϕ) in $\mathbb{S}^3$:

$$\begin{aligned} v_1 &= \sin\chi \sin\theta \cos\phi \\ v_2 &= \sin\chi \sin\theta \sin\phi \\ v_3 &= \sin\chi \cos\theta \\ v_4 &= \cos\chi \end{aligned}$$

and choose $\mathbf{w} = (0, 0, 0, 1)$. We get the following equation:

$$\int_0^{2\pi} d\chi \int_0^{\pi} d\theta \int_0^{\pi} d\phi \frac{(\sin\chi \cos\theta + i\cos\chi)^{n-1}}{2(1-\cos\chi)} \sin^2\chi \sin\theta = \lambda_n i^{n-1}$$

Doing the integration with respect to θ, ϕ we get

$$\frac{2\pi}{n} \int_0^{2\pi} d\chi \frac{\sin\chi}{1-\cos\chi} \sin(n\chi) = \lambda_n,$$

which finally yields

$$\lambda_n = \frac{2\pi^2}{n}.$$

Comparing with the Eq. (9.22) we obtain $p_0 = \frac{1}{n}$ and thus

$$E_n = -\frac{1}{2n^2}$$

Thus we recover the point spectrum of the Hydrogen atom with its degeneracy: $n^2 = \dim(\mathscr{H}_4^{(n-1)})$.

We shall now use Eq. (9.20) to show that the eigenfunctions of $\hat{H}$ map onto the spherical harmonics defined on $\mathbb{S}^3$.

Recall that $\hat{\mathbf{L}}$ is the quantum angular momentum operator, and

$$\hat{\mathbf{L}}^2 = \hat{L}_1^2 + \hat{L}_2^2 + \hat{L}_3^2.$$

Consider the normalized eigenfunctions of the Hydrogen atom $\Psi_{n,\ell,m}$ satisfying:

$$\hat{H}\Psi_{n,\ell,m} = -\frac{1}{n^2}\Psi_{n,\ell,m} \tag{9.23}$$

$$\hat{\mathbf{L}}^2\Psi_{n,\ell,m} = \ell(\ell+1)\Psi_{n,\ell,m}, \quad \ell = 0, 1, \ldots, n-1 \tag{9.24}$$

$$\hat{L}_3\Psi_{n,\ell,m} = m\Psi_{n,\ell,m}, \quad m = -\ell, -\ell+1, \ldots, \ell-1, \ell. \tag{9.25}$$

Recall that $\mathscr{E}_n$ is the n^2-dimensional vector space spanned by the functions $\{\Psi_{n,\ell,m} \mid 0 \le l \le n-1, \ -l \le m \le l\}$.

Let $\mathscr{H}_{pp}$ be the subspace of $L^2(\mathbb{R}^3)$ spanned by the eigenfunctions of $\hat{H}$. We define the operator $U : \mathscr{H}_{pp} \mapsto L^2(\mathbb{S}^3)$ using Eq. (9.20):

$$(U\hat{\Psi}_{n,\ell,m})(\mathbf{w}) = \frac{1}{\sqrt{p_0}}\left(\frac{p^2(\mathbf{w}) + p_0^2}{2p_0}\right)^2 \hat{\Psi}_{n,\ell,m}(\mathbf{p}(\mathbf{w}))$$

and extend U linearly to all the space $\mathscr{H}_{pp}$. Since $U\hat{\Psi}_{n,\ell,m}$ satisfies Eq. (9.22) $U\hat{\Psi}_{n,\ell,m}$ must belong to the space $\mathscr{H}_4^{(n-1)}$, thus one has

$$\Delta_3(U\hat{\Psi}_{n,\ell,m}) = n^2 U\hat{\Psi}_{n,\ell,m},$$

where Δ_3 is the modified Laplacian on $\mathbb{S}^3$ with eigenvalue n^2.

Since U preserves the L^2 norm and the spaces $\mathscr{E}_n$ and $\mathscr{H}_4^{(n-1)}$ have the same finite dimension U is an unitary operator from $\mathscr{E}_n$ onto $\mathscr{H}_4^{(n-1)}$. Furthermore it is a unitary operator from $\mathscr{H}_{pp} = \bigoplus_{n\in\mathbb{N}} \mathscr{E}_n$ onto $L^2(\mathbb{S}^3) = \bigoplus_{n\in\mathbb{N}} \mathscr{H}_4^{(n-1)}$.

Let Π_{pp} be the orthogonal projector onto the pure-point spectrum space of $\hat{H}$. One has

$$-\frac{1}{2}\hat{H}^{-1}\Pi_{pp} = \sum_{n=1}^{\infty} n^2 \Pi_n$$

where Π_n is the projector onto $\mathscr{E}_n$. One has:

Proposition 116

$$U(\hat{H}^{-1}\Pi_{pp})U^{-1} = -2\Delta_3.$$

Proof Let Π_n' be the projector onto $\mathscr{H}_4^{(n-1)}$. One has

$$\Delta_3 = \sum_{n=1}^{\infty} n^2 \Pi_n'.$$

Since

$$U\Pi_n U^{-1} = \Pi_n'$$

this yields the result. □

9.3 The Coherent States of the Hydrogen Atom

Using the unitary operator already introduced in Sect. 9.2, $U : \mathscr{H}_{pp} \to L^2(\mathbb{S}^3)$, we define the coherent states of the hydrogen atom as

$$\hat{\Psi}_{\alpha,k} = U^{-1}\Psi_{\alpha,k}. \tag{9.26}$$

From now on we define $\Psi_{\alpha,k}(\mathbf{w}) = c_k(\alpha \cdot \mathbf{w})^k$ where the constant c_k is chosen so that $\|\Psi_{\alpha,k}\|_{L^2(\mathbb{S}^3)} = 1$. It was computed in Sect. 9.1: $c_k^2 = \frac{k+1}{2\pi^2}$.

Equation (9.26) reads using (9.20):

$$\hat{\Psi}_{\alpha,k}(\mathbf{p}) = \sqrt{p_0}\left[\frac{2p_0}{\mathbf{p}^2 + p_0^2}\right]^2 \Psi_{\alpha,k}(\mathbf{w}_k(\mathbf{p})) = c_k\sqrt{p_0}\left[\frac{2p_0}{\mathbf{p}^2 + p_0^2}\right]^2 (\alpha \cdot \mathbf{w}_k(\mathbf{p}))^k$$

(we recall that $\mathbf{w}_k(\mathbf{p})$ is the stereographic projection (9.17) for $p_0 = k$).

Let us define the dilation operator D_k in $L^2(\mathbb{R}^3)$ as

$$(D_k\Psi)(\mathbf{x}) = k^{3/2}\Psi(k\mathbf{x}),$$

or in momentum space

$$(D_k\hat{\Psi})(\mathbf{p}) = k^{-3/2}\hat{\Psi}\left(\frac{\mathbf{p}}{k}\right).$$

Defining J to be the multiplication operator by $\left(\frac{2}{\mathbf{p}^2+1}\right)^2$ we get

$$\hat{\Psi}_{\alpha,k}(\mathbf{p}) = (D_k J\Psi_{\alpha,k})(\mathbf{p}) = c_k D_k\left[\left(\frac{2}{\mathbf{p}^2+1}\right)^2 (\alpha \cdot \mathbf{w}_1(\mathbf{p}))^k\right]. \tag{9.27}$$

Taking the Fourier transform we see that the coherent state for the hydrogen atom in configuration space equals

$$\Psi_{\alpha,k}(\mathbf{x}) = \frac{c_k}{(2\pi k)^{3/2}}\int_{\mathbb{R}^3} \exp\left(\frac{i\mathbf{p}\cdot\mathbf{x}}{k}\right)\left(\frac{2}{\mathbf{p}^2+1}\right)^2 (\alpha \cdot \mathbf{w}_1(\mathbf{p}))^k d\mathbf{p}.$$

Now we shall consider the state $\Psi_{\alpha,k}(\mathbf{x})$ dilated by k^2:

$$\begin{aligned}\Phi_{\alpha,k}(\mathbf{x}) :=& (D_{k^2}\Psi_{\alpha,k})(\mathbf{x})\\ =& c_k\left(\frac{k}{2\pi}\right)^{3/2}\int_{\mathbb{R}^3}\left(\frac{2}{\mathbf{p}^2+1}\right)^2 \exp(ik\mathbf{p}\cdot\mathbf{x} + k\log(\alpha\cdot\mathbf{w}_1(\mathbf{p})))d\mathbf{p}.\end{aligned}$$

Note that $\log(\alpha \cdot \mathbf{w}_1(\mathbf{p}))$ is well defined: if $\alpha_4 = 0$ then $|\alpha \cdot \mathbf{w}_1(\mathbf{p})| \to 0$ as $|\mathbf{p}| \to \infty$ and one thus gets a decrease outside a compact K of $\mathbb{R}^3$. On K the logarithm is defined locally. If $\alpha_4 \neq 0$ then

$$\alpha \cdot \mathbf{w}_1(\mathbf{p}) = \alpha_4 + O(\frac{1}{|\mathbf{p}|})$$

where $\alpha_4 \in \mathbb{C}$. Then there exists $R > 0$ such that for $|\mathbf{p}| > R$ the logarithm is well defined.

The aim is now to show that $\Phi_{\alpha,k}(\mathbf{x})$ concentrates in the neighborhood of a Kepler orbit when k becomes large. This is a semiclassical result since k plays the role of $\frac{1}{\hbar}$. For doing this we use complex stationary phase estimates applied to the integral:

$$\Phi_{\alpha,k}(\mathbf{x}) = c_k \left(\frac{k}{2\pi}\right)^{3/2} \int_{\mathbb{R}^3} d\mathbf{p} \left(\frac{2}{\mathbf{p}^2+1}\right)^2 \exp(kf(\mathbf{x},\mathbf{p}))$$

with

$$f(\mathbf{x},\mathbf{p}) = i\mathbf{x}\cdot\mathbf{p} + \log(\alpha \cdot \mathbf{w}_1(\mathbf{p})). \tag{9.28}$$

The stationary phase condition reads:

$$\Re f(\mathbf{x}\cdot\mathbf{p}) = 0, \tag{9.29}$$

$$\nabla_{\mathbf{p}} f(\mathbf{x},\mathbf{p}) = 0. \tag{9.30}$$

But

$$\Re f(\mathbf{x},\mathbf{p}) = \log(|\alpha \cdot \mathbf{w}_1(\mathbf{p})|)$$

We have that $|\alpha \cdot \mathbf{w}_1(\mathbf{p})| \leq 1$ with equality only when $\mathbf{w}_1(\mathbf{p})$ is in the plane generated by $a = \Re\alpha,\ b = \Im\alpha$. Take for simplicity

$$\alpha = \hat{e}_1 + i(\hat{e}_2 \cos\gamma + \hat{e}_4 \sin\gamma).$$

We assume $\gamma \neq \frac{\pi}{2}, \frac{3\pi}{2}$. The vectors $\hat{e}_i$ are unit vectors in the direction of the components w_i. Thus the first stationary phase condition (9.29) imposes that $\mathbf{w}_1(\mathbf{p})$ must satisfy a parametric equation of the form:

$$\mathbf{w}_1(\mathbf{p}) = \hat{e}_1 \cos\beta + \sin\beta(\hat{e}_2 \sin\gamma + \hat{e}_4 \cos\gamma)$$

The corresponding conditions for the momentum components are

$$\begin{aligned} p_1 &= \frac{\cos\beta}{1-\sin\beta\sin\gamma}, \\ p_2 &= \frac{\sin\beta\cos\gamma}{1-\sin\beta\sin\gamma}, \\ p_3 &= 0. \end{aligned}$$

So $\mathbf{p}$ describes the circle $\mathscr{C}_\gamma$ in the plane $p_3 = 0$: $p_1^2 + (p_2 - \tan\gamma)^2 = \frac{1}{\cos^2\gamma}$.

One sees easily that $\alpha \cdot \mathbf{w}_1(\mathbf{p}) = e^{i\beta}$ which is a complex number of modulus 1 as required. Now we consider condition (9.30): it reads

$$ix_j + \frac{2}{(\mathbf{p}^2 + 1)(\alpha \cdot \mathbf{w}_1(\mathbf{p}))}[\alpha_j + (\alpha_4 - \alpha \cdot \mathbf{w}_1(\mathbf{p}))p_j] = 0. \tag{9.31}$$

Thus $\mathbf{x}$ must satisfy the parametric equations:

$$\begin{aligned} x_1(\beta) &= \sin\beta - \sin\gamma \\ x_2(\beta) &= -\cos\beta\cos\gamma \\ x_3 &= 0 \end{aligned}$$

Thus $\mathbf{x}$ must belong to the ellipse $\mathscr{E}(\gamma)$ of equation

$$(x_1 + \sin\gamma)^2 + \left(\frac{x_2}{\cos\gamma}\right)^2 = 1. \tag{9.32}$$

of energy $-1/2$:

$$\frac{\mathbf{p}^2}{2} - \frac{1}{|\mathbf{x}|} = -\frac{1}{2}$$

To apply the saddle point method (see Appendix A.4) one needs to show that the Hessian matrix is non singular at the critical point $(\mathbf{x}(\beta), \mathbf{p}(\mathbf{x}(\beta)))$. One calculates the Hessian matrix H_β

$$H_\beta = \partial^2_{p,p} f\,(\mathbf{x}(\beta), \mathbf{p}(\mathbf{x}(\beta)))\,.$$

First of all we easily see that if $\mathbf{x}$ is not on the ellipse $\mathscr{E}(\gamma)$ then we have

$$\Phi_{\alpha,k}(\mathbf{x}) = O(k^{-\infty}).$$

A tedious calculation sketched in Appendix A.3 shows that on $\mathscr{E}(\gamma)$ we have

$$|\det H_\beta| = (1 - \sin\beta\sin\gamma)^4\sqrt{(\sin^2\gamma + 2\sin\beta\sin\gamma + 1)(\sin^2\gamma - 2\sin\beta\sin\gamma + 1)}. \tag{9.33}$$

So we have $\det H_\beta \neq 0$.

From now on it is enough to consider a neighborhood $\mathscr{V}$ of a fixed point $\mathbf{x}_0 := \mathbf{x}(\beta)$ on $\mathscr{E}(\gamma)$. $f(x, p)$ being holomorphic in a complex neighborhood of $(\mathbf{x}_0, \mathbf{p}(\mathbf{x}_0))$ in $\mathbb{C}^3 \times \mathbb{C}^3$, the saddle point method (see Appendix A.4) can be applied and gives for every $\mathbf{x} \in \mathscr{V}$,

$$|\Phi_{\alpha,k}(\mathbf{x})| = Cstk^{1/2}\left|\frac{2}{1+\mathbf{p}^2(\mathbf{x})}\right|^2 \exp(kf(\mathbf{x},\mathbf{p}(\mathbf{x})))\frac{1}{\sqrt{|\det H_\beta|}}(1+O(k^{-1/2}). \tag{9.34}$$

Let us remark that the exponent in (9.34) is fast decreasing in k for $\mathbf{x} \notin \mathscr{E}(\gamma)$.

To analyze more carefully the behaviour of $|\Phi_{\alpha,k}(\mathbf{x})|$ nearby $\mathbf{x}_0$ we choose convenient coordinates. Let $\mathbf{x}(\beta, t, s) = \mathbf{x}(\beta) + t\nu_\beta + s\hat{e}_3$ where $\beta \in [0, 2\pi[$, ν_β is the normal vector to $\mathscr{E}(\gamma)$ at $\mathbf{x}(\beta)$: $\nu_\beta = \sin\beta\cos\gamma\hat{e}_1 - \cos\beta\hat{e}_2$. (t, s) are such that $t^2 + s^2 < \delta^2$ with $\delta > 0$ small enough.

We use the shorter notations $\mathbf{x}_0 = \mathbf{x}(\beta)$, $\mathbf{x}_{t,s} = \mathbf{x}(\beta, t, s)$, $\mathbf{p}_0 = \mathbf{p}(\mathbf{x}_0)$. Using the Taylor expansion we get

$$\begin{aligned} f(\mathbf{x}_{t,s}, \mathbf{p}(\mathbf{x}_{t,s})) - f(\mathbf{x}_0, \mathbf{p}_0) &= \partial_{\mathbf{x}} f(\mathbf{x}_0, \mathbf{p}_0)\cdot(\mathbf{x}_{t,s} - \mathbf{x}_0) \\ &+\frac{1}{2}\partial^2_{\mathbf{x},\mathbf{x}} f(\mathbf{x}_0, \mathbf{p}_0)(\mathbf{x}_{t,s} - \mathbf{x}_0)\cdot(\mathbf{x}_{t,s} - \mathbf{x}_0) \\ &+\frac{1}{2}\partial^2_{\mathbf{x},\mathbf{p}} f(\mathbf{x}_0, \mathbf{p}_0)(\mathbf{x}_{t,s} - \mathbf{x}_0)\cdot(\mathbf{p}(\mathbf{x}_{t,s}) - \mathbf{p}_0) + \\ \frac{1}{2}\partial^2_{\mathbf{p},\mathbf{p}} f(\mathbf{x}_0, \mathbf{p}_0)&(\mathbf{p}(\mathbf{x}_{t,s}) - \mathbf{p}_0)\cdot(\mathbf{p}(\mathbf{x}_{t,s}) - \mathbf{p}_0) + O(|\mathbf{x}_{t,s} - \mathbf{x}_0|^3) \end{aligned} \tag{9.35}$$

and

$$\mathbf{p}(\mathbf{x}_{t,s}) - \mathbf{p}_0 = \partial_{\mathbf{x}} p(\mathbf{x}_0)(\mathbf{x}_{t,s} - \mathbf{x}_0) + O((t^2+s^2)^{3/2}). \tag{9.36}$$

But we have $H_\beta = -i\partial_{\mathbf{p}}\mathbf{x}$ so $\partial_{\mathbf{x}} p(\mathbf{x}_0) = iH_\beta^{-1}$ and we get, taking the real part in the Taylor expansion,

$$\Re f(\mathbf{x}_{t,s}, \mathbf{p}(\mathbf{x}_{t,s})) = \frac{1}{2}\Re H_\beta^{-1}(t\nu_\beta + s\hat{e}_3)\cdot(t\nu_\beta + s\hat{e}_3) + O((t^2+s^2)^{3/2}). \tag{9.37}$$

Here we have used that $\partial_x f$ is imaginary and $\partial^2_{x,x}(x_0, p_0) = 0$, $\partial^2_{x,p}(x_0, p_0) = i$.

The matrix $\Re H^{-1}$ (real part of H^{-1}) has the following form (see computations in Appendix A.3):

$$\Re H^{-1} = -\frac{1}{h(\beta,\gamma)}\begin{pmatrix} \sin^2\beta\cos^2\gamma & -\sin\beta\cos\beta\cos\gamma & 0 \\ -\sin\beta\cos\beta\cos\gamma & \cos^2\beta & 0 \\ 0 & 0 & \frac{h(\beta,\gamma)}{\sin^2\gamma - 2\sin\beta\sin\gamma + 1} \end{pmatrix}$$

where

$$h(\beta,\gamma) = (1 - \sin\beta\sin\gamma)^2(\sin^2\gamma + 2\sin\beta\sin\gamma + 1).$$

The eigenvectors of $\Re H^{-1}$ are

$$\begin{aligned} v_1 &= (\cos\beta, \sin\beta\cos\gamma, 0) \\ v_2 &= v_\beta = (\sin\beta\cos\gamma, -\cos\beta, 0) \\ v_3 &= \hat{e}_3 = (0, 0, 1) \end{aligned} \tag{9.38}$$

with corresponding eigenvalues:

$$\begin{aligned} \lambda_1 &= 0, \\ \lambda_2 &= -\frac{\sin^2\beta\cos^2\gamma + \cos^2\beta}{h(\beta, \gamma)}, \\ \lambda_3 &= -\frac{1}{\sin^2\gamma - 2\sin\beta\sin\gamma + 1}. \end{aligned} \tag{9.39}$$

Since $\lambda_2, \ \lambda_3 < 0$ we see that for k large $|\Phi_{\alpha,k}(\mathbf{x})|^2$ behaves like a Gaussian highly concentrated around the ellipse at the point $\mathbf{x}_0$. Furthermore it decreases in the direction of the eigenvectors $v_2, \ v_3$ namely in the direction perpendicular to the plane of the ellipse and in the plane of the ellipse in the direction normal to the ellipse (note that $\mathbf{p}(\mathbf{x}_0) \cdot v_2 = 0$). More precisely we have

$$\left|\Phi_{\alpha,k}(\mathbf{x}_{t,s})\right|^2 = C^{st} \left|\frac{2}{\mathbf{p}(\mathbf{x}_{t,s})^2 + 1}\right|^4 \frac{1}{|\det H_{\beta,t,s}|} e^{kQ(t,s)} k + O(k^{-1}) \tag{9.40}$$

where

$$Q(t, s) = \frac{1}{2}\Re H_\beta^{-1}(tv_\beta + s\hat{e}_3) \cdot (tv_\beta + s\hat{e}_3) + O((t^2 + s^2)^{3/2}).$$

The quadratic form $Q_0(t, s) := \frac{1}{2}\Re H_\beta^{-1}(tv_\beta + s\hat{e}_3) \cdot (tv_\beta + s\hat{e}_3)$ is definite-negative in the plane (t, s).

Now we shall see that as $k \to +\infty$ the density probability $\left|\Phi_{\alpha,k}\right|^2$ converges to a probability measure supported in the ellipse $\mathscr{E}(\gamma)$. The following statement is close to [251] (Thesis, Prop. 4.1).

Proposition 117 *For every continuous and bounded function ψ in $\mathbb{R}^3$ we have*

$$\lim_{k\to+\infty} \int_{\mathbb{R}^3} \left|\Phi_{\alpha,k}(\mathbf{x})\right|^2 \psi(\mathbf{x}) d\mathbf{x} = C^{st} \int_{\mathscr{E}(\gamma)} \psi(e) d(e, O) d\ell(e), \tag{9.41}$$

where $d\ell(e)$ is the length measure on the ellipse $\mathscr{E}(\gamma)$, $d(e, O)$ is the distance of $e \in \mathscr{E}(\gamma)$ to its focus O, C^{st} is a normalization constant.

Proof From computations already done we have

$$\left|\frac{2}{\mathbf{p}(\mathbf{x}_0)^2 + 1}\right|^4 \frac{1}{|\det H_\beta|} = (\sin^2\gamma + 2\sin\gamma\sin\beta + 1)^{-1/2}(\sin^2\gamma - 2\sin\gamma\sin\beta + 1)^{-1/2}$$

and

$$\det Q_0 = \lambda_2\lambda_3$$
$$= \frac{1-\sin^2\gamma\sin^2\beta}{(1-\sin\gamma\sin\beta)^2(\sin^2\gamma+2\sin\gamma\sin\beta+1)(\sin^2\gamma-2\sin\gamma\sin\beta+1)}$$

So we can apply the stationary phase theorem in variables (t,s) to get

$$\lim_{k\to+\infty}\int_{\mathbb{R}^3}\left|\Phi_{\alpha,k}(\mathbf{x})\right|^2\psi(\mathbf{x})d\mathbf{x} = Cst\int_0^{2\pi}(1-\sin\gamma\sin\beta)(1-\sin^2\gamma\sin^2\beta)^{1/2}\psi(\mathbf{x}(\beta))d\beta.$$

But for $e=\mathbf{x}(\beta)$ we have $(1-\sin^2\gamma\sin^2\beta)^{1/2}\psi(\mathbf{x}(\beta))d\beta = d\ell(e)$ and $(1-\sin\gamma\sin\beta)$ is the distance between $\mathbf{x}(\beta)$ and O.

Let us remark that the speed $v(\beta)$ of a classical particule travelling on $\mathscr{E}(\gamma)$ is

$$v(\beta) = \frac{\sqrt{1-\sin^2\gamma\sin^2\beta}}{1-\sin\gamma\sin\beta}.$$

□

Kepler ellipse

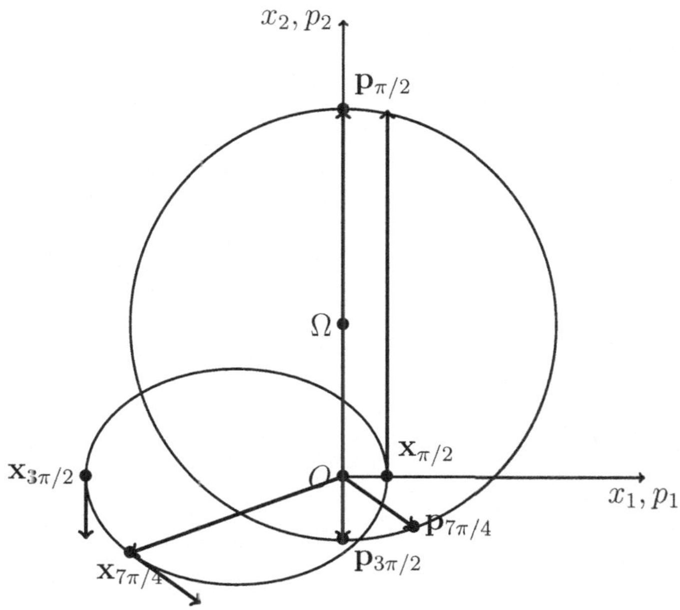

Part III
More Advanced Results on Harmonic Coherent States and Applications

Chapter 10
Characterizations of Harmonic Pseudodifferential Operators

Abstract There are many connections between Weyl quantization of classical observables and coherent states as we have already seen in Chap. 2. Here we put emphasis on necessary and sufficient conditions to characterize some classes of classical observables (called symbols). In particular, natural classes of symbols are associated with the Harmonic oscillator. They are defined as follows: for any $m \in \mathbb{R}$, $\rho \in [0, 1]$,

$$S^m_\rho := \{A : \mathbb{R}^{2n} \to \mathbb{C}, \quad |\partial^\alpha_X A(X)| \le C_\alpha \langle X\rangle^{m-\rho|\alpha|}\}.$$

These classes were introduced by Shubin [235] (see also [139]). Our aim here is to give several criteria to decide when a linear continuous operator $\hat{A} : \mathscr{S}(\mathbb{R}^n) \to \mathscr{S}'(\mathbb{R}^n)$ has its Weyl symbol in some S^m_ρ for some $m \in \mathbb{R}$. In particular we shall revisit the Beals charaterization [21] and the Chodosh characterization [54]. Some applications are given.

10.1 Operators Classes Associated with the Harmonic Oscillator

10.1.1 The Harmonic Classes

These classes are defined by estimates on the size of the derivatives of the Weyl symbols. For any $m \in \mathbb{R}$, let us denote by S^m_ρ the space of smooth functions $A : \mathbb{R}^{2n} \mapsto \mathbb{C}$ such that for any multiindex $\alpha \in \mathbb{N}^{2n}$ we have

$$|\partial^\alpha_X A(X)| \le C_\alpha \langle X\rangle^{m-\rho|\alpha|}, \; \forall X \in \mathbb{R}^{2n}. \tag{10.1}$$

In Chap. 2, we have considered the class S^m_0. Here we shall consider the case $\rho > 0$. Another difference with Chap. 2 is that in this chapter the Planck constant is fixed: $\hbar = 1$ and the orders of operators are measured by powers of the weight $\lambda(X) := \langle X\rangle$.

So we put emphasis on smoothness and behavior at infinity in the configuration space. So the symbolic calculus rule to compute the Moyal product is formulated as follows. Let be $A \in S^m_\rho$, $B \in S^{m'}_\rho$. We have seen that $\hat{A} \cdot \hat{B} = \hat{C}$ where

D. Robert and M. Combescure, *Coherent States and Applications in Mathematical Physics*, Theoretical and Mathematical Physics,
https://doi.org/10.1007/978-3-030-70845-0_10

$$C(X) = (\pi)^{-2n} \iint_{\mathbb{R}^{4n}} e^{2i\sigma(Y,Z)} A(X+Z)B(X+Y)dYdZ. \tag{10.2}$$

Following the same method as in Chap. 2 see also [153, 217], we get that the Weyl symbol $C(x, \xi)$ is in $S_\rho^{m+m'}$, more precisely we have the following

Theorem 47 *For every $A \in S_\rho^m$ and $B \in S_\rho^{m'}$, there exists a unique $C \in S_\rho^{m+m'}$ such that $\hat{A} \cdot \hat{B} = \hat{C}$. Moreover we have an asymptotic formula to compute C: $C \asymp \sum_{j\geq 0} C_j$ and $C_j \in S_\rho^{m+m'-2j\rho}$ is computed by the formula:*

$$C_j(x,\xi) = \frac{1}{2^j} \sum_{|\alpha+\beta|=j} \frac{(-1)^{|\beta|}}{\alpha!\beta!} (D_x^\beta \partial_\xi^\alpha A).(D_x^\alpha \partial_\xi^\beta B)(x,\xi),$$

and for any $N \geq 1$ we have $\left(C - \sum_{0\leq j\leq N-1} C_j\right) \in S_\rho^{m+m'-2\rho N}$.

Corollary 30 *For every $A \in S_\rho^m$ and $B \in S_\rho^{m'}$ the commutator $[\hat{A}, \hat{B}]$ is in $\hat{S}_{2\rho}^{m+m'}$.*

Operators in $\hat{S}_\rho^m$ acts in a nice way in the Harmonic weighted Sobolev spaces $\mathcal{H}^r$ defined by the powers of the harmonic oscillator $\hat{H}_{\text{os}} = \frac{1}{2}(-\Delta + |x|^2)$. For $r \geq 0$, $\mathcal{H}^r = \text{Dom}(\hat{H}_{\text{os}}^{r/2})$. $\mathcal{H}^r$ is an Hilbert space for the graph norm $\|\varphi\|_r := \left(\|\hat{H}_{\text{os}}^{r/2}\varphi\|^2 + \|\varphi\|^2\right)^{1/2}$.

For $r < 0$, $\mathcal{H}^r$ is defined by duality. In other words, $f \in \mathcal{H}^r$ if f is a temperate distribution (i.e. $f \in \mathcal{S}'(\mathbb{R}^n)$) such that

$$|\langle\varphi, f\rangle| \leq C_f \|\varphi\|_{-r}, \ \forall \varphi \in \mathcal{S}(\mathbb{R}^n)$$

In Chap. 1 of the book we have introduced an orthonormal basis (usually called the tensorial basis) $\{\phi_{\mathbf{k}}, \ \mathbf{k} \in \mathbb{N}^n\}$. $\phi_{\mathbf{k}}$ is a normalized eigenfunction for $\hat{H}_{\text{os}}$, with eigenvalue $E_{\mathbf{k}} = k_1 + k_2 + \cdots + k_n + \frac{n}{2}$. So for any integer $j \in \mathbb{N}$ the multiplicity ν_j of the eigenvalue $j + \frac{n}{2}$ is the number of multiindex $\mathbf{k}$ such that $|\mathbf{k}| = j$. Now we can easily deduce a characterization of the harmonic Sobolev space $\mathcal{H}^r$ on the above orthornormal basis.

Proposition 118 *Let $\psi \in \mathcal{S}'(\mathbb{R}^n)$ and denote $c_{\mathbf{k}}(\psi) = \langle\phi_{\mathbf{k}}, \psi\rangle$. Then $\psi \in \mathcal{H}^r$ if and only if $\sum_{\mathbf{k}\in\mathbb{N}^n} |c_{\mathbf{k}}|^2(1+|\mathbf{k}|)^r < +\infty$. Moreover the norm*

$$\|\psi\|_r^* := \left(\sum_{\mathbf{k}\in\mathbb{N}^n} |c_{\mathbf{k}}|^2(1+|\mathbf{k}|)^r\right)^{1/2}$$

is equivalent to the norm $\|\psi\|_r$.

An important property of harmonic pseudodifferential operators is boundness on the scale of harmonic Sobolev spaces.

Theorem 48 *Let $A \in S^m_\rho$, $m \in \mathbb{R}$. Then for any $r \in \mathbb{R}$, $\hat{A} \in \mathcal{L}(\mathcal{H}^{r+m}, \mathcal{H}^r)$.*

Notice that for $m = r = 0$ Theorem 48 is the Calderon-Vaillancourt Theorem proved in Chap. 2. We shall deduce the general case from this particular case and a functional calculus result for the harmonic oscillator $\hat{H}_{\rm os}$.

Definition 21 A smooth function f from $\mathbb{R}$ to $\mathbb{C}$ is a symbol of order $\nu \in \mathbb{R}$ if for any $j \in \mathbb{N}$ we have

$$|\frac{d^j f(\lambda)}{d\lambda^j}| \le C_j \langle\lambda\rangle^{\nu-j}, \ \forall\lambda \in \mathbb{R}.$$

The space of symbols of order ν is denoted S^ν.

A typical example is $f(\lambda) = (1 + \lambda^2)^{\nu/2}$.

Definition 22 A operator A from $\mathcal{S}'(\mathbb{R}^n)$ into $\mathcal{S}(\mathbb{R}^n)$ is called an N-smoothing operator if for any $r \in \mathbb{R}$ we have $A \in \mathcal{L}(\mathcal{H}^r, \mathcal{H}^{r+N})$.

Example: $e^{-tH_{osc}}$ is N-smoothing for any N and $H_{\rm os}^{-j}$ is $2j$-smoothing.

Proposition 119 *Let f be a symbol of order $\nu \in \mathbb{R}$. Then $f(\hat{H}_{\rm os})$ is in $\hat{S}^{2\nu}_1$ modulo a smoothing operator. That means here that for any $N \ge 1$ there exist $A_{f,N} \in S^{2\nu}_H$ and a smooth function g_N such that $|g_N(\lambda)| \le C_N\langle\lambda\rangle^{-N}$, and*

$$f(\hat{H}_{\rm os}) = \hat{A}_{f,N} + g_N(\hat{H}_{\rm os}). \tag{10.3}$$

Proof of Theorem 48 The idea is to apply the Calderon-Vaillancourt Theorem to show that $\hat{H}^r_{\rm os}\hat{A}\hat{H}_{\rm os}^{-(m+r)/2}$ is bounded on $L^2(\mathbb{R}^n)$. This would be clear if we had proved that $\hat{H}^r_{\rm os}$ is a pseudodifferential operator without a smoothing error (we shall prove later that this is true). To tackle here with this technical problem, using the product rule (47), it is not difficult to prove that for any integers M, N we have

$$\mathbb{1} = \hat{B}_{M,N}H^M_{\rm os} - \hat{R}_{M,N}, \tag{10.4}$$

where $B_{M,N} \in S^{2M}_H$ and $R_{M,N} \in S^{-N}_H$.

We begin by proving the theorem for $m = 0$. So we have to prove that $\hat{H}^r_{\rm os}\hat{A}\hat{H}^{-r}_{\rm os}$ is bounded on $L^2(\mathbb{R}^n)$. Using Proposition 119 we have

$$\hat{H}^r_{\rm os}\hat{A}\hat{H}^{-r}_{\rm os} = \hat{E} + \hat{H}^r_{\rm os}\hat{A}\hat{R}_N + \hat{R}'_N\hat{A}\hat{H}^{-r}_{\rm os} + \hat{R}'_N\hat{A}\hat{R}_N$$

where $E \in S^0_H$, $R_N, R'_N \in S^{-N}_H$. $\hat{E}$ and $\hat{R}'_N\hat{A}\hat{R}_N$ are bounded on L^2 by the Calderon-Vaillancourt Theorem.

Now we have $\hat{H}_{\mathrm{os}}^r \hat{A} \hat{R}_N = \hat{A}_{r,N} + \hat{R}_{r,N}$, where $A_{r,N} \in S_H^{2r}$ and $\hat{R}_{r,N}$ is an N-smoothing operator. So it is enough to prove that $r > 0$ and for any $F \in S_H^{2r}$, $\hat{F}\hat{R}_{r,N}$ is bounded. But we have, using (10.4), is

$$\hat{F}\hat{R}_{r,N} = (\hat{F}\hat{B}_{M,N})(\hat{H}_{\mathrm{os}}^M \hat{R}_N) - (\hat{F}\hat{R}_{M,N})\hat{R}_N.$$

Then we get that $\hat{H}_{\mathrm{os}}^r \hat{A} \hat{H}_{\mathrm{os}}^{-r}$ is L^2 bounded by choosing M, N large enough.

Using the same techniques (we leave the details to the reader) we can extend the result to every $m \in \mathbb{R}$. □

Proof of Proposition 119 We shall use an explicit expression for the Weyl symbol of $\mathrm{e}^{itH_{\mathrm{os}}}$ which is a particular case of the formulas obtained in Chap. 3 (Sect. 5).

Let $t \in]-\pi, \pi[$. The Weyl symbol $W(t, X)$ of $\mathrm{e}^{itH_{\mathrm{os}}}$ is given by the well known Mehler formula

$$W(t, X) = (\cos(t/2))^{-n} \exp\left(i \tan(t/2)|X|^2\right). \tag{10.5}$$

By inverse Fourier transform we have

$$f(H_{\mathrm{os}}) = (2\pi)^{-n} \int_{\mathbb{R}} \tilde{f}(t)\mathrm{e}^{itH_{\mathrm{os}}} dt.$$

Let χ be a cut-off C^∞ function such that $\chi(t) = 1$ for $|t| \le \pi/2$ and $\chi(t) = 0$ for $|t| \ge 2\pi/3$. Denote f_0, f_1 defined by their Fourier transforms: $\tilde{f_0}(t) = \chi(t)f(t)$ and $\tilde{f_1}(t) = 1 - \tilde{f_0}(t)$. Using the definition of S^ν and standard properties of the Fourier transform, we get that $f_1 \in \mathscr{S}(\mathbb{R})$ and $f_0 \in S^\nu$. Notice that we also have

$$f_1(H_{\mathrm{os}}) = \sum_{j \in \mathbb{N}} f_1\left(j + \frac{n}{2}\right) \Pi_j, \tag{10.6}$$

where Π_j denotes the spectral projector on eigenvalue $j + \frac{n}{2}$. Hence using the characterization given in Proposition 118 we get easily that $f_1(H_{\mathrm{os}})$ is an N-smoothing operator for any $N \in \mathbb{N}$.

Now we use (10.5) to compute the Weyl symbol of $f_0(\hat{H}_{\mathrm{os}})$. Denoting $\lambda = 2|X|^2$, we have to compute

$$\sigma_f(\lambda) = (2\pi)^{-1} \iint_{\mathbb{R}^2} \frac{\chi(t)}{\cos^n t} \lambda f(\lambda s) \mathrm{e}^{i\lambda(\tan t - ts)} dt ds. \tag{10.7}$$

In order to simplify the phase we write $\mathrm{e}^{i\lambda(\tan t - ts)} = \mathrm{e}^{i\lambda(\tan t - t)}\mathrm{e}^{i\lambda t(1-s)}$ and we expand

$$\mathrm{e}^{i\lambda(\tan t - t)} = \sum_{0 \le j \le N-1} \frac{\tan t - t}{j!} + r_N(\tan t - t)$$

where r_N satisfies

$$\left|\frac{d^j}{ds^j} r_N(s)\right| \leq \frac{|s|^{N-j}}{(N-j)!}, \quad \text{for } 0 \leq j \leq N.$$

Now using that $\tan t - t$ vanishes at the order 3 at $t = 0$ and that the critical point of the quadratic form $t(1-s)$ is at $t = 0, s = 1$ we get that according the stationary phase principle,

$$\sigma_f(\lambda) = f(\lambda) + \sum_{3k \leq j \leq N-1} \gamma_{j,k} \lambda^k \frac{d^j}{d\lambda^j} f(\lambda) + R_{f,N}(\lambda), \tag{10.8}$$

and for any M there exists N large enough such that

$$\left| \lambda^k \frac{d^j}{d\lambda^j} R_{f,N}(\lambda)\right| \leq C, \text{ for } k + j \leq M.$$

In particular we have proved Proposition 119. □

Remark 56 The proof given here for Proposition 119 is adapted to the harmonic oscillator. More general methods are known, in particular the Helffer-Sjöstrand formula which is explain in Appendix C.

10.1.2 The Beals Commutator Classes

This class was introduced by R. Beals [21] using conditions on commutators Recall the notation $\mathrm{ad}\hat{A} \cdot \hat{B} := [\hat{A}, \hat{B}]$ where $\hat{A}, \hat{B}$ are arbitrary operators. Let be $\{L_j\}_{1\leq j \leq J}$ any finite family of linear forms on the phase space $\mathbb{R}^{2n}$. Using Theorem 47 we see easily that if $A \in S^m_\rho$ then $(\mathrm{ad}\hat{L}_1 \mathrm{ad}\hat{L}_2 \ldots \mathrm{ad}\hat{L}_J)\hat{A} \in \hat{S}_H^{m-\rho J}$. Hence using Theorem 48 we get that

$$(\mathbf{com})_\rho \quad (\mathrm{ad}\hat{L}_1 \ldots \mathrm{ad}\hat{L}_J).\hat{A} \in \mathscr{L}(\mathscr{H}^{r+m-\rho J}, \mathscr{H}^r), \quad \forall J \geq 1.$$

for any $r \in \mathbb{R}$. For the class S^0_0 of pseudodifferential operators (see the definition in Chap. 3), Beals [21] proved that $\hat{A} \in \hat{\mathrm{S}}^0_0$ if and only if for any $J \geq 0$, $\mathrm{ad}\hat{L}_1 \mathrm{ad}\hat{L}_2 \cdots \mathrm{ad}\hat{L}_J \hat{A}$ is bounded on $L^2(\mathbb{R}^n)$.

J. M. Bony get a simpler proof in [35]. A proof is also given in the book [139] for the harmonic class S^m_1 and in [36] in a more general setting. We shall give here a different proof for the classes S^m_ρ with $\rho \in [0, 1]$, following [217].

Theorem 49 *Let be $\hat{A} \in \mathscr{L}(\mathscr{S}(\mathbb{R}^n), \mathscr{S}'(\mathbb{R}^n))$. Then $\hat{A} \in \hat{\mathrm{S}}^m_\rho$ if and only if the condition $(\mathbf{com})_\rho$ is satisfied.*

Proof It is clear that, using the Moyal product rule and Theorem 48, if $A \in S^m_\rho$ then $\hat{A}$ satisfies the condition $(\mathbf{com})_\rho$. Assume now that $(\mathbf{com})_\rho$ is satisfied. In a first step we assume that $m = \rho = 0$. We shall see that the general case will be a consequence of this particular case.

We begin by some remarks concerning the duality bracket $< \cdot, \cdot >$ between $\mathscr{L}(\mathscr{S}'(\mathbb{R}^n))$ and $\mathscr{L}(\mathscr{S}(\mathbb{R}^n))$. Ee have, for any $A \in \mathscr{L}(\mathscr{S}(\mathbb{R}^n), \mathscr{S}'(\mathbb{R}^n))$ and any $B \in \mathscr{L}(\mathscr{S}'(\mathbb{R}^n), \mathscr{S}(\mathbb{R}^n))$,

$$< A, B >= (2\pi)^n \mathrm{Tr}(\hat{A}\hat{B}). \tag{10.9}$$

On the l.h.s the bracket is the usual duality between $\mathscr{S}'(\mathbb{R}^{2n})$ and $\mathscr{S}(\mathbb{R}^{2n})$. On the r.h.s the trace is an extension by a regularisation procedure of the usual trace for class trace operators in $L^2(\mathbb{R}^n)$ (for details see [217]). Using standard properties of the trace (cyclicity), we have the following properties.

$$\mathrm{Tr}((\mathrm{ad}\hat{L} \cdot \hat{A})\hat{B}) = -\mathrm{Tr}(\hat{A}(\mathrm{ad}\hat{L} \cdot \hat{B})), \tag{10.10}$$

for any differential operator $\hat{L}$ with polynomials coefficients. In particular if L is a linear form defined by: $L(X) = \sigma(Z, X)$, we have

$$\mathrm{Tr}(\hat{A}.\mathrm{ad}(\hat{L} \cdot \hat{B})) = (2\pi)^{-n} \int_{\mathbb{R}^{2n}} A(X) D_Z B(X) dX, \tag{10.11}$$

where D_Z is the derivative in the direction Z.
So for J linear forms $L_j(X) = \sigma(Z_j, X)$ we have

$$< D_{Z_1} \dots D_{Z_J} A, B >= \mathrm{Tr}(\mathrm{ad}\hat{L}_1 \dots \mathrm{ad}\hat{L}_J \cdot \hat{A}).\hat{B}). \tag{10.12}$$

So using standard estimates of traces, in particular Theorem 5 of Chap. 2, and condition $(\mathbf{com})_0$, we get

$$| < D_{Z_1} \dots \partial D_{Z_J} A, B > | \leq (2\pi)^n \|\mathrm{ad}\hat{L}_1 \dots \mathrm{ad}\hat{L}_J.\hat{A}\| \|B\|_{W^{1,2n+1}(\mathbb{R}^{2n})} \tag{10.13}$$

where $W^{p,m}(\mathbb{R}^{2n})$ denotes the usual Sobolev space.

From (10.13) we get that $\partial^\gamma A \in W^{\infty,-2n-1}(\mathbb{R}^{2n})$ for any multiindex γ. To achieve the proof we apply the following result.

Let Δ be the Laplace operator on $\mathbb{R}^M$ and denote $G_k = (1 - \Delta)^{-k}$. G_k is a convolution operator by the kernel

$$G_k(x) = (2\pi)^{-M} \int_{\mathbb{R}^M} (1 + \xi^2)^{-k} e^{ix \cdot \xi} d\xi.$$

So $G_k \in L^1(\mathbb{R}^M)$ if $k > M/2$ and $\partial^\gamma G_k \in L^1(\mathbb{R}^M)$ if $k > \frac{M+|\gamma|}{2}$.

Hence we have

$$F(u) = \int_{\mathbb{R}^M} G_k(u-v)(1-\Delta)^k F(v)dv. \tag{10.14}$$

So let $N_0 \in \mathbb{N}$ such that $k > \frac{M+N_0}{2}$. Then we get

$$|F(u)| \le \|G_k\|_{W^{1,N_0}} \| \|(1-\Delta)^k F\|_{W^{\infty,-N_0}} \tag{10.15}$$

Applying this estimate to $\partial^\gamma F$ we get the following

Lemma 63 *Let be $F \in \mathscr{S}'(\mathbb{R}^M)$ such that for any multiindex γ, $\partial^\gamma F \in W^{\infty,-N_0}$ $(\mathbb{R}^M)$ for some $N_0 > 0$. Then for any multiindex β we have $\partial^\beta F \in L^\infty(\mathbb{R}^M)$.*

Applying the above Lemma with $M = 2n$ and $N_0 = 2n+1$ we have proved the Beals characterization for $m = \rho = 0$.
For later use we notice that we have got the following universal estimates for the symbol:

Lemma 64 *There exists universal constants $C_{\alpha,n}$ such that*

$$\|\partial_X^\alpha A\|_{L^\infty(\mathbb{R}^{2n})} \le C_{\alpha,n} \sum_{|\gamma|\le 2n+1+|\alpha|} \|\mathrm{Op}_1^w(\partial_X^\gamma A)\|_{L^2,L^2} \tag{10.16}$$

Assume now that $m = 0$ and $\rho \in]0,1]$.

We prove that $\langle X\rangle^{\rho|\gamma|}\partial^\gamma A \in \mathrm{S}_0^0$ by induction on $|\gamma|$.

For any linear form L we have that $(\mathrm{ad}\hat{L}\cdot\hat{A})\hat{H}_0^{\rho/2}$ is bounded on $L^2(\mathbb{R}^n)$. Using the symbolic calculus we get that the operator $\hat{B}_1 := (\mathrm{ad}\hat{L}\cdot\hat{A})\hat{H}_0^{\rho/2}$ satisfies $(\mathbf{com})_0$. So applying the first part of the proof we get $B_1 \in \mathrm{S}_0^0$. Using the symbolic calculus it results that $\langle X\rangle^\rho D_Z A \in \mathrm{S}_0^0$, for any $Z \in \mathbb{R}^{2n}$.

Assume that for $K \le J-1$, for any $Z_1, \ldots, Z_K \in \mathbb{R}^{2n}$, we have

$$\langle X\rangle^{k\rho} D_{Z_1} \ldots D_{Z_K} \in \mathrm{S}_0^0.$$

But

$$(\mathrm{ad}\hat{L}_1 \cdots \mathrm{ad}\hat{L}_J)\cdot\hat{A})\hat{H}_{\mathrm{os}}^{J\rho/2} = \mathrm{ad}\hat{L}_1\cdot(\mathrm{ad}\hat{L}_2\cdots\mathrm{ad}\hat{L}_J)\cdot\hat{A})\hat{H}_{\mathrm{os}}^{(J-1)\rho/2}\hat{H}_{\mathrm{os}}^{\rho/2}$$

So we get by induction hypothesis and the first steps, that $\langle X\rangle^{J\rho} D_{Z_1}\cdots D_{Z_J} A \in \mathrm{S}_0^0$ for any $Z_1, \cdots, Z_J \in \mathbb{R}^{2n}$. □.

Corollary 31 *[Wiener property] Let be $A \in \mathrm{S}_0^0$ and assume that $\hat{A}^{-1}$ is invertible in $L^2(\mathbb{R}^n)$. Then $\hat{A}^{-1} \in \hat{\mathrm{S}}_0^0$*

Proof One check the condition $(\mathbf{com}_0)$ for $\hat{A}^{-1}$ starting with the formula $[\hat{L}, \hat{A}^{-1}] = -\hat{A}^{-1}[\hat{L}, \hat{A}]\hat{A}^{-1}$. □

10.1.3 Coherent States Classes

If $\hat{A} \in \mathcal{L}(\mathcal{S}(\mathbb{R}^n), \mathcal{S}'(\mathbb{R}^n))$ its covariant symbol is defined by $A_c(z, z') := \langle \varphi'_z, \hat{A}\varphi_z \rangle$. In this section we assume that $\hbar = 1$.

Definition 23 Let $\rho \in [0, 1]$ and $m \in \mathbb{R}$. $A_c : \mathbb{R}^{2n} \times \mathbb{R}^{2n} \to \mathbb{C}$ is a covariant symbol in the class $S^m_{CS,\rho}$ if for any $N \geq 0$, any multiindices $\alpha, \alpha' \in \mathbb{N}^{2n}$ we have

$$|\partial_z^\alpha \partial_{z'}^{\alpha'} A_c(z, z')| \leq C_{N,\alpha,\alpha'} \langle z + z' \rangle^{m-\rho(|\alpha|+|\alpha'|)} (1 + |z - z'|)^{-N}. \tag{10.17}$$

We shall prove that the classes S^m_{CS0} and S^m_0 define the same operator classes.

Theorem 50 *For any $\rho \in [0, 1]$, if $\hat{A} \in \hat{S}^m_\rho$ then $A_c \in S^m_{CS,\rho}$.*
Conversely, for $\rho = 0, 1$, if $A_c \in S^m_{CS,\rho}$ then $\hat{A} \in \hat{S}^m_\rho$.

Proof Let us recall that $W_{z',z}(X)$ denotes the Wigner function of the pair $(\varphi_{z'}, \varphi_z)$. We have the following formulas

$$A_c(z', z) = (2\pi)^{-n} \int_{\mathbb{R}^{2n}} A(X) W_{z',z}(X) dX, \tag{10.18}$$

$$A(X) = (2\pi)^{-2n} \iint_{\mathbb{R}^{2n} \times \mathbb{R}^{2n}} A_c(z', z) W_{z',z}(X) dz dz'. \tag{10.19}$$

Recall that $W_{z',z}(X) = c_0 e^{\Phi(z,z',X)}$ where c_0 is a universal constant and

$$\Phi(z, z', X) = -\left|X - \frac{z' + z}{2}\right|^2 - i\sigma(X - \frac{z'}{2}, z - z').$$

It is convenient to change the variables z', z into $u = \frac{z'+z}{2}$ and $v = z - z'$.

Assume that $A \in S^m_\rho$. We have $\nabla_X \Phi(z, z', X) = -2(X - u) - iJv$. So if $v \neq 0$ we can integrate by parts with the differential operator $M := \frac{\overline{\nabla\Phi}}{|\nabla\Phi|^2} \cdot \nabla_X$ in the integral (10.18). After N iterations we get the estimate in $O(\langle v \rangle^{-N})$.

To gain the estimate in u, we use the Peetre inequality: for any $r \in \mathbb{R}$ we have:

$$\langle u \rangle^{-r} \leq 2^{-r} \langle X \rangle^{-r} \langle u - X \rangle^{|r|}. \tag{10.20}$$

Hence we get estimate (10.17) for $\alpha = \alpha'$. In the same way we can get (10.17) for any derivatives in α, α'. Assume now that (10.17) is satisfied with $\rho = 0$. We use here formula (10.19). First using (10.20) we get that $\langle X \rangle A^m \in L^\infty(\mathbb{R}^{2n})$. So we can assume that $m = 0$. Then computing $\partial_X A$, using again (10.20) and estimates on A_c, we get that $\partial_X A \in L^\infty(\mathbb{R}^{2n})$. More generally we can prove that $\partial_X^\gamma A \in L^\infty(\mathbb{R}^{2n})$ for any γ.

Now consider the case $\rho = 1$. We have to prove that $\langle X\rangle^{|\gamma|}\partial^\gamma A$ is bounded in $\mathbb{R}^{2n}$ for any γ. We explain the idea of the proof for $|\gamma| = 1$. It is easy to extend the the estimate to any γ.

In coordinates (u, v) the phase Φ is

$$\Phi(z, z', X) = \Phi(u, v, X) = -|X - u|^2 + iJ(X - \frac{u}{2}) \cdot v$$

So we have

$$\partial_X \Pi(u, v, X) = -\partial_u \Pi(u, v, X) = -2(X - u) - iJv.$$

So integrating by parts we get

$$\partial_X A(X) = 2^{1-n}(2\pi)^{-2n} \iint_{\mathbb{R}^{2n}\times\mathbb{R}^{2n}} \partial_u A_c(u - v/2, u + v/2)) \cdot \cdot W_{u-v/2,u+v/2}(X)dudv. \tag{10.21}$$

Hence using (10.17) and (10.20) we get that $\langle X\rangle\partial_X A$ is in $L^\infty(\mathbb{R}^{2n})$. By iterations we can prove in the same way that, for any γ, $\langle X\rangle^{|\gamma|}\partial_X^\gamma A$ is in $L^\infty(\mathbb{R}^{2n})$ □

10.1.4 The Matrix Class

This class was introduced by Chodosh [54] for giving a condition on the matrix elements $A_{\mathbf{j},\mathbf{k}} := \{\langle\phi_\mathbf{j}, \hat{A}\phi_\mathbf{k}\rangle$ to characterize a class of pseudodifferential operators.

Definition 24 Let us introduce the following discrete weighted derivatives on the matrix elements $A_{\mathbf{j},\mathbf{k}}$.

$$\delta_\ell(A)_{\mathbf{j},\mathbf{k}} = \sqrt{j_\ell + 1}A_{\mathbf{j}+\mathbf{e}_\ell,\mathbf{k}} - \sqrt{k_\ell}A_{\mathbf{j},\mathbf{k}-\mathbf{e}_\ell}, \tag{10.22}$$

$$\delta_\ell^\dagger(A)_{\mathbf{j},\mathbf{k}} = \sqrt{j_\ell}A_{\mathbf{j}-\mathbf{e}_\ell,\mathbf{k}} - \sqrt{k_\ell + 1}A_{\mathbf{j},\mathbf{k}+\mathbf{e}_\ell}. \tag{10.23}$$

$\hat{A}$ is in the matrix class $\hat{\mathrm{S}}^m_{\rho,\mathrm{Mat}}$, $m \in \mathbb{R}$ if for any $\alpha, \beta \in \mathbb{N}^n$ and any $N \geq 1$ there exists $C_{\alpha,\beta,N}$ such that

$$|(\delta^\dagger)^\beta\delta^\alpha(A)_{\mathbf{j},\mathbf{k}}| \leq C_{\alpha,\beta,N}(1 + |\mathbf{j} + \mathbf{k}|)^{m-\rho(|\alpha|+|\beta|)/2}(1 + |\mathbf{j} - \mathbf{k}|)^{-N}, \tag{10.24}$$

where $\delta^\alpha = \delta_1^{\alpha_1}\cdots\delta_n^{\alpha_n}$ and $(\delta^\dagger)^\alpha = (\delta_1^\dagger)^{\alpha_1}\cdots(\delta_n^\dagger)^{\alpha_n}$.

Theorem 51 $\hat{A} \in \hat{\mathrm{S}}^m_1$ *if and only if the condition* (10.24) *is satisfied for* $\rho = 1$.

Proof It is enough to assume that $m = 0$ by applying the result to $\hat{H}_{\rm os}^{-m/4}\hat{A}\hat{H}_{\rm os}^{-m/4}$. First assume that $\hat{A} \in \hat{\mathbb{S}}_1^0$. Then $[\hat{H}_{{\rm os},\ell}, \hat{A}] \in \hat{\mathbb{S}}_1^0$ where $\hat{H}_{{\rm os},\ell} = \frac{1}{2}\left(-\partial_{x_\ell}^2 + x_\ell^2\right)$ and

$$\langle\phi_{\mathbf{j}}, [\hat{H}_{{\rm os},\ell}, \hat{A}]\phi_{\mathbf{k}}\rangle = (j_\ell - k_\ell)A_{j,k}.$$

By iteration we get (10.24) for any $N \geq 1$ for $\alpha = \beta = 0$. So now we can assume that $N = 0$.

Recall the following formulas (see Chap. 1)

$$a_\ell\phi_{\mathbf{j}} = \sqrt{j_\ell}\,\phi_{\mathbf{j}-e_\ell}, \tag{10.25}$$

$$a_\ell^\dagger\phi_{\mathbf{j}} = \sqrt{j_\ell + 1}\,\phi_{\mathbf{j}+e_\ell}, \tag{10.26}$$

where $\{e_1, \cdots, e_n\}$ is the standard basis of $\mathbb{R}^n$.

We easily see that we have

$$(\delta^\dagger)^\beta\delta^\alpha(A)_{\mathbf{j},\mathbf{k}} = \langle\phi_{\mathbf{j}}, {\rm ad}^{\beta_1}a_1^\dagger\cdots{\rm ad}^{\beta_n}a_n^\dagger.{\rm ad}^{\alpha_1}a_1\cdots{\rm ad}^{\alpha_n}a_n.\hat{A}\phi_{\mathbf{k}}\rangle. \tag{10.27}$$

But ${\rm ad}^{\beta_1}a_1^\dagger\cdots{\rm ad}^{\beta_n}a_n^\dagger{\rm ad}^{\alpha_1}a_1\cdots{\rm ad}^{\alpha_n}a_n.\hat{A}$ is in $\hat{\mathbb{S}}_1^{-|\alpha|-|\beta|}$ so we get (10.24) as explain before.

Conversely if (10.24) is satisfied using (10.27) and the Beals characterization Theorem 49 we get that $\hat{A} \in \hat{\mathbb{S}}_1^m$. □

Remark 57 It is not difficult to prove that if $A \in \mathbb{S}_\rho^m$, $\rho \in [0, 1]$, then (10.24) is satisfied. It is not clear if the reverse claim is true if $\rho \in [0, 1[$.

Remark 58 For $n = 1$ it is not difficult to prove that condition (10.24) is equivalent to the Chodosh condition

$$|\Delta^\ell A(j, k)| \leq C_{\ell,N}(1 + |j - k|)^{-N}(1 + j + k)^{m-\ell}, \tag{10.28}$$

where $\Delta A(j, k) = A_{j+1,k+1} - A_{j,k}$.

Corollary 32 *Let be $f : \mathbb{R} \to \mathbb{R}$ a symbol of order $\nu \in \mathbb{R}$. Then $f(\hat{H}_{\rm os}) \in \hat{\mathbb{S}}_1^{2\nu}$. In other words there exists a symbol $H_{{\rm os}f}$ such that $f(\hat{H}_{\rm os}) = \hat{H}_{{\rm os}f}$. Moreover H_f is a radial function: $H_{{\rm os}f}(X) = f^\sharp(|X|^2)$, where $f^\sharp$ is a smooth function on* $[0, \infty[$.

Proof Notice that Proposition 119 states that the result is true modulo smoothing operators. Here we get a result without error term using Theorem 51.
We have the matrix elements computation:

$$F_{\mathbf{j},\mathbf{k}} = \langle\phi_{\mathbf{j}}; f(\hat{H}_{\rm os})\phi_{\mathbf{k}}\rangle = \delta_{\mathbf{j},\mathbf{k}}\langle\phi_{\mathbf{j}}; f(\hat{H}_{\rm os})\phi_{\mathbf{j}}\rangle = \delta_{\mathbf{j},\mathbf{k}}f\left(|\mathbf{j}| + \frac{n}{2}\right).$$

Hence it is not difficult to check that condition (10.24) is satisfied for the matrix $F_{\mathbf{j},\mathbf{k}}$. So $f(\hat{H}_{\rm os})$ has an exact Weyl symbol in the right class. But we know (see Theorem 2, Chap. 3) that

$$e^{it\hat{H}_{os}}\hat{A}e^{it\hat{H}_{os}} = \widehat{A \circ F^t_{os}},$$

where F^t_{os} is the classical flow of H_{os}. Hence we get that $H_{os\,f}(F^t_{os}(X)) = H_{os\,f}(X)$ for every $t \in \mathbb{R}$, $X \in \mathbb{R}^2$. So $H_{os\,f}$ is rotation invariant. □

10.1.5 An Application to Time Dependent Perturbations of Harmonic Oscillator

Let be $\hat{V}(t)$ a time dependent operator such that $V(t) \in S^0_1(\mathbb{R}^{2n})$, continuous and periodic in t. We consider the Schrödinger equation

$$i\partial_t\psi(t) = \hat{H}_{os}\psi(t) + \hat{V}(t)\psi(t), \quad \psi(0) = \psi_0. \tag{10.29}$$

The following Lemma claims that the Cauchy problem (10.29) is well posed in $L^2(\mathbb{R}^n)$. The proof is left to the reader.

Proposition 120 *(i) For any $\psi_0 \in L^2(\mathbb{R}^n)$, (10.29) has a unique solution $\psi(t) \in L^2(\mathbb{R}^n)$ and $\|\psi(t)\|_0 = \|\psi_0\|_0$. Moreover this solution solved the fixed point equation*

$$\psi(t) = \mathscr{U}_{os}(t)\psi_0 + \int_{t_0}^t \mathscr{U}_{os}(t-s)\hat{V}(s)\psi(s)ds, \tag{10.30}$$

where $\mathscr{U}_{os}(t) = e^{-itH_{os}}$.
(ii) Moreover the Cauchy problem (10.29) is well posed in $\mathscr{H}^r$ for any $r \in \mathbb{R}$ and there exist $C_r > 0$ such that

$$\|\psi(t)\|_r \le C_r\langle t\rangle^r\|\psi_0\|_r, \quad \forall t \in \mathbb{R}. \tag{10.31}$$

Proof Recall the following Duhamel formula. Denote $\hat{H}(t) = H_{os} + \hat{V}(t)$ and $\psi(t) = \mathscr{U}(t, t_0)\psi_0$ the solution defined in (10.30). Consider the Cauchy problem

$$i\partial_t\psi(t) = \hat{H}(_{os})t)\psi(t) + g(t), \quad \psi(0) = \psi_0, \tag{10.32}$$

where $g : \mathbb{R} \to L^2(\mathbb{R})$ is continuous. Then we have

$$\psi(t) = \mathscr{U}(t, t_0)\psi_0 + \int_{t_0}^t \mathscr{U}(t, s)g(s)ds. \tag{10.33}$$

For simplicity we assume that $r = 2k$, $k \in \mathbb{N}$.We shall prove (ii) by induction on k. Consider the case $k = 1$. We have

$$i\partial_t H_{os}\psi(t) = \hat{H}(t)\hat{H}_0\psi(t) + [\hat{H}_0, \hat{V}(t)]\psi(t).$$

So using (10.33) and that $[\hat{H}_0, \hat{V}(t)] \in \hat{S}_1^0$ we have

$$\|H_{os}\psi(t)\|_0 \leq \|H_{os}\psi_0\|_0 + C|t - t_0|\|\psi_0\|_0 \tag{10.34}$$

It is not difficult now to achieve the induction by computing $i\partial_t H_{os}^{k+1}\psi(t)$ and using again (10.33). □.

Now we shall see that the growth estimate is sharp. This is a non trivial result. It was first proved by Delort [80]. We shall prove it here following the rather elementary and illuminating proof given by Maspero [190].

Theorem 52 *There exists a time dependent pseudo-differential operator $\hat{V}(t) \in S_1^0(\mathbb{R}^{2n})$, continuous and periodic in time, such that for any $r > 0$ and any $\psi \in \mathscr{S}(\mathbb{R}^n)$, $\psi \neq 0$, there exists $c_r > 0$ such that*

$$\|\psi(t)\|_r \geq c_r\langle t\rangle^r, \quad \forall t \in \mathbb{R}. \tag{10.35}$$

Proof We are looking for a time dependent perturbation $\hat{V}_A(t) := \mathrm{e}^{it\hat{H}_{os}}\hat{A}\mathrm{e}^{-it\hat{H}_{os}}$, with $A \in S_1^0$, real. It is clear that $\psi(t)$ is solution of (10.29) for $\hat{V}(t) = \hat{V}_A(t)$ if and only if $\eta(t) := \mathrm{e}^{-it\hat{H}_{os}}\psi(t)$ satisfies the equation $i\partial_t\eta(t) = \hat{A}\eta(t)$. So $\eta(t) = \mathrm{e}^{-it\hat{A}}\psi_0$ and $\|\psi(t)\|_r = \|\eta(t)\|_r$. So it is enough to prove that

$$\|\eta(t)\|_r \geq c_r\langle t\rangle^r. \tag{10.36}$$

Let $r = 2k$, $k \in \mathbb{N}$.
Recall the Liebniz rule for derivatives

$$(\mathrm{ad}\hat{A})^m \cdot \hat{H}^k = \sum_{j_1+\cdots+j_k=m} \binom{m}{j_1, \cdots j_k} (\mathrm{ad}\hat{A})^{j_1} \cdot \hat{H} \cdots (\mathrm{ad}\hat{A})^{j_k} \cdot \hat{H}. \tag{10.37}$$

We shall use the following Lemma :

Lemma 65 *There exists a real symbol $A \in \mathrm{S}_1^0$ such that for every $j \geq 1$ we have $(\mathrm{ad}\hat{A})^j \cdot \hat{H}_{os}^j \neq 0$ and $(\mathrm{ad}\hat{A})^{j+1} \cdot \hat{H}_{os}^j = 0$.*

This Lemma will be proved later.
We have

$$\|\eta(t)\|_k^2 = \langle \mathrm{e}^{it\hat{A}}\hat{H}_{os}^k\mathrm{e}^{-it\hat{A}}\psi_0, \psi_0\rangle$$

and

$$\partial_t^\ell(\mathrm{e}^{it\hat{A}}\hat{H}_{os}^k\mathrm{e}^{-it\hat{A}}) = \mathrm{e}^{it\hat{A}}(\mathrm{ad}\hat{A})^\ell \cdot \hat{H}_{os}^k\mathrm{e}^{-it\hat{A}}.$$

Choosing A like in Lemma 65, we get that

$$\partial_t^\ell(\mathrm{e}^{it\hat{A}}\hat{H}_{\mathrm{os}}^k\mathrm{e}^{-it\hat{A}}) = 0 \text{ if } \ell \geq k+1, \text{ and} \tag{10.38}$$

$$\partial_t^k(\mathrm{e}^{it\hat{A}}\hat{H}_{\mathrm{os}}^k\mathrm{e}^{-it\hat{A}}) = \frac{(it)^k}{k!}(\mathrm{ad}\hat{A}\cdot\hat{H}_{\mathrm{os}})^k. \tag{10.39}$$

But there exists $\psi_0 \in \mathscr{S}(\mathbb{R})$ such that $(\mathrm{ad}\hat{A}\cdot\hat{H}_{\mathrm{os}})^k\psi_0 \neq 0$ (notice that $i^{-1}\mathrm{ad}\hat{A}\cdot\hat{H}_{\mathrm{os}})$ is self-adjoint). For example we can take $\psi_0 = \phi_{j_0}$, for any $j_0 \geq 0$.
So we get

$$\liminf_{t\to+\infty}\frac{\|\eta(t)\|_k}{t^k} \geq \frac{1}{k!}\|(\mathrm{ad}\hat{A}\cdot\hat{H}_{\mathrm{os}})^k\psi_0\|_0,$$

and (10.36) follows. □

Proof of Lemma 65 Following [190] we define $\hat{A}$ on the basis of Hermite functions. Let $\delta > 0$. Define $\hat{A}$ by $\hat{A}\phi_j = \delta(\phi_{j+1}+\phi_{j-1})$ ($\phi_j = 0$ if $j < 0$). A direct computation gives $(\mathrm{ad}\hat{A}\cdot\hat{H}_{\mathrm{os}})\phi_j = \delta(\phi_{j-1}-\phi_{j+1})$ and $((\mathrm{ad}\hat{A})^2\cdot\hat{H}_{\mathrm{os}})\phi_j = 0$ for any $j \in \mathbb{N}$. Moreover it is not difficult to see that the matrix $A_{j,k} = \langle\phi_j, \hat{A}\phi_k\rangle$ satisfies the condition (10.28). So by the remark 58 we can apply Theorem 51 hence we get that $\hat{A} \in \hat{\mathrm{S}}_1^0$. Notice that here $(\mathrm{ad}\hat{A}\cdot\hat{H}_{\mathrm{os}})$ is injective in $L^2(\mathbb{R})$ so we can start with any initial data in the Theorem 52. □

Remark 59 The lower bound (10.35) shows that as time t is increasing a large part of the energy $E(t) := \langle\psi(t), \hat{H}_{\mathrm{os}}\psi(t)\rangle$ is transferred from low modes to higher modes. This can be observed on the lower bound estimate:

$$E(t) = \sum_{j\geq 0}(j+1/2)|c_j(t)|^2 \geq \mu\langle t\rangle, \quad \mu > 0. \tag{10.40}$$

The contribution of the mode ϕ_j to the energy is given by $|c_j(t)|^2$ which can be very large for large time according (10.40).

Notice that without time perturbations the energy is conserved or almost conserved.
Let us consider a stationary real perturbation $V \in \mathrm{S}_1^m$ with $m < 2$ and $\hat{H} = \hat{H}_{\mathrm{os}} + \hat{V}$ a time independent Hamiltonian.

$\hat{H}$ is a self-adjoint operator on the domain $\mathrm{Dom}(\hat{H}) = \mathscr{H}^2$. Furthermore we have, for every $k \in \mathbb{N}$, $\mathscr{H}^{2k} = \mathrm{Dom}(H^k)$ with a norm equivalent to the graph norm of H^k. Hence for any $k \geq 0$ there exists $C_k > 0$ such that

$$\|\psi(t)\|_{2k}^2 \leq C_k\|H^k\psi_0\|^2 + \|\psi_0\|^2), \quad \forall t \in \mathbb{R}. \tag{10.41}$$

So the upper-bound (10.41) shows that for perturbations of order < 2 of H_{os} the lower bound (10.40) is a purely dynamical effect, due to the time oscillations of the perturbation $\hat{V}(t)$ and not to its size.

This is a rather subtle phenomenum analyzed in several papers, [16, 190, 191] and the references in these papers.

10.2 The Semi-classical Setting

In quantum mechanics the quantization of a classical observable A depends on an effective small Planck constant $\hbar$ i.e. $\hat{A} = \mathrm{Op}_\hbar^W A)$ with the Weyl quantization. We consider now the problem of establishing uniform estimates as $\hbar \searrow 0$.

We are revisiting the characterization of semi-classical pseudo-differential operators with the Planck constant $\hbar$. Let L_j linear forms on $\mathbb{R}^{2n}$. Now we denote $\hat{L}_j = L_j(\hat{Q}, \hat{P})$, $1 \le j \le J$.

Let m be a weight function on $\mathbb{R}^{2n}$. Denote $\mathrm{S}_{0,u}^{\log m}$ the class of $\hbar$-dependent symbols uniformly in S_0^m which means

$$|\partial_X^\gamma A^\hbar(X)| \le C_\gamma m(X), \ \ \forall \hbar \in]0, 1], \forall X \in \mathbb{R}^{2n}.$$

The constant weight $m = 1$ corresponds to the pseudo-differential operators of order 0.

Recall that the covariant Weyl-symbol of an operator $\hat{A}$ is $A_c(z, z') := \langle \varphi_{z'}, \hat{A}\varphi_z \rangle$, where φ_z is the coherent state defined in Chap. 1.

Proposition 121 *Let $\{\hat{A}^\hbar\}_{0<\hbar\le 1}$ be a family of operators in $\mathscr{L}(\mathscr{S}(\mathbb{R}^n), \mathscr{S}'(\mathbb{R}^n))$. Then the three following conditions are equivalent:*
(1) $A^\hbar \in \mathrm{S}_{0,u}^0$
(2) For any family of linear forms $\{L_j\}_{1\le j\le J}$ on the phase space $\mathbb{R}^{2n}$ we have

$$\|(\mathrm{ad}\hat{L}_1 \cdots \mathrm{ad}\hat{L}_J)(\hat{A}^\hbar)\|_{L^2,L^2} \le C_J \hbar^J, \ \ \forall \hbar \in]0, 1].$$

(3)

$$|\partial_z^\alpha \partial_{z'}^{\alpha'} A_c(z, z')| \le C_{N,\alpha,\alpha'} \left(1 + \frac{|z - z'|}{\sqrt{\hbar}}\right)^{-N}, \ \ \forall \hbar \in]0, 1]. \tag{10.42}$$

Proof Assume (1) satisfied. We know that (2) is satisfied using the $\hbar$-pseudodifferential calculus.
Using integrations by parts like for the proof of (10.17) we get that (3) is satisfied. Assume now that (3) is satisfied. In (10.19) the Wigner function has the following $\hbar$-dependence.

$$W_{z',z}(X) = c_n \mathrm{e}^{\hbar^{-1}\Phi(z,z',X)}.$$

So using integrations by parts like in the proof of Theorem 50 we get property (3).

Assume now that (2) is satisfied. It is convenient here to notice the following dilation property.

$$\mathrm{Op}_\hbar^w(A) = \Lambda_\hbar \mathrm{Op}_1^w(A)\Lambda_\hbar^{-1}$$

So we consider now the family of operators $\hat{S}_\hbar = \Lambda_\hbar^{-1}\hat{A}^\hbar\Lambda_\hbar$ and the linear forms $\tilde{L}_j = \mathrm{Op}_1^w L_j$. Using our assumption we have easily

$$\|\mathrm{ad}\tilde{L}_1 \cdots \mathrm{ad}\tilde{L}_J(S_{\tilde{\hbar}})\|_{L^2,L^2} \le C_J \hbar^{J/2}.$$

So using (10.16) we get that $\hat{S}_\hbar$ has a 1-symbol satisfying (1). But the $\hbar$-symbol of $\hat{A}^\hbar$ is equal to $S_\hbar$. So (1) is satisfied. □

In applications we need to consider that $\hat{A}^\hbar$ has an $\hbar$-Weyl symbol with an asymptotic expansion in $\hbar$.

Definition 25 A $\hbar$-dependent symbol $A^\hbar \in \mathrm{S}^0_{0,u}$ is in $\mathrm{S}^0_{0,sc}$ there exists a serie of symbols $A_j \in S^0_0$, $j \in \mathbb{N}$, such that for any N we have

$$A^\hbar - \sum_{0 \le j \le N} \hbar^j A_j = \hbar^{-N-1} R_{\hbar,N}.$$

where $R_N \in \mathrm{S}^0_{0,u}$. Then we say that $A^\hbar$ is a semiclassical symbol.

We have the same definition for symbols of weight $m > 0$.

Remark 60 In Chap. 2 (and Appendix C) we have given asymptotic expansions for $f(\hat{H})$ with remainder estimates in L^2-norms of operators. Here we shall prove that the asymptotic expansions hold true for the semi-norms of the full symbol of $f(\hat{H})$.

Theorem 53 *Let $\hat{H}$ a uniformly elliptic semi-classical Hamiltonian of weight m and f as in Proposition 119 with $r = 0$ (see Chap. 2). Then $f(\hat{H}) = \mathrm{Op}^w_\hbar(H_f)$ where $H_f \in \mathrm{S}^0_{0,u}$.*

Proof We give here a sketched proof, leaving the details to the reader.
The first step is to prove the theorem for the resolvent $\hat{B}_\lambda = (\hat{H} - \lambda)^{-1}$ and in a second step to use the Helffer-Sjöstrand formula (see Appendix C).

We want to apply Proposition 121 to $\hat{B}_\lambda$ for $\lambda \in \mathbb{C}\backslash\mathbb{R}$. Let L be a linear form on $\mathbb{R}^{2n}$. Using the equality

$$\mathrm{ad}\hat{L}(\hat{B}_\lambda) = -\hat{B}_\lambda \mathrm{ad}L(\hat{H})\hat{B}_\lambda$$

we get

$$\|\mathrm{ad}\hat{L}(\hat{B}_\lambda)\|_{L^2,L^2} \le C_1 \frac{\hbar}{|\Im\lambda|^2}(1 + |\lambda|)$$

Hence by induction on $J \ge 1$ we can prove that for any linear forms L_j, $j \le J$,

$$\|(\mathrm{ad}\hat{L}_1 \cdots \mathrm{ad}\hat{L}_J)(\hat{B}_\lambda)\|_{L^2,L^2} \le C_J \left(\frac{\hbar^J}{|\Im\lambda^{J+1}}\right)(1 + |\lambda|^J),\ \forall \hbar \in]0, 1].$$

Then applying Proposition 121 we get that the full symbol B_λ of $(\hat{H} - \lambda)^{-1}$ is in $S^0_{0,u}$, with control in λ:

$$|\partial^\gamma_X B_\lambda(X)| \le C_\gamma \left(\frac{<\lambda>^{|\gamma|}}{(\Im\lambda)^{|\gamma|+1}}\right). \tag{10.43}$$

Then plugging (10.43) in the Helffer-Sjöstrand formula we get the Theorem.

Corollary 33 *Let be H a classical symbol of weight m (see Chap. 2) and f like in Theorem 53. Then for $\lambda \in \mathbb{C}\backslash\mathbb{R}$, the full symbol $B_\lambda^\hbar$ of $(\hat{H}-\lambda)^{-1}$ is a semi-classical symbol of weight m^{-1}. Moreover the symbol H_f of $f(\hat{H})$ is a semi-classical symbol of order 0 (see Definition 25).*

Proof For simplicity we give the proof for the resolvent. A similar argument can be use for $f(\hat{H})$.
In Chap. 2 and Appendix C we have seen that for any $N \geq 1$ we have a parametrix $B_\lambda^{(N)} = \sum_{0\leq j\leq N} \hbar^j b_{\lambda,j}$ such that

$$(H-\lambda) \circledast B_\lambda^{(N)} = \mathbb{1} + \hbar^{N+1} R_\lambda^{(N)}$$

where the symbols $b_{\lambda,j}$ are computed and $R_\lambda^{(N)}$ satisfies for some fixed n_1,

$$|\partial_X^\gamma R_\lambda^{(N)}(X)| \leq C_{\gamma,N} \left|\frac{<\lambda>}{\Im\lambda}\right|^{2N+|\gamma|+n_1}.$$

Hence from the Calderon-Vaillancourt we deduce that there exists $c(\lambda) > 0$ (which can be estimated) such that we have, for N large enough,

$$\|R_\lambda^{(N)}\|_{L^2,L^2} \leq c(\lambda)^{3(N+1)}.$$

So for $\hbar$ small enough: $0 < \hbar \leq 1/2c(\lambda)^{-3})$ we get

$$\left(\mathbb{1} + \hbar^{N+1} R_\lambda^{(N)}\right)^{-1} - \mathbb{1} = \hbar^{N+1} T_\lambda^{(N)}.$$

Now applying Proposition 121 we get that $T_\lambda^{(N)} \in S^0_{0,u}$. So using the composition formula for symbols

$$B_\Lambda = B_\lambda^{(N)} \circledast (1 + \hbar^{N+1} T_\lambda^{(N)}).$$

we get that $B_\lambda \in S^{m^{-1}}_{0,u}$.

Chapter 11
Herman-Kluk Approximation for the Schrödinger Propagator

Abstract In this Chapter our goal is to study the Schrödinger propagator associated with a time dependent sub-quadratic Hamiltonian $H(t)$ on the phase space $\mathbb{R}^n \times \mathbb{R}^n = T^*(\mathbb{R}^n)$. The solution $\psi(t)$ of the Schrödinger equation

$$i\hbar\frac{\partial\psi(t)}{\partial t} = \hat{H}(t)\psi(t), \quad \psi(t=s) = \psi_s$$

defined a propagator $\mathcal{U}^\hbar(t,s)\psi_s := \psi(t)$. When $\hat{H}(t) = H(t, x, \hbar D_x)$ ($\hbar$-Weyl quantization of $H(t)$), we analyze the Schwartz kernel of $\mathcal{U}^\hbar(t,s)$, describing it by a quadratic phase and an amplitude. At every time t, when $\hbar$ is small enough, we get that it is "essentially supported" in a neighborhood of the graph of the classical flow generated by $H(t)$. This is proved using the Fourier-Bargman representation of the propagator $\mathcal{U}^\hbar(t,s)$. Furthermore we get that the amplitude has a full uniform asymptotic expansion in $\hbar$ in the phase space. One of the simplest choice of the phase, related with frozen Gaussians, is known in chemical physics as Herman-Kluk formula. Moreover we prove that the semiclassical expansion obtained for the propagator is valid below an Ehrenfest time, $|t| << \frac{1}{8\delta}|\log\hbar|$, where $\delta > 0$ is a stability parameter for the classical system. Most of the results can be extended to multi-component systems with constant multiplicities, like Dirac operators (see Chap. 14). Finally the results are applied to recover the Van-Vleck formula and to revisit the Aharonov-Bohm effect.

11.1 Sub-quadratic Hamiltonians

In this Chapter we consider time dependent sub-quadratic Hamiltonians $H(t, X)$ where $t \in \mathbb{R}$, $X = (x, \xi) \in \mathbb{R}^n \times \mathbb{R}^n$. We assume that $H(t, X)$ is continuous in time t, C^∞-smooth in X and sub-quadratic, that means:

$$|\partial_X^\gamma H(t, X)| \le C_\gamma(1+|X|)^{(2-|\gamma|)_+}, \ \forall t \in \mathbb{R}\ , \ \forall X \in \mathbb{R}^{2n}, \ C_\gamma > 0, \qquad (11.1)$$

D. Robert and M. Combescure, *Coherent States and Applications in Mathematical Physics*, Theoretical and Mathematical Physics,
https://doi.org/10.1007/978-3-030-70845-0_11

Notice that we could consider the case where t is in some bounded real interval. For simplicity we assume that $H(t)$ is defined for $t \in \mathbb{R}$ even if our estimates will be valid only locally uniformly in time.

As before we denote $\hat{H}(t)$ the $\hbar$-Weyl quantization of $H(t)$ and the time dependent Schrödinger equation:

$$i\hbar\frac{\partial\psi(t)}{\partial t} = \hat{H}(t)\psi(t), \quad \psi(t = s) = \psi_s. \tag{11.2}$$

A nice property of sub-quadratic Hamiltonians is that classical and quantum dynamics are well defined at any time.

The classical motion for the Hamiltonian $H(t)$ is given by the Hamilton equation

$$\dot{X}(t) = J\nabla_X H(t, X(t)), \quad X(t_0) = X_0. \tag{11.3}$$

Proposition 122 *The solution $X(t)$ of* (11.3). *is well defined for any $t \in \mathbb{R}$. Moreover there exits a constant $C > 0$ (depending on the constants C_γ of* (11.1)*) such that*

$$|X(t)|^2 \le C|X(t_0)|^2 e^{C|t-t_0|}, \quad \forall t \in \mathbb{R}. \tag{11.4}$$

Proof Let us consider the quadratic Hamiltonian $K_0(X) := |X|^2$. Using (11.1), we have

$$\frac{d}{dt}|X(t)|^2 = \{K_0, H(t)\}(X(t) \le C|X(t_0)|^2. \tag{11.5}$$

□

Remark 61 The proof of Proposition 122 uses a weaker property than subquadratic. It is enough to assume that

$$|\{K_0, H(t)\}(X)| \le C|X|^2, \forall t \in \mathbb{R}, \; X \in \mathbb{R}^{2n}.$$

Proposition 123 *For any $\psi_s \in L^2(\mathbb{R}^n)$* (11.2) *has a unique solution $\psi(t) \in L^2(\mathbb{R}^n)$ determined by a propagator: $\psi^\hbar(t) = \mathscr{U}^\hbar(t, s)\psi_s$ where $\mathscr{U}^\hbar(t, s)$ is a unitary operator in $L^2(\mathbb{R}^n)$. Moreover $\mathscr{U}^\hbar(t, s)$ is a bounded operator in the Harmonic weighted Sobolev spaces $\Sigma^k_\hbar$ for any $k \in \mathbb{N}$ and there exists $C_k > 0$ such that*

$$\|\mathscr{U}^\hbar(t, s)\psi_s\|_k \le C_k e^{C_k|t-s|}\|\psi_s\|_k, \quad \forall t, s \in \mathbb{R}. \tag{11.6}$$

We use the harmonic oscillator $K_0 := \hat{H}_{os}$ and the smoothing operators

$$R_N := \left(1 + \frac{K_0}{N}\right)^{-1}.$$

The following lemma describes the main properties of the smoothing operator R_N. Its proof is left to the reader.

Let us denote $\|R\|_{k,\ell}$ the operator norm $\|R\|_{\mathcal{L}(\Sigma_\hbar^k,\Sigma_\hbar^\ell)}$.

Lemma 66 *For every $N > 0$ the following holds true:*

$$\|R_N\|_{k,k+2} \leq N. \tag{11.7}$$
$$\|R_N\|_{k,k} \leq C. \tag{11.8}$$
$$\|R_N - \mathbb{1}\|_{k+2,k} \leq \tfrac{C}{N}. \tag{11.9}$$

The constant C depending only on k.

Now we regularize the operator $\hat{H}(t)$ by defining

$$\hat{H}_N(t) := R_N\,\hat{H}(t)\,R_N\,.$$

For every N the propagator for $\hat{H}_N(t)$ exists because $\hat{H}_N(t)$ is a bounded operator. We shall prove that a propagator for $\hat{H}(t)$ can be obtained by controlling the limit $N \to +\infty$.

Lemma 67 *For every $N \geq 1$ $H_N(t)$ is symmetric on $L^2(\mathbb{R}^n)$ and bounded on $\Sigma_\hbar^k$ for any $k > 0$. Furthermore there exists a positive C such that for $0 \leq k \leq k_0$ we have:*

$$\|\hat{H}_N(t) - \hat{H}(t)\|_{k+4,k} \leq \frac{C}{N},\quad N \geq 1, t \in \mathbb{R}\,. \tag{11.10}$$

We prove only the estimate. By Lemma 67 one has

$$\begin{aligned}\|\hat{H}_N(t) - \hat{H}(t)\|_{k+4,k} &\leq \|R_N\hat{H}(t)(R_N - \mathbb{1})\|_{k+4,k}\\ +\|(R_N - \mathbb{1})\hat{H}(t)(t)\|_{k+4,k} &\leq \frac{C}{N}\,.\end{aligned} \tag{11.11}$$

Consider the regularized Schrödinger equation

$$\begin{cases} i\partial_t\psi = \hat{H}_N(t)\psi \\ \psi|_{t=s} = \psi_s\,, \quad s \in \mathbb{R}. \end{cases} \tag{11.12}$$

Since the operator $\hat{H}_N(t)$ is bounded, it generates a flow $\mathcal{U}_N(t,s) \in C(\mathbb{R}\times\mathbb{R}, \mathcal{L}(\Sigma_\hbar^k))$ which is unitary in $L^2(\mathbb{R}^n)$.

Lemma 68 *Fix $0 < k$. There exists a positive constant C_k, independent of N, such that*

$$\|\mathcal{U}_N(t,s)\|_{k,k} \leq e^{C_k|t-s|}\,,\quad \forall N > 0\,.$$

Proof We must show that $\hat{H}_{\mathrm{os}}^k\,\mathcal{U}_N(t,s)\,\hat{H}_{\mathrm{os}}^{-k}$ is bounded uniformly in N as an operator from $L^2(\mathbb{R}^n)$ to itself. Remark that, due to the unitarity of $\mathcal{U}_N(t,s)$ in $L^2(\mathbb{R}^n)$, one has

$$\|H_{\mathrm{os}}^k\,\mathcal{U}_N(t,s)\,H_{\mathrm{os}}^{-k}\|_{0,0} = \|\mathcal{U}_N(t,s)^*\,H^k\,\mathcal{U}_N(t,s)\,\hat{H}_{\mathrm{os}}^{-k}\|_{0,0}\,.$$

Now one has

$$\begin{aligned}\mathscr{U}_N(t,s)^*\,H_{\rm os}^k\,\mathscr{U}_N(t,s)\,H_{\rm os}^{-k} &= \mathbb{1} + \int_s^t \mathscr{U}_N(r,s)^*\,[\hat{H}_N(r),\hat{H}_{\rm os}^k]\,\mathscr{U}_N(r,s)\,\hat{H}_{\rm os}^{-k}dr \\ &= \mathbb{1} + \int_s^t \mathscr{U}_N(r,s)^*\,R_N\,[\hat{H}(r),\hat{H}_{\rm os}^k] \\ &\quad\times H_{\rm os}^{-k}\,R_N\,H_{\rm os}^k\,\mathscr{U}_N(r,s)\,H_{\rm os}^{-k}dr \qquad (11.13)\end{aligned}$$

where we used that $[R_N, H_{\rm os}^k]=0$ and

$$[\hat{H}_N(t), H_{\rm os}^k] = R_N\,[H(t), H_{\rm os}^k]\,R_N\,.$$

But we can prove easily that $\|[\hat{H}(t),\hat{H}_{\rm os}^k]\,\hat{H}_{\rm os}^{-k}\|_{0,0}\le C_k$ for some positive constant C_k, thus it follows (using also Lemma 66 (ii)) that uniformly in N. we have

$$\|R_N\,[\hat{H}(t),\hat{H}_{\rm os}^k]\,\hat{H}_{\rm os}^{-k}\,R_N\|_{0,0}\le C_k\,,\quad \forall N>0\,.$$

Then such estimate combined with the unitarity of $\mathscr{U}_N(t,s)$ in $\mathscr{L}(L^2(\mathbb{R}^n))$ give

$$\begin{aligned}\|\hat{H}^k\,\mathscr{U}_N(t,s)\,\hat{H}^{-k}\|_{0,0} &\le \mathbb{1} + \int_s^t \|\mathscr{U}_N(r,s)^*\,R_N\,[\hat{H}(r),\hat{H}_{\rm os}^k]\,\hat{H}_{\rm os}^{-k}\,R_N \\ &\quad\times\hat{H}_{\rm os}^k\,\mathscr{U}_N(r,s)\,\hat{H}_{\rm os}^{-k}\|_{0,0}\,dr \\ &\le \mathbb{1} + C_k\int_s^t \|\hat{H}^k\,\mathscr{U}_N(r,s)\,\hat{H}^{-k}\|_{0,0}\,dr \qquad (11.14)\end{aligned}$$

which by Gronwall allows us to conclude that

$$\|\hat{H}_{\rm os}^k\,\mathscr{U}_N(t,s)\,\hat{H}_{\rm os}^{-k}\|_{0,0}\le e^{C_k|t-s|}\,,$$

thus proving the claim. □

Proof of Proposition 123 Fix arbitrary $t,s\in\mathbb{R}$. The first step is to show that for every $\psi\in\Sigma_\hbar^{k+4}$, the sequence $\{\mathscr{U}_N(t,s)\psi\}_N$ is a Cauchy sequence in the space $\Sigma_\hbar^k$. One has

$$\begin{aligned}\|\mathscr{U}_N(t,s)\psi - \mathscr{U}_{N'}(t,s)\psi\|_k &= \|\int_s^t \partial_\tau(\mathscr{U}_{N'}(t,\tau)\,\mathscr{U}_N(\tau,s)\psi)\,d\tau\|_k \\ &= \|\int_s^t \mathscr{U}_{N'}(t,\tau)\,\left(\hat{H}_N(\tau)-\hat{H}_{N'}(\tau)\right)\mathscr{U}_N(\tau,s)\psi\,d\tau\|_k \\ &\le C\left(\frac{1}{N}+\frac{1}{N'}\right)\,|t-s|\,e^{(C_k+C_{k+4})|t-s|}\,\|\psi\|_{k+4}.\end{aligned}$$

For any t, s in a bounded interval, and $\psi \in \Sigma_\hbar^{k+2\tau+2m}$, the sequence $\{\mathcal{U}_N(t,s)\psi\}_N$ is a Cauchy sequence in $\Sigma_\hbar^{k+4}$. Since $\Sigma_\hbar^{k+4}$ is dense in $\Sigma_\hbar^k$ and $\|\mathcal{U}_N(t,s)\|_{k,k} \le e^{C_k|t-s|}$ uniformly in N, by an easy density argument one shows that $\{\mathcal{U}_N(t,s)\psi\}_N$ is also a Cauchy sequence in $\Sigma_\hbar^k$. Thus for every $\psi \in \Sigma_\hbar^k$ the limit

$$\mathcal{U}^\hbar(t,s)\psi := \lim_{N\to\infty} \mathcal{U}_N(t,s)\psi$$

exists in $\Sigma_\hbar^k$. Moreover we have the following error estimate, for $C > 0$ large enough,

$$\|\mathcal{U}^\hbar(t,s) - \mathcal{U}_N(t,s)\psi\|_k \le \frac{C}{N}|t-s|\mathrm{e}^{C|t-s|}\|\psi\|_{k+4}. \tag{11.15}$$

By the principle of uniform boundedness, $\mathcal{U}^\hbar(t,s) \in \mathcal{L}(\Sigma_\hbar^k)$. Since $\mathcal{U}_N(t,s)$ is an isometry in $L^2(\mathbb{R}^n)$,

$$\|\mathcal{U}^\hbar(t,s)\psi\|_0 := \lim_{N\to\infty} \|\mathcal{U}_N(t,s)\psi\|_0 = \|\psi\|_0$$

which implies that $\mathcal{U}^\hbar(t,s)$ is an isometry on $L^2(\mathbb{R}^n)$ satisfying the Schrödinger equation (11.2)). Denote $\psi^N(t) = \mathcal{U}_N(t,s)\psi_s$. Then we have

$$\psi^N(t) = \psi_s + i^{-1}\int_s^t \hat{H}_N(r)\psi^N(r)dr. \tag{11.16}$$

Using Lemma 67 and estimate (11.15) there exists $C > 0$, depending on a, b, k such that for $a \le s \le r \le t \le b$ we have

$$\|\hat{H}_N(r)\psi^N(r) - \hat{H}(r)\psi(r)\|_k \le C\left(\|\psi(r) - \psi^N(r)\|_{k+2\tau+2m} + \frac{1}{N}\|\psi\|_{k+2\tau+2m}\right).$$

So we can pass to the limit in (11.16) and we get

$$\psi(t) = \psi_s + i^{-1}\int_s^t \hat{H}(r)\psi(r)dr. \tag{11.17}$$

In particular if $\psi_s \in \Sigma_\hbar^{k+4}$ then $t \mapsto \psi(t)$ is strongly derivable from $\mathbb{R}$ into $\Sigma_\hbar^k$ and satisfies the Schrödinger equation ((11.2)).

Furthermore

$$\mathcal{U}^\hbar(t,s)\psi = \lim_{N\to\infty} \mathcal{U}_N(t,s)\psi = \lim_{N\to\infty} \mathcal{U}_N(t,\tau)\mathcal{U}_N(\tau,s)\psi = \mathcal{U}^\hbar(t,\tau)\mathcal{U}^\hbar(\tau,s)\psi$$

where the limits are in the $\Sigma_\hbar^k$ topology. This shows the group property.
Estimate (11.6) is deduced from Lemma 68.

Finally we have shown that $(t,s) \mapsto \mathscr{U}^\hbar(t,s) \in \mathscr{L}(\Sigma_\hbar^{k+4}, \Sigma_\hbar^k)$ is strongly continuously differentiable with strong derivatives

$$i\hbar\partial_t \mathscr{U}^\hbar(t,s) = \hat{H}(t)\mathscr{U}^\hbar(t,s)\,.$$

With the same proof we get also

$$-i\hbar\partial_s \mathscr{U}^\hbar(t,s) = \mathscr{U}^\hbar(t,s)\hat{H}(s)\,.$$

□

Remark 62 More general results and applications to growth of Sobolev norms can be found in the paper [191]. More abstract results are also proved in [191] which can be applied to subquadratic multi-component systems in $L^2(\mathbb{R}^n, \mathbb{C}^m)$ considered in Chap. 14.

Moreover the conclusion of Proposition 123 is still valid assuming that the commutator $[\hat{K}_0, \hat{H}(t)]\hat{K}_0^{-1}$ is bounded on the spaces $\Sigma_1^k(\mathbb{R}^n)$, $k \in \mathbb{N}$.

11.2 Semi-classical Fourier-Integral Operators

The Schwartz kernel of the Schrödinger propagator $\mathscr{U}^\hbar(t,s)$ is conveniently describe as a particular case of a class of operators known in P.D.E theory as **Fourier Integral operators**. They are obtained by a suitable quantization of canonical transformations of the phase space.

Recall that a canonical transformation of the phase space is a map which preserves the symplectic two form on $\mathbb{R}^n_x \times \mathbb{R}^n_\xi$:

$$\sigma = \mathrm{d}x \wedge \mathrm{d}\xi.$$

Definition 1 A map $\chi : \mathbb{R}^{2n} \to \mathbb{R}^{2n}$ is called a smooth bilipschitz canonical transformation if χ is a smooth transformation of the phase space: $(x,\xi) \mapsto \chi(x,\xi)$ such that

- $\partial_x^\alpha \partial_\xi^\beta \chi(x,\xi)| \le C_{\alpha,\beta}, \quad |\alpha| + |\beta| \ge 1.$
- χ is a diffeomorphism.
- Both χ and χ^{-1} are Lipschitz continuous.
- χ preserves the symplectic form, i.e. if $(y,\eta) = \chi(x,\xi)$ then $\mathrm{d}y \wedge \mathrm{d}\eta = \mathrm{d}x \wedge \mathrm{d}\xi$.

To any bilipschitz canonical transformation χ we can associate a class of operators, named semiclassical Fourier integral operators, (F.I.O) by using the Fourier-Bargmann transform. Recall the definition of the Fourier-Bargman transform (Chaps. 1 and 2 for properties) :

$$\mathscr{F}^{\mathscr{B}}\psi(X) = \psi^{\sharp}(X) = (2\pi\hbar)^{-n/2}\langle\varphi_X, \psi\rangle.$$

Definition 2 Let $\chi : \mathbb{R}^{2n} \to \mathbb{R}^{2n}$ be a smooth bilipschitz canonical transformation. A family of operators $\mathscr{U}^{\hbar} : \mathscr{S}(\mathbb{R}^n) \to \mathscr{S}'(\mathbb{R}^n)$, $\hbar \in (0, 1]$, is a semiclassical Fourier integral operator of order[1] $m \in \mathbb{R}$ if the kernel $K^{\hbar}$ of $\mathscr{F}^{\mathscr{B}}\mathscr{U}^{\hbar}\mathscr{F}^{\mathscr{B}*}$ satisfies the following condition, for every $N \geq 0$, there exists C_N such that,

$$|K^{\hbar}(X, Y)| \leq C_N\,\hbar^{m-n}\left(1 + \frac{|X - \chi(Y)|}{\sqrt{\hbar}}\right)^{-N}, \ \forall X, Y \in \mathbb{R}^{2n},\ \hbar \in (0, 1]. \tag{11.18}$$

The above definition describes a natural way to quantize canonical transformations. For example the identity operator is associated with the trivial canonical transformation. Any pseudo-differential operator $\mathrm{Op}^w_{\hbar}$ is also a Fourier integral with the trivial canonical transformation. A metaplectic transformation in $L^2(\mathbb{R}^n)$ is a F.I.O associated with a linear symplectic transformation χ (see Chap. 2). Conversely to any canonical transformation satisfying Definition 2 one can associate ($\hbar$-uniformly) bounded transformations in $L^2(\mathbb{R}^n)$ by the following formula. Let us introduce the kernel

$$K^{\hbar}_{\chi}(X, Y) = \hbar^{-n}a(X, Y)\mathrm{e}^{-\frac{|X-\chi(Y)|^2}{\hbar}}$$

where $a \in L^{\infty}(\mathbb{R}^{4n})$ and define:

$$\mathscr{I}^{\hbar}_{\chi}(u)(x) = \int_{\mathbb{R}^{4n}} K^{\hbar}_{\chi}(X, Y)\varphi_X(x)\langle\varphi_Y, u\rangle dY dX,$$

where φ_Z is the normalized coherent state. Using over-completness property of coherent states one can prove that $\mathscr{I}_{\chi}$ is a uniformly bounded operator.

As an example we have already computed in Chap. 3 (Sect. 3.1) the Schrödinger propagator $\mathscr{U}(t, 0)$ when $H(t, X)$ is a quadratic Hamiltonian. In this case we get using computations of Chap. 3 (see Chap. 2, Wigner functions)

$$|\langle\varphi_X, \mathscr{U}^{\hbar}(t, 0)\varphi_Y\rangle| \leq C\hbar^{-n}\mathrm{e}^{-c\frac{|X-F_tY|^2}{\hbar}}, \ \forall t \in [0, T], \tag{11.19}$$

where F_t is the linear flow defined by $H(t)$ such that $F_0 = \mathbb{1}$ and C, c may depend on $0 \leq T < +\infty$.

An easy consequence of the definition and Chap. 2, is:

Proposition 124 *A semiclassical F.I.O of order* 0 *is uniformly bounded in* $L^2(\mathbb{R}^n)$, *as* $\hbar \in]0, 1]$.

We prove now that this class of operators is closed by composition. We have

---

[1] Here the order is defined according the behavior of operators in $\hbar$.

Theorem 54 *Let $\mathscr{U}_j^\hbar$ be a semiclassical F.I.O of order m_j associated to the bilipschitz canonical transformation χ_j, $j = 1, 2$. Then $\mathscr{U}_1^\hbar \circ \mathscr{U}_2^\hbar$ is a semiclassical FIO of order $m_1 + m_2$ associated to $\chi_1 \circ \chi_2$.*

Proof Let $K_j^\hbar$ be the kernel of $\mathscr{F}^{\mathscr{B}} \mathscr{U}_j^\hbar \mathscr{F}^{\mathscr{B}*}$, $j = 1, 2$. Then, the kernel $K_\hbar$ of

$$\mathscr{F}^{\mathscr{B}} \mathscr{U}_1^\hbar \mathscr{U}_2^\hbar \mathscr{F}^{\mathscr{B}*} = \mathscr{F}^{\mathscr{B}} \mathscr{U}_1^\hbar \mathscr{F}^{\mathscr{B}*} \, \mathscr{F}^{\mathscr{B}} \mathscr{U}_2^\hbar \mathscr{F}^{\mathscr{B}*}$$

is given by

$$K^\hbar(X, Y) = \int_{\mathbb{R}^{2n}} K1^\hbar(X, Z) K_2^\hbar(Z, Y) dZ.$$

Using the definition of semiclassical F.I.O, we have for all $N, N' \in \mathbb{N}$, such that $N' \geq N$,

$$|K^\hbar(X, Y)| \leq C_{N,N'} \hbar^{m_1+m_2-2n} \int_{\mathbb{R}^{2n}} dZ \left(1 + \frac{|X - \chi_1(Z)|}{\sqrt{\hbar}}\right)^{-N} \left(1 + \frac{|Z - \chi_2(Y)|}{\sqrt{\hbar}}\right)^{-N'} \tag{11.20}$$

On the other hand, one has by the triangle inequality

$$|X - \chi_1 \circ \chi_2(Y)| \leq |X - \chi_1(Z)| + |\chi_1(Z) - \chi_1 \circ \chi_2(Y)|,$$

and since χ_2 is Lipschitz, we obtain

$$|X - \chi_1 \circ \chi_2(X)| \leq C \left(|X - \chi_1(Z)| + |Z - \chi_2(Y)|\right), \; C > 0.$$

Thus, Peetre's inequality yields

$$1 + \frac{|X - \chi_1 \circ \chi_2(Y)|}{\sqrt{\hbar}} \leq C \left(1 + \frac{|X - \chi_1(Z)|}{\sqrt{\hbar}}\right) \left(1 + \frac{|Z - \chi_2(Y)|}{\sqrt{\hbar}}\right). \tag{11.21}$$

Now, insert (11.21) into (11.20), we obtain

$$|K\hbar(X, Y)| \leq C_{N,N'} \hbar^{m_1+m_2-2n} \left(1 + \frac{|X - \chi_1 \circ \chi_2(Y)|}{\sqrt{\hbar}}\right)^{-N} \times \int_{\mathbb{R}^{2n}} \left(1 + \frac{|Z - \chi_2(Y)|}{\sqrt{\hbar}}\right)^{N-N'} dZ.$$

Hence, after a change of variables, for $N' - N > 2n$, we deduce

$$|K^{\hbar}(X,Y)| \le C_N \hbar^{m_1+m_2-n}\left(1+\frac{|X-\chi_1\circ\chi_2(Y)|}{\sqrt{\hbar}}\right)^{-N}, \quad \forall N \ge 0.$$

This achieves the proof of the Theorem. □

One of the main result of this chapter is the following one [93].

Theorem 55 *Let $H(t,X)$ be a real subquadratic Hamiltonian (2.82), continuous in time for $t\in\mathbb{R}$, C^∞ in X. Then for any $t, t_0 \in \mathbb{R}$, the propagator $\mathscr{U}_H^{\hbar}(t,t_0)$ for the Schrödinger equation* (11.2) *is a semiclassical FIO of order* 0 *associated to the Hamilton flow ϕ^{t,t_0}.*

Proof Let us first remark that we know that the propagator $\mathscr{U}_H^{\hbar}(t,t_0)$ is well defined according Proposition 123 and the classical flow Φ^{t,t_0} is also well defined according Proposition 122. It is clearly uniformly Lipschitz for $t, t_0 \in [-T,T]$ and bi-Lipchitz for $(\Phi^{t,t_0})^{-1} = \Phi^{t_0,t}$.

Step 1. We assume here that the initial data is a coherent state: $\psi_{t_0} = \varphi_z$, $z\in\mathbb{R}^{2n}$. We transform the problem (11.2) into a problem independent of $\hbar$ as in Chap. 4 (Sect. 4.3).

Let $\psi(t) = \mathscr{T}_\hbar(z_t)\Lambda_\hbar u(t)$. Then the function $u(t)$ satisfies the following equation

$$\begin{cases} i\hbar\partial_t u(t) = \Lambda_\hbar^{-1}\mathscr{T}_\hbar(z_t)^{-1}\left(\hat{H}(t)\mathscr{T}_\hbar(z_t) - i\hbar\partial_t\mathscr{T}_\hbar(z_t)\right)\Lambda_\hbar u(t) \\ u(t_0) \quad = g. \end{cases} \tag{11.22}$$

We have the useful formula

$$\Lambda_\hbar^{-1}\mathscr{T}_\hbar(z_t)^{-1}\hat{H}(t)\mathscr{T}_\hbar(z_t)\Lambda_\hbar = Op_1^w H(t, z_t+\sqrt{\hbar}X). \tag{11.23}$$

Taylor's expansion yields

$$H(t,z_t+\sqrt{\hbar}X) = H(t,z_t)+\sqrt{\hbar}\partial_q H(t,z_t)x+\sqrt{\hbar}\partial_p H(t,z_t)\xi+\hbar P_2(t,X)+\hbar R_3(t,X), \tag{11.24}$$

where $P_2(t)$ is a homogeneous polynomial of degree 2 in X:

$$P_2(t,X) = \frac{1}{2}\sum_{|\alpha|=2}\partial_X^\alpha H(t,z_t)X^\alpha,$$

and R_3 is the remainder term given by

$$\begin{aligned} R_3(t,X) = \hbar^{-1}\Big(&H(t,z_t+\sqrt{\hbar}X) - H(t,z_t) - \\ &\hbar\partial_q H(t,z_t)x - \sqrt{\hbar}\partial_p H(t,z_t)\xi - \hbar P_2(t,X)\Big). \end{aligned} \tag{11.25}$$

Consider the function $v(t) = \mathrm{e}^{-\frac{i}{\hbar}\delta_t}u(t)$ where δ_t is a classical action function defined in Chap. 4. By (11.23)–(11.25), the problem (11.22) becomes

$$\begin{cases} i\partial_t v(t) = Op_1^w[P_2(t)]v(t) + Op_1^w[R_3(t)]v(t) \\ v(t_0) \ = g. \end{cases}$$

Since $P_2(t)$ is quadratic, by the metaplectic representation (see Chap. 3), we use another change of function $v(t) = \mathscr{R}[F_t]b(t)g$. The new unknown function $w(t,x) = b(t,x)g(x)$ satisfies the following Cauchy problem

$$\begin{cases} i\partial_t w(t,x) = Op_1^w[R_3(t, F_t(x,\xi))](w(t,x)) \\ w(t_0,x) \ = g(x). \end{cases} \tag{11.26}$$

Step 2. We claim that problem (11.26) is well defined. This is a consequence of Proposition 123 and of the next lemma. Let us introduce a bounded interval $I_T = [t_0 - T, t_0 + T]$, for any $T > 0$. In what follows all our estimates are uniform for $t \in I_T$ and the constants depend only on T.

Lemma 69 *Under Hypothesis of Theorem 11.113, $R_3(t)$ is a subquadratic symbol, with uniform estimates with respect to $h \in (0,1]$ and $t \in I_T$.*

Proof Let $\gamma \in \mathbb{N}^{2n}$. If $|\gamma| \geq 2$, one has by (11.25)

$$|\partial_X^\gamma R_3(t,X)| \leq C_{\gamma,T}, \quad C_{\gamma,T} \geq 0, \tag{11.27}$$

for every $\hbar \in (0,1]$ and $t \in I_T$. Then equation (11.27) and the mean value Theorem applied on $\partial_X R_3(t)$ and $R_3(t)$ respectively, show that

$$|\partial_X^\gamma R_3(t,X)| \leq C_{\gamma,T}|X|^{2-|\gamma|}, \quad |\gamma| \leq 1,$$

for all $\hbar \in (0,1]$ and $t \in I_T$. □

Taking into account that F_t is linear on $\mathbb{R}^{2n}$, it results from Lemma 69 and the chain rule that $R_3(t, F_t)$ is also subquadratic. Hence, Proposition 123 ensures that problem (11.2) is well defined in weighted Sobolev spaces $\mathscr{H}^k$, $k \in \mathbb{N}$. Denote by $V(t)$ its quantum propagator. Then, the evolution operator of the Schrödinger equation (11.2) has the following form

$$\psi(t) = \mathscr{U}_H^\hbar(t,t_0)\varphi_z = \mathscr{T}_\hbar(z_t)\Lambda_\hbar u(t) = \mathscr{T}_\hbar(z_t)\Lambda_\hbar \mathscr{R}[F_t](V(t)g)\,\mathrm{e}^{\frac{i}{\hbar}\delta_t}. \tag{11.28}$$

Last step. Using (11.28) we shall deduce that that $\mathscr{U}_H^\hbar(t,t_0)$ is a semiclassical FIO of order 0, associated to the flow ϕ^{t,t_0}. We have, from the previous computations, denoting $z'_t = \phi^{t,t_0}z'$,

$$\begin{aligned} \langle \varphi_z, \mathscr{U}_H^\hbar(t,t_0)\varphi_{z'}\rangle &= \langle \varphi_z \mathscr{T}_\hbar(z'_t)\Lambda_\hbar u(t)\rangle \\ &= \left\langle \Lambda_\hbar^{-1}\mathscr{T}_\hbar(-z'_t)\mathscr{T}_\hbar(z)\Lambda_\hbar g, u(t)\right\rangle \\ &= \exp\left(\frac{1}{2i\hbar}\sigma(z,z'_t)\right)\left\langle \mathscr{T}\left(\frac{z-z'_t}{\sqrt{\hbar}}\right)g, u(t)\right\rangle. \end{aligned} \tag{11.29}$$

From Proposition 123 we know that that $u(t) \in \mathscr{S}(\mathbb{R}^n)$. Hence estimate (11.18) for $\mathscr{U}_H^{\hbar}(t, t_0)$ is a consequence of the following Lemma:

Lemma 70 *Let $u, v \in \mathscr{S}(\mathbb{R}^n)$. Then for all $N \in \mathbb{N}$, there exists $C_N > 0$ such that*

$$|\langle \mathscr{T}(z)u, v\rangle| \le C_N(1+|z|)^{-N}\|u\|_{\mathscr{H}^N}\|v\|_{\mathscr{H}^N}.$$

Proof The case $N = 0$ is obvious by Cauchy-Schwarz inequality. For $N \ge 1$, we proceed by induction. Let $\alpha \in \mathbb{N}^n$ such that $|\alpha| = N$. One has

$$\begin{aligned}
|q^\alpha\langle \mathscr{T}(z)uv\rangle| &\le \int |q^\alpha u(x-q)v(x)|\, dx \\
&\le C_\alpha \sum_{\beta\le\alpha}\int |(x-q)^\beta u(x-q)|\,|x^{\alpha-\beta}v(x)|dsx \\
&\le C_\alpha \sum_{\beta\le\alpha}\|x^\beta u\|_{L^2(\mathbb{R}^n)}\|x^{\alpha-\beta}v\|_{L^2(\mathbb{R}^n)}.
\end{aligned} \tag{11.30}$$

In the other hand,

$$\begin{aligned}
|p^\alpha\langle \mathscr{T}(z)u, v\rangle| &= \left|\int p^\alpha e^{ipx}u(x-q)v(x)\, dx\right| \\
&= \left|\int \partial_x^\alpha e^{ipx}u(x-q)v(x)\, dx\right| \\
&\le C_\alpha \sum_{\gamma\le\alpha}\int \left|\partial_x^\gamma u(x-q)\right|\left|\partial_x^{\alpha-\gamma}v(x)\right|\, dx \\
&\le C_\alpha \sum_{\gamma\le\alpha}\|\partial_x^\gamma u\|_{L^2(\mathbb{R}^n)}\|\partial_x^{\alpha-\gamma}v\|_{L^2(\mathbb{R}^n)}.
\end{aligned} \tag{11.31}$$

We have $|z|^N \le 2^{N-1}(|q|^N + |p|^N)$.
Consequently, (11.30) and (11.31) imply for all $N \ge 1$:

$$\begin{aligned}
|z|^N|\langle \mathscr{T}(z)uv\rangle| &\le C_N\Bigg(\sum_{\beta\le\alpha}\|x^\beta u\|_{L^2(\mathbb{R}^n)}\|x^{\alpha-\beta}v\|_{L^2(\mathbb{R}^n)} \\
&\quad + \sum_{\gamma\le\alpha}\|\partial_x^\gamma u\|_{L^2(\mathbb{R}^n)}\|\partial_x^{\alpha-\gamma}v\|_{L^2(\mathbb{R}^n)}\Bigg) \\
&\le C_N\|u\|_{\mathscr{H}^N}\,\|v\|_{\mathscr{H}^N}.
\end{aligned}$$

This Lemma simultaneously with (11.29) provide that

$$|K_\hbar(z, z')| \le (2\pi\hbar)^{-n}C_N\left(1+\frac{|z-z'_t|}{\sqrt{\hbar}}\right)^{-N}, \quad \forall N \in \mathbb{N}.$$

Finally, using the definition of the flow Φ^{t,t_0} we deduce estimate (11.18) and therefore Theorem 11.113. □

Remark 63 The above definition given for F.I.O is in agreement with the correspondence principle between classical mechanics and quantum mechanics as it was formulated by N. Bohr.

11.3 The Herman-Kluk Semi-classical Approximation

We have seen that the time dependent propagator for a subquadratic Hamiltonian is a semi-classical F.I.O. Notice that the result is easily extended for $\hbar$-dependent classical Hamiltonians $H(t, X) \asymp \sum_{j\geq 0} \hbar^j H_j(t, X)$ where $H_j(t)$ is in $S_0^{(2-j)_+}$ with uniformly bounded semi-norms as $t \in \mathbb{R}$ and

$$\forall N \geq 1, \hbar^{-N-1}\big(H(t, X) - \sum_{0\leq j\leq N} \hbar^j H_j(t, X)\big) \text{ is bounded in } \mathrm{S}_0^0, \tag{11.32}$$

with uniformly bounded semi-norms as $t \in \mathbb{R}$ and $\hbar \in]0, 1]$. In applications the contribution of the sub-principal symbol $H_1(t)$ may have a non trivial effect (this will be more obvious for systems considered later.

In this section our aim is to analyze with more accuracy the structure of the Schwartz kernel of the propagator $\mathscr{U}_H^\hbar(t, t_0)$. This can be done using coherent states and the following lemma.

Lemma 71 *Let $\hat{A}$ be a continuous linear operator $\hat{A} : \mathscr{S}(\mathbb{R}^n) \to \mathscr{S}'(\mathbb{R}^n)$. Then its Schwartz kernel $\mathscr{K}^A(x, y)$, as a distribution in $\mathscr{S}'(\mathbb{R}^n_x \times \mathbb{R}^n_y)$, satisfies*

$$\mathscr{K}^A(x, y) = (2\pi\hbar)^{-n} \int_{\mathbb{R}^{2n}} \overline{\hat{A}^*\varphi_z(y)}\, \varphi_z(x) dz. \tag{11.33}$$

In particular let $\mathscr{K}_H^\hbar(t, t_0; x, y)$ be the Schwartz kernel of a quantum propagator $\mathscr{U}_H^\hbar(t, t_0)$ such that in Theorem 56 then we have

$$\boxed{\mathscr{K}_H^\hbar(t, t_0; x, y) = (2\pi\hbar)^{-n} \int_{\mathbb{R}^{2n}} \mathscr{U}_H^\hbar(t, t_0)\varphi_z(x)\overline{\varphi_z(y)} dz}. \tag{11.34}$$

Proof Recall here that we compute kernels as distributions.
71 follows from the inverse Bargmann transform applied to $\hat{A}\psi$.
We get (11.34) using that $\mathscr{U}_H^\hbar(t, t_0)^* = \mathscr{U}_H^\hbar(t_0, t)^{-1}$. □

We wish now to use formula (11.34) in order to apply the results obtained in Chap. 4. concerning semi-classical propagation of coherent states.

As it is expected, and suggested by Sect. 11.2, the semi-classical approximation depends stongly on the classical flow generated by the principal term $H_0(t, X)$ of the Hamiltonian. As in Chap. 4, Let us introduce the flow ϕ^{t,t_0} for $H_0(t, X)$ and the stability matrix along this flow:

$$F_{t,t_0} = \begin{pmatrix} A_{t,t_0} & B_{t,t_0} \\ C_{t,t_0} & D_{t,t_0} \end{pmatrix}$$

Recall that if $\phi^{t,t_0}(q, p) = q_{t,t_0}(q, p), p_{t,t_0}(q, p) = z_{t,t_0}(z)$ then we have

$$A_{t,t_0} = \frac{\partial q_{t,t_0}}{\partial q}, \quad B_{t,t_0} = \frac{\partial q_{t,t_0}}{\partial p}, \quad C_{t,t_0} = \frac{\partial p_{t,t_0}}{\partial q}, \quad D_{t,t_0} = \frac{\partial p_{t,t_0}}{\partial p} \tag{11.35}$$

We also introduce the classical action

$$S(t, t_0, z) = \int_{t_0}^{t} (p_{s,t_0} \cdot \dot{q}_{s,t_0} - H(s, z_{s,t_0}))ds, \tag{11.36}$$

and the symmetric complex matrix

$$\Gamma_{t,t_0} = (C_{t,t_0} + iD_{t,t_0})(A_{t,t_0} + iB_{t,t_0})^{-1}. \tag{11.37}$$

Notice that Γ_{t,t_0} depends also on $z \in \mathbb{R}^{2n}$ the initial data of the classical trajectory z_t.

Using that the coherent state family $\{\varphi_z\}_{z\in\mathbb{R}^{2n}}$ is overcomplete in $L^2(\mathbb{R}^n)$ and the propagation Theorem proved in Chap. 4, we get the following analysis of the Schwartz kernel of the propagator $\mathscr{U}(t, t_0)$.

Theorem 56 *Let $H(t)$ be a time dependent subquadratic Hamiltonian satisfying (11.32). Let $\mathscr{K}_H^\hbar(t, t_0; x, y)$ be the Schwartz kernel of its quantum propagator $\mathscr{U}_H^\hbar(t, t_0)$.*

$$\mathscr{K}_H^\hbar(t, t_0, x, y) \asymp \int_{\mathbb{R}^{2n}} e^{\frac{i}{\hbar}\Phi_\Gamma(t,t_0,z;x,y)} c(\hbar; t, t_0, z)dz, \tag{11.38}$$

where $c(\hbar; t, t_0, \cdot)$ is a semi-classical symbol of order 0 and

$$\Phi_{\Gamma_{t,t_0}}(t, t_0, z; x, y) = S(t, t_0, z) + p_t \cdot (x - q_t) - p \cdot (y - q) + \frac{i}{2}|y - q|^2 + \Gamma_{t,t_0}(x - q_t) \cdot (x - q_t).$$

More precisely, if $\mathscr{U}_H^{\hbar,N}(t, t_0)$ denotes the operator with the Schwartz kernel

$$\mathscr{K}_H^{\hbar,N}(t, t_0, x, y) = (2\pi\hbar)^{-3n/2} \int_{\mathbb{R}^{2n}} e^{\frac{i}{\hbar}\Phi_{\Gamma_{t,t_0}}(t,z;x,y)} \left(\sum_{j=0}^{N} \hbar^j c_j(t, t_0, z)\right) dz,$$

then for every t_0, T, and every $N \geq 1$, there exists $C_{N,T}$ such that

$$\|\mathscr{U}_H^{\hbar}(t, t_0) - \mathscr{U}_H^{\hbar,N}(t, t_0)\|_{L^2 \to L^2} \leq C_{N,T}\, \hbar^{N+1}, \quad \forall \hbar \in (0, 1],\ |t - t_0| \leq T. \tag{11.39}$$

The leading term is given by the following formula

$$c_0(t, t_0; z) = \det{}^{-1/2}(A_{t,t_0} + iB_{t,t_0}) \exp\Big(-i \int_{t_0}^{t} H_1(z_s)ds\Big) \tag{11.40}$$

where the square root is defined by continuity starting from $t = t_0$ ($c_0(t_0; z) = 2^{n/2}$).

Proof The asymptotic formula (11.38) is a direct application of results proved in Chap. 4 (Sect. 4.3) concerning the time evolution coherent states, using Lemma 71. Hence we get easily the quadratic phase $\Phi_\Gamma(t, t_0, z; x, y)$. The only new thing we have to prove is the remainder estimate (11.39) using the sub-quadratic assumption. It is easier to prove this estimate in the Fourier-Bargmann representation. Recall the following formulas obtained in Chap. 4 (4.3.1).

$$i\hbar\partial_t \psi_z^{(N)} = \widehat{H}(t)\psi_z^{(N)} + R_z^{(N)}(t) \tag{11.41}$$

where

$$\psi_z^{(N)}(t) = \mathrm{e}^{i\delta_t/\hbar}\hat{T}(z_t)\Lambda_\hbar \hat{R}[F_t]\left(\sum_{0\leq j\leq N} \hbar^{j/2} b_j(t) g\right) \tag{11.42}$$

and

$$R_z^{(N)}(t, x) = \mathrm{e}^{i\delta_t/\hbar}\left(\hbar^{j/2} \sum_{\substack{j+k=N+3\\ k\geq 3}} \hat{T}(z_t)\Lambda_\hbar \hat{R}[F_t] Op_1^w[R_k(t) \circ F_t](b_j(t)g)\right). \tag{11.43}$$

So have

$$i\hbar\partial_t \mathscr{U}^{\hbar,(N)}(t, t_0) = \hat{H}(t)\mathscr{U}^{\hbar,(N)}(t, t_0) + \mathscr{R}^{(N)}(t, t_0) \tag{11.44}$$

Denote $\mathscr{B}R^{(N)}(t, t_0, z, z') = \langle \varphi_{z'}, \mathscr{R}^{(N)}(t, t_0)\varphi_z\rangle$.
We have $\mathscr{B}R^{(N)}(t, t_0, z, z') = \langle \varphi_{z'}, R_z^N(s)\rangle$. Now, using Lemma 70, we get that for any $M > 0$. there exists $C_{M,T}$ such that

$$|\langle \varphi_{z''}, R_z^N(s)\rangle| \leq C_{M,T}\hbar^{(N+3)/2}\left(1 + \frac{|z'' - z_{s,t_0}|}{\sqrt{\hbar}}\right)^{-M}. \tag{11.45}$$

Notice that (11.45) means that the operator $\mathscr{R}^{(N)}(t, t_0)$ is semi-classical F.I.O of order $\frac{N+1}{2} + n$.
We know that $\mathscr{U}(t_0, t)$ is a semi-classical F.I.O.

Using the Duhamel formula we get for $\mathscr{B}E^{(N)}(z, z') = \langle \varphi_z, (\mathscr{U}_H^\hbar(t, t_0) - \mathscr{U}^{\hbar,(N)}(t, t_0))\varphi_{z'}\rangle$,

$$\mathscr{B}E^{(N)}(z, z') = \frac{1}{i\hbar}\int_{t_0}^{t} \langle \varphi_z, \mathscr{U}_H^\hbar(t, s)\mathscr{R}_z^{(N)}(s, t_0)\varphi_{z'}\rangle ds \tag{11.46}$$

Using Lemma 70 we get that $\mathscr{U}_H^\hbar(t, s)\mathscr{R}_z^{(N)}(s)$ is a semi-classical F.I.O of order $\frac{N+3}{2} + n$ associated with the canonical transform ϕ^{t,t_0}. This entails the estimate:

$$\|\mathscr{U}_H^\hbar(t, t_0) - \mathscr{U}^{\hbar,(M)}(t, t_0)\|_{L^2\to L^2} \le C_{M,T}\, \hbar^{\frac{M+1}{2}}, \quad \forall\, \hbar \in (0, 1]. \tag{11.47}$$

But it is not difficult to see that $\mathscr{U}^{\hbar,(M)}(t, t_0)$ is a semiclassical F.I.O of order 0 and for $M > N$, $\mathscr{U}^{\hbar,(M)}(t, t_0) - \mathscr{U}^{\hbar,(N)}(t, t_0)$ is of order $N + 1$. Taking M large enough we get (11.39). □

In the following theorem we get that in the phase function $\Phi_{\Gamma_{t,t_0}}$ we can replace the matrix Γ_{t,t_0} by any matrix $\Theta \in \mathfrak{S}(n)^+$ by modifying the amplitude. In particular we can choose $\Theta = i\mathbb{1}$. This property is known as the Hermann-Kluk approximation of the propagator.

Theorem 57 *([219]) Let $H(t)$ be a time dependent subquadratic Hamiltonian satisfying* (11.32)*. Let $K_\hbar(t, t_0; x, y)$ be the Schwartz kernel of its quantum propagator $\mathscr{U}_H^\hbar(t, t_0)$. Let be $\Theta \in \mathfrak{S}(n)^+$, a smooth family of complex matrices. dependending smoothly in $t, z) \in \mathbb{R} \times \mathbb{R}^{2n}$. Then we have*

$$K^\hbar(t, t_0, x, y) \asymp \int_{\mathbb{R}^{2n}} e^{\frac{i}{\hbar}\Phi_\Theta(t,t_0,z;x,y)} a(\hbar; t, t_0, z) dz, \tag{11.48}$$

where $a(\hbar; t, t_0)$ is a semi-classical symbol of order 0 *and*

$$\Phi_\Theta(t, z; x, y) = S(t, z) + p_t \cdot (x - q_t) - p \cdot (y - q) + \frac{1}{2}\big(\Theta(x - q_t) \cdot (x - q_t) + i|y - q|^2\big). \tag{11.49}$$

More precisely, if $\mathscr{U}^{\hbar,(N)}(t, t_0)$ denotes the operator with the Schwartz kernel

$$K^{\hbar,(N)}(t, t_0, x, y) = (2\pi\hbar)^{-3n/2} \int_{\mathbb{R}^{2n}} e^{\frac{i}{\hbar}\Phi_\Theta(t,z;x,y)} \left(\sum_{j=0}^{N} \hbar^j a_j(t, t_0, z)\right) dz,$$

then for every t, $|t - t_0| \le T$ and every $N \ge 1$, there exists $C_{N,T}$ such that

$$\|\mathscr{U}_H^\hbar(t) - \mathscr{U}^{\hbar,(N)}(t)\|_{L^2\to L^2} \le C_{N,T}\, \hbar^{N+1}, \quad \forall\, \hbar \in (0, 1].$$

The leading term is given by the following formula

$$a_0(t, t_0; z) = \det^{1/2}(C_{t,t_0} - iD_{t,t_0} - \Theta(A_{t,t_0} - iB_{t,t_0}))\exp\Big(-i\int_{t_0}^{t} H_1(z_s)ds\Big) \tag{11.50}$$

where the square root is defined by continuity starting from $t = t_0$ ($a_0(t_0, t_0; z) = 2^{d/2}$).
Moreover, the amplitudes a_j are smooth functions defined by transport equations and for every $T > 0$ they are uniformly bounded in S_0^0 for $|t| \leq T$.

The previous result is valid on bounded time intervals. Now we give a result with a control on large time intervals. Let us assume that conditions on $\hat{H}(t)$ are satisfied uniformly for $t \in \mathbb{R}$. Moreover assume that there exists a positive real function $\mu(T) \geq 1$, $T > 0$, such that the classical flow ϕ^t satisfies, for every multiindex γ, $|\gamma| \geq 1$, we have for some $C_\gamma > 0$,

$$|\partial_z^\gamma \phi^{t,t'}(z)| \leq C_\gamma \mu(T)^{|\gamma|}, \quad \text{for } |t| + |t'| \leq T, \ \forall z \in \mathbb{R}^{2n}. \tag{11.51}$$

The condition (11.51) is discussed in [40] . In particular this condition is fulfilled with $\mu(T) = e^{\delta T}$ for $\delta = \sup_{X \in \mathbb{R}^{2n}, t \in \mathbb{R}} \|J\partial_{X,X}^2 H(t, X)\|$.

Theorem 58 *Choosing the phase as in theorem 57, for $j \geq 0$ the amplitudes $a_j(t, z)$ satisfy the following estimates, for every multiindex γ there exist a constant $C_{j,\gamma}$ such that*

$$|\partial_z^\nu a_j(t, z)| \leq C_{j\gamma}|\det^{1/2} M(t, z)|\mu(T)^{8j+2|\nu|}, \quad \forall T > 0, |t| \leq T, \ \forall z \in \mathbb{R}^{2n}. \tag{11.52}$$

In particular for $\mu(T) = e^{\delta T}$ we have the following Ehrenfest type estimate. For every $N \geq 1$ and every $\varepsilon > 0$ there exists $C_{N,\varepsilon}$ such that we have

$$\|\mathscr{U}_H^\hbar(t, t_0) - \mathscr{U}^{\hbar,(N)}(t, t_0)\| \leq C_{N,\varepsilon}\hbar^{\varepsilon(N+1)}, \quad \forall t, |t| \leq \frac{1-\varepsilon}{8\delta}|\log \hbar|, \ \forall \hbar \in]0, 1]. \tag{11.53}$$

For quadratic Hamiltonians the Herman-Kluk formula is exact (without error term). So assume that $H(t, X)$ is quadratic in $X \in \mathbb{R}^{2n}$ then we have the following representation formula for the propagator generated by $\hat{H}(t)$.

$$K^\hbar(t, t_0, x, y) = (2\pi\hbar)^{-3n/2} a(t, t_0) \int_{\mathbb{R}^{2n}} e^{\frac{i}{\hbar}\Phi_\Theta(t,z;x,y)}\, dz, \tag{11.54}$$

where $a(t, t_0)$ is a number depending only on the coefficient of $H(t)$ and which can be computed.

Remark 64 In [219] more explicit and exact computations are given for quadratic Hamiltonians.

In previous works an Ehrenfest time $T_E = c|\log \hbar|$, $c > 0$, was estimated for propagation of Gaussians in [65] and for propagation of observables in [40]. For

Gaussians we got $c = \frac{1}{6\delta}$, for observables $c = \frac{1}{2\delta}$ and here for the Hermann-Kluk approximation $c = \frac{1}{8\delta}$.
Notice that if $\mu(T) \approx T^\kappa$, $\kappa > 0$ then we get an algebraic critical time T_{HK} for the validity of the Herman-Kluk approximation with $T_{HK} \approx \hbar^{\frac{1}{8\kappa}}$.

We give two different proofs for Theorem 57 and Theorem 58. For Theorem 57 we start from Theorem 56 and we use a deformation argument from the phase Φ_Γ (with $\Gamma = \Gamma_{\cdot,t_0}$ depends also on $z \in \mathbb{R}^{2n}$, to the phase $\Phi_{i\mathbb{1}}$.
For proving Theorem 58 we directly use an ansatz to compute the amplitudes $a_j(t, t_0, z)$ solving transport equations.

11.3.1 *Proof of Theorem 57 by Deformation*

Our aim here is to prove Theorem 57 from Theorem 56. For that purpose let us consider a C^1 curve in $\mathfrak{S}(n)^+$, $s \mapsto \Theta(s)$, $0 \leq s \leq 1$, $\Theta(0) = \Gamma_{t,t_0}$, $\Theta(1) = \Theta$. For simplicity we take $\Theta(s) = s\vartheta + (1-s)\Gamma_{t,t_0}$.

We shall prove that we can get a family of semiclassical symbols: $a^{(s)}(\hbar, t, t_0, z) = \sum_{j\geq 0} \hbar^j a_j^{(s)}(t, t_0, z)$ such that

$$\partial_s \left(\int_0^1 a^{(s)}(\hbar, t, t_0, z) e^{\frac{i}{\hbar}\Phi_{\Theta(s)}(t,z;x,y)} dz \right) \approx 0. \tag{11.55}$$

We have

$$\partial_s \Phi_{\Theta(s)} = \frac{1}{2}(\Theta - \Gamma_{t,t_0})(x - q_t) \cdot (x - q_t)$$

We shall compute (11.55) using integration by parts with the formula

$$M(s)^T (x - q_t) e^{\frac{i}{\hbar}\Phi_{\Theta(s)}} = \frac{\hbar}{i}(\partial_q - \partial_p) e^{\frac{i}{\hbar}\Phi_{\Theta(s)}} \tag{11.56}$$

where $M(s) = C_{t,t_0} - iD_{t,t_0} - \Theta(s)(A_{t,t_0} - iB_{t,t_0})$.

To explain the details it is convenient here to consider a more general setting. Let us consider a class of operators (of Fourier-integral type) defined as follows.
χ is a bilipchitz symplectic diffeomorphism in $\mathbb{R}^{2n}$. We denote $z = (q, p) \in \mathbb{R}^{2n}$, $\chi(z) = (Q(z), P(z)) \in \mathbb{R}^d \times \mathbb{R}^d$ and S an action for χ, that means that it is a primitive on $\mathbb{R}^{2n}$ of the closed 1-form $PdQ - pdq$. We consider the following phases

$$\begin{aligned}\Psi^{(\chi,\Theta,\Gamma)}(z; x, y) &= S(z) + P \cdot (x - Q) - p \cdot (y - q) \\ &+ \frac{1}{2}\big(\Theta(x - Q) \cdot (x - Q) - \overline{\Gamma}(y - q).(y - q)\big).\end{aligned} \tag{11.57}$$

Let us denote $\mathscr{I}(a, \Gamma, \Theta)$ the operator in $\mathscr{S}(\mathbb{R}^n)$ having the following Schwartz kernel

$$\mathscr{K}_{\chi,a}(x,y) = (2\pi\hbar)^{-3n/2}\int_{\mathbb{R}^{2n}} e^{\frac{i}{\hbar}\Psi^{(\chi,\Theta,\Gamma)}(z;x,y)}a(z)dz \tag{11.58}$$

where $a \in \mathfrak{A}_0$, and $\Gamma, \Theta \in \mathfrak{S}^+(n)$. We assume that the matrix Γ is constant (pratically $\Gamma = i\mathbb{1}$) but Θ may depends on z and satisfies

$$\Im\Theta(z)v\cdot v \geq c_0|v|^2,\ \forall z \in \mathbb{R}^{2d},\ \forall v \in \mathbb{C}^d,\quad \text{with } c_0 > 0. \tag{11.59}$$

$$|\partial_z^\alpha \Theta(z)| \leq C_\alpha,\ \forall z \in \mathbb{R}^{2d}. \tag{11.60}$$

Pratically if χ is an Hamiltonian flow at time t: $z \mapsto \Phi^t(z)$ then $\Theta = (C(t,z) + \Gamma_0 D(t,z))(A(t,z) + \Gamma_0 B(t,z))^{-1}$, where we have

$$\begin{pmatrix} A(t,z) & B(t,z) \\ C(t,z) & D(t,z) \end{pmatrix} := \partial_z \Phi^t(z).$$

Remark 65 We know that $(C + D\Gamma)$ and $(A + B\Gamma)$ are invertible and that the matrix $(C + D\Gamma)(A + B\Gamma)^{-1} \in \mathfrak{S}^+(n)$.

Proposition 125 *Let be Γ, Θ as above and $a \in S_0^0$. Then there exists a semi-classical symbol $b^{(\hbar)} \approx \sum_{j\geq 0}\hbar^j b_j$, in S_0^0 such that we have for the L^2 operator norm,*

$$\mathscr{I}(a, \Gamma, \Theta) = \mathscr{I}(b^{(\hbar)}, \Gamma, \Gamma_0) + O(\hbar^\infty) \tag{11.61}$$

Moreover we have $b^\hbar = b_0 + \hbar b_1 + \mathscr{O}(\hbar^2)$ where

$$b_0(z) = a_0(z)\frac{\det^{1/2}(M(1))}{\det^{1/2}(M(0))} \tag{11.62}$$

and

$$M(s) := C + D\bar{\Gamma} - \Big((1-s)\Theta + s\Gamma_0\Big)(A + B\bar{\Gamma}). \tag{11.63}$$

The proof consists in an evolution argument by considering the path in $\mathfrak{S}^+(n)$,

$$s \mapsto \Theta(s) := (1-s)\Theta(z) + s\Gamma_0,\quad 0 \leq s \leq 1$$

and by looking for $s \mapsto a(s) \in \mathfrak{A}_0$ such that

$$\mathscr{I}(a, \Gamma, \Theta) = \mathscr{I}(a(s), \Gamma, \Theta(s)) + O(\hbar^\infty)$$

for any $s \in [0, 1]$. This is done thanks to the matrix $M(s)$ which has good properties according to the next remark.

Remark 66 Note that we have

$$M(s) = C + D\bar{\Gamma} - \Theta(s)(A + B\bar{\Gamma})(C + D\Gamma) = (A + B\overline{\Gamma})^{-1} - \Theta(s))(A + B\overline{\Gamma}).$$

Therefore, by Remark 65, $C + D\Gamma)(A + B\Gamma)^{-1} \in \mathfrak{S}^+(n)$ and so $(C + D\Gamma)(A + B\Gamma)^{-1} - \bar{\Theta}) \in \mathfrak{S}^+(n)$. Therefore, $M(s)$ is invertible for all $s \in [0, 1]$.

We shall also use the following Lemma

Lemma 72 *For every $b \in \mathrm{S}^0_0$ and $\beta \in \mathbb{N}^n$ we have the L^2 estimate, uniformly in $s \in [0, 1]$,*

$$\|\mathscr{I}((x - Q)^\beta b, \Gamma, \Theta)\| = \mathscr{O}\big(\hbar^{|\beta|/2}\big) \tag{11.64}$$

Proof It follows from an elementary Gaussian estimate and the use of the Bargman transform. Indeed, the Schwartz kernel of the operator $\mathscr{I}((x - Q)^\beta b, \Gamma, \Theta)$ is the function

$$(x, y) \mapsto (2\pi\hbar)^{-n} \int_{\mathbb{R}^{2n}} (x - Q)^\beta e^{\frac{i}{\hbar}S(z)} \overline{g_z^{\Gamma_0}(y)} g_Z^{\Theta,\hbar}(x) b(z) dz.$$

Its Bargman transform is the function

$$k_{\mathscr{B}}(X, Y) = \hbar^{\frac{|\beta|}{2}} (2\pi\hbar)^{-2n} \int_{\mathbb{R}^{2n}} e^{\frac{i}{\hbar}p_Y \cdot q_Y - p_X \cdot q_X + \frac{i}{\hbar}S(z)} \langle g_Y^\hbar, g_z^{\Gamma,\hbar}\rangle \cdot$$
$$\cdot \left\langle \left(\frac{x - Q}{\sqrt{\hbar}}\right)^\beta g_Z^{\Theta,\hbar}, g_X^\hbar \right\rangle b(z) dz. \tag{11.65}$$

and it is not difficult to see that for all $N \in \mathbb{N}$, there exists $C_N > 0$ such that

$$\left(\langle X\rangle^N + \langle Y\rangle^N\right) |k_{\mathscr{B}}(X, Y)| \le C_N \hbar^{\frac{|\beta|}{2}}.$$

The conclusion then follows from Lemma 5.5 of our paper or 3.4 of [219].

Proof We set

$$\Psi^{(s)} = \Psi^{(\chi, \Theta(s), \Gamma)}$$

and we look for a C^1 family symbol $a^{(s)}$, $0 \le s \le 1$ such that $a^{(0)} = a$, $a^{(1)} = b$ and

$$\partial_s \left(\int_{\mathbb{R}^{2d}} e^{\frac{i}{\hbar}\Psi^{(s)}(z; x, y)} a^{(s)}(z) dz \right) = O(\hbar^\infty), \quad \forall s \in [0, 1] \tag{11.66}$$

We have

$$\partial_s \Psi^{(s)} = \frac{1}{2} \partial_s \Theta(s)(x - Q) \cdot (x - Q), \tag{11.67}$$

and we claim that

$$(\partial_q + \bar{\Gamma}\partial_p)\Psi^{(s)} = \frac{i}{\hbar} M(s)^T(x-Q)\Psi^{(s)} + \frac{i}{2\hbar}\left(\partial_q\Theta + \bar{\Gamma}\partial_p\Theta\right)(x-Q)\cdot(x-Q)\Psi^{(s)} \tag{11.68}$$

This relation comes from a direct computation.

Finally, using that $M(s)$ is invertible according to Remark 66, we can write

$$(\partial_q + \bar{\Gamma}\partial_p)\Psi^{(s)} = \frac{i}{\hbar}\left(M^T(s,z)(x-Q) + N(s,z)(x-Q, x-Q)\right)\Psi^{(s)}, \tag{11.69}$$

$$(x-Q)\mathrm{e}^{\frac{i}{\hbar}\Psi^{(s)}} = \frac{\hbar}{i}(M^T(s))^{-1}(s,z)\big(\partial_q + \bar{\Gamma}\partial_p\big)\mathrm{e}^{\frac{i}{\hbar}\Psi^{(s)}} - N_0(s,z)(x-Q, x-Q)\mathrm{e}^{\frac{i}{\hbar}\Psi^{(s)}} \tag{11.70}$$

where $N_0(s, z)$ is the smooth tensor of order 2 given by $M(s) := C + D\bar{\Gamma} - \Theta^{(s)}(A + B\bar{\Gamma})$

$$N_0(s,z) := \frac{1}{2}(M^T(s))^{-1}\left(\partial_q\Theta(s) + \bar{\Gamma}\partial_p\Theta(s)\right).$$

We are now going to use the latter relation for finding an equation on $s \mapsto a^{(s)}$. Using (11.67), we have

$$\partial_s\mathscr{I}(a^{(s)}, \Gamma, \Theta(s)) = \mathscr{I}(\partial_s a^{(s)} + \frac{i}{2\hbar}\partial_s\Theta(s)(x-Q)\cdot(x-Q)a^{(s)}, \Gamma, \Theta^{(s)}).$$

We then use (11.70) to write

$$\begin{aligned}
&\frac{i}{2\hbar}\partial_s\Theta(s)(x-Q)\cdot(x-Q)a^{(s)}\mathrm{e}^{\frac{i}{\hbar}\Psi^{(s)}}\\
&\quad = a^{(s)}(M(s)^{-1}\partial_s\Theta(s)(x-Q))\cdot\big(\partial_q + \bar{\Gamma}\partial_p\big)\mathrm{e}^{\frac{i}{\hbar}\Psi^{(s)}} + N_1(s,z)[x-Q]^3\mathrm{e}^{\frac{i}{\hbar}\Psi^{(s)}}
\end{aligned}$$

where $N_1(s, z)$ is a smooth tensor of order 3 (independent of $\hbar$). We then observe

$$\begin{aligned}
(\partial_q + \overline{\Gamma}\partial_p)\cdot\big(a^{(s)}M(s)^{-1}\partial_s\Theta(s)(x-Q)\big)
&= a^{(s)}\mathrm{Tr}\left(M(s)^{-1}\partial_s\Theta(s)(A + B\overline{\Gamma})\right) + n_1(s,z)\cdot(x-Q)\\
&= -a^{(s)}\mathrm{Tr}\left(M(s)^{-1}\partial_s M(s)\right) + n_1(s,z)\cdot(x-Q)
\end{aligned}$$

for some vector n_1 which is independent from $\hbar$, and where we have used

$$\partial_s\Theta(s) = \Gamma_0 - \Theta(z) = -\partial_s M(s)(A + B\overline{\Gamma})^{-1}.$$

Therefore, integration by part in q and p shall generate powers of $(x - Q)$. However, if we stop here the transformations, we shall only have a term of order 1 in $(x - Q)$ which only generates a remainder of order $\sqrt{\hbar}$. Therefore, we push the transformation one step higher by using again (11.70) and writing

$$n_1(s,z)\cdot(x-Q)e^{\frac{i}{\hbar}\Psi^{(s)}} = \frac{\hbar}{i}M(s)^{-1}n_1(s,z)\cdot\left(\partial_q+\bar{\Gamma}\partial_p\right)e^{\frac{i}{\hbar}\Psi^{(s)}}$$
$$-n_1(s,z)\cdot N_0(s,z)[x-Q]^2e^{\frac{i}{\hbar}\Psi^{(s)}}. \quad (11.71)$$

Then, performing a first integration by parts and using Lemma 72, we obtain

$$\frac{d}{ds}\mathscr{I}(a^{(s)},\Gamma,\Theta(s)) = \mathscr{I}(\partial_s a^{(s)} - f(s,z)a^{(s)},\Gamma,\Theta(s)) + O(\hbar) \quad (11.72)$$

and

$$f(s,z) = \mathrm{Tr}((M(s))^{-1}\partial_s M(s)) = (\det M(s))^{-1}\partial_s(\det M(s)).$$

Note that the second equality is the Liouville formula. At this stage of the proof, by solving the equation $\partial_s a^{(s)} - f(s,z)a^{(s)} = 0$ we get the first term b_0 of the asymptotic expansion by the formula (11.62).

Note that the structure of the rest term in (11.72) is known, it is of the form $I(r^{\hbar}(s),\Gamma,\Theta(s))$, where $s\mapsto r^{\hbar}(s)$ is a smooth map from $[0,1]$ to the set S_0^0 and $r^{\hbar}$ admits an asymptotic expansion in $\hbar$ at any order. This opens the way to a recursive argument based on (11.72) for concluding the proof by constructing the coefficients of the asymptotic series $\sum\hbar^j b_j$. Therefore, let us assume that we have constructed $\mathscr{C}^1$ maps $t\mapsto b_j(s)$ such that

$$\frac{d}{ds}\mathscr{I}\left(\sum_{j=1}^{j_0}\hbar^j b_j(s),\Gamma,\Theta(s)\right) = \hbar^{j_0+1}I(r_{j_0}(s),\Gamma,\Theta(s)) + O(\hbar^{j_0+2})$$

and let us look for $s\mapsto b_{j_0+1}(s)$ for pushing the estimate one step further. The analysis above and the recursive assumption leads to

$$\frac{d}{ds}\mathscr{I}\left(\sum_{j=1}^{j_0+1}\hbar^j b_j(s),\Gamma,\Theta(s)\right) = \hbar^{j_0+1}\mathscr{I}(\partial_s b_{j_0+1}(s) - f(s,z)b_{j_0+1}(s),\Gamma,\Theta(s))$$
$$+\hbar^{j_0+1}\mathscr{I}(r_{j_0+1}(s),\Gamma,\Theta(s)) + O(\hbar^{j_0+2}).$$
$$(11.73)$$

We then conclude the recursive argument by choosing $b_{j_0+1}(s)$ such that $b_{j_0+1}(0)=0$ and

$$\partial_s b_{j_0-1}(s) - f(s,z)b_{j_0+1}(s) = -r_{j_0}(s),$$

whence the result of the proposition. □

11.3.2 A Proof by Solving Transport Equations

This alternative proof seems more convenient to control the estimates in the large time regime.
Let us denote $\mathscr{I}(a, \Phi)$ the formal operator having the Schwartz kernel

$$K_a^{\hbar}(x, y) = (2\pi\hbar)^{-3n/2}\int_{\mathbb{R}^{2n}} e^{\frac{i}{\hbar}\Phi(t,z;x,y)}a(t, z)dz. \tag{11.74}$$

For simplicity we take $t_0 = 0$ in all the proof of Theorem 57.

As usual for this kind of problems there are two steps :① Determine the amplitudes a_j solving by induction transport differential equations, ② Estimate the error between the approximated propagator and the exact one. It is convenient to write

$$e^{\frac{i}{\hbar}\Phi} = (\pi\hbar)^{n/2}\varphi_{z_t}(x)\bar{\varphi}_z(y)e^{\frac{i}{\hbar}(S(t,z)+(p\cdot q-p_t\cdot q_t)/2)}. \tag{11.75}$$

Then we have to compute $\hat{H}^{\hbar}(t)\varphi_{z_t}$. It is not difficult to add contributions of the lower order terms of the Hamiltonian, so we shall assume for simplicity that $H^{\hbar}(t) = H_0(t) := H(t)$.

Lemma 73 *For every $N \geq 2$ we have*

$$\hat{H}(t)\varphi_{z_t}(x) = \sum_{|\gamma|\leq N}\frac{\hbar^{|\gamma|/2}}{\gamma!}\partial_X^{\gamma}H(t, z_t)\Pi_{\gamma}\Big(\frac{x-q_t}{\sqrt{\hbar}}\Big)\varphi_{z_t}(x) + \hbar^{(N+1)/2}T(z_t)\Lambda_{\hbar}Op_1^w[R_N(t, z_t)]g(x) \tag{11.76}$$

where

$$R_N(t, z_t, X) = \int_0^1\frac{(1-s)^N}{N!}\sum_{|\gamma|=N+1}\partial_X^{\gamma}H(t, z_t + s\sqrt{\hbar}X)X^{\gamma}ds. \tag{11.77}$$

and Π_{γ} is a universal polynomial of degree $\leq |\gamma|$ which is even or odd according $|\gamma|$ is even or odd.

Proof Let us recall that $\varphi_z = \hat{T}(z)\Lambda_{\hbar}g$. In this proof we put $z_t = z$. An easy property of Weyl quantization gives

$$\Lambda_{\hbar}^{-1}\hat{T}(z)\hat{H}(t)\hat{T}(z)\Lambda_{\hbar} = Op_1^w[H(\sqrt{\hbar}X + z)], \tag{11.78}$$

where the quantization is in $X = (x, \xi) \in \mathbb{R}^{2n}$. So the Lemma follows easily from the Taylor formula with integral remainder. □

In this first step we don't take care of remainder estimates, this will be done in the next step.

From the Lemma 73 we can write

$$\hat{H}(t)\mathscr{I}(a,\Phi) \sim \mathscr{I}(b,\Phi), \quad \text{where}$$
$$b \sim \sum_{\gamma} \frac{\hbar^{|\gamma|/2}}{\gamma!} \partial_X^{\gamma} H(t,z_t)\Pi_{\gamma}\Big(\frac{x-q_t}{\sqrt{\hbar}}\Big)a. \tag{11.79}$$

We have

$$\Pi_{\gamma}(x) = \sum_{\beta\leq\gamma} h_{\gamma,\beta}x^{\beta}. \tag{11.80}$$

A direct computation, using Hamilton equation for the classical flow, gives

$$\begin{aligned} b \sim H(t,z_t)a + (\partial_q H(t,z_t) + i\partial_p H(t,z_t))\cdot(x-q_t)a + \\ \hbar\Big(E\big(\frac{x-q_t}{\sqrt{\hbar}}\big)\cdot\big(\frac{x-q_t}{\sqrt{\hbar}}\big) + \mathrm{Tr}(K-iL)\Big)a \end{aligned} \tag{11.81}$$

where we denote $\partial^2_{X,X}H(t,X)$ the Hessian matrix of $H(t)$,

$$\partial^2_{X,X}H(t,z_t) = \begin{pmatrix} G & L \\ L & K \end{pmatrix}, \quad E = G + 2iL - K. \tag{11.82}$$

with $G := \partial^2_{q.q}H(t,z_t)$, $L := \partial^2_{q,p}H(t,z_t)$, $K := \partial^2_{p,p}H(t,z_t)$.
Reacall that the stability matrix $F_t = \begin{pmatrix} A_t & B_t \\ C_t & D_t \end{pmatrix}$ satisfies
$\dot{F}_t = J\partial^2_{X,X}H(t,z_t)F_t$, $F_{t=0} = \mathbb{1}$.
It turns out that the power of $(x-q_t)$ can be transformed into power of $\hbar$, using the following Lemma:

Lemma 74 *Let us denote $M(t,z) = (C_t - B_t) - i(A_t + D_t)$. We have*

$$\hbar(\partial_q - i\partial_p)e^{\frac{i}{\hbar}\Phi} = iM(t,z)^T(x-q_t)e^{\frac{i}{\hbar}\Phi}, \quad \text{with} \tag{11.83}$$

$$\big|\det M(t,z)\big| \geq 2^{-n}. \tag{11.84}$$

In particular $M(t,z)$ is invertible.

Proof For simplicity, let us forget the lower index t.
Let us consider the $2d \times 2d$ matrix

$$\begin{aligned} \mathbb{1} + F + iJ(\mathbb{1} - F) &= \begin{pmatrix} \mathbb{1} + A - iC & B + i(\mathbb{1} - D) \\ C - i(\mathbb{1} - A) & \mathbb{1} + D + iB \end{pmatrix} \\ &= \begin{pmatrix} \mathbb{1} + A - iC & -i(D+iB) + i \\ i(A - iC) & \mathbb{1} + D + iB \end{pmatrix}. \end{aligned} \tag{11.85}$$

A linear algebra computation gives

$$\det(\mathbb{1} + F + iJ(\mathbb{1} - F)) = 2^n \det\big(A + D + i(B - C)\big). \tag{11.86}$$

Using that F is symplectic, we get

$$\begin{aligned}&(\mathbb{1} + F + iJ(\mathbb{1} - F))^*(\mathbb{1} + F + iJ(\mathbb{1} - F))\\ &= (\mathbb{1} + F^\tau)(\mathbb{1} + F) + (1 - F^\tau)(\mathbb{1} - F) \geq \mathbb{1}_{2n}\end{aligned} \tag{11.87}$$

hence (11.84) follows.
Let us recall classical computations for the derivatives of the action

$$\partial_q S = (\partial_q q_t)^T p_t - p, \tag{11.88}$$
$$\partial_p S = (\partial_p q_t)^T p_t. \tag{11.89}$$

Then we can compute $\partial_q \Phi$, $\partial_p \Phi$ and we get (2.45).
Let us prove now (11.115).
We know that $\Im \Gamma_t(s)$ is an Hermitian non negative matrix. □

Integrating by parts using Lemma 74, we get

$$(i\hbar\partial_t - \hat{H}(t))\mathscr{I}(a, \Phi) \asymp \mathscr{I}(f, \Phi) \tag{11.90}$$

where

$$\begin{aligned}f &\sim i\hbar\big(\partial_t a - \frac{1}{2}\mathrm{Tr}\big(\dot{M}M^{-1}\big)a\\ &+ \sum_{|\gamma|\geq 3} \frac{\hbar^{|\gamma|/2-1}}{\gamma!}\partial_X^\gamma H(t, z_t)\Pi_\gamma\Big(\frac{x - q_t}{\sqrt{\hbar}}\Big)a\end{aligned} \tag{11.91}$$

Hence using the Liouville formula, we get the first term

$$a_0(t, z) = 2^{n/2}\det{}^{1/2}\big(iM(t, z)\big) \tag{11.92}$$

We shall obtain the next terms a_j by successive integrations by parts. This is solved more explicitly with the following Lemma.

Lemma 75 *For any symbol $b \in S_0^0$, and every multiindex $\alpha \in \mathbb{N}^{2d}$ we have*

$$\begin{aligned}&\int_{\mathbb{R}^{2n}} (x - q_t)^\alpha e^{\frac{i}{\hbar}\Phi} b(z)dz =\\ &\sum_{\frac{|\alpha|}{2}\leq|\beta|\leq|\alpha|} \hbar^{|\beta|} \sum_{|\sigma|\leq 2|\beta|-|\alpha|} \int_{\mathbb{R}^{2n}} f_{\alpha,\beta,\sigma}(t, z)e^{\frac{i}{\hbar}\Phi}\partial_z^\sigma b(z)dz\end{aligned} \tag{11.93}$$

where $f_{\alpha,\beta,\sigma}(t,z)$ are symbols of order 0, uniformly bounded in S^0_0 on bounded time intervals. They only depend on the classical flow $\phi^t(z)$ and its derivatives.

The Lemma is easily obtained by induction on $|\alpha|$ using Lemma 74 □

Now, to determine the transport equation, we solve inductively on $j \geq 0$, the equations

$$(i\hbar\partial_t - \hat{H}(t))\mathscr{I}\Big(\sum_{0\leq k\leq j+1} \hbar^k a_k(t), \Phi\Big) = O(\hbar^{j+2}). \tag{11.94}$$

Reasoning by induction on $j \geq 0$, we get the transport equation for $a_{j+1}(t)$ by cancellation of the coefficient of $\hbar^{j+1}$ in (11.118).

$$\partial_t a_{j+1}(t,z) = \frac{1}{2}\mathrm{Tr}\big(\dot{M}M^{-1}\big)a_{j+1}(t,z) + b_j(t,z), \quad a_{j+1}(0,z) = 0, \tag{11.95}$$

where

$$b_j(t,z) = \sum_{|\alpha|+2k\leq 2(j+2)} F_{j,k,\alpha}(t,z)\partial_z^\alpha a_k(t,z). \tag{11.96}$$

Moreover, $F_{j,k,\alpha}(t,z)$ depends only on the classical flow $\phi^t(z)$ and its derivatives and satisfies, for all $z \in \mathbb{R}^{2n}$, $|t - t_0| \leq T$,

$$|\partial_z^\gamma F_{j,k,\alpha}(t,z)| \leq C_{j,k,\alpha,\gamma}(T). \tag{11.97}$$

So we get, for every $j \geq 0$,

$$a_{j+1}(t,z) = \int_0^t \det{}^{1/2}\big(M(t,z)M(s,z)^{-1}\big)b_j(s,z)ds. \tag{11.98}$$

11.3.3 Error Estimates on Finite Time Intervals

Let us denote

$$\mathscr{R}_N(t) = (i\hbar\partial_t - \hat{H}(t))\mathscr{I}\Big(a^{(N)}(t), \Phi\Big) \tag{11.99}$$

where $a^{(N)}(t) = \sum_{0\leq k\leq N} \hbar^k a_k$. Using Duhamel formula we have

$$\|\mathscr{U}^\hbar(t) - \mathscr{U}^{N,\hbar}(t)\| \leq \hbar^{-1}\int_0^t \|\mathscr{R}_N(s)\|ds \tag{11.100}$$

where $t_0 = 0$, $\mathscr{U}^\hbar(t) = \mathscr{U}^\hbar(t,0)$, $\mathscr{U}^{N,\hbar}(t) = \mathscr{I}\Big(a^{(N)}(t), \Phi\Big)$.

So we have to estimate $\|\mathscr{R}_N(t)\|$. Let us denote $K^{\hbar,(N)}(x,y)$ the Schwartz kernel of $\mathscr{U}^{N,\hbar}(t)$ and $\tilde{K}_R^{(N)}(X,Y)$ the Schwartz kernel of $\mathscr{R}_N(t)$ in the Fourier-Bargmann representation :

$$\tilde{K}^{(N)}(X,Y) = \iint_{\mathbb{R}^d\times\mathbb{R}^d} K^{(N)}(x,y)\varphi_X(y)\overline{\varphi_Y(x)}dxdy. \tag{11.101}$$

Let $\tilde{\mathscr{R}}_N(t)$ be the operator with Schwartz kernel $\tilde{K}^{(N)}(X,Y)$. The following Lemma is well known.

Lemma 76 *We have the following L^2 norm estimate*

$$\|\mathscr{R}_N(t)\|_{L^2(\mathbb{R}^n)} \le (2\pi\hbar)^{-n}\|\tilde{\mathscr{R}}_N(t)\|_{L^2(\mathbb{R}^{2n})}. \tag{11.102}$$

In particular we have

$$\|\mathscr{R}_N(t)\|_{L^2(\mathbb{R}^n)} \le$$
$$(2\pi\hbar)^{-n}\max\left\{\sup_Y\int|\tilde{K}_R^{(N)}(X,Y)|dX, \sup_X\int|\tilde{K}_R^{(N)}(X,Y)|dY\right\}. \tag{11.103}$$

Proof For inequality (11.102) we use that the Fourier-Bargmann transform is an isometry.
Inequality (11.103) is known as Carleman (or Schur) L^2 estimate. □

Using Lemma 73 we get

$$\tilde{K}^{(N)}(X,Y) = 2^{-3n/2}(\pi\hbar)^{-n}\int_{\mathbb{R}^{2n}}\left\langle\hat{T}(z_t)\Lambda_\hbar \mathrm{Op}_1^w[R_N(t)]g,\varphi_Y\right\rangle\langle\varphi_X,\varphi_z\rangle\cdot$$
$$\cdot a^{(N)}(t,z)e^{\frac{i}{\hbar}\delta(t,z}dz,$$

where $\delta(t,z) = S(t,z) + \frac{p\cdot q - p_t\cdot q_t}{2}$.
Using Weyl commutation formula we have

$$\langle\varphi_X,\varphi_z\rangle = \exp\Big(-\frac{|X-z|^2}{4\hbar} + \frac{i}{2\hbar}\sigma(X,z)\Big) \tag{11.104}$$

$$\left\langle\hat{T}(z_t)\Lambda_\hbar Op_1^w[R_N(t)]g,\varphi_Y\right\rangle = \left\langle Op_1^w[R_N(t)]g, g_{\frac{Y-z_t}{\sqrt{\hbar}}}\right\rangle. \tag{11.105}$$

We know the Wigner function $W_{0,Z}$ of the pair (g, g_Z), $Z\in\mathbb{R}^{2n}$ (see Chap. 2)

$$W_{0,Z}(X) = 2^{2n}\exp\Big(-|X-\frac{Z}{2}|^2 - i\sigma(X,Z)\Big) \tag{11.106}$$

By a well known property of Weyl quantization [104], for any symbol s, we have

$$\langle Op_1^w[s]g, g_Z\rangle = (2\pi)^{-n}\int_{\mathbb{R}^{2n}} s(X)W_{0,Z}(X)dX \tag{11.107}$$

We shall use the following Lemma

Lemma 77 *Let f be a symbol, $f \in S^0_0(2n)$. For every $\gamma \in \mathbb{N}^{2n}$ and $m > 0$ there exists $C_{\gamma,m}$ such that*

$$\left|\int_{\mathbb{R}^{2n}} X^\gamma f(X) e^{-|X-Z|^2 - iJZ\cdot X} dX\right| \le C_{\gamma,m}(1+|Z|)^{-m} \sup_{|\alpha|\le m+|\gamma|;\ Y\in\mathbb{R}^{2d}} |\partial_Y^\alpha f(Y)| \tag{11.108}$$

Proof It is enough to assume $|Z| \ge 1$. We integrate m times by parts with the differential operator

$$\mathscr{L} = \frac{2(X-Z) - iJZ\cdot X}{4|X-z|^2 + |JZ|^2}\partial_X \tag{11.109}$$

using that $(\mathscr{L}^\tau)^m = \sum_{|\alpha|\le m} l_{m,\alpha}\partial_X^\alpha$, with $|l_{m,\alpha}| \le C_{m,\alpha}(|Z| + |X-Z|)^{-m}$, where $\theta(X) = -|X-Z|^2 - iJZ\cdot X$. □

So using Lemma 77 we get the following estimate : for every N, N' there exists $C_{N,N'}$ (depending only on semi-norms $|H(t)|_{\infty,\gamma}$, $2 \le |\gamma| \le N+N'$, such that for $X, Y \in \mathbb{R}^{2d}$ and $|t| \le T$ we have

$$|\tilde{K}^{(N)}(X,Y)| \le C_{N,N'}(\mu(T))^{N+N'}\hbar^{\frac{N+1}{2}-n}\int_{\mathbb{R}^{2n}} e^{-\frac{|X-z|^2}{4\hbar}}\left(1+\frac{|Y-z_t|}{\sqrt{\hbar}}\right)^{-N'} \times |a^{(N)}(t,z)|dz. \tag{11.110}$$

Let us denote $\phi^{*t} = \phi^{0,t} = (\phi^t)^{-1}$. We have the Lipchitz estimate, for $|t| \le T$,

$$|\phi^{*,t}Y - z| \le \mu(T)|Y - z_t| \tag{11.111}$$

So we get

$$\left|\int_{\mathbb{R}^{2n}} e^{-\frac{|X-z|^2}{4\hbar}}\left(1+\frac{|Y-z_t|}{\sqrt{\hbar}}\right)^{-N'} dz\right| \le C_{N'}\left(1+\frac{|\phi^{t*}Y - X|}{\mu(T)\sqrt{\hbar}}\right)^{-N'} \tag{11.112}$$

and

$$|\tilde{K}^{(N)}(X,Y)| \le C_{N,N'}(\mu(T))^{N+N'}\hbar^{\frac{N+1}{2}}\left(1+\frac{|\phi^{t*}Y-X|}{\mu(T)\sqrt{\hbar}}\right)^{-N'} \times \sup_{z\in\mathbb{R}^{2n},|t|\le T}|a^{(N)}(t,z)| \tag{11.113}$$

Then using Lemma 76 and choosing $N' > 2d$, we get the following uniform L^2 estimate for the remainder term, for $|t| \le T$,

$$\|\mathscr{R}_N(t)\| \le C_N(\mu(T))^{N+1}\hbar^{(N+1)/2} \sup_{z\in\mathbb{R}^{2n},|t|\le T}|a^{(N)}(t,z)|. \tag{11.114}$$

If T is fixed, pushing the expansion up to $2N$ instead of N we get easily Theorem 57 using Duhamel formula. □

11.4 Large Time Estimates

In this section we revisit the estimates proved in the previous two sections under the assumptions of Theorem 58, in particular using assumption (11.51).

Lemma 78 *Let* $L(t,z) = \left(iM(t,z)^T\right)^{-1}$. *For any* $\sigma \in \mathbb{N}^n$ *we have*

$$\|\partial_z^\sigma L(t,z)\| \le C_\sigma \mu(T)^{(n+1)|\sigma|+n-1}, \quad \forall t \in [-T,T], \forall T > 0. \tag{11.115}$$

Proof We use Lemma 22 of the book [67]. We have with the notation of [67], $M = iY$. So we have $M^*M \ge 4\mathbb{1}$. Then we get

$$\|M^{-1}(t,z)\| \le \frac{1}{2}$$

So for any multi-index σ we have

$$\|\partial_z^\sigma M^{-1}\| \le C_\alpha \mu(T)^{2|\sigma|}, \quad \forall |t| \le T.$$

□

Lemma 79 *For any symbol* $b \in S_0^0$, *and every multiindex* $\alpha \in \mathbb{N}^{2n}$ *we have*

$$\int_{\mathbb{R}^{2n}} (x-q_t)^\alpha e^{\frac{i}{\hbar}\Phi} b(z)dz = \sum_{\frac{|\alpha|}{2}\le|\beta|\le|\alpha|} \hbar^{|\beta|} \sum_{|\sigma|\le 2|\beta|-|\alpha|} \int_{\mathbb{R}^{2n}} f_{\alpha,\beta,\sigma}(t,z) e^{\frac{i}{\hbar}\Phi} \partial_z^\sigma b(z)dz. \tag{11.116}$$

Then we have for any $T > 0$ *and* $|t| \le T$, *we have*

$$|\partial_z^\nu f_{\alpha,\beta,\sigma}(t,z)| \le C_{\alpha,\beta,\sigma,\nu} \mu(T)^{2(|\alpha|-|\sigma|+|\nu|)}. \tag{11.117}$$

Proof The Lemma is easily obtained by induction on $|\alpha|$ using Lemma (11.115).

□

Now, to determine the transport equation, we solve inductively on $j \ge 0$, the equation

$$(i\hbar\partial_t - \hat{H}(t))\mathscr{I}\Big(\sum_{0\le k\le j+1} \hbar^k a_k(t), \Phi\Big) = O(\hbar^{j+2}) \tag{11.118}$$

Reasoning by induction on $j \ge 0$, we get the transport equation for $a_{j+1}(t)$ by cancellation of the coefficient of $\hbar^{j+1}$ in (11.118).

$$\partial_t a_{j+1}(t,z) = \frac{1}{2}\mathrm{Tr}\big(\dot{M}M^{-1}\big)a_{j+1}(t,z) + b_j(t,z), \quad a_{j+1}(0,z) = 0, \tag{11.119}$$

where

$$b_j(t,z) = \sum_{|\alpha|+2k\le 2(j+2)} F_{j,k,\alpha}(t,z)\partial_z^\alpha a_k(t,z). \tag{11.120}$$

Moreover, $F_{j,k,\alpha}(t,z)$ depends only on the classical flow $\phi^t(z)$ and its derivatives and satisfies

$$|\partial_z^\gamma F_{j,k,\alpha}(t,z)| \le C_{j,k,\alpha,\gamma}\mu(T)^{2(j-k+2)+|\gamma|-|\alpha|} \tag{11.121}$$

where $C_{j,k,\alpha,\gamma}$ only depends on $\sup_{|t|\le T} |H(t)|_{\infty,\gamma}$, $2 \le |\gamma| \le j+2$.

So we get, for every $j \ge 0$,

$$a_{j+1}(t,z) = \int_0^t \det{}^{1/2}\big(M(t,z)M(s,z)^{-1}\big)b_j(s,z)ds \tag{11.122}$$

Moreover, from (11.120) and (11.121), we get, by induction on j, the following estimate, for every $j \ge 0$, $|t| \le T$, $z \in \mathbb{R}^{2n}$,

$$|\partial_z^\nu a_j(t,z)| \le C_{j,\gamma}|\det{}^{1/2} M(t,z)|\mu(T)^{8j+2|\nu|} \tag{11.123}$$

with the same remark as in (11.121) for the constant $C_{j,\gamma}$. Now revisiting Sects. 11.3.2 and 11.4 we get Theorem 58. □

11.5 Application to the Aharonov-Bohm Effect

11.5.1 The Van-Vleck Formula

This is one of the oldest mathematical result (1928) in quantum mechanics. It is often deduced using Feynman integral. Here we deduce it using Theorem 56. The goal is to get a point-wise approximation for the Schwartz kernel $\mathscr{K}^\hbar(t,t_0,x,y)$, as in Theorem 56. In order to apply the stationary phase theorem with complex phase ([153], Theorem 7.7.5) we introduce the following assumption.

(V-V). There exist a finite number m of classical trajectories of the Hamilton flow ϕ^{t,t_0} connecting in the configuration space the initial data (t_0, y) to the final data (t, x) such that (t_0, y) and (t, x) are not conjugate. That means there exist m momenta $p^{(j)}$, $1 \le j \le m$ such that

$$q_{t,t_0}(y, p^{(j)}) = x, \quad \text{and}$$
$$\det\left(\frac{\partial q_{t,t_0}}{\partial p}(y,p^{(j)})\right) \neq 0\,, \quad 1 \le j \le m. \tag{11.124}$$

Theorem 59 *(i) If* $q_{t,t_0}(y, p^{(j)}) \neq 0$, $\forall p \in \mathbb{R}^n$ *then*

$$\mathscr{K}^\hbar(t, t_0, x, y) = O(\hbar^\infty).$$

If assumption **(V-V)** *is fulfilled then we have the following Van-Vleck formula*

$$\mathscr{K}(t, t_0, x, y) = (2i\pi\hbar)^{-n/2} \sum_{1\le j\le m} \mathrm{e}^{-i\frac{\pi}{2}\sigma_j} |\mathrm{det}\partial^2_{x,y} S(t, t_0; y, p^{(j)})|^{1/2} +$$
$$\sum_{1\le j\le m, k\ge 0} \hbar^{1+k/2} c_{j,k}(t, t_0; x, y) + O(\hbar^\infty). \tag{11.125}$$

The asymptotic expansion is in the sense of L^2*-norm of operators.* S *is the classical action between the data* (t_0, y) *and* (t, x). $\sigma_j \in \mathbb{Z}$ *is the Maslov index of the path* (j). *Moreover if* $\hat{H}(t)$ *is an elliptic operator (like for* $\hat{H}(t) = -\hbar^2\Delta + V(t, x)$ *then the asymptotic expansion is true in the* L^∞*-norm in the variables* (x, y).

Proof This is a direct application of the stationary phase theorem ([153], Theorem 7.7.5) to the oscillating integral (11.38) or (11.48). We have to compute the leading term, which is known as the Van-Vleck semi-classical approximation.
In (11.38) we have

$$\Im\Phi_\Gamma = 0 \Longleftrightarrow \{q = y,\ q_t = x\}.$$

Then we have to compute $\partial_z\Phi_\Gamma$ and the Hessian matrix $\partial^2_{zz}\Phi_\Gamma$ at the critical points.
We compute the first derivatives of the classical action S in (90).

$$\partial_p S(t, t_0; q, p) = \left(\frac{\partial q_{t,t_0}}{\partial p}\right)^T p_{t,t_0}, \tag{11.126}$$

$$\partial_q S(t, t_0; q, p) = \left(\frac{\partial q_{t,t_0}}{\partial q}\right)^T p_{t,t_0} - p. \tag{11.127}$$

Then we get

$$\partial_q\Phi_\Gamma = C^T(x - q_{t,t_0} - A^T\Gamma(x - q_{t,t_0}) - i(y - q) \tag{11.128}$$
$$\partial_q\Phi_\Gamma = D^T(x - q_{t,t_0} - B^T\Gamma(x - q_{t,t_0}) - (y - q) \tag{11.129}$$

where

$$A = \frac{\partial q_{t,t_0}}{\partial q},\ B = \frac{\partial q_{t,t_0}}{\partial p}, C = \frac{\partial p_{t,t_0}}{\partial q},\ D = \frac{\partial p_{t,t_0}}{\partial p}. \tag{11.130}$$

Applying the implicit function theorem in (11.124), we get that the phase Φ_Γ has a finite number m of critical points $z^{(j)} = (y, p^{(j)})$, with $q = y$, $p = p^{(j)}(t, t_0; x, y)$,

$1 \le j \le m$. Now we show that these critical points are non degenerate by computing $\partial^2_{zz}\Phi_\Gamma$ for $z = z^{(j)}$.
Using (11.126) and that F_{t,t_0} is symplectic, we get

$$\partial^2_{zz}\Phi_\Gamma(t, t_0; z^{(j)}, x, y) = \begin{pmatrix} 2i\mathbb{1} + (A+iB)^{-1}B & i(A+iB)^{-1}B \\ i(A+iB)^{-1}B & -(A+iB)^{-1}B \end{pmatrix}. \quad (11.131)$$

Notice that A, B are computed for $(t, t_0, z = z^{(j)})$.
The determinant is easily computed on these points:

$$\det(\partial^2_{zz}\Phi_\Gamma) = \det(-2i\partial^2_{zz}\Phi_\Gamma(A+iB)^{-1}B). \quad (11.132)$$

Using assumption **(V-V)** we get that $\det(\partial^2_{zz}\Phi_\Gamma) \neq 0$. Hence the stationary phase theorem can be applied.
Moreover we have to compute the correct phase for the square root $\det^{-1/2}(\partial^2_{zz}\Phi_\Gamma)$ on critical points (analytic continuation and continuity (see [153], section VII).
Using the implicit function theorem we get on the critical points

$$\partial^2_{x,y}S = -B^{-1}.$$

□

Remark 67 If $\partial_p H(q,p)$ is invertible (like for $H(q,p) = |p|^2 + V(q)$) then for $|t - t_0| \le \delta$ with $\delta > 0$ small enough then **(V-V)** is fulfilled with $m = 1$ and then the Maslov index is 0.

11.5.2 A Setting for the Time-Dependent Aharonov-Bohm Effect

The Aharonov-Bohm effect is a purely quantum effect in the sense that a quantum particle may feel the effect of a magnetic potential even if the field vanishes in a neighborhood of the classical trajectory of the particle. This can be checked on the propagator as a consequence of the Van-Vleck formula proved above.

Let us consider a two dimensional Schrödinger Hamiltonian

$$\hat{H}_{\kappa\vec{a},V}(t) = \frac{1}{2}(\hbar D_x - \kappa\vec{a}(x))^2 + V(t,x), \ \vec{a} = (a_1, a_2).$$

The magnetic field $\vec{b}$ is perpendicular to the plane $\mathbb{R}^2_x$, $x = (x_1, x_2)$ and so is identified with the scalar function

$$b(x) = \partial_{x_2}a_1(x) - \partial_{x_1}a_2(x).$$

For simplicity we assume that, for $|\alpha| \geq 1$, $\partial_x a(x)$ is bounded on $\mathbb{R}^2$ and for $|\alpha| \geq 2$, $\partial_x V(t,x)$ is bounded on $\mathbb{R}^2$. Under these assumptions the propagator for $\hat{H}_{\vec{a},V}(t)$ is well defined as a unitary operator on $L^2(\mathbb{R}^2)$ (see the proof of Proposition 123).

Our geometric setting is the following (see also the picture).

(**GAB**). There exist exactly two classical trajectories in the configuration space connecting the initial data (s, y) to the final data (t, x). More precisely, the classical flow $(y, p) \mapsto q(t, s, y, p), p(t, s, y, p)$ is such that the equation $q(t, s, y, p) = x$ has exactly two solutions in $p = p^{1,2} = p^{(1,2)}(t, s, x, y)$.

Thus in the plane configuration space the two paths $\tau \mapsto q(\tau, s, y, p^{1,2}, s \leq \tau \leq t$ are the oriented boundary $\gamma := \gamma_1 \cup \gamma_2$ of a bounded open domain Ω.

We assume that the support of the magnetic field b is inside Ω. We denote by $\varphi(b)$ the magnetic flux through the domain Ω. Moreover we put a filter Ψ to prevent any classical particle starting from y to penetrate in supp(b). For that we assume that $\Psi \in C_0^\infty(\mathscr{V})$ where $\mathscr{V}$ is a small open set such that $p^1, p^2 \in \mathscr{V}$ and $\Psi(p^{1,2}) = 1$.

Let us denote $\mathscr{K}^{\hbar,\Psi}_{H_{\kappa a,V}}(t, s; x, y)$ the Schwartz kernel of $\mathscr{U}^{\hbar}_{H_{\kappa a,V}}(t, s)\Psi(\hbar D_x)$ and by $S_{\kappa a,V}(\gamma)$ the classical action along the oriented path γ for the Hamiltonian $\hat{H}_{\kappa\vec{a},V}(t)$.

The magnetic Aharonov-Bohm effect

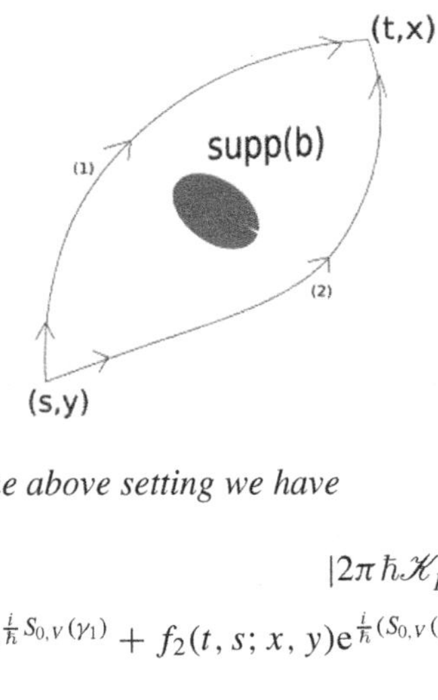

Theorem 60 *Under the above setting we have*

$$|2\pi\hbar\mathscr{K}^{\hbar,\Psi}_{H_{\kappa a,V}}(t, s; x, y)|^2 =$$
$$|f_1(t, s; x, y)e^{\frac{i}{\hbar}S_{0,V}(\gamma_1)} + f_2(t, s; x, y)e^{\frac{i}{\hbar}(S_{0,V}(\gamma_2)-\kappa\varphi(b))}|^2 + \mathscr{O}(\hbar) \quad (11.133)$$

where $f_{1,2}(t, s; x, y) = |\det \partial^2_{x,y}S(t, s; y, p^{1,2})|^{1/2}e^{-i\frac{\pi}{2}\sigma_{1,2}}$ *and* $\sigma_{1,2}$ *a Maslov index like in Sect. 11.5.1.*

Proof This is a direct consequence of Theorem 11.5.1 by computing carefully the classical actions. Let us recall here this well known computation.

Let us introduce the Lagrangian $\mathscr{L}_{\kappa a,V}$ for the electric potential $(\kappa a, V)$. We have

$$\mathscr{L}_{\kappa a,V}(q, \dot{q}) = \mathscr{L}_{0,V}(q, \dot{q}) + \kappa a(q) \cdot \dot{q}.$$

The equation of motion of the classical particle is

$$\ddot{a} = \partial_q V(q) + \kappa b(q) \cdot \dot{q}.$$

Then we have for the actions

$$S_{\kappa a, V}(\gamma) = S_{0,V}(\gamma) + \kappa \int_\gamma a dq.$$

Then using Stokes formula we get the flux:

$$\int_{\gamma_1} a dq = \int_{\gamma_2} a dq - \varphi(b).$$

□

Remark 68 We can compute more explicitly the Aharonov-Bohm effect as follows. Denote $\rho_{1,2} = |f_{1,2}|$ (they are not null). Then we have

$$|2\pi\hbar \mathscr{K}_{H_{\kappa a,V}}^{\hbar,\Psi}(t,s;x,y)|^2 - |2\pi\hbar \mathscr{K}_{H_{0,V}}^{\hbar,\Psi}(t,s;x,y)|^2 = 2\rho_1\rho_2(\sin\left(\theta - \frac{\kappa\varphi(b)}{\hbar}\right) - \sin\theta) \tag{11.134}$$

where $\theta = \frac{\pi}{2}(\sigma_2 - \sigma_1) + \frac{S_1 + -S_2}{\hbar}$.
To be more specific assume for example that $\sigma_1 = \sigma_2$ and $S_1 = S_2$ and that the magnetic flux is quantized: $\kappa\varphi(b) = \frac{\pi\hbar}{2}$. Then we have

$$|2\pi\hbar \mathscr{K}_{H_{\kappa a,V}}^{\hbar,\Psi}(t,s;x,y)|^2 = |2\pi\hbar \mathscr{K}_{H_{0,V}}^{\hbar,\Psi}(t,s;x,y)|^2 + 2\rho_1\rho_2.$$

So even if the classical particle never meet the magnetic field the motion of corresponding quantum particle is modified by the field.

A similar phenomenon can be observed by perturbing a quantum system by a *time dependent electric potential*.

We construct a perturbation of the electric potential $V(t,q)$ as follows. Let

$$W(s,q) = \chi_1(s)Y_1(q) + \chi_2(s)Y_2(q).$$

Let us consider thin tubular open neighborhoods $\mathbf{T}_{1,2}$, $\widetilde{\mathbf{T}_{1,2}}$, of the pieces $\{q(\tau,s,y,p^\pm);\ t_1 \le \tau \le t_2\}, s < t_1 < t_2 < t$, of path $\gamma_\pm$, such that $\overline{\mathbf{T}_{1,2}} \subseteq \widetilde{\mathbf{T}_{1,2}}$. Let us choose $\chi_{1,2} \in C_0^\infty]t_1,t_2]$, $Y_{1,2} \in C_0^\infty(\widetilde{\mathbf{T}_{1,2}})$, $Y_{1,2} = 1$ on $\mathscr{T}_{1,2}$. Then we have:

Theorem 61 *Under the above assumptions we have*

$$|2\pi\hbar\mathcal{K}^{\hbar,\Psi}_{H_{V+\kappa W}}(t,s;x,y)|^2 =$$
$$\left|f_+ e^{\frac{i}{\hbar}S_{0,V}(\gamma_1)} + f_- e^{\frac{i}{\hbar}S_{0,V}(\gamma_2)-\kappa\delta}\right|^2 + \mathcal{O}(\hbar), \tag{11.135}$$

where $\delta = \int_s^t(\chi_2(\tau)-\chi_1(\tau))d\tau$.

The electric Aharonov-Bohm effect

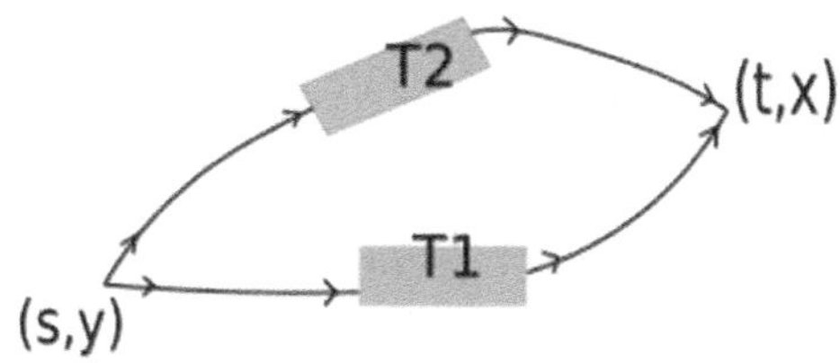

Notice that $\partial_q W(s,q) = 0$ in a neighborhood of $\gamma_{1,2}$. So the classical trajectories never see the perturbation of the electric field. Nevertheless we also have here an electric Aharonov-Bohm effect, where the magnetic flux is replaced by the difference of voltage $\kappa\delta$ between the tubes $\mathbf{T}_1$ and $\mathbf{T}_2$.

11.6 Application to the Spectrum of Periodic Hamiltonians Flow

In this section we shall give a sketchy alternative proof of the Theorem 32, Chap. 5., using coherent states and the Herman-Kluk approximation.

We assume for simplicity that H is subquadratic (after cut-off it is not a restriction), that $H^{-1}[E-\eta, E+\eta]$ is compact and the Hamiltonian flow Φ_H^t is 2π-periodic and 2π is a primitive period on $\Sigma_e := H^{-1}(e)$ for $E-\eta \le e \le E+\eta$. Moreover Σ_e is supposed to be connected.

Let us choose an energy cutoff $\chi \in C_0^\infty(]E-\eta, E+\eta[)$, $\chi = 1$ on $[E-\eta/2, E+\eta/2]$.

Theorem 62 *Let* $\sigma(\hbar) = \frac{\mu}{4} + \frac{\lambda_E}{\hbar}$ *(*μ *is the Maslov index of the periodic orbits* $\gamma_E \subset \Sigma_E$ *and* $\lambda_E = \frac{1}{2\pi}\oint_{\gamma_E} pdq$*). Then we have*

$$\chi(\hat{H})e^{-i\frac{2\pi}{\hbar}(\hat{H}-\sigma(\hbar))} = \chi(\hat{H})(\mathbb{1} + \hbar R_1^\hbar) \tag{11.136}$$

where $\|R_1^\hbar\| = \mathcal{O}(1)$ *as* $\hbar \to 0$.

Proof This is a direct application of Theorem 57 using that $\Phi_H^{2\pi} = \mathbb{1}_{\mathbb{R}^{2n}}$. The Maslov index is obtained by computing the argument of a complex determinant associated to the monodromy matrix of the flow. □

A s a consequence we deduce the following well known clustering property for the spectrum of periodical classical systems[142].

Corollary 34

$$\boxed{\mathrm{Spect}[\hat{H}] \cap [E-\eta/2, E+\eta/2] \subseteq \bigcup_{k\in\mathbb{Z}} I(k,\hbar)},$$

where

$$I(k,\hbar) = [(k+\frac{\mu}{4})\hbar + \lambda - C\hbar^2, (k+\frac{\mu}{4})\hbar + \lambda + C\hbar^2] \text{ with } C>0 \text{ large enough.}$$

Proof For $\hbar > 0$ small enough we can define the bounded operator $\hat{L} = \frac{\log(\mathbb{1}+\hbar\hat{R}_1^{\hbar})}{2i\pi}$. So we have

$$\chi(\hat{H})\mathrm{e}^{-i\frac{2\pi}{\hbar}(\hat{H}-\sigma(\hbar)+\hat{L}^{\hbar})} = \chi(\hat{H}). \qquad (11.137)$$

The two compact self adjoint operators $\chi(\hat{H})\hat{H}$ and $\chi(\hat{H})\hat{L}$ commutes so they have a joint basis $\{\psi_j\}_{1\le j\le M_\hbar}$ of eigenfunctions in the spectral subspace Range $[\mathbb{1}_{[E-\eta,E+\eta]}(\hat{H})$:

$$\hat{H}\psi_j = \lambda_j\psi_j$$
$$\hat{L}\psi_j = \nu_j\psi_j.$$

Now using (11.137) and $\|\hat{L}\| = \mathscr{O}(\hbar)$ we get for $\lambda_j \in [E-\eta/2, E+\eta/2]$ that there exists $k \in \mathbb{Z}$ such that

$$\lambda_j - \hbar\sigma(\hbar) + \mathscr{O}(\hbar^2) = k\hbar.$$

□

Remark 69 If the period depends on the energy e we can apply Proposition 49 of Chap. 5 to reduce to the case of constant period.

Notes

Proposition 123 is a particular case of a more general result proved in [191]. Theorem 11.113 was first proved without the small parameter $\hbar$ in [242]. The case with $\hbar \searrow 0$ consider here follows [93].

We could easily extent the results of this Chapter to multi-components systems with constant multiplicities (see Chap. 14). This is an exercise for the interested reader.

The Herman-Kluk approximation was discovered by chemical-physicists [138, 149, 164]. The first mathematical proof appeared in [227] and in a more general setting in [219].

Chapter 12
Semi-classical Measures: Stationary and Large Time Behavior

Abstract In these chapter our aim is to develop some applications of the theory of semi-classical measures (introduced in Chap. 2) to study quantum properties of ergodic classical system and also for some periodical classical systems. We have already seen that there are several connections between semi-classical measures and coherent states. We first consider quantization of classical ergodic systems and the consequence in the quantum world, a part of a domain much studied in the physicist literature under the name "quantum chaos". We also consider some quantum properties for quantizations of classical systems with periodic classical flow. We shall see that semi-classical measures are well adapted to study quantum systems in the semi-classical regime at large time, much beyond the well known Ehrenfest time.

12.1 Semi-classical Measures for Bound States

In Chap. 2 we have discussed some basic properties of the semi-classical measures associated with a family $\psi^{\hbar}$ of normalized L^2 states on $\mathbb{R}^n$.
Recall that a probability measure μ on the phase space $\mathbb{R}^{2n}$ is a semi-classical measure (for the family $\psi^{\hbar}$) if there exists a sequence $\hbar_\ell \to 0$ such that for any $A \in S_0^0$ we have

$$\lim_{\ell\to+\infty} \langle \psi^{\hbar_\ell}, \mathrm{op}^w_{\hbar_\ell}(A)\psi^{\hbar_\ell}\rangle = \int_{\mathbb{R}^{2n}} A(X)\mu(dX).$$

Recall that the mesure μ tell us where the state $\psi^{\hbar}$ is concentrate in the phase space is the semi-classical regime. In particular we shall consider this problem for bound states.

12.1.1 More on the Weyl Law

Let H be a classical Hamiltonian satisfying the assumptions (H.0) and (H.1) of Chap. 5. In particular the Liouville measure on $\Sigma_E := H^{-1}\{E\}$ is well defined

D. Robert and M. Combescure, *Coherent States and Applications in Mathematical Physics*, Theoretical and Mathematical Physics,
https://doi.org/10.1007/978-3-030-70845-0_12

$$dL_E := \frac{d\Sigma_E}{\|\nabla_z H(z)\|}.$$

Moreover there exists some $\epsilon_0 > 0$, small enough, such that the spectrum of $\hat{H}$ in the interval $I_E(\eta) := [E-\eta, E+\eta]$ is discrete (only eigenvalues of finite multiplicities). Let $N(E,\eta)$ be the number (with multiplicities) of eigenvalues of $\hat{H}$ in $I_E(\eta)$ and $\{\psi_j\}_{1\leq j\leq N(E,\eta)}$ an orthonormal family of eigenfunctions. So we have $\hat{H}\psi_j = E_j\psi_j$. Notice that E_j, ψ_j depend on $\hbar$. Moreover we have the following basic result known as the extended semi-classical Weyl law. The proof follows easily from the proof given in Chap. 5 for the Weyl law and the Gutzwiller trace formula.

Theorem 63 *With the previous notations, for any observable $A \in \mathrm{S}^0_0$, we have:*

$$\sum_{1\leq j\leq N(E,\eta)} \langle\psi_j, \hat{A}\psi_j\rangle = (2\pi\hbar)^{-n}\int_{H^{-1}[E-\eta,E+\eta]} A(X)dX + O(\hbar^{1-n}). \quad (12.1)$$

In particular the Weyl law read:

$$N(E,\eta) = (2\pi\hbar)^{-n}\mathrm{Vol}_{\mathbb{R}^{2n}}(H^{-1}[E-\eta, E+\eta]) + O(\hbar^{1-n}). \quad (12.2)$$

The following corollary will be useful. Let us consider $\hbar$-dependent energy intervals: $I(\hbar) = [E-\delta(\hbar), E+\delta(\hbar)]$, with $\lim_{\hbar\to 0}\delta(\hbar) = 0$ and $\delta(\hbar) \geq \hbar^\tau$, for some $\tau < 1$.

Corollary 35 *We have the following consequence:*

$$\lim_{\hbar\searrow 0}\frac{1}{N(E,\delta(\hbar))}\sum_{1\leq j\leq N(E,\delta(\hbar))}\langle\psi_j, \hat{A}\psi_j\rangle = \int_{\Sigma_E} A(X)L_E(dX). \quad (12.3)$$

Proof Let $F_A(\lambda) = \int_{H^{-1}[E-\eta_0,\lambda]} A(X)dX$, for $\eta_0 > 0$, $E-\eta_0 < \lambda < E+\eta_0$. F_A is smooth in a neighborhood of E and

$$\frac{d}{d\lambda}F_A(\lambda) = \int_{\Sigma_\lambda} A(X)L_\lambda(dX). \quad (12.4)$$

Using (12.4), (12.1) and (12.2) we get (12.3). □

Remark 70 In the Cesaro mean we have seen that the quantum averages of the observable $\hat{A}$ converges toward the classical average of A for the Liouville probability measure.

The behaviour of individual quantum averages $\langle\psi_j, \hat{A}\psi_j\rangle$ is much more delicate to understand. A way to do that is to consider the mapping family $A \mapsto \langle\psi_j, \hat{A}\psi_j\rangle$ as a distribution, or better if possible, as a Radon measure depending on $\hbar$. This can be done by modifying the Weyl quantization in the positive preserving quantization i.e. the anti-Wick quantization (see Chap. 2). Afterwards it is possible to study the statistic of the real numbers $\langle\psi_j, \hat{A}\psi_j\rangle$ as we shall see in this Chapter.

12.1.2 More About Semi-classical Measures

In Chap. 2 we have introduced the anti-Wick quantization $\hat{A}^{\mathrm{aw}}$ of an observable A and the Husimi function of a L^2-normalized family of states $\psi^\hbar \in L^2(\mathbb{R}^n)$: $\mathscr{H}_\psi^\hbar(X) = (2\pi\hbar)^{-n}|\langle\varphi_X, \psi\rangle|^2$.
Recall the two following useful properties (Proposition 27):
(1) For any $A \in \mathrm{S}_0^0$ we have

$$\|\mathrm{Op}_\hbar^w(A) - \mathrm{Op}_\hbar^{aw}(A)\| = \mathscr{O}(\hbar).$$

(2) For any $\psi^\hbar \in L^2(\mathbb{R}^n)$,

$$\langle\psi^\hbar, \mathrm{Op}_\hbar^{aw}(A)\psi^\hbar\rangle = (2\pi\hbar)^{-n}\int_{\mathbb{R}^{2n}} A(X)\mathscr{H}_{\psi^\hbar}(X)dX.$$

Corollary 36 *For any $\psi \in L^2(\mathbb{R}^n)$, $A \mapsto \langle\psi^\hbar, \hat{A}^{aw}\psi^\hbar\rangle$ extends to a positive Radon measure $\mu_\psi^\hbar$ on $\mathbb{R}^{2n}$. Moreover $\mu_\psi^\hbar$ has a density with respect to the Lebesgue measure on $\mathbb{R}^{2n}$ equal to the Husimi function $\mathscr{H}_\psi$ of ψ.*

Let us denote $\mu_j^\hbar = \mu_{\psi_j}^\hbar$ where $\{\psi_j\}_{1\le j\le N(E,\delta(\hbar))}$.

Then for the weak convergence of Radon measures we have

$$\lim_{\hbar\to 0}\frac{\sum_{j=1}^{N(E,\delta(\hbar))}\mu_j^\hbar}{N(E,\delta(\hbar))} = \nu_E,$$

where ν_E is the Liouville probability measure: $\nu_E = \frac{L_E}{L_E(\Sigma_E)}$.

This Corollary shows that the arithmetical averages of the quantum probability measures $\mu_j^\hbar : A \mapsto \langle\psi_j^\hbar, \hat{A}^{aw}\psi_j^\hbar\rangle$ converge towards the Liouville probability on Σ_E. We shall see in this chapter that the behavior of the individual measures $\mu_j^\hbar$ is much more intricate and depends heavily on dynamical properties of the classical flow Φ_H^t on Σ_E when the frequency set of $\psi^\hbar$ is concentrated on Σ_E. Existence of semi-classical measures follows from the following statement.

Proposition 126 *(i) Let $\{\psi^\hbar\}_{0<\hbar\le 1}$ be a bounded family in the Hilbert space $L^2(\mathbb{R}^n)$. Then there exists a subsequence of $\hbar \to 0$ and a finite positive measure μ on $\mathbb{R}^{2n}$ such that for any $A \in \mathrm{S}_0^0$ we have*

$$\lim_{\hbar\to 0}\langle\psi^\hbar, \hat{A}\psi^\hbar\rangle = \int_{\mathbb{R}^{2n}} A(X)\mu(dX).$$

Furthermore assume that $\{\psi^\hbar\}_{0<\hbar\leq 1}$ *is a family of* L^2*-normalized states in* $L^2(\mathbb{R}^n)$ *and the following condition (*$\hbar$-tightness*)*

$$\lim_{R\to+\infty}\limsup_{\hbar\to 0}\left(\int_{|X|\geq R}|\langle\varphi_X,\psi^\hbar\rangle|^2 dX\right)=0. \tag{12.5}$$

Then for all sequence of $\hbar\to 0$ *there exists a subsequence* $\hbar_\ell$ *and a* probability measure ν, *such that for all* $A\in S^0_0$ *we have*

$$\lim_{\ell\to+\infty}\langle\psi^{\hbar_\ell},\hat{A}\psi^{\hbar_\ell}\rangle=\int A(X)\nu(dX).$$

Proof (a sketch) (i) We have

$$\langle\psi^\hbar,\hat{A}\psi^\hbar\rangle=\int A(X)\mathscr{H}_{\psi^\hbar}(X)dX+\mathscr{O}(\hbar^\eta),\ \eta>0.$$

Hence $A\mapsto\int_{\mathbb{R}^{2n}}A(X)\mathscr{H}_{\psi^\hbar}(X)dX$ defines a norm bounded family of linear forms on the Banach space $C_b(\mathbb{R}^{2n})$ of bounded and continuous real functions on $\mathbb{R}^{2n}$. So this family is compact for the $\star$-weak topology on the dual space $C'_b(\mathbb{R}^{2n})$ (Alaoglou theorem). Then there exists an accumulation point μ for the measures family $\mathscr{H}_{\psi^\hbar}(X)dX$.
(ii) Using (12.5) it is not difficult to prove that ν is a probability measure.This proves the proposition. □

Remark 71 It is not difficult to prove that condition (12.5) is equivalent to the following, using localization with the harmonic oscillator $\hat{H}^\hbar_{\mathrm{os}}$,

$$\lim_{R\to+\infty}\limsup_{\hbar\to 0}\|\mathbb{1}_{[R,+\infty[}(\hat{H}^\hbar_{\mathrm{os}})\psi^\hbar\|_{L^2(\mathbb{R}^n)}=0. \tag{12.6}$$

In the literature, in different contexts, condition (12.5) or equivalently (12.6), is named tightness-condition or non-oscillating condition. This condition garanties that no mass escape to infinity.

Corollary 37 *Assume that* $\psi^\hbar$ *satisfies the Schrödinger equation*

$$\hat{H}\psi=E(\hbar)\psi_\hbar+O_{L^2}(\hbar),\ \|\psi^\hbar\|=1,$$

with $E(\hbar)\in[a,b]$ *such that*

$$\liminf_{|X|\to\infty}\mathrm{dist}(H(X),[a,b])>0.$$

Then $\psi^\hbar$ *has a semi-classical probability measure* ν. *Moreover we have* $\mathrm{supp}(\nu)\subseteq H^{-1}[a,b]$.

Other examples:

Let us consider the coherent state $\psi_\hbar:=\varphi_X$, $X\in\mathbb{R}^{2n}$. The only semi-classical mea-

sure is the Dirac mass δ_X.
If $\psi_\hbar(x) = f(x)e^{i\frac{S(x)}{\hbar}}$ (WKB states), with $f \in L^2(\mathbb{R}^n)$ with compact support and $S \in C^1(\mathbb{R}^n)$. The only semi-classical measure is the measure: $\nu = |f(x)|^2 dx \otimes \delta(\xi - \nabla S(x))$.

12.2 Semi-Classical Quantum Ergodicity

The following semi-classical version of the Schnirelmann theorem was given in [140] following the proof by Colin de Verdière [56] and Zelditch [267] on compact Riemannian manifolds with negative curvature.
Assume: **(ERG)** *The dynamical system* $\left(\Sigma_E, dL_E, \Phi_H^t\right)$ *is ergodic which means: for every continuous function* A *on* Σ_E*, we have, for almost all* $z \in \Sigma_E$:

$$\lim_{T\nearrow+\infty} \frac{1}{T}\int_0^T A(\Phi_{H_0}^t(z))dt = \int_{\Sigma_E} A d\nu_E.$$

Let us recall that ν_E is the Liouville probability measure on Σ_E.

Remark 72 Well known examples of ergodic dynamical systems are compact manifolds with strictly negative curvature.
A. Knauf (1987) has given examples of ergodic systems on the Euclidean space with Hamiltonian $H(q, p) = \frac{1}{2}(p_1^2 + p_2^2) + V(q_1, q_2)$.

Let us consider now $\hbar$-dependent energy intervals: $I_E(\hbar) = [E - \delta(\hbar), E + \delta(\hbar)]$, with $\lim_{\hbar\to 0} \delta(\hbar) = 0$, $\delta(\hbar) \geq \epsilon_2 \hbar$, for some $\epsilon_2 > 0$. Let us denote: $\Lambda(\hbar) = \{j,\ E_j(\hbar) \in I(\hbar)\}$ and $N(\hbar) = \#\Lambda(\hbar)$. Notice that the eigenvalues $E_j(\hbar)$ are repeated according their multiplicity.

The quantum signature of classical chaos is given by the following :

Theorem 64 *(*Ergodic Semi-Classical Theorem*)[140]) Under the ergodic assumption* **(ERG)***, for every* $\hbar > 0$*, there exists a subset of density one:*
$M(\hbar) \subseteq \Lambda(\hbar)$*, depending only on the Hamiltonian* $\hat{H}$*, such that :*

$$\lim_{\hbar\searrow 0}\left(\frac{\#M(\hbar)}{\#\Lambda(\hbar)}\right) = 1, \quad \text{and} \quad \lim_{[\hbar\searrow 0,\ j\in M(\hbar)]} \langle\psi_j, \hat{A}\psi_j\rangle = \int_{\Sigma_E} A d\nu_E, \quad \forall A \in \mathrm{S}_0^0. \tag{12.7}$$

In other words almost all eigenfunctions are uniformly distributed on the energy shell Σ_E.

Remark 73 The following question is still open : can we take $M(\hbar) = \Lambda(\hbar)$ in the conclusion of the above lemma, if $n \geq 2$?(it is true for $n = 1$).

We shall give two different proofs of Theorem 64. The first one is close to the proof given in [56, 267]. The second one will be given in the next section.

Proof Let us introduce the Schrödinger unitary group : $\mathcal{U}_H^\hbar(t) = \mathrm{e}^{-it\hbar^{-1}\hat{H}}$. We have

$$\langle \hat{A}\psi_j, \psi_j\rangle = \frac{1}{T}\int_0^T \langle \mathcal{U}_H^\hbar(-t)\hat{A}\,\mathcal{U}_H^\hbar(t)\psi_j, \psi_j\rangle dt. \tag{12.8}$$

Denote $\bar{A}_T(X) = \frac{1}{T}\int_0^T A(\Phi^t(X))dt$, $\hat{A}_T = \frac{1}{T}\int_0^T \mathcal{U}_H^{\hbar,*}(t)\hat{A}\,\mathcal{U}_H^\hbar(t)dt$, and $\langle A\rangle_E := \int_{\Sigma_E} AdL_E$. From Cauchy-Schwarz inequality we have:

$$\begin{aligned}\varepsilon(\hbar) &:= \frac{1}{N(E,\delta(\hbar))}\sum_{1\le j\le N(E,\delta(\hbar))} |\langle\psi_j, \hat{A}\psi_j\rangle - \langle A\rangle_E|^2 = \\ &\frac{1}{N(E,\delta(\hbar))}\sum_{1\le j\le N(E,\delta(\hbar))} |\langle\psi_j, \hat{A}_T\psi_j\rangle - \langle A\rangle_E|^2 \le \\ &\frac{1}{N(E,\delta(\hbar))}\sum_{1\le j\le N(E,\delta(\hbar))} \langle\psi_j, (\hat{A}_T - \langle A\rangle_E)^2\psi_j\rangle \end{aligned}\tag{12.9}$$

Let $B_T(X) = (\bar{A}_T(X) - \langle A\rangle_E)^2$. Using using Egorov theorem and the Moyal product rule (see Chap. 2) we get

$$\varepsilon(\hbar) \le \frac{1}{N(E,\delta(\hbar))}\sum_{1\le j\le N(E,\delta(\hbar))} \langle\psi_j, (\mathrm{Op}_\hbar^w B_T)\psi_j\rangle + \mathcal{O}_T(\hbar). \tag{12.10}$$

$$\varepsilon(\hbar) \le \int_{\Sigma_E} B_T(X)L_E(dX) + \mathcal{O}_T(\hbar). \tag{12.11}$$

Now using the ergodicity assumption we have

$$\lim_{T\to+\infty}\int_{\Sigma_E} B_T(X)L_E(dX) = 0.$$

Hence taking first T large enough then $\hbar$ small enough we get

$$\lim_{\hbar\to 0}\left(\frac{1}{N(E,\delta(\hbar))}\sum_{1\le j\le N(E,\delta(\hbar))} |\langle\psi_j, \hat{A}_T\psi_j\rangle - \langle A\rangle_E|^2\right) = 0. \tag{12.12}$$

(12.7) will be deduced from (12.12) and the Tchebichev inequality following [271].
Denote $M(\hbar) = \{j \in \Lambda(\hbar),\ |\langle\psi_j, (A - \langle A\rangle_E)\psi_j\rangle| \le \alpha(\hbar)\}$ for some suitable $0 < \alpha(\hbar)$.
Let $\Delta_j := \langle\psi_j, \hat{A}\psi_j\rangle - \langle A\rangle_E$. Then we have

$$\frac{1}{N(\hbar)}\sum_{1\le j\le N(\hbar)}\Delta_j^2 \ge \frac{\#\left(\Lambda(\hbar)\backslash M(\hbar)\right)}{N(\hbar)}\alpha(\hbar)^2$$

and we get

$$\frac{\#M(\hbar)}{N(\hbar)} \ge 1 - \frac{\varepsilon(\hbar)}{\alpha(\hbar)^2}$$

Then to conclude we choose $\alpha(\hbar)$ such that

$$\lim_{\hbar\searrow 0}\alpha(\hbar) = 0, \quad \text{and} \quad \lim_{\hbar\searrow 0}\frac{\epsilon(\hbar)}{\alpha(\hbar)^2} = 0.$$

We shall see now some consequences of ergodicity on non diagonal matrices elements: $A_{j,k} = \langle\psi_j, \hat{A}\psi_k\rangle$.
Classical ergodicity implies that almost all classical trajectory is dense in Σ_E. This implies also that the quantum spectrum is asymptotically dense in $\mathbb{R}$. More precisely we have

Theorem 65 *Assume that there exists on Σ_E a non periodical trajectory for the flow Φ^t. Then for every $c > 0$, $\delta > 0$ and every $\hbar_0 > 0$ the set:*

$$\mathcal{T}_{E,\delta} := \{\omega_{jk}(\hbar) : \ E_j(\hbar), E_k(\hbar) \in [E - c\hbar^{1-\delta}, E + c\hbar^{1-\delta}], \ 0 < \hbar \le \hbar_0\}$$

is dense in $\mathbb{R}$, where $\omega_{jk}(\hbar) = \frac{E_j(\hbar)-E_k(\hbar)}{\hbar}$.

Proof Let $f \in C_0^\infty(\mathbb{R})$ be such that $f = 0$ on $\mathcal{T}_{E,\delta}$. We have to show that $f \equiv 0$ on $\mathbb{R}$. Let us introduce $\chi \in C_0^\infty(]-c, c[)$ with $\chi(0) = 1$. Following Helton[146] we consider the operator:

$$\hat{A}_{E,f}(\hbar) = \int \hat{f}(t)\mathcal{U}(-t,\hbar)\hat{A}_E(\hbar)\mathcal{U}(t,\hbar)dt, \tag{12.13}$$

with $A_E(\hbar) = \chi(\frac{\hat{H}-E}{h^{1-\delta}})\hat{A}\chi(\frac{\hat{H}-E}{h^{1-\delta}})$, $A \in C_0^\infty(\mathbb{R}^{2n})$. By inverse Fourier transform, we also have:

$$\hat{A}_{E,f}(\hbar) = 2\pi\sum_{j,k} f(\omega_{jk}(\hbar))\chi\left(\frac{E_j(\hbar)-E}{\hbar^{1-\delta}}\right)\chi\left(\frac{E_k(\hbar)-E}{\hbar^{1-\delta}}\right)\Pi_k\hat{A}\Pi_j, \tag{12.14}$$

where Π_k is the projection on the eigenstate ψ_k. From (12.13), (12.14) we have:

$$\int \hat{f}(t)U(-t,\hbar)\hat{A}_E(\hbar)U(t,\hbar)dt = 0, \ \forall A \in S_0^0. \tag{12.15}$$

From (12.15) we would like to prove that $f \equiv 0$ by going to the classical limit $\hbar \searrow 0$. To do that we first use the semi-classical propagation theorem (Sect. 1) and functional calculus with parameters for pseudodifferential operators [40]. We test (12.15) by computing the trace of the product with any operator $\hat{B}$, $B \in S^0_0$. Using (12.15) and Egorov theorem we get easily, $\forall B \in S^0_0$,

$$\lim_{\hbar \searrow 0} (2\pi\hbar)^n . Tr\left(\hat{A}_{E,f}(\hbar).\hat{B}\right) = \int\int \hat{f}(t) A\left(\Phi^t_H(z)\right) B(z) dz dt = 0. \quad (12.16)$$

So we get

$$\int \hat{f}(t) A\left(\Phi^t_H(z)\right) dt = 0, \ \forall z \in \Sigma_E. \quad (12.17)$$

Now, choose $z_0 \in \Sigma_E$ such that $t \to \Phi^t(z_0)$ is not periodic, we should like to deduce from (12.17) that $\hat{f} \equiv 0$. We use now the following result to finish the proof. □

Lemma 80 *For $T > 0$ we can find $\rho_T > 0$ such that the mapping $\Phi : (t, z) \overset{F}{\mapsto} (t, \Phi^t(z))$ is a diffeomorphism from $]-T, T[\times D_{\rho_T}(z_0)$ onto an open neighborhood $\mathcal{N}_T$ of the curve: $\{\Phi^t(z_0), \ -T < t < T\}$; where $D_{\rho_T}(z_0)$ is the euclidean ball with center z_0 and radius ρ_T in the orthogonal plane to the curve at time 0.*
Furthermore, for every g in $C^\infty_0(]-T, T[)$ we can construct some $A \in C^\infty_0(\mathbb{R}^{2n})$ such that:
$g(t) = A(\Phi^t(z_0)) + h(t), \ \forall t \in \mathbb{R}$ with:
(i) $Supp(h) \cap]-T, T[= \emptyset$,
(ii) $\sup_{\mathbb{R}} |h| \leq \sup_{\mathbb{R}} |g|$.

Proof Starting with the diffeomorphism F, let us choose $u \in C^\infty_0(D_{\rho_T}(z_0))$, $u(z_0) = 1$, $0 \leq u \leq 1$. We define $B(t, z) := g(t)u(z)$ for $(t, z) \in D_{\rho_T}(z_0)$ and $A(z) := B(F^{-1}(z))$ for $z \in \mathcal{N}_T$. We have clearly $A(\Phi^t(z_0)) = g(t)$ if $|t| \leq T$. For $|t| \geq T$ and $\Phi^t(z_0) \in \mathcal{N}_T$ we have $\Phi^t(z_0) = \Phi^t_1(z)$ with $|t_1| \leq T$, $z \in D_{\rho_T}(z_0)$ so $h(t) = -g(t_1)u(z)$ and we get the announced properties for h.

So, with the above notation, using the above lemma and (12.17) we get:

$$\int \hat{f}(t) g(t) dt = \int \hat{f}(t) h(t) dt \leq \sup_{\mathbb{R}} |g| \int_{|t| \geq T} |\hat{f}(t)| dt,$$

taking T large, we have clearly $\hat{f} \equiv 0$ hence $f \equiv 0$. □

Remark 74 Theorem 65 has to be compared with the clustering property of the spectra when the Hamiltonian flow is periodic (see Chap. 5). It is remarkable that one non periodical trajectory at energy E is enough to perturbed dramatically the spectrum of $\hat{H}$ close to E.

Concerning the non diagonal matrix elements we have the following result.

Theorem 66 *Assume that the classical system is ergodic on Σ_E. Then for all $A \in S^0_0$ such that $\langle A \rangle_E = 0$, we have*

$$(\star)\quad \lim_{\hbar\to 0,\delta\to 0}\left(\frac{1}{N_\hbar}\sum_{E_k\in I(\hbar),\ |\omega_{jk}(\hbar)|\le\delta}|A_{jk}|^2\right)=0.$$

Moreover, the condition ($\star$) *holds if and only if the classical system is ergodic on the energy level* E.

Proof Let be θ a bounded with compact support function such that $\int_{\mathbb{R}}\theta(t)dt=1$. Denote $\hat{A}_\theta=\int\theta(t)\mathscr{U}(-t)\hat{A}\,\mathscr{U}(t)dt$. Let J_E be a fix neighborhood of E. Using Parseval equality and localization of eigenfunctions in a neighborhood of Σ_E we have

$$\frac{1}{N_\hbar}\sum_{E_j\in I(\hbar),\ E_k\in J_E}|\langle\psi_j,\hat{A}_\theta\psi_k\rangle|^2=\frac{1}{N_\hbar}\sum_{E_j\in I(\hbar)}|\hat{\theta}\left(\omega_{jk}\right)|^2|A_{jk}|^2+o(1).\tag{12.18}$$

But we have

$$\frac{1}{N_\hbar}\sum_{E_j\in I(\hbar),\ E_k\in J_E}|\langle\psi_j,\hat{A}_\theta\psi_k\rangle|^2=\frac{1}{N_\hbar}\sum_{E_j\in I(\hbar)}\langle A_\theta^*A_\theta\psi_j,\psi_j\rangle+o(1).\tag{12.19}$$

Using Corollary 35 we have

$$\lim_{\hbar\to 0}\frac{1}{N_\hbar}\sum_{E_j\in I(\hbar)}\langle A_\theta^*A_\theta\psi_j,\psi_j\rangle=\int_{\Sigma_E}\left|\int\theta(t)A(\Phi_H^t(X))dt\right|^2\nu_E(dX).$$

The results follows by choosing $\theta(t)=\frac{1}{2T}\mathbb{1}_{[-T,T]}$, so $\hat{\theta}(\lambda)=\frac{\sin(T\lambda}{T\lambda})$ with T large enough. □

Remark 75 We shall see later that quantum ergodicity, in the sense of Theorem 64, can occurs for non classically ergodic systems (even completely integrable systems). For these systems ($\star$) cannot be satisfied. These probability transitions were considered in [172] in section 51 for $n=1$.
Classical ergodicity brings information not only on diagonal matrix elements but also on nearby non diagonal matrix elements (see [64] for more details, in particular for considering classical mixing systems).

12.3 The Quantum Ergodic Birkhoff Theorem

This section is inspired from a paper by G. Rivière [216]. We adapt it to a semi-classical setting.
Let us recall the classical Birkhoff ergodic theorem for the Hamiltonian flow Φ_H^t on Σ_E with the Liouville invariant measure ν_E (it is well defined, only assuming that E is a non-critical energy).

Theorem 67 (see reference[252]) *There exists a Borel subset $\Gamma_E \subseteq \Sigma_E$, $\nu_E(\Gamma_E) = 1$ such that for all $X \in \Gamma_E$ and all $A \in C^0(\Sigma_E)$ we have*

$$\lim_{T \to +\infty} \frac{1}{T} \int_0^T A(\Phi_H^t(X))dt = \mu_X^B(A)$$

where μ_X^B is a Radon invariant (for the flow Φ_H^t) probability on Σ_E.

We follow the method of [216] to prove the following quantum consequence.

Denote by μ_j the Schwartz distribution $A \mapsto \langle \psi_j, \hat{A}\psi_j \rangle$, $A \in C^\infty(K_E)$ where $K_E := H^{-1}[E - \varepsilon, E + \varepsilon]$.

The Weyl quantization is not positive, so it is more convenient to replace it, as before, by the positive quantization $\hat{A}^{aw}$ such that $\|\hat{A} - \hat{A}^{aw}\| = O(\hbar)$ (anti-Wick quantization). Hence $A \mapsto \langle \psi_j, \hat{A}^{aw}\psi_j \rangle$ are Radon measures on K_E.

Let us introduce the Sobolev space $W^{s,p}(K_E)$, $s > 0$, $p > 1$.

By the Sobolev theorem we have $W^{s,p}(K_E)) \subseteq C^0(K_E)$ if $s > \frac{2d}{p}$. So we have $\mu_j \in (W^{s,p}(K_E))' = W^{-s,p'}(K_E)$ where $\frac{1}{p} + \frac{1}{p'} = 1$ (B' denotes the dual space of the Banach space B) .

Theorem 68 *For any $\hbar \in]0, 1]$ there exists a subset $\tilde{\Lambda}_\hbar \subseteq \Lambda_\hbar$ of density one in $\Lambda_\hbar$ i.e.:*

$$\lim_{\hbar \searrow 0} \frac{\#\tilde{\Lambda}_\hbar}{N_\hbar} = 1$$

and for any map $\hbar \mapsto j(\hbar) \in \tilde{\Lambda}_\hbar$ any accumulation point μ of $\mu_{j(\hbar)}$ in the Banach space $W^{-s,p'}(\Sigma_E)$ is in the closed convex hull of the Birkhoff measures $\{\mu_X^B,\ X \in \Sigma_E\}$ (for the weak convergence).

When the Hamiltonian flow of H is ergodic on Σ_E, we have essentially one Birkhoff measure: it is the Liouville measure, so Theorem 68 entails Theorem 64 (Quantum Ergodic Theorem).

In the opposite case, if the Hamiltonian flow is periodic on Σ_E we have much more Birkhoff measures. Let be T_0 a period for the flow. Then for any $X \in \Sigma_E$ we have

$$\lim_{T \to +\infty} \frac{1}{T} \int_0^T A(\Phi^t(X))dt = \frac{1}{T_0} \int_0^{T_0} A(\Phi^t(X))dt := \delta_\gamma(A)$$

i.e. the Birkhoff measures are the Dirac measures supported by the periodic trajectories γ of the flow. The following Proposition applies in particular to the Harmonic oscillator: $\hat{H} = \frac{1}{2}(-\hbar^2\Delta + |x|^2)$ with $E > 0$.

Proposition 127 *Let us assume that the Hamiltonian H satisfies the periodicity assumptions of Theorem 35 (Chap. 5). Moreover we assume that $\hat{H}$ has only one eigenvalue (with multiplicity) in each cluster $I(k, \hbar)$.Then we have:*

(i) for any 2π-periodical trajectory $\gamma \subseteq \Sigma_E$ its Dirac measure δ_γ is the Wigner measure for eigenstates associated with the eigenvalue $E(k, \hbar) \in I(k, \hbar)$ as $\hbar \to 0$.

(ii) More generally any probability measure μ on Σ_E, invariant for the flow Φ_H^t, is the Wigner measure for a sequence $\psi_{\hbar_\ell}$ of normalized bound states with $\hat{H}\psi_\hbar = E_\hbar\psi_\hbar$ for $\hbar = \hbar_\ell$ and $\lim_{\ell\to+\infty} E_{\hbar_\ell} = E$.

Let us recall here the shape of the clusters in the completely periodic case, where μ is the Maslov index and λ is a geometric constant,

$$I(k,\hbar) = [(k+\frac{\mu}{4})\hbar + \lambda - C\hbar^2, (k+\frac{\mu}{4})\hbar + \lambda + C\hbar^2] \text{ with } C > 0 \text{ large enough.}$$

Remark 76 Notice that in part (i) of this proposition the conclusion is valid for any subsequence $\hbar_\ell \to 0$. It is not needed to extract a specific subsequence.

The proof of the Theorem 68 is a consequence of the following lemma plus the Hahn-Banach Theorem

Lemma 81 *There exists $\tilde{A}_\hbar$ and $\hbar \mapsto j(\hbar)$ like in Theorem 68 such that*

$$\text{essinf}_{X\in\Sigma_E}\mu_X^B(A) \le \liminf_{\hbar\searrow 0}\mu_{j(\hbar)}(A) \le \limsup_{\hbar\searrow 0}\mu_{j(\hbar)}(A) \le \text{esssup}_{X\in\Sigma_E}\mu_X^B(A) \tag{12.20}$$

Proof (Theorem 68 using Lemma 81)
By contradiction. Assume that μ is not in the closed convex hull $\mathscr{C}^B$ of $\{\mu_X^B, X \in \Sigma_E\}$. From the Hahn-Banach Theorem and reflexivity of the Banach space $W^{s,p}(K_E)$ there exists $f \in W^{s,p}(K_E)$ such that

$$\nu(f) \le a < \mu(f), \quad \forall \nu \in \mathscr{C}^B.$$

In particular

$$\sup_{X\in\Sigma_E}\mu_X^B(f) \le a < \mu(f)$$

which cannot be true because we have (12.20). □

Proof (Lemma 81)
Let $A_T(X) = \frac{1}{T}\int_0^T A(\Phi^t(X))dt$. We have

$$\mu_j(A) = \mu_j(A_T) + o_T(1) = \langle\psi_j, \hat{A}^{aw}\psi_j\rangle + o_T(1).$$

Denote $R_0 = \text{esssup}\mu_X^B(A)$, $\tilde{A}_T$ is a smooth observable such that

(i) $\tilde{A}_T(X) \le A_T(X), \forall X \in \Sigma_E$,
(ii) $\tilde{A}_T(X) \le R_0 + \sqrt{\varepsilon}, \forall X \in \Sigma_E$,
(iii) $\tilde{A}_T(X) = A_T(X)$ if $A_T(X) \le R_0 + \frac{\sqrt{\varepsilon}}{2}$.

We can choose $\tilde{A}_T = \theta_\varepsilon(A_T)$ where θ_ε is a an increasing function on $\mathbb{R}$ such that $\theta_\epsilon(x) = x$ for $|x| \le R_0 + \frac{\sqrt{\varepsilon}}{2}$ and $|\theta_\varepsilon(x)| \le R_0 + \sqrt{\varepsilon}$ for every $x \in \mathbb{R}$.

From the Weyl law we get

$$\frac{1}{N_\hbar}\sum_{1\le j\le N_\hbar}\langle\psi_j,\mathrm{Op}_\hbar^{aw}(A_T-\tilde{A}_T)\psi_j\rangle=\int_{\Sigma_E}(A_T-\tilde{A}_T)dL_E+o_T(1). \quad (12.21)$$

To estimate this term we use that

$$\int_{\Sigma_E}(A_T-\tilde{A}_T)dL_E=\int_{\Omega_\varepsilon}(A_T-\tilde{A}_T)dL_E\le\|A\|_{L^\infty}L_E(\Omega_{T,\varepsilon}).$$

where $\Omega_{T,\varepsilon}=\{X\in\Sigma_E,\ |A_T(X)|\ge R_0+\frac{\sqrt{\varepsilon}}{2}\}$.
But from definition of R_0 we have

$$\lim_{T\to+\infty}L_E(\Omega_{T,\varepsilon})=0.$$

So, using (12.21), for any $\eta>0$ there exists $T_{\varepsilon\eta}$ and $\hbar_T$ such that for $T\ge T_{\varepsilon\eta}$ and $0<\hbar\le\hbar_T$ we have

$$\frac{1}{N_\hbar}\sum_{1\le j\le N_\hbar}\langle\psi_j,\mathrm{Op}_\hbar^{aw}(A_T-\tilde{A}_T)\psi_j\rangle\le\eta\varepsilon.$$

Then by Tchebichev inequality we get

$$\frac{1}{N_\hbar}\#\{j\in\Lambda_\hbar,\ \langle\psi_j,\mathrm{Op}_\hbar^{aw}(A_T-\tilde{A}_T)\psi_j\rangle\ge\sqrt{\varepsilon}\}\le\eta\sqrt{\varepsilon}$$

We also have for $\hbar$ small enough

$$\langle\psi_j,\mathrm{Op}_\hbar^{aw}\tilde{A}_T\psi_j\rangle\le R_0+2\sqrt{\varepsilon}$$

then writing

$$\mu_j(\hbar)=\langle\psi_j,\mathrm{Op}_\hbar^{aw}\tilde{A}_T\psi_j\rangle+\langle\psi_j,\mathrm{Op}_\hbar^{aw}(A_T-\tilde{A}_T)\psi_j\rangle,$$

and denoting $S_\varepsilon(\hbar)=\{j\in\Lambda_\hbar,\ \mu_{j(\hbar)}\le R_0+3\sqrt{\varepsilon}\}$, from the above estimates we get

$$\liminf_{\hbar\searrow0}\frac{\#S_\varepsilon(\hbar)}{N_\hbar}\ge1-\eta\sqrt{\varepsilon}.$$

This is true for any $\eta>0$ so $S_\epsilon(\hbar)$ has density one.
Finally, by a diagonal argument we get, in a subset $\tilde{\Lambda}(\hbar)$ of $\Lambda_\hbar$, of density one, (independent on A), such that

$$\limsup_{\hbar\searrow0}\mu_{j(\hbar)}(A)\le\mathrm{esssup}_{X\in\Sigma_E}\mu_X^B(A).$$

The other part of inequality (12.20) is obtained by applying the above inequality to $-A$. □

Proof (Proof of Proposition 127) We recall that in Chap. 5 we have obtained quasi-modes ψ_{γ_E} with energy $E^\hbar$ such that

$$\hat{H}\psi_{\gamma_E} = E_k^\hbar \psi_{\gamma_E} + \mathscr{O}(\hbar^{7/4}).$$

Moreover $\|\psi_{\gamma_E}\| = m_E \hbar^{1/4} + \mathscr{O}(\hbar^{1/2})$ with $m_E > 0$ and $E_k^\hbar \in I(k, \hbar)$.
The quasi-mode ψ_{γ_E} is obtained by propagating coherent states according the formula:

$$\psi_{\gamma_E} = \int_0^{2\pi} \mathrm{e}^{i\frac{t}{\hbar}(E^\hbar - \hat{H})} \varphi_z dt, \quad z \in \gamma. \tag{12.22}$$

In particular ψ_{γ_E} is microlocalized on γ which means $\mathrm{FS}[\psi_{\gamma_E}] \subseteq \gamma$.
Now we defined the L^2-normalized quasi-mode $\psi_\gamma^\hbar := c_\hbar \psi_{\gamma_E}$ with $c_\hbar > 0$, $c_\hbar \approx \hbar^{-1/4}$.
Using a stationary phase argument as in Chap. 5 we get for any $A \in S_0^0$,

$$\langle \psi_\gamma^\hbar, \hat{A}\psi_\gamma^\hbar \rangle = \frac{1}{2\pi} \int_0^{2\pi} A(\Phi^t(z))dt + \mathscr{O}(\sqrt{\hbar}). \tag{12.23}$$

To prove the part (i) of the proposition our strategy consists to prove that we can replace the quasi-mode $\psi_\gamma^\hbar$ by a genuine eigenfunction of $\hat{H}$. Let be $\eta > 0$ small enough. Let us denote $\Pi_k^\hbar$ the spectral projector of $\hat{H}$ on the interval $I(k, \hbar)$. We shall always assume that $I(k, \hbar) \subseteq]E - \eta, E + \eta]$. Let us choose an energy cutoff $\chi \in C_0^\infty(]E - \eta, E + \eta[)$, $\chi = 1$ on $[E - \eta/2, E + \eta/2]$. From our assumptions we know that $\hat{H}\Pi_k = E(k, \hbar)\Pi_k$ for some $E(k, \hbar) \in I(k, \hbar)$. Hence we have

$$\chi(\hat{H})\mathrm{e}^{-i\frac{t\hat{H}}{\hbar}} = \sum_{k\in\mathbb{Z}} \chi(E(k, \hbar))\mathrm{e}^{-i\frac{tE(k,\hbar)}{\hbar}} \Pi_k^\hbar.$$

Using that $E(k, \hbar) = k\hbar - \frac{\mu}{4}\hbar - \lambda + \mathscr{O}(\hbar^2)$ we get the Fourier series

$$\sum_{k\in\mathbb{Z}} \chi(E(k, \hbar))\mathrm{e}^{-itk} \Pi_k^\hbar = \chi(\hat{H})\mathrm{e}^{-i\frac{t\hat{H}}{\hbar}} \mathrm{e}^{it(\frac{\mu}{4} + \frac{\lambda}{\hbar})} + \mathscr{O}(\hbar^{7/4}).$$

So we have

$$\chi(E(k, \hbar))\Pi_k^\hbar = \frac{1}{2\pi} \int_0^{2\pi} \chi(\hat{H})\mathrm{e}^{-i\frac{t}{\hbar}\left(\hat{H} - E^\hbar\right)} dt + \mathscr{O}(\hbar^{7/4}).$$

Now we can choose $c(\hbar)$ of order $\hbar^{-1/4}$ such that $\|c(\hbar)\Psi_{\gamma_E}\| = 1$ and for $z \in \gamma$ we have

$$c(\hbar)\Pi_k^\hbar \varphi_z = \frac{c(\hbar)}{2\pi}\psi_{\gamma_E} + \mathscr{O}(\hbar^{3/2}). \tag{12.24}$$

So we get the part (i) of the proposition using (12.24) and (12.23).
Let us now prove (ii). Let γ_j, $1 \le j \le N$ different periodical trajectories on Σ_E. Let us consider the measure $\mu = \alpha_1^2\delta_{\gamma_1} + \cdots + \alpha_N^2\delta_{\gamma_N}$, $\alpha_j \ge 0$, $\alpha_1^2 + \cdots + \alpha_N^2 = 1$. We prove that μ is a semi-classical mesure on Σ_E.
From (12.24) we have obtained quasi-modes $\{\psi_j^\hbar\}_{1\le j\le N}$ with frequency set in γ_j and such that $\Pi_k^\hbar \psi_j^\hbar = \psi_j^\hbar + \mathscr{O}(\hbar^{3/2})$. So they are almost orthogonal.
Let us denote $\psi_\alpha^\hbar = \sum_{1\le j\le N} \alpha_j \psi_j^\hbar$ where $\psi_j^\hbar = 2\pi c(\hbar)\Pi_k^\hbar \varphi_{z_j}$, $z_j \in \gamma_j$.

To conclude we use that any invariant measure μ on Σ_E is the $\star$-weak-limit of finite convex sums of Dirac measures δ_{γ_j} for periodical orbits γ_j. $\square$

12.4 Time Dependent Semi-classical Measures

Let us consider now a sub-quadratic time dependent semi-classical Hamiltonian $H(t) \approx \sum_{j\ge 1} \hbar^j H_j(t, X)$ satisfying the assumptions of Theorem 12 (Chap. 2). Recall that

$$|\partial_X^\gamma H_j(t, X)| \le C_\gamma, \text{ for } (t, X) \in \mathbb{R} \times \mathbb{R}^{2n} \text{ and } |\gamma| + j \ge 2.$$

We know that $\hat{H}(t)$ generates a unitary Schrödinger propagator in $L^2(\mathbb{R}^n)$ which will be denoted $\mathscr{U}^\hbar(t, t_0)$ and $\mathscr{U}_t^\hbar = \mathscr{U}^\hbar(t, 0)$.
We denote $\Phi_{H_0}^{t,t_0}$ the classical Hamiltonian flow associated with the principal symbol H_0.
Let us consider a normalized initial data $\psi_0^\hbar$ satisfying (12.5) and let be μ_0 a corresponding semi-classical measure.
Let $\{\psi_0^\hbar\}_{0<\hbar\le 1}$ be a bounded set in $L^2(\mathbb{R}^n)$. Then $A \mapsto \langle \psi_t^\hbar, \hat{A}^{aw}\psi_t^\hbar\rangle$ defines a finite positive Radon measure $\mu_t^\hbar$ on $\mathbb{R}^{2n}$.
Recall that if $A \in S_0^0$ then $\langle \psi_t^\hbar, \hat{A}^w\psi_t^\hbar\rangle = \langle \psi_t^\hbar, \hat{A}^{aw}\psi_t^\hbar\rangle + \mathscr{O}(\hbar)$.
But $\mu_t^\hbar \in L^\infty(\mathbb{R}_t, C_b'(\mathbb{R}^{2n}))$ and is a uniformly bounded set for the $\star$-weak topology as $0 < \hbar \le 1$. So by a compactness argument there exists $\mu_t \in L^\infty(\mathbb{R}_t, C_b'(\mathbb{R}^{2n})$ such that for a sequence of $\hbar \to 0$ and any $A \in C_b(\mathbb{R}^{2n})$, for almost every $t \in \mathbb{R}$ we have

$$\lim_{\hbar\to 0} \langle \psi_t^\hbar, \hat{A}^{aw}\psi_t^\hbar\rangle = \int_{\mathbb{R}^{2n}} A(X)\mu_t(dX).$$

Moreover μ_t is a positive finite Radon measure.

Proposition 128 *Assume that* $\|\psi_0^\hbar\| = 1$ *and satisfies the tightness condition* (12.5). *Then for any* $t \in \mathbb{R}$, $\psi_t^\hbar$ *satisfies* (12.5) *and any semi-classical measure* μ_t *for* $\psi_t^\hbar$, *is a probability measure on* $\mathbb{R}^{2n}$.

Proof We have, using the propagation rule for coherent states (Chap. 4),

$$|\langle\varphi_X, \psi_t^\hbar\rangle| = |\langle \mathscr{U}_t^{\hbar,\star}\varphi_X, \psi_0^\hbar\rangle| = |\langle\varphi_{Y_t}, \psi_t^\hbar\rangle| + \mathscr{O}(\sqrt{\hbar}),$$

where $Y_t = \Phi_{H_0}^{0,t}(X)$. Hence the conclusion follows from the following lemma. □

Lemma 82 *For any $T > 0$ there exists C_T such that for any $t \in [0, T]$ we have*

$$\|\Phi_{H_0}^{t,0}X\| \le C_T(\|X\| + 1), \ \forall X \in \mathbb{R}^{2n}. \tag{12.25}$$

Proof We begin by proving the inequality for $T = t_0 > 0$ small enough.
Let us denote $X_t = \Phi_{H_0}^{t,0}X$. From the Hamilton equation we get a constant $C > 0$ such that for all $t \ge 0$,

$$\| X_t\| \le \|X\| + Ct \sup_{0\le s\le t} \langle X_s\rangle.$$

So choosing $t_0 = \frac{1}{2C}$ we get, for $t \in [0, t_0]$, $\|X_t\| \le 2\langle X\rangle$. Then by iteration, writing for $kt_0 \le t \le (k+1)t_0$, $\Phi^{t,0} = \Phi^{t,kt_0} \circ \Phi^{kt_0,(k-1)t_0} \circ \cdot \circ \Phi^{t_0,0}$, we get (12.25). □

Now we consider large time behavior of the measures $\mu_t^\hbar$. Following [184, 185] we introduce a time scale $\tau_\hbar$ such that $\limsup_{\hbar\to 0} \tau_\hbar = +\infty$.

Recall that $\psi_t^\hbar$ satisfies a semi-classical time dependent Schrödinger equation

$$i\hbar\partial_t\psi_t^\hbar = \hat{H}\psi_t^\hbar, \quad \psi_{t=0}^\hbar = \psi_0^\hbar. \tag{12.26}$$

$\{\psi_0^\hbar\}$ is supposed to be bounded in $L^2(\mathbb{R}^n)$ and $\mu_t^\hbar$ is the Husimi measure defined by $\psi_t^\hbar$.
We define $\tilde{\mu}_t^\hbar = \mu_{t\tau_\hbar}^\hbar$. As above, a compactness argument shows that $\tilde{\mu}_t^\hbar$ has accumulation points $\tilde{\mu}_t \in L^\infty(\mathbb{R}_t, C_b'(\mathbb{R}^{2n}))$. Moreover, for a.e $t \in \mathbb{R}$, $\tilde{\mu}_t$ is a finite measure if $\psi_0^\hbar$ is uniformly bounded in $L^2(\mathbb{R}^n)$.

$$\int_{\mathbb{R}^{2n}} \tilde{\mu}_t(dX) \le \limsup_{\hbar\to 0} \|\psi_0^\hbar\|^2.$$

We shall use the following formula, where $\theta \in L^1(\mathbb{R}_t, dt)$, for a sequence of $\hbar \to 0$, we have

$$\lim_{\hbar\to 0} \int \theta(t)\langle\psi_{t\tau_\hbar}^\hbar, \hat{A}^w\psi_{t\tau_\hbar}^\hbar\rangle dt = \int_{\mathbb{R}_t} \theta(t)\left(\int_{\mathbb{R}_X^{2n}} A(X)\tilde{\mu}_t(dX)\right) dt/ \tag{12.27}$$

Proposition 129 *Recall that $\lim_{\hbar\to 0} \tau_\hbar = +\infty$.*
(i) *a.e $t \in \mathbb{R}$, the measure $\tilde{\mu}_t$ is invariant under the Hamiltonian flow $\Phi_{H_0}^t$.*
(ii) *Assume that the initial data $\psi_0^\hbar$ has a semi-classical measure μ_0 and that $\tau_\hbar = o(\hbar^{-2})$. Let $A \in S_0^0$ invariant under the Hamiltonian flow $\Phi_{H_0}^t$. Then for a*

subsequence $\hbar \to 0$, *for any* $t \in \mathbb{R}$, *we have:*

$$\lim_{\hbar\to 0}\langle \psi^{\hbar}_{t\tau_\hbar}, \hat{A}^w \psi^{\hbar}_{t\tau_\hbar}\rangle = \int_{\mathbb{R}^{2n}} A(X)\mu_0(dX). \tag{12.28}$$

Proof We follow [185] (see also [186] for time dependent semi-classical measures for large times).
(i) Let $A_s(X) = A(\phi^{-s}(X))$. Using the propagation of observables we have

$$\mathcal{U}_s^{\hbar}\hat{A}_s\,\mathcal{U}_s^{\hbar\star} = \hat{A} + \mathcal{O}(\hbar).$$

Hence we get

$$\int_{\mathbb{R}}\theta(t)\langle\psi_0^\hbar, \mathcal{U}^{\hbar,\star}_{t\tau_\hbar}\hat{A}^w_s\mathcal{U}^{\hbar}_{t\tau_\hbar}\psi_0^\hbar\rangle dt = \int_{\mathbb{R}}\theta(t)\langle\psi_0^\hbar, \mathcal{U}^{\hbar,\star}_{(t-\frac{s}{\tau_\hbar})\tau_\hbar}\hat{A}^w\mathcal{U}^{\hbar}_{(t-\frac{s}{\tau_\hbar})\tau_\hbar}\psi_0^\hbar\rangle dt + \mathcal{O}(\hbar). \tag{12.29}$$

Hence

$$\int_{\mathbb{R}}\theta(t)\langle\psi_0^\hbar, \mathcal{U}^{\hbar,\star}_{t\tau_\hbar}\hat{A}^w_s\mathcal{U}^{\hbar}_{t\tau_\hbar}\psi_0^\hbar\rangle dt = \int_{\mathbb{R}}\theta\left(t+\frac{s}{\tau_\hbar}\right)\langle\psi_0^\hbar, \mathcal{U}^{\hbar,\star}_{t\tau_\hbar}\hat{A}^w\mathcal{U}^{\hbar}_{t\tau_\hbar}\psi_0^\hbar\rangle dt + \mathcal{O}(\hbar). \tag{12.30}$$

Using that $\lim_{\hbar\to 0}\tau_\hbar = +\infty$ and passing to the limit in $\hbar \to 0$ we get for any $\theta \in L^1(\mathbb{R})$,

$$\int_{\mathbb{R}}\theta(t)\int_{\mathbb{R}^{2n}} A_s(X)\tilde{\mu}_t(dX) = \int_{\mathbb{R}}\theta(t)\int_{\mathbb{R}^{2n}} A(X)\tilde{\mu}_t(dX).$$

(ii) We have $\{A, H_0\} = 0$. So using property of Weyl quantization we get that $[\hat{H}, \hat{A}] = \mathcal{O}(\hbar^3)$ (a little computation shows that the coefficient of $\hbar^2$ vanishes).
We have:

$$i\frac{d}{dt}\langle\psi^\hbar_{t\tau_\hbar}, \hat{A}^w\psi^\hbar_{t\tau_\hbar}\rangle = \tau_\hbar\langle\psi_0^\hbar, \mathcal{U}^{\hbar,\star}_{t\tau_\hbar}[\hat{H}, \hat{A}]\,\mathcal{U}^{\hbar}_{t\tau_\hbar}\psi_0^\hbar\rangle.$$

Then after integration between 0 and t and using that $\lim_{\hbar\to 0}\hbar^2\tau_\hbar = 0$, we get (12.28). □

Corollary 38 *Assume that the conditions of Proposition 129 are satisfied and that the Hamiltonian flow of* H_0 *is* 2π *periodic in the compact set* $H_0^{-1}[E-\eta, E+\eta]$ ($\eta > 0$ *small enough) with* E *non critical for* H_0. *Then for any semi-classical measure* $\tilde{\mu}_t$ *defined by* (12.27) *and any* $A \in C_0^\infty(H_0^{-1}]E-\eta, E+\eta[)$, *we have a.e* $t \in \mathbb{R}$,

$$\int_{\mathbb{R}^{2n}} A(X)\tilde{\mu}_t(dX) = \int_{\mathbb{R}^{2n}}\langle A\rangle(X)\mu_0(dX), \tag{12.31}$$

where $\langle A\rangle(X) = \frac{1}{2\pi}\int_0^{2\pi} A(\phi^s(X))ds$.

Proof $\langle A\rangle(X)$ is invariant by the Hamiltonian flow so from (12.28) we have

$$\int_{\mathbb{R}^{2n}} \langle A\rangle(X)\tilde{\mu}_t(dX) = \int_{\mathbb{R}^{2n}} \langle A\rangle(X)\mu_0(dX).$$

Then using that $\tilde{\mu}_t$ is invariant by the flow and Fubini Theorem we get (12.31). □

Let us give now an application of time dependent semi-classical measures to estimate non diagonal matrix elements.

Corollary 39 *We assume that the assumptions of the Corollary 38 are satisfied. Let $\psi_j^\hbar$ normalized eigenfunctions of $\hat{H}$ for eigenvalues $E_j^\hbar$ in clusters $I(j,\hbar)$. Let $\omega_{jk}(\hbar) = \frac{E_j(\hbar)-E_k(\hbar)}{\hbar}$.*
Assume that, for a subsequence, $\lim_{\hbar\to 0}\frac{\omega_{j,k}(\hbar)}{\hbar} = c \neq 0$. Then we have $\lim_{\hbar\to 0}\langle\psi_j^\hbar, \hat{A}\psi_k^\hbar\rangle = 0$ for a subsequence $\hbar \to 0$.

Proof Following [185] again we consider $\psi^\hbar = \psi_j^\hbar + \psi_k^\hbar$ and its large time evolution.
We have

$$\langle\psi^\hbar(s), \hat{A}\psi^\hbar(s)\rangle = \langle\psi_j^\hbar, \hat{A}\psi_j^\hbar\rangle + \langle\psi^\hbar, \hat{A}\psi^\hbar\rangle + 2\Re(e^{is\omega_{i,k}(\hbar)}\langle\psi_j^\hbar, \hat{A}\psi_k^\hbar\rangle).$$

Using this egality for $s = \frac{t}{\hbar}$ and using (12.31) with $\tau_\hbar = \hbar^{-1}$, , there exists a finite real measure μ and and complex finite measure ν on $\mathbb{R}^{2n}$ such that for any $\theta \in L^1(\mathbb{R})$s we have

$$\tilde{\theta}(0)\int A(X)\mu(dX) = 2\Re\left(\tilde{\theta}(c)\int A(X)\nu(dX)\right).$$

Then we get that for a subsequence of $\hbar$,

$$\lim_{\hbar\to 0}\langle\psi_j^\hbar, \hat{A}\psi_k^\hbar\rangle = \int A(X)\nu(dX) = 0.$$

□

Remark 77 Notice that the transition amplitudes $\langle\psi_j^\hbar, \hat{A}\psi_k^\hbar\rangle$ considered here concerns energies which are in the same cluster $I(k,\hbar)$, so the result is not intuitive. A similar phenomenon was proved for Zoll manifolds [184].
An example of such a situation holds true for a perturbation of the 2-dimensional harmonic oscillator: $\hat{H} = \hat{H}_{\rm os} + \hbar\hat{L}$ where $\hat{L}$ is the generator of rotations in the plane: $\hat{L} = \hbar(x\partial_y - y\partial_x)$. Its spectrum is $\hbar\mathbb{Z}$. We know that $[\hat{H}_{\rm os}, \hat{L}] = 0$. So in each spectral subspace $\mathscr{E}_j$ for the eigenvalue $j+1$ of $\hat{H}_{\rm os}$ we can find a common basis of eigenfunctions $\{\psi_{j,\ell}^\hbar\}_{0\le\ell\le j}$ for $\hat{H}_{\rm os}$ and $\hat{L}$. So the eigenvalues of $\hat{H}$ in the cluster are $(j+1)\hbar + \ell\hbar^2$. Then applying the Corollary for this basis we have for $\ell \neq \ell'$,

$$\lim_{\hbar\to 0}\langle\psi_{j,\ell}^\hbar, \hat{A}\psi_{j,\ell'}^\hbar\rangle = 0.$$

12.5 Quantum Unique Ergodicity for a Random Hermite Basis

We consider now the harmonic oscillator $\hat{H}_{os} = \frac{1}{2}(-\Delta + |x|^2)$ in $L^2(\mathbb{R}^n)$, $n \geq 2$. The corresponding classical Hamiltonian flow is surely not ergodic on the energy shell $|\xi|^2 + |x|^2 = 2$. Nevertheless we shall see that most of orthonormal basis of eigenfunctions of $\hat{H}_{os}$ are quantum ergodic (meaning that in these basis the matrix diagonal elements for a class of observables $\mathrm{Op}_1^w(A)$ converge towards the average of A for the Liouville probability. This rather surprising phenomenon was discovered by Zelditch for the Laplace operator on the sphere $\mathbb{S}^n$ [268]. In [222] the authors have proved that the same phenomenon holds true for the harmonic oscillator. As we have seen classical ergodicity is related to the semi-classical behaviour of the non-diagonal elements $\langle \psi_j^\hbar, \hat{A}\psi_k^\hbar \rangle$, $j \neq k$, close to the diagonal elements.

The eigenvalues of $\hat{H}_{os}$ are $\lambda_j = j + \frac{n}{2}$, $j \in \mathbb{N}$, with mutiplicities $m_j \approx j^{n-1}$. Let be $\mathscr{E}_j$ the corresponding eigenspace , $\mathbb{S}^{m_j}$ the unit sphere of $\mathscr{E}_j$ and $\{\varphi_{j,k},\ 1 \leq k \leq m_j\}$ a fixed orthonormal basis of $\mathscr{E}_j$ (Hermite functions).

We can identify the space of the orthonormal basis of $\mathscr{E}_j$ with the unitary group $U(m_j)$. We consider on $U(m_j)$ its Haar probability measure ρ_j. Then the set $\mathscr{B}$ of the Hilbertian basis of eigenfunctions of H_{os} in $L^2(\mathbb{R}^n)$ can be identified with the infinite product

$$\mathscr{B} = \times_{j\in\mathbb{N}} U(N_{m_j}),$$

which can be endowed with the product of probability measures:

$$\mathrm{d}\rho = \otimes_{j\in\mathbb{N}}\, \mathrm{d}\rho_j.$$

Denote by $B = (\psi_{j,\ell})_{j\in\mathbb{N},\, 1\leq \ell \leq m_j} \in \mathscr{B}$ a typical orthonormal basis of $L^2(\mathbb{R}^n)$ so that for all $j \in \mathbb{N}$, $(\psi_{j,\ell})_{1\leq\ell\leq m_j} \in U(m_j)$ is an orthonormal basis of $\mathscr{E}_{m_j}$.
Denote S_0 the space of symbols on $\mathbb{R}^n \times \mathbb{R}^n$ smooth and homogeneous of degree 0 (outside a neighborhood of 0). Then we have the following result.

Theorem 69 *For $B \in \mathscr{B}$ and $A \in S_0$ let us denote*

$$D_j(B) = \max_{1\leq \ell \leq m_j} \left| \langle \psi_{j,\ell}, \hat{A}\psi_{j,\ell} \rangle - \nu_1(A) \right|,$$

where ν_1 is here the uniform probability on the sphere $\sqrt{2}\mathscr{S}_{2n-1}$.
Then we have

$$\lim_{j\to+\infty} D_j(B) = 0, \quad \rho - \text{a.s on } \mathscr{B}.$$

In other words, a random orthonormal basis of Hermite functions is Quantum Unique Ergodique (QUE).

A main step in the proof of Theorem 69 is to prove a semi-classical probabilistic result for the quantum oscillator $\hat{H} = \frac{1}{2}(-\hbar^2\Delta + |x|^2)$. More generally we assume

that $\hat{H}$ is a rather general quantum Hamiltonian with a discrete spectrum in some small interval $I(\hbar) = [E - \delta(\hbar), E + \delta(\hbar)]$. We assume that H satisfies the assumptions (H.0) and (H.1) of Chap. 5 and $\delta(\hbar)$ is chosen like in Sect. 1 of this Chapter, $\delta(\hbar) \approx \hbar^{1-\tau}$, with $\tau > 0$.
Moreover H satisfies the periodicity conditions of Proposition 35 (Chap. 5) close to the energy E with $0 < c < 1$. Let us introduce the following notations.
$\Pi^\hbar$ is the spectral projector of $\hat{H}$ on the interval $I(\hbar)$. $\{\psi_j\}_{1\leq j\leq N_\hbar}$ an orthonormal basis of eigenfunctions of the spectral subspace $\mathscr{E}_\hbar := \Pi^\hbar(L^2(\mathbb{R}^n))$ and $\hat{H}\psi_j = E_j\psi_j$ where $E_j \in I(\hbar)$. $\mathbf{S}_\hbar$ is the unit sphere of $\mathscr{E}_\hbar$. (Notice that $\dim(\mathscr{E}_\hbar) := N_\hbar \approx \hbar^{-n+\tau}$.)
Moreover if H satisfies the periodicity condition of Proposition 35 (Chap. 5) close to the energy E we shall choose $\delta(\hbar) = \hbar$ and $\Pi^\hbar$ is the spectral projector in $[E, E + \hbar[$ with $E > 0$. Notice that there is only one eigenvalue in $[E, E + \hbar[$ (with high multiplicity).

Lemma 83 *Assume that H satisfies the periodicity conditions of Proposition 35 (Chap. 5) close to the energy E. Then choosing $I(\hbar) = [E, E + \hbar[$ we have*
(i) $N_\hbar = (2\pi\hbar)^{1-n}L_E(\Sigma_E) + o(\hbar^{1-n})$
(ii)

$$\lim_{\hbar\rightarrow} \frac{\mathrm{Tr}(\Pi^\hbar \hat{A})}{N_\hbar} = \int_{\Sigma_E} A(X)\nu_E(dX).$$

Proof We can use the same technicalities as in Chap. 5 for proving the Gutzwiller trace formula (see also the Chapter about Herman-Kluk approximation).
For simplicity we shall assume that $\hat{H} = H^\hbar_{\mathrm{os}} := -\frac{\hbar^2}{2}\triangle + \frac{|x|^2}{2}$.
So we know from Chap. 1 that

$$\mathrm{e}^{-i\frac{t}{\hbar}H^\hbar_{\mathrm{os}}}\varphi_z = \mathrm{e}^{-it\frac{n}{2}}\varphi_{z_t}$$

where $z \mapsto z_t$ is the classical harmonic flow.
Denote $\Pi^\hbar_k$ the spectral projector for the eigenvalue $E^\hbar_j = (j + \frac{n}{2})\hbar$.

Let us consider a random wave $V(\omega) = N_\hbar^{-1/2} \sum_{1\leq j\leq N_\hbar} X_j(\omega)\psi_j$ where $X_j(\omega)$ are iid random variables of mean 0 and variance 1 (Gaussians or Bernoulli variables). Then we get a Borelian probability measure $\mathbf{P}_\hbar$ on $\mathbf{S}_\hbar$ such that for any continuous function $F : \mathbf{S}_\hbar \longrightarrow \mathbb{R}$ we have

$$\int_{\mathbf{S}_\hbar} F(u)\mathbf{P}_\hbar(du) = \int_\Omega F\left(\frac{V(\omega)}{\|V(\omega)\|}\right)\mathbb{P}(d\omega),$$

where $\{\Omega, \mathbb{P}\}$ is an abstract probability space.
For simplicity we assume here that X_j follows the complex normal law. This means that $(\Re X_j, \Im X_j)$ are independent real random variables and the pair follows the two

dimensional normal law. Then it can be shown that $\mathbf{P}_\hbar(du)$ is the uniform probability density on the sphere $\mathbf{S}_\hbar$.

Proposition 130 *Under the previous assumptions there exist $c, C > 0$ so that for all $r > 0$, $\hbar \in]0, 1]$ and $A \in \mathrm{S}_0^0$*

$$\mathbf{P}_h\Big[u \in \mathbf{S}_h : |\langle u, \hat{A}(h)u\rangle - \nu_E(A)| > r\Big] \leq Ce^{-cN_\hbar r^2}. \tag{12.32}$$

As a consequence, there exists $C > 0$ so that for all $p \geq 2$, $h \in]0, 1]$ and $A \in \mathrm{S}_0^0$ we have

$$\|\langle u, \hat{A}(h)u\rangle - L_E(A)\|_{L^p_{\mathbf{P}_h}} \leq CN_h^{-1/2}\sqrt{p}.$$

Recall that ν_E is the Liouville probability on Σ_E. It may be surprising that this result is true without any assumption on the dynamical system. Why this is true? This is a consequence of the concentration property of probability measures on spheres of large dimension.

For simplicity we assume that $\mathbf{P}_\hbar$ is the uniform probability on $\mathbf{S}_\hbar$.

Proposition 131 (concentration of measure, P. Levy, M. Ledoux *[174]) There exist constants $K > 0$, $\kappa > 0$ such that for every Lipschitz function $F : \mathbf{S}_\hbar \longrightarrow \mathbb{R}$ satisfying*

$$|F(u) - F(v)| \leq \|F\|_{Lip}\|u - v\|_{L^2(\mathbb{R}^d)}, \quad \forall u, v \in \mathbf{S}_h,$$

we have

$$\mathbf{P}_\gamma\Big[u \in \mathbf{S}_\hbar : |F - \mathscr{M}_F| > r\Big] \leq Ke^{-\frac{\kappa N_\hbar r^2}{\|F\|^2_{Lip}}}, \quad \forall r > 0, \quad h \in]0, 1], \tag{12.33}$$

where $\mathscr{M}_F$ is a median for F.

Recall that a median $\mathscr{M}_F$ for F is defined by

$$\mathbf{P}_\gamma\big[u \in \mathbf{S}_h : F \geq \mathscr{M}_F\big] \geq \frac{1}{2}, \quad \mathbf{P}_\gamma\big[u \in \mathbf{S}_h : F \leq \mathscr{M}_F\big] \geq \frac{1}{2}.$$

The factor N_h in the exponential of r.h.s of (12.33) will be crucial in applications to get large deviation estimates.

Proof (Proposition 130).
Let $\mathscr{M}_A$ be a median for the map $F_A(u) := \langle u, \hat{A}u\rangle$. We have

$$|F_A(u) - F_A(v)| \leq 2\|\hat{A}\|\|u - v\|.$$

So we can apply Proposition 131. To go further we have to estimate the median $\mathscr{M}_A$ by using the expectation $\mathbf{E}_\hbar$ associated with the probability law $\mathbf{P}_\hbar$.

The Proposition 130 is a consequence of the three following lemmas.

Lemma 84 *There exists $C > 0$ such that*

$$|\mathbf{E}_\hbar[F_A - \mathscr{M}_A]| \leq C N_\hbar^{-1/2}, \ \forall \hbar \in]0, 1].$$

Lemma 85

$$\lim_{\hbar \to 0} \mathbf{E}_\hbar[F_A] = \int A(X)\nu_E(dX).$$

Lemma 86 *Let us denote the random variable $\mathbf{v}(\omega) = \|V(\omega)\|^2$. Then there exists $s_0 > 0$, small enough, such that for any $\epsilon > 0$ we have*

$$\mathbb{P}[|\mathbf{v} - 1| \geq \varepsilon] \leq \mathrm{e}^{-\varepsilon \frac{s_0 N_\hbar}{2}}, \ \forall \hbar \in]0, 1].$$

We get easily Proposition 130 using Proposition 131 and lemmas 85, 84
Let us prove now the three lemmas.

Proof (Lemma 84). We have, using Proposition 131,

$$\begin{aligned} |\mathbf{E}_\hbar[F_A - \mathscr{M}_A]| &\leq \int_0^{+\infty} \mathbf{P}_\hbar[|F_A - \mathscr{M}_A| > r]dr \\ &\leq C N_\hbar^{-1/2}. \end{aligned} \tag{12.34}$$

□

Proof (Lemma 85). Using Corollary 1 of Theorem 63 it is enough to prove

$$\lim_{\hbar \to 0} \left(\mathbf{E}_\hbar[F_A] - \frac{\mathrm{Tr}(\hat{A}\Pi^\hbar)}{N_\hbar} \right) = 0.$$

We have

$$\mathbf{E}_\hbar[F_A] = \int_\Omega \langle \frac{V(\omega)}{\|V(\omega)\|}, \hat{A}\frac{V(\omega)}{\|V(\omega)}\rangle \mathbb{P}(d\omega).$$

Using properties of the random variables, it is clear that we have

$$\int_\Omega \langle V(\omega), \hat{A}V(\omega)\rangle \mathbb{P}(d\omega) = \frac{\mathrm{Tr}(\hat{A}\Pi^\hbar)}{N_\hbar}.$$

Define $\Omega_\hbar = \{\omega : | \ \|(\omega)\|^2 - 1| \leq N_\hbar^{-1/2}\}$ and

$$I_\hbar := \int_{\Omega_\hbar} \frac{1 - \|V\|^2}{\|V\|^2} \langle V, \hat{A}V\rangle \mathbb{P}(d\omega), \ II_\hbar := \int_{\Omega_\hbar^c} \frac{1 - \|V\|^2}{\|V\|^2} \langle V, \hat{A}V\rangle \mathbb{P}(d\omega)$$

From Proposition 131 we have

$$\mathbb{P}(\Omega_\hbar^c) \le C e^{-cN_\hbar^{-1/2}}.$$

And using the Cauchy-Schwarz inequality gives

$$II_\hbar \le \|\hat{A}\|\mathbb{P}(\Omega_\hbar^c)\left(\int_\Omega (1+|V\|)^2)^2\mathbb{P}(d\omega)\right).$$

Using that the integral is $\mathscr{O}(1)$ as $\hbar \to 0$ we have $\lim_{\hbar\to 0} II_\hbar = 0$.
We also have $I_\hbar \le N_\hbar^{-1/2}\|\hat{A}\|$ so the proof of Lemma 85 is completed. □

Proof (Lemma 86). We reproduce here a usual method to prove a large deviation estimate. Let $f(s) = \mathbb{E}(e^{sX_1^2})$ and $g(s) = \log f(s)$, for $s > 0$ close to 0. For $\tau > 0$ we have

$$\mathbb{P}[\mathbf{v} > \tau] = \mathbb{P}[e^{sN_\hbar \mathbf{v}} > e^{sN_\hbar \tau}] \le \mathbb{E}(e^{sN_\hbar \mathbf{v}})e^{-sN_\hbar \tau}.$$

Using the definition of $\mathbf{v}$ we have $\mathbb{E}(e^{sN_\hbar \mathbf{v}}) = e^{-(s\tau - g(s))N_\hbar}$. Hence if $\tau \ge 1+\varepsilon$ here $s > 0$ small enough such that we have for any $\varepsilon > 0$,

$$s\tau - g(s) = s(\tau - 1) + \mathscr{O}(s^2) \ge \frac{\varepsilon s}{2}.$$

Then we get $\mathbb{P}[\mathbf{v} > 1+\varepsilon] \le e^{-\frac{\varepsilon}{2}sN_\hbar}$. By the same argument we get $\mathbb{P}[\mathbf{v} < 1-\varepsilon] \le e^{-\frac{\varepsilon}{2}sN_\hbar}$. □

Now we can prove Theorem 69 as a consequence of Proposition 130.

Proof Let $\hat{H}_{\text{os}}^\hbar = -\frac{\hbar^2}{2}\Delta + \frac{|x|^2}{2}$.

Every $B \in \mathscr{B}$ can be identified with $\{B_j\}_{j\ge 1}$ where $B_j \in U(N_{h_j})$. The random variables D_j are independent and D_j depends only on B_j. So for every $r > 0$ we have

$$\begin{aligned}
\rho\big[D_j(B) > r\big] &= \rho_j\big[D_j(B) > r\big] \\
&= \rho_j\Big[\exists \ell \in \mathbb{N}\cap[1, N_{h_j}],\ \big|\langle \varphi_{j,\ell}, \hat{A}\varphi_{j,\ell}\rangle - \nu_1(A)\big| > r\Big] \\
&\le \sum_{1\le \ell \le N_{h_j}} \rho_j\Big[\big|\langle \varphi_{j,\ell}, \hat{A}\varphi_{j,\ell}\rangle - \nu_1(A)\big| > r\Big] \\
&\le N_{h_j}\mathbf{P}_{h_j}\Big[\big|\langle u, \hat{A}(h_j)u\rangle - \nu_1(A)\big| > r\Big].
\end{aligned}$$

In the last line we have used that the probability $\mathbf{P}_h$ is the push-forward of the Haar measure of $U(M_h)$ by the maps: $U(N_h) \ni M \mapsto Mv \in \mathbf{S}_h$, for any $v \in \mathbf{S}_h$.
So using Proposition 130 we get, with positive constants C_1, C_2,

$$\rho\big[D_j(B) > r\big] \le C_1 j^{n-1}\exp\Big[-C_2 j^{n-1} r^2\Big].$$

In particular for any $n \geq 2$ we get

$$\sum_{j \geq 1} \rho\big[D_j(B) > r\big] < +\infty$$

and the result is a consequence of the Borel-Cantelli Lemma. □

Notes

There are many papers studying properties of semi-classical mesures in different geometry settings for the Laplace-Beltrami operator on compact Riemannian manifolds [4, 184, 185]. In this chapter we have adapted, in the context of confining Hamiltonians on the Euclidian space $\mathbb{R}^n$, some ideas taken in these papers.
Section 5 is taken from a paper by Robert-Thomann [222].

Chapter 13
Open Quantum Systems and Coherent States

Abstract In this chapter we study some properties of systems of coupled quadratic quantum Hamiltonians. The well known example is coupled harmonic oscillators appearing in the Quantum Brownian Motion. An open system is a physical (S) interacting with the environment (E) such that the total system $(S \cup E)$ is isolated from the rest of the universe. So we have two independent Hamiltonians $\hat{H}_S$, $\hat{H}_E$ and an interaction $\hat{V}$ between (S) and (E). The main problem discussed for open systems is to understand the dynamical action of the interaction $\hat{V}$ with the environment on the time dependent evolution of (S) which is now irreversible because a part of its energy escapes to the environment. An important physical effect of the interaction of the environment on a quantum system is *decoherence* which give an explanation for the mechanism driving the quantum-to-classical transition. We restrict ourself to few mathematical problems related with this very active field of research in physics [155, 200, 231, 270]. In a first part we study the purity of a state of (S) (or the linear entropy) in a short time approximation. In a second part we discuss the separability of bipartite systems. In a third part we establish a master equation (or a Fokker-Planck type equation) for the time evolution of the reduced matrix density for coupled quadratic Hamiltonians by a bilinear inter-action. We give an explicit formula for the solution of this master equation so that the time evolution of the reduced density at time t is written as a convolution integral for the reduced density at initial time $t_0 = 0$, with a Gaussian kernel, for $0 \leq t < t_c$ where $t_c \in]0, \infty]$ is a critical time such that reversibility is lost for $t > t_c$ if $t_c < +\infty$. Some manifestations of *decoherence*, for $t > 0$ small enough, are proved on our simple model which is similar to a representation of the Schrödinger cat.

13.1 Introduction to Open Quantum Systems

13.1.1 A General Setting

The general setting considered here is a quantum system (S) interacting with an environment (E). The total system $(S \cup E)$ is assumed to be an isolated quantum system and we are interested in dynamical properties of (S) alone, which is an open

D. Robert and M. Combescure, *Coherent States and Applications in Mathematical Physics*, Theoretical and Mathematical Physics,
https://doi.org/10.1007/978-3-030-70845-0_13

system because of its interactions with (E). In particular, during the time evolution, the energy of (S) is not preserved and its evolution is not determined by a Schrödinger or Liouville–von Neumann equation unlike for the total system $(S \cup E)$. A detailed presentation concerning the quantum dynamics of open systems can be found in the book [43].

The Hilbert space of the total system $(S \cup E)$ is the tensor product $\mathscr{H} = \mathscr{H}_S \widehat{\otimes} \mathscr{H}_E$,[1] and its Hamiltonian $\widehat{H}$ is decomposed as follows:

$$\widehat{H}_V(t) = \underbrace{\widehat{H}_S \otimes \mathbb{1} + \mathbb{1} \otimes \widehat{H}_E}_{\widehat{H}_0} + \widehat{H}_I(t) = \widehat{H}_0 + \widehat{V}(t). \tag{13.1}$$

A basic example is the quantum Brownian motion, coupling two harmonic oscillators with bilinear potential $V_g(t)$:

$$\widehat{H}_S = \frac{1}{2}(-\partial^2_{x^2} + \omega_0^2 x^2), \quad \widehat{H}_E = \frac{1}{2} \sum_{1 \le j \le N} (-\partial^2_{y_j^2} + \omega_j^2 y_j^2), \tag{13.2}$$

$$V_g(t, x, y) = x \cdot \sum_{1 \le j \le N} g_j(t) y_j, \quad x, y_j, g_j(t) \in \mathbb{R}, \tag{13.3}$$

where $\mathscr{H}_S = L^2(\mathbb{R})$, $\mathscr{H}_E = L^2(\mathbb{R}^N)$.

The whole system $(S \cup E)$ is isolated by assumption and its evolution obeys the Schrödinger (or Liouville–von Neumann) equation with the Hamiltonian $\widehat{H}$. Assuming that at time $t = 0$ the system (S) and the environment (E) are decorrelated, the problem is to understand the evolution of the system (S) alone in the Hilbert space $\mathscr{H}_S$ when an interaction with the environment is switched on.

All Hamiltonians here are selfadjoint operators on their natural domain, $\widehat{H}$ and $\widehat{H}_I = \widehat{V}$ are defined in $\mathscr{H}$, $\widehat{H}_S$ in $\mathscr{H}_S$, $\widehat{H}_E$ in $\mathscr{H}_E$. Quantum observables are denoted with a hat accent, the corresponding classical observables (also named Wigner functions or Weyl-symbols) are written without hat accent.

We shall assume that the interacting potential $\widehat{V}$ has the following form

$$\widehat{V}_g(t) = \sum_{1 \le j \le N} g_j(t) \widehat{S}_j \otimes \widehat{E}_j \tag{13.4}$$

where $\widehat{S}_j$ and $\widehat{E}_j$ are selfadjoint operators in $\mathscr{H}_S$ and $\mathscr{H}_E$ respectively. Assumptions on their domains will be given later.

Recall that a density matrix $\widehat{\rho}$ is a positive trace class operator with trace one ($\mathrm{Tr}(\widehat{\rho}) = 1$), see Chap. 1. Let $\widehat{\rho}_0$ be a state of the total system $(S \cup E)$ in the Hilbert space $\mathscr{H}$ at the initial time. The time evolution of $\widehat{\rho}$ obeys the following Liouville–von Neumann equation:

[1] Here $\widehat{\otimes}$ means that the tensor product is the Hilbert space obtained by completion of the algebraic tensor product.

$$\dot{\widehat{\rho}} = i^{-1}[\widehat{H}, \widehat{\rho}], \quad \widehat{\rho} : t \mapsto \widehat{\rho}(t), \quad \widehat{\rho}(0) = \widehat{\rho}_0. \tag{13.5}$$

Time derivatives are denoted by a dot, $[\,\cdot\,,\,\cdot\,]$ denotes the commutator of two observables.

We assume that for $t = 0$ the system and the environment are decoupled, which means that

$$\widehat{\rho}(0) = \widehat{\rho}_S(0) \otimes \widehat{\rho}_E(0), \; \widehat{\rho}_{S,E}(0) \text{ are density matrices in } \mathscr{H}_{S,E}. \tag{13.6}$$

The time evolution $\widehat{\rho}(t)$ is given by the propagator $\mathscr{U}_V(t, s)$ generated by the Hamiltonian $H_V(t)$. For $V = 0$ we have $\mathscr{U}_0(t, s) = \mathrm{e}^{-i(t-s)\widehat{H}_S} \otimes \mathrm{e}^{-i(t-s)\widehat{H}_E}$. So we have $\widehat{\rho}(t) = \widehat{\rho}_S(t) \otimes \widehat{\rho}_E(t)$, $t \in \mathbb{R}$, where $\rho_{S,E}(t) = \mathscr{U}^*_{S,E}\rho_{S,E}(0)\mathscr{U}_{S,E}$. In particular, if $\rho_S(0)$ is a pure state, then $\rho_S(t)$ is pure for any $t \in \mathbb{R}$. When the interaction is switched on, $\rho_S(t)$ is defined as the partial trace of $\widehat{\rho}(t)$ over the environment degrees of freedom and one question is to check if $\rho_S(t)$ is pure or not for $t \neq 0$.

Let us recall some definitions concerning some properties of density matrix states.

Definition 28 Let $\widehat{\rho}$ be a state in the Hilbert space $\mathscr{H}$, i.e., a nonnegative operator in $\mathscr{H}$ with trace 1.
(i) $\widehat{\rho}$ is a pure state if and only if $\widehat{\rho}$ is the orthogonal projector on a unit vector ψ of $\mathscr{H}$.
(ii) $\widehat{\rho}$ is a mixed state if it is not a pure state.
(iii) Let $\widehat{\rho}$ be a state in $\mathscr{H}_S \otimes \mathscr{H}_E$; $\widehat{\rho}$ is a separable state if there exist a probability space $(\Omega, \mathbf{P})$ and a mesurable family of states: $\omega \mapsto \widehat{\rho}_\omega^{(S,E)}$ from Ω into $\mathfrak{S}^1(\mathscr{H}_{S,E})^2$ such that[3]

$$\widehat{\rho} = \int\limits_{\Omega} \widehat{\rho}_\omega^{(S)} \otimes \widehat{\rho}_\omega^{(E)} P(d\omega) \tag{13.7}$$

(iv) $\widehat{\rho}$ is entangled if it is not separable.

It is easy to see that a density matrix $\widehat{\rho}$ is a mixed state if and only if $\widehat{\rho}$ has an eigenvalue $\lambda \in]0, 1[$. To decide if a state is pure or not we consider its purity function

$$\mathscr{P}_\rho(t) = \mathrm{Tr}(\widehat{\rho}(t)^2).$$

We clearly have $0 \leq \mathscr{P}_\rho(t) \leq 1$. It is not difficult to see that $\rho(t)$ is pure if and only if $\mathscr{P}_\rho(t) = 1$. Suppose that at time $t = 0$ the state $\widehat{\rho}(t)$ is pure. So 0 is a maximum for $\mathscr{P}_\rho$, hence we have $\ddot{\mathscr{P}}_\rho(t)(0) \leq 0$. If furthermore $\ddot{\mathscr{P}}_\rho(0) < 0$, then for some $\varepsilon > 0$, the state $\widehat{\rho}(t)$ is mixed whenever $0 < t < \varepsilon$.

For isolated (closed) systems, if the initial state is pure, then it stays pure at every time, as it can easily be seen by using Eq. (13.5).

---

[2]Let us recall that $\mathfrak{S}^1(\mathscr{H})$ denotes the space of class-trace operators in the Hilbert space $\mathscr{H}$.

[3] In physics literature only discrete probability laws are considered, but it seems necessary to extend the definition to any probability law.

Remark 78 The function $S_\ell(\rho) := 1 - \mathscr{P}_\rho$ is called the linear entropy. Recall that in quantum mechanics the Boltzman-von Neumann entropy is defined as $S_{BN}(\rho) =: -\mathrm{Tr}(\hat\rho\log(\hat\rho))$. Using the linear approximation $\log(\hat\rho) \approx \hat\rho - \mathbb{1}$ one get $S_{BN}(\rho) \approx S_\ell(\rho)$.

Another quantum entropy, with better properties than S_{BN}, was introduced by Wehrl [254] by using the covariant Berezin symbol $\rho_c(z) = \langle\varphi_z, \hat\rho\varphi_z\rangle$ (see Chap. 2). We assume here that $\mathscr{H} = L^2(\mathbb{R})$.
The Wehrl entropy is

$$S_W(\rho) = -\int_{\mathbb{R}^2} \rho_c(z)\log(\rho_c(z))dz.$$

For interesting properties of $S_W(\rho)$ we refer to [179, 254]. In particular it is proved in [179] that $\min_\rho S_W(\rho) \geq 1$ and that $S_W(\rho) = 1$ if and only if $\hat\rho$ is the orthogonal projection on a coherent state φ_z ("Wehrl conjecture").

So our first step is to compute the second derivative $\ddot{\mathscr{P}}_S(t)(0)$ for the purity of the reduced system $\widehat{\rho}_S$ when $\widehat{\rho}_S(0)$ is pure. In particular, for a large class of models, including the Quantum Brownian Motion with N oscillators, we shall prove that for the reduced system we have $\ddot{\mathscr{P}}_S(0) < 0$. So for these models there is an indication that the state of the reduced system may be immediately decoherent and is entangled with the environment.

Our second goal is to consider the separability (or entanglement) property of the mixed state $\widehat{\rho}(t)$. We explain here results obtained by physicsts: the necessary Horodecki condition [156] which turns out to be sufficient for Gaussian states and this can be applied to the quantum Brownian motion [92, 236].

Our third goal is to compute the time evolution $\widehat{\rho}_S(t)$ of the state of the small system (S). We know that $\widehat{\rho}_S(t)$ does not satisfy a Liouville–von Neumann equation because the system is open so diffusion terms appear. We shall give a proof that for any time dependent quadratic systems, $\widehat{\rho}_S(t)$ obeys an exact "master equation" similar to a Fokker–Planck type equation, with time dependent coefficients, and from this equation we can get an explicit formula for $\widehat{\rho}_S(t)$. Similar results were obtained by different methods in [85, 86, 103].

Let us now recall a mathematical definition for the reduced evolution $\rho_S(t)$ and related properties. This can be done by introducing partial trace (or relative trace) over the environment for a state $\widehat{\rho}$ of the global system.

Definition 29 Let $\widehat{A}$ be a trace class operator in $\mathscr{H}_S \otimes \mathscr{H}_E$. We denote by $\mathrm{Tr}_E\widehat{A}$ the unique trace class operator on $\mathscr{H}_S$ satisfying, for every bounded operator $\widehat{B}$ on $\mathscr{H}_S$,

$$\mathrm{Tr}_S\Big((\mathrm{Tr}_E\widehat{A})\widehat{B}\Big) = \mathrm{Tr}\Big(\widehat{A}(\widehat{B}\otimes\mathbb{1}_E)\Big), \tag{13.8}$$

We have denoted by "Tr" the trace in the total space $\mathscr{H}$, and by $\mathrm{Tr}_{S,E}$ the trace in $\mathscr{H}_{S,E}$, respectively. Of course we have the "Fubini property:" $\mathrm{Tr}_S(\mathrm{Tr}_E\widehat{A}) = \mathrm{Tr}\widehat{A}$.

Moreover, we have $\mathrm{Tr}_E(\widehat{B} \otimes \widehat{C}) = \widehat{B} \cdot (\mathrm{Tr}_E \widehat{C})$ if $\widehat{B}$ and $\widehat{C}$ are trace class operators in $\mathscr{H}_S$ and $\mathscr{H}_E$ respectively.

If $\widehat{\rho}$ is a density matrix in $\mathscr{H}$, then $\mathrm{Tr}_E \widehat{\rho}$ is a density matrix in $\mathscr{H}_S$, called the reduced density matrix or the reduced state. From (13.8) we can compute the matrix element of $\mathrm{Tr}_E \widehat{A}$ by the formula

$$\langle \psi, (\mathrm{Tr}_E \widehat{A})\varphi \rangle = \mathrm{Tr}\left(\widehat{A}(\Pi_{\psi,\varphi} \otimes \mathbb{1}_E)\right), \quad \psi, \varphi \in \mathscr{H}_S.$$

The partial trace was first introduced by L. Landau (1927) and then more recently in quantum mechanics to explain *entanglement* and *decoherence* (see [200, 231] for more details).

Let

$$\Psi \in \mathscr{H}_S \otimes \mathscr{H}_E, \quad \Psi = \sum_{1 \le j \le N} \psi_j \otimes \eta_j,$$

where Ψ is a pure state in $\mathscr{H} = \mathscr{H}_S \otimes \mathscr{H}_E$. Let us compute the partial trace of Π_Ψ in $\mathscr{H}_E$. Applying Definition 29, we easily get

$$\mathrm{Tr}_E(\Pi_\Psi) = \sum_{1 \le j,k \le N} \langle \eta_j, \eta_k \rangle \Pi_{\psi_j,\psi_k} \tag{13.9}$$

where Π_{ψ_j,ψ_k} is a rank one operator in $\mathscr{H}_S : \Pi_{\psi_j,\psi_k}(\varphi) = \langle \psi_j, \varphi \rangle \psi_k$. The decoherence phenomenon [162, 270] means that there exists a independent vectors $\{\eta_j\}$ such that $|\langle \eta_j, \eta_k \rangle|$ becomes negligible for $j \neq k$, so that $\mathrm{Tr}_E(\Pi_\Psi)$ is very close to

$$\sum_{1 \le j \le N} p_j \Pi_{\psi_j}, \quad p_j = \langle \eta_j, \eta_j \rangle.$$

Hen ce quantum interferences between the states ψ_j of the system (S) are destroyed (13.9).

13.1.2 Coupled Quadratic Hamiltonians

Formula (13.8) is an operator version of the Fubini integration theorem as we can see in the Weyl quantization setting.

Let $\mathscr{H}_S = L^2(\mathbb{R}^d)$, $\mathscr{H}_E = L^2(\mathbb{R}^N)$. Denote $z = (x, \xi) \in \mathbb{R}^{2d}$, $u = (y, \eta) \in \mathbb{R}^{2N}$, $A = \sigma_w \widehat{A}$ the Weyl symbol of $\widehat{A}$ in the Schwartz space $\mathscr{S}(\mathbb{R}^{2(d+N)})$ (see Chap. 2 for more details); then the Weyl symbol of $\mathrm{Tr}_E \widehat{A}$ is

$$\sigma_w(\mathrm{Tr}_E \widehat{A})(z) = (2\pi)^{-N} \int_{\mathbb{R}^{2N}} A(z, u) du. \tag{13.10}$$

Our main applications here concern the Weyl-quadratic case where:

- H_S and H_E are real quadratic Hamiltonians (generalized harmonic oscillators) respectively in the phase spaces $\mathbb{R}^{2d}$, $\mathbb{R}^{2N}$;
- $\widehat{H}_S$ and $\widehat{H}_E$ are quantum Hamiltonians (Weyl quantization of H_S, respectively H_E) in $L^2(\mathbb{R}^d)$, respectively $L^2(\mathbb{R}^N)$;
- $\widehat{\rho}_S$ is the projector to $\psi \in \mathscr{S}(\mathbb{R}^d)$;
- ρ_E is a Gaussian with 0 mean;
- S_j, respectively E_j is a linear form on $\mathbb{R}^{2d}$, respectively $\mathbb{R}^{2N}$ (bilinear coupling).

In applications one considers thermal equilibrium states for the environment: $\widehat{\rho}_E = Z(\beta)^{-1}\mathrm{e}^{-\beta\widehat{H}_E}$, where H_E is a quadratic form, positive definite, and $\beta = \frac{1}{T}$, $T > 0$, is the temperature, $Z(\beta) = \mathrm{Tr}(\mathrm{e}^{-\beta\widehat{H}_E})$. In this setting it is possible to find the exact time dependent master equation satisfied by $\widehat{\rho}_S(t)$, as we shall see. This model was analyzed in several papers [102, 103].

13.1.3 A Schrödinger Cat Model

A coherent state model for the famous Schrödinger cat is built as follows. Let be $z^\pm \in \mathbb{R}^2$, $z\pm \neq (0,0)$. For simplicity we choose $z^- = -z^+$.

φ_{z^+} is the alive cat state, φ_{z^-} is the dead cat state. $\chi_\pm$ is an orthonormal basis of $\mathbb{C}^2$, for the atomic states (if χ_+ the cat is alive, dead if χ_-). The states space of the system "cat+atom" is the Hilbert space $L^2(\mathbb{R}) \otimes \mathbb{C}^2$ and the state of the system is

$$\psi_{\text{cat+at}} = \frac{1}{N}\left(\varphi_{z^+} \otimes \chi_+ + \varphi_{z^-} \otimes \chi_-\right) \tag{13.11}$$

N is a normalization constant.

Notice that $\varphi_{z^\pm}$ are almost orthogonal ($N \approx \sqrt{2}$ as $\hbar \to 0$).

The cat state is the partial trace :

$$\hat{\rho}_{\text{cat}} = \mathrm{Tr}_{\mathbb{C}^2}(\hat{\rho}_{\text{cat+at}}),$$

where $\hat{\rho}_{cat+at}$ is the orthogonal projector on $\psi_{\text{cat+at}}$. We find

$$\hat{\rho}_{\text{cat}} = \frac{1}{N}\left(\Pi_+ + \Pi_-\right),$$

$\Pi_\pm$ is the orthogonal projector on $\varphi_{\pm z}$.

So the cat state in entangled. Moreover we get

$$\text{Probability[alive cat]} = \mathrm{Tr}(\hat{\rho}_{\text{cat}}\Pi_+) \approx \frac{1}{2},$$

$$\text{Probability[dead cat]} = \mathrm{Tr}(\hat{\rho}_{\text{cat}}\Pi_-) \approx \frac{1}{2}.$$

For a finer analysis of the situation we compute the Wigner function of the state ψ_{cat+at} using the spin coherent state (Chap. 7). We have seen that coherent 1/2-spin states of an atom are parametrized by the sphere $\mathbb{S}^2$. Let $\mathbf{n} \in \mathbb{S}^2$, $\mathbf{n} = (n_1, n_2, n_3)$: $n_3 = \cos\theta$, $n_2 = \sin\theta\cos\phi$, $n_1 = \sin\theta\sin\phi$. Let the spin coherent state $\psi_{\mathbf{n}} = \sin(\theta/2)e^{i\phi}\chi_- + \cos(\theta/2)\chi_+$. If $\chi \in \mathbb{C}^2$, norm 1, its Wigner-Husimi function is $\mathcal{W}_\chi(\mathbf{n}) = |\langle\chi, \psi_{\mathbf{n}}\rangle|^2$. Now we have the Wigner function $\mathcal{W}_{\text{cat+at}}$ of $\psi_{\text{cat+at}}$

$$\frac{N^2}{2}\mathcal{W}_{\text{cat+at}}(X, \mathbf{n}) = \cos^2(\theta/2)\mathcal{W}_{\varphi_z}(X) + \sin^2(\theta/2)\mathcal{W}_{\varphi_{-z}}(X) + \sin(\theta)\cos\left(\frac{2\sigma(X, z)}{\hbar} - \phi\right)\mathcal{W}_{\varphi_0}(X) \tag{13.12}$$

$X \in \mathbb{R}^2$ and $\mathbf{n} \in \mathbb{S}^2$ with coordinates (θ, ϕ).

If $\mathbf{n}$ is in the north or south pole, the state of the cat is known. Taking the average on the sphere $\mathbb{S}^2$ for $\mathcal{W}_{cat+at}(X, \mathbf{n})$ we find the Wigner function of $\hat{\rho}_{cat}$: $\frac{1}{N^2}(\mathcal{W}_{\varphi_{z+}} + \mathcal{W}_{\varphi_{z-}})$.

Choosing $\mathbf{n}$ on the equator with $\phi = 0$ we find

$$\frac{N^2}{2}\mathcal{W}_{\psi^{\text{cat}}}(X) = e^{-\frac{|X-z^+|^2}{\hbar}} + e^{-\frac{|X-z^-|^2}{\hbar}} + 2e^{-\frac{|X|^2}{\hbar}}\cos\left(\frac{2\sigma(X, z^+)}{\hbar}\right). \tag{13.13}$$

corresponding with the state $\psi^{\text{cat}} = \frac{1}{N}(\varphi_{z^+} + \varphi_{z^-})$. This is a superposition of two states usually called "states of the Schrödinger cat". The first two terms in (13.13) correspond to the Wigner functions of $\varphi_{z^\pm}$. The third term is an oscillating term between life and death which is responsible of the coherence of the cat state (here is the *Schrödinger paradox*).

13.2 Computation of the Purity

13.2.1 Assumption and Statements

Let us consider the following abstract setting. We fix a selfadjoint and positive operator K_0 in $\mathcal{H}$ and we define the scale of abstract Sobolev space $\mathcal{H}^s = \mathrm{Dom}(K_0^s)$ if $s \geq 0$ and $\mathcal{H}^s$ is the dual of $\mathcal{H}^{-s}$ if $s \leq 0$. Define $\mathcal{H}^\infty = \bigcap_{s\in\mathbb{R}}\mathcal{H}^s$ and $\mathcal{H}^{-\infty} = \bigcup_{s\in\mathbb{R}}\mathcal{H}^s$. Moreover, we assume that K_0^{-1} is in the Schatten class $\mathfrak{S}^{p_0}(\mathcal{H})$. For definitions and properties of Schatten classes we refer to [120]. We introduce the following definition.

Definition 30 A linear operator $A : \mathscr{H}^{\infty} \to \mathscr{H}^{-\infty}$ is said to be of order $m \in \mathbb{R}$ if, for any $k \in \mathbb{R}$, we have $A \in \mathscr{L}(\mathscr{H}^{k+m}, \mathscr{H}^{k})$.

Now we introduce our assumptions on the Hamiltonian (13.1).

- [A0] H_S and H_E are selfadjoint operators in $\mathscr{H}_S$, respectively $\mathscr{H}_E$. Moreover, $\widehat{H}_0 := \widehat{H}_S \otimes \mathbb{1} + \mathbb{1} \otimes \widehat{H}_E$ is an operator of order m_0 in $\mathscr{H}$, is a symmetric operator in $\mathscr{H}^{\infty}$ and is such that $[H_0, K_0]K_0^{-1}$ is of order 0.
- [A1] $\widehat{V} = \widehat{V}(t)$ is of order m_0 such that $\widehat{V} \in \mathbb{C}^0(I_0, \mathscr{L}(\mathscr{H}^{k+m_0}, \mathscr{H}^{k}))$, for every k and $\widehat{V}$ is symmetric on $\mathscr{H}^{\infty}$, for $t \in I_0 :=] - T_0, T_0[$ for some $T_0 > 0$. Moreover $[\widehat{V}(t), K_0]K_0^{-1} \in \mathbb{C}^0(I_0, \mathscr{L}(\mathscr{H}^{k}, \mathscr{H}^{k}))$, for all k.
- [A2] The initial state $\widehat{\rho}(0)$ is of order r, with $r < -p_0 - km_0$, for some $k \geq 3$.

We have the following preliminary results (see [191]).

Proposition 132 *Assume that conditions* [A0] *and* [A1] *are satisfied. Then* $\widehat{H}_V(t)$ *has a unique selfadjoint extension in* $\mathscr{H}$. *In particular,* $\mathscr{H}_0$ *is essentially selfadjoint. Moreover, the time dependent Hamiltonian* $\widehat{H}_V(t)$ *defines a unitary propagator* $\mathscr{U}_V(t, s)$ *in* $\mathscr{H}$ *and* $(t, s) \mapsto \mathscr{U}_V(t, s)$ *is continuous from* $I_0 \times I_0$ *into* $\mathscr{L}(\mathscr{H}^{k})$, *for any* $k \in \mathbb{N}$.

Then the time evolution of an initial matrix density $\widehat{\rho}(0)$ is well defined and is given by

$$\rho(t) = \mathscr{U}_V(t, 0)^* \widehat{\rho}(0) \mathscr{U}_V(t, 0). \tag{13.14}$$

It is useful to compute the second derivative of the purity of the state ρ_S of the system. Assume that $\rho_S(0)$ is pure and that its purity $\mathscr{P}_S(t)$ is C^3 in a neighborhood of 0. Then we have

$$\mathscr{P}_S(t) = 1 + \frac{\ddot{\mathscr{P}}_S(0)}{2} t^2 + O(t^3), \quad t \searrow 0.$$

So if $\ddot{\mathscr{P}}_S(0) < 0$, then $\widehat{\rho}(t)$ becomes a mixed state for t close to 0, $t \neq 0$.

Theorem 70 *Assume that conditions* [A0], [A1], *and* [A2] *are satisfied. Then* $t \mapsto \widehat{\rho}(t)$ *is* C^k *from* I_0 *into* $\mathfrak{S}^1(\mathscr{H})$. *Assume furthermore that* $k \geq 2$ *and that* $\rho_S(0)$ *is a pure state in* $\mathscr{H}_S$. *Then we have the following formula for the second derivative of* $\mathscr{P}_S(t)$ *at* 0:

$$\ddot{\mathscr{P}}_S(0) = -4\mathrm{Tr}\left(\widehat{V}\widehat{\rho}(0)\widehat{V}(\widehat{\rho}_S^{\perp}(0) \otimes \mathbb{1})\right), \tag{13.15}$$

where $\widehat{\rho}_S^{\perp} = \mathbb{1} - \widehat{\rho}_S$

Now we get a more explicit formula for the separable interactions (13.4).

Corollary 40 *Assume that we have* (13.4) *and that* $\mathrm{Tr}_E(\widehat{\rho}_E(0)\widehat{E}_j\widehat{E}_{j'}) = \delta_{j,j'}d_j$, $1 \leq j, j' \leq N$, *then we have*

$$\ddot{\mathscr{P}}_S(0) = -4 \sum_{1 \leq j \leq N} \left(\|\widehat{S}_j\psi_S\|^2 - \langle\psi_S, \widehat{S}_j\psi_S\rangle^2\right) d_j, \tag{13.16}$$

where $\rho_S(0)$ is the projection to ψ_S. In particular, if there exists j_0 such that $d_{j_0} > 0$ and ψ_S is not an eigenfunction of $\widehat{S}_{j_0}$, then $\ddot{\mathscr{P}}_S(0) < 0$.

We can apply this result to the Quantum Brownian Motion (13.2), (13.3). Denote $\pi_S(z) := 2\pi^{-d}\rho_S(z)\,dz$ and $\pi_E(u) := (2\pi)^{-N}\rho_E(u)du$ (they are quasiprobabilities in $\mathbb{R}^{2d}$, respectively $\mathbb{R}^{2N}$). Here we assume that $\widehat{\rho}_S(0)$ is a pure state $\psi_S \in \mathscr{S}(\mathbb{R}^d)$. We introduce the commutators:

$$s_{j'j} = [\widehat{S}_{j'}, \widehat{S}_j], \quad e_{j'j} = [\widehat{E}_{j'}, \widehat{E}_j],$$

and the classical (symmetric) covariance matrices:

$$\Sigma^{(S)}_{jj'} = \int_{\mathbb{R}^{2d}} \pi_S(z) S_{j'}(z) S_j(z)\,dz - \Big(\int_{\mathbb{R}^{2d}} \pi_S(z) S_j(z)\,dz\Big)\Big(\int_{\mathbb{R}^{2d}} \pi_S(z) S_{j'}(z)\,dz\Big)$$

and

$$\Sigma^{(E)}_{jj'} = \int_{\mathbb{R}^{2N}} \pi_E(u) E_{j'}(u) E_j(u)\,du.$$

So we get the following statement.

Corollary 41

$$\ddot{\mathscr{P}}_S(0) = -4\mathrm{Tr}_{\mathbb{R}^{2m}}(\Sigma^{(S)}\Sigma^{(E)}) + \sum_{1\le j,j'\le m} s_{j'j} e_{j'j}. \tag{13.17}$$

We remark that on the right-hand side in formula (13.17), the first term is classical and the second term is a quantum correction, which vanishes if the coupling of the system and its environment is only in position (or momentum) variables (as it is usually in the physics literature). Assume now that

$$S_j(z) = z_j \quad \text{and} \quad E_j(u) = \sum_{1\le \ell \le 2N} G_{j,\ell} u_\ell\,,$$

where $1 \le j \le 2d$. Then $G_{j,\ell}$ is a $(2d \times 2N)$-matrix with real coefficients. With these notations, for the quantum correction we have

$$q_c = \sum_{1\le j,j'\le 2d} s_{j'j} e_{j'j} = \mathrm{Tr}_{\mathbb{R}^{2d}}(J_S G J_E G^\top), \tag{13.18}$$

where J_S, J_E are, respectively, the matrix of the canonical symplectic form in $\mathbb{R}^{2d}$ and in $\mathbb{R}^{2N}$: $\sigma_S(z, z') = z \cdot J_S z'$, $\sigma_E(u, u') = u \cdot J_E u'$ (the scalar products are denoted by a dot).

For the "classical part" we have

$$-4\mathrm{Tr}_{\mathbb{R}^{2m}}\big(\Sigma^{(S)}\Sigma^{(E)}\big) = -4\mathrm{tr}\big(G^{\top}\mathrm{Cov}_{\rho_S}G\mathrm{Cov}_{\rho_E}\big). \tag{13.19}$$

Corollary 42

$$\boxed{\ddot{\mathscr{P}}_S(0) = -4\mathrm{tr}_{\mathbb{R}^{2d}}\big(\mathrm{Cov}_{\pi_S}G\mathrm{Cov}_{\pi_E}G^{\top}\big) + \mathrm{Tr}_{\mathbb{R}^{2d}}\big(J_S G J_E G^{\top}\big)} \tag{13.20}$$

where Cov_{π} *is the covariance matrix for the quasiprobability* π.

Recall here that the Wigner function of a pure state ψ is nonnegative if and only if ψ is a Gaussian by the Hudson theorem (see Chap. 2).

We get now a more explicit formula for the quantum Brownian motion or a variant of the quantum Brownian motion. Let us assume that

$$\begin{aligned} V(x,\xi;y,\eta) = x\cdot&\Big(\sum_{1\le j\le N} c_j y_j + \sum_{1\le j\le N} c_{j+N}\eta_j\Big) \\ &+\xi\cdot\Big(\sum_{1\le j\le N} d_j y_j + \sum_{1\le j\le N} d_{j+N}\eta_j\Big). \end{aligned}$$

Then the quantum correction q_c is obtained by the formula

$$q_c = \big(2(d_1c_{N+1}+\cdots+d_Nc_{2N}) - 2(c_1d_{N+1}+\cdots+c_Nd_{2N})\big).$$

In the quantum Brownian motion, it is assumed that the initial data for the environment is a thermal state:

$$\rho_{\tau,E}(y,\eta) = (2\tau)^N\exp\Big(-\tau\sum_{1\le j\le N}(y_j^2+\eta_j^2)\Big),\quad 0<\tau<1.$$

Corollary 43 *Assume that* $d=1$, $c_j=0$ *for* $j\ge N+2$, *and* $d_j=0$ *for* $j\ge 2$, $\psi_S\in\mathscr{S}(\mathbb{R})$, *and that* $\rho_E(0)=\rho_{\tau,E}$ *is a thermal state. Then we have*

$$\begin{aligned} \ddot{\mathscr{P}}_S(0) &= \frac{2}{\tau}\left((c_1^2+\cdots+c_N^2+c_{N+1}^2)\mathrm{Cov}_{x,x}(\psi_0)\right. \\ &\quad+\big(2c_1d_1\mathrm{Cov}_{x,\xi}(\psi_0)+d_1^2\mathrm{Cov}_{\xi,\xi}(\psi_0)\big)+\underbrace{c_{N+1}d_1}_{\text{quantum correction}} \end{aligned} \tag{13.21}$$

where $\mathrm{Cov}(\psi_S)$ *is the* (2×2) *covariance matrix of the Wigner function of* ψ_S *and*

$$\mathrm{Cov}(\psi_S) = \begin{pmatrix}\mathrm{Cov}_{x,x}(\psi_S) & \mathrm{Cov}_{x,\xi}(\psi_S)\\ \mathrm{Cov}_{x,\xi}(\psi_S) & \mathrm{Cov}_{\xi,\xi}(\psi_S)\end{pmatrix}.$$

So we see a quantum correction term appearing when the coupling mixes positions and momenta variables. When $d_1 = 0$, we recover a known formula for the quantum Brownian motion.

13.2.2 Proofs of the Statements of Sect. 13.2.1

13.2.2.1 Proof of Theorem 70

It is easier here to consider the interaction formulation of quantum mechanics to "eliminate" the "free" (noninteracting) evolution: $\mathcal{U}_0(t) = e^{-it\widehat{H}_0}$. We introduce the inter-acting evolution

$$\widehat{\rho}^{(I)}(t) = \mathcal{U}_0(t)^*\widehat{\rho}(t)\mathcal{U}_0(t),$$

We easily see that

$$\widehat{\rho}_S^{(I)}(t) := \mathrm{Tr}_E(\widehat{\rho}^{(I)}(t)) = e^{it\widehat{H}_S}\widehat{\rho}_S(t)e^{-it\widehat{H}_S},$$

and $\mathscr{P}_S(t) = \mathrm{Tr}_S(\widehat{\rho}_S^I(t)^2)$. We shall use the simpler notation $\widehat{r}(t) = \widehat{\rho}_S^{(I)}(t)$. Using cyclicity of the trace, we obtain

$$\dot{\mathscr{P}}_S(t) = 2\mathrm{Tr}_S(\widehat{r}(t)\dot{\widehat{r}}(t)), \tag{13.22}$$

$$\ddot{\mathscr{P}}_S(t) = 2\left(\mathrm{Tr}_S(\widehat{r}(t)\ddot{\widehat{r}}(t)) + \mathrm{Tr}_S(\dot{\widehat{r}}(t))^2\right). \tag{13.23}$$

We have, at time $t = 0$, where $V = V(0)$,

$$\dot{\widehat{r}}(0) = i^{-1}\mathrm{Tr}_E[\widehat{V}, \widehat{\rho}_S \otimes \widehat{\rho}_E], \tag{13.24}$$

$$\ddot{\widehat{r}}(0) = i^{-1}\mathrm{Tr}_E[\dot{\widehat{V}}, \widehat{\rho}_S \otimes \widehat{\rho}_E] - \mathrm{Tr}_E[\widehat{V}, [\widehat{V}, \widehat{\rho}_S \otimes \widehat{\rho}_E]]. \tag{13.25}$$

Notice that

$$\mathrm{Tr}_S\left(\mathrm{Tr}_E[\dot{\widehat{V}}, \widehat{\rho}_S \otimes \widehat{\rho}_E]\widehat{\rho}_S\right) = 0. \tag{13.26}$$

This follows from the definition of the relative trace Tr_E and the cyclicity of the trace. Hence we get

$$\mathrm{Tr}_S\left(\mathrm{Tr}_E[\dot{\widehat{V}}, \widehat{\rho}_S \otimes \widehat{\rho}_E]\widehat{\rho}_S\right) = \mathrm{Tr}\left([\dot{\widehat{V}}, \widehat{\rho}_S \otimes \widehat{\rho}_E]\widehat{\rho}_S \otimes 1\right) = 0.$$

Finally we get formula (13.15) using the identity

$$[\widehat{V}, [\widehat{V}, \widehat{\rho}]] = \widehat{V}^2\widehat{\rho} + \widehat{\rho}\widehat{V}^2 - 2\widehat{V}\widehat{\rho}\widehat{V}.$$

□

13.2.2.2 Proof of the Corollaries

They are easily obtained after standard computations for traces of operators starting with formula (13.15).

13.3 Separability and Entanglement

13.3.1 Statements

We assume in this section that

$$\mathscr{H}_S = L^2(\mathbb{R}^d) \quad \text{and} \quad \mathscr{H}_E = L^2(\mathbb{R}^N),$$

and we use here properties of Weyl quantization (see Chap. 2 for properties used here). We have to consider partial transposition.

In the complex Hilbert space $L^2(\mathbb{R}^m)$, we define the bilinear form

$$u \cdot v = \int_{\mathbb{R}^m} u(x)v(x)\,dx.$$

If $\widehat{A}$ is a bounded operator on $L^2(\mathbb{R}^m)$, its transpose $\widehat{A}^\top$ is defined by the formula

$$\widehat{A}u \cdot v = u \cdot \widehat{A}^\top v, \quad u,\, v \in L^2(\mathbb{R}^m).$$

We easily see that the Weyl symbol $A^\top$ of $\widehat{A}^\top$ is given by $A^\top(x,\xi) = A(x,-\xi)$ and if $\widehat{A}$ is a quantum state (a density matrix), then $\widehat{A}^\top$ is also a quantum state.

Now we define the partial transpose in the environment space. Let $\psi_S^{(j)} \in \mathscr{H}_S$ and $\psi_E^{(j)} \in \mathscr{H}_E$, $j = 1, 2$. There exists a unique bounded operator $\widehat{A}_{\psi_S^{(1)},\psi_S^{(2)}}$ in $\mathscr{H}_E$ such that

$$(\psi_S^{(2)} \otimes \psi_E^{(2)}) \cdot \widehat{A}(\psi_S^{(1)} \otimes \psi_E^{(1)}) = \psi_E^{(2)} \cdot \widehat{A}_{\psi_S^{(1)},\psi_S^{(2)}} \psi_E^{(1)}.$$

Definition 31 The partial transposition of $\widehat{A}$ is a unique bounded operator $\widehat{A}^{\top E}$ in $\mathscr{H}_S \widehat{\otimes} \mathscr{H}_E$ satisfying

$$\begin{aligned} \left(\psi_S^{(2)} \otimes \psi_E^{(2)}\right) \cdot \widehat{A}^{\top E}\left(\psi_S^{(1)} \otimes \psi_E^{(1)}\right) &= \psi_E^{(1)} \cdot \widehat{A}_{\psi_S^{(1)},\psi_S^{(2)}} \psi_E^{(2)}, \\ \psi_S^{(j)} \in \mathscr{H}_S, \quad \psi_E^{(j)} &\in \mathscr{H}_E. \end{aligned} \tag{13.27}$$

Proposition 133 *([156]) Let $\widehat{\rho}$ be a separable density matrix in $\mathscr{H} = \mathscr{H}_S \otimes \mathscr{H}_E$. Then the partial transpose $\widehat{\rho}^{\top E}$ is again a density matrix in $\mathscr{H}$.*

Using anti-Wick quantization (see Chap. 2), we get a sufficient condition for separability.

Proposition 134 *Assume that $\mathscr{H}_S = L^2(\mathbb{R}^d)$ and $\mathscr{H}_E = L^2(\mathbb{R}^N)$. Let $\widehat{\rho}$ be a mixed state in $L^2(\mathbb{R}^{d+N})$ such that $\widehat{\rho}$ admits an anti-Wick symbol defining a Radon probability measure in $\mathbb{R}^{2(d+N)}$. Then $\widehat{\rho}$ is separable. In particular, if $\widehat{\rho} = \rho^{(\Gamma,m)}$ is a Gaussian mixed state with covariance matrix Γ, mean m, and such that $\Gamma \geq \frac{1}{2}$, then $\rho^{(\Gamma,m)}$ is separable.*

A remarquable result is the following reverse statement for Proposition 133 in [156, 236].

Theorem 71 *Assume that $\mathscr{H}_S = L^2(\mathbb{R})$, $\mathscr{H}_E = L^2(\mathbb{R}^N)$. Let $\widehat{\rho}$ be a density matrix with a Gaussian Weyl symbol $\rho = G^{(\Gamma,m)}$. Then $\widehat{\rho}$ is separable if and only if its partial transpose $\widehat{\rho}^{\top E}$ is a density matrix.*

Theorem 71 was used in [92] to prove separability results for propagation of Gaussian states in the quantum Brownian motion (13.2), (13.3). We shall give later the details for a mathematical proof of these results.

Recall that a thermal state at temperature $\frac{1}{\beta}$ for the Hamiltonian H_E is the mixed state defined as

$$\widehat{\mathscr{T}_{\beta,E}} = \frac{\mathrm{e}^{-\beta H_E}}{\mathrm{Tr}(\mathrm{e}^{-\beta H_E})}. \tag{13.28}$$

We denote $\widehat{\rho_{g,\beta}}(t) = \mathscr{U}_g^*(t,0)\widehat{\rho}(0)\,\mathscr{U}_g(t,0)$, the time evolution of the initial density matrix $\widehat{\rho}(0) = \widehat{\rho_S}(0) \otimes \mathscr{T}_\beta$, by the Hamiltonian (13.2), (13.3).

Theorem 72 *For any quantum Brownian motion* (13.2), (13.3) *with $\omega_S > 0$, $\omega_j > 0$, $1 \leq j \leq N$, we have the following results.*

(A) *Assume that the coupling g is time independent. Then for any quantum Brownian motion model with $\omega_j > 0$, $1 \leq j \leq N+1$, sufficiently small (see* (13.32)*), there exists a Gaussian initial density matrix $\widehat{\rho_S}(0)$ and a temperature $\frac{1}{\beta} > 0$ such that $\widehat{\rho_{g,\beta}}(t)$ is separable for any $t \in \mathbb{R}$.*

(B) *If furthermore $g(0) \neq 0$ then, for any pure state $\widehat{\rho_S}(0)$ and for any temperature $\frac{1}{\beta} > 0$ there exists $\varepsilon > 0$ such that $\widehat{\rho_{g,\beta}(t)}$ is not separable for any $t \in]0, \varepsilon[$.*

Remark 79 The results (A) and (B) of Theorem 72 were stated in [92]. A mathematically rigorous proof of (B) was given in [220].

13.3.2 Proofs of the Separability Results

First we recall some properties concerning Gaussian density matrices. We recall the definition.

Definition 32 An operator $\widehat{G}$ in the Hilbert space $L^2(\mathbb{R}^n)$ is said to be Gaussian if its Weyl symbol G is a Gaussian in the phase space $\mathbb{R}^{2n}$, denoted by $G^{\Gamma,m}$ where Γ

is the covariance matrix (a positive definite $(2n \times 2n)$-matrix) and $m \in \mathbb{R}^{2d}$ is the mean of G. So we have

$$G^{(\Gamma,m)}(z) = c_\Gamma e^{-\frac{1}{2}(z-m)\cdot\Gamma^{-1}(z-m)}, \quad z = (x,\xi) \in \mathbb{R}^{2n}, \tag{13.29}$$

where $c_\Gamma = \det(\Gamma)^{-1/2}$.

Some condition is needed on the covariance matrix Γ such that (13.29) defines a nonnegative operator in $L^2(\mathbb{R}^n)$ (see Remark 80).

Here J is the matrix of the symplectic form on $\mathbb{R}^n \times \mathbb{R}^n$ and $X = (x,\xi) \in \mathbb{R}^{2n}$.

Proposition 135 *$G^{\Gamma,m}(X)$ defines a density matrix if and only if $\Gamma + i\frac{J}{2} \geq 0$, or equivalently, if and only if all the symplectic eigenvalues of Γ are greater than $\frac{1}{2}$. Moreover, up to conjugation by a unitary metaplectic transform in $L^2(\mathbb{R}^n)$, $\rho^{\Gamma,0}$ is a product of one degree of freedom thermal states $\mathcal{T}_\beta$ (see Remark 80). Next, $\rho^{\Gamma,m}(X)$ determines a pure state if and only if 2Γ is positive and symplectic or equivalently if $2\Gamma = FF^\top$, where F is a linear symplectic transformation.*

For an elementary proof of Proposition 135 we refer to [220].
From these results we can compute the purity (hence the linear entropy):

$$\mathcal{P}_{\widehat{G}^{\Gamma,m}} = (2\pi)^{-n} \int_{\mathbb{R}^{2d}} G^{\Gamma,m}(z)^2\, dz = 2^{-n}(\det\Gamma)^{-1/2}. \tag{13.30}$$

Remark 80 From Proposition 135 we claim that the environment state $\widehat{\rho}_{\tau,E}$ is nonnegative if and only if $\tau \leq 1$. We can give a direct proof of this. It suffices to prove the claim for $N = 1$. By a holomorphic continuation argument in τ, we can see that, for $\tau > 1$, $\widehat{\rho}_{\tau,E}$ is negative on the subspace of $L^2(\mathbb{R})$ spanned by the odd Hermite functions.

13.3.2.1 Proof of Propositions 133 and 134

To prove Proposition 133 (and the necessity part of Theorem 71), we remark that $(\widehat{\rho}_S \otimes \widehat{\rho}_E)^{\top E} = \widehat{\rho}_S \otimes (\widehat{\rho}_E)^\top$ and that for any $\psi_E \in \mathcal{H}_E$ we have $\langle \psi_E, \widehat{\rho}_E^\top \psi_E\rangle = \langle \bar{\psi}_E, \widehat{\rho}_E \bar{\psi}_E\rangle$. So we see that $\widehat{\rho}^{\top E}$ is a nonnegative operator. It is easy to check that it is in trace class so it is a density matrix. □

The proof of Proposition 134 results from properties of coherent states. Recall the notation of Chap. 1 for coherent states φ_z. Here we denote $z = (z_S, z_E)$, $z_S \in \mathbb{R}^{2d}$, $z_E \in \mathbb{R}^{2N}$. So from the definition of coherent states we have $\varphi_z = \varphi_{z_S} \otimes \varphi_{z_E}$ and $\Pi_z = \Pi_{z_S} \otimes \Pi_{z_E}$, where Π_z is the orthogonal projection to φ_z. If ρ^{aw} is the anti-Wick symbol of $\widehat{\rho}$, then we have

$$\widehat{\rho} = \int_{\mathbb{R}^{2(d+N)}} (\Pi_{z_S} \otimes \Pi_{z_E})\rho^{aw}(dz_S, dz_E).$$

Usually it is assumed that an anti-Wick symbol is L^∞. But the definition can be extended to finite Radon measures on the phase space.

Assume now that $\rho = \rho^\Gamma$. The Fourier transform $\widetilde{\rho}$ of ρ satisfies

$$\widetilde{\rho}(Y) = \mathrm{e}^{-\frac{\Gamma}{2}Y\cdot Y} = \widetilde{\rho}^{aw}(Y)\mathrm{e}^{-\frac{1}{2}Y\cdot Y}.$$

So ρ^{aw} defines a Schwartz distribution if and only if $\Gamma \geq \frac{1}{2}$. If this condition is satisfied, ρ^{cw} is a finite Radon measure. Hence we have proved Proposition 134. □

13.3.2.2 Proof of Theorem 71

Here $\rho = G^{(\Gamma,m)}$. It suffices to assume that $m = 0$ and denote $G^{(\Gamma)} = G^{(\Gamma,0)}$. If $G^{(\Gamma)}$ is separable, we saw in Proposition 133 that $G^{(\Gamma),\top E}$ is a density matrix.

The reverse property is more difficult to prove. For simplicity, we give the details only for $N = 1$ following [236]. The general case is proved by induction on N (see [258])

The coordinates in the phase space for $X \in \mathbb{R}^4$ are denoted as follows:

$$X = (x_1, \xi_1, x_2, \xi_2),$$

and the matrices are written as 4 blocks of (2×2)-matrices:

$$\Gamma = \begin{pmatrix} A & C \\ C^\top & B \end{pmatrix}, \quad J = \begin{pmatrix} J_1 & 0 \\ 0 & J_1 \end{pmatrix},$$

where A, B are symmetric, $J_1 = \begin{pmatrix} 0 & 1 \\ -1, & 0 \end{pmatrix}$.

We know that $\Gamma + \frac{i}{2}J \geq 0$ (Proposition 135) so we have $A \geq \frac{\mathbb{1}}{2}$ and $B \geq \frac{\mathbb{1}}{2}$. To prove that $G^{(\Gamma)}$ is separable, we have to prove that $\Gamma \geq \frac{1}{2}$ (Proposition 134). From the assumption we know that $G^{(\Gamma),\top E}$ is a mixed Gaussian state with the covariance matrix

$$\Gamma^{\top E} = \begin{pmatrix} A & C\sigma \\ \sigma C^\top & \sigma B\sigma \end{pmatrix}; \quad \sigma = \begin{pmatrix} 1 & 0 \\ 0 & -1 \end{pmatrix}.$$

So we also have $\Gamma^{\top E} + \frac{i}{2}J \geq \frac{1}{2}$.

We remark here that separability for states defined in $L^2(\mathbb{R}^2)$ is preserved by the metaplectic transformations associated with the symplectic transformations of the form $S = \begin{pmatrix} S_1 & 0 \\ 0 & S_2 \end{pmatrix}$, where S_1, S_2 are (2×2)-symplectic matrices. Matrices like these constitute a subgroup, denoted by $Sp_{1,1}(\mathbb{R})$, of the symplectic group $Sp_2(\mathbb{R})$ of $\mathbb{R}^2_{x_1,\xi_1} \times \mathbb{R}^2_{x_2,\xi_2}$. So it suffices to prove that there exists $S \in Sp_{1,1}(\mathbb{R})$ such that $S^\top \Gamma S \geq \frac{1}{2}$. This is proved in two steps. A, B being real and symmetric, they can be diagonalized by rotations R_1, R_2. So there exits $S_1 = \begin{pmatrix} R_1 & 0 \\ 0 & R_2 \end{pmatrix}$ such that

$$S_1^\top \Gamma S_1 = \Gamma_1 := \begin{pmatrix} a\mathbb{1}_2 & C' \\ C'^{,\top} & b\mathbb{1}_2 \end{pmatrix}.$$

But we have $C' = E R_3$, where E is symmetric and R_3 is a rotation. Then there exist rotations R_1', R_2' such that if $S_2 = \begin{pmatrix} R_1' & 0 \\ 0 & R_2' \end{pmatrix}$, we have

$$S_2^\top \Gamma S_2 = \Gamma_1 := \begin{pmatrix} a & 0 & c_1 & 0 \\ 0 & a & 0 & c_2 \\ c_1 & 0 & b & 0 \\ 0 & c_2 & 0 & b \end{pmatrix}$$

with $a, b \geq \frac{1}{2}$, $c_1, c_2 \in \mathbb{R}$.

Assume that $\det C > 0$. We may assume that $c_1 > 0$, $c_2 > 0$.

At the second step we conjugate Γ_1 by a family of squeezing symplectic transformations $\Sigma(r, s) = \Sigma(r) \otimes \Sigma(s)$, where $\Sigma(r) = \begin{pmatrix} r & 0 \\ 0, & \frac{1}{r} \end{pmatrix}$, $r, s \in]0, \infty[$. Choosing $\frac{s}{r} = \theta := \left(\frac{ac_2+bc_2}{ac_1+bc_2}\right)^{1/2}$, we can find a symplectic rotation R in $\mathbb{R}^4$ such that $\Gamma_2 := \Sigma(\theta s, s)^\top \Gamma_1 \Sigma(\theta s, s)$ can be diagonalized. Then we can choose s such that the smallest eigenvalue of Γ_2 is greater than $\frac{1}{2}$. So we get Theorem 71 if $\det C > 0$. A slight modification is needed if $\det C = 0$. Now if $\det C < 0$, we have $\det(C^{\top E}) > 0$, then $G^{(\Gamma)\top E}$ is separable so $G^{(\Gamma)}$ is separable. □

13.3.3 Proof of Theorem 72

(A) Assuming the coupling constants g and β sufficiently small, we have to find an initial Gaussian density matrix G_S such that the time evolution $\widehat{\rho}_{\gamma,\beta}(t)$ of the initial state $G_S \otimes \mathcal{T}_{\beta,E}$ stays separable at any time. It is well known that $\widehat{\rho}_{V,\beta}(t)$ is a Gaussian state (for computations see Sect. 13.4). Denote by $\Gamma_{g,\beta}(t)$ the covariance matrix of $\widehat{\rho}_{g,\beta}(t)$ and by $\Gamma_{g,\beta}^{\top E}(t)$ the covariance matrix of $\widehat{\rho}_{g,\beta}^{\top E}(t)$.

If Λ_E denotes the $(2N+2) \times (2N+2)$-matrix defined by the reflection

$$(y_1, y', \eta_1, \eta') \mapsto (y_1, y', \eta_1, -\eta'),$$

for $y_1, \eta_1 \in \mathbb{R}$, $y', \eta' \in \mathbb{R}^N$, denote

$$M^{\top E} = \Lambda_E M \Lambda_E,$$

where M is any $(2N+2)\times(2N+2)$-matrix. According to Theorem 71 and Proposition 135, we have to prove that

$$\Gamma^{\top E}_{g,\beta}(t)+\frac{i}{2}J\geq 0,\quad \text{or}\quad \Gamma_{g,\beta}(t)+\frac{i}{2}J^{\top E}\geq 0. \tag{13.31}$$

Let us introduce the full potential, where $y_1 = x$ and $y' = (y_2, y_3, \ldots, y_{N+1})$,

$$W_g(y_1, y') = \frac{1}{2}\Big(\sum_{1\leq j\leq N+1}\omega_j^2 y_j^2\Big) + y_1\sum_{2\leq j\leq N+1} g_j y_j.$$

W_g is a quadratic form on $\mathbb{R}^{N+1}$. Its matrix is denoted by M_g. It is positive definite if the following condition is satisfied:

$$\max_{2\leq j\leq N+1}\frac{|g_j|}{2\omega_1\omega_j} < 1. \tag{13.32}$$

Hence there exists a rotation R in $\mathbb{R}^{1+N}$ such that $RM_gR^\top := D_g$ is a diagonal matrix with eigenvalues $\widetilde{\omega}_j^2$, $1\leq j\leq N+1$.

So if $\mathbf{R} := \begin{pmatrix} R & 0\\ 0 & R\end{pmatrix}$, the quadratic form H_V is transformed in a sum of independent oscillators:

$$\widetilde{H}(x,\xi) := H(\mathbf{R}(x,\xi)) = \frac{1}{2}\Big(\sum_{1\leq j\leq N+1}\xi_j^2+\widetilde{\omega}_j^2 x_j^2\Big). \tag{13.33}$$

The thermal state at temperature $\frac{1}{\delta}$ of $\widetilde{H}_V$ ($V=V_g$) is denoted by $\mathscr{T}_{\delta\widetilde{H}}$ and can be computed by using Remark 80. Denote $\tau(u)=\tanh(u/2)$, $u\in\mathbb{R}$. We have

$$\mathscr{T}_{\delta\widetilde{H}}(x,\xi) = 2^{-1-N}\prod_{1\leq jk\leq N+1}\tau(\delta\widetilde{\omega}_j)\exp\Big(-\sum_{1\leq j\leq N+1}\tau(\delta\widetilde{\omega}_j)\Big(\widetilde{\omega}_j x_j^2+\frac{1}{\widetilde{\omega}_j}\xi_j^2\Big)\Big). \tag{13.34}$$

If $\Gamma_{\delta\widetilde{H}}$ denotes the covariance matrix of the Gaussian state $\widehat{\mathscr{T}}_{\delta\widetilde{H}}$, elementary estimates show that we have

$$2\Gamma_{\delta\widetilde{H}}\geq \mathbb{1}_{2+2N}\quad\text{if}\quad 0<\delta\leq\ell(\widetilde{\omega}_j),\quad\text{for}\quad 1\leq j\leq 1+N, \tag{13.35}$$

where $\ell(u)=\frac{1}{u}\ln\left(\frac{1+u}{|1-u|}\right)$ if $u\neq 1$ and $\ell(1)=\infty$. Using the relation

$$\Gamma_{\delta H} = \mathbf{R}\Gamma_{\delta\widetilde{H}}\mathbf{R}^\top,$$

we get (13.35) for $\Gamma_{\delta H}$.

Denote by $\Phi^t_{H_V}$ the linear flow defined by the quadratic Hamiltonian H_V. Then we have $\Gamma_{g,\beta}(t) = \Phi^t_{H_V}\Gamma_{g,\beta}(0)\Phi^{t,\top}_{H_V}$, where

$$\Gamma_{g,\beta}(0) = \begin{pmatrix} \Gamma_0^S & 0 \\ 0 & \Gamma_{\beta H_E} \end{pmatrix}.$$

So we write

$$\Gamma_{g,\beta}(t) + \frac{i}{2}J^{\top E} = \Phi^t_{H_V}(\Gamma_{g,\beta}(0) - \Gamma_{\delta H_V})\Phi^{t,\top}_{H_V} + \Gamma_{\delta H_V} + \frac{i}{2}J^{\top E}. \qquad (13.36)$$

To conclude, we see that it is possible to choose the (2×2)-symmetric matrix Γ_0^S and β such that the symmetric matrix

$$E_{g,\beta,\delta} := \Gamma_{g,\beta}(0) - \Gamma_{\delta H}$$

is diagonally dominant [30]. Hence, using the Gersgorin Theorem [30] and (13.35), we see that $E_{g,\beta,\delta}$ is nonnegative hence we get (13.31). □

(B) Now we assume that $\widehat{\rho}_S(0)$ is a pure state. Then Γ_0^S is a symplectic symmetric matrix. Here it suffices to assume that $N = 1$. With the notation of part (A), let us denote $\Gamma_t := \Gamma_{g,\beta}(t)$, we have to prove that the Hermitian matrix $2\Gamma_t + iJ^{\top E}$ has at least one negative eigenvalue for $0 < t < \varepsilon$ if $g(0) \neq 0$. Equivalently, we have to prove that the smallest eigenvalue, $\lambda_{\min}(t)$, of $4|iJ^{\top E}\Gamma_t|^2$ is smaller than 1 for $t \in]0, \varepsilon[$.

Recall that $\Gamma_t = \begin{pmatrix} A_t & C_t \\ C_t^\top & B_t \end{pmatrix}$. The eigenvalues of $4|iJ^{\top E}\Gamma_t|^2$ satisfiy

$$\lambda^4 - 4\lambda^2(\det A_t + \det B_t - 2\det C_t) + 16\det\Gamma_t = 0.$$

So we have

$$\lambda_{\min}(t)^2 = 2\left(d_t - \sqrt{d_t^2 - 4\gamma_t}\right) =: 2v(t), \qquad (13.37)$$

where $d_t = a_t + b_t - 2c_c$, $a_t = \det A_t$, $b_t = \det B_t$, $c_t = \det C_t$, $\gamma_t = \det\Gamma_t$. In particular, $\lambda_{\min}(0) = 1$.

The initial data Γ_0 is chosen here as the diagonal matrix

$$\Gamma_0 = \mathrm{Diag}\left(\frac{r}{2}, \frac{1}{2r}, \frac{s}{\omega_2}, s\omega_2\right)$$

where $r > 0$, $s > 1$, $\omega_2 > 0$. The parameter s is related to the temperature of the thermal state: $s = \coth(\beta\omega_2/2)$.

We are going to prove that $\dot{v}(0) = 0$ and $\ddot{v}(0) < 0$. In [92] it was claimed that $\dot{v}(0) < 0$, which is not correct.

Recall here the Liouville formula to compute the derivative of the determinant for an invertible matrix $M(t)$, smooth in time t:

$$\frac{d}{dt}\det(M(t)) = \det(M(t))\mathrm{Tr}\Big(M(t)^{-1}\frac{d}{dt}(M(t))\Big). \tag{13.38}$$

The time evolution of Γ_t is given by the classical flow determined by the quadratic classical Hamiltonian $H_V(X) = \frac{1}{2}X \cdot S(t)X$, where $S(t)$ is a symmetric matrix. Denote by Φ^t this flow. We have

$$\dot{\Phi}^t = JS(t)\Phi^t \quad \text{and} \quad \Gamma_t = (\Phi^t)^\top \Gamma_0 \Phi^t. \tag{13.39}$$

As a consequence of (13.37) and (13.39), we can compute $\dot{v}(0)$ and $\ddot{v}(0)$. Denote $\check{S} := JS$, $\Gamma = \Gamma(0)$, $S = S(0)$. We compute

$$\dot{\Gamma} = -\check{S}\Gamma - \Gamma\check{S}^\top, \tag{13.40}$$
$$\ddot{\Gamma} = 2\check{S}\Gamma\check{S}^\top + \check{S}^2\Gamma + \Gamma(\check{S}^\top)^2. \tag{13.41}$$

For the first derivative we get

$$\dot{\Gamma} = \begin{pmatrix} 0 & \frac{r}{2}\omega_1^2 - \frac{1}{2r} & 0 & \kappa\frac{r}{2} \\ \frac{r}{2}\omega_1^2 - \frac{1}{2r} & 0 & \frac{\kappa s}{2\omega_2} & \\ 0 & \frac{\kappa s}{2\omega_2} & 0 & 0 \\ \kappa\frac{r}{2} & 0 & 0 & 0 \end{pmatrix}. \tag{13.42}$$

From (13.42) using (13.38) and (13.37), we deduce easily that $\dot{d} = \dot{\gamma} = 0$, hence $\dot{v} = 0$.

For the second derivatives similar computations give:

$$\ddot{a} = 2\kappa^2\frac{rs}{\omega_2}, \quad \ddot{b} = \kappa^2\frac{rs}{2\omega_2}, \quad \ddot{c} = \kappa^2\frac{rs}{2\omega_2}, \quad \ddot{\gamma} = -\kappa^2\frac{rs}{16\omega_2}(s^2 - 2s - 1). \tag{13.43}$$

We get finally

$$\ddot{v} = \kappa^2\frac{rs}{\omega_2}\left(\frac{3}{2} - \frac{4s^2 - 2s + 2}{s^2 - 1}\right) \le -3.9\frac{rs}{\omega_2}\kappa^2. \tag{13.44}$$

Hence $t \mapsto \lambda_{\min}(t)$ is strictly concave at $t = 0$ (see the following figure). Morever, in this figure we see that $v(t)$ oscillates: above 0.5 the state is separable, below 0.5 the state is entangled. □

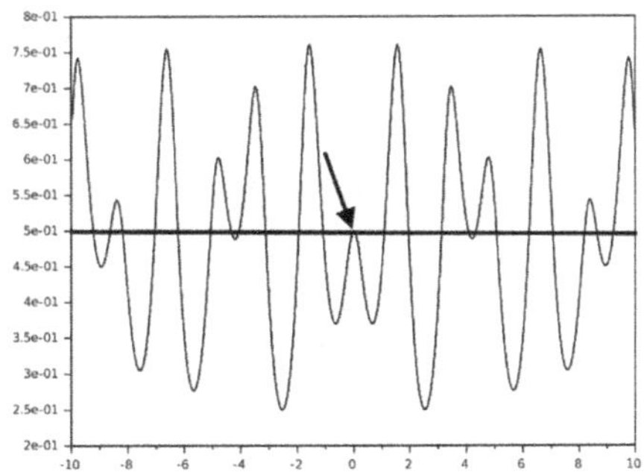

Time dependent oscillations between separability and entanglement.

13.4 The Master Equation in the Quadratic Case

13.4.1 General Quadratic Hamiltonians

In this section we get a time dependent partial differential equation (often called the master equation) satisfied by the reduced density matrix $\widehat{\rho}_S$ of the open system (S). Moreover, this equation can be solved explicitly using the well-known characteristics method.

The quantum Brownian motion model with bilinear position coupling was considered in particular in [103, 106, 136, 158]. Most of papers use the path integral method. In [85, 86] the authors put emphasis on a Fokker–Planck type equation.

Our aim here is to give a different mathematical presentation, with straightforward general computations for the time evolution of the reduced density matrix $\widehat{\rho}_S(t)$.

Here we assume that H_S and H_E are the quadratic Weyl symbols of $\widehat{H}_S$ and $\widehat{H}_E$, respectively. We denote by Φ_0^t the Hamiltonian flow generated by H_0, and by Φ^t the Hamiltonian flow generated by $H = H_0 + V(t)$. The symbol $V(t)$ stands for a C^1-time dependent bilinear coupling between the system and the environment:

$$V(t, z, u) = z \cdot G(t)u,$$

where $G(t)$ is a linear map from $\mathbb{R}^{2N}$ to $\mathbb{R}^{2d}$ (note that the coupling can mix positions and momenta variables).

Recall that $\mathbb{R}^{2d}$ (respectively $\mathbb{R}^{2N}$) is the classical phase space of the system (respectively, the environment). Next, $\Psi^t = \Phi_0^{-t}\Phi^t$ is the interacting flow in the global phase space $\mathbb{R}^{2(d+N)}$. The phase space of the global system is identified with the direct sum $\mathbb{R}^{2d} \oplus \mathbb{R}^{2N}$ and in this decomposition the flows are represented by 4 matrix blocks

$$\Psi^t = \begin{pmatrix} \Psi^t_{ii} & \Psi^t_{ie} \\ \Psi^t_{ei} & \Psi^t_{ee} \end{pmatrix} \quad \text{and} \quad \Psi^t_0 = \begin{pmatrix} \Phi^t_S & 0 \\ 0 & \Phi^t_E \end{pmatrix}, \quad \text{with} \quad \Psi^0_0 = \Psi^0 = \mathbb{1}_{\mathbb{R}^{2d+2N}}.$$

The interacting classical Hamiltonian is

$$H_I(t, z, u) = V(t, z, u) = z \cdot G^{(I)}(t)u, \quad \text{where} \quad G^{(I)}(t) = (\Phi^t_S)^\top \cdot G(t) \cdot \Phi^t_E.$$

The classical interacting evolution is given by the equation

$$\dot{\Psi}^t = J\nabla^2_{z,u} V(t)\Psi^t, \quad \text{with } J = \begin{pmatrix} J_S & 0 \\ 0 & J_E \end{pmatrix},$$

where $\nabla^2_{z,u} V(t)$ is the Hessian matrix in variables $(z, u) \in \mathbb{R}^{2d} \times \mathbb{R}^{2N}$. So for the block components of the interacting dynamics we have

$$\dot{\Psi}^t_{ii} = J_S G(t)\Psi^t_{ei}, \quad \dot{\Psi}^t_{ie} = J_S G(t)\Psi^t_{ee}, \tag{13.45}$$

$$\dot{\Psi}^t_{ei} = J_E G(t)^\top \Psi^t_{ii}, \quad \dot{\Psi}^t_{ee} = J_E G(t)^\top \Psi^t_{ie}. \tag{13.46}$$

Since all the Hamiltonians considered here are quadratic (eventually time dependent), they generate well-defined quantum dynamics in Hilbert spaces $L^2(\mathbb{R}^n)$ where $n = d$ for the system (S), $n = N$ for the environment (E)), and $n = d + N$ for the global system $(S \cup E)$.

Recall the notations

$$\mathscr{U}(t) = \mathrm{e}^{-it\widehat{H}}, \quad \mathscr{U}_0(t) = \mathrm{e}^{-it\widehat{H}_0}, \quad \mathscr{U}_S(t) = \mathrm{e}^{-it\widehat{H}_S}, \quad \mathscr{U}_E(t) = \mathrm{e}^{-it\widehat{H}_E}.$$

We have $\mathscr{U}_0(t) = \mathscr{U}_S(t) \otimes \mathscr{U}_E(t)$ and the quantum interacting dynamics $\mathscr{U}_I(t)$ is $\mathscr{U}_0(t)^*\mathscr{U}(t)$.

At time $t = 0$ we assume that the states of the system and of the environment are decoupled: $\widehat{\rho}(0) = \widehat{\rho}_S(0) \otimes \widehat{\rho}_E(0)$, where $\widehat{\rho}_S$ (respectively, $\widehat{\rho}_E$) is a density matrix in the Hilbert space $\mathscr{H}_S = L^2(\mathbb{R}^d)$ (respectively in $\mathscr{H}_E = L^2(\mathbb{R}^N)$);
ρ, ρ_S, ρ_E are the Weyl symbols (i.e., the Weyl–Wigner functions, with $\hbar = 1$, of the corresponding density matrices).

The coefficients of the quadratic forms H_S, H_E, V may be time dependent. In this case $\mathscr{U}(t)$ means $\mathscr{U}(t, 0)$, where $\mathscr{U}(t, s)$ is the propagator solving

$$i\frac{\partial}{\partial t}\mathscr{U}(t, s) = \widehat{H}(t)\mathscr{U}(t, s), \quad \mathscr{U}(s, s) = \mathbb{1}_{\mathscr{H}}.$$

We denote $\mathscr{U}(t) := \mathscr{U}(t, 0)$. It is well known that for quadratic Hamiltonians, every mixed state $\widehat{\rho}$ of $(S \cup E)$ propagates in accordance with the classical evolution (see Chap. 3):

$$\widehat{\rho}(t) := \mathscr{U}^*(t)\widehat{\rho}(0)\mathscr{U}(t), \quad \rho(t, X) = \rho(0, \Phi^t X), \quad X \in \mathbb{R}^{2(d+N)}. \tag{13.47}$$

Our aim is to compute $\widehat{\rho}_S(t) = \mathrm{Tr}_E(\widehat{\rho}(t))$. The Weyl symbol of $\widehat{\rho}_S(t)$ is given by the following integral:

$$\rho_S(t, z) = \int_{\mathbb{R}^{2N}} \rho(t, z, u)\, du. \tag{13.48}$$

In particular, if $\rho(0)$ is Gaussian in all the variables,

$$X = (z, u) \in \mathbb{R}^{2d} \times \mathbb{R}^{2N},$$

then $\rho_S(t, z)$ is Gaussian in z. Nevertheless, a direct computation on the formula (13.48) seems not easy to perform, so a different strategy will be used.

13.4.2 Time Evolution of Reduced Mixed States

Here we state and prove the main results of this section. It is convenient to work in the interacting setting. Recall that we have

$$\widehat{\rho}_S^{(I)}(t) = \mathcal{U}_S^*(t)\widehat{\rho}_S(0)\, \mathcal{U}_S(t)), \text{ hence } \rho_S^{(I)}(t, z) = \rho_S(t, \Phi_S^t z). \tag{13.49}$$

Theorem 73 *Let $t_c > 0$ be the largest time t_c such that Ψ_{ii}^t is invertible for every $t \in [0, t_c[$. Assume that the environment density matrix $\widehat{\rho}_E(0)$ is a Gaussian with mean m_E. Then there exist two time-dependent $(2d \times 2d)$-matrices $A^{(I)}(t)$ and $B^{(I)}(t)$, and a time dependent vector $v^{(I)}(t) \in \mathbb{R}^{2d}$ such that for $t \in [0, t_c[$ and for every $\rho_S(0)$ in $\mathscr{S}(\mathbb{R}^{2d})$, the Weyl symbol $\rho_S^{(I)}(t, z)$ of the interacting evolution $\widehat{\rho}_S^{(I)}(t)$ of the system (S) satisfies the following master equation (a Fokker–Planck type equation):*

$$\begin{aligned}\frac{\partial}{\partial t}\rho_S^{(I)}(t, z) = (A^{(I)}(t)\nabla_z) \cdot z\rho_S^{(I)}(t, z) + (B^{(I)}(t)\nabla_z) \cdot \nabla_z\rho_S^{(I)}(t, z)\\ + v^{(I)}(t) \cdot \nabla_z\rho_S^{(I)}(t, z).\end{aligned} \tag{13.50}$$

Moreover, we have the following formula to compute $A^{(I)}(t)$ and $B^{(I)}(t)$:

$$A^{(I)}(t)^\top = -J_S G(t)\Psi_{ei}^t(\Psi_{ii}^t)^{-1}, \quad B^{(I)}(t) = \frac{L(t) + L(t)^\top}{2}, \tag{13.51}$$

where

$$L(t) = J_S \cdot G(t)\left(\Psi_{ee}^t - \Psi_{ei}^t(\Psi_{ii}^t)^{-1}\Psi_{ie}^t\right)\mathrm{Cov}_{\rho_E}(\Psi_{ie}^t)^\top, \tag{13.52}$$

$$v^{(I)}(t) = -J_S G(t)\Psi_{ee}^t m_E. \tag{13.53}$$

Proof In this proof (and only here) we shall erase the upper index (I) for the interacting dynamics. Taking the partial trace in Eq. (13.5), we have

$$\dot{\rho}_S(t,z) = \int_{\mathbb{R}^{2N}} \{V(t), \rho(t)\}(z,u)\, du.$$

For simplicity we shall assume that $m_E = 0$. It is not difficult to take this term into account.

The Poisson bracket is taken in the variables (z, u) but due to integration in u we have only to consider the Poisson bracket in z. Hence we get

$$\dot{\rho}_S(t,z) = -\int_{\mathbb{R}^{2N}} J_S G(t)u \cdot \nabla_z \rho(0, \Psi^t(z,u))\, du. \tag{13.54}$$

Denote by $\widetilde{f}(\zeta)$ the Fourier transform of f in the variable z. Then we have

$$\widetilde{\dot{\rho}}_S(t,\zeta) = i\zeta \cdot \iint_{\mathbb{R}^{2(d+N)}} J_S G(t)u\rho(0, \Psi^t(z,u) e^{-iz\cdot\zeta}\, dz\, du.$$

Now we perform the symplectic change of variables $(z', u') = \Psi^t(z, u)$,

$$\begin{aligned}\widetilde{\dot{\rho}}_S(t,\zeta) &= i\zeta \\ &\times \int_{\mathbb{R}^{2(d+N)}} dz'\, du'\, J_S G(t)(\Psi^t_{ei} z' + \Psi^t_{ee} u')\rho_S(z')\rho_E(u') e^{-i\zeta\cdot(\Psi^t_{ii} z' + \Psi^t_{ie} u')}.\end{aligned} \tag{13.55}$$

Denoting $\varphi = \Psi^t_{ii} z' + \Psi^t_{ie} u'$ and using the identity

$$i(\Psi^t_{ii})^{-1}\nabla_\zeta e^{-i\zeta\cdot\varphi} = (z' + (\Psi^t_{ii})^{-1}\Psi^t_{ie} u') e^{-i\zeta\cdot\varphi},$$

we get

$$\begin{aligned}\widetilde{\dot{\rho}}_S(t,\zeta) =& -\zeta \cdot J_S G(t)\Psi^t_{ei}(\Psi^t_{ii})^{-1}\nabla_\zeta \widetilde{\rho}_S(t,\zeta) \\ &- i\zeta \cdot \int_{\mathbb{R}^{2(d+N)}} dz'\, du'\, J_S G(t)\Psi^t_{ei}(\Psi^t_{ii})^{-1}\Psi^t_{ie}(u')\rho_E(u')\rho_S(z') e^{-i\zeta\cdot\varphi} \\ &+ i\zeta \cdot \int_{\mathbb{R}^{2(d+N)}} dz'\, du'\, J_S G(t)(\Psi^t_{ee} u')\rho_E(u')\rho_S(z') e^{-i\zeta\cdot\varphi}.\end{aligned} \tag{13.56}$$

To absorb the linear terms in u', we use the fact that ρ_E is Gaussian, $\rho_E = c_\Lambda e^{-1/2u\cdot\Lambda u}$, where Λ is a positive definite $(2N \times 2N)$-matrix and c_Λ a normalization constant. So we have $\Lambda^{-1}\nabla_u \rho_E(u) = u\rho_E(u)$, and integrating by parts

we have

$$\dot{\widetilde{\rho}}_S(t,\zeta) = -\zeta \cdot A^{(I)}(t)^\top \nabla_\zeta \widetilde{\rho}_S(t,\zeta) - (\zeta \cdot L(t)\zeta)\widetilde{\rho}_S(t,\zeta). \tag{13.57}$$

We get (13.50) by taking inverse Fourier transforms.

Corollary 44 *In the notation of Theorem 73, the reduced density matrix for the system satisfies the following master equation*:

$$\begin{aligned}\frac{\partial}{\partial t}\rho_S(t,z) &= \{H_S, \rho_S(t)\}(z) \\ &\quad + (A(t)\nabla_z)\cdot z\rho_S(t,z) + (B(t)\nabla_z)\cdot\nabla_z\rho_S(t,z) \\ &\quad + v(t)\cdot\nabla_z\rho_S(t,z), \quad \textit{for} \quad t \in [0, t_c[.\end{aligned} \tag{13.58}$$

Remark 81 The function $\rho_S(t,z)$ is of course well defined for every time $t \in \mathbb{R}$ but the coefficients of the master equation (13.50) may have singularity at $t = t_c$.

Denote $\gamma = \sup_{t\geq 0}\|G(t)\|$ and $\varphi(t) = \|\Phi_S^t\| + \|\Phi_E^t\|$.

Proposition 136 *If* $\sup\limits_{t\geq 0}\varphi(t) < +\infty$ *then there exists* $c > 0$ *such that* $t_c \geq \frac{c}{g}$.

If there exist C, $\delta > 0$ *such that* $\varphi(t) \leq C\mathrm{e}^{\delta t}$ *for every* $t > 0$, *then there exists* $C_1 \in \mathbb{R}$ *such that*

$$t_c \geq \frac{1}{\delta}\log(\frac{1}{g}) + C_1.$$

The constants δ *and* C_1 *are independant of the coupling strength* g.

Proof It suffices to work in the interaction representation. Using interacting time evolution Ψ^t for the total classical system $(S \cup E)$, we get

$$\|\Psi^t - \mathbb{1}\| \leq g\int_0^t \varphi(s)ds + g\int_0^t \varphi(s)\|\Psi^s - \mathbb{1}\|ds.$$

Denote $\phi(t) = \int_0^t \varphi(t)ds$. Using the Gronwall lemma and integrating by parts, we get

$$\|\Psi^t - \mathbb{1}\| \leq g\phi(t) + \frac{g^3}{6}\phi(t)^3\mathrm{e}^{g\phi(t)}. \tag{13.59}$$

From inequality (13.59), we easily get the proposition.

Remark 82 Explicite computations on the model of coupled harmonic oscillators show that we may have $t_c < +\infty$ or $t_c = \infty$. Assume that $d = N = 1$, $V(x,y) = gxy$, $\widehat{H} = \widehat{H}_S + \widehat{H}_E + V$,

$$\widehat{H}_S = \frac{1}{2}(-\partial^2_{x^2} + \omega_S{}^2 x^2), \quad \widehat{H}_E = \frac{1}{2}(-\partial^2_{y^2} + \omega_E{}^2 y^2).$$

If $\omega_S^2 = 1$, $\omega_E^2 = -1$ and $g \neq 0$, then $0 < t_c < +\infty$. If $\omega_S^2 = 1 = \omega_E^2$ and $0 \leq g \leq \frac{1}{\sqrt{2}}$, then $t_c = +\infty$.

Remark 83 The coefficients of equation (13.50) are related to the first and second moments of the reduced density matrix $\rho_S(t)$. We denote

$$m_j^{(I)}(t) = \int_{\mathbb{R}^{2d}} z_j \rho_S^{(I)}(t, z)\, dz, \quad \mu_{jk}^{(I)}(t) = \int_{\mathbb{R}^{2d}} z_j z_k \rho_S^{(I)}(t, z)\, dz.$$

From (13.50), after computations using (13.46), we get

$$m^{(I)}(t) = \Psi_{ii}^t m^{(I)}(0). \tag{13.60}$$

Computations of the second moments, by using (13.54), (13.46), and (13.50), give

$$\mu^{(I)}(t) = \frac{1}{2}\left(\Psi_{ii}^{t\top} \mu^{(I)}(0) \Psi_{ii}^t + \Psi_{ie}^{t\top} \mu_E(0) \Psi_{ie}^t\right), \tag{13.61}$$

where $\mu_E(0)$ is the second moments matrix of $\rho_E(0)$.

Remark 84 Suppose that the initial state of the total system $(S \cup E)$ is Gaussian:

$$\rho(0, z, u) = c_\Gamma \mathrm{e}^{-(1/2)\Gamma^{-1}(z,u)\cdot(z,u)}, \quad c_\gamma > 0$$

is a normalization constant. Then we can compute $\rho_S(t, z)$. We get first the Fourier transform:

$$\widetilde{\rho}_S(t, \zeta) = \mathrm{e}^{-(1/2)\Gamma(t)(\zeta,0)\cdot(\zeta,0)},$$

where $\Gamma(t) = (\Phi^t)^\top \Gamma(\Phi^t)$. Using inverse Fourier transforms, we obtain

$$\rho_S(t, z) = c_{\gamma,t} \mathrm{e}^{-(1/2)\Gamma_S^{-1}(t) z\cdot z} \tag{13.62}$$

where $\Gamma_S(t)$ is the matrix of the positive definite quadratic form

$$\mathcal{Q}_t(\zeta, \zeta) := \Gamma(t)(\zeta, 0) \cdot (\zeta, 0)$$

on $\mathbb{R}^{2d}$ and $c_{\Gamma,t} = (2\pi)^{-d} \det^{1/2} \Gamma_t$.

We can see on this example what is the meaning of the critical times t_c. Assume that $\Gamma = \Gamma_S \oplus \Gamma_E$, where the $\Gamma_{S,E}$ are positive definite quadratic forms on $\mathbb{R}^{2d}$, respectively, on $\mathbb{R}^{2N}$. We have easily: $\mathcal{Q}_0(\zeta, \zeta) = \Gamma_S \zeta \cdot \zeta$ and

$$\mathcal{Q}_t(\zeta, \zeta) = \Gamma_S(\Phi_{ii}^t)\zeta \cdot (\Phi_{ii}^t)\zeta + \Gamma_E(\Phi_{ei}^t)\zeta \cdot (\Phi_{ei}^t)\zeta. \tag{13.63}$$

We see from (13.63) that the initial state $\widehat{\rho}_S(0)$ cannot be recovered by its evolution at time t_c: only the restriction of $\mathcal{Q}_0$ to $(\ker \Phi_{ii})^{\perp}$ is recovered by $\mathcal{Q}_{t_c}$. The physical interpretation is that a part of information contained in $\widehat{\rho}_S(t_c)$ has escaped in the environment represented here by Γ_E.

Now we consider more general initial $\rho_S \in \mathscr{S}(\mathbb{R}^d)$ for the system (S). The master equation (13.50) can easily be solved by the characteristics method after a Fourier transform. As in the proof of Theorem 73, $\widetilde{\rho}_S^{(I)}(t, \zeta)$ denotes the Fourier transform of $\rho_S^{(I)}(t)$.

Theorem 74 *Assume (for simplicity) that the mean of the environment state is* 0. *With the notation of Theorem* 73, *we have*:

$$\widetilde{\rho}_S^{(I)}(t, \zeta) = \widetilde{\rho}_S(0, (\Psi_{ii}^t)\zeta) \exp\left(-\frac{1}{2}\zeta \cdot \Theta^{(I)}(t)\zeta\right), \quad t \in \mathbb{R}, \tag{13.64}$$

where $\Theta^{(I)}(t) = \Psi_{ie}^t \mathrm{Cov}_{\rho_E(0)}(\Psi_{ie}^{t,\mathsf{T}})$.

Remark 85 The interpretation of the right-hand side in formula (13.64) is the following: the first factor is a transport term. The second term is a dissipation term due to the influence of the environment which is controlled by the nonnegative matrix $\Theta^{(I)}(t)$.
In formula (74) for the Brownian quantum motion model the matrix $\Theta^{(I)}(t)$ is named the "thermal covariance" when the environment is in a thermal equilibrium state. This term determines a dumping factor for the state of the small system S exponentially decreasing in the Fourier variable like $\mathrm{e}^{-\gamma|\Psi_{ie}^{t,\mathsf{T}}\zeta|^2}$ where $\gamma > 0$ is the smallest eigenvalue of the covariance matrix of ρ_E.

As far as Φ_{ii}^t is invertible, from (13.64) we see that the time evolution of $\widehat{\rho}_S$ is reversible but this is no longer true for $t \geq t_c$. From (13.64) we get an explicit representation formula for the reduced density of the system as a convolution integral.

Corollary 45 *With the notation of Theorem* 74, *we have*

$$\begin{aligned}
\rho_S^{(I)}(t, z) &= c_S^{(I)}(t) \int_{\mathbb{R}^{2d}} \rho_S\left(0, \Psi_{ii}^{(t,0)} z'\right) \exp\left(-\frac{1}{2}(z-z') \cdot \Theta^{(I)}(t)^{-1}(z-z')\right) dz', \\
\rho_S(t, z) &= c_S(t) \int_{\mathbb{R}^{2d}} \rho_S\left(0, \Phi_{ii}^{(t,0)} z'\right) \exp\left(-\frac{1}{2}(z-z') \cdot \Theta(t)^{-1}(z-z')\right) dz',
\end{aligned} \tag{13.65}$$

where

$$c_S^{(I)}(t) = (2\pi)^{-d/2} \det(\Psi_{ii}^t)^{-1} \det\left(\Theta^{(I)}(t)\right)^{-1/2}$$

and

$$c_S(t) = (2\pi)^{-d/2} \det(\Phi_{ii}^t)^{-1} \det(\Theta(t))^{-1/2}.$$

If $\det(\Phi^t_{ii}) = 0$ *or* $\det((\Phi^t_{ie}(\Phi^t_{ie})^\top) = 0$, *the integrals in* (13.65) *are defined as convolutions of distributions.*

Formula (13.65) shows clearly the damping influence of the environment on the system because $\Theta(t)$ is a non negative matrix under our assumptions. If $\Theta(t)$ is singular, then the Fourier transform of $\exp\left(-\frac{1}{2}\zeta \cdot \Theta(t)\zeta\right)$ is a distribution supported in some linear subspace of $\mathbb{R}^{2d}$ where the damping occurs.

13.4.3 About the Schrödinger Cat

Let us come back to the Schrödinger cat model 13.1.3: $\psi^{\mathrm{cat}} = \frac{1}{N}(\varphi_{z^+} + \varphi_{z^-})$.
Here we consider the quantum evolution of the cat state in a system of quantum oscillators like in the quantum Brownian motion:

$$\widehat{H}_S = \frac{1}{2}(-\partial^2_{x^2} + \omega_0{}^2 x^2),\quad \widehat{H}_E = \frac{1}{2}\sum_{1\le j\le N}(-\partial^2_{y_j^2} + \omega_j{}^2 y_j^2),$$

with the bilinear coupling defined by the Weyl symbol

$$V(x,\xi,y,\eta) = (x,\xi)\cdot G\cdot (y,\eta)^\top,$$

where G is a real $2\times 2N$ matrix, $(x,\xi)\in\mathbb{R}^2$, $(y,\eta)\in\mathbb{R}^{2N}$.
The Hilbert spaces are here $\mathscr{H}_{cat} = L^2(\mathbb{R})$, $\mathscr{H}_E = L^2(\mathbb{R}^N)$ and for the whole system $(S\cup E)$, $\mathscr{H} = L^2(\mathbb{R}\times\mathbb{R}^N)$.

The initial data for the system $(\mathrm{cat}\cup E)$ is $\hat\rho(0) := \hat\rho_{\mathrm{cat}} \otimes \hat{\mathscr{T}}_{\omega,\beta}$ for the inverse temperature $\beta > 0$ and frequences $\omega = (\omega_1,\cdots,\omega_N)$.
Notice that the Fourier transform of the Wigner function of the initial cat state is

$$\tilde\rho_{\mathrm{cat}}(\zeta) = \pi\left(2\mathrm{e}^{-\frac{|\zeta|^2}{4}}\cos(\zeta\cdot z^+) + \mathrm{e}^{-\frac{|\zeta - J\zeta z^+|^2}{4}} + \mathrm{e}^{-\frac{|\zeta + J\zeta z^+|^2}{4}}\right).$$

The thermal state $\hat{\mathscr{T}}_{\omega,\beta}$ associated with the Hamiltonian $\widehat{H}_E$ has the following Weyl symbol

$$\mathscr{T}_{\omega,\beta}(Y) = Z(\beta,\omega)^{-1}\exp\left(-\sum_{j=1}^{j=N}\tanh(\beta\omega_j/2)(\eta_j^2 + \omega_j^2 y_j^2)\right), \tag{13.66}$$

where $Z(\beta,\omega)$ is a normalisation constant ($\int_{\mathbb{R}^{2N}}\mathscr{T}_\beta(Y)dY = (2\pi)^N$). From (13.66) we see that the covariance matrice $\Gamma_{\omega,\beta}$ of the Gaussian state $\mathscr{T}_{\omega,\beta}$ the block-diagonal matrix with the 2×2 blocks

$$\Gamma_{\omega_j,\beta} := \frac{1}{2\tanh(\beta\omega_j/2)} \begin{pmatrix} \omega_j^{-2} & 0 \\ 0 & 1 \end{pmatrix}.$$

Using Theorem 74 we can estimate the purity of the quantum superposition of the cat state.

Proposition 137 *Let us assume for simplicity that the rank of the coupling G is one and $N = 1$.*

$$G = \begin{pmatrix} g & 0 \\ 0 & 0 \end{pmatrix}, \quad g > 0.$$

Then there exists $t_1 > 0$, $C_1 > 0$ such that for any $t \in]0, t_1]$, $\beta \neq 0$, $\omega_1 \neq 0$, $g > 0$, we have

$$\mathrm{Tr}(\hat{\rho}_{\mathrm{cat}}(t)^2) \leq C_1 \frac{\omega_1 \tanh(\beta\omega_1^2/2)}{tg} \tag{13.67}$$

In particular for $t > 0$, small enough, the purity of $\rho_{\mathrm{cat}}(t)$ can be arbitrary small if the coupling strength or the temperature are chosen large enough.

Proof Using Theorem 74 we need to compute Ψ_{ie}^t for t close the 0. From (13.45) we have $\Psi_{ie}^t = tJ_S G + \mathcal{O}(t^2)$. So the thermal covariance is approximate as

$$\Theta^{(I)}(t) = -t^2 J_S G^\top \Gamma_{\omega_1,\beta} G J_S + \mathcal{O}(t^3).$$

Hence

$$\Theta(t) = -\frac{t^2}{2\tanh(\beta\omega_1/2)} \begin{pmatrix} 0 & 0 \\ 0 & -g^2\omega_1^2 \end{pmatrix} + \mathcal{O}(t^3). \tag{13.68}$$

Using Plancherel Theorem we have

$$\mathrm{Tr}(\hat{\rho}_{\mathrm{cat}}(t)^2) = \int_{\mathbb{R}^2} \tilde{\rho}_{\mathrm{cat}}(t,\zeta)^2 d\zeta$$

Hence, using Theorem 74, the expression of G and that $\hat{\rho}_S(0)$ is Gaussian, we get easily (13.67). □.

Remark 86 (i) From (13.67) we can get an estimate for the rank of $\hat{\rho}_{\mathrm{cat}}(t)$:

$$\mathrm{Rank}[\hat{\rho}_{\mathrm{cat}}(t)] \geq \frac{\log \tau^{-1}}{\log 2}$$

where τ is the r.h.s of (13.67) (using that $\hat{\rho}_{\mathrm{cat}}(t)$ is a density matrix).
In other words the initial coherent state of the cat is transformed into an incoherent mixture of a large number of independent states under the influence of the environment.
(ii) It is not difficult to adapt the proof for any coupling $G \neq 0$.

13.4.3.1 Decoherence on the Density Matrix

Now we analyze the integral kernel $\mathscr{K}_{\rm cat}(t,x,x')$ of the reduce density matrix $\hat\rho_{\rm cat}(t)$ outside the diagonal for $t>0$ small. A manifestation of decoherence is the fast decreasing of $\mathscr{K}_{\rm cat}(t,x,x')$ for $x\neq x'$ due to the influence of the environment (see many papers in the physics literature [155, 162, 231]).

Here we fix $\hbar=1$ but we introduce coherent states with a small variance $\sigma^2>0$ in position $x\in\mathbb{R}$.

So we introduce the coherent states $\varphi_z^\sigma=\hat T(z)g^\sigma$ where $g^\sigma(x)=\pi^{-1/4}\sigma^{-1/2}\mathrm{e}^{-\frac{x^2}{2\sigma^2}}$. We consider here the following superposition of states of the Schrödinger states:

$$\psi_{\sigma,a}(x):=\frac{\varphi_\alpha^\sigma(x)+\varphi_{-\alpha}^\sigma(x)}{\sqrt{N_{\sigma,a}}},$$

where $\alpha=(a,0)\in\mathbb{R}^2$ and $N_{\sigma,a}$ is the normalization constant: $N_{\sigma,a}=2(1-\mathrm{e}^{-a^2/\sigma^2})$.

Denote by $W_{\sigma,a}$ the Wigner transform of $\psi_{\sigma,a}$. We have

$$W_{\sigma,a}(X)=\mathscr{W}_{\varphi_\alpha^\sigma,\varphi_\alpha^\sigma}+\mathscr{W}_{\varphi_{-\alpha}^\sigma,\varphi_{-\alpha}^\sigma}+\mathscr{W}_{\varphi_\alpha^\sigma,\varphi_{-\alpha}^\sigma}.$$

We need to compute the Wigner function of the pairs $(\varphi_z^\sigma,\varphi_{z'}^\sigma)$. Using the computations of Chap. 2 (Proposition 16) it is not difficult to prove that, with $X=(x,\xi)\in\mathbb{R}^2$,

$$\mathscr{W}_{\varphi_z^\sigma,\varphi_{z'}^\sigma}(X)=\mathscr{W}_{z_\sigma,z'_\sigma}(X_\sigma)=$$

$$2\exp\left(-\frac{1}{\sigma^2}\left|x-\frac{q+q'}{2}\right|^2-\sigma^2\left|\xi-\frac{p+p'}{2}\right|^2-i\sigma(X-\frac{z'}{2},z-z')\right).\tag{13.69}$$

where $z_\sigma=(\sigma^{-1}q,\sigma p)$ if $z=(q,p)$.

Hence we get

$$W_{\sigma,a}(x,\xi)=\frac{1}{N_{\sigma,a}}\left(\mathrm{e}^{-\frac{x-a|^2}{\sigma^2}-\sigma^2\xi^2}+\mathrm{e}^{-\frac{|x+a|^2}{\sigma^2}-\sigma^2\xi^2}+2\mathrm{e}^{-\frac{|x|^2}{\sigma^2}-\sigma^2\xi^2}\cos(2a\xi)\right).\tag{13.70}$$

And the Fourier transform is:

$$\tilde W_{\sigma,a}(u,v)=C_{\sigma,a}\left(2\mathrm{e}^{-\frac{v^2}{4\sigma^2}-\frac{\sigma^2}{4}u^2}\cos(au)+\mathrm{e}^{-\frac{\sigma^2}{4}u^2}\left(\mathrm{e}^{-\frac{(v+2a)^2}{4\sigma^2}}+\mathrm{e}^{-\frac{(v-2a)^2}{4\sigma^2}}\right)\right),\tag{13.71}$$

where $C_{\sigma,a}=\frac{1}{2(1+\mathrm{e}^{-a^2/\sigma^2})}$. Now we compute the integral kernel of the density matrix. Let $\hat\rho$ be an arbitrary density matrix. Using computations of Chap. 2 and playing with the Fourier transform $\tilde\rho$ of ρ, we have the following formula for the integral kernel $K_\rho(x,x')$ of $\hat\rho$:

$$K_\rho(x,x') = (2\pi)^{-2}\int_{\mathbb{R}} \tilde{W}(u, x-x')e^{iu\cdot\frac{x+x'}{2}}du. \qquad (13.72)$$

Plugging (13.64) in (13.72) and using (13.68) we get that for $t>0$ the kernel $K_{\mathrm{cat}}(t,x,x')$ of $\hat{\rho}_{\mathrm{cat}}(t)$ has now a dumping multiplicative factor :

$$D(t,x,x') = e^{-t^2 A|x-x'|^2}$$

where $A = \frac{g^2\omega_1^2}{2\tanh(\beta\omega_1/2)}$. In particular the dumping is stronger at high temperature $T=\frac{1}{\beta}$ or for large coupling strength.

To illustrate the off-diagonal dumping phenomenon let us compute $K_{\mathrm{cat}}(0,x,x')$

$$\begin{aligned} K_{\mathrm{cat}}(0,x,x') &= \frac{\sqrt{\pi}}{\sigma}e^{-\frac{(x-x')^2}{4\sigma^2}}\left(e^{-\frac{1}{4\sigma^2}(\frac{x+x'}{2}+a)^2} + e^{-\frac{1}{4\sigma^2}(\frac{x+x'}{2}-a)^2}\right) \\ &+ \frac{\sqrt{\pi}}{\sigma}e^{-\frac{(x+x')^2}{16\sigma^2}}\left(e^{-\frac{1}{4\sigma^2}(x-x'+2a)^2} + e^{-\frac{1}{4\sigma^2}(x-x'-2a)^2}\right) \\ &= K_{\mathrm{cat}}^{\mathrm{diag}}(0,x,x') + K_{\mathrm{cat}}^{\mathrm{andiag}}(0,x,x'). \end{aligned} \qquad (13.73)$$

So, for $\sigma>0$ small enough, $K_{\mathrm{cat}}^{\mathrm{diag}}(0,x,x')$ is concentrated in a neighborhood of $\{(a,a),(-a,-a)\}$ and $K_{\mathrm{cat}}^{\mathrm{andiag}}(0,x,x')$ is concentrated in a neighborhood of $\{(-a,a),(a,-a)\}$. So the dumping factor $D(t,x,x')$ strongly decreases for $t>0$, T large, the anti-diagonal part $K_{\mathrm{cat}}^{\mathrm{andiag}}(0,x,x')$ which is responsible of the quantum interferences for the cat state and modifies weakly the diagonal part. Notice that the diagonal part is the kernel of the decoherent state $\frac{1}{2}(\Pi_{a,0}+\Pi_{-a,0})$ (Π_z is the projector on the coherent state φ_z). This is a toy model for the decoherence phenomenom [198].

13.4.3.2 Decoherence on the Wigner Function

We compute here the Wigner transform $W_{\sigma,a,t}(x,p)$ of the density matrix $\hat{\rho}_{\mathrm{cat}}(t)$, which is nothing else than $\rho_{\mathrm{cat}}(t,x,p)$. (here p is the momentum, denoted ξ before). We introduce the following notations.

- $N(\sigma,a) = 2(1+e^{-a^2/\sigma^2})$
- $f(\sigma,t) = 1+4At^2\sigma^2)$
- $\sigma(t) = \sqrt{2}\sigma f(\sigma,t)^{-1/2}$
- $N(\sigma,a,t) = N(\sigma,a)\sqrt{f(\sigma(t)}$.

Then we have

$$W(\sigma, a, t, x, p) = \frac{1}{N(\sigma, a, t)} \Bigg((e^{-\frac{|x-a|^2}{\sigma^2}} + e^{-\frac{|x+a|^2}{\sigma^2}}) e^{-\frac{\sigma^2 p^2}{f(\sigma,t)}}$$

$$+ 2e^{-\frac{x^2}{\sigma^2}} e^{-\frac{\sigma(t)^2}{2} p^2} \cos\left(\frac{\sigma(t)^2}{\sigma^2} ap\right) \underbrace{\exp\left[-\frac{a^2}{\sigma^2}\left(1 - \frac{\sigma(t)^2}{2\sigma^2}\right)\right]}_{\text{decoherence factor}} \Bigg). \quad (13.74)$$

Let us make more explicit the decoherence factor. We have

$$\exp\left[-\frac{a^2}{\sigma^2}\left(1 - \frac{\sigma(t)^2}{2\sigma^2}\right)\right] = \exp\left[-\frac{2Aat^2}{1 + 4At^2\sigma^2}\right].$$

So we see that the decoherence factor is exponentially small as Aat^2 is large and $At^2\sigma^2$ is small. In this regime the oscillations of the cat state are strongly damped as we see on the following pictures.

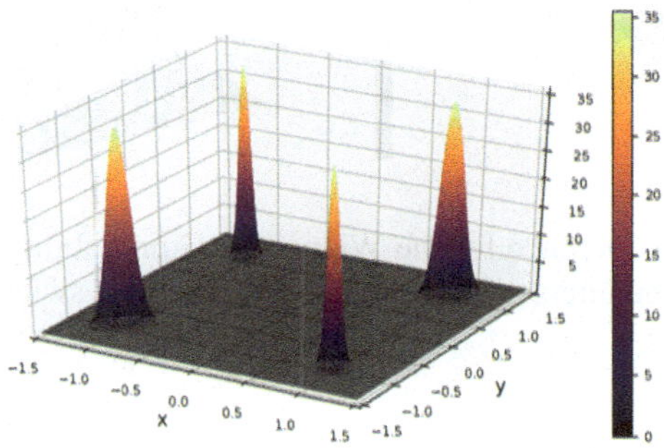

t=0: full coherence for the density matrix, diagonal and non-diagonal peaks

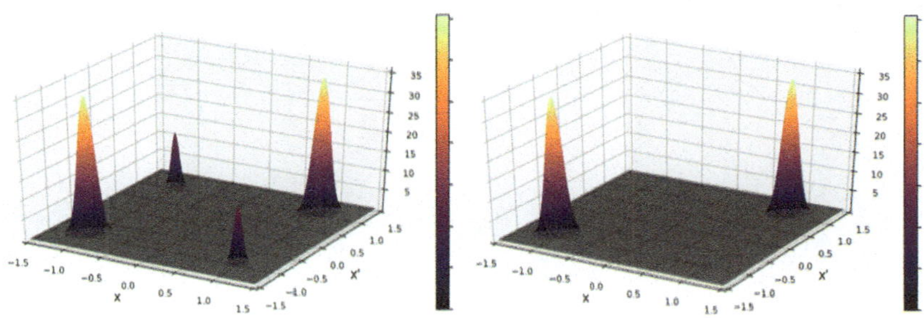

t>0: the non-diagonal of the density matrix are decreasing when the interaction with the environment is switched on

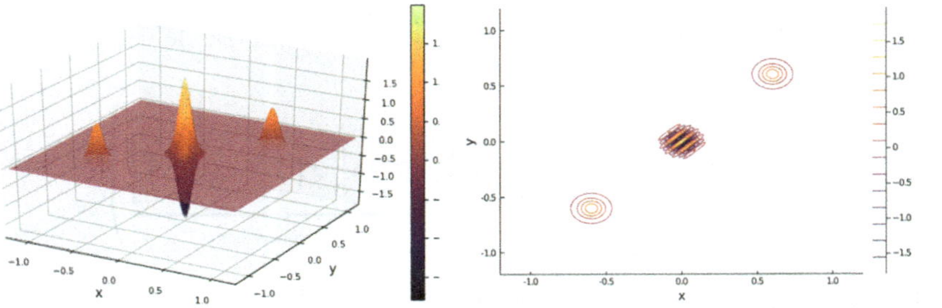

t=0: full coherence for the Wigner quasi-probability with strong oscillations in the middle part

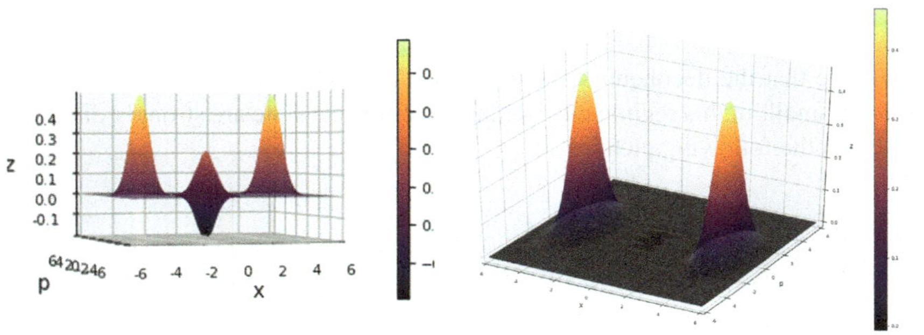

t>0: the quantum coherence for the Wigner quasi-probability in the middle part is decreasing when the interaction with the environment is switched on

Chapter 14
Adiabatic Decoupling and Time Evolution of Coherent States for Multi-component Systems

Abstract Coherent states are well adapted to understand adiabatic decoupling for multi-componenents systems like the Dirac operator or systems obtained by reduction in the Born-Oppenheimer approximation [58]. Hagedorn [130] has done an accurate analysis of this using a variant of coherent states. The new fact for systems is that now states are vectors valued in $\mathbb{C}^m$ for $m \geq 2$ and the eigenvalues of the matrix of the system may cross so that one cannot diagonalize it smoothly. In the case when the eigenvalues are not crossing one can construct an adiabatic modification of the spectral projectors on the modes and compute the time evolution of coherent states following the same strategy as in the scalar case. Here we shall consider only non crossing cases where we can prove an adiabatic decoupling between the modes. For the more difficult crossing case we refer to the literature [130] where non-adiabatic transitions can be computed [101, 129] for more recent results.

14.1 Semi-classical Analysis for Multi-components Systems

Until now we have assumed that the classical Hamiltonian $H(t)$ is real valued. For systems with spin degrees of freedom we have to consider multi-components systems and so their classical counterparts are Hermitian matrix valued Hamiltoninan, depending on the semi-classical parameter $\hbar$. So we shall consider in this section more general classical Hamiltonians $H(t, X)$, taking their values in the space $\mathbb{C}^{m,m}$ of $m \times m$ complex matrices. $|\bullet|$ denotes a matrix norm on $\mathbb{C}^{m,m}$. Let us introduce a suitable class of matrix observables of size m. Let recall that $\mathscr{Z}$ denotes the phase space $\mathbb{R}^n_x \times \mathbb{R}^n_\xi$ and $X = (x, \xi)$ a generic point. Sometimes it will be denoted $z = (q, p)$.

Our aim in this Chapter is to analyze solutions of the Schrödinger equation

$$i\hbar\partial_t\psi(t) = \hat{H}(t)\psi(t), \quad \psi(t_0) = \varphi_{z_0}\vec{v}_0, \tag{14.1}$$

where φ_{z_0} is a coherent states living at $z_0 \in \mathscr{Z}$, $\vec{v}_0 \in \mathbb{C}^m$, $\hat{H}(t)$ is the $\hbar$-Weyl quantization of $H(t)$, $\psi(t) \in L^2(\mathbb{R}^n, \mathbb{C}^m)$.

D. Robert and M. Combescure, *Coherent States and Applications in Mathematical Physics*, Theoretical and Mathematical Physics,
https://doi.org/10.1007/978-3-030-70845-0_14

Definition 33 We say that $A \in S^{m,\nu}$, $\nu \in \mathbb{R}$, if and only if A is a C^∞-smooth function on $\mathscr{X}$ with values in $\mathbb{C}^{m,m}$ such that for every multiindex γ there exists $C > 0$ such that

$$|\partial_X^\gamma A(X)| \le C\langle X\rangle^\nu, \ \forall X \in \mathscr{X}.$$

Let us denote $S^{m,+\infty} = \bigcup_\nu S^{m,\nu}$. We have obviously $\bigcap_\nu S^{m,\nu} = \mathscr{S}(\mathscr{X}, \mathbb{C}^{m,m})$ (the Schwartz space for matrix valued functions).

These symbol spaces are equipped with the natural family of semi-norms (Fréchet spaces).

For every $A \in S^{m,\nu}$ and $\psi \in \mathscr{S}(\mathscr{X}, \mathbb{C}^m)$ we can define $\widehat{A}\psi = (Op_\hbar^w A)\psi$ by the same formula as for the scalar case ($m = 1$) considered in Chap. 2. Now $A(\frac{x+y}{2}, \xi)\psi(y)$ means the action of the matrix $A(\frac{x+y}{2}, \xi)$ on the vector $\psi(y)$. Most of general properties already seen in the scalar case can be easily extended in the matrix case:

1. $A \mapsto \widehat{A}\psi$ is a linear continuous mapping from $S^{m,\nu}$ into $\mathscr{S}(\mathbb{R}^n, \mathbb{C}^m)$.
2. $\widehat{A^\star} = \widehat{A}^\star$ and $\widehat{A}$ is a linear continuous mapping on $\mathscr{S}'(\mathscr{X}, \mathbb{C}^{m,m})$.

We have an operational calculus defined by the *product rule for quantum observables*. Let be $A, B \in \mathscr{S}(\mathscr{X}, \mathbb{C}^{m,m})$. We look for a classical observable C such that $\widehat{A} \cdot \widehat{B} = \widehat{C}$. Some computations with the Fourier transform on the phase space give the following formula (see [153, 217]),

$$C(x, \xi) = \exp\left(\frac{i\hbar}{2}\sigma(D_x, D_\xi; D_y, D_\eta)\right) A(x, \xi)B(y, \eta)|_{(x,\xi)=(y,\eta)}, \tag{14.2}$$

where σ is the symplectic bilinear form introduced above. By expanding the exponent we get a formal series in $\hbar$:

$$C(x, \xi) = \sum_{j\ge 0} \frac{\hbar^j}{j!}\left(\frac{i}{2}\sigma(D_x, D_\xi; D_y, D_\eta)\right)^j A(x, \xi)B(y, \eta)|_{(x,\xi)=(y,\eta)}. \tag{14.3}$$

We see that in general C is not a classical observable because of the $\hbar$ dependence of the series. It can be proved that it is a *semi-classical observable* in the following sense. We say that A is a semi-classical observable of weight ν and size m if there exists $A_j \in S^{m,\nu}$, $j \in \mathbb{N}$, so that A is a map from $]0, \hbar_0]$ into $S^{m,\nu}$ satisfying the following asymptotic condition : for every $N \in \mathbb{N}$ and every $\gamma \in \mathbb{N}^{2n}$ there exists $C_N > 0$ such that for all $\hbar \in]0, 1[$ we have

$$\sup_{\mathscr{X}}\langle X\rangle^{-\nu}\left|\frac{\partial^\gamma}{\partial X^\gamma}\left(A(\hbar, X) - \sum_{0\le j\le N} \hbar^j A_j(X)\right)\right| \le C_N \hbar^{N+1}, \tag{14.4}$$

A_0 is called the principal symbol, A_1 the sub-principal symbol of $\hat{A}$.

The set of semi-classical observables of weight ν and size m is denoted by $S_{sc}^{m,\nu}$.

Now we can state the product rule.

Theorem 75 *For every $A \in S^{m,\nu}$ and $B \in S^{m,\mu}$, there exists a unique $C \in S_{sc}^{m,\nu+\mu}$ such that $\widehat{A} \cdot \widehat{B} = \widehat{C}$ with $C \asymp \sum_{j\geq 0} \hbar^j C_j$. We have the computation rule*

$$C_j(x,\xi) = \frac{1}{2^j} \sum_{|\alpha+\beta|=j} \frac{(-1)^{|\beta|}}{\alpha!\beta!} (D_x^\beta \partial_\xi^\alpha A).(D_x^\alpha \partial_\xi^\beta B)(x,\xi), \tag{14.5}$$

where $D_x = i^{-1}\partial_x$.

In particular we have

$$C_0 = A_0 B_0, \quad C_1 = A_0 B_1 + A_1 B_0 + \frac{1}{2i}\{A_0, B_0\} \tag{14.6}$$

where $\{R, S\}$ is the Poisson bracket of the matrix observables $R = \{R_{j,k}\}$ and $S = \{S_{j,k}\}$ defined by the matrices equality

$$\{R, S\} = \left(\{R, S\}_{j,k}\right)_{j,k}, \quad \{R, S\}_{j,k} = \sum_{1\leq \ell \leq m} \{R_{j,\ell}, S_{\ell,k}\}$$

Let us recall that the Poisson bracket of two scalar observables F, G is defined by $\{F, G\} = \partial_\xi F \cdot \partial_x G - \partial_x F \cdot \partial_\xi G$.

Notations. We shall denote $C := A \circledast B$ (Moyal product) and $A^{2\circledast} = A \circledast A$.

The Moyal bracket $[A, B]_\circledast = A \circledast B - B \circledast A$ is the full symbol of the commutator operator $[\hat{A}, \hat{B}]$.

The sub-principal term of the Moyal bracket is given by the formula

$$[A, B]_{\circledast,1} = [A_1, B_0] + [A_0, B_1] + \frac{1}{2i}(\{A_0, B_0\} - \{B_0, A_0\}). \tag{14.7}$$

In particular if the matrices A, B are Hermitian then $[A, B]_{\circledast,1}$ is anti-Hermitian.

A proof of this theorem in the scalar case, with an accurate remainder estimate, is given in the Appendix of [40]. This rather technical proof could be easily extended to the matrix case considered here.

Let us recall also some other useful properties concerning Weyl quantization of observables. Detailed proofs can be found in [153, 217] for the scalar case and the extension to the matricial one is easy.

- if $A \in S^{m,0}$ then $\widehat{A}$ is bounded in $L^2(\mathscr{Z}, \mathbb{C}^m)$ (Calderon-Vaillancourt theorem).
- if $A \in L^2(\mathscr{Z}, \mathbb{C}^{m,m})$ then $\hat{A}$ is an Hilbert–Schmidt operator in $L^2(X, \mathbb{C}^m)$ and its Hilbert–Schmidt norm is

$$\|\widehat{A}\|_{HS} = (2\pi\hbar)^{-n/2} \left(\int_{\mathscr{Z}} \|A(z)\|^2 dz \right)^{1/2},$$

where $\|A(z)\|$ is the Hilbert–Schmidt norm for matrices.

- if $A \in S^{m,\nu}$ with $\nu < -2n$ then $\widehat{A}$ is a trace-class operator. Moreover we have

$$\mathrm{tr}(\hat{A}) = (2\pi\hbar)^{-n} \int_{\mathscr{X}} \mathrm{tr}(A(z))dz. \tag{14.8}$$

- If $A, L \in L^2(\mathscr{X}, \mathbb{C}^{m,m}$, then $\hat{A} \cdot \hat{L}$ is a trace class operator in $L^2(X, \mathbb{C}^m)$ and

$$\mathrm{tr}(\hat{A} \cdot \hat{B}) = (2\pi\hbar)^{-n} \int_{\mathscr{X}} \mathrm{tr}(A(z)B(z))dz.$$

The extension to the matrix case of the propagation of coherent states may be difficult if the principal symbol $H_0(t, X)$ has crossing eigenvalues. We shall not consider this case here (accurate results have been obtained in [132, 249]). We shall consider here two cases: firstly the principal symbol $H_0(t, X)$ is scalar and we shall write $H_0(t, X) = H_0(t, X)\mathbb{1}_m$ where we have identified $H_0(t, X)$ with a scalar Hamiltonian; secondly $H_0(t, X)$ is a matrix with two distinct eigenvalues of constant multiplicity (like Dirac Hamiltonians). The general case of eigenvalues of constant multiplicities is no more difficult.

Our goal in the following sections is to construct asymptotic solutions for the Schrödinger system

$$i\hbar\frac{\partial\psi(t)}{\partial t} = \widehat{H}(t)\psi(t), \quad \psi(t = t_0) = \psi_0, \tag{14.9}$$

where $\psi_0 \in L^2(\mathbb{R}^n, \mathbb{C}^m)$. Later we shall give more details when the initial state is a wave-packet.

14.2 Systems with a Scalar Leading Term

We shall see that the coherent states analysis of the scalar case can be extended to multicomponent systems as far as the classical Hamiltonian for the principal symbol has eigenvalues of constant multiplicity (without crossing). A state $\psi \in L^2(\mathbb{R}^n, \mathbb{C}^m)$ can be seen as a map $(x, s) \mapsto \psi(x, s)$ from $\mathbb{R}^d \times \{1, \dots, m\}$ into $\mathbb{C}$. The variable $s \in \{1, \dots, m\}$ corresponds to a spin variable in quantum mechanics. The Fourier-Bargmann is defined for $u \in L^2(\mathbb{R}^n, \mathbb{C}^m)$, $u = (u_1, \dots, u_m)$, using the more convenient notation $u_s(x) = u(x, s)$, $s \in \{1, 2, \dots, m\}$,

$$\mathscr{F}^{\mathscr{B}}[u](z, s) = (2\pi)^{-n/2}\langle g_z, u_s\rangle.$$

It defines an isometry from $L^2(\mathbb{R}^n, \mathbb{C}^m)$ into $L^2(\mathbb{R}^{2n}, \mathbb{C}^m)$ (injective but not surjective). Obviously we have the inversion formula

$$u_s(x) = \int_{\mathscr{Z}} \mathscr{F}_{\mathscr{B}}[u](z,s)\varphi_z(x)dz, \quad \text{in the } L^2-\text{sense}, \tag{14.10}$$

where $\langle \cdot, \cdot \rangle$ is the scalar product in $L^2(\mathbb{R}^n, \mathbb{C}^m)$. We shall also use the notation $\varphi_{z,v} = \varphi_z v$ for $(z, v) \in \mathscr{Z} \times \mathbb{C}^m$.

Let us introduce the following assumptions.

(**SQ**) $H(t)$ is a semi-classical sub-quadratic matrix Hamiltonian (continuously time dependent and $\hbar$-dependent) if $H(t, X) \asymp \sum_{j \geq 0} H_j(t, X)\hbar^j$ and satisfies, for $j + |\alpha| \geq 2$ and $t \in [t_0 - T, t_0 + T]$:

$$|\partial_X^\alpha H_j(t, X)| \leq C_{\alpha,T}, \quad \forall X \in \mathbb{R}^{2n}$$

and for any $N \geq 1$ and $|\alpha| \geq 2$,

$$\left| \partial_X^\alpha \left(H(t, X) - \sum_{j \leq N} \hbar^j H_j(t, X) \right) \right| \leq C_{N,\alpha,T}\hbar^{N+1}, \quad \forall X \in \mathbb{R}^{2n}, \hbar \in [0, 1].$$

Recall here that $H_j(t, X)$ is a Hermitian complex matrix. It follows that the $\hbar$-Weyl quantization of $H(t)$ has a unique propagator $\mathscr{U}_H^\hbar(t, t_0)$. Moreover in a first step we assume that the principal symbol H_0 is a scalar matrix.

(**SCA**) $H_0(t, X) = \lambda(t, X)\mathbb{1}_m$ such that $\lambda(t, X)$ is a scalar subquadratic Hamiltonian in X. Hence the classical flow for $\lambda(t, X)$ exists for every initial data $(t_0, z_0) \in \mathbb{R} \times \mathbb{R}^{2n}$.

We shall denote this flow $\Phi_\lambda^{t,t_0} z_0 := z(t, t_0, z_0)$ or with the shorter notation $z_t := z(t, t_0, z_0)$.

The first new fact in this case is the contribution of the subprincipal term $H_1(t, X)$ in the propagation of states by the unitary propagator $\mathscr{U}_H^\hbar(t, t_0)$ (spin-orbit interaction).

The same remark has been done concerning propagation of observables for systems [9]. Let us begin with this last problem.

Let us consider a matrix observable of size m, $A \in S^{m,0}$ and its quantum time evolution: $\hat{A}(t, t_0) := \mathscr{U}_H^\hbar(t_0, t)\hat{A}\, \mathscr{U}_H^\hbar(t, t_0)$.

Theorem 76 *Assume hypothesis* (**SQ**) *and* (**SCA**) *are satisfied and let* $A \in S^{m,0}$. *Then at any time* t *the observable*

$$\hat{A}(t, t_0) := \mathscr{U}_H^\hbar(t_0, t)\hat{A}\, \mathscr{U}_H^\hbar(t, t_0), \tag{14.11}$$

has a semi-classical symbol $A(t, t_0) \in S_{sc}^{m,0}$. *Its principal symbol* $A_0(t, t_0)$ *is given by*

$$A_0(t, t_0; X) = \mathscr{R}^*(t, t_0, X)A(\Phi_\lambda^{t,t_0}(X))\mathscr{R}(t, t_0, X), \tag{14.12}$$

where $\mathscr{R}(t, t_0, X)$ *solves the equation*

$$\partial_t \mathscr{R}(t, t_0, X) = -iH_1(t, (\Phi_\lambda^{t,t_0}(X))\mathscr{R}(t, t_0, X), \ \mathscr{R}(t_0, t_0, X) = \mathbb{1}_m. \tag{14.13}$$

Proof The proof is similar to the proof given in Chap. 2 for the scalar case. We only compute here the principal symbol. Otherwise the details can be reproduced following the Chap. 2 (The semi-classical propagation Theorem).

Using (14.6) we get for $A_0(t, t_0)$ the equation

$$\partial_t A_0(t, t_0) = \{A_0(t, t_0), \lambda(t)\} + i^{-1}[A_0(t, t_0), H_1(t)]. \tag{14.14}$$

Using (14.13) we get easily that Eq. 14.14 is solved by formula (14.12). □

Here the dynamics of the observable is driven by the classical flow distorted by a precession $\mathscr{R}$ acting on the components of the states (Larmor precession). This dynamics is written as

$$B(X) \mapsto \mathscr{R}^*(t, t_0, X)B(\Phi_\lambda^{t,t_0}(X))\mathscr{R}(t, t_0, X). \tag{14.15}$$

Notice that if A is scalar then $A(t)$ is also scalar at any time.

Concerning the propagation of coherent states, we also perform the same analysis as for the scalar case (Chap. 4) with natural adaptations. Let us denote by $k_2(t, t_0)X \cdot X$ the quadratic part of the Taylor expansion of $\lambda(t, z)$ around the trajectory z_t. Almost nothing is changed, except that now the coefficients $K_j(t, t_0)$ of the Taylor expansion of $H(t, X)$ around the classical trajectory z_t are time dependent matrices in $\mathbb{C}^{m,m}$. Moreover here we have to add to the scalar Hamiltonian $\hbar k_2(t, t_0, X)\mathbb{1}_m$ the matrix $\hbar H_1(t, z_t)$. We shall see later that a sub-principal term (of order 1 in $\hbar$) appears after diagonalizing of non-crossing systems. So the spin of the system will be modified along the time evolution according to the "rotation" matrix $\mathscr{R}(t, t_0)$ defined by (14.13). This phenomenon is related with the Berry phase "factors accompanying adiabatic changes". (Berry 1984).

The following lemma is easy to prove and is standard for differential equations.

Lemma 87 *Let us denote respectively by $\mathscr{U}_2(t, t_0)$ and $\mathscr{U}_2'(t, t_0)$ the quantum propagators defined by the Hamiltonians $\widehat{k_2(t, t_0)}\mathbb{1}_m$ respectively $\widehat{k_2(t)}\mathbb{1}_m + H_1(t, z_t)$. Then we have*

$$\mathscr{U}_2'(t, t_0) = \mathscr{R}(t, t_0, z_0)\mathscr{U}_2(t, t_0). \tag{14.16}$$

Let us recall that if $F(t, t_0)$ denotes the linear classical flow associated with $k_2(t, t_0)$(stability or Jacobi matrix). The corresponding quantum flow $\mathscr{U}_2(t, t_0)$ is also the metaplectic transformations $\mathscr{M}(F(t, t_0))$ (see Chap. 3).

The wave packet transform is a convenient way to built coherent states:

Definition 34 Let $z = (q, p) \in \mathbb{R}^{2n}$ and $f \in \mathscr{S}(\mathbb{R}^n, \mathbb{C}^m)$ (a profile) . We set

$$\mathscr{WP}_z^\hbar f(x) = \hbar^{-n/4} e^{\frac{i}{\hbar}p\cdot(x-q)} f\left(\frac{x-q}{\sqrt{\hbar}}\right). \tag{14.17}$$

The operator

$$f \mapsto \mathcal{W}\mathcal{P}_z^\hbar f$$

is a unitary map in $L^2(\mathbb{R}^n, \mathbb{C}^m)$ which maps continuously $\Sigma_1^{m,k}(\mathbb{R}^n)$ into $\Sigma_\hbar^{m,k}(\mathbb{R}^n)$, where

$$\Sigma_\hbar^{m,k}(\mathbb{R}^n) = \{f \in L^2(\mathbb{R}^n, \mathbb{C}^m),\ (-\hbar^2\Delta + |x|^2)^{k/2} f \in L^2(\mathbb{R}^n, \mathbb{C}^m)\},$$

equipped with the $\hbar$-dependent graph-norm. Gaussian coherent states are the Wave Packet transforms of Gaussian profiles, up to a phase factor:

$$\varphi_z^\Gamma = \mathrm{e}^{-\frac{i}{2\hbar}p\cdot q}\mathcal{W}P_z^\hbar(g^\Gamma), \tag{14.18}$$

where $g^\Gamma(x) = \pi^{-n/4}\det(\Im\Gamma)^{1/4}\mathrm{e}^{\frac{i}{2}\Gamma x\cdot x}$, $\Gamma \in \mathfrak{S}^+(n)$.

Moreover we have the following properties whose proofs are left to the reader

- $\mathcal{W}\mathcal{P}_z^\hbar = \mathrm{e}^{-\frac{i}{2\hbar\hbar}p\cdot q}\,\hat{T}(z)\,\Lambda_\hbar$
- $\varphi_z = \mathrm{e}^{\frac{i}{2\hbar}p\cdot q}\mathcal{W}\mathcal{P}_z^\hbar g$ where g is the standard Gaussian.
- For all z, z' in $\mathbb{R}^{2d}$,

$$\mathcal{W}\mathcal{P}_{z+z'}^\hbar = \mathrm{e}^{-\frac{i}{\hbar}p\cdot q'}\mathcal{W}\mathcal{P}_z^\hbar\,\Lambda_\hbar^{-1}\mathcal{W}\mathcal{P}_{z'}^\hbar.$$

The following Lemma concerns the action of an $\hbar$-Weyl quantized observable on wave packets.

Lemma 88 *If A is any classical observable with values in $\mathbb{C}^{m,m}$,*

$$\mathrm{op}_\hbar^w(A)\mathcal{W}\mathcal{P}_z^\hbar = \mathcal{W}\mathcal{P}_z^\hbar\,\mathrm{Op}_1^w(A_{\hbar,z}), \tag{14.19}$$

$A_{\hbar,z}$ denotes the $\mathbb{C}^{m,m}$ matrix $A_{\hbar,z}(X) = A(\sqrt{\hbar}X + z)$, $X \in \mathbb{R}^{2n}$. Moreover for any $f \in \mathcal{S}(\mathbb{R}^n, \mathbb{C}^m)$ we have the following expansion at any order:

$$\mathrm{op}_\hbar^w(A)\mathcal{W}\mathcal{P}_z^\hbar f = \sum_{|\alpha|\le N}\frac{\hbar^{\frac{|\alpha|}{2}}}{\alpha!}\partial_X^\alpha\mathcal{W}\mathcal{P}_z^\hbar\mathrm{op}_1^w(X^\alpha)f + \mathcal{O}(\hbar^{N+1}). \tag{14.20}$$

Proof (14.19) follows from known properties of Weyl quantization. Equation (14.20) follows from (14.19) and the Taylor formula. □

As in the scalar case, for every $N \geq 1$, we can get approximate solutions $\psi_{z_0,v}^{(N)}(t, t_0)$ for the exact solution $\psi_{z_0,v}(t, t_0)$ of the Eq. (14.9) with initial data $\psi_{z_0,v}(t_0, t_0) = \mathcal{W}\mathcal{P}_{z_0}^\hbar f\mathbf{v}$, where $\mathbf{v} \in \mathbb{C}^m$ and $f \in \mathcal{S}(\mathbb{R}^n)$(a wave packet polarized along the direction $\mathbf{v}$).

We do not repeat here all the details given in Chap. 4 for the scalar case.

First, we get a family of differential operators $\mathbf{b}_j(t, t_0)$, polynomials in (x, ∂_x) of degree smaller than $3j$ and with coefficients in $\mathbb{C}^m$, uniquely defined by the following induction formula for $j \geq 1$, starting with $\mathbf{b}_0(t, t_0)f = \mathbf{v}f$,

$$\partial_t \mathbf{b}_j(t, t_0) f(x) = \sum_{k+\ell=j+2,\ \ell\geq 3} Op_1^w[K_\ell^\#(t, t_0)](\mathbf{b}_k(t, t_0) f \tag{14.21}$$

$$\mathbf{b}_j(t_0, t_0) f = 0, \tag{14.22}$$

where $K_\ell^\#(t, t_0, X) = K_\ell(t, t_0, F(t_0, t)X)$. Recall that the matrix symbols $K_\ell(t, t_0)$ are obtained by Taylor expanding the full Hamiltonian $H(t)$ around the classical flow $\Phi_\lambda^{t,t_0}(z_0)$.

Moreover, coming back to the Schrödinger equation, we get, in the same way as for the scalar case, for every $N \geq 0$,

$$i\hbar\partial_t \psi_{z,v}^{(N)}(t, t_0) = \widehat{H}(t)\psi_{z,v}^{(N)}(t, t_0) + R_{z,v}^{(N)}(t, t_0) \tag{14.23}$$

where

$$\psi_{z_0,v}^{(N)}(t, t_0, \cdot) = \mathrm{e}^{iS(t,t_0,z_0)/\hbar} \mathscr{W}\mathscr{P}_{z_t}^\hbar \left(\mathscr{R}(t, t_0)\mathscr{M}[F(t, t_0)] \sum_{0\leq j\leq N} \hbar^{j/2} \mathbf{b}_j(t, t_0) f \right) \tag{14.24}$$

and

$$R_{z_0,v}^{(N)}(t, t_0) = \mathrm{e}^{iS(t,t_0,z_0)/\hbar} \sum_{\substack{j+k=N+2\\ k\geq 3}} \hbar^{j/2} \mathscr{W}\mathscr{P}_{z_t}^\hbar \Big(\mathscr{R}(t, t_0)\mathscr{M}[F(t, t_0)]$$
$$\mathrm{Op}_1^w[R_k(t, t_0) \circ F(t, t_0)](\mathbf{b}_j(t, t_0) f) \Big) = \mathscr{O}_{L^2(\mathbb{R}^n, \mathbb{C}^m)}(\hbar^{\frac{N+3}{2}}), \tag{14.25}$$

where $R_k(t, t_0)$ is a remainder term in the Taylor expansion of $H(t, z)$ around z_t.

Putting all these things together we get the following result:

Theorem 77 *Let us assume that the assumptions (***SQ***) and (***SCA***) are satisfied. Then we have, for every $(z, v) \in \mathscr{Z} \times \mathbb{C}^m$, for every $t \in I_T = \{t,\ |t - t_0| \leq T\}$,*

$$\|\psi_{z_0,v}(t, t_0) - \psi_{z_0,v}^{(N)}(t, t_0)\|_{L^2(\mathbb{R}^d, \mathbb{C}^m)} \leq C_{N,T,z} \hbar^{(N+1)/2}. \tag{14.26}$$

In particular we have for the approximation mod $\mathscr{O}(\hbar)$, in $L^2(\mathbb{R}^n, \mathbb{C}^m)$,

$$\psi_{z_0,v}^{(1)}(t, t_0) = \mathrm{e}^{\frac{i}{\hbar}S(t,t_0,z_0)} \mathscr{W}\mathscr{P}_{z_t}^\hbar \mathscr{R}(t, t_0)\mathscr{M}[F(t, t_0)] \left(f\mathbf{v} + \sqrt{\hbar}\mathbf{b}_1(t, t_0) f\mathbf{v} \right), \tag{14.27}$$

where

$$\mathbf{b}_1(t,t_0)f\mathbf{v} = \sum_{|\alpha|=3}\frac{1}{\alpha!}i^{-1}\int_{t_0}^{t}\partial_z^{\alpha}\lambda(s,z_s)\mathrm{Op}_1^w((F(t_0,s)X)^{\alpha}\mathbf{v}f ds$$
$$+i^{-1}\int_{t_0}^{t}\partial_X H_1(s,z_s)\mathrm{Op}_1^w(F(t_0,s)X)f\mathbf{v}ds. \quad (14.28)$$

Remark 87 Using results obtained in [101, 191] we could get estimates in $\Sigma_{\hbar}^k$-norms.

Remark 88 We can apply Theorem 77 with a Gaussian profile $f = g^{\Gamma}$ or a finite sum of excited state of the standard harmonic oscillator: $f = \sum_{|\mathbf{k}|\leq\nu} c_{\mathbf{k}}\phi_{\mathbf{k}}^1$. (Here $\{\phi_{\mathbf{k}}^1\}_{\mathbf{k}\in\mathbb{N}^n}$ denotes the usual orthonormal basis of Hermite functions for $\hbar = 1$). Using creation and annihilation operators we get that

$$\psi_{z_0,v}^{(N)}(t,t_0) = \mathrm{e}^{\frac{i}{\hbar}S(t,t_0,z_0)}$$
$$\mathcal{WP}_{z_t}^{\hbar}\left(\mathcal{R}(t,t_0)\mathcal{M}[F(t,t_0)]\sum_{|k|\leq J+3N} c_{\mathbf{k}}(t,t_0,\hbar)\phi_{\mathbf{k}}^1\right). \quad (14.29)$$

One can compute the vector coefficients $c_{\mathbf{k}}(t,t_0,\hbar)$ following the method used in Chap. 4 (Theorem 21).

In the next section we can consider the particular case of **spin-orbit** interaction.

14.3 Spin-Orbit Interaction

We have already studied spin-coherent states in Chap. 7. In this section we consider a quantum Hamiltonian where the motion of a particle is coupled with its spin motion. This is a particular case of Sect. 14.2 with the subprincipal term $H_1(t,X) = \sum_{1\leq k\leq 3}\mathfrak{C}_k(t,X)\hat{S}_k = \mathfrak{C}(t,X)\cdot\mathbf{S}$.

More precisely $\hat{S}_k$ are the spin operators in the complex linear space $\mathcal{V}^{(j)}$, $\dim\mathcal{V}^{(j)} = 2j+1$, $j\in\frac{1}{2}\mathbb{N}$ is the spin number and $\hat{S}_k = D^{(j)}(\sigma_k/2)$ are defined by the irreducible representation $D^{(j)}$ of the Lie algebra $\mathfrak{su}(2)$. Recall that we have the commutation relations

$$[S_k, S_\ell] = i\varepsilon_{k,\ell,m}S_m. \quad (14.30)$$

$\mathfrak{C}_k(t)$ are real scalar symbols satisfying the assumption **(SQ)**. So the full Hamiltonian here is $H(t,X) = H_0(t,X) + \hbar\mathfrak{C}(t,X)\cdot\mathbf{S}$, with the spin operator $\mathbf{S} = (\hat{S}_1,\hat{S}_2,\hat{S}_3)$.

Hence the quantum Hamiltonian $\hat{H}(t)$ generates a propagator $\mathcal{U}(t,t_0)$ in the Hilbert space $L^2(\mathbb{R}^n,\mathbb{C}^{2j+1})$.

We have seen in Chap. 7 that spin coherent states are labelled by the sphere $\mathbb{S}^2$. So it is natural to study the semi-classical limit of the propagator in the phase space $T^*(\mathbb{R}^n) \times \mathbb{S}^2$. In particular we can reformulate the propagation for coherent states $\varphi_z \otimes \psi_{\mathbf{n}}$ labelled by $(z, \mathbf{n}) \in \mathbb{R}^{2n} \times \mathbb{S}^2$ where φ_z was defined in Chap. 1 and $\psi_{\mathbf{n}}^{(j)} = \psi_{\mathbf{n}}$ in Chap. 7.

Our goal here is to apply Theorem 77 for $f = g^{\Gamma_0}$ (Gaussian with covariance matrix Γ_0), $v = \psi_{\mathbf{n}_0}$ a spin coherent state where $\mathbf{n}_0 \in \mathbb{S}^2$. So in particular for the leading term we have to compute the time evolution $\mathscr{R}(t, t_0)\psi_{\mathbf{n}_0}$ with $H_1 = \mathfrak{C}(t, X)$. It is enough to solve the equation for $\mathscr{R}(t, t_0)$ for the spin number $j = \frac{1}{2}$. Let $G(t, t_0) \in SU(2)$ satisfying

$$\partial_t G(t, t_0) + \frac{i}{2}(\mathfrak{C}(t) \cdot \sigma)G(t, t_0) = 0, \quad G(t_0, t_0) = \mathbf{1}.$$

Using the notations of Chap. 7, let

$$\phi(t) = \mathscr{R}(t, t_0)\psi_{\mathbf{n}_0}^{(j)} = T^{(j)}(G(t, t_0))T^{(j)}(g_{\mathbf{n}_0})\psi_0^{(j)}. \tag{14.31}$$

Lemma 89 *The adjoint action of $G(t, t_0)$ has the following property*

$$G(t, t_0)(\mathbf{n}_0 \cdot \sigma)G(t, t_0)^{-1} = \mathbf{n}(t) \cdot \sigma$$

where $\mathbf{n}(t) = R_{G(t,t_0}\cdot \mathbf{n}_0$ satisfies

$$\partial_t \mathbf{n}(t) = \mathfrak{C}(t) \wedge \mathbf{n}(t).$$

Proof Using the canonical cover of $SO(3)$ by $SU(2)$ we can compute to get the Lemma. □

Lemma 90 *The spin states (14.31) coîncides with the coherent state $\psi_{\mathbf{n}(t)}^{(j)}$ modulo a phase factor*

$$\phi(t) = \mathrm{e}^{ij\alpha(t)}\psi_{\mathbf{n}(t)}^{(j)}, \text{ where } \alpha(t) \in \mathbb{R}. \tag{14.32}$$

Proof We have

$$\mathbf{n}(t) = R_{G(t,t_0)g_{\mathbf{n}_0}} e_3 = R_{g_{\mathbf{n}}(t)} e_3,$$

hence there exists $\alpha(t) \in \mathbb{R}$ such that

$$g_{\mathbf{n}(t)}^{-1} G(t, t_0) g_{\mathbf{n}_0} = \mathrm{e}^{\frac{i}{2}\alpha(t)\sigma_3}.$$

So we get (14.32) for $j = \frac{1}{2}$ hence for any $j \in \frac{\mathbb{N}}{2}$ by considering the representation $T^{(j)}$. □

In a last step we compute the phase $\alpha(t)$.

Lemma 91 *In polar coordinates* (θ, φ), $\mathbf{n} = (\sin\theta\cos\varphi, \sin\theta\sin\varphi, \cos\theta)$ *we have*

$$\alpha(t) = \int_{t_C}^{t} (\mathfrak{C}(\tau)\cdot\mathbf{n}(\tau) + (1-\cos\theta(\tau)\dot{\varphi}(\tau))\,d\tau. \tag{14.33}$$

Proof Applying the equation satisfied by $G(t, t_0)$ to $G(t, t_0) = e^{\frac{i}{2}\alpha(t)} g_{\mathbf{n}(t)}$ we get

$$\dot{\alpha}(t) g_{\mathbf{n}(t)} = \Lambda(t)\cdot\frac{\sigma(t)}{2} - 2i\partial_t g_{\mathbf{n}(t)}.$$

Computing the average of this equation in the initial $\frac{1}{2}$-spin coherent state ψ_0, a direct computation gives

$$\dot{\alpha}(t) = \Lambda(t)\cdot\mathbf{n}(t) + (1-\cos\theta(t))\dot{\varphi}(t).$$

Here we have used that

$$\langle g_{\mathbf{n}}\psi_0, \sigma_k g_{\mathbf{n}}\psi_0\rangle = -\mathbf{n}_k, \quad 1 \le k \le 3.$$

□

So from Theorem 77 we get the following result (spin-orbit motion)

Theorem 78 ([33]) *For the initial state* $\Psi_{z_0,\mathbf{n}_0} = \varphi_{z_0}\otimes\psi^{(j)}_{\mathbf{n}_0}$ *we have*

$$\Psi_{z_0,\mathbf{n}_0}(t, t_0) = e^{\frac{i}{\hbar}S(t,t_0)+ij\alpha(t)}\varphi^{\Gamma(t)}_{z(t)}\otimes\psi^{(j)}_{\mathbf{n}(t)} + O(\sqrt{\hbar})$$

where the covariance matrix $\Gamma(t)$ *is computed in Chaps. 3 and 4 for the dynamics generated by the Hamiltonian* $H_0(t, X)$.

Remark 89 This Theorem could be extended to an approximation modulo any power of $\hbar$ like in Theorem 77 (see also [33]).

14.4 Systems with Constant Multiplicities Eigenvalues

Let us consider now the case where the principal part $H_0(t, X)$ has several different eigenvalues. For simplicity we assume that $H_0(t, X)$ has two eigenvalues $\lambda_\pm(t, X)$ such that $\lambda_\pm(t,\cdot) \in C^\infty(\mathbb{R}^{2n})$. Moreover we assume that $\lambda_\pm(t, X)$ have a constant multiplicity $\mu_\pm$ for any $(t, X) \in [t_0, t_0+T]\times\Omega_2$, where Ω_2 is an open bounded subset of the phase space $\mathscr{X}$.

In the semi-classical limit wave-packets are localized in the phase space and we know that their time evolution follows the classical trajectories, in the scalar case. So we shall see that it is enough to decouple the considered system in a neighborhood of classical trajectories for the different modes.

More precisely we introduce the following non-crossing assumptions:
(**NCC**) There exists $\delta > 0$ such that $|\lambda_+(t, X) - \lambda_-(t, X)| \geq \delta$, $\forall (t, X) \in [t_0, t_0 + T] \times \Omega_2$.

Let us introduce a local existence property for the flows $\Phi_\pm^{t,t_0}$ generated by the Hamiltonians $\lambda_\pm(t, X)$: for a given $T > 0$ there exist bounded open set Ω_j, $1 \leq j \leq 2$, such that $\bar{\Omega}_1 \subset \Omega_2$ and for any $X \in \Omega_1$ and $t \in [t_0, t_0 + T]$, the solution $X(t)$ of the Hamilton equations[1]

$$\partial_t X_\pm(t) = J \nabla_X \lambda_\pm(t, X_\pm(t)), \quad X(t_0) = X$$

exists and stays in Ω_2 $\forall t \in [t_0, t_0 + T]$., where Ω_2 is like in (**NCC**).

Let us remark that $\lambda_\pm(t, X)$ may collide for some $X \in \mathbb{R}^{2n} \backslash \Omega_2$ which is outside the accessible classical area. Moreover the only global assumption we shall use concerns the quantum propagator for $\hat{H}(t)$. In particular this holds true if $H(t)$ is sub-quadratic.

Let us denote by $\pi_\pm(t, X)$ the spectral projectors on $\ker(H_0(t, X) - \lambda_\pm(t, X)\mathbb{1}_m)$. All these functions are C^∞-smooth in $X \in \Omega_2$ and continuous in $t \in [t_0, t_0 + T]$.

To construct asymptotic solutions of equation (14.9) we shall show that the time evolution of the vector states $(t, x) \mapsto \psi(t, x) \in \mathbb{C}^m$ splits into two parts: one part coming from the eigenvalue $\lambda_+(t, X)$ the other part coming from $\lambda_-(t, X)$. We shall see that at the leading order the two quantum evolutions $\pm$ are not mixing but they can mix up at the order one or greater in $\hbar$. In physics this phenomenon is called *adiabaticity*. This will be demonstrated for the propagation of wave-packets (or coherent states) and more generally for initial data with compact frequency set.

In a first step we work with formal semi-classical series in $\hbar$ whose coefficients are matrix symbols in $\mathbb{C}^{m,m}$. Let us denote by $\mathfrak{A}_{\Omega_1}$ the set of formal series $A = \sum_{j \geq 0} A_j \hbar^j$ where A_j is a C^∞-smooth application from Ω_1 in $\mathbb{C}^{m,m}$. $\mathfrak{A}_{\Omega_1}$ is an algebra for the Moyal product defined by $C = A \circledast B$ where $C = \sum_{j \geq 0} C_j \hbar^j$ with the notations used in (14.5). This product is associative but non commutative. The Moyal bracket will be denoted $\{A, B\}_\circledast = A \circledast B - B \circledast A$ where $A, B \in \mathfrak{A}_{\Omega_1}$.

Definition 35 A formal self-adjoint observable A is a formal series in Ω_1: $A = \sum_{j \geq 0} \hbar^j A_j$ $A \in \mathfrak{A}_{\Omega_1}$ and $A_j(X)$ is an Hermitian matrix for any $X \in \Omega_1$. Moreover A is continuous in time $t \in [t_0, t_0 + T]$ iff $A_j(t, \cdot)$ is continuous in $t \in [t_0, t_0 + T]$.

A is a formal projection in Ω_1 if A is self-adjoint and satisfies $A \circledast A = A$ in Ω_1.

In a first step we construct a formal semi-classical projector following the quantum evolution driven by $\hat{H}(t)$. It will be convenient to write the $m \times m$ complex matrices

[1] Notice that for a general subquadratic matrix Hamiltonian $H(t, X)$ it is not always true that $\lambda_\pm(t, \cdot)$ are subquadratic in $\mathbb{R}^{2n}$. We have easily $\lambda_\pm(t, X) = O(< X >^2)$ and $\partial_X \lambda_\pm(t, X) = O(< X >)$ but $\partial^2_{X,X} \lambda_\pm(t, \cdot)$ is not necessarily bounded in $\mathbb{R}^{2n}$. This is why we introduce a local assumption which allows us to work with the subquadratic Hamiltonians $\tilde{\lambda}_\pm(t, X) = \chi(X)\lambda_\pm(t, X)$ for some $\chi \in C_0^\infty(\Omega_2)$, $\chi(X) = 1$ if $X \in \Omega_1$.

symbols in $\mathbb{C}^{m,m}$ in the splitting $\mathbb{C}^m = \Im\pi_+(t,X) \oplus \Im\pi_-(t,X)$ as a 2×2 matrix with matrix-valued coefficients. We shall use the following notation. Let be $A(t,X) \in \mathbb{C}^{m,m}$ then

$$A(t,X) = \begin{pmatrix} A^{++}(t,X) & A^{+-}(t,X) \\ A^{-+}(t,X) & A^{--}(t,X) \end{pmatrix}$$

where $A^{++}(t,X) = \pi_+(t,X)A(t,X)\pi_+(t,X)$, $A^{--}(t,X) = \pi_-(t,X)A(t,X)$ $\pi_-(t,X)$, $A^{+-}(t,X) = \pi_+(t,X)A(t,X)\pi_-(t,X)$, $A^{-+}(t,X) = \pi_-t,X)A(t,X)$ $\pi_+(t,X)$.

Notice that $A(t,X)$ is Hermitian if and only if $A(t,X)^{++}$, $A(t,X)^{--}$ are Hermitian and $A(t,X)^{-+} = (A(t,X)^{+-})^*$.

We shall say that $A(t,X)$ is diagonal if $A(t,X)^{+-} = A(t,X)^{-+} = 0$ and that it is off-diagonal if $A(t,X)^{++} = A(t,X)^{--} = 0$.

Notice that $\partial_t\pi_\pm(t,X)$ is off-diagonal (uses that $\pi_\pm(t,X)^2 = \pi_\pm(t,X)$) as well as $[\pi_\pm(t,X), A(t,X)]$.

In the following Theorem we also denote $\pi(t,X) = \pi_+(t,X)$ and $\Pi(t,X) = \Pi^+(t,X)$ and $\pi^\perp(t) = \mathbb{1}_m - \pi(t) = \pi_-$.

Theorem 79 (Semiclassical projector-adiabatic evolution) *There exists a unique self-adjoint formal projection $\Pi(t,X) = \sum_{j\geq 0}\hbar^j\Pi_j(t,X)$, continuous in $t \in [t_0, t_0 + T]$, C^∞-smooth in $X \in \Omega_1$, such that $\Pi_0(t,X) = \pi(t,X)$ and*

$$i\hbar\partial_t\Pi(t) = \{H(t), \Pi(t)\}_\circledast \tag{14.34}$$

Moreover the sub-principal term $\Pi_1(t)$ (Hermitian matrix) is given by the following formulas:

$$\begin{aligned} \Pi_1(t)^{++} &= \frac{-1}{2i}\pi(t)\{\pi(t),\pi(t)\}\pi(t) \\ \Pi_1(t)^{--} &= \frac{-1}{2i}\pi(t)^\perp\{\pi(t),\pi(t)\}\pi(t)^\perp. \end{aligned} \tag{14.35}$$

$$\Pi_1(t)^{-+} = i(H_0(t) - \lambda(t))^{-1}\pi(t)^\perp S_1(t)^{-+}, \tag{14.36}$$

and $\Pi_1(t)^{+-} = (\Pi_1(t)^{-+})^*$, where

$$S_1(t) = \partial_t\pi(t) + \frac{1}{2}(\{H_0(t),\pi(t)\} - \{\pi(t), H_0(t)\}) - [H_1(t),\pi(t)].$$

Proof Using (14.6) the result will follow by an induction argument.

Let us remark that in formula (14.36) $(H_0(t) - \lambda(t))^{-1}$ is well defined on the range of $\pi(t)^\perp$.

We use the following convention: we say that two formal series L, M are equal modulo $\mathcal{O}(\hbar^{N+1})$ if $L_j - M_j = 0$ for $0 \leq j \leq N$ in Ω_1 and we shall write $L = M + \mathcal{O}(\hbar^{N+1})$.

Starting with $\Pi_0(t) = \pi(t)$ we see that it is a Moyal projector modulo $[\hbar]$. For the next term we are looking for $\Pi_1(t)$ such that $\Pi^{(1)} := \pi(t) + \hbar\Pi_1(t)$ is a projector modulo $\mathscr{O}(\hbar^2)$ and such that Eq. (14.34) is satisfied modulo $\mathscr{O}(\hbar^2)$. So computing the term in $\hbar$ we find the equations

$$\Pi_1(t) - (\pi(t)\Pi_1(t) + \Pi_1(t)\pi(t)) = \frac{1}{2i}\{\pi(t), \pi(t)\} \tag{14.37}$$

and

$$[\Pi_1(t), H_0(t)] = \frac{1}{2i}(\{H_0(t), \pi(t)\} - \{\pi(t), H_0(t)\}) + [H_1(t), \pi(t)] - i\partial_t\pi(t). \tag{14.38}$$

Then the formula (14.35) follows from the first equation. Now we can check that the l.h.s in (14.38) is off-diagonal so we can compute the off-diagonal terms of $\Pi_1(t, X)$ and (14.36) follow by elementary algebra. Notice that we have used the gap assumption so that $(H_0(t) - \lambda(t))^{-1}\pi(t)^\perp$ is well defined. Hence we have found that $\Pi_1(t, X)$ exists and is unique.

Now, reasoning by induction on $N \geq 1$, assume that $\Pi_0(t), \ldots, \Pi_N(t)$ are known such that $\Pi^{(N)}(t) = \sum_{0\leq k\leq N} \hbar^k\Pi_k(t)$ is a projector modulo $\mathscr{O}(\hbar^{N+1})$ satisfying (14.34) modulo $[\hbar^{N+1}]$ then we have to compute $\Pi_{N+1}(t)$ such that $\Pi^{(N+1)}(t) := \Pi^{(N)}(t) + \Pi_{N+1}(t)$ is a Hermitian projector and satisfies Eq. (14.34) modulo $\mathscr{O}(\hbar^{N+2})$.

Let us denote $R_N(t)$ the coefficient of $\hbar^{N+1}$ in the $\hbar$-expansion of the matrix symbol $\Pi^{(N)}(t) - (\Pi^{(N)})^{2\circledast}(t)$ and $S_N(t)$ the coefficient of $\hbar^{N+1}$ in the $\hbar$-expansion of $\hbar\partial_t\Pi^{(N)}(t) - \frac{1}{i}[H(t), \Pi^{(N)}(t)]_\circledast$.

Hence $\Pi^{(N+1)}(t)$ must satisfy the equations

$$\Pi_{N+1}(t) - (\pi(t)\Pi_{N+1}(t) + \Pi_{N+1}(t)\pi(t)) = R_N(t) \tag{14.39}$$

$$[H_0(t), \Pi_{N+1}(t)] = iS_N(t) \tag{14.40}$$

As for the computation of $\Pi_1(t)$ we have to prove compatibility conditions.

We remark first that $R_N(t, X)$ and $S_N(t, X)$ ($X \in \mathbb{R}^{2d}$) are Hermitian matrices. To check (14.39) and (14.40) we have to prove that $R_N(t)$ is diagonal and $S_N(t)$ is off-diagonal.

We have, for the Moyal product,

$$\begin{gathered}\Pi^{(N)}(t)\left(\Pi^{(N)}(t)^{2\circledast} - \Pi^{(N)}(t)\right)(\mathbb{1} - \Pi^{(N)}(t)) = \\ \left(\Pi^{(N)}(t)^{2\circledast} - \Pi^{(N)}(t)\right)\left(\Pi^{(N)}(t) - \Pi^{(N)}(t)^{2\circledast}\right) = \mathscr{O}(\hbar^{2N+2}).\end{gathered} \tag{14.41}$$

Then computing the coefficient of $\hbar^{N+1}$ in (14.41) we get that $\pi(t)R_N(t)\pi^\perp(t) = 0$, hence taking the adjoint, $\pi^\perp(t)R_N(t)\pi(t) = 0$.

Notice that taking the time derivative of

$$\Pi^{(N)}(t)^{2\circledast} - \Pi^{(N)}(t) = \mathcal{O}(\hbar^{N+1})$$

we get

$$\Pi^{(N)}(t) \circledast \partial_t \Pi^{(N)}(t) \circledast \Pi^{(N)}(t) = \mathcal{O}(\hbar^{N+1}). \tag{14.42}$$

We have the same properties for $\mathbb{1} - \Pi^{(N)}(t)$. Now let us prove that S_N is off-diagonal.

Using (14.42) we have

$$\Pi^{(N)}(t) \left(i\hbar\partial_t \Pi^{(N)}(t) - [H(t), \Pi^{(N)}(t)]_{\circledast}\right) \Pi^{(N)}(t) =$$
$$\left(\Pi^{(N)}(t)^{2\circledast} - \Pi^{(N)}(t)\right) H(t)\Pi^{(N)}(t) - \Pi^{(N)}(t)H(t)\left(\Pi^{(N)}(t)^{2\circledast} - \Pi^{(N)}(t)\right).$$

Computing the coefficient of $\hbar^{N+1}$ in the last equation we get

$$i\pi(t)S_N(t)\pi(t) = R_N(t)H_0(t)\pi(t) - \pi(t)H_0(t)R_N(t) = \lambda(t)[R_N(t), \pi(t)] = 0.$$

By the same argument applied with $(\mathbb{1} - \Pi^{(N)}(t))$ instead of $\Pi^{(N)}(t)$ we also have $\pi^{\perp}(t)S_N(t)\pi^{\perp}(t) = 0$.

So now we can solve equations (14.39), (14.40) and we get for $\Pi_{N+1}(t)$ the following formula

$$\Pi_{N+1}(t) = \pi(t)^{\perp}R_N(t)\pi(t)^{\perp} - \pi(t)R_N(t)\pi(t) +$$
$$\frac{i}{\lambda(t) - \lambda^{\perp}(t)}\left(\pi(t)S_N(t)\pi^{\perp}(t) - \pi^{\perp}(t)S_N(t)\pi(t)\right). \tag{14.43}$$

Moreover the proof gives also uniqueness for the series $\{\Pi_j(t)\}_{j\geq 0}$ such that $\Pi_0(t) = \pi$. □

Corollary 46 *We have the following formal series equality in* $[t_0, t_0 + T] \times \Omega_2$*:*

$$\Pi^{+}(t, X) + \Pi^{-}(t, X) = \mathbb{1}_m.$$

In particular $\Pi^{+}(t) \circledast \Pi^{-}(t) = \Pi^{-}(t) \circledast \Pi^{+}(t) = 0$.

Corollary 47 *Assuming that condition* **(NCC)** *is satisfied then in* $[t_0, t_0 + T] \times \Omega_2$ *we have for any* $j \in \mathbb{N}$ *and* $\alpha \in \mathbb{N}^{2n}$*, the estimates:*

$$\|\partial_X^{\alpha}\Pi_j^{\pm}(t, X)\| \leq C_{j,\alpha}\delta^{-j-|\alpha|} \tag{14.44}$$

where $\|\cdot\|$ *is a fixed norm in* $\mathbb{C}^{m,m}$.

Proof Consider the case +. We only give an idea of the proof for $j = 0, 1$. The general case can be checked by an induction argument.

For $j = 0$ we use the integral formula

$$\pi_+(t, X) = \frac{i}{2\pi} \oint_{\gamma(t,X)} (H_0(t, X) - z)^{-1} dz, \tag{14.45}$$

where $\gamma(t, X)$ is a loop around $\lambda_+(t, X)$ with no other eigenvalue of $H(t, X)$ inside the loop. But we have

$$(H_0(t, X) - z)^{-1} = \frac{\pi_+(t, X)}{\lambda_+(t, X) - z} + \frac{\pi_-(t, X)}{\lambda_-(t, X) - z}. \tag{14.46}$$

So computing $\partial_X^\alpha \pi_+(t, X)$ with (14.45) then plugging (14.46) in (14.45) and using the residue Theorem we get easily (14.44) for $j = 0$. Then using Theorem 79. we deduce (14.60) for $j = 1$. For $j \geq 2$ we have to use an inductive argument as in the proof of Theorem 79.

As above in the following statement we denote $\lambda(t, X) = \lambda_\pm(t, X)$, $\pi(t, X) = \pi_\pm(t, X)$ and so on. The assumptions **(NCC)** are supposed to be satisfied.

Theorem 80 (formal adiabatic decoupling) *There exists a formal semi-classical time dependent Hermitian Hamiltonian in* $\mathfrak{A}_{\Omega_2}$, $H^{\mathrm{adia}}(t) = \sum_{j\geq 0} \hbar^j H_j^{\mathrm{adia}}(t)$, *smooth in X, continuous in t, such that* $H_0^{\mathrm{adia}}(t, X) = \lambda(t, X)\mathbb{1}_m$ *and*

$$\Pi(t) \circledast H^{\mathrm{adia}}(t) = \Pi(t) \circledast H(t). \tag{14.47}$$

Moreover, the subprincipal term $H_1^{\mathrm{adia}}(t)$ *is a Hermitian matrix, such that* $\pi(t)$ *satisfies in* Ω_1 *the following adiabatic transport equation along the classical flow for the Hamiltonian* $\lambda(t)$.

$$\partial_t \pi(t) + \{\lambda(t), \pi(t)\} = \frac{1}{i}[H_1^{\mathrm{adia}}(t), \pi(t)]. \tag{14.48}$$

The diagonal part of H_1^{adia} *satisfies*

$$\pi(t) H_1^{\mathrm{adia}}(t) \pi(t) = \pi(t) H_1(t) \pi(t) + \frac{1}{2}\pi(t)\{\pi(t), H_0(t)\}\pi(t). \tag{14.49}$$

Moreover for $j \geq 1$, *the blocks* $H_j^{\mathrm{adia}+\pm}$ *are unique and the block* $H_j^{\mathrm{adia}--}$ *can be chosen arbitrary.*

Proof As for the last Theorem we can compute $H_j^{\mathrm{adia}}(t)$ by induction on j starting with $H_0^{\mathrm{adia}}(t) = \lambda(t)\mathbb{1}_m$. Let compute $H_1^{\mathrm{adia}}(t)$ such that Eq. (14.47) is satisfied modulo $\mathscr{O}(\hbar^2)$. This gives

$$\pi(t)(H_1^{\mathrm{adia}}(t) - H_1(t)) = \Pi_1(t)\,(H_0(t) - \lambda(t)) + \frac{1}{2i}\{\pi(t), H_0(t) - \lambda(t)\}. \tag{14.50}$$

First of all we have to prove that (14.50) has at least a solution $H_1^{\mathrm{adia}}(t)$. For that we check that we have

$$\pi(t)^{\perp}\left(\Pi_1(t)\,(H_0(t)-\lambda(t))+\frac{1}{2i}\{\pi(t),H_0(t)-\lambda(t)\}\right)=0. \tag{14.51}$$

Using the following identity for the Poisson bracket of matrix symbols:

$$\{A,BC\}-\{AB,C\}=\{A,B\}C-A\{B,C\}, \tag{14.52}$$

we get that

$$\{\pi,\lambda\pi\}=\{\pi,\pi\}\lambda+\pi^{\perp}\{\pi,\lambda\},\quad \{\pi,\lambda^{\perp}\pi^{\perp}\}=-\{\pi,\pi\}\lambda+\pi^{\{}\pi,\lambda-\perp\}.$$

We apply these formulas to $H_0=\lambda\pi+\lambda^{\perp}\pi^{\perp}$ to prove (14.51).

Then to have uniqueness of $H_1^{\mathrm{adia}}(t)$ we can impose the (arbitrary) condition

$$(\mathbb{1}-\pi(t))(H_1^{\mathrm{adia}}(t)-H_1(t))(\mathbb{1}-\pi(t))=0.$$

So using (14.36), (14.35) and (14.50), $H_1^{\mathrm{adia}}(t)$ is well determined.

Then by induction on j we can compute $H_{j+1}^{\mathrm{adia}}(t)$ such that the Eq. (14.50) is satisfied modulo $\mathscr{O}(\hbar^{j+2})$.

Let us prove now Eq. (14.48). For simplicity let us assume here that $H_1(t)=0$. This term is not important.

We have

$$i[H_1^{\mathrm{adia}}(t),\pi(t)]=-i\left(\pi(t)H_1^{\mathrm{adia}}(t)\pi(t)^{\perp}-\pi(t)^{\perp}H_1^{\mathrm{adia}}(t)\pi(t)\right).$$

So from (14.50) we get

$$\pi(t)H_1^{\mathrm{adia}}(t)\pi(t)^{\perp}=\Pi_1(t)(H_0(t)-\lambda(t))\pi(t)^{\perp}+\frac{1}{2i}\{\pi(t),H_0(t)-\lambda(t)\}\pi(t)^{\perp}. \tag{14.53}$$

Using (14.36) we get

$$\pi(H_1^{\mathrm{adia}}\pi^{\perp}=-\pi\left(i\partial_t\pi-\frac{1}{2i}(\{H_0,\pi\}-\{\pi,H_0\})\right)\pi^{\perp}+\frac{1}{2i}\pi\{\pi;H_0-\lambda\}\pi^{\perp}. \tag{14.54}$$

We compute $\{H_0,\lambda\}=\{\lambda\pi+\lambda^{\perp}\pi^{\perp},\lambda\}$ using (14.52) and that $\{\pi(t),\pi(t)\}$ is diagonal and using the adjoint of (14.54) we get (14.48).

For the diagonal term using (14.50) we have

$$\pi(t)H_1^{\mathrm{adia}}(t)\pi(t)=\pi(t)H_1(t)\pi(t)+\frac{1}{2}\pi(t)\{\pi(t),H_0(t)\}\pi(t). \tag{14.55}$$

Then by induction on $N \geq 0$, we can check $H^{\mathrm{adia}}_{N+1}(t)$ such that (14.47) is satisfied mod. $\mathscr{O}(\hbar^{N+2})$.

More precisely, assume H^{adia}_j well defined for $0 \leq j \leq N$ such that

$$\Pi^{(N)}(t) \circledast H^{\mathrm{adia}(N)}(t) = \Pi^{(N)}(t) \circledast H(t) + \mathscr{O}(\hbar^{N+1}), \tag{14.56}$$

where $\Pi^{(N)} = \sum_{0\leq k\leq N} \hbar^k \Pi_k, \; H^{\mathrm{adia}(N)}(t) = \sum_{0\leq k\leq N} H^{\mathrm{adia}}_k(t)$.

We have to determine $H^{\mathrm{adia}}_{N+1}(t)$ such that (14.56) is satisfied for $N \to N+1$.

For simplicity we always assume that $H = H_0$ (the extension with sub-principal terms can be easily done).

Let us denote $T_N(t)$ the coefficient of $\hbar^{N+1}$ in $\Pi^{(N+1)}(t) \circledast (H(t) - H^{\mathrm{adia}(N)}(t))$. So we have to find H^{adia}_{N+1} such that

$$\pi(t) H^{\mathrm{adia}}_{N+1}(t) = T_N(t). \tag{14.57}$$

Hence we need to check that $\pi^{\perp}(t) T_N(t) = 0$. This is done by computing the coefficient of $\hbar^{N+1}$ in the expansion of

$$\left(\mathbb{1}_m - \Pi^{(N+1)}(t)\right) \Pi^{(N+1)}(t)(H(t) - H^{\mathrm{adia}(N)}(t)) = \mathscr{O}(\hbar^{N+2}).$$

So existence for the H^{adia}_j ($j \geq 1$) is proven. From the proof we get that the blocks $H^{\mathrm{adia},--}_j$ can be choose arbitrary, the other blocks are uniquely determined by our algorithm. □

From Theorem 80 we get that the time evolution $\mathscr{U}_H(t, t_0)$ for the Hamiltonian $\hat{H}(t)$ is formally decoupled between the modes $+$ and $-$. From Theorem 80 we get two formal quantum Hamiltonians $H^{\mathrm{adia}}_{\pm} = \sum_{j\geq 0} \hbar^j H^{\mathrm{adia},\pm}_j$ with $H^{\mathrm{adia},\pm}_0 = \lambda_{\pm}$.

Corollary 48 *Assume that condition* (**NCC**) *is satisfied in* $[t_0, t_0 + T] \times \mathbb{R}^{2n}$ *and that* $\hat{H}(t)$, $H^{\mathrm{adia}}_{\pm}(t)$ *generate propagators* $\mathscr{U}_H(t, t_0)$, $\mathscr{U}^{\mathrm{adia}}_{\pm}(t, t_0)$. *Then for any* $\psi_0 \in L^2(\mathbb{R}^d, \mathbb{C}^m)$ *we have, at the formal level for Weyl quantization,*

$$\boxed{\mathscr{U}_H(t, t_0)\psi_0 = \hat{\Pi}^+(t)\mathscr{U}^{\mathrm{adia}}_+(t, t_0)\hat{\Pi}^+(t_0)\psi_0 + \hat{\Pi}^-(t)\mathscr{U}^{\mathrm{adia}}_-(t, t_0)\hat{\Pi}^-(t_0)\psi_0}. \tag{14.58}$$

Proof Let us denote $\tilde{\psi}(t)$ the r.h.s of (14.58) and let us compute $i\hbar\partial_t\tilde{\psi}(t)$. Using Theorems 79 and 80 we get easily

$$(i\hbar\partial_t - \hat{H}(t))\hat{\Pi}^{\pm}(t) = \hat{\Pi}(t)(i\hbar\partial_t - \hat{H}^{\mathrm{adia}}_{\pm}(t)), \tag{14.59}$$

then we compute $i\hbar\partial_t\tilde{\psi}(t) = H(t)\tilde{\psi}(t)$ and $\tilde{\psi}(t_0) = \psi(t_0) = \psi_0$. Hence by uniqueness we get that $\psi(t) = \tilde{\psi}(t)$. □

Corollary 49 *Assuming that condition* (**NCC**) *is satisfied then in* $[t_0, t_0 + T] \times \Omega_2$ *we have for any* $j \in \mathbb{N}$ *and* $\alpha \in \mathbb{N}^{2n}$*, the estimates:*

$$\|\partial_X^\alpha H_j^{\text{adia},\pm}(t, X)\| \le C_{j,\alpha}\delta^{1-j-|\alpha|} \tag{14.60}$$

where $\|\cdot\|$ *is a fixed norm in* $\mathbb{C}^{m,m}$.

Our aim in the following step is to give an asymptotic meaning as $\hbar \searrow 0$ for solutions of (14.58) under the local non-crossing assumption (**NCC**) concerning the Hamilton flow of $\lambda^\pm(t, X)$. To do that we shall assume:

(CFS) The frequency set of the initial state $\psi_0^\hbar$ is localized in a compact set $K_0 \subset \mathbb{R}^{2n}$.

Recall that the frequency set for $\mathcal{U}_\pm^\hbar(t, t_0)\psi_0^\hbar$ propagates with the classical flow $\Phi_\pm^{t,t_0}$ according to the modified dynamics (14.15). Using this property and assumption **(CFS)** we shall see that we can give an $\hbar$-asymptotic meaning to (14.58) by localization in bounded set in the phase space.

We shall see by applying Theorems (79) and (80) that we can get asymptotic solutions for (14.9) with $\psi_0^\hbar(x) = \mathcal{WP}_{z_0}^\hbar g(x)\vec{v}$, $z_0 \in \Omega_0$ and $\vec{v}$ is an eigenvector of $H_0(t_0, z_0)$. More generally we can assume that the frequency set $\mathrm{FS}[\psi_0^\hbar]$ is in a compact K_C.

To do that we have to transform formal expansion in $\hbar$ into semiclassical quantum observables with error estimates.

Assuming property (**NCC**) let us introduce cutoff functions $\chi_j \in C_0^\infty(\Omega_j)$ such that $\chi_1 = 1$ in K_0, $\chi_2 \in C_0^\infty(\Omega_2)$ such that $\chi_2 = 1$ on $\{\Phi_\pm^{t,t_0}X,\ t \in [T_0, t_0 + T], X \in K_0\} \subseteq \Omega_2$.

We have to give a meaning as an equality between operators after cutting off. We forget here the sign $\pm$.

Proposition 138 *Let* $\chi_3(t, X)$ *be a time dependent cut-off, continuous in* $t \in [t_0, t_0 + T]$, C^∞ *with a compact support in* $X \in \Omega_2$ *such that* $\chi_2 = 1$ *on* $\mathrm{supp}(\chi_3)$. *Then we have*

$$(i\hbar\partial_t - \hat{H}(t))\widehat{\chi_2\Pi^{(N)}}(t)\hat{\chi}_3 = \widehat{\chi_2\Pi^{(N)}}(t)(i\hbar\partial_t - \widehat{\chi_2 H^{\text{adia},(N)}}(t))\hat{\chi}_3 + \mathcal{O}(\hbar^{N+1}), \tag{14.61}$$

Proof It is easier to compute at the symbol level. We start with the equality:

$$(i\hbar\partial_t - H(t)) \circledast \chi_2\Pi^{(N)}(t) = \chi_2\Pi^{(N)}(t) \circledast (i\hbar\partial_t - \chi_2 H^{\text{adia},(\mathrm{N})}(t))$$
$$+ i\hbar\partial_t(\chi_2\Pi^{(N)}(t)) + \chi_2\Pi^{(N)}(t) \circledast \chi_2 H^{\text{adia},(N)}(t) - H(t) \circledast \chi_2\Pi^{(N)}(t)$$

where $H^{\text{adia},(\mathrm{N})}(t) = \sum_{0\le j\le N} \hbar^j H_j^{\text{adia}}$. Using (14.34) we get

$$i\hbar\partial_t(\chi_2\Pi^{(N)}(t) = \chi_2\{H(t), \Pi^{(N)}(t)\}_\circledast^{(N)} + \chi_2\Pi(t)^{(N)} i\hbar\partial_t + \mathcal{O}(\hbar^{N+1}).$$

But we have

$$H(t) \circledast \chi_2 \Pi^{(N)}(t)) = (\chi_2 \Pi^{(N)}) \circledast H(t) + \{H(t), \chi_2 \Pi^{(N)}\}_{\circledast}.$$

Then using (14.47) (mod $\mathcal{O}(\hbar^{N+1})$), local properties of the Moyal product and that $\chi_2 = 1$ on supp(χ_3), we get

$$(i\hbar\partial_t - H(t)) \circledast \chi_2 \Pi^{(N)}(t) \circledast \chi_3 = \chi_2 \Pi^{(N}(t) \circledast (i\hbar\partial_t - \chi_2 H^{\text{adia},(N)}) \circledast \chi_3 + \mathcal{O}(\hbar^{N+1}). \tag{14.62}$$

□

Let us denote by $\mathcal{U}_{\pm}^{(N)}(t, t_0)$ the Schrödinger propagator for the Hamiltonian $\chi_2 H^{\text{adia},(N)}(t)$

Let us consider the Schrödinger equation (14.1) with the initial data $\psi(t_0) = \psi_0^\hbar$ with a compact frequency set in K_0. Let us denote

$$\psi_{\pm}^{(N)}(t) = \widehat{\chi_2 \Pi_{\pm}^{(N)}}(t) \mathcal{U}_{\pm}^{(N)}(t, t_0) \widehat{\chi_2 \Pi_{\pm}^{(N)}}(t_0) \psi_0^\hbar. \tag{14.63}$$

Notice that with the previous notations we have $\hat{\chi}_1 \psi_0^\hbar = \psi_0^\hbar + \mathcal{O}(\hbar^{+\infty})$. Using (14.61) and Egorov theorem we get that

$$(i\hbar\partial_t - \hat{H}(t)) \underbrace{(\psi_+^{(N)}(t) + \psi_-^{(N)}(t))}_{\psi^{(N)}(t)} = \mathcal{O}_{L^2}(\hbar^{N+1}). \tag{14.64}$$

Notice that we have applied (14.62) with $\hat{\chi}_3 = \mathcal{U}_{\pm}^{(N)}(t, t_0) \hat{\chi}_1 \mathcal{U}_{\pm}^{(N)}(t_0, t)$.

Using again Theorem 79, and that $\chi_2 = 1$ on K_0, we have

$$\psi^{(N)}(t_0) = \widehat{\chi_2 \Pi_+^{(N)}(t_0)}^2 \psi_0^\hbar + \widehat{\chi_2 \Pi_-^{(N)}(t_0)}^2 \psi_0^\hbar = \psi_0^\hbar + \mathcal{O}_{L^2}(\hbar^{N+1}).$$

Hence from the Duhamel formula we deduce the asymptotic adiabatic reduction.

Theorem 81 *Assume that the condition* **(NCC)** *is satisfied, that the initial data* $\psi_0^\hbar$ *satisfies* **(CFS)** *and that the cut-off* χ_1, χ_2 *are fixed as above. Then for* $t \in [t_0, t_0 + T]$, *the Schrödinger equation*

$$i\hbar\partial_t \psi(t) = \hat{H}(t)\psi(t), \quad \psi(t_0) = \psi_0^\hbar + \mathcal{O}(\hbar^\infty), \tag{14.65}$$

satisfies

$$\psi(t) = \psi^{(N)}(t) + \mathcal{O}_{L^2}(\hbar^N), \tag{14.66}$$

where $\psi^{(N)}(t)$ *is defined in (14.63) and (14.64).*

Remark 90 At the beginning of the Chapter we have assumed the $H(t)$ is sub-quadratic for simplicity. Moreover with this assumption the estimates can be extended to the harmonic spaces. In fact for proving it is enough to assume:

(i) Existence of the quantum propagator for $\hat{H}(t)$.

(ii) $H(t, X)$ is in a reasonable class of Hermitian symbols defined for some $M \in \mathbb{R}$, by estimates as

$$|\partial_X^\alpha H(t, X)| \le C_\alpha \langle X \rangle^M, \quad \forall t \in [t_0, t_0 + T], \ \forall X \in \mathbb{R}^{2n}.$$

Remark 91 The adiabatic reduction allows us to apply the results proved in Sect. 14.2. In particular many results known for scalar Hamiltonian can be extended to Hamiltonian systems with constant multiplicity eigenvalues, in the semi-classical regime.

We now apply the general adiabatic reduction results when the initial data is a coherent state.

Corollary 50 *Let* $\mathbf{v}_0$ *be a unitary eigenvector for* $H(t_0, z_0)$ *for the eigenvalue* $\lambda_+(t_0, z_0)$. *Then the solution* $\psi(t)$ *for (14.1) satisfies*

$$\psi(t) = \mathrm{e}^{\frac{i}{\hbar} S_+(t, t_0, z_0)} \vec{v}_+(t) \varphi_{z_t}^{\Gamma_t^+} \left(1 + \sqrt{\hbar} u^+(t) \cdot \frac{(x - q_t^+(t))}{\sqrt{\hbar}}\right) + \mathcal{O}(\hbar)$$

where $\vec{v}_+(t) = \mathscr{R}_+(t, t_0)\mathbf{v}_0$ *and* $u^+(t) = (u_1^+(t), \ldots, u_n^+(t))$ *can be computed using Theorem 77.*

Moreover $\vec{v}_+(t)$ *solves the differential equation*

$$\frac{d}{dt}\vec{v}_+(t) + i H_1^{\mathrm{adia}}(t)\vec{v}_+(t) = 0,$$

and

$$H(t)\vec{v}_+(t) = \lambda_+(t)\vec{v}_+(t), \quad \|\vec{v}_+(t)\|_{\mathbb{C}^m} = \|\vec{v}_+(t_0)\|_{\mathbb{C}^m}.$$

.

Proof Using Theorem 79 and (80) we have modulo $\mathcal{O}(\hbar^\infty)$,

$$\mathscr{U}(t, t_0)\varphi_{z,v} \asymp \widehat{\Pi^+}(t)\mathscr{U}_+(t, t_0)\widehat{\Pi^+}(t_0)\varphi_{z,v} + \widehat{\Pi^-}(t)\mathscr{U}_-(t, t_0)\widehat{\Pi^-}(t_0)\varphi_{z,v}$$

Therefore it is enough to prove the following formula

$$\pi_\pm(t, z_t^\pm)\mathscr{R}^\pm(t, t_0)\pi_\pm(t_0, z) = \mathscr{R}^\pm(t, t_0)\pi_\pm(t_0, z). \tag{14.67}$$

Let us denote $v(t) = \mathscr{R}^\pm(t, t_0)\pi_\pm(t_0, z)$. Computing derivatives in t of $v(t)$ and $\pi_\pm(t, z_t^\pm)v(t)$, and using (14.48), we can see that they satisfy the same differential equation and we conclude that $v(t) = \pi_\pm(t, z_t^\pm)v(t)$ for every $t \in I_T$. □

Remark 92 It is not difficult to extend the proof of Theorem 77 to systems such that the principal term has any number of constant multiplicities eigenvalues. We left to the reader the study of the remainder term for large T (estimation of the

Ehrenfest time) which can be obtained following the method already used for scalar Hamiltonians.

To finish this section let us apply the results to the Dirac operator.

Let us recall that the Dirac Hamiltonian is defined in the following way. Its symbol is

$$H(t,x,\xi) = c\sum_{j=1}^{3}\alpha_j\cdot(\xi_j - A_j(t,x)) + \beta mc^2 + V(t,x),$$

where $\{\alpha_j\}_{j=1}^3$ and β are the 4×4-matrices of Dirac satisfying the anti-commutation relations

$$\alpha_j\alpha_k + \alpha_k\alpha_j = 2\delta_{jk}I_4, \quad 1\le j,k\le 4,$$

($\alpha_4 = \beta$, $\mathbb{1}_4$ is the 4×4 identity matrix). $A = (A_1, A_2, A_3)$ is the magnetic vector potential and

$$V = \begin{pmatrix} V_+\mathbb{1}_2 & 0 \\ 0 & V_-\mathbb{1}_2 \end{pmatrix}$$

where $V_\pm$, is a scalar potential (1_2 is the identity matrix on $\mathbb{C}^2$). The physical constant m (mass) and c (velocity) are fixed so we can assume $m = c = 1$. We assume for simplicity that the potentials are C^∞ and there exists $\mu\in\mathbb{R}$ such that for very $k\in\mathbb{N}$, $\alpha\in\mathbb{N}^d$ there exists $C>0$ such that for all $(t,x)\in I_T\times\mathbb{R}^d$ we have

$$|\partial_t^k\partial_x^\alpha A(t,x)| + |\partial_t^k\partial_x^\alpha V(t,x) \le C < x >^\mu .$$

$H(t,X)$ has two eigenvalues $\lambda_\pm = \pm\sqrt{1+|\xi - A(t,x)|^2} + V_\pm(t,x)$. To apply our results it could be sufficient to assume that $V_+(t,x) - V_-(t,x) \ge -2+\delta$ for some $\delta>0$. But to make computations easier we shall assume $V_+ = V_- = V$. Then the spectral projections are given by the formula

$$\pi_\pm(t,x) = \frac{1}{2} \pm \frac{\alpha\cdot(\xi - A(t,x)) + \beta}{2\sqrt{1+|\xi - A(t,x)|^2}}. \tag{14.68}$$

Then we compute the subprincipal terms $H^\pm$ by the general formula (14.50). After a computation left to the reader, we get the following formula

$$H_1^\pm(t,x,\xi) = \frac{\pm i}{2(1+|p|^2)}(p\mathbb{1}_4 - \alpha\cdot H(p))\cdot(\partial_t A + \partial_x V) - \frac{1}{\sqrt{1+|p|^2}}B\cdot S, \tag{14.69}$$

where $p = \xi - A(t,x)$, $B_{j,k} = \partial_{x_j}A_k - \partial_{x_k}A_j$, $S_{j,k} = -i\alpha_j\alpha_k$. Let us recall the notation $S\cdot B = \sum_{1\le j<k\le 3} B_{j,k}S_{j,k}$ for the coupling of the spin with the magnetic field.

Let us now consider the simpler case where V and B are constant in time and space variables. Then we have $H_1^{\pm}(t,x,\xi) = -\frac{1}{\sqrt{1+|p|^2}} B \cdot S$. But in this case the magnetic kinetic energy $|p|^2$ is constant along the time evolution, so we have, if $E = \sqrt{1+|p|^2}$,

$$\mathscr{R}^{\pm}(t,0) = e^{\frac{it}{E} B\cdot S}. \tag{14.70}$$

The unitary matrix $\mathscr{R}^{\pm}(t,0)$ measures explicitly the effect of the spin on the trajectory of the particle in this particular case. In general the effect of the spin is computed by solving an ODE along the classical trajectories.

14.5 Figure

Non-crossing energy surfaces λ_+ and λ_-

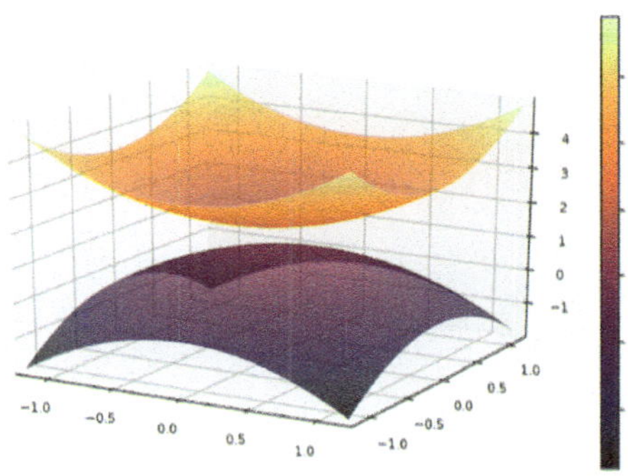

Notes

Similar mathematical results explained in this section were proved in [32]. In [9] another presentation concerning decoupling of multicomponent systems is given as well as application on the spectra.

Adiabatic decoupling setting follows Sjöstrand [94, 238] and the references there. For a different setting see Hagedorn-Joye [132] or Nenciu [199].

For more details concerning physical interpretation of the term $B \cdot S$ we refer to the book [243].

In the paper [33] one find more details concerning the spin-orbit coupling. But unfortunately the Lemma 4.5 in [33], is not correct. In our Chap. 7 an explicite computation reveals a term of order $\sqrt{j}$ in the expansion ($j = s$ in [33]) so the proof of Theorem 4.7 [33] has to be modified accordingly.

Concerning systems with crossing eigenvalues and the Landau-Zener formula there are several papers giving an accurate semi-classical analysis of the behavior of solutions close to the crossings after the pioneer works [129, 130], we refer the reader to the following advanced mathematical papers [97–100, 189, 250].

Part IV
Coherent States with Infinitely Many Degrees of Freedom

Chapter 15
Bosonic Coherent States

Abstract In a first part we give a brief presentation of general Fock space setting to describe quantum field theory. Bosons are quantum particles with integer spin and have symmetric wave functions; Fermions are quantum particles with half-integer spin and are represented by anti-symmetric wave functions. The functional setting is given by a symmetric or anti-symmetric tensor product of Hilbert spaces. We describe these spaces and transformations between these spaces. We shall follow the references [23, 42]. Coherent states are defined by translating the vacuum states with the Weyl operators. This is easily done here for bosons. We shall see in the next chapter how to deal with fermions. In a second part we give an interesting application of bosonic coherent states to the study of the classical limit as $\hbar \searrow 0$ of non-relativistic boson systems with two body interaction in the neighborhood of a solution of the classical system (here the Hartree equation). The classical limit corresponds here to the mean-field limit as the number of particles goes to infinity. As we have done for finite systems, we use here Hepp's method, which is a linearization procedure of the quantum Hamiltonian around the classical field. The fluctuations around this solution are controlled by a purely quadratic Hamiltonian. In a series of several important papers [114–116] Ginibre and Velo have proven an asymptotic expansion and remainder estimates for these quantum fluctuations. Finally, following the paper [225] one can show that, in the limit $\hbar \to 0$, the marginal distribution of the time-evolved coherent states tends in trace-norm to the projector onto the solution of the classical fields equation (Hartree equation) with a uniform remainder estimates in time.

15.1 Introduction

This chapter is very different from the others in this book. Until now we have considered coherent states depending on a parameter living in a finite dimensional space (typically a phase space for a classical mechanical system or more generally a Lie group). But coherent states may also be a useful tool to analyze quantum systems with an infinite number of particles (this was the main motivation for the founder of coherent states, R. J. Glauber). Systems with large number of particles are stud-

D. Robert and M. Combescure, *Coherent States and Applications in Mathematical Physics*, Theoretical and Mathematical Physics,
https://doi.org/10.1007/978-3-030-70845-0_15

ied in many domains of physics: statistical mechanics, quantum field theory, quantum optics for example. There are many books and papers in the physical literature (Weinberg [256]). There exists also books more rigorous from the mathematical point of view [42] and for a discussion concerning physical and mathematical aspects see the book [105].

15.2 Fock Spaces

15.2.1 Bosons and Fermions

Let us start with a quantum system of identical particles. Each particle has its states in the Hilbert space $\mathfrak{h}$. The states of systems of k particles are in the Hilbert space $\mathfrak{h}^{\otimes k} = \mathfrak{h} \otimes \cdots \otimes \mathfrak{h}$ and if the number of particles is not fixed (like in quantum field theory) the total Hilbert space is the Fock space

$$\mathscr{F}(\mathfrak{h}) := \bigoplus_{k\geq 0} \mathfrak{h}^{\otimes k} \tag{15.1}$$

where $\mathfrak{h}^{\otimes 0} = \mathbb{C}$ ("no-particle" space).

Let us recall here that if $\{e_j\}_{j\in J}$ is an orthonormal basis of $\mathfrak{h}$ then $\{e_{i_1} \otimes e_{i_2} \otimes \cdots \otimes e_{i_k}, |i_1, \dots i_k \in J\}$ is an orthonormal basis of $\mathfrak{h}^{\otimes k}$. So if we denote $\psi^{(k)}_{i_1,i_2,\dots,i_k} = \langle e_{i_1} \otimes e_{i_2} \otimes \cdots \otimes e_{i_k}, \psi^{(k)}\rangle$ then we have

$$\|\psi\|^2 = \sum_{k\geq 0, i_1,\dots,i_k\in J} |\psi^{(k)}_{i_1,i_2,\dots,i_k}|^2. \tag{15.2}$$

Recall that the differences between bosons and fermions are determined by their behaviour under permutations (Pauli exclusion principle for fermions). Let us denote $\mathfrak{S}_k$ the group of permutations of $\{1, 2, \dots, k\}$ and by ε_π the signature of $\pi \in \mathfrak{S}_k$.

The following equalities can be extended in two projections in $\mathscr{F}(\mathfrak{h})$

$$\begin{aligned} \Pi_{\mathrm{B}}(\psi_1 \otimes \cdots \otimes \psi_k) &= \frac{1}{k!} \sum_{\pi\in\mathfrak{S}_k} \psi_{\pi 1} \otimes \psi_{\pi 2} \otimes \cdots \otimes \psi_{\pi k} \\ \Pi_{\mathrm{F}}(\psi_1 \otimes \cdots \otimes \psi_k) &= \frac{1}{k!} \sum_{\pi\in\mathfrak{S}_k} \varepsilon_\pi \psi_{\pi 1} \otimes \psi_{\pi 2} \otimes \cdots \otimes \psi_{\pi k} \end{aligned} \tag{15.3}$$

where $\psi_1, \dots, \psi_k \in \mathfrak{h}$. The following notations will be used:

$$\psi_1 \wedge \psi_2 \wedge \cdots \wedge \psi_k = \Pi_F(\psi_1 \otimes \cdots \otimes \psi_k), \quad \psi_1 \vee \psi_2 \cdots \vee \psi_k = \Pi_B(\psi_1 \otimes \cdots \otimes \psi_k).$$

Definition 36 The subspace $\mathscr{F}_{\mathrm{B}}(\mathfrak{h}) := \Pi_{\mathrm{B}}\mathscr{F}(\mathfrak{h})$ is the Fock space of bosons and the subspace $\mathscr{F}_{\mathrm{F}}(\mathfrak{h}) := \Pi_{\mathrm{F}}\mathscr{F}(\mathfrak{h})$ is the Fock space of fermions.

$\mathfrak{h}^{\otimes k}_{\mathrm{B,F}} = \Pi_{\mathrm{B,F}}(\mathfrak{h}^{\otimes k})$ are the k-particles subspaces for bosons (B) or fermions (F).

The number operator $\mathbf{N}$ is defined as follows:

$$\mathbf{N}\psi^{(k)} = k\psi^{(k)},\ \psi^{(k)} \in \mathfrak{h}^{\otimes k}. \tag{15.4}$$

$\mathbf{N}$ can be extended as a self-adjoint operator in $\mathscr{F}(\mathfrak{h})$ with domain

$$D(\mathbf{N}) = \{\psi \in \mathscr{F}(\mathfrak{h}),\ \sum_{k\geq 0} k^2\|\psi^{(k)}\|^2 < +\infty\} \tag{15.5}$$

Moreover $\mathbf{N}$ commutes with $\Pi_{B,F}$ so $\mathbf{N}$ is a self-adjoint operator in the spaces $\mathscr{F}_{F,B}(\mathfrak{h})$.

We can define operators in $\mathscr{F}_{\mathrm{B,F}}(\mathfrak{h})$ starting from an Hamiltonian H in $\mathfrak{h}$ by a method known as *second quantization* as follows.

Define $H^{(0)} = 0$ and for $k \geq 1$,

$$H^{(k)}(\Pi_{\mathrm{B,F}}(\psi_1 \otimes \cdots \otimes \psi_k)) =$$
$$\Pi_{\mathrm{B,F}}\left(\sum_{1\leq j\leq k} \psi_1 \otimes \cdots \otimes \psi_{j-1} \otimes H\psi_j \otimes \psi_{j+1} \otimes \cdots \otimes \psi_k\right). \tag{15.6}$$

By linearity the direct sum of $H^{(k)}$ defines an operator $\mathbf{H} := \bigoplus_k H^{(k)}$ in $\mathscr{F}_{\mathrm{B,F}}(\mathfrak{h})$. $\mathbf{H}$ is the second quantization of H. It is convenient to introduce the dense subspace defined as

$$\mathscr{F}_0(\mathfrak{h}) = \{\psi \in \mathscr{F}(\mathfrak{h}) | \psi^{(k)} = 0 \ \text{ if } \ k \text{ large enough}\}$$

$\bigoplus_k H^{(k)}$ is well defined in $\mathscr{F}_0(\mathfrak{h})$. More precisely we have the following easy to prove lemma.

Lemma 92 *If H is a self-adjoint operator in $\mathfrak{h}$ then $\mathbf{H}$ can be extended as a unique self-adjoint operator (with dense domain) in $\mathscr{F}_{\mathrm{B,F}}(\mathfrak{h})$. This operator is also denoted $d\Gamma(H)$ or $\mathbf{H}$.*

If U is a unitary operator then $\bigoplus_k U^{(k)}$ can be extended in a unique unitary operator in $\mathscr{F}_{\mathrm{B,F}}(\mathfrak{h})$. This operator is denoted $\Gamma(U)$ or $\mathbf{U}$.

Remark 93 If $H = \mathbb{1}$ then we see that $d\Gamma(\mathbb{1}) = \mathbf{N}$, the number operator.

If $U_t = \mathrm{e}^{-itH}$ with H self-adjoint in $\mathfrak{h}$ then we have $\Gamma(U_t) = \mathrm{e}^{-itd\Gamma(H)}$ in other words the infinitesimal generator of $\Gamma(U_t)$ is the second quantization of the generator of U_t.

In quantum field theory the number of particles of the system is not constant so we have to define two kind of observables : annihilation operators and creation operators (other names are absorption and emission operators).

Definition 37 For every $f \in \mathfrak{h}$ we define the operators $a(f)$ and $a^*(f)$ by the following conditions

$$\begin{aligned} a(f)\psi^{(0)} = 0, \quad a^*(f)\psi^{(0)} = f, \\ a(f)(\psi_1 \otimes \cdots \otimes \psi_k) = (k+1)^{1/2}\langle f, \psi_1\rangle \psi_2 \otimes \cdots \otimes \psi_k, \\ a^*(f)(\psi_1 \otimes \cdots \otimes \psi_k) = k^{-1/2} f \otimes \psi_1 \otimes \psi_2 \otimes \cdots \otimes \psi_k. \end{aligned}$$

Remark that $f \mapsto a(f)$ is antilinear and $f \mapsto a^*(f)$ is linear on $\mathfrak{h}$.

Lemma 93 *For every* $\psi^{(k)} \in \mathfrak{h}^{\otimes k}$, $f \in \mathfrak{h}$, *we have*

$$\|a(f)\psi^{(k)}\| \le k^{1/2}\|f\|\|\Psi^{(k)}\|, \quad \|a^*(f)\psi^{(k)}\| \le (k+1)^{1/2}\|f\|\|\Psi^{(k)}\| \tag{15.7}$$

$a(f)$ *and* $a^*(f)$ *are defined on the linear space* $D(\mathbf{N}^{1/2})$ *and satisfy,*

$$\|a(f)\psi\| \le \|f\|\|(\mathbf{N}+1)^{1/2}\psi\|, \tag{15.8}$$

$$\|a^*(f)\psi\| \le \|f\|\|(\mathbf{N}+1)^{1/2}\psi\|, \quad \forall \psi \in D(\mathbf{N}^{1/2}). \tag{15.9}$$

$a(f)$ *and* $a^*(f)$ *leave the subspaces* $\mathscr{F}_{\mathrm{B,F}}(\mathfrak{h})$ *invariant. So the annihilation and creation operators for bosons (B) and fermions (F) are defined as follows*

$$\begin{aligned} a_{B,F}(f) = a(f)\Pi_{\mathrm{B,F}} = \Pi_{\mathrm{B,F}}a(f) \\ a^*_{B,F}(f) = a^*(f)\Pi_{\mathrm{B,F}} = \Pi_{\mathrm{B,F}}a^*(f). \end{aligned} \tag{15.10}$$

Remark 94 Starting from the vacuum state: $\Omega = (1, 0, \ldots, 0, \ldots)$ we create a particle with state $a^*_{B,F}(f)\Omega = (0, f, 0, \ldots)$. More generally if $h_1, \ldots, h_k \in \mathfrak{h}$ we get k particles in the state $a^*(h_1)a^*(h_2)\ldots a^*(h_k)\Omega$. It is not difficult to prove that Ω is cyclic, that means that the family $\{a^*(h_1)a^*(h_2)\ldots a^*(h_k)\Omega, |, h_j \in \mathfrak{h}, k \in \mathbb{N}\}$ is dense in $\mathscr{F}(\mathfrak{h})$ and the same property holds true for bosons and fermions.

In the Fock spaces $\mathscr{F}_{\mathrm{B,F}}(\mathfrak{h})$ we have the canonical commutation relations (CCR) for bosons and anticommutation relations (CAR) for fermions. More explicitly, If H, K are two operators , we denote the commutator $[H, K] := HK - KH$ and the anticommutator $[H, K]_+ := HK + KH$. In what follows operators are defined on $\mathscr{F}_{\mathrm{B,F}}(\mathfrak{h}) \cap \mathscr{F}_0(\mathfrak{h})$, $h_1, h_2 \in \mathfrak{h}$. We have for bosons the **(CCR)** relations :

$$[a_{\mathrm{B}}(h_1), a^*_B(h_2)] = \langle h_1, h_2\rangle \mathbb{1}, \quad [a_B(h_1), a_B(h_2)] = [a^*_B(h_1), a^*_{\mathrm{B}}(h_2)] = 0,$$

and for fermions the **(CAR)** relations :

$$[a_{\mathrm{F}}(h_1), a^*_{\mathrm{F}}(h_2)]_+ = \langle h_1, h_2\rangle \mathbb{1}, \quad [a_{\mathrm{F}}(h_1), a_{\mathrm{F}}(h_2)]_+ = [a^*_{\mathrm{F}}(h_1), a^*_{\mathrm{F}}(h_2)]_+ = 0.$$

Remark 95 If an orthonormal basis $\{\varphi_j\}_{i\in I}$ of $\mathfrak{h}$ is given the annihilation/creation operators are determined by $a_i^{(*)} := a^{(*)}(\varphi_i)$ (the subscript F, B is erased when the context is clear). In particular the number operator can be written as

$$\mathbf{N} = \sum_{i\in I} a_i^* a_i \tag{15.11}$$

We shall now detail some consequences of relations (CCR) and (CAR).

15.2.2 Bosons

First of all we remark that the Bargmann–Fock realization of quantum mechanics for n particles is isomorphic to the bosonic Fock realization with $\mathfrak{h} = \mathbb{C}^n$.

Recall that we have seen in Chap. 1 that in the Bargmann space $\mathscr{F}(\mathbb{C}^n)$ we have

$$\left[\zeta_j, \frac{\partial}{\partial \zeta_k}\right] = \delta_{j,k}$$

and an orthonormal basis

$$\phi_\alpha^\#(\zeta) = (2\pi\hbar)^{-n/2}(\alpha!)^{-1/2}\zeta^\alpha.$$

If $\{e_j\}_{1\leq j\leq n}$ is the canonical basis of $\mathbb{C}^n$, we get a unitary map Φ_B from $\mathscr{F}(\mathbb{C}^n)$ onto $\mathscr{F}_B(\mathbb{C}^n)$ by the property

$$\Phi_B(\phi_\alpha^\#) = \Phi_B(e_{\alpha_1} \otimes e_{\alpha_2} \otimes \cdots \otimes e_{\alpha_k}).$$

Note that $\Phi_B(e_{\alpha_1} \otimes e_{\alpha_2} \otimes \cdots \otimes e_{\alpha_k})$ is the symmetric tensor product of the e_{α_j}. We have easily

$$a_j = \Phi_B \frac{\partial}{\partial \zeta_j} \Phi_B^{-1}, \quad a_j^* = \Phi_B \hat{\zeta}_j \Phi_B^{-1} \tag{15.12}$$

which proved that bosonic Fock realization and Bargmann–Fock realization of quantum mechanics are equivalent.

In quantum field theory the one particle space is usually the infinite dimensional Hilbert space $\mathfrak{h} = L^2(\mathbb{R}^n)$. In physical applications it is convenient to consider field operators depending on a point $x \in \mathbb{R}^n$ (each particle has n degree of freedom). They are operator valued distributions on $L^2(\mathbb{R}^n)$. Let $\mathscr{F}_n$ be the bosonic Fock space :

$$\mathscr{F}_n = \bigoplus_{k\geq 0} \overset{k}{\vee} L^2(\mathbb{R}^n)$$

with

$$\overset{0}{\vee} L^2(\mathbb{R}^n) = \mathbb{C}, \quad \overset{k}{\vee} L^2(\mathbb{R}^n) = \otimes_s^k L^2(\mathbb{R}^n).$$

The subscript s indicates the symmetric tensor product. $\otimes_s^k L^2(\mathbb{R}^n)$ is the subspace of $\otimes^k L^2(\mathbb{R}^n)$ of symmetric functions on $\underbrace{\mathbb{R}^n \times \mathbb{R}^n \ldots \mathbb{R}^n}_{k \text{ times}}$. So $\overset{k}{\vee} L^2(\mathbb{R}^n)$ is the k-particles space for bosons.

Recall that a vector $\psi \in \mathscr{F}_n$ is a sequence

$$\psi = \{\psi^{(k)}\}_{k \geq 0}$$

of k-particle wavefunctions $\psi^{(k)} \in \overset{k}{\vee} L^2(\mathbb{R}^n)$. The scalar product in $\mathscr{F}_n$ of two functions $\psi_\alpha,\ \psi_\beta$ is given by

$$\langle \psi_\alpha, \psi_\beta \rangle = \sum_{k \geq 0} \langle \psi_\alpha^{(k)}, \psi_\beta^{(k)} \rangle_{L^2(\mathbb{R}^{nk})}.$$

Recall that the state $\Omega = \{1, 0, \ldots 0, \ldots\}$ is called the vacuum.

The creation and annihilation operators $a^*(x),\ a(x)$ are defined as operator-distribution by

$$\begin{aligned} (a^*(x)\psi)^{(k)}(x_1, \ldots x_k) &= \frac{1}{\sqrt{k}} \sum_{j=1}^{k} \delta(x - x_j) \psi^{(k-1)}(x_1, \ldots, \hat{x}_j, \ldots, x_k), \\ (a(x)\psi)^{(k)}(x_1, \ldots x_k) &= \sqrt{k+1}\, \psi^{(k+1)}(x, x_1, \ldots x_k), \end{aligned} \tag{15.13}$$

where $\hat{x}_j$ means that x_j is absent.

The canonical commutation relations assume the form:

$$[a(x), a^*(y)] = \delta(x - y), \quad [a(x), a(y)] = [a^*(x), a^*(y)] = 0$$

For $f \in L^2(\mathbb{R}^n)$ we recover the definitions:

$$a^*(f) = \int dx\ f(x) a^*(x) \tag{15.14}$$

$$a(f) = \int dx\ \bar{f}(x) a(x) \tag{15.15}$$

The number operator $\mathbf{N}$ has the form

$$\mathbf{N} = \int dx a^*(x) a(x) \tag{15.16}$$

Later we shall consider the Hamiltonian of a bosons system with pairwise interactions described by a potential $V = V(x - y)$. V is supposed to be an even, real function on $\mathbb{R}^n$. This Hamiltonian can be written as follows in the Fock space $\mathscr{F}$. We assume here that $\hbar = 1$.

$$\mathbf{H} = \frac{1}{2}\int dx\, \nabla a^*(x) \cdot \nabla a(x) + \frac{1}{2}\int dx\, dy\, V(x - y)a^*(x)a^*(y)a(y)a(x). \tag{15.17}$$

Formula (15.17) needs to be interpreted in the distribution sense. A direct computation shows that restriction $H^{(k)}$ of $\mathbf{H}$ to the k-particles space is, as it is expected:

$$H^{(k)} = -\frac{1}{2}\sum_{1\leq j\leq k} \Delta_j + \sum_{1\leq i<j\leq k} V(x_i - x_j) \tag{15.18}$$

where

$$\Delta_j = \underbrace{\mathbb{1}\otimes\cdots\otimes\mathbb{1}}_{j-1}\otimes\Delta\otimes\underbrace{\mathbb{1}\cdots\otimes\mathbb{1}}_{k-j}$$

and Δ_j is the Laplace operator in variables $x_j \in \mathbb{R}^n$.

Let us explain why representations (15.17) and (15.18) formally coincide. We have to understand the meaning of the r.h.s in (15.17). We have two terms, the first is the kinetic energy the second is the potential energy. If $\hat{A}$ is an operator in the one particle space $L^2(\mathbb{R}^n)$ its second quantization $\mathbf{A}$ can be written as

$$\mathbf{A} = \sum_{j,k\in I} \langle e_j, \hat{A}e_k\rangle a^*(e_j)a(e_k). \tag{15.19}$$

Introduce the distributions $a^*(x)$ and $a(y)$ and the Schwartz integral kernel K_A of $\hat{A}$, we have

$$K_A(x, y) = \sum_{j,k} \langle e_j, \hat{A}e_k\rangle e_j(x)\bar{e}_k(y)$$

so we get

$$\mathbf{A} = \int dxdy K_A(x, y)a^*(x)a(y).$$

In particular if K_A is null outside the diagonal we can write

$$\mathbf{A} = \int dx A(x)a^*(x)a(x)$$

where $A(x)$ is some function or distribution depending in one particle variable. In particular this is true if $\hat{A}$ is a multiplication operator in $L^2(\mathbb{R}^n)$. Its kernel is $K_A(x, y) = A(x)\delta(x - y)$. If $\hat{A}$ is a convolution operator (like Laplace operator) we have the same interpretation using Fourier transform in variable x.

For the potential energy term we have the same interpretation with $L^2(\mathbb{R}^n)$ replaced by $L_s^2(\mathbb{R}^n \times \mathbb{R}^n) = L^2(\mathbb{R}^n) \vee L^2(\mathbb{R}^n)$ and using that $\{e_j \vee e_k\}_{j,k\in I}$ is an orthonormal basis. $V = V(x-y)$ acting as a multiplication operator in $L_s^2(\mathbb{R}^n \times \mathbb{R}^n)$. So we have the formal equality

$$\int dxdya^*(y)a^*(x)V(x-y)a^*(x)a^*(y)a(x)a(y)$$
$$= \sum_{\substack{j,k\in I\\ j',j'}} \langle e_j \vee e_k, V(e_{j'} \vee e_{k'})\rangle a^*(e_j)a^*(e_k)a(e_{k'})a(e_{j'}). \tag{15.20}$$

So we get a rigorous interpretation for (15.17) for $\mathbf{H}$ acting in the finite number dense subspace $\mathscr{F}_0(L^2(\mathbb{R}^n))$. More assumptions will be needed later on V.

15.3 The Bosons Coherent States

As above, here we forget the subscript B, considering only bosons.

We want to extend for infinite number of bosons the coherent states already defined for finite systems (Chap. 1).

The idea is the same: translate by Weyl operators the vacuum (the ground state for coherent states of harmonic oscillator).

We first define the field operator (or Segal operator) :

$$\Phi(f) = \frac{a(f) + a^*(f)}{\sqrt{2}}, \quad f \in \mathfrak{h}. \tag{15.21}$$

Remark that if we define $\Pi(f) = \Phi(if)$ then we have

$$a(f) = \frac{\Phi(f) + i\Pi(f)}{\sqrt{2}}, \quad a^*(f) = \frac{\Phi(f) - i\Pi(f)}{i\sqrt{2}}.$$

Lemma 94 *(i) For every $f \in \mathfrak{h}$, $\Phi(f)$ is symmetric on the subspace $\mathscr{F}_0(\mathfrak{h})$ and is essentially self-adjoint.*

(ii) $D(\mathbf{N}^{1/2})$ is in the domain of $\Phi(f)$ and for every $\psi \in D(\mathbf{N}^{1/2})$, $f \mapsto \Phi(f)\psi$ is continuous on $\mathfrak{h}$.

(iii) If $\psi \in D(\mathbf{N})$ and $f, g \in \mathfrak{h}$ then we have for commutators:

$$[\Phi(f), \Phi(g)]\psi = i\Im\langle f, g\rangle\psi.$$

Proof The only non trivial statement is that $\Phi(f)$ is essentially self-adjoint. Using the Nelson criterium [215] it is enough to prove that each $\psi \in \mathscr{F}_0(\mathfrak{h})$ is an analytic vector for $\Phi(f)$. This a consequence of the estimate

$$\|\Phi(f)^n\psi^{(k)}\| \le 2^{n/2}(k+n)^{1/2}(k+n-1)^{1/2}\dots(k+1)^{1/2}\|\psi^{(k)}\|\|f\|^n$$

It follows that for every $r > 0$ we have

$$\sum_{n\geq 0}\frac{r^n}{n!}\|\Phi(f)^n\psi^{(k)}\| < +\infty.$$

□

The Weyl operators and their compagnons coherent states are defined as follows:

Definition 38 For $f \in \mathfrak{h}$ we define the Weyl translation operators:

$$T(f) = \exp\left(a^*(f) - a(f)\right) = \exp\left(\int dx\ (f(x)a^*(x) - \bar{f}(x)a(x))\right). \quad (15.22)$$

The coherent state $\Psi(f)$ for every $f \in \mathfrak{h}$ is then defined as

$$\boxed{\Psi(f) = T(f)\Omega}.$$

The bosonic coherent states have the following expression (analogue of an expression already given in Chap. 1 for finite systems of bosons)

Proposition 139 *For every $f \in \mathfrak{h}$ we have*

$$\Psi(f) = \mathrm{e}^{-\frac{\|f\|^2}{2}}\sum_{k\geq 0}\frac{1}{\sqrt{k!}}f^{\otimes k}.$$

In particular the probability to have k particles in $\psi(f)$ is equal to $\mathrm{e}^{-\|f\|^2}\|f\|^{2k}/k!$, where we recognize the Poisson law with mean $\|f\|^2$.

Proof We give formal argument from which it is not difficult to supply rigourous proofs.

One has the useful formula:

$$T(f) = e^{-\frac{\|f\|^2}{2}}\exp(a^*(f))\exp(-a(f)) \quad (15.23)$$

since the commutator $[a(f), a^*(f)] = \|f\|^2$ commutes with $a(f)$, $a^*(f)$. We deduce

$$\Psi(f) = \mathrm{e}^{-\frac{\|f\|^2}{2}}\sum_{k\geq 0}\frac{(a^*(f))^k}{k!}\Omega = \mathrm{e}^{-\frac{\|f\|^2}{2}}\sum_{k\geq 0}\frac{f^{\otimes k}}{\sqrt{k!}}$$

where $f^{\otimes k}$ is the Fock-vector $\{0, 0, \dots, f^{\otimes k}, 0\dots\}$.

Let π_k be the orthogonal projector onto the k-particle space $\otimes_s^k\mathfrak{h}$. We have found

$$\pi_k(\Psi(f)) = \mathrm{e}^{-\frac{\|f\|^2}{2}} \frac{f^{\otimes k}}{\sqrt{k!}}$$

so the probability to have k particles in $\psi(f)$ is equal to $\|\pi_k(\Psi(f))\|^2 = e^{-\|f\|^2}\|f\|^{2k}/k!$. □

The main properties of Weyl operators and coherent states are given in the following proposition.

Proposition 140 *Let $f,\ g \in \mathfrak{h}$.*

(i) *$T(f)$ is a unitary operator and one has*

$$T(f)^* = T(f)^{-1} = T(-f).$$

(ii) *The Weyl operator satisfies the commutation relations:*

$$T(f)T(g) = T(g)T(f)\exp\left(-2i\Im\langle f, g\rangle\right) = T(f+g)\exp\left(-i\Im\langle f, g\rangle\right).$$

In particular we have

$$T(g)\Psi(f) = \mathrm{e}^{-i\Im\langle g, f\rangle}\Psi(g+f)$$

(iii) *We have*

$$T^*(f)a(g)T(f) = a(g) + \langle g, f\rangle \mathbb{1}, \quad T^*(f)a^*(g)T(f) = a^*(g) + \langle f, g\rangle \mathbb{1}.$$

(iv) *The coherent states are eigenfunctions of the annihilation operators:*

$$a(g)\Psi(f) = \langle g, f\rangle\Psi(f).$$

(v) *The expectation of the number operator* **N** *in the coherent state $\Psi(f)$ is*

$$\langle\Psi(f), \mathbf{N}\Psi(f)\rangle = \|f\|^2,$$

also we have for the variance:

$$\langle\Psi(f), \mathbf{N}^2\Psi(f)\rangle - \langle\Psi(f), \mathbf{N}\Psi(f)\rangle^2 = \|f\|^2.$$

(vi) *The coherent states are normalized but not orthogonal to each other:*

$$\langle\Psi(f), \Psi(g)\rangle = \exp\left(-\frac{1}{2}(\|f-g\|^2 - i\Im\langle f, g\rangle)\right)$$

which implies that

$$|\langle \Psi(f), \Psi(g)\rangle| = \exp\left(-\frac{1}{2}\|f-g\|^2\right).$$

(vii) *The set of operators* $\{T(f),\ f \in \mathfrak{h}\}$ *is irreducible on* $\mathscr{F}(\mathfrak{h})$*: the only bounded operators* $\mathbf{B}$ *in* $\mathscr{F}(\mathfrak{h})$ *commuting with* $T(f)$ *for all* $f \in \mathscr{F}(\mathfrak{h})$ *are the scalar* $\mathbf{B} = \lambda\mathbb{1}$, $\lambda \in \mathbb{C}$.

In particular the set of coherent states $\{\Psi(f),\ f \in \mathfrak{h}\}$ *is total in* $\mathscr{F}(\mathfrak{h})$.

Proof Properties (i) to (ii) are left to the reader. For (iii) we compute

$$\frac{d}{dt}T^*(tf)a(g)T(tf) = T^*(tf)[a(g), a^*(f)]T(tf) = T^*(tf)[\langle g, f\rangle]T(tf)$$

and we get (iii) integrating in t between 0 and 1. (iv) are (v) are consequences of (iii).

For (vi) we write, using (15.23): $T(f)\Omega = \mathrm{e}^{-\frac{1}{2}\|f\|}\mathrm{e}^{a^*(f)}\Omega$. So we get

$$\begin{aligned}\langle T(f)\Omega, T(g)\Omega\rangle &= \langle \Omega, T^*(f)T(g)\Omega\rangle \\ &= \langle \Omega, T(g-f)\Omega\rangle \mathrm{e}^{i\Im\langle f,g\rangle} \\ &= \mathrm{e}^{-\frac{1}{2}\|g-f\|^2}\mathrm{e}^{i\Im\langle f,g\rangle}. \end{aligned} \tag{15.24}$$

Let us now prove (vii). Remark first that $\mathbf{B}$ commutes with $\Phi(f)$ for every $f \in \mathfrak{h}$ hence with $a(f)$ and $a^*(f)$. In particular $\mathbf{B}$ commutes with the number operator $\mathbf{N}$. Let $\{e_i\}_{i\in I}$ be an orthonormal basis for $\mathfrak{h}$ and $a_i = a(e_i)$. Denote $\psi_{k_1,\ldots,k_n} = (k_1!\ldots k_n!)^{-1/2}(a_1^*)^{k_1}\ldots(a_n^*)^{k_n}\Omega$. This is an orthonormal basis for the Fock space $\mathscr{F}_B(\mathfrak{h})$ with the obvious index set. Compute $\langle \psi_{k_1,\ldots,k_n}, \mathbf{B}\psi_{j_1,\ldots,j_m}\rangle$. This is 0 if the sets $\{k_1,\ldots,k_n\}$. $\{j_1,\ldots,j_m\}$are not equal. Finally we have, using (CCR),

$$\langle \psi_{i_1,\ldots,i_n}, \mathbf{B}\psi_{i_1,\ldots,i_n}\rangle = \langle \Omega, a_{i_1}\ldots a_{i_n}a_{i_1}^*\ldots a_{i_n}^*\Omega\rangle = \langle \Omega, \mathbf{B}\Omega\rangle$$

hence $\mathbf{B} = \langle \Omega, \mathbf{B}\Omega\rangle\mathbb{1}$. □

The two following lemmas will be useful later. Let us introduce the operators family $\langle \mathbf{N}\rangle^r = (\mathbb{1} + \mathbf{N}^2)^{r/2}$ for $r \in \mathbb{R}$.

Lemma 95 *For every* $r \in \mathbb{R}$ *and every* $f \in \mathfrak{h}$, $\langle \mathbf{N}\rangle^r T(f)\langle \mathbf{N}\rangle^{-r}$ *extends in a bounded operator in the Fock space* $\mathscr{F}_B(\mathfrak{h})$. *In other words for every* $r \geq 0$, $T(f)$ *is bounded on the Hilbert space* $D(\langle \mathbf{N}\rangle^r)$ *for the norm* $\|\psi\|_r := \|\langle \mathbf{N}\rangle^r\psi\|_{\mathscr{F}_B(\mathfrak{h})}$.

Proof From the previous proposition (iii) we have

$$T^*(f)\mathbf{N}T(f) = \mathbf{N} + a(f) + a^*(f) + \|f\|^2\mathbb{1}. \tag{15.25}$$

We prove the lemma for $r = \frac{1}{2}$. It is not difficult by iteration and interpolation to prove the result for every r.

We have

$$\|\langle\mathbf{N}\rangle^{1/2}T(f)\langle\mathbf{N}\rangle^{-1/2}\Psi\|^2 = \langle\Psi, \langle\mathbf{N}\rangle^{-1/2}T^*(f)\langle\mathbf{N}\rangle T(f)\langle\mathbf{N}\rangle^{-1/2}\Psi\rangle$$

We can replace $\langle\mathbf{N}\rangle$ by $\mathbf{N}$, use (15.25) and Cauchy–Schwarz inequality to get

$$\|\langle\mathbf{N}\rangle^{1/2}T(f)\langle\mathbf{N}\rangle^{-1/2}\|^2 \le C\|f\|^2.$$

□

Lemma 96 *Let $\alpha \in \mathscr{C}^1(\mathbb{R}, \mathfrak{h})$, $t \to \alpha(t)$. Then $T(\alpha(t))$ is strongly differentiable in t from $D(\mathbf{N}^{1/2})$ to $\mathscr{F}(\mathfrak{h})$. The derivative is given by*

$$\frac{d}{dt}T(\alpha(t)) = T(\alpha(t))[a^*(\dot\alpha(t)) - a(\dot{\bar\alpha}) + i\Im(\bar\alpha\cdot\dot\alpha)] \tag{15.26}$$

where $\dot\alpha = \frac{d}{dt}\alpha$.

Proof It is enough to compute the derivative for $t = 0$. Using Lemma 3 the computations can be done for $\psi \in D(\langle\mathbf{N}\rangle^{1/2})$. In what follows ψ will be omitted.

We have $T(\alpha(t)) - T(\alpha(0)) = T(\alpha(0))T^*(\alpha(0))(T(\alpha(t)) - \mathbb{1})$ and

$$T^*(\alpha(0))T(\alpha(t)) = T(\alpha(t) - \alpha(0))e^{i\Im\langle\alpha(0),\alpha(t)\rangle}.$$

Using Duhamel formula on $D(\mathbf{N}^{1/2})$ we get

$$T(\alpha(t) - \alpha(0)) = \mathbb{1} + t(a^*(\dot\alpha(0)) - a(\dot\alpha(0))) + O(t^2)$$

hence

$$\frac{d}{dt}T^*(\alpha(0))T(\alpha(t))|_{t=0} = a^*(\dot\alpha(0)) - a(\dot\alpha(0)) + i\Im\langle\alpha(0), \dot\alpha(0)\rangle.$$

The formula (15.26) follows. □

In the following section we shall study the mean field behavior of large systems of bosons with weak two particles interactions. For that purpose we introduce one particle density operator $\Gamma_\Psi^{(1)}$ for every $\Psi \in \mathscr{F}(\mathfrak{h})$, as follows. It is defined as a sesquilinear form in $\mathfrak{h}$:

$$(f, g) \mapsto \frac{1}{\langle\Psi, \mathbf{N}\Psi\rangle}\langle\Psi, a^*(f)a(g)\Psi\rangle := \langle f, \Gamma_\Psi^{(1)}g\rangle, \quad f, g \in \mathfrak{h}.$$

If $\mathfrak{h} = L^2(\mathbb{R}^3)$ the Schwartz kernel of $\Gamma_\Psi^{(1)}$ satisfies

$$\Gamma_\Psi^{(1)}(x, y) = \frac{1}{\langle\psi, \mathbf{N}\Psi\rangle}\langle\Psi, a^*(x)a(y)\Psi\rangle \tag{15.27}$$

Moreover if Ψ is a k-particle state then $\Gamma_{\Psi}^{(1)}$ is the relative trace in $\mathfrak{h} = L^2(\mathbb{R}^n)$ of the projector $|\Psi\rangle\langle\Psi|$ and we have

$$\Gamma_{\Psi}^{(1)}(x, y) = \int_{\mathbb{R}^{n(k-1)}\times\mathbb{R}^{n(k-1)}} dx'dy'\Psi(x', x)\bar{\Psi}(y', y)$$

We shall be interested to considering the one particle density for $\Psi(t)$ being a time evolution of a coherent state $\Psi(\varphi_{\hbar}(t))$, where $\varphi_{\hbar}(t) = \hbar^{-1/2}\varphi(t)$, depending on a small (semi-classical) parameter $\hbar$. We call it $\Gamma_{\hbar,t}^{(1)}$. We shall see that, under some conditions, $\Gamma_{\hbar,t}^{(1)}$ converges in trace norm operator to $|\varphi(t)\rangle\langle\varphi(t)|$ when $\hbar \searrow 0$ where $\varphi(t)$ follows a classical evolution.

15.4 The Classical Limit for Large Systems of Bosons

15.4.1 Introduction

We have already considered the classical limit problem for systems with a finite number of bosons in Chap. 1. It has been a natural question since the early days of quantum mechanics to compare the classical and quantum mechanical descriptions of physical systems. One of the oldest and by now best known relation between the two theories goes back to Ehrenfest [91]. This has been put on a firm mathematical basis by Hepp [148]. He proved that in the limit where the Planck constant $\hbar$[1] tends to zero, the matrix elements of quantum observables between suitable $\hbar$-dependent coherent states tend to the classical values evolving according to the appropriate equation. Moreover he proved that the quantum mechanical fluctuations evolve according to the equation obtained by linearizing the quantum mechanical evolution around the classical solution. Hepp's approach covers the case of quantum mechanics (that we have studied in detail in Chap. 4) of boson field theories, both relativistic and nonrelativistic, and more generally of all quantum theories which can be expressed in terms of observables satisfying the Canonical Commutation Relations (CCR). One is led to study a perturbation problem for the evolution of a set of operators satisfying the (CCR) in a suitable representation. The small parameter which characterizes the perturbation theory is $\hbar^{1/2}$. In the most favourable cases this evolution is implemented by a unitary group of operators $W(t, s)$. The solution of the unperturbed problem is given by a unitary group $\mathscr{U}_2(t, s)$, the infinitesimal generator of which is quadratic in the field operators and depends on the classical solution around which one is considering the classical limit. The operator $\mathscr{U}_2(t, s)$ describes the evolution of the quantum fluctuations. Hepp's result consists in proving strong convergence of $W(t, s)$ towards $\mathscr{U}_2(t, s)$ when $\hbar$ goes to zero. We shall explain the results obtained by

[1] In physics $\hbar$ is a constant equal to 1.055×10^{-34}J.S. As it is usual in quantum mechanics we consider here $\hbar$ as an effective Planck constant obtained by scaling, for example $\hbar \mapsto \frac{\hbar}{\sqrt{2m}}$, where m is the mass and $\hbar \searrow 0$ means $m \nearrow +\infty$.

Ginibre-Velo [114–116] to estimate the error term in the Hepp's results. We also explain results obtained more recently by Rodnianski-Schlein [225] using Hepp's approach to get the convergence of the one particle marginal for evolved coherent states towards a classical field. Note that this result is somehow an extension to the quantum field context of the semi-classical expansion considered before in Chap. 4 for time evolution of Gaussian coherent states for a fixed number of bosons.

15.4.2 The Hepp Method

We follow here the presentation given in [115]. We start with the abstract setting of a general Fock space $\mathscr{F}(\mathfrak{h})$ and an orthonormal basis $\{e_i\}_{i\in I}$ in $\mathfrak{h}$. $I = \{1, 2, \ldots, \nu\}$, $\nu \le +\infty$. Recall that $a_i := a(e_i)$.

So the system is described by the family of quantum operators $\mathbf{a} = (a_i)_{i\in I}$ satisfying the canonical commutation rules:

$$[a_i, a_j] = 0, \quad [a_i, a_j^*] = \delta_{i,j}.$$

The variables $\mathbf{a}_\hbar$ expected to have a classical limit are related to $\mathbf{a}$ by

$$\mathbf{a}_\hbar = \hbar^{1/2}\mathbf{a} \ \text{ or } \ a_\hbar(f) = \hbar^{1/2}a(f), \quad f \in \mathfrak{h} \ \text{ or } \ a_i^\hbar = \hbar^{1/2}a_i.$$

Consider a self adjoint Hamiltonian $\mathbf{H}$ in $\mathscr{F}_B(\mathfrak{h})$, suitably regular (for instance polynomial in $\mathbf{a}^*$, $\mathbf{a}$), with no explicit $\hbar$-dependence :

$$\mathbf{H} = \sum C(n_1, \ldots, n_k|m_1, \ldots, m_\ell)(a_1^*)^{n_1} \ldots (a_1^*)^{n_k} a_1^{m_1} \ldots a_\ell^{m_\ell}, \ C(\bullet|\bullet) \in \mathbb{C}.$$

In the Heisenberg picture the time evolution $\mathbf{a}_\hbar(t)$ of $\mathbf{a}_\hbar$ is given by the following equation

$$i\hbar\frac{d}{dt}\mathbf{a}_\hbar(t) = [\mathbf{a}_\hbar(t), H(\mathbf{a}_\hbar)], \quad \mathbf{a}_\hbar(0) = \mathbf{a}_\hbar. \tag{15.28}$$

We want to relate the time dependent operators $\mathbf{a}_\hbar(t)$ with a family of $\hbar$-independent c-number variables[2]:

$$\varphi(t) = \{\varphi_i(t)\}_{i\in I}$$

which will appear to be the classical limits. $\varphi(t)$ can be identified with a classical trajectory in the Hilbert space $\mathfrak{h}$ writing $\varphi(t) = \sum_{i\in I} \varphi_i(t)e_i$.

[2]a c-number here is a family of time dependent complex numbers indexed by I. It be can identified with a vector in $\mathfrak{h}$. In the language of quantum mechanics c-numbers are the opposite of $\Gamma^{(1)}_{\hbar,t}$, operators in an Hilbert space.

We thus expand $\mathbf{H}$ in power series of $a_i^\hbar - \varphi_i,\ (a_i^\hbar)^* - \bar{\varphi}_i$ in a neighborhood of $\varphi,\ \bar{\varphi}$:

$$\mathbf{H}(\mathbf{a}_\hbar) = H(\varphi) + H_1(\mathbf{a}_\hbar - \varphi) + H_2(\mathbf{a}_\hbar - \varphi) + H_{\geq 3}(\mathbf{a}_\hbar - \varphi) \tag{15.29}$$

where the functions $H_1,\ H_2,\ H_{\geq 3}$ are polynomials in $\mathbf{a}_\hbar - \varphi,\ \mathbf{a}_\hbar^* - \bar{\varphi}$ with total degree $1, 2$ and ≥ 3 respectively, with time dependent coefficients. Defining

$$H_k'(\mathbf{a}) = [\mathbf{a}, H_k(\mathbf{a})], \quad k = 1, 2, \geq 3$$

we see that H_k' are $\hbar$-independent, that $H_1'(\mathbf{a})$ is a c-number and that H_2' is linear in $\mathbf{a},\ \mathbf{a}^*$. Equation (15.28) can be rewritten as

$$i\hbar\frac{d}{dt}\varphi + i\hbar\frac{d}{dt}(\mathbf{a}_\hbar - \varphi) = H_1'(\mathbf{a}_\hbar) + H_2'(\mathbf{a}_\hbar - \varphi) + H_{\geq 3}'(\mathbf{a}_\hbar - \varphi). \tag{15.30}$$

But we have $H_1'(\mathbf{a}_\hbar) = \hbar H_1'(\mathbf{a})$. So we choose φ to be a solution of the classical evolution equation associated with Hamiltonian H, namely

$$i\frac{d}{dt}\varphi = H_1'(\mathbf{a}). \tag{15.31}$$

We define

$$\varphi_\hbar = \hbar^{-1/2}\varphi.$$

Then Eq. (15.30) becomes

$$i\frac{d}{dt}(\mathbf{a} - \varphi_\hbar) = H_2'(\mathbf{a} - \varphi_\hbar) + \hbar^{-1/2}H_{\geq 3}'(\hbar^{1/2}(\mathbf{a} - \varphi_\hbar)).$$

Let us now introduce the Weyl operator for any $(\alpha_i)_{i\in I}$

$$T(\alpha) = \exp\left[\sum_{i\in I}(\alpha_i a_i^* - \alpha_i^* a_i)\right] = \exp(a^*(\alpha) - a(\alpha)) \quad \text{where } \alpha = \sum_{i\in I}\alpha_i e_i, \tag{15.32}$$

where we have chosen initial time $s = 0$ for solving equation (15.28). Recall that $T(\alpha)$ are unitary and obey

$$T(\alpha)^*\mathbf{a}T(\alpha) = \mathbf{a} + \alpha.$$

Note that here we have chosen coordinates in $\mathfrak{h}$. If $f = \sum_{i\in I}\alpha_i e_i$ we have $T(f) = T(\alpha)$.

We define a new variable $\mathbf{b}(t)$ as

$$\mathbf{b}(t) = T(\varphi_\hbar(s))^*(\mathbf{a}(t) - \varphi_\hbar(t))T(\varphi_\hbar(s)).$$

The initial value problem for Eq. (15.28) then reduces to finding a family $\mathbf{b}(t)$ of operators satisfying the **(CCR)** and the relations

$$\begin{aligned}\mathbf{b}(0) &= \mathbf{a}\\ i\frac{d}{dt}\mathbf{b} &= H_2'(\mathbf{b}) + \hbar^{-1/2}H_{\geq 3}'(\hbar^{1/2}\mathbf{b})\end{aligned} \quad (15.33)$$

Note that the second term in the right hand side of equation (15.33) is $O(\hbar^{1/2})$ since $H_{\geq 3}'$ has degree at least two. Therefore $\mathbf{b}(t)$ is expected to converge towards the solution of the linearized equation

$$i\frac{d}{dt}\mathbf{b}' = H_2'(\mathbf{b}'),\quad \mathbf{b}'(0) = \mathbf{a} \quad (15.34)$$

which governs the quantum fluctuations around the classical equation. We introduce the propagator $\mathscr{U}_2(t,s)$ defined by the quadratic Hamiltonian $\hat{H}_2(t)$:

$$i\frac{d}{dt}\mathscr{U}_2(t,s) = \hat{H}_2(t)\mathscr{U}_2(t,s),\quad \mathscr{U}_2(s,s) = \mathbb{1}.$$

So we have

$$\mathbf{b}'(t) = \mathscr{U}_2(t,0)^*\mathbf{a}\mathscr{U}_2(t,0).$$

In the same way, the time evolution of operators $\mathbf{b}(t)$ will be implemented by a unitary group $W(t,s)$ such that:

$$\mathbf{b}(t) = W(t,0)^*\mathbf{a}W(t,0)$$

where $W(t,s)$ obeys the differential equation

$$i\frac{d}{dt}W(t,s) = \left\{H_2(\mathbf{a}) + \hbar^{-1/2}H_{\geq 3}(\hbar^{1/2}\mathbf{a})\right\}W(t,s),\quad W(s,s) = \mathbb{1}. \quad (15.35)$$

One has the following result, which is easily proved using Lemma 96.

Proposition 141 *One has:*

$$W(t,s) = \exp(i\omega_\hbar(t,s))\hat{T}(\varphi_\hbar(t))\mathscr{U}(t-s)\hat{T}(\varphi_\hbar(s))$$

with

$$\mathscr{U}(t-s) = \exp\left\{-i\hbar^{-1}(t-s)H(\mathbf{a}_\hbar)\right\}$$

and

$$\omega_\hbar(t,s) = \hbar^{-1}\int_s^t d\tau\left\{H(\varphi(\tau)) - \Re\langle\varphi(\tau), H_1'(\varphi(\tau))\rangle\right\}.$$

Therefore we have proven at least formally the following result.

Proposition 142

$$T(\varphi_\hbar(s))^*\mathscr{U}(t-s)^*(\mathbf{a}(s)-\varphi_\hbar(t))\mathscr{U}(t-s)T(\varphi_\hbar(s)) = W(t,s)^*\mathbf{a}(s)W(t,s)$$

with $\mathscr{U}(t)$ *and* $W(t,s)$ *given by Proposition 141.*

The difficult mathematical problem is to analyze the unitary propagators $\mathscr{U}_2(t,s)$ and $W(t,s)$.

One has the following result (see [148]):

Let $\varphi(t,x)$ be a solution of the classical Eq. (15.31) with initial data φ at $t=s$. For $f \in L^2(\mathbb{R}^n)$ let us define

$$\varphi^\sharp(f,t) = \int dx f(x)\varphi^\sharp(t,x)$$

Similarly consider the solutions $\mathbf{b}'(t)$ of the linearized problem (15.34). Let

$$(b')^\sharp(t,f) = \sum_{i\in I} \bar{f}_i b_i^\sharp(t), \text{ if } f = \sum_{i\in I} f_i e_i.$$

As usual define the operator valued distributions $(b')^\sharp(t,x)$ such that

$$(b')^\sharp(t,f) = \int dx f(x)(b')^\sharp(t,x).$$

Here $\sharp$ denotes either nothing or $*$.

It is possible to apply the strategy described above to prove semi-classical limit results for boson systems when $\hbar \searrow 0$. The Hamiltonian $\mathbf{H}$ is defined as

$$\mathbf{H} = \frac{\hbar}{2}\int dx\ \nabla a^*(x)\cdot\nabla a(x) + \frac{\hbar^2}{2}\int dx\ dy\ V(x-y)a^*(x)a^*(y)a(y)a(x). \tag{15.36}$$

Note that the limit $\hbar \searrow 0$ is equivalent to the mean-field limit $N \nearrow +\infty$ considered in [225] ($N = \hbar^{-1}$) for the Hamiltonian

$$\mathbf{H_N} = \frac{1}{2}\int dx\ \nabla a^*(x)\cdot\nabla a(x) + \frac{1}{2N}\int dx\ dy\ V(x-y)a^*(x)a^*(y)a(y)a(x). \tag{15.37}$$

Here we have

$$H_1' = -\frac{1}{2}\Delta\varphi(x) + \varphi(x)\int dy V(x-y)|\varphi(y)|^2 \tag{15.38}$$

so that one requires that the classical evolution $\varphi_t(x) = \varphi(t,x)$ is a solution of Hartree equation:

$$i\frac{d}{dt}\varphi_t = -\frac{1}{2}\Delta\varphi_t + \varphi_t(V * |\varphi_t|^2). \tag{15.39}$$

To state rigorous results some technical assumptions are needed for the potential V. Recall the following definition (see [151] for more details). For simplicity we only consider the case $\mathfrak{h} = L^2(\mathbb{R}^3)$.

Definition 39 A potential $V(x)$, $x \in \mathbb{R}^3$ is called a Hardy potential if V is real and there exists $C > 0$ such that

$$\|V\varphi\|_{L^2(\mathbb{R}^3)} \le C\|\varphi\|_{H^1(\mathbb{R}^3)}, \ \forall\varphi \in H^1(\mathbb{R}^3). \tag{15.40}$$

It is well known that if $V(x) = \frac{c}{|x|}$ $c \in \mathbb{R}$ then V is a Hardy potential (by the usual Hardy inequality). Hardy class potentials included the Kato class potentials. (see [151] for details).

The following proposition is a particular case of more general ones concerning Hartree equation [116]. The following proposition is sketched in [225], Remark 1.3. In [114–116] more refined results are given concerning solutions for the Hartree equation with singular potentials.

Proposition 143 *Let $V(x)$ be a Hardy potential. Let $\varphi_0 \in H^1(\mathbb{R}^3)$. Then Eq. (15.39) has a unique solution $\varphi \in \mathscr{C}(\mathbb{R}, H^1(\mathbb{R}^3))$. Furthermore one has the property of conservation of the L^2-norm and of the energy:*

$$\|\varphi_t\| = \|\varphi_0\|, \quad \forall t \in \mathbb{R}$$

$$\mathscr{E}(\varphi_t) = \mathscr{E}(\varphi_0), \quad \forall t \in \mathbb{R}$$

where

$$\mathscr{E}(\varphi) = \frac{1}{2}\|\nabla\varphi\|^2 + \frac{1}{2}\int dx\, dy\ V(x-y)|\varphi(x)|^2\,|\varphi(y)|^2.$$

Proposition 144 *Let $V(x)$ be an even Hardy potential. Assume that the initial data for the Hartree equation satisfies $\varphi_s \in H^1(\mathbb{R}^3)$. Then one has for $|t-s| < T$:*

(i) The Hamiltonian **H** *is essentially self-adjoint on the dense subspace $\mathscr{F}_{0,0}$ of the finite particle states with compact supported Fourier transform. In particular the time evolutions $\mathscr{U}(t)$, $\mathscr{U}_2(t,s)$ and $W(t,s)$ are well defined and are unitary operators in the Fock space $\mathscr{F}_{\mathrm{B}}(L^2(\mathbb{R}^3))$.*

(ii) We have the following limit result for the fluctuation operator around the classical solution $\varphi(t)$

$$\mathrm{s}-\lim_{\hbar\to 0} W(t,s) = \mathscr{U}_2(t,s). \tag{15.41}$$

(iii) One also has

$$\begin{aligned}\mathrm{s}-\lim_{\hbar\to 0} T(\varphi_\hbar)^*\mathscr{U}(t-s)^*\exp[(a^*(f) - \overline{\varphi_\hbar}(f,t)) - \mathrm{h.c.}]\mathscr{U}(t-s)T(\varphi_\hbar)\\ = \exp[(b')^*(f,t) - b'(f^*,t)]\end{aligned} \tag{15.42}$$

where $\varphi_\hbar = \hbar^{-1/2}\varphi$.

Proof This proposition is essentially due to Hepp [148].

It is well known that for every $k \in \mathbb{N}$, $\mathbf{H}$ is essentially self-adjoint in $\overset{k}{\vee} L^2(\mathbb{R}^3)$ so we get that $\mathbf{H}$ is essentially self-adjoint.

(ii) is proved with Duhamel formula and the following weight estimate for $\mathscr{U}_2(t,s)$ proved in [115]. In this case the generator H_2 of $\mathscr{U}_2(t,s)$ is given by

$$H_2 = \frac{1}{2}\int dx\, \nabla a^*(x)\cdot \nabla a(x) + \frac{1}{2}\int dxdy\; V(x-y)|\varphi_t(y)|^2 a^*(x)a(x) + \\ \frac{1}{2}\int dxdy\; \varphi_t(x)V(x-y)|\bar{\varphi}_t(y)|^2 a^*(x)a(y) + L^* + L \tag{15.43}$$

where

$$L = \frac{1}{2}\int dxdy\; \bar{\varphi}_t(x)V(x-y)\bar{\varphi}_t(y)a(x)a(y). \tag{15.44}$$

Lemma 97 *For every $\delta > 0$ and every $T > 0$ there exists $C(\delta, T)$ such that*

$$\|\langle \mathbf{N}\rangle^{\delta}\mathscr{U}_2(t,s)\langle \mathbf{N}\rangle^{-\delta}\| \le C(\delta, T), \text{ for } |t-s| \le T \tag{15.45}$$

The Duhamel formula gives

$$W(t,s) = \mathscr{U}_2(t,s) - i\int_s^t W(t,\tau)(\hbar^{1/2}H_3(\tau,\mathbf{a}) + \hbar H_4(\mathbf{a}))\mathscr{U}_2(\tau,s)d\tau \tag{15.46}$$

where $H_3(t.\,\mathbf{a}) = A_3(t) + A_3(t)^*$ and

$$A_3(t) = \int dydx V(x-y)\bar{\varphi}_t(x)a^*(y)a(y)a(x) \tag{15.47}$$

and

$$H_4 = \int dydx V(x-y)a^*(x)a^*(y)a(y)a(x). \tag{15.48}$$

Using that $\mathscr{F}_{0,0}$ is dense in $\mathscr{F}(L^2(\mathbb{R}^3))$ and that $W(t,s)$ is unitary in $\mathscr{F}(L^2(\mathbb{R}^3))$, it is enough to prove that

$$\lim_{\hbar\to 0} W(t,s)\Psi = \mathscr{U}_2(t,s)\Psi, \text{ for every } \Psi \in \mathscr{F}_{0,0}. \tag{15.49}$$

This result is proved using (15.46)–(15.48), assumptions on V and φ and standard estimates.

(iii) is proven using Proposition 142 and (i). □

Corollary 51 *Let* **A** *be a smooth and bounded function of* **a** *and* **a** *(see [23]). Then we have the following semi-classical limit evolution for quantum expectations in coherent states:*

$$\lim_{\hbar\to 0}\langle \mathcal{U}(t)\Psi(\varphi_\hbar), \mathcal{U}(t)\mathbf{A}(\mathbf{a}-\varphi_\hbar, \mathbf{a}^*-\bar{\varphi}_\hbar)\mathcal{U}(t)\Psi(\varphi_\hbar)\rangle = \langle \Omega, A(\mathbf{b}'(t), (\mathbf{b}'(t)^*)\Omega\rangle \tag{15.50}$$

where $\mathbf{b}'(t) = \mathcal{U}_2(t,0)^*\mathbf{a}\,\mathcal{U}_2(t,0)$ *is the linear evolution at the classical evolution* $\varphi(t)$.

Remark 96 In [116] the authors proved a full asymptotic expansion in $\hbar^{1/2}$ for a dense subset of states ψ. We shall see later that we have a better result if V is bounded.

15.4.3 Remainder Estimates in the Hepp Method

Hepp's method was revisited by Ginibre-Velo [116, 117] to extend it to singular potentials and to get quantum correction in $\hbar$ at any order. The spirit of the works of [116, 117] is to exploit the differential equation (15.35) (and the formula (15.46)) to write a Dyson series expansion for $W(t,s)$. The authors obtain a power series in $\hbar^{1/2}$ and study its analyticity properties in $\kappa := \hbar^{1/2}$. For bounded potentials V this series can be shown to be Borel summable in vector norm when applied to fixed $\hbar$-independent vectors taken from a suitable dense set including coherent states. The potential V is also assumed to be stable, which means that there exists a constant $B \geq 0$ such that

$$H_4 + B\mathbf{N} \geq 0 \tag{15.51}$$

($\mathbf{N}$ is the number operator (15.16)).

Roughly speaking it means that the potential is sufficiently repulsive near the origin if attractive somewhere else. More explicitly we can find in [229] a sufficient condition of stability for a potential V. H_4 has the following expression

$$(H_4\Psi)^{(k)}(x_1,\ldots,x_k) = \sum_{i<j} V(x_i - x_j)\Psi^{(k)}(x_1,\ldots,x_k).$$

Let V be such that

$$\begin{aligned} V(x) &\geq \varphi_1(|x|), \text{ for } |x| \leq r_1 \\ V(x) &\geq -\varphi_2(|x|), \text{ for } |x| \geq r_2 \end{aligned} \tag{15.52}$$

where φ_1, φ_2 are positive decreasing on $]0, r_1[$, $[r_2, +\infty[$ respectively and for some $\nu > 3$ we have

$$\int \varphi_1(t) t^{\nu-1} dt = +\infty, \quad \int \varphi_2(t) t^{\nu-1} dt < +\infty.$$

Then V is stable ([229], Proposition 3.2.8).

Remark that the stability condition is a restriction on the negative part of V.

Before to state the summability result let us recall some definitions concerning asymptotic series.

Consider a formal power complex series $f^\sharp(\kappa) = \sum_{j \in \mathbb{N}} \alpha_j \kappa^j$.

Definition 40 (i) $f^\sharp$ is a Gevrey series of order $1/s$, $s > 0$, if there exists $C > 0$, $\rho > 0$ such that

$$|\alpha| \le C\rho^j (j!)^{1/s}, \quad \forall j \in \mathbb{N}. \tag{15.53}$$

(ii) The s-Borel transform of the Gevrey series $f^\sharp$ is defined as

$$\mathscr{B}_s f(\tau) = \sum_{j \in \mathbb{N}} \frac{\alpha_j}{\Gamma(1 + \frac{j}{s})} \tau^j \tag{15.54}$$

where the series converges for $\tau \in \mathbb{C}$, $|\tau| < \rho^{-1}$.

(iii) The Gevrey series $f^\sharp$ is said Borel s-summable if
$(iii)_1$ its s-Borel transform $\mathscr{B}_s f$ has an analytic extension to a neighborhood of the positive real axis,
$(iii)_2$ the following integral

$$\int_0^\infty du \mathscr{B}_s f(u) u^{s-1} \mathrm{e}^{-(\frac{u}{\kappa})^s}$$

converges for $0 < \kappa < \kappa_0$. If this holds then we say that $f^\sharp$ has a s-Borel sum $f(\kappa)$ defined by

$$f(\kappa) = s\kappa^{-s} \int_0^\infty du \mathscr{B}_s f(u) u^{s-1} \mathrm{e}^{-(\frac{u}{\kappa})^s}. \tag{15.55}$$

Remark 97 (i) For $s = 1$ the above definition corresponds to the usual Borel summability.

(ii) At the formal level, formula (15.55) is easy to check using definition of Γ function and changes of variables.

(iii) $(j!)^{j/s}$ and $\Gamma(1 + \frac{j}{s})$ have the same order as $j \to$ as a consequence of the Stirling formula

$$\Gamma(1 + u) = \sqrt{2\pi u} \left(\frac{u}{\mathrm{e}}\right)^u (1 + o(1)) \text{ as } u \to +\infty.$$

We state now a sufficient condition for Borel summability due to Watson, Nevanlinna and Sokal (see [214, 239] and references with extension to any $s > 0$).

Theorem 82 *Let f be an holomorphic function in the complex domain $D_{s,R} := \{\kappa \in \mathbb{C},\ \Re\kappa^{-s} > R^{-s}\}$ for some $s > 0$. Assume that there exist $C > 0$, $\rho > 0$ such that in this domain we have*

$$|f(\kappa) - \sum_{0\leq j<N} \alpha_j \kappa^j| \leq C\rho^N (N!)^{1/s} |\kappa|^N. \tag{15.56}$$

Then the power series $\sum_{j\in\mathbb{N}} \alpha_j \kappa^j$ is s-Borel summable and its s-Borel sum is equal to $f(\kappa)$ in $D_{s,R}$.

In particular on the interval $]0, \kappa_0]$, $f(\kappa)$ is uniquely determined by its asymptotic expansion.

Remark 98 Estimate (15.56) on the interval $]0, \kappa_0]$ entails that $f(\kappa)$ is determined by its asymptotic expansion up to an exponentially small error like $\mathrm{e}^{-c\kappa^{-s}}$ $c > 0$. This is easily seen by stopping the series at the order $N \approx \frac{\delta}{\kappa^s}$ with δ small enough. Adding an analytic condition in a suitable domain as in the above theorem erase this error term so that $f(\kappa)$ is is uniquely determined.

The following result gives an accurate asymptotic description for the quantum fluctuation operator $W(t, s)$, improving Proposition 144.

Theorem 83 *Let V be stable and bounded potential. Let $\varphi \in \mathscr{C}^1(\mathbb{R}, L^2(\mathbb{R}^3))$ be a solution of Hartree equation (15.39). Let $\beta > 0$ and $\Phi \in \mathscr{D}(\exp(\beta N))$. Then there exists $\theta > 0$ such that for all $s, t \in \mathbb{R}$, $t \geq s$ such that $t - s \leq \theta$, $W(t, s)\Phi$ is analytic in κ in the sector $-\frac{\pi}{2} < \mathrm{Arg}\kappa < 0$, $|z|$ small and has an asymptotic expansion at $\kappa = 0$ which is 2-Borel summable to $W(t, s)\Phi$ itself. The constant θ depends on V, β, φ but can be taken independent of B and uniform in V for $\|V\|_\infty$ and B bounded, and uniform in φ for $\|\varphi\|$ bounded.*

Proof We only give here the main steps for the proof. We refer to the paper [116] for a detailed proof.

A first step is to consider the Duhamel formula (15.46). By iterating it we get a Dyson series. Then by reordering the Dyson series we get a formal power series in $\hbar^{1/2}$:

$$W^\sharp(t, s) = \sum_{j\in\mathbb{N}} \hbar^{j/2} W_j(t, s) \tag{15.57}$$

where the coefficient $W_j(t, s)$ are operators. $W_0(t, s) = \mathscr{U}_2(t, s)$.

In a second step, 2-Gevrey type estimates [3] are obtained for $W_j(t, s)\Phi$ where $\Phi \in \mathscr{D}(\mathrm{e}^{\varepsilon N})$, $\varepsilon > 0$ (Proposition 3.1 in [116]).

In a third step it is proved that $W(t, s)\Phi$ has an analytic expansion in $\kappa := \hbar^{1/2}$ in a domain like $D_{2,R}$. Here the stability condition is used.

[3] Notations for Borel summability have been defined before for complex valued series, the extension to vector valued series is straightforward.

The last step consists in estimating the remainder term of the asymptotic expansion of $W(t,s)\Psi$ in κ small. To do that the Dyson series expansion is performed in two steps, introducing an auxiliary evolution operator $\mathscr{U}_4(t,s)$ obeying

$$i\frac{d}{dt}\mathscr{U}_4(t,s) = (H_2(t) + \hbar H_4)\mathscr{U}_4(t,s).$$

Then if $\mathscr{U}_2(t,s)$ is the propagator for $H_2(t)$ (which describes the evolution of the quantum fluctuations), one has

$$\mathscr{U}_4(t,s) = \mathscr{U}_2(t,s) - i\int_s^t d\tau\ \mathscr{U}_2(t,\tau)\hbar H_4\mathscr{U}_4(\tau,s)$$

$$W(t,s) = \mathscr{U}_4(t,s) - i\int_s^t d\tau\ \mathscr{U}_4(t,\tau)\hbar^{1/2}H_3W(\tau,s)$$

Then it can be proved that the series $\sum_{j\in\mathbb{N}}\hbar^{j/2}W_j(t,s)\Phi$ is 2-Borel summable and its 2-Borel sum is $W(t,s)\Phi$. □

Using similar methods the case of unbounded potentials has been considered in [117]. The authors obtain the same analyticity domain as in [116] of $W(t,s)$ with respect to $\kappa = \hbar^{1/2}$ and obtain that the series is still asymptotic (in the Poincaré sense) and Gevrey of order 2. It is not known that the series is still 2-Borel summable for singular potentials. Note that $\mathscr{D}(\exp(\beta N))$ contains the coherent states $\Psi(f)$, $f \in L^2(\mathbb{R}^3)$.

15.4.4 Time Evolution of Coherent States

We have seen that the Hepp's method concerns the quantum fluctuations in a neighborhood of a classical trajectory. It does not give informations for the quantum motion itself, in particular when the initial state is a coherent state. In Chap. 4 this problem was considered for finite bosons systems. Our goal here is to explain an extension of this result for large systems of bosons, following the paper [225].

According to the previous section one has to study the evolution operator $\mathbf{U}(t) = e^{-i\mathbf{H}_\hbar t}$ associated with the Hamiltonian

$$\mathbf{H} = \frac{1}{2}\int dx\ \nabla a^*(x)\cdot\nabla a(x) + \frac{\hbar}{2}\int dx\,dy\ V(x-y)a^*(x)a^*(y)a(y)a(x)$$

by the evolution equation

$$i\frac{d}{dt}\mathbf{U}(t) = \mathbf{H}\mathbf{U}(t)$$

and apply it to the coherent state $\Psi(\hbar^{-1/2}\varphi)$ where $\varphi_\hbar = \hbar^{-1/2}\varphi$, φ is a solution of the Hartree equation (15.39). By definition the Hamiltonian $\mathbf{H}$ leaves sectors $\overset{k}{\vee}L^2(\mathbb{R}^3)$

with fixed number of particles invariant. One thus have

$$\mathbf{U}(t)^*\mathbf{N}\mathbf{U}(t) = \mathbf{N} \tag{15.58}$$

One has the following result. $\Gamma^{(1)}_{\hbar,t}$ (defined at the end of Sect. 15.3.) is the marginal operator in the one particle space $L^2(\mathbb{R}^3)$ deduced from the quantum evolution $\mathbf{U}(t)\Psi(\varphi_\hbar)$ of the coherent state $\Psi(\varphi_\hbar)$. Recall that in an Hilbert space $\mathfrak{h}$ the trace norm of an operator A is defined as $\|A\|_{\mathrm{Tr}} = \sqrt{\mathrm{Tr}(A^*A)}$ (see Chap. 1).

Theorem 84 *Suppose that V is a Hardy potential (see (15.40). Then there exists constants $C,\ K > 0$ (only depending on the $H^1(\mathbb{R}^3)$ norm of φ and on C) such that*

$$\left\| \Gamma^{(1)}_{\hbar,t} - |\varphi_t\rangle\langle\varphi_t| \right\|_{\mathrm{Tr}} \leq C\hbar e^{Kt}, \quad t \in \mathbb{R}. \tag{15.59}$$

φ_t is the solution of Hartree equation (15.39) at time t with $\|\varphi_0\| = 1$.

Proof We shall give here the main ideas of the proof. The details are in the paper [225].

We now write the function $\Gamma^{(1)}_{\hbar,t}(x, y)$ using (15.27). One has to calculate the denominator:

$$\langle \mathbf{U}(t)\psi(\varphi_\hbar)\Omega, \mathbf{N}\mathbf{U}(t)\psi(\varphi_\hbar)\Omega\rangle$$

Using equation (15.58) we are left with

$$\langle \psi(\varphi_\hbar)\Omega, \mathbf{N}\psi(\varphi_\hbar)\Omega\rangle.$$

We now use the translation property of the Weyl operator:

$$T(\varphi_\hbar)^*\mathbf{N}T(\varphi_\hbar) = \int dx\ (a^*(x) - \bar{\varphi}_\hbar(x))(a(x) - \varphi_\hbar(x)) \tag{15.60}$$

But using that $a(x)\Omega = 0$ we see that the expectation value of (15.60) in the vacuum simply equals

$$\hbar^{-1}\|\varphi\|^2 = \hbar^{-1}.$$

Therefore $\Gamma^{(1)}_{\hbar,t}(x, y)$ has the following decomposition.

$$\Gamma^{(1)}_{\hbar,t}(x, y) = \hbar\langle \Omega, T(\hbar^{-1/2}\varphi)^*U(t)^*a^*(y)a(x)U(t)T(\hbar^{-1/2}\varphi)\Omega\rangle$$

$$= \bar{\varphi}_t(y)\varphi_t(x) + \hbar^{1/2}\bar{\varphi}(y)\langle \Omega, T(\varphi_\hbar)^*U(t)^*(a(x) - \varphi_{t,\hbar}(x))U(t)T(\varphi_\hbar)\Omega\rangle$$

$$+\hbar^{1/2}\varphi_t(x)\langle \Omega, T(\varphi_\hbar)^*U(t)^*(a^*(x) - \varphi_{t,\hbar}(y))T(\varphi_\hbar)\Omega\rangle$$

$$+\hbar\langle \Omega, T(\varphi_\hbar)^*U(t)^*(a^*(y) - \bar{\varphi}_{t,\hbar}(y))(a(x) - \varphi_{t,\hbar}(x))U(t)T(\varphi_\hbar)\Omega\rangle$$

Now we use the fact demonstrated in the previous section that

$$T(\varphi_{s,\hbar})^* U(t)^* (a(x) - \varphi_{t,\hbar}(x)) U(t) T(\varphi_{s,\hbar}) = W(t,s)^* a(x) W(t,s)$$

Thus we get the equality between the two kernels:

$$\begin{aligned} \Gamma^{(1)}_{\hbar,t}(x,y) - \varphi_t(x)\bar{\varphi}_t(y) = \hbar\langle \Omega, W(t,0)^* a^*(y) a(x) W(t,0)\Omega\rangle + \\ \hbar^{1/2}\varphi_t(x)\langle \Omega, W(t,0)^* a^*(y) W(t,0)\Omega\rangle + \\ \hbar^{1/2}\bar{\varphi}_t(y)\langle \Omega, W(t,0)^* a(x) W(t,0)\Omega\rangle. \quad (15.61) \end{aligned}$$

It is remarked in [225] that here it is enough to estimate the Hilbert-Schmidt norm of $\Gamma^{(1)}_{\hbar,t} - |\varphi_t\rangle\langle\varphi_t|$ instead of its trace-norm. So the main technical part of the paper [225] is to show that the L^2 norm in (x, y) of the right hand side of (15.61) is bounded above by $C\hbar e^{Kt}$, using suitable approximation of the dynamics $W(t, s)$. As in [116, 117] $W(t, s)$ is compared with the dynamics $\mathcal{U}_4(t, s)$ generated by $H_2(t) + \hbar H_4$ (without the $H_3(t)$ term), namely

$$i\frac{d}{dt}\mathcal{U}_4(t,s) = (H_2(t) + \hbar H_4)\mathcal{U}_4(t,s).$$

In [225] the following lemmas are proven.

Lemma 98 *One has*

$$\|(W(t,0) - \mathcal{U}_4(t,0))\Omega\| \le C\hbar^{1/2}e^{Kt}.$$

Lemma 99

$$\langle W(t,0)\Omega, \mathbf{N}W(t,0)\Omega\rangle \le Ce^{Kt}.$$

Now, introducing $\mathcal{U}_4(t, s)$, the right hand side of equation (15.61) is rewritten as

$$\begin{aligned} \Gamma^{(1)}_{\hbar,t}(x,y) - \varphi_t(x)\bar{\varphi}_t(y) = \hbar\langle \Omega, W(t,0)^* a^*(y) a(x) W(t,0)\Omega\rangle + \\ \hbar^{1/2}\varphi_t(x)\big(\langle \Omega, W(t,0)^* a^*(y)(W(t,0) - \mathcal{U}_4(t,0))\Omega\rangle + \\ \langle \Omega, (W(t,0)^* - \mathcal{U}_4(t,0)^*) a^*(y)\mathcal{U}_4(t,0)^*\Omega\rangle\big) + \\ \hbar^{1/2}\bar{\varphi}_t(y)\big(\langle \Omega, W(t,s)^* a(x)(W(t,s) - \mathcal{U}_4(t,s))\Omega\rangle + \\ \langle \Omega, (W(t,s)^* - \mathcal{U}_4(t,s)^*) a(x)\mathcal{U}_4(t,s)\Omega\rangle\big) \quad (15.62) \end{aligned}$$

where we use that $\mathcal{U}_4(t, s)$ preserves the parity of the number of particles:

$$\langle \Omega, \mathcal{U}_4(t,s)^* a^*(y)\mathcal{U}_4(t,s)\Omega\rangle = \langle \Omega, \mathcal{U}_4(t,s)^* a(x)\mathcal{U}_4(t,s)\Omega\rangle = 0.$$

Then we get from Lemmas 98 and 99 the Hilbert-Schmidt estimate

$$\int_{\mathbb{R}^3\times\mathbb{R}^3} dxdy|\Gamma^{(1)}_{\hbar,t}(x,y) - \varphi_t(x)\bar{\varphi}_t(y)|^2 \le C\hbar^2 e^{2Kt}, \quad \forall t \ge 0.$$

□

Remark 99 In [53] the authors have extended the previous result to the case of arbitrary factorized initial data.

Chapter 16
Fermionic Coherent States

Abstract This chapter is an introduction to some computation techniques for fermionic states. After defining Grassmann algebras it is possible to get a classical analogue for the fermionic degrees of freedom in a quantum system. Following the basic work of Berezin [23, 26], we show that we can compute with Grassmann numbers as we do with complex numbers: derivation, integration, Fourier transform. After that we show that we have quantization formula for fermionic observables. In particular there exists a Moyal product formula. As an application we consider explicit computations for propagators with quadratic Hamiltonians in annihilation and creation operators.

16.1 Introduction

We have seen in the Chapter on Bosons 15 that the canonical commutation relations (CCR) can be realized with real or complex numbers. We shall see here that the anti-commutation relations (CAR) require the introduction of new kind of numbers, nilpotents, called Grassmann numbers. In some sense, nilpotence is classically equivalent to the Pauli exclusion principle which characterizes fermions. This appears to be strange from a physicist point of view because the Pauli exclusion principle is a purely quantum property, without a classical analogue. Nevertheless such a mathematical model exists. Even if "classical fermions" do not exist in Nature, they are a convenient mathematical tool for computations and allow to put bosons and fermions on the same footing. This is important to elaborate supersymmetric models which will be considered in more details in the next Chapter.

Our main goal here is to introduce fermionic coherent states and to describe their properties. The main difference from the bosonic case considered in Chap. 1 is that we have to replace complex numbers by Grassmann numbers. So the constructions have to be revisited with some care. So we shall study in more details Grassmann algebras and we shall see that many properties and constructions well known for usual numbers can be extended to Grassmann algebras. Our construction of fermionic

D. Robert and M. Combescure, *Coherent States and Applications in Mathematical Physics*, Theoretical and Mathematical Physics,
https://doi.org/10.1007/978-3-030-70845-0_16

coherent states mainly follows the paper [47]. We also consider in more details the propagation of Fermions for quadratic evolutions.

In [25] (see the book [208], Chap. 9) the author introduced Fermionic coherent states from another point of view, without Grassmann algebras. He considered an irreducible representation of the rotation group $SO(2n, \mathbb{R})$ (for a system of n fermions) called the spin representation. We shall see that the two points of views are mathematically equivalent.

16.2 From Fermionic Fock Spaces to Grassmann Algebras

It follows from the Chapter on bosons that the fermionic Fock space is the Hilbert space $\mathscr{F}_F(\mathfrak{h}) = \oplus_{k\in\mathbb{N}} \wedge^k \mathfrak{h}$, where $\mathfrak{h}$ is the one fermion space and $\wedge^k\mathfrak{h}$ is the anti-symmetric tensor spanned by the states

$$\frac{1}{k!}\sum_{\pi\in\mathfrak{S}_k} \varepsilon_\pi \psi_{\pi 1}\otimes\psi_{\pi 2}\otimes\cdots\times\psi_{\pi k},\quad \psi_j\in\mathfrak{h}.$$

For any $f \in \mathfrak{h}$, the annihilation and creation operators $a(f)$, $a^*(f)$ are bounded operators in $\mathscr{F}_F(\mathfrak{h})$. This is a consequence of the Canonical Anticommutation Relation

$$a(f_1)a^*(f_2) + a^*(f_1)a(f_2) = \langle f_1, f_2\rangle \mathbb{1}. \tag{16.1}$$

Let us begin by an explicit model to realize the CAR relations (16.1) which is called the spin model.

We start with $\mathscr{H}_F = \mathbb{C}^2$ and the matrices

$$\sigma_+ = \begin{pmatrix} 0 & 1\\ 0 & 0\end{pmatrix},\ \sigma_- = \begin{pmatrix} 0 & 0\\ 1 & 0\end{pmatrix}.$$

So we have (CAR) for $a = \sigma_-$ and $a^* = \sigma_+$:

$$[\sigma_+, \sigma_-]_+ = \mathbb{1}_2,\ \sigma_+^2 = \sigma_-^2 = 0.$$

This is a model for the states of one spin. We have the number operator $N = a^*a = \begin{pmatrix} 1 & 0\\ 0 & 0\end{pmatrix}$ and the ground state is $e_0 = \begin{pmatrix} 0\\ 1\end{pmatrix}$.

We get a model for n spin states in the Hilbert space $\mathscr{H}_H = (\mathbb{C}^2)^{\otimes n} = \mathbb{C}^{2^n}$. The annihilation and creation operators are

$$a_k = \underbrace{\sigma_3 \otimes \cdots \sigma_3}_{k-1} \otimes a \otimes \underbrace{\mathbb{1}_2 \otimes \cdots \otimes \mathbb{1}_2}_{n-k},$$

$$a_k^* = \underbrace{\sigma_3 \otimes \cdots \sigma_3}_{} \otimes a * \otimes \underbrace{\mathbb{1}_2 \otimes \cdots \otimes \mathbb{1}_2}_{n-k}.$$

The ground state is here $\Omega_0 = \underbrace{e_0 \otimes \cdots \otimes e_0}_{n}$.

This model is unitary equivalent to the fermionic Fock model with $\mathfrak{h} = \mathbb{C}^n$. We have

$$\mathscr{H}_F = \mathbb{C} \oplus \mathbb{C}^n \oplus (\wedge^2\mathbb{C}^n) \oplus \cdots \oplus (\wedge^n\mathbb{C}^n).$$

Notice that $\dim(\mathscr{H}_F) = 2^n$.

Let us give now some specific properties for the general fermionic Fock model (for detailled proofs see [42]).

Proposition 145 *(1) For every $f \in \mathfrak{h}$, $a(f)$ and $a^*(f)$ are bounded operators in $\mathscr{F}_F(\mathfrak{h})$:*

$$\|a(f)\| = \|a^*(f)\| = \|f\|, \forall f \in \mathfrak{h}. \tag{16.2}$$

(2) Let $\{\varphi_j\}_{j\in J}$ be an orthonormal basis for $\mathfrak{h}$. The family of states defined as

$$\psi_{j_1,j_2,\ldots,j_k} = a^*(\varphi_{j_1})\ldots a^*(\varphi_{j_n})\Omega, \quad j_k \in J, \ n \geq 0,$$

is an orthonormal basis for $\mathscr{F}_F(\mathfrak{h})$.

(3) If T is a bounded operator in $\mathscr{F}_F(\mathfrak{h})$ commuting with all the operators $a(f)$ and $a^*(g)$, $f, g \in \mathfrak{h}$, then $T = \lambda\mathbb{1}$ for some $\lambda \in \mathbb{C}$.

Property (3) of the Proposition means that the Fock representation for Fermion is irreducible.

When the system has n identical particles the creation/annihilation operators are denoted $a_j^{(*)} := a^{(*)}(\varphi_j)$, $1 \leq j \leq n$.

A natural problem is to represent the anti-commutation relations (CAR) with derivative and multiplication operators as it can be done for (CCR) with q and $\frac{d}{dq}$ or in the Fock–Bargmann representation (see Chap. 1). In other word we would like to represent anticommutation relations like

$$\frac{\partial}{\partial\theta}\theta + \theta\frac{\partial}{\partial\theta} = \mathbb{1}. \tag{16.3}$$

This is not possible in the naive sense. This problem is equivalent to build a classical analogue for Fermionic observables. In order to satisfy (16.3) it is necessary to replace the usual real or complex numbers by Grassmann variables as it was discovered by Berezin [26]. Let us define the Grassmann algebras.

Definition 41 The Grassmann algebra $\mathscr{G}_n$ with n generators $\{\theta_1, \ldots, \theta_n\}$ is an algebra, with unit 1, a product and a $\mathbb{K}$-linear space ($\mathbb{K} = \mathbb{R}$ or $\mathbb{K} = \mathbb{C}$) such that

$$\theta_j\theta_k + \theta_k\theta_j = 0, \ \forall j, k = 1, \ldots, n.$$

and every $g \in \mathcal{G}_n$ can be written as

$$g = c_0(g) + \sum_{k\geq 1}\sum_{j_1,\ldots,j_k} c_{j_1,\ldots,j_k}(g)\theta_{j_1}\ldots\theta_{j_k}. \tag{16.4}$$

where $c_0(g)$ and $c_{j_1,\ldots,j_k}(g)$ are $\mathbb{K}$-numbers.

This definition has some direct algebraic consequences given below and easy to prove

1. In equality (16.4) the decomposition is not unique. We get a unique decomposition if we add in the sum the condition $j_1 < j_2 < \cdots < j_k$. So we get a basis of $\mathcal{G}_n$, $\{\theta_{j_1}\ldots\theta_{j_k},\ \ j_1 < j_2 < \cdots j_k;\, 1 \geq k \geq, n\}$ and its dimension is 2^n.
   It is sometimes convenient to introduce $\mathcal{E}[n] = \{0, 1\}^n$, $\varepsilon = (\varepsilon_1, \ldots, \varepsilon_n)$ where $\varepsilon_k = 0, 1$, and $\theta^\varepsilon = \theta_1^{\varepsilon_1}\ldots\theta_n^{\varepsilon_n}$. So the above basis can be written as $\{\theta^\varepsilon,\ \varepsilon \in \mathcal{E}[n]\}$. $|\varepsilon| := \varepsilon_1 + \cdots + \varepsilon_n$ is the number of fermionic states occupied. In particular (16.4) can be written as
$$g = \sum_{\varepsilon\in\mathcal{E}[n]} c_\varepsilon(g)\theta^\varepsilon \tag{16.5}$$
2. $g \in \mathcal{G}_n$ is invertible if and only if $c_0(g) \neq 0$.
3. Equality (16.4) can be interpreted as a generating function for fermionic states where $c_{j_1,\ldots,j_k}(g)$ are the coefficients of a states in the basis $\{\psi_{j_1,j_2,\ldots,j_k}\}$ of Proposition 145.
4. Derivatives are defined as follows. If $g \in \mathcal{G}_n$, for $1 \leq j \leq n$, we have $g = g_0 + \theta_j g_1$ where g_0 and g_1 are independent on θ_j. So we define the (left derivative)
$$\frac{\partial g}{\partial\theta_j} = g_1$$
   We also a right derivative denoted $g\frac{\partial}{\partial\theta_j}$ and defined by writing $g = g_0 + g_2\theta_j$, where g_0 and g_2 are θ_j-independent, so $g_2 =: g\frac{\partial}{\partial\theta_j}$.
5. We have (CAR) relations
$$\left[\frac{\partial}{\partial\theta_j}, \hat{\theta}_k\right]_+ = \delta_{j,k} \tag{16.6}$$
   where $\hat{\theta}_k$ is left multiplication by θ_k in $\mathcal{G}_n$ and $[\bullet]_+$ is the anticommutator in $\mathcal{G}_n$: $[f, g]_+ = fg + gf$.
   Let us remark that (16.6) are analogue for fermions of the commutation relation for bosons in the holomorphic representation (see Chap. 1).

It is possible to write the relations (CAR) in a real form. For a finite system we introduce the self-adjoint operators:

$$\hat{Q}_j = \sqrt{\frac{\hbar}{2}}(a_{F,j} + a^*_{F,j}), \quad \hat{P}_j = i^{-1}\sqrt{\frac{\hbar}{2}}(a_{F,j} - a^*_{F,j}). \tag{16.7}$$

So (16.6) is transformed to

$$[\hat{Q}_j, \hat{Q}_k]_+ = [\hat{P}_j, \hat{P}_k]_+ = \hbar\delta_{j,k}\mathbb{1} \ \ [\hat{Q}_j, \hat{P}_k]_+ = 0, \ 1 \le j, k \le n \tag{16.8}$$

Equation (16.8) can be compared to the relations (CCR) : anti-commutators replace commutator. We have seen in Chap. 1 that (CCR) is a representation of the Weyl–Heisenberg Lie algebra. (CAR) (16.8) is a representation of the Clifford algebra $\mathrm{Cl}(\mathbb{R}^n)$ which is defined below.

Definition 42 Let Φ be a symmetric bilinear form on a linear space V (over $K = \mathbb{R}$ or $\mathbb{C}$). The Clifford algebra is the associative algebra with unit 1 denoted, $\mathrm{Cl}(V, \Phi)$, generated by V and such that for every u, v in V we have

$$u \cdot v + v \cdot u = \Phi(u, v) \cdot 1. \tag{16.9}$$

Taking $V = \mathbb{R}^n$ with the canonical basis $\{e_k\}_{1\le k\le n}$ and Φ the usual scalar product the Clifford algebra $\mathrm{Cl}(\mathbb{R}^n)$ is the algebra defined by the relations

$$e_j \cdot e_k + e_k \cdot e_j = \delta_{j,k}. \tag{16.10}$$

Hence we see that the anti-commutation relations (16.8) for $\hbar = 1$ define a representations of the Clifford algebra $\mathrm{Cl}(\mathbb{R}^{2n})$ as the commutation relations (CAR) are a representation of the Weyl–Heisenberg algebra. We shall see later that $\mathrm{Cl}(\mathbb{R}^{2n})$ is not a Lie algebra but a graded Lie algebra or super-Lie algebra.

On the other side the representation (16.6) of (CAR) is equivalent to the representation given in Proposition 145.

Proposition 146 *Let us consider a system of n identical fermions ($\mathfrak{h} = \mathbb{C}^n$). The linear map Φ from $\mathscr{F}_F(\mathbb{C}^n)$ onto $\mathscr{G}_n$ is defined by*

$$\Phi(\psi_{j_1,\dots,j_k}) = \theta_{j_1}\dots\theta_{j_k}.$$

We have

$$a^*_j = \Phi^{-1}\hat{\theta}_j\Phi, \ \text{ and } \ a_j = \Phi^{-1}\frac{\partial}{\partial\theta_j}\Phi. \tag{16.11}$$

16.3 Integration on Grassmann Algebra

The construction follows the bosonic construction with the important difference that complex numbers are replaced by Grassmann variables which are anticommuting. In particular the classical-quantum correspondence has many differences from the bosonic case. The main tool is derivation-integration over Grassmann variables introduced by Berezin and which can appear to be sometimes strange. But it is the right algebra to preserve similarities with Bosons and to have a classical analogue of Fermions.

16.3.1 More Properties of Grassmann Algebras

A Grassmann algebra can be defined on a field $\mathbb{K} = \mathbb{R}$ or $\mathbb{K} = \mathbb{C}$. The Grassmann algebra with n generators $\{\theta_1, \ldots, \theta_n\}$ will be denoted $\mathcal{G}_n$ or $\mathbb{K}[\theta_1, \ldots, \theta_n]$.

In applications, we have to replace $\mathbb{K}$ itself by a Grassmann algebra $\mathcal{G}_m := \mathbb{K}[\zeta_1, \ldots, \zeta_m]$ where $[\zeta_j, \zeta_k]_+ = 0$. In this case we shall consider the right $\mathcal{G}_m$-module: $\mathcal{G}_m[\theta_1, \ldots, \theta_n]$ with generators $\{\theta_1, \ldots, \theta_n\}$. It will be sometimes denoted $\mathcal{G}_n^m$.

More explicitly a generic element g of $\mathcal{G}_n^m$ can be written as

$$g = \sum_{i_1 < \cdots i_k} \theta_{i_1} \ldots \theta_{i_k} \cdot c_{i_1 \ldots i_k}$$

where $c_{i_1 \ldots i_k} \in \mathbb{K}[\zeta_1, \ldots, \zeta_m]$. Multiplication of a vector v in $\mathcal{G}_n^m := \mathcal{G}_m[\theta_1, \ldots, \theta_n]$ by a Grassmann number $\lambda \in \mathcal{G}_m$ is $v \cdot \lambda$ where $\cdot$ is multiplication in $\mathbb{K}[\theta_1, \ldots, \theta_n, \zeta_1, \ldots, \zeta_m]$ We also have a left multiplication $\lambda \cdot v$ which may be different of $v \cdot \lambda$.

Recall here that this rule is important because Grassmann algebras are not commutative.

Because of properties of Grassmann variables the (right)-module $\mathcal{G}_n^m$ has a basis $\{\theta^\varepsilon\}_{\varepsilon \in \mathcal{E}[n]}$.

The parity operator is also very useful; it is the linear operator defined by $\mathbf{P}(\theta^\varepsilon) = (-1)^{|\varepsilon|}\theta^\varepsilon$ where $|\varepsilon| = \varepsilon_1 + \cdots \varepsilon_n$. So we have the direct sum decomposition

$$\mathcal{G}_n^m = \mathcal{G}_{n,+}^m \oplus \mathcal{G}_{n,-}^m$$

where $\mathbf{P} = \mathbb{1}$ on $\mathcal{G}_{n,+}^m$ and $\mathbf{P} = -\mathbb{1}$ on $\mathcal{G}_{n,-}^m$.

Elements in $\mathcal{G}_{n,+}^m$ are said **even** and elements in $\mathcal{G}_{n,-}^m$ are said **odd**.

Let A be a linear operator in $\mathcal{G} := \mathcal{G}_n^m$. A is said even if A and $\mathbf{P}$ commute: $[A, \mathbf{P}] = 0$. In other words that means that A has the following matrix representation

$$A = \begin{pmatrix} A_+ & 0 \\ 0 & A_- \end{pmatrix}$$

where $A_\pm : \mathcal{G}_{n,\pm} \to \mathcal{G}_{n,\pm}$. The following lemmas are useful in computations

Lemma 100 *Even elements commute with every element: if $\psi \in \mathcal{G}_{n,+}^m$ and $\varphi \in \mathcal{G}_n^m$ then $\psi\varphi = \varphi\psi$.*

Odd elements anticommute: if $\psi \in \mathscr{G}^m_{n,-}$ *and* $\varphi \in \mathscr{G}^m_{n,-}$ *then we have* $[\psi, \varphi]_+ = \psi\varphi + \varphi\psi = 0$

Proof Exercise.

The Liebnitz rule for derivative of products depends on parity:

$$\frac{\partial}{\partial \theta_k}(\psi\varphi) = (\mathbf{P}\psi)\left(\frac{\partial}{\partial \theta_k}\varphi\right) + \left(\frac{\partial}{\partial \theta_k}\psi\right)\varphi. \tag{16.12}$$

If f is a smooth function around 0 and $\mathscr{G}$ a Grassmann algebra we can define $f(\psi)$ for every $\psi \in \mathscr{G}$ by the Taylor formula

$$f(\psi) = \sum_{k \in \mathbb{N}} \frac{f^{(k)}(0)}{k!}\psi^k.$$

We remark that the sum is finite because $\psi^k = 0$ if $k > n$. In particular if $\theta = (\theta_1, \ldots, \theta_n), \gamma = (\gamma_1, \ldots, \gamma_n)$ are $2n$ Grassmann generators then $\mathrm{e}^{\theta \cdot \gamma}$ is well defined where $\theta \cdot \gamma = \sum_{1 \le k \le n} \theta_k \gamma_k$.

Lemma 101 *If* $k \geq 2$ *we have*

$$\theta_1\theta_2 \ldots \theta_k = (-1)^{\mu(k)}\theta_k\theta_{k-1} \ldots \theta_1. \tag{16.13}$$

where

$$\mu(k) = \begin{cases} 1 & \text{if } k \equiv 2, 3 \bmod(4) \\ 0 & \text{if } k \equiv 0, 1 \bmod(4) \end{cases} \tag{16.14}$$

$$\mathrm{e}^{\theta \cdot \gamma} = 1 + \theta \cdot \gamma + \sum_{\varepsilon \in \mathscr{E}[n], |\varepsilon| \geq 2} (-1)^{\nu(\varepsilon)}\theta^\varepsilon \gamma^\varepsilon \tag{16.15}$$

where the integer $\nu(\varepsilon)$ *is defined as follows*

$$\nu(\varepsilon) = \begin{cases} 1 & \text{if } |\varepsilon| \equiv 2, 3 \bmod(4) \\ 0 & \text{if } |\varepsilon| \equiv 0, 1 \bmod(4) \end{cases} \tag{16.16}$$

Proof Equation (16.13) is true for $k = 1, 2$. We get the general case by induction on k using Lemma 100.

Equation (16.15) is proved by induction on n using the identity

$$\mathrm{e}^{\theta \cdot \gamma} = \prod_{1 \le k \le n} \mathrm{e}^{\theta_k \gamma_k} = \prod_{1 \le k \le n} (1 + \theta_k \gamma_k).$$

□

16.3.2 Calculus with Grassmann Numbers

We have already defined derivatives in Grassmann algebras in the previous section. To preserve analogy with bosons it is useful to define integration in Grassmann variables.

Definition 43 Consider a Grassmann algebra $\mathscr{G}_n$ with generators $\{\theta_1, \ldots, \theta_n\}$.
Let $\psi \in \mathscr{G}_n$ and $1 \le j_1, \ldots, j_k \le n$.
$\int \psi d\theta_{j_1} \ldots d\theta_{j_k} \in \mathscr{G}_n$ is defined by the following properties.

(i) $\psi \mapsto \int \psi d\theta_{j_1} \ldots d\theta_{j_k} \in \mathscr{G}_n$ is linear
(ii) $\int d\theta_j = 0, \forall j = 1, \ldots, n.$
(iii) $\int \theta_j d\theta_k = \delta_{j,k}, \forall j, k.$
(iv) $\int d\theta_j d\theta_k = 0, \forall j, k.$
(v) $\int\int \psi(\theta_j)\varphi(\theta_k) d\theta_j d\theta_k = \int \psi(\theta_j) d\theta_j \int \varphi(\theta_k) d\theta_k, \ \forall j, k$

We see that for fermions integration coincides with differentiation; we have easily

$$\int \psi d\theta_{j_1} \ldots d\theta_{j_k} = \frac{\partial}{\partial \theta_{j_k}} \cdots \frac{\partial}{\partial \theta_{j_1}} \psi(\theta_1, \ldots, \theta_n). \tag{16.17}$$

This formula will be used to compute integrals.

Lemma 102 (change of variables) *Let A be a real invertible $n \times n$ matrix (A may have its coefficient in a Grassmann algebra such that A is even and invertible). Then we have:*

$$\int \psi(A\theta) d\theta = \det(A) \int \psi(\theta) d\theta. \tag{16.18}$$

Proof If $\theta' = \theta A$ we have

$$\theta'_k = \sum_{1 \le j \le n} \theta_j A_{j,k}.$$

Using definition of determinant we get

$$\theta'_n \ldots \theta'_1 = \theta_n \ldots \theta_1 (\det A).$$

But we have $\int \psi d\theta = \psi_{(1,\ldots,1)}$ where $\psi_{(1,\ldots,1)}$ is the component of ψ in its expansion $\psi(\theta) = \sum_{\varepsilon \in \mathscr{E}[n]} \theta^{\varepsilon} \psi_{\varepsilon}$. Hence we deduce (16.18). □

From the Leibnitz rule we deduce an integration by parts formula:

$$\int \left(\frac{\partial \psi}{\partial \theta_k}\right) \varphi d\theta^* d\theta = \int \mathbf{P}\psi \left(\frac{\partial \varphi}{\partial \theta_k}\right) d\theta^* d\theta. \tag{16.19}$$

Remark on notation: here $d\theta^* d\theta$ means $\prod_{1\leq j\leq n} d\theta_j^* d\theta_j$. Sometimes it will be denoted $d^2\theta$.

16.3.3 Gaussian Integrals

In order to preserve analogy with the Bargmann-Fock realization with holomorphic states (see Chap. 1) we need to have a complex structure on our Grassmann algebra. So we consider the Grassmann algebra $\mathcal{G}_{2n}$ with generators $\{\theta_1, \ldots, \theta_n; \theta_1^*, \ldots, \theta_n^*\}$. In $\mathcal{G}_{2n}$ we can define a complex anti- linear involution such that $\theta_k \mapsto \theta_k^*$. We impose that $\psi \mapsto \psi^*$ is $\mathbb{R}$-linear and satisfies

(i) $(z\alpha)^* = \bar{z}\alpha^*$ if $z \in \mathbb{C}$ and α is a generator
(ii) $(\alpha\beta)^* = \beta^*\alpha^*$ for every $\alpha, \beta \in \mathcal{G}_{2n}$

Let us denote $\theta^* = (\theta_1^*, \ldots, \theta_n^*)$. The Grassmann algebra $\mathcal{G}_{2n}$ with its complex structure will be denoted $\mathcal{G}_n^c$.
Integration on $\mathcal{G}_n^c$ is defined as follows

$$\int \psi d\theta d\theta^* = \int \psi d\theta_1 d\theta_1^* \ldots d\theta_n d\theta_n^*.$$

We have the following property.

Lemma 103 *The integral $\int \psi d\theta d\theta^*$ is invariant under unitary change of variable: $\theta = U\zeta$, $U \in SU(n)$. So we have*

$$\int \psi d\theta d\theta^* = \int \psi(\bar{U}\zeta, U\zeta) d\zeta d\zeta^*.$$

Proof From the proof of the change of variable lemma we have: $\theta_n \ldots \theta_1 = (\det U)\zeta_n \ldots \zeta_1$. Then we get

$$\theta_n^*\theta_n \ldots \theta_1^*\theta_1 = (\zeta_n^*\zeta_n \ldots \zeta_1^*\zeta_1)\overline{\det U} \det U = \zeta_n^*\zeta_n \ldots \zeta_1^*\zeta_1.$$

The lemma follows. □

Proposition 147 *Let B be a $n \times n$ Hermitian matrix. Then we have*

$$\int \mathrm{e}^{-\theta^* \cdot B\theta} d\theta^* \cdot d\theta = \det B. \tag{16.20}$$

Proof There exists a unitary matrix U such that $UBU^* = D$ where D is the diagonal matrix $D = (b_1, \ldots, b_n)$. Using the change of variables $\eta = U\theta$ we get

$$\int e^{-\theta^*\cdot B\theta}d\theta^*\cdot d\theta = \int e^{-\eta^*\cdot D\eta}d\eta^*\cdot d\eta = \prod_{1\le j\le n}\int e^{-b_j\eta_j^*\eta_j}d\eta_j^* d\eta_j = \det B.$$

□.

We can extend the above proposition to more general Gaussian integrals.

Proposition 148 *Let $\mathbb{C}[\gamma,\gamma^*]$ be an other copy of the complex Grassmann algebra $\mathcal{G}_n^c$ and B a $n\times n$ Hermitian matrix. Then we have*

$$\int e^{-\theta^*\cdot B\theta}e^{\gamma^*\cdot\theta+\theta^*\cdot\gamma}d\theta^*\cdot d\theta = (\det B)e^{\gamma^*\cdot B^{-1}\gamma}. \tag{16.21}$$

Proof As in the proof of Proposition 147 we begin by a unitary change of variable to diagonalize B. Then we get the result after some computations (reduction to the case $n=1$) left to the reader. □

Another useful Gaussian integral computation is

Proposition 149 *Let us consider a quadratic form in (γ,γ^*):*

$$\Phi(\gamma,\gamma^*) = \frac{1}{2}(\gamma\cdot K\gamma + \gamma^*\cdot L\gamma^* + 2\gamma^*\cdot M\gamma)$$

where K, L are antisymmetric matrices. So the following matrix Λ of Φ is antisymmetric, where

$$\Lambda = \begin{pmatrix} K & -M^T \\ M & L \end{pmatrix}.$$

Assume Λ is non degenerate. Then we have the Fourier transform result

$$\int e^{\Phi(\gamma,\gamma^*)}e^{\gamma^*\cdot\xi-\xi^*\cdot\gamma}d\gamma^* d\gamma = \mathrm{Pf}\,\Lambda e^{\Phi^{(-1)}(\xi^*,\xi)} \tag{16.22}$$

where $\mathrm{Pf}\,\Lambda$ is the Pfaffian of Λ and $\Phi^{(-1)}$ is the quadratic form with the matrix Λ^{-1}.

Proof For $\xi=0$ it is well known that the integral is equal to $\mathrm{Pf}\,\Lambda$ which is the Pfaffian of Λ (see [269]). Recall that $\mathrm{Pf}\,\Lambda^2 = \det\Lambda$.

Then by a usual trick we can eliminate the linear terms in the integral. We do that by performing a change of Grassmann variables $\gamma = \theta - \alpha_\xi$, $\gamma^* = \zeta - \beta_\xi$. θ and ζ are new Grassmann integration variables (not necessarily conjugated) and α_ξ, β_ξ are computed such that the linear terms are eliminated. So we find

$$\begin{pmatrix}\alpha_\xi\\ \beta_\xi\end{pmatrix} = \Lambda^{-1}\begin{pmatrix}\xi^*\\ \xi\end{pmatrix}.$$

The formula (16.22) follows. □

16.4 Super-Hilbert Spaces and Operators

As in the bosonic case we want to find a space to represent fermionic states as functions in some L^2-space.

16.4.1 A Space for Fermionic States

Let us define $\mathscr{H}^{(n)}$ the subspace of $\mathscr{G}_{2n}$ of holomorphic functions in $\theta = (\theta_1, \ldots, \theta_n)$. $\psi \in \mathscr{H}^{(n)}$ means $\frac{\partial}{\partial \theta_k^*}\psi = 0$ for $1 \leq k \leq n$.

$\mathscr{H}^{(n)}$ is a complex vector space of dimension n with the basis $\{\theta^\varepsilon,\ \varepsilon \in \mathscr{E}[n]\}$.

Moreover $\mathscr{H}^{(n)}$ is a Hilbert space for the scalar product

$$\langle \psi, \varphi \rangle := \int \mathrm{e}^{\theta^* \cdot \theta} \psi(\theta)^* \varphi(\theta) d\theta d\theta^*.$$

and $\{\theta^\varepsilon,\ \varepsilon \in \mathscr{E}[n]\}$ is an orthonormal basis. We shall denote $e_\varepsilon(\theta) = \theta^\varepsilon$ for $\varepsilon \in \mathscr{E}[n]$. The space $\mathscr{H}^{(n)}$ has dimension 2^n over $\mathbb{C}$ and is isomorphic to the Fock space $\mathscr{H}_F^{(n)}$.

Remark 100 The notation ψ^* for the complex involution is sometimes replaced by the more suggestive notation $\bar{\psi}$ which may be sometimes confusing.

The creation/annihilation operators in $\mathscr{H}^{(n)}$ are

$$(a_j^* \psi)(\theta) = \theta_j \psi(\theta), \quad a_j \psi(\theta) = \frac{\partial}{\partial \theta_j} \psi(\theta).$$

They satisfy anti-commutation rules (CAR) and a_k^* is the hermitian adjoint of a_k for every $1 \leq k \leq n$.

The Hilbert space $\mathscr{H}^{(n)}$ is not large enough to represent Fermionic states and to compute with them. As usual in Grassmann calculus we need to add new Grassmann generators (γ, γ^*), $\gamma = (\gamma_1, \ldots, \gamma_n)$. Let us denote $\Gamma_n^c := \mathbb{C}[\gamma, \gamma^*]$.

As above we introduce $\tilde{\mathscr{H}}^{(n)}$ the sub-Grassmann algebra of $\mathscr{G}_n^c \otimes \Gamma_n^c$ of holomorphic elements in θ.

$\tilde{\mathscr{H}}^{(n)}$ will be seen as a module[1] over the algebra Γ_n^c with basis $\{\theta^\varepsilon, ; \varepsilon \in \mathscr{E}[n]\}$. $\tilde{\mathscr{H}}^{(n)}$ is a kind of linear space where the complex field number is replaced by the Grassmann algebra numbers: Γ_n^c. The following operations make sense because they are well defined in $\mathscr{G}_n^c \otimes \Gamma_n^c$, with usual properties easy to state: if $\psi, \varphi \in \tilde{\mathscr{H}}^{(n)}$, $\lambda \in \Gamma_n^c$, then $\psi + \varphi \in \tilde{\mathscr{H}}^{(n)}$ and $\lambda\psi$, $\psi\lambda \in \tilde{\mathscr{H}}^{(n)}$. In particular every $\tilde{\psi} \in \tilde{\mathscr{H}}^{(n)}$ can be decomposed as

[1] A module over some ring or algebra is like a linear space where the field number $\mathbb{R}$ or $\mathbb{C}$ is replaced by a ring or an algebra (see any textbook in advanced algebra).

$$\psi = \sum_{\varepsilon \in \mathscr{E}[n]} \theta^{\varepsilon} c_{\varepsilon}(\psi)$$

where $c_{\varepsilon}(\psi) \in \Gamma_n^c$.

The scalar product in $\mathscr{H}^{(n)}$ can be extended as sesquilinear form to $\tilde{\mathscr{H}}^{(n)}$ by the formula

$$\langle \psi, \varphi \rangle := \int \psi(\theta)^* \varphi(\theta) \mathrm{e}^{\theta^* \cdot \theta} d\theta d\theta^*$$

but now $\langle \psi, \varphi \rangle$ is a Grassmann number in Γ_n^c and not always a complex number. The map $(\psi, \varphi) \mapsto \langle \psi, \varphi \rangle$ has usual properties: it is sesquilinear and non negative. But it is degenerate and only its restriction to the Hilbert space $\mathscr{H}^{(n)}$ is non-degenerate. Here we do not use the general theory of super-space, on our examples direct computations can be done (see the book [76] for more details concerning super-Hilbert spaces).

Let us remark that every linear operator in $\mathscr{H}^{(n)}$ can be extended as a linear operator in $\tilde{\mathscr{H}}^{(n)}$. In particular the parity operator $\mathbf{P}$ in variables θ is well defined.

As a first application let us consider the Fermionic Dirac distribution at θ. It is not difficult to establish the following identity for every $\psi \in \mathscr{H}^{(n)}$,

$$\psi(\theta) = \int \psi(\gamma) \mathrm{e}^{-(\theta - \gamma) \cdot \gamma^*} d\gamma^* d\gamma. \tag{16.23}$$

In other words the identity is an integral operator with a kernel

$$E(\theta, \gamma) := \mathrm{e}^{-(\theta - \gamma) \cdot \gamma^*} = \prod_{1 \le k \le n} (1 - (\theta_k - \gamma_k)\gamma_k^*). \tag{16.24}$$

It is usual to denote $E(\theta, \gamma) = \delta(\theta - \gamma)$. We have also the more symmetric form

$$\delta(\theta - \gamma) = \prod_{1 \le k \le n} (\theta_k - \gamma_k)(\theta^* - \gamma_k^*)$$

This could be used to prove that every linear operator in $\mathscr{H}^{(n)}$ has a kernel.

16.4.2 Integral Kernels

Proposition 150 *For every linear operator $\hat{H} : \mathscr{H}^{(n)} \to \mathscr{H}^{(n)2}$ there exists $K_H \in \mathbb{C}[\theta, \gamma^*]$ such that*

[2]As for bosons, when we consider quantization of observables, it is convenient to denote operators with a hat accent to make a difference between classical and quantum observables. Sometimes this rule is not applied when the context is clear.

$$\hat{H}\psi(\theta) = \int K_H(\theta, \gamma^*)\psi(\gamma)e^{\gamma^* \cdot \gamma} d\gamma d\gamma^*, \quad \forall \psi \in \mathscr{H}^{(n)}. \tag{16.25}$$

Moreover $K_H \in \mathbb{C}[\theta, \gamma^*]$ *can be computed as follows:*

$$K_H(\theta, \gamma^*) = \sum_{\varepsilon', \varepsilon \in \mathscr{E}[n]} \langle e_{\varepsilon'}, \hat{H} e_\varepsilon \rangle e_{\varepsilon'}(\theta) e_\varepsilon(\gamma)^*. \tag{16.26}$$

Proof Using that $\{e_\varepsilon\}$ is an orthonormal system it is not difficult to prove that Eq. (16.25) is satisfied for $\psi = e_\varepsilon$ with K_H given by (16.26). So get the result for every ψ. □

Corollary 52 *Every linear operator* $\hat{H} : \mathscr{H}^{(n)} \to \mathscr{H}^{(n)}$ *has a unique decomposition like*

$$\hat{H} = \sum_{\varepsilon', \varepsilon \in \mathscr{E}[n]} H_{\varepsilon, \varepsilon'} a^{*\varepsilon'} a^\varepsilon \tag{16.27}$$

where $H_{\varepsilon, \varepsilon'} \in \mathbb{C}$.

These results can be easily extended to linear operators in the super-Hilbert space $\tilde{\mathscr{H}}^{(n)}$.

Proposition 151 *For every linear operator* $\hat{H} : \tilde{\mathscr{H}}^{(n)} \to \tilde{\mathscr{H}}^{(n)}$ *there exists* $K_H \in \mathbb{C}[\theta, \gamma, \gamma^*]$ *such that*

$$\hat{H}\psi(\theta) = \int K_H(\theta, \gamma^*)\psi(\gamma)e^{\gamma^* \cdot \gamma} d\gamma d\gamma^*, \quad \forall \psi \in \tilde{\mathscr{H}}^{(n)}. \tag{16.28}$$

Moreover $K \in \mathbb{C}[\theta, \gamma, \gamma^*]$ *can be computed as follows:*

$$K(\theta, \gamma^*) = \sum_{\varepsilon', \varepsilon \in \mathscr{E}[n]} e_{\varepsilon'}(\theta) \langle e_{\varepsilon'}, \hat{H} e_\varepsilon \rangle e_\varepsilon(\gamma)^*, \tag{16.29}$$

and every linear operator $\hat{H}$ *in* $\tilde{\mathscr{H}}^{(n)}$ *has a unique decomposition*

$$\hat{H} = \sum_{\varepsilon', \varepsilon \in \mathscr{E}[n]} H_{\varepsilon, \varepsilon'} a^{*\varepsilon'} a^\varepsilon \tag{16.30}$$

where $H_{\varepsilon, \varepsilon'} \in \Gamma_n^c$. *This representation, with annihilation operators, first is call the* **normal representation** *of* $\hat{H}$.

We denote $\mathrm{End}(\tilde{\mathscr{H}}^{(n)})$ the space of linear operators in $\tilde{\mathscr{H}}^{(n)}$.

Corollary 53 *Every linear operator* $\hat{H}$ *in* $\tilde{\mathscr{H}}^{(n)}$ *has an hermitian conjugate* $\hat{H}^*$*:*

$$\langle \psi, \hat{H}\varphi \rangle = \langle \hat{H}^* \psi, \varphi \rangle, \quad \forall \psi, \varphi \in \tilde{\mathscr{H}}^{(n)}. \tag{16.31}$$

Moreover $\hat{H} \mapsto \hat{H}^*$ *is a linear complex involution in* $\mathrm{End}(\tilde{\mathscr{H}}^{(n)})$*, satisfying for every linear operator* $\hat{H}, \hat{F}$ *in* $\tilde{\mathscr{H}}^{(n)}$ *and* $\lambda \in \Gamma_n^c$*,*

- $(\hat{H}\hat{F})^* = \hat{F}^*\hat{H}^*$.
- $(\lambda\hat{H})^* = \hat{H}^*\lambda^*$, $(\hat{H}\lambda)^* = \lambda^*\hat{H}^*$.
- a_k^* *is the hermitian conjugate of* a_k *for* $1 \leq k \leq n$.

It is convenient to define a fermionic Fourier transform which is a kind of Fourier-symplectic transform. This will be used for fermionic quantization.

16.4.3 A Fourier Transform

Definition 44 For any $\psi \in \tilde{\mathscr{H}}^{(n)}$ the Fourier transform is defined as

$$\psi^{\mathscr{F}}(\alpha) = \int e^{\xi^*\cdot\alpha - \alpha^*\cdot\xi}\psi(\xi)d\xi^*d\xi.$$

The following properties are easy to prove

1. $\psi(\xi) = \int e^{(\alpha^*\cdot\xi - \xi^*\cdot\alpha)}\psi^{\mathscr{F}}(\alpha)d\alpha^*d\alpha$ (inverse formula)
   The Fourier transform is idempotent: $(\psi^{\mathscr{F}})^{\mathscr{F}} = \psi$.
2. $1^{\mathscr{F}}(\alpha) = \alpha^*\cdot\alpha$, $\xi^{\mathscr{F}}(\alpha) = \alpha$, $\xi^{*\mathscr{F}}(\alpha) = -\alpha^*$.
3. $\int \psi^{\mathscr{F}}(\alpha)(\varphi^{\mathscr{F}}(\alpha))^*d\alpha^*d\alpha = \int \psi(\xi)(\varphi(\xi))^*d\xi^*d\xi$ (Parseval's relation 1)
4. $\int \psi^{\mathscr{F}}(\alpha)\varphi^{\mathscr{F}}(-\alpha)d\alpha^*d\alpha = \int \psi(\xi)\varphi(\xi)d\xi^*d\xi$ (Parseval's relation 2)
5. $(\psi\varphi)^{\mathscr{F}}(\alpha) = \int \psi^{\mathscr{F}}(\alpha - \beta)\varphi^{\mathscr{F}}(\beta)d\beta^*d\beta$ (Fourier-convolution)
6. $(\tau_\zeta\psi)^{\mathscr{F}}(\alpha) = \psi^{\mathscr{F}}(\alpha)e^{\zeta^*\cdot\alpha - \alpha^*\cdot\zeta}$, (translation-modulation), where $\tau_\zeta\psi(\xi) = \psi(\xi - \zeta)$.

Let us remark that these properties are also satisfied if ψ, φ are replaced by linear operators in $\mathrm{End}(\tilde{\mathscr{H}}^{(n)})$ depending on Grassmann variables.

16.5 Coherent States for Fermions

As in the previous section we consider two complex Grassmann algebras $\mathscr{G}_n^c = \mathbb{C}[\theta, \theta^*]$ and $\Gamma_n^c = \mathbb{C}[\gamma, \gamma^*]$. $\mathscr{H}^{(n)}$ (and $\tilde{\mathscr{H}}^{(n)}$) are spaces of holomorphic states in variables θ.

16.5.1 Weyl Translations

The Weyl translations are defined as usual:

$$\hat{T}(\gamma) = \mathrm{e}^{a^*\cdot\gamma - \gamma^*\cdot a}, \quad \gamma = (\gamma_1, \ldots, \gamma_n). \tag{16.32}$$

Remark that $\hat{T}(\gamma)$ depends on the $2n$ independent Grassmann variables (γ, γ^*). Nevertheless we simply note $\hat{T}(\gamma)$ and sometimes $\hat{T}(\gamma, \gamma^*)$ if necessary. They are translations in the phase space and in particular we have

$$\hat{T}(0, \gamma^*)\psi(\theta) = \psi(\theta - \gamma^*), \quad \hat{T}(\gamma, 0)\psi(\theta) = \mathrm{e}^{\theta\cdot\gamma}\psi(\theta).$$

Recall that $\gamma \cdot a = \sum_{1\le k\le n} \gamma_k a_k$ and $a^* \cdot \gamma = \sum_{1\le k\le n} a_k^*\gamma_k$ (beware of the order!).
Using anti-commutations relations (CAR) we have the commutation relations:

$$\begin{aligned} \alpha_j a_j &= -a_j\alpha_j \text{ for } \alpha_j = \gamma_j, \gamma_j^* \\ [a_j^*\gamma_j, \gamma_k^* a_k] &= \gamma_j\gamma_k^*\delta_{j,k} \\ [a_j, a_k^*\gamma_k] &= \gamma_k\delta_{j,k} \\ [a_j, a_k\gamma_k] &= 0. \end{aligned} \tag{16.33}$$

We have similar relation inverting a and a^* and using $[A, B]^* = -[A^*, B^*]$. From these relations we get:

$$\hat{T}(\gamma) = \prod_{1\le k\le n} \mathrm{e}^{a_k^*\gamma_k - \gamma_k^* a_k} = \prod_{1\le k\le n} \left(1 + a_k^*\gamma_k - \gamma_k^* a_k + (a_k^* a_k - \frac{1}{2})\gamma_k^*\gamma_k\right). \tag{16.34}$$

The following properties are easily obtained with little algebraic computations similar to the bosonic case (see Chap. 1). In particular we also have a Baker–Campbell–Hausdorff formula.

Lemma 104 *Let A, B, be Γ_n^c-linear operators in $\tilde{\mathscr{H}}^{(n)}$ such that $[A, B]$ commutes with A, B. Then*

$$\mathrm{e}^A\mathrm{e}^B\mathrm{e}^{-\frac{1}{2}[A,B]} = \mathrm{e}^{A+B}.$$

Proof Let us remark first that e^A is well defined by the Taylor series because $\dim[\tilde{\mathscr{H}}^{(n)}] < +\infty$ and e^A is a Γ_n^c-linear operator. Hence the result follows as in Lemma 1 of Chap. 1. □.

Now we state mains properties of the translation operators $\hat{T}(\gamma)$.

1. $\hat{T}(\gamma)$ is Γ_n^c-linear in $\tilde{\mathscr{H}}^{(n)}$ (on right and left)
2. $\hat{T}(\alpha + \gamma) = \hat{T}(\alpha)\hat{T}(\gamma)\exp\left(\frac{1}{2}(\alpha^*\cdot\gamma + \alpha\cdot\gamma^*)\right)$
3. $(\hat{T}(\gamma))^{-1} = \hat{T}(-\gamma) = \hat{T}(\gamma)^*$. In particular $\hat{T}(\gamma)$ is a unitary operator in the super-Hilbert space, that means that the super-inner product is preserved:

$$\langle \hat{T}(\gamma)\psi, \hat{T}(\gamma)\varphi\rangle = \langle\psi, \varphi\rangle$$

4. Translation property:

$$\hat{T}(\gamma)^* a \hat{T}(\gamma) = a + \gamma \tag{16.35}$$

$$\hat{T}(\gamma)^* a^* \hat{T}(\gamma) = a^* + \gamma^*. \tag{16.36}$$

16.5.2 Fermionic Coherent States

Now we can give a definition for Fermionic coherent states.

Definition 45 For every Grassmann generator γ we associate a state in the super-Hilbert space $\tilde{\mathscr{H}}^{(n)}$ denoted $\psi_\gamma = |\gamma\rangle$ by the formula

$$\psi_\gamma = \hat{T}(\gamma)\psi_\emptyset \tag{16.37}$$

where $\psi_\emptyset(\theta) = e_{0,\ldots,0}(\theta)$ is the vacuum state, usually denoted $|0\rangle$.

First properties of coherent states ψ_γ are easily deduced from properties of the translation $\hat{T}(\gamma)$.

1. ψ_γ are eigenvectors of annihilation operators a: $a\psi_\gamma = \gamma\psi_\gamma$ where $a = (a_1, \ldots, a_n)$ and $\gamma = (\gamma_1, \ldots, \gamma_n)$.
2. The inner product of two coherent states satisfies

$$\langle\psi_\gamma, \psi_\alpha\rangle = \exp\left(\gamma^* \cdot \alpha - \frac{1}{2}(\gamma^* \cdot \gamma + \alpha^* \cdot \alpha)\right) \tag{16.38}$$

3.
$$\hat{T}(\gamma)\psi_\alpha = \exp\left(\frac{1}{2}(\alpha^* \cdot \gamma - \gamma^* \cdot \alpha)\right)\psi_{\alpha+\gamma}. \tag{16.39}$$

We can give a more explicit formula for ψ_γ which is very similar to that obtained in the Bargmann representation for bosons (Chap. 1).

From above computations we have

$$\begin{aligned}
\psi_\gamma(\theta) &= \mathrm{e}^{-\gamma^*\cdot\gamma/2} \prod_{1\le k\le n} (1 + a_k^*\gamma_k)\psi_\emptyset \\
&= \mathrm{e}^{-\gamma^*\cdot\gamma/2} \sum_{k_1<k_2<\cdots k_j} a_{k_1}^*\gamma_{k_1}\ldots a_{k_j}^*\gamma_{k_j}\psi_\emptyset \\
&= \mathrm{e}^{-\gamma^*\cdot\gamma/2} \sum_{\varepsilon\in\mathscr{E}[n]} (-1)^{\nu(\varepsilon)}\theta^\varepsilon\gamma^\varepsilon \\
&= \mathrm{e}^{(\theta-\frac{\gamma^*}{2})\cdot\gamma}
\end{aligned}$$

where $\nu(\varepsilon)$ was defined in Eq. (16.15). In particular we have proved that

$$\langle e_\varepsilon, \psi_\gamma\rangle = (-1)^{\nu(\varepsilon)}\gamma^\varepsilon \mathrm{e}^{-\gamma^*\cdot\gamma/2}. \tag{16.40}$$

Then for every $\psi \in \mathscr{H}^{(n)}$ we have

$$\langle \psi_\gamma, \psi \rangle = \mathrm{e}^{-\gamma^* \cdot \gamma/2} \sum_{\varepsilon \in \mathscr{E}[n]} \gamma^{*\varepsilon} \psi_\varepsilon. \tag{16.41}$$

Remark 101 Let us remark that the chirality operator[3] χ on coherent state ψ_γ gives $\chi \psi_\gamma = \psi_{-\gamma}$.

One of the most useful property of coherent states is their completeness :

Proposition 152 *For every* $\psi \in \tilde{\mathscr{H}}^{(n)}$ *we have*

$$\psi(\theta) = \int \langle \psi_\gamma, \psi \rangle \psi_\gamma(\theta) d\gamma^* d\gamma. \tag{16.42}$$

Proof It is enough to prove (16.42) for $\psi = e_\varepsilon$ for every $\varepsilon \in \mathscr{E}[n]$. Computing the integrand in the right hand side of (16.42) using (16.40) and (16.41) we get

$$\langle \psi_\gamma, e_\varepsilon \rangle \psi_\gamma(\theta) = \sum_{\varepsilon' \in \mathscr{E}[n]} (-1)^{\nu(\varepsilon)+\nu(\varepsilon')} \mathrm{e}^{\gamma \cdot \gamma^*} \theta^{\varepsilon'} \gamma^{\varepsilon'} (\gamma^\varepsilon)^*. \tag{16.43}$$

Then using orthogonality relations in Grassmann variables γ we get

$$\int \langle \psi_\gamma, e_\varepsilon \rangle \psi_\gamma(\theta) d\gamma^* d\gamma = \theta^\varepsilon. \tag{16.44}$$

□

Remark 102 Coherent states are obtained by translations from a fixed state. Instead of the vacuum $\psi_\emptyset$ we could start from the state $\psi_{\mathbb{1}}(\theta) = \theta_1 \ldots \theta_n$ (all modes are occupied). So we define another family of coherent states:

$$\psi'_\gamma = \hat{T}(\gamma)\psi_{\mathbb{1}}. \tag{16.45}$$

Any property of ψ_γ can be translated in a property of ψ'_γ:

- $a_k^* \psi'_\gamma = \gamma_k^* \psi'_\gamma$ (ψ'_γ is an eigenstate of the creation operators)
- Expansion in the orthonormal basis:

$$\begin{aligned} \psi'_\gamma(\theta) &= \mathrm{e}^{\frac{\gamma^* \cdot \gamma}{2}} \sum_{\varepsilon \in \mathscr{E}[n]} \theta_1^{1-\varepsilon_1} (\gamma_1^*)^{\varepsilon_1} \ldots \theta_n^{1-\varepsilon_n} (\gamma^*)^{\varepsilon_n} \\ &= \mathrm{e}^{\frac{\gamma^* \cdot \gamma}{2}} \sum_{\varepsilon \in \mathscr{E}[n]} (-1)^{\nu'(\varepsilon)} \theta^{\varepsilon^c} (\gamma^*)^\varepsilon. \end{aligned} \tag{16.46}$$

where $\varepsilon^c = (1 - \varepsilon_1, \ldots, 1 - \varepsilon_n)$ and $\nu'(\varepsilon)$ is 0 or 1.

[3] A definition of the chirality operator is given in the next section.

- Completeness: for every $\psi \in \tilde{\mathscr{H}}^{(n)}$,

$$\psi(\theta) = \int \psi'_\gamma(\theta)\langle\psi'_\gamma, \psi\rangle d\gamma d\gamma^*. \quad (16.47)$$

We shall see in the next section that fermionic coherent states are related with a fermionic harmonic oscillator.

16.6 Representations of Operators

Besides their integral kernels, operators can be represented by several other functions or (Schwartz distributions) often called "symbols" in the same way as for finite bosons systems Weyl (covariant, contravariant), Wick, anti-Wick, Wigner functions (see Sect. 12.1). We shall see now that this can be also defined for finite systems of fermions.

A useful tool to compute with operators (quantum observables) is the trace which is an important spectral invariant.

16.6.1 *Trace*

The trace of any linear operator $\hat{H}$ in $\tilde{\mathscr{H}}^{(n)}$ is defined as usual by the sum of diagonal elements

$$\mathrm{Tr}\hat{H} = \sum_{\varepsilon\in\mathscr{E}[n]} \langle e_\varepsilon, \hat{H} e_\varepsilon\rangle. \quad (16.48)$$

The parity operator $\mathbf{P}$ defined a Γ_n^c linear operator in the super-space $\tilde{\mathscr{H}}^{(n)}$ and another useful invariant is the super-trace defined as follows. Let us denote $\mathbf{P}_\pm$ the orthogonal projection on even/odd states in $\tilde{\mathscr{H}}^{(n)}$. So every operator $\hat{H}$ can be decomposed as

$$\hat{H} = \mathbf{P}_+\hat{H}\mathbf{P}_+ + \mathbf{P}_-\hat{H}\mathbf{P}_- + \mathbf{P}_+\hat{H}\mathbf{P}_+ + \mathbf{P}_-\hat{H}\mathbf{P}_+.$$

In a matricial form this is written as

$$\hat{H} = \begin{pmatrix} \hat{H}_{++} & \hat{H}_{+-} \\ \hat{H}_{-+} & \hat{H}_{--} \end{pmatrix}.$$

So we have $\mathrm{Tr}\hat{H} = \mathrm{Tr}(\hat{H}_{++}) + \mathrm{Tr}(\hat{H}_{--})$.

Definition 46 (*super-trace*) The super-trace of $\hat{H}$ is defined as

$$\mathrm{Str}\hat{H} = \mathrm{Tr}(\hat{H}_{++}) - \mathrm{Tr}(\hat{H}_{--}).$$

Or if we introduce the chirality operator $\chi = \begin{pmatrix} \mathbb{1} & 0 \\ 0 & -\mathbb{1} \end{pmatrix}$ we have

$$\mathrm{Str}\hat{H} = \mathrm{Tr}(\hat{H}\chi).$$

Remark 103 If $\hat{A}$ is a linear invertible operator in $\tilde{\mathscr{H}}^{(n)}$ and if $\hat{A}$ is even: $\hat{A}\mathbf{P} = \mathbf{P}\hat{A}$ then

$$\mathrm{Str}(\hat{A}^{-1}\hat{H}\hat{A}) = \mathrm{Str}\hat{H}.$$

In particular the super-trace is invariant under even unitary transformations of $\tilde{\mathscr{H}}^{(n)}$.

More generally we have

$$\mathrm{Str}(\hat{A}\hat{B}) = (-1)^{\varepsilon}\mathrm{Str}(\hat{B}\hat{A}).$$

where $\varepsilon = 1$ if $\hat{A}$ and $\hat{B}$ are odd, $\varepsilon = 1$ otherwise.

Recall that the trace and determinant are related through the usual trace-determinant relation: $\mathrm{e}^{\mathrm{Tr}H} = \det(\mathrm{e}^{H})$. To the super-trace is associated a super-determinant $\mathrm{Ber}(\hat{H})$ (found by Berezin [24]) such that

$$\exp\left(\mathrm{Str}\hat{H}\right) = \mathrm{Ber}(\exp\hat{H}).$$

Assume for simplicity that $\hat{H}$ has complex coefficients.

Definition 47 Let $\hat{H} = \begin{pmatrix} \hat{H}_{++} & \hat{H}_{+-} \\ \hat{H}_{-+} & \hat{H}_{--} \end{pmatrix}$ be such that $\hat{H}_{--}$ is invertible.

The Berezian $\mathrm{Ber}(\hat{H})$ of $\hat{H}$ is defined by

$$\mathrm{Ber}(\hat{H}) = \det(\hat{H}_{++} - \hat{H}_{+-}\hat{H}_{--}^{-1}\hat{H}_{-+})\det(\hat{H}_{--})^{-1}.$$

The following property is an easy consequence of the definition: if $\hat{H}$ and $\hat{L}$ are super-operators then

$$\mathrm{Ber}(\hat{H}\hat{L}) = \mathrm{Ber}(\hat{H})\mathrm{Ber}(\hat{L}).$$

Proposition 153 *Let $\hat{H}$ be as above. Then we have*

$$\exp\left(\mathrm{Str}\hat{H}\right) = \mathrm{Ber}(\exp\hat{H}). \tag{16.49}$$

Proof If $\hat{H}_{+-} = \hat{H}_{-+} = 0$ (16.49) is a consequence of the trace-determinant relation. The general case can be deduced from this case. □

Proposition 154 *Let $\hat{H}$ be a linear operator in $\tilde{\mathscr{H}}^{(n)}$ and $K_H(\theta, \gamma^*)$ its integral kernel. We assume that the element of matrix of $\hat{H}$ are even in Γ_n^c. Then we have*

$$\mathrm{Tr}\hat{H} = \int K_H(\theta, -\theta^*)\mathrm{e}^{\theta^*\cdot\theta}d\theta d\theta^* \tag{16.50}$$

$$\mathrm{Str}\hat{H} = \int K_H(\theta, \theta^*)\mathrm{e}^{\theta^*\cdot\theta}d\theta d\theta^*. \tag{16.51}$$

Proof From assumption $\langle e_{\varepsilon'}, \hat{H}e_\varepsilon\rangle$ are even Grassmann numbers, so we have

$$\int K_H(\theta, \theta^*)\mathrm{e}^{\theta^*\cdot\theta}d\theta d\theta^* = \sum_\varepsilon \langle e_\varepsilon, \hat{H}e_\varepsilon\rangle e_\varepsilon(\theta)e_\varepsilon(\theta)^*\mathrm{e}^{\theta^*\cdot\theta}d\theta d\theta^*$$

Splitting the sum according the parity of ε we get

$$\int K_H(\theta, \theta^*)\mathrm{e}^{\theta^*\cdot\theta}d\theta d\theta^* = \sum_{|\varepsilon|\mathrm{even}} \langle e_\varepsilon, \hat{H}e_\varepsilon\rangle - \sum_{|\varepsilon|\mathrm{odd}} \langle e_\varepsilon, \hat{H}e_\varepsilon\rangle \tag{16.52}$$

$$= \mathrm{Str}\hat{H}. \tag{16.53}$$

The same computation gives the expression for the trace formula. □

The integral kernel of operators are closely related with the matrix elements on coherent states (see Chap. 1 concerning Bargmann representation):

Proposition 155 *Under the same assumption on* $\hat{H}$ *as in proposition 154 we have*

$$\langle\psi_\alpha, \hat{H}\psi_\beta\rangle = \mathrm{e}^{\frac{1}{2}(\alpha\cdot\alpha^*+\beta\cdot\beta^*)}K_H(\alpha^*, \beta) \tag{16.54}$$

where α, β *are independent Grassmann generators with their complex conjugate* (α^*, β^*).

In particular the integral kernel of the identity operator is the following expression of the delta function

$$\delta(\theta - \gamma^*) = \mathrm{e}^{\theta\cdot\gamma^*}. \tag{16.55}$$

Proof Using Parseval relation we have

$$\langle\psi_\alpha, \hat{H}\psi_\beta\rangle = \sum_\varepsilon \langle e_\varepsilon, \psi_\alpha\rangle^*\langle e_\varepsilon, \hat{H}\psi_\beta\rangle$$

$$= \mathrm{e}^{\frac{1}{2}(\alpha\cdot\alpha^*+\beta\cdot\beta^*)}\left(\sum_{\varepsilon',\varepsilon}(-1)^{\nu(\varepsilon)+\nu(\varepsilon')}\langle\varepsilon, \hat{H}e_{\varepsilon'}\rangle\alpha^{*\varepsilon}\beta^{\varepsilon'}\right). \tag{16.56}$$

Putting $\alpha = \theta^*$, $\beta = \gamma^*$ we recognize the integral kernel of H.

Formula (16.55) is easily obtained using properties of coherent states. Note that this expression for Dirac function is different of that found before because the measure on the space is different. □

$H_c(\alpha, \beta) := \langle\psi_\alpha, \hat{H}\psi_\beta\rangle$ plays the role of a covariant-Wick symbol for $\hat{H}$. We shall see later that we also have contravariant Wick symbols.

Corollary 54 *We have*

$$\mathrm{Tr}\hat{H} = \int H_c(\theta, -\theta)d\theta^* d\theta \quad (16.57)$$

$$\mathrm{Str}\hat{H} = \int H_c(\theta, \theta)d\theta^* d\theta. \quad (16.58)$$

In particular for rank one operators $\Pi_{\eta,\varphi}\psi = \eta\langle\varphi, \psi\rangle$ *we have*

$$\mathrm{Str}(\Pi_{\eta,\varphi}) = \langle\varphi, \eta\rangle \quad (16.59)$$

$$\mathrm{Tr}(\Pi_{\eta,\varphi}) = \langle\chi\varphi, \eta\rangle = \langle\varphi, \chi\eta\rangle, \quad (16.60)$$

where χ *is the chirality operator.*

Proof Let us prove (16.59). The super-trace formula follows from completeness relation for coherent states and (16.57). So we have

$$\mathrm{Str}\Pi_{\eta,\varphi} = \int \langle\psi_\alpha, \eta\rangle\langle\varphi, \psi_\alpha\rangle = \langle\varphi, \eta\rangle.$$

We get the trace formula using the $\mathrm{Tr}(\Pi_{\eta,\varphi}) = \mathrm{Str}(\chi\Pi_{\eta,\varphi})$. □

Remark 104 As usual we have the following relation between the trace and matrix-elements

$$\mathrm{Tr}(\hat{H}\Pi_{\beta,\gamma}) = \langle\psi_{-\gamma}, \hat{H}\psi_\beta\rangle,$$

where $\Pi_{\beta,\gamma}(\psi) = \psi_\beta\langle\psi_\gamma, \psi\rangle$. Furthermore if $\hat{H}$ is even, we have

$$\mathrm{Tr}(\hat{H}\Pi_{\beta,\gamma}) = \langle\psi_\gamma, \hat{H}\psi_{-\beta}\rangle.$$

Concerning the super-trace we have

$$\mathrm{Str}(\hat{H}\Pi_{\beta,\gamma}) = \langle\psi_\gamma, \hat{H}\psi_\beta\rangle.$$

Now we define Weyl symbols for fermionic operators.

16.6.2 Representation by Translations and Weyl Quantization

Definition 48 The operator $\hat{H}$ has a covariant Weyl symbol H_w if we have the operator equality:

$$\hat{H} = \int H_w(\xi)\hat{T}(-\xi)d\xi^* d\xi \quad (16.61)$$

where the symbol $H_w \in \Gamma_n^c$, complex Grassmann algebra with generators (ξ, ξ^*).

Let us remark that H_w depends on the $2n$ independent Grassmann variables $(\xi_1, \ldots, \xi_n; \xi^*, \ldots, \xi_n^*)$. Nevertheless for simplicity we note $H_w(\xi)$.

Proposition 156 *Every linear operator $\hat{H}$ in $\tilde{\mathscr{H}}^{(n)}$ has a covariant Weyl symbol. This symbol is unique and given by the formula*

$$H_w(\xi) = \mathrm{Str}(\hat{H}\hat{T}(\xi)). \tag{16.62}$$

Proof The proof is done in three steps.

First we remark that identity operator $\mathbb{1}$ and creation operators have Weyl symbols. Direct computations give

$$\mathbb{1}_w(\xi) = \xi \cdot \xi^*, \quad a_{k,w}(\xi) = \xi_k, \quad a_{k,w}^*(\xi) = -\xi_k^*.$$

We remark that if $\hat{H}, \hat{L}$ have Weyl symbols, H_w, L_w then $\hat{H}\hat{L}$ and $\hat{H}^*$ have Weyl symbols (it is possible to give formula, this is left to the reader).

It results that every operator $\hat{H}$ has a Weyl symbol because $\hat{H}$ is a sum of products of $\hat{a}_k^*, \hat{a}_j$.

To compute the symbol of $\hat{H}$ we first compute the diagonal matrix element of translation operators. We get easily:

$$\langle \psi_\alpha, \hat{T}(\gamma)\psi_\alpha \rangle = \exp\left(\alpha^* \cdot \gamma - \gamma^* \cdot \alpha - \frac{1}{2}\gamma^* \cdot \gamma\right). \tag{16.63}$$

Remark that we have an analogous formula for anti-diagonal element

$$\langle \psi_\alpha, \hat{T}(\gamma)\psi_{-\alpha} \rangle = \exp\left(\alpha^* \cdot \gamma + \gamma^* \cdot \alpha - 2\alpha^* \cdot \alpha - \frac{1}{2}\gamma^* \cdot \gamma\right) \tag{16.64}$$

Using formula (16.61) we get after straightforward computations

$$\mathrm{Str}\left(\hat{H}\hat{T}(\alpha)\right) = \int H_w(\xi)$$
$$\exp\left((\beta^* + \frac{\xi^*}{2}) \cdot (\alpha - \xi) - (\alpha^* - \xi^*) \cdot (\beta + \frac{\xi}{2})\right) d^2\xi d^2\beta. \tag{16.65}$$

Computing first integral in β we have

$$\mathrm{Str}\left(\hat{H}\hat{T}(\alpha)\right) = \int H_w(\xi)(\xi - \alpha) \cdot (\xi^* - \alpha^*)$$
$$\exp\left(\frac{\xi^*}{2} \cdot (\alpha - \xi) - (\alpha^* - \xi^*) \cdot \frac{\alpha}{2}\right) d^2\xi. \tag{16.66}$$

Hence a computation using Grassmann algebra rules gives that

$$\mathrm{Str}[\hat{H}\hat{T}(\alpha)] = H_w(\alpha). \tag{16.67}$$

□

Remark 105 A direct consequence of our computations is that

$$\mathrm{Str}\left(\hat{T}(-\xi)\hat{T}(\gamma)\right) = \delta(\gamma - \xi).$$

As usual we have the following relation between the trace and matrix-elements

$$\mathrm{Tr}(\hat{H}\Pi_{\beta,\gamma}) = \langle \psi_\gamma, \hat{H}\psi_{-\beta}\rangle$$

where $\Pi_{\beta,\gamma}(\psi) = \psi_\beta\langle\psi_\gamma, \psi\rangle$.

Now we get the contravariant Weyl symbol as Fourier transform of the covariant symbol

Proposition 157 *For every linear operator $\hat{H}$ we have*

$$\hat{H} = \int H^w(\alpha)\hat{T}^{\mathscr{F}}(\alpha)d^2\alpha \tag{16.68}$$

where $H^w(\alpha) = H_w^{\mathscr{F}}(\alpha)$ and

$$\hat{T}^{\mathscr{F}}(\alpha) = \int \mathrm{e}^{(a^* - \alpha^*)\cdot\gamma + (a-\alpha)\cdot\gamma^*} d^2\gamma.$$

In particular we have the following formula

$$1^w = 1, \quad a^w(\alpha) = \alpha, \quad a^{*w}(\alpha) = \alpha^*. \tag{16.69}$$

Proof Formula (16.68) is a direct application of the Parseval relation 2 for Fourier transform.

Corollary 55 *The supertrace of $\hat{H}$ is equal to the integral of the contravariant Weyl symbol H^w:*

$$\mathrm{Str}\hat{H} = \int H^w(\alpha)d^2\alpha. \tag{16.70}$$

More generally we have the following super trace product formula

$$\mathrm{Str}(\hat{H}\hat{G}) = \int H^w(\alpha)G^w(\alpha)d^2\alpha. \tag{16.71}$$

Proof Using the Fourier transform we get

$$\mathrm{Str}\hat{H} = H_w(0) = \int H^w(\alpha)d^2\alpha.$$

so we have (16.70).

We have for the product

$$\hat{H}\hat{G} = \int H^w(\alpha)\hat{T}^{\mathcal{F}}(\alpha)\hat{G}d^2\alpha \Rightarrow \text{Str}(\hat{H}\hat{G}) = \int H^w(\alpha)\text{Str}(\hat{T}^{\mathcal{F}}(\alpha)\hat{G})d^2\alpha.$$

But $\text{Str}(\hat{T}^{\mathcal{F}}(\alpha)\hat{G})$ is the Fourier transform of $\text{Str}(\hat{T}(\alpha)\hat{G})$ which is the Weyl symbol G^w. So we get (16.71). □

Weyl symbols and integral kernel are closely related.

Proposition 158 *For every operator $\hat{H}$ its integral kernel K_H and its covariant Weyl symbol H_w satisfy*

$$K_H(\theta, \gamma^*) = e^{\theta\cdot\gamma^*} \int H_w(\xi)e^{\frac{\xi\cdot\xi^*}{2}} e^{\xi^*\cdot\gamma^* - \theta\cdot\xi} d\xi^* d\xi \tag{16.72}$$

$$H_w(\xi) = e^{\frac{\xi^*\cdot\xi}{2}} \int e^{\alpha\cdot\alpha^*} K_H(\alpha^*, \alpha)e^{\alpha^*\cdot\xi - \xi^*\cdot\alpha} d\alpha^* d\alpha. \tag{16.73}$$

Proof Remark that the two formulas are clearly equivalent using Fourier and inverse Fourier transform. To prove the first formula we compute the integral kernel of any translation operator $\hat{T}(\zeta)$.

From the relation

$$\psi(\theta) = \int e^{\theta\cdot\gamma^*}\psi(\gamma)e^{\gamma^*\cdot\gamma} d\gamma d\gamma^*.$$

we get that

$$K_{T(\zeta)}(\theta, \gamma^*) = e^{\zeta\cdot\zeta^*} e^{(\theta - \zeta^*)\cdot\gamma^* + \theta\cdot\zeta}.$$

Using that $\hat{H}$ is a superposition of translations with density $H_w(\zeta)$we get the first formula (16.72). □

One of the more useful tool to analyze mean values of observables for a n-bosons system are Wigner–Weyl functions because of the following well known Chap. 2

$$\text{Tr}(\rho\hat{O}) = (2\pi\hbar)^{-n} \int_{\mathbb{R}^{2n}} W_\rho(X)O(X)dX. \tag{16.74}$$

where ρ is a density operator, $\hat{O}$ a quantum observable, $W_\rho(X)$ and $O(X)$ are the Wigner–Weyl symbols, defined in the phase space $\mathbb{R}^{2n}$, for ρ and $\hat{O}$. Our goal now is to find fermionic analogues for formula (16.74). Here we have to overcome algebraic problems to deal with Grassmann variables. We again follow the paper [47].

16.6.3 Wigner–Weyl Functions

Ordering is always a problem for computations in non commutative algebras in particular for quantization of observables. With fermions even classical variables do not commute.

To prepare further computations we introduce now two families of operators from which we can get different quantization procedures with interesting connexion between them.

$$\hat{T}_N(\gamma) = \hat{T}_1(\gamma) = \mathrm{e}^{a^*\cdot\gamma}\mathrm{e}^{-\gamma^*\cdot a}(\text{normal ordering}) \tag{16.75}$$

$$\hat{T}_A(\gamma) = \hat{T}_{-1}(\gamma) = \mathrm{e}^{-\gamma^*\cdot a}\mathrm{e}^{a^*\cdot\gamma}(\text{anti-normal ordering}). \tag{16.76}$$

We have

$$\hat{T}_N(\gamma) = \hat{T}(\gamma)\mathrm{e}^{\frac{\gamma^*\cdot\gamma}{2}} = \hat{T}_1(\gamma) \tag{16.77}$$

$$\hat{T}_A(\gamma) = \hat{T}(\gamma)\mathrm{e}^{-\frac{\gamma^*\cdot\gamma}{2}} = T_{-1}(\gamma) \quad \text{where} \tag{16.78}$$

$$\hat{T}_s(\gamma) = \hat{T}(\gamma)\mathrm{e}^{s\frac{\gamma^*\cdot\gamma}{2}}, \quad s \in \mathbb{R}. \tag{16.79}$$

We also define

$$\hat{R}_A(\xi) = \hat{R}_{-1}(\xi) = \int \mathrm{e}^{\xi\cdot\gamma^* - \gamma\cdot\xi^*}\Pi_{\gamma,-\gamma}d\gamma^* d\gamma \tag{16.80}$$

$$\hat{R}_N(\xi) = \hat{R}_1(\xi) = \hat{R}_{-1}(\xi)\mathrm{e}^{\xi^*\cdot\xi}. \tag{16.81}$$

Remark that $\hat{R}_{-1}(\xi)$ is the Fourier transform in γ of $\Pi_{\gamma,-\gamma}$. We also introduce

$$\hat{R}_s(\xi) = \exp\left(\frac{1+s}{2}\xi^*\xi\right)\hat{R}_{-1}(\xi), \quad s \in \mathbb{R}. \tag{16.82}$$

We now show that $\hat{R}_s$ generates a quantization for each s as well as the translations $\hat{T}_s$ do and that there exists a kind of duality between them.

Proposition 159 *For every linear operator $\hat{H}$ in $\tilde{\mathscr{H}}^{(n)}$ and every real number s there exist unique symbols $H_{\mathscr{G},s}$, $H_{w,s}$ in Γ_n^c such that*

$$\hat{H} = \int H_{\mathscr{G},s}(\xi)\hat{R}_{-s}(-\xi)d\xi^* d\xi, \tag{16.83}$$

$$\hat{H} = \int H_{w,s}(\xi)\hat{T}_{-s}(-\xi)d\xi^* d\xi. \tag{16.84}$$

Moreover we have

$$H_{\mathcal{G},s}(\xi) = \mathrm{Tr}\left(\hat{T}_s(\xi)\hat{H}\right), \tag{16.85}$$

$$H_{w,s}(\xi) = \mathrm{Tr}\left(\hat{R}_s(\xi)\hat{H}\right). \tag{16.86}$$

Proof The proof will be done following the method of Sect. 16.5.2. For simplicity we only consider the case $s = 1$ and the symbol $H_{\mathcal{G},1}$. The general case is not more difficult.

First compute the action of $\hat{R}_{-1}(-\xi)$ on coherent states. Using properties of coherent state, we get

$$\hat{R}_{-1}(-\xi)\psi_\alpha(\theta) = \mathrm{e}^{(\xi+\alpha)\cdot(\theta+\xi^*)}. \tag{16.87}$$

Hence we get the matrix elements

$$\langle\psi_\beta, \hat{R}_{-1}(-\xi)\psi_\alpha\rangle = \exp\left((\xi+\alpha)\cdot(\xi^*+\beta^*) - \frac{1}{2}(\beta^*\cdot\beta + \alpha^*\cdot\alpha)\right). \tag{16.88}$$

In particular we get the following representation formula for operators $\mathbb{1}$, a_k, a_k^*:

$$\mathbb{1} = \int (1 - \xi\cdot\xi^*)\hat{R}_{-1}(\xi)d\xi^* d\xi \tag{16.89}$$

$$a_k = \int (-\xi_k)\hat{R}_{-1}(\xi)d\xi^* d\xi \tag{16.90}$$

$$a_k^* = \int \xi_k^*\hat{R}_{-1}(\xi)d\xi^* d\xi. \tag{16.91}$$

From this, using that every operator is a polynomial in a and a^*, we get that for any operator $\hat{H}$ there exists $H_{\mathcal{G},-1}$ satisfying (16.83).

To get (16.85) we apply the following lemma where we see a duality between quantization by translations and quantization by coherent states. □

Lemma 105 *For every $s \in \mathbb{R}$ we have*

$$\mathrm{Tr}(\hat{T}_s(\xi)\hat{R}_{-s}(-\zeta)) = \delta(\xi - \zeta). \tag{16.92}$$

Proof Using the trace formula with coherent states, we have

$$\mathrm{Tr}(\hat{T}_s(\xi)\hat{R}_{-s}(-\zeta)) = \int \langle\hat{T}_s(-\xi)\psi_{-\gamma}, \hat{R}_{-s}(-\zeta)\psi_\gamma\rangle d^2\gamma. \tag{16.93}$$

In (16.93) we use the properties (16.38), (16.39), (16.87) and after a Berezin integral computation we get (16.92). □

Let us consider a density operator $\hat{\rho}$ which means that $\hat{\rho}$ is a linear, non negative, hermitian operator in $\tilde{\mathcal{H}}^{(n)}$ such that $\mathrm{Tr}\rho = 1$;

For physical reasons (see [47]) we assume that $\hat{\rho}$ is even (it commutes with the parity operator $\mathbf{P}$).

We introduce the characteristic function χ_ρ defined as

$$\chi_\rho(\xi) := \mathrm{Tr}\left(\hat{\rho}\mathrm{e}^{\xi\cdot a^* - a\cdot\xi^*}\right). \tag{16.94}$$

We can easily see that from properties of trace and translation operators we also have

$$\chi_\rho(\xi) = \mathrm{Tr}(\hat{\rho}\hat{T}(\xi)) = \mathrm{Tr}(\hat{\rho}\hat{T}(-\xi)).$$

In particular χ_ρ is even. The normal and anti-normal characteristic functions are

$$\chi_{\rho,1}(\xi) = \mathrm{Tr}\left(\hat{\rho}\mathrm{e}^{\xi\cdot a^*}\mathrm{e}^{-a\cdot\xi^*}\right) \tag{16.95}$$
$$\chi_{\rho,-1}(\xi) = \mathrm{Tr}\left(\hat{\rho}\mathrm{e}^{-a\cdot\xi^*}\mathrm{e}^{\xi\cdot a^*}\right). \tag{16.96}$$

Using a trace computation with coherent states we have:

$$\chi_{\rho,-1}(\xi) = \int \mathrm{e}^{\xi\cdot\beta^* - \beta\cdot\xi^*}\langle\psi_\beta, \hat{\rho}\psi_{-\beta}\rangle d\xi^* d\xi. \tag{16.97}$$

So $\chi_{\rho,-1}(\xi)$ is the Fourier transform of the matrix element $\langle\psi_\beta, \hat{\rho}\psi_{-\beta}\rangle$.

Introduce now the "Fourier transform operator":

$$\hat{T}_s^{\mathscr{F}}(\gamma) = \int \mathrm{e}^{\xi\cdot\gamma^* - \gamma\cdot\xi^*}\hat{T}_s(\xi)d\xi^* d\xi \tag{16.98}$$
$$\hat{R}_s^{\mathscr{F}}(\gamma) = \int \mathrm{e}^{\gamma\cdot\xi^* - \xi\cdot\gamma^*}\hat{R}_s(\xi)d\xi^* d\xi. \tag{16.99}$$

In particular we can easily compute:

$$\hat{R}_{-1}^{\mathscr{F}}(\gamma) = \Pi_{\gamma,-\gamma}, \quad \hat{R}_1^{\mathscr{F}}(\gamma) = \Pi'_{\gamma,-\gamma} \tag{16.100}$$

where

$$\Pi'_{\gamma,\beta}(\psi) = \psi'_\gamma\langle\psi'_\beta, \psi\rangle.$$

From Parseval's relation we get the following representation formula.

$$\hat{H} = \int H^{w,-s}(\alpha)\hat{T}_s^{\mathscr{F}}(\alpha)d\alpha^* d\alpha \tag{16.101}$$
$$\hat{H} = \int H^{\mathscr{G},-s}(\alpha)\hat{R}_s(\alpha)^{\mathscr{F}} d\alpha^* d\alpha \tag{16.102}$$

where

$$H^{w,-s}(\alpha) = \int \mathrm{e}^{\xi\cdot\alpha^* - \alpha\cdot\xi^*}\mathrm{Tr}(\hat{H}\hat{R}_{-s}(\xi))d\xi^* d\xi \tag{16.103}$$
$$H^{\mathscr{G},-s}(\alpha) = \int \mathrm{e}^{\xi\cdot\alpha^* - \alpha\cdot\xi^*}\mathrm{Tr}(\hat{H}\hat{T}_{-s}(\xi))d\xi^* d\xi. \tag{16.104}$$

Recall that

$$\delta(\xi - \zeta) = \mathrm{Tr}(\hat{T}_s(\xi)\hat{R}_{-s}(-\zeta)).$$

So we get

$$\mathrm{Tr}(\hat{H}\hat{G}) = \int \mathrm{Tr}\left(\hat{H}\hat{R}_{-s}(\xi))\mathrm{Tr}(\hat{G}\hat{T}_{-s}(-\xi)\right) d\xi^* d\xi \tag{16.105}$$

for any operators H, G. Applying Parseval relation we get

$$\mathrm{Tr}(\hat{H}\hat{G}) = \int H^{R,-s}(\gamma)G^{w,s}(\gamma)d\gamma^* d\gamma. \tag{16.106}$$

Applying this to the density operator $\hat{\rho}$ we have

$$\hat{\rho} = \int \chi_{\rho,s}(\xi)\hat{R}_{-s}(-\xi)d\xi^* d\xi. \tag{16.107}$$

Now we introduce the Wigner–Weyl function or quasi-probability as the Fourier transform of the characteristic function

$$W_{\rho,s}(\gamma) = \int \chi_{\rho,s}(\xi)\mathrm{e}^{\gamma\cdot\xi^* - \xi\cdot\gamma^*} d\xi^* d\xi. \tag{16.108}$$

We see that $W_{\rho,s}(\gamma) = \rho^{\mathscr{G},s}(\gamma)$ hence

$$\hat{\rho} = \int W_{\rho,s}(\gamma)\hat{R}^{\mathscr{F},-s}(\gamma)d\gamma^* d\gamma \tag{16.109}$$

and for every observable $\hat{A}$,

$$\mathrm{Tr}(\hat{\rho}\hat{A}) = \int W_{\rho,s}(\gamma)A^{w,-s}d^2\gamma. \tag{16.110}$$

This is why $W_{\rho,s}(\gamma)$ is called a quasi-probability.

In particular for $s = 1$ (normal ordering) and $s = -1$ (anti-normal ordering) we have

$$\hat{\rho} = \int P_\rho(\gamma)\Pi_{\gamma,-\gamma}d\gamma^* d\gamma = \int Q_\rho(\gamma)\Pi'_{\gamma,-\gamma}d\gamma^* d\gamma \tag{16.111}$$

where $P_\rho(\gamma) = W_{\rho,1}(\gamma)$ and $Q_\rho(\gamma) = W_{\rho,-1}(\gamma)$. These functions are quasi-probabilities and are similar to anti-Wick symbols or contravariant Wick symbols.

A main application of above fermionic symbolic calculus is computation of mean values of observables. We know that every observable G on the super-Hilbert space $\tilde{\mathscr{H}}^{(n)}$ is a linear expression in $(a^*)^\varepsilon a^{\varepsilon'}$ $(\varepsilon, \varepsilon' \in \mathscr{E}[n])$.

Theorem 85 *For any $\varepsilon, \varepsilon' \in \mathscr{E}[n]$ we have*

$$\mathrm{Tr}\left(\hat{\rho}(a^*)^{\varepsilon}a^{\varepsilon'}\right) = \int P_{\rho}(\gamma)(\gamma^*)^{\varepsilon}\gamma^{\varepsilon'}d\gamma^* d\gamma \tag{16.112}$$

$$\mathrm{Tr}\left(\hat{\rho}a^{\varepsilon}(a^*)^{\varepsilon'}\right) = \int Q_{\rho}(\gamma)\gamma^{\varepsilon}(\gamma^*)^{\varepsilon'}d\gamma^* d\gamma. \tag{16.113}$$

Proof In the normal case we easily compute the matrix elements of $G = (a^*)^{\varepsilon}a^{\varepsilon'}$ on coherent states then we apply the trace formula with coherent state. The proof is analogous in the anti-normal case. □

Concerning the Weyl symbols ($s = 0$) we have the following useful result

Proposition 160 *Let $\hat{H}, \hat{G}$ be two operators in $\tilde{\mathscr{H}}_n$. Then we have*

$$\mathrm{Tr}(\hat{H}\hat{G}) = 2^{-n}\int H^w(\zeta)G_w(2\zeta)d^2\zeta, \quad \text{where } \zeta \in \mathscr{G}_n^c. \tag{16.114}$$

Proof From definition of covariant and contravariant Weyl symbol we have

$$\mathrm{Tr}(\hat{H}\hat{G}) = \int d^2\gamma d^2\zeta H^w(\gamma)G_w(\zeta)\mathrm{Tr}\left(\hat{T}(-\zeta)\hat{T}^{\mathscr{F}}(\gamma)\right).$$

A computation using coherent states gives

$$\mathrm{Tr}\left(\hat{T}(-\zeta)\hat{T}(\eta)\right) = \int d^2\theta\langle\hat{T}(\zeta)\psi_{\theta}, \hat{T}(\eta)\psi_{-\theta}\rangle = 2^n \mathrm{e}^{1/2(\zeta^*\cdot\eta-\eta^*\cdot\zeta)}. \tag{16.115}$$

Using the Fourier transform we get

$$\mathrm{Tr}(\hat{H}\hat{G}) = 2^n\int H^w\left(\frac{\zeta}{2}\right))G_W(\zeta)d^2\zeta.$$

and a change of fermionic variables gives (16.114). □

Notice the 1/2 in H^w, so there is a difference with the bosonic case concerning the trace of a product.

16.6.4 The Moyal Product for Fermions

We have seen before several phase space representations for operators acting in fermionic spaces. That means that if $\hat{G}$ is a quantum (fermionic) observable then $\hat{G}$ has an integral representation on the phase space, with some symbol G, for example $G = G_w$ or $G = G^w$ for Weyl quantizations.

As we have already seen for bosons, these representations give what it is called a symbolic calculus. In particular there exists a formula to compute the symbol of the product of two operators $\hat{G}\hat{H}$. That means that we have $\hat{G}\hat{H} = \widehat{G \circledast H}$, where $G \circledast H$ is a non commutative product of G and H (it can be interpreted as a twist convolution). Moyal [195] gave an explicit integral formula to compute $G \circledast H$ for bosons (see Sect. 12.1). Our aim here is to give an analogous formula for fermions. This kind of formula were discussed in many places, see for examples the book [23] and the papers [107, 150] and their bibliographies.

Let us start with two operators $\hat{G}, \hat{H}$ in the super Hilbert space $\mathscr{H}^n$, with their covariant and contravariant Weyl symbols G_w, G^w, H_w, H^w.

We have the following lemma

Lemma 106 *For every Grassmann variables* $\xi = (\xi_1, \ldots, \xi_n)$ *we have the following properties*

$$\begin{aligned}
\hat{G}\hat{T}(\xi) &= \widehat{G^{\xi}_{w}}, \quad \text{where} \quad G^{\xi}_{w}(\beta) = G_w(\beta + \xi)\mathrm{e}^{1/2(\beta^*\cdot\xi + \beta\cdot\xi^*)} \\
\hat{T}(\eta)\hat{G} &= \widehat{G_{w,\eta}}, \quad \text{where} \quad G_{w,\eta}(\beta) = G(\beta + \eta)\mathrm{e}^{-1/2(\beta^*\cdot\eta + \beta\cdot\eta^*)} \\
\hat{T}(-\xi)\hat{G}\hat{T}(\xi) &= \widehat{(\mu_\xi G)}, \quad \text{where} \quad (\mu_\xi G)_w(\beta) = \mathrm{e}^{\beta^*\cdot\xi + \beta\cdot\xi^*} G_w(\beta).
\end{aligned}$$

Proof We have

$$\begin{aligned}
\hat{G}\hat{T}(\xi) &= \int G_w(\alpha)\hat{T}(-\alpha)\hat{T}(\xi)d^2\alpha \\
&= \int G_w(\alpha)\hat{T}(\xi - \alpha)\mathrm{e}^{1/2(\alpha^*\cdot\xi + \alpha\cdot\xi^*)}d^2\alpha \\
&= \int G_w(\beta + \xi)\hat{T}(-\beta)\mathrm{e}^{1/2(\beta^*\cdot\xi + \beta\cdot\xi^*)}d^2\beta. \qquad (16.116)
\end{aligned}$$

We get the first formula. The two other formulas are easily deduced. □

Let us now give a formula for the Moyal product for covariant Weyl-symbols

Lemma 107 *Here* $G_w \circledast H_w$ *denotes the covariant Weyl symbol of* $\hat{G}\hat{H}$. *Then we have*

$$(G_w \circledast H_w)(\gamma) = \int G_w(\gamma - \xi)H_w(\xi)\mathrm{e}^{-1/2(\gamma\cdot\xi^* + \gamma^*\cdot\xi)}d^2\xi. \qquad (16.117)$$

Proof From Lemma 106 we have

$$\hat{G}\hat{T}(-\xi) = \int G_w(\alpha - \xi)\mathrm{e}^{-1/2(\alpha\cdot\xi^* + \alpha^*\cdot\xi)}\hat{T}(-\alpha)d^2\alpha \qquad (16.118)$$

We have

$$\hat{G}\hat{H} = \int \hat{G}\hat{T}(-\xi)H_w(\xi)d^2\xi.$$

So we get

$$\hat{G}\hat{H} = \iint G_w(\gamma - \xi)H_w(\xi)e^{-1/2(\gamma\cdot\xi^* + \gamma^*\cdot\xi)}\hat{T}(-\gamma)d^2\xi d^2\gamma. \tag{16.119}$$

The Lemma follows. □

Now we shall deduce the Moyal formula for contravariant Weyl symbols by applying a Fourier transform on the covariant Moyal formula.

We need some preliminaries. The left derivative ∂_θ was defined putting the fermionic variable θ on the left. It will be denoted $\overrightarrow{\partial_\theta}$ and its action on a function G is denoted $\overrightarrow{\partial_\theta}G$. We have also a right derivative defined by putting θ on the right and its action on a function ψ is denoted $G\overleftarrow{\partial_\theta}$. We have the following relationship between these derivatives

$$G\overleftarrow{\partial_\theta} = (-1)^{1+\pi(G)}\overrightarrow{\partial_\theta}G, \tag{16.120}$$

where $\pi(G)$ is the parity of G. In particular the signs are opposite if G is even.

We also need a fermionic analogue of Fourier multipliers. Let M be an even polynomial in Grassmann variables (ζ, ζ^*). It defines a Fourier multiplier by the usual formula

$$\hat{M}G(\xi) = \int G^{\mathcal{F}}(\zeta, \zeta^*)M(\zeta, \zeta^*)e^{\zeta\cdot\xi^* + \zeta^*\cdot\xi}d^2\zeta. \tag{16.121}$$

We have the following formula

$$\hat{M}G = M(\partial_{\xi^*}, \partial_\xi)G. \tag{16.122}$$

Now we can state the following result.

Proposition 161 *Let $\hat{G}, \hat{H}$ be two linear operators in the super-Hilbert space $\tilde{\mathcal{H}}^{(n)}$. Then the contravariant symbol $G^w \circledast H^w$ of $\hat{G}\hat{H}$ is given by the following Moyal formula*

$$(G^w \circledast H^w)(\alpha) = G^w(\alpha)e^{\frac{1}{2}\left(\overleftarrow{\partial_{\alpha^*}}\cdot\overrightarrow{\partial_\alpha} + \overleftarrow{\partial_\alpha}\cdot\overrightarrow{\partial_{\alpha^*}}\right)}H^w(\alpha). \tag{16.123}$$

To compute with formula (16.123) it is useful to note that $\overleftarrow{\partial_{\alpha_j^*}}\cdot\overrightarrow{\partial_{\alpha_j}} = -\overrightarrow{\partial_{\alpha_j}}\cdot\overleftarrow{\partial_{\alpha_j^*}}$

Proof Using (16.117) we have

$$\begin{aligned}(G^w \circledast H^w)(\alpha) &= \int G_w(\gamma - \xi)H_w(\xi)e^{-\frac{1}{2}(\gamma\cdot\xi^* + \gamma^*\cdot\xi)}e^{\alpha\cdot\gamma^* + \alpha^*\cdot\gamma}d^2\xi d^2\gamma \quad (16.124)\\ &= \int G_w(\beta)H_w(\xi)e^{-\frac{1}{2}(\beta\cdot\xi^* + \beta^*\cdot\xi)}e^{\xi\cdot\alpha^* + \xi^*\cdot\alpha + \beta\cdot\alpha^* + \beta^*\cdot\alpha}d^2\xi d^2\beta.\end{aligned}$$

Now we see that the right hand side of this formula can be seen as a Fourier multiplier in the variables $\zeta = (\beta, \xi)$ with the multiplier $M(\zeta, \zeta^*) = e^{-\frac{1}{2}(\beta\cdot\xi^* + \beta^*\cdot\xi)}$, using that $G_w(\beta)H_w(\xi) = (G^w \otimes H^w)^{\mathcal{F}}(\beta, \xi)$ (β and ξ are independent Grassmann variables and $\otimes$ is the associated tensor product). Hence we get (16.123) using properties (16.120)–(16.122). □

Remark 106 From the formula (16.123) we can compute the Weyl symbol of the commutator $[\hat{G}, \hat{H}]$. Expanding the exponential in a Taylor series we get that the first term for the symbol of $\frac{1}{i}[\hat{G}, \hat{H}]$ is the Poisson bracket $\{G, H\}$ as expected where

$$\{G, H\} = i(\partial_\alpha G \cdot \partial_{\alpha^*} H - \partial_\alpha H \cdot \partial_{\alpha^*} G)$$

if we assume G, H even.

16.7 Examples

16.7.1 The Fermi Oscillator

We shall see in the next chapter that the Fermionic analogue of the harmonic oscillator is $\hat{H}_{\mathrm{fos}} = \omega a^* a$, $\omega \in \mathbb{R}$. We can compute the Weyl symbol of $\hat{H}_{\mathrm{fos}}$ using the Moyal formula. We get

$$(a^* \circledast a)(\alpha) = \alpha^* (1 + \frac{1}{2} \overleftarrow{\partial_{\alpha^*}} \cdot \overrightarrow{\partial_\alpha}) \alpha = \alpha^* \alpha + \frac{1}{2}. \tag{16.125}$$

Let us compute the Weyl symbol of $\mathrm{e}^{za^* a}$ for $z \in \mathbb{C}$.

Using that $(a^* a)^2 = a^* a$ and a Taylor expansion we have

$$\mathrm{e}^{za^* a} = 1 + (\mathrm{e}^z - 1) a^* a.$$

The Weyl symbol of $a^* a$ is $\alpha^* \alpha + \frac{1}{2}$. So we get a formula analogous to the bosonic case

$$(\mathrm{e}^{za^* a})^w (\alpha) = \frac{\mathrm{e}^z + 1}{2} \mathrm{e}^{2 \tanh(z/2) \alpha^* \alpha}. \tag{16.126}$$

Remark that here the right hand side is everywhere smooth, even if $z = i\frac{\pi}{2}$, this is not true for the bosonic case.

16.7.2 The Fermi-Dirac Statistics

The most popular example is an ensemble of independent fermions in thermal equilibrium with a reservoir. This system is described by means of the grand canonical ensemble with the density operator

$$\hat{\rho} = Z(\beta, \mu)^{-1} \mathrm{e}^{-\beta(\hat{H} - \mu \hat{N})} \tag{16.127}$$

where

$$Z(\beta,\mu)=\operatorname{Tr}\left(e^{-\beta(\hat{H}-\mu\hat{N})}\right),\ \hat{H}=\sum_{1\le k\le n}\omega_k a_k^* a_k,\ \hat{N}=\sum_{1\le k\le n}a_k^* a_k.$$

To compute $Z(\beta,\mu)$ we first compute the action of $e^{-\beta(\hat{H}-\mu\hat{N})}$ on coherent states.

Introduce the following notations for $z_k\in\mathbb{C}$, $\gamma_k\in\Gamma_n^c$: $z_k=-\beta(\omega_k-\mu)$, $Z_k=e^{z_k}-1$, $z=(z_1,\dots,z_n)$, $\gamma(z)=(e^{z_1}\gamma_1,\dots,e^{z_n}\gamma_n)$, $\hat{H}(z)=\sum_{1\le k\le n}z_k a_k^* a_k$, we have

Lemma 108

$$e^{\hat{H}(z)}\psi_\gamma(\theta)=\psi_{\gamma(z)}(\theta)e^{\frac{1}{2}(\gamma(z)^*\cdot\gamma(z)-\gamma^*\cdot\gamma)}. \tag{16.128}$$

Proof It is a simple Grassmann computation.

First using that $(a_k^* a_k)^2=a_k^* a_k$ and Taylor expansion we have

$$e^{z_k a_k^* a_k}=1+(e^{z_k}-1)a_k^* a_k=1+Z_k a_k^* a_k.$$

Using that ψ_γ is an eigenvector for a_k:

$$\begin{aligned}
e^{\hat{H}(z)}\psi_\gamma(\theta) &= \prod_{k=1}^n(1+Z_k a_k^* a_k)\psi_\gamma(\theta)\\
&= \sum_{k_1<k_2<\cdots<k_m} Z_{k_1}\dots Z_{k_m}(\theta_{k_1}\gamma_{k_1})\dots(\theta_{k_m}\gamma_{k_m})\psi_\gamma(\theta)\\
&= e^{\theta\cdot\gamma(z)}e^{-\frac{1}{2}\gamma^*\cdot\gamma}\\
&= \psi_{\gamma(z)}(\theta)e^{\frac{1}{2}(\gamma(z)^*\cdot\gamma(z)-\gamma^*\cdot\gamma)}.
\end{aligned} \tag{16.129}$$

□

Remark 107 Equality (16.128) gives evolution by quadratic Hamiltonian of coherent states with $z_k=-it\omega_k$.

Then using the overlap formula for coherent states we get:

$$\langle\psi_{-\gamma},e^{\hat{H}(z)}\psi_\gamma\rangle=e^{\gamma(z)\cdot\gamma^*+\gamma\cdot\gamma^*}. \tag{16.130}$$

So we get for the trace

$$\operatorname{Tr}(e^{\hat{H}(z)})=\int\langle\psi_{-\gamma},e^{\hat{H}(z)}\psi_\gamma\rangle d\gamma^* d\gamma=\prod_{1\le k\le n}(1+e^{z_k}). \tag{16.131}$$

In particular we have

$$\boxed{Z(\beta,\mu)=\prod_{1\le k\le n}(1+e^{-\beta(\omega_k-\mu)})}. \tag{16.132}$$

An analogous computation gives the mean number $<\hat{N}>$ of Fermions:

$$<\hat{N}>=\mathrm{Tr}(\hat{\rho}\hat{N})=\sum_{1\le k\le n}\frac{1}{1+\mathrm{e}^{\beta(\omega_k-\mu)}}. \tag{16.133}$$

16.7.3 Quadratic Hamiltonians and Coherent States

A general quadratic self-adjoint fermionic Hamiltonian has the following form

$$\hat{H}=\frac{1}{2}(a^*\cdot Ma+a^*\cdot La^*-a\cdot\bar{L}a)-\frac{\mathrm{Tr}M}{4} \tag{16.134}$$

where $a=(a_1,\ldots,a_n)$, L, M are $n\times n$ matrices, M is hermitian, L is antisymmetric ($\bar{L}$ is the complex conjugate of L, so $\bar{L}=(L^*)^T$ where A^T is the transpose matrix of A). We assume for simplicity that L, M are time independent, extension to time dependent Hamiltonians is straightforward.

The Weyl symbol of $\hat{H}$ is

$$H(\alpha)=\frac{1}{2}(\alpha^*\cdot M\alpha+\alpha^*\cdot L\alpha^*-\alpha\cdot\bar{L}\alpha).$$

We also introduce the matrix of the quadratic form H:

$$\Xi=\begin{pmatrix}-\bar{L} & -\frac{M^T}{2}\\ \frac{M}{2} & L\end{pmatrix}.$$

Our goal here is to compare the quantum evolution of $\hat{H}$ and the pseudo-classical evolution for the Hamiltonian $H=H^w$.

The equations of motion are

$$\dot{\alpha}=\{\alpha,H\}=i\partial_{\alpha_j^*}H, \tag{16.135}$$

$$\dot{\alpha}^*=\{\alpha^*,H\}=i\partial_{\alpha_j}H. \tag{16.136}$$

We have to remark here that the Poisson bracket $\{,\}$ is defined according the usual quantization rule

$$\widehat{\{G,H\}}=\frac{1}{i}[\hat{G},\hat{H}].$$

To solve this problem we introduce another set of $2n$ Grassmann variables (η,η^*) so that $\alpha=\alpha_t=A_t\eta+B_t\eta^*$, $\alpha_t^*=\bar{B}_t\eta+\bar{A}_t\eta^*$ where A_t, B_t are complex $n\times n$ matrices depending on the time t.

Using that $\{\alpha,H\}=i(L\alpha^*+\frac{M}{2})$ and $\{\alpha^*,H\}=-i(\bar{L}\alpha+\frac{\bar{M}}{2})$ the equations of motion (16.135) are

$$\dot{A}_t = i\left(L\bar{B}_t + \frac{M}{2}A_t\right), \quad \dot{B}_t = i\left(L\bar{A}_t + \frac{M}{2}B_t\right). \tag{16.137}$$

Or, equivalently,

$$\frac{d}{dt}\begin{pmatrix} A_t & B_t \\ \bar{B}_t & \bar{A}_t \end{pmatrix} = i\begin{pmatrix} \frac{M}{2} & L \\ -\bar{L} & -\frac{\bar{M}}{2} \end{pmatrix} \cdot \begin{pmatrix} A_t & B_t \\ \bar{B}_t & \bar{A}_t \end{pmatrix}. \tag{16.138}$$

The matrix $\begin{pmatrix} \frac{M}{2} & L \\ -\bar{L} & -\frac{\bar{M}}{2} \end{pmatrix}$ is hermitian. So the "classical evolution" F_t is unitary (in fermionic pseudoclassical mechanics unitarity replaces symplecticity for evolution in the phase space).

The $2n \times 2n$ matrix $F_t = \begin{pmatrix} A_t & B_t \\ \bar{B}_t & \bar{A}_t \end{pmatrix}$ is a representation of the classical flow because we have

$$\frac{d}{dt}\begin{pmatrix} \alpha_t \\ \alpha_t^* \end{pmatrix} = i\begin{pmatrix} \frac{M}{2} & L \\ -\bar{L} & -\frac{\bar{M}}{2} \end{pmatrix} \cdot \begin{pmatrix} \alpha_t \\ \alpha_t^* \end{pmatrix}, \quad \text{with } \alpha_0 = \eta, \alpha_0^* = \eta^*.$$

Note that F_t has a natural action in the Grassmann algebra $\mathbb{C}[\eta, \eta^*]$. We shall see in the next chapter that $\mathbb{C}[\eta, \eta^*]$ can be seen as the space of smooth functions on the fermionic phase space so any element $g \in \mathbb{C}[\eta, \eta^*]$ is a classical observable in the fermionic sense. So the natural action of F_t in $\mathbb{C}[\eta, \eta^*]$ is $(F_t g)(\eta, \eta^*) = g(F_t(\eta, \eta^*))$.

But F_t has a natural action in $\mathbb{C}^n$: $\zeta \mapsto A_t\zeta + B_t\zeta^*$ and $\mathbb{C}^n$ is identified with $\mathbb{R}^{2n}$ by $\zeta = x + iy$, $(x, y) \in \mathbb{R}^n \times \mathbb{R}^n$. It is not difficult to prove that F_t is an isometry in $\mathbb{C}^n$ if and only if the matrix F_t is a unitary $2n \times 2n$ matrix. So F_t can be considered as a rotation in $\mathbb{R}^{2n}$ and belongs to the classical Lie group $SO(2n)$. Conversely for any $G \in SO(2n)$ there exists a flow F_t such that $F_1 = G$. This can be proved using a complex diagonal form for G.

Now we compute the quantum evolution for quadratic Hamiltonians as we did in the bosonic case. Let us first consider the quantum evolution for the annihilation/creation operators a and a^*:

$$\begin{pmatrix} a_t \\ a_t^* \end{pmatrix} = \mathrm{e}^{-it\hat{H}} \begin{pmatrix} a \\ a^* \end{pmatrix} \mathrm{e}^{it\hat{H}}. \tag{16.139}$$

Then we find

$$\begin{pmatrix} a_t \\ a_t^* \end{pmatrix} = F_t \begin{pmatrix} a \\ a^* \end{pmatrix}. \tag{16.140}$$

As an application, we can define a fermionic analogue for the metaplectic representation which is called the **spin representation**. For more details about the spin group we refer to the book [248], Chap. 5. We comment here about its fermionic interpretation. In [82] the authors compare bosonic and fermionic gaussian states.

To any $F \in SO(2n)$ we can associate a unitary (even) operator $\hat{S}(F)$ in the super-Hilbert space $\mathscr{H}^{(n)}$ such that

$$\hat{S}^*(F)\begin{pmatrix} a \\ a^* \end{pmatrix}\hat{S}(F) = F\begin{pmatrix} a \\ a^* \end{pmatrix}. \tag{16.141}$$

Formula (16.141) is known in physics literature as a Bogoliubov transformation and in mathematics as a Shale–Weil representation formula.

The map $F \mapsto \hat{S}(F)$ is a projective representation: if $F \in SO(2n)$ and if $\hat{S}_1(F)$ and $\hat{S}_2(F)$ are to operators satisfying (16.141) then there exists $\lambda \in \mathbb{C}$, $|\lambda| = 1$ such that $\hat{S}_1^*(F) = \lambda \hat{S}_2^*(F)$. This is true because the Fock representation is irreducible.

As in the bosonic case it is possible to determine the spin representation up to a sign and we have $\hat{S}(F^1F^2) = \pm\hat{S}(F^1)\hat{S}(F^2)$. This can be proved using propagation of coherent states which will now be studied.

We have seen (16.128) that the Fermi oscillator propagates a coherent state into a coherent state following the classical motion. We prove now that this is true for any quadratic Hamiltonian if we replace coherent states by squeezed coherent states as in the bosonic case.

Let us denote $\psi_{\gamma,t} = \mathrm{e}^{-it\hat{H}}\psi_\gamma$. It is enough to consider the case $\psi_\gamma = \psi_0 = 1$ because we have

$$\mathrm{e}^{-it\hat{H}}\psi_\gamma = \mathrm{e}^{-it\hat{H}}\hat{T}(\gamma)\mathrm{e}^{it\hat{H}}\mathrm{e}^{-it\hat{H}}\psi_0.$$

From the first part we have $\mathrm{e}^{-it\hat{H}}\hat{T}(\gamma)\mathrm{e}^{it\hat{H}} = \hat{T}(F_t\gamma)$. So it is enough to compute $\psi_{0,t} = \mathrm{e}^{-it\hat{H}}\psi_0$. Imitating the bosonic case we consider the following ansatz

$$\psi_{0,t}(\theta) = s(t)\mathrm{e}^{\frac{1}{2}\theta\cdot\Gamma_t\theta}$$

where Γ_t is an antisymmetric complex $n \times n$ matrix. We compute

$$\mathrm{e}^{-\frac{1}{2}\theta\cdot\Gamma_t\theta}\frac{d}{dt}\psi_{0,t} = \dot{s}(t) + s(t)\theta\cdot\frac{\dot{\Gamma}_t}{2}\theta \tag{16.142}$$

and

$$\mathrm{e}^{-\frac{1}{2}\theta\cdot\Gamma_t\theta}\hat{H}\psi_{0,t} = \frac{1}{2}\left[\theta\cdot(M\Gamma_t + L)\theta + \theta\cdot\Gamma_t\bar{L}\Gamma_t\theta - \mathrm{Tr}\bar{L}\Gamma_t - \frac{1}{2}\mathrm{Tr}M\right]s(t). \tag{16.143}$$

Identification between antisymmetric matrices gives the following Ricatti equation for Γ_t:

$$i\dot{\Gamma}_t = \frac{1}{2}(M\Gamma_t + \Gamma_t\bar{M}) + L + \Gamma_t\bar{L}\Gamma_t, \quad \Gamma_0 = 0, \tag{16.144}$$

and the Liouville equation for the prefactor s:

$$i\dot{s}(t) = -\frac{1}{2}\mathrm{Tr}\left(\bar{L}\Gamma_t + \frac{M}{2}\right)s(t),\ s(0) = 1. \tag{16.145}$$

If $\Gamma_t = V_t W_t^{-1}$ we see that Γ_t solves (16.144) if the matrix $\begin{pmatrix} V_t & \bar{W}_t \\ W_t & \bar{V}_t \end{pmatrix}$ solves

$$i\frac{d}{dt}\begin{pmatrix} V_t & \bar{W}_t \\ W_t & \bar{V}_t \end{pmatrix} = \begin{pmatrix} \frac{M}{2} & L \\ -\bar{L} & -\frac{\bar{M}}{2} \end{pmatrix} \cdot \begin{pmatrix} V_t & \bar{W}_t \\ W_t & \bar{V}_t \end{pmatrix}.$$

We have to solve this linear differential equation with the initial data $\begin{pmatrix} 0 & \mathbb{1} \\ \mathbb{1} & 0 \end{pmatrix}$. So we find

$$\begin{pmatrix} V_t & \bar{W}_t \\ W_t & \bar{V}_t \end{pmatrix} = F_t^* \begin{pmatrix} 0 & \mathbb{1} \\ \mathbb{1} & 0 \end{pmatrix} = \begin{pmatrix} B_t^T & A_t^* \\ A_t^T & B_t^* \end{pmatrix}.$$

So that we get $V_t = B_t^T$ and $W_t = A_t^T$ hence $\Gamma_t = B_t^T (A_t^T)^{-1}$ as far as A_t is invertible.

Concerning the prefactor s we get

$$\dot{s}(t) = \frac{i}{2}\mathrm{Tr}\left(\bar{L}\Gamma_t + \frac{M}{2}\right)s(t) = \frac{1}{2}\mathrm{Tr}(\dot{W}_t W_t)s(t) = \frac{1}{2}\mathrm{Tr}(\dot{A}_t A_t^{-1})s(t),$$

hence using the Liouville formula, we have

$$s(t) = \det(A_t)^{\frac{1}{2}}. \tag{16.146}$$

Finally we get the following result:

Proposition 162 *For any Grassmann generators $\alpha = (\alpha_1, \ldots, \alpha_n)$ the evolution of the coherent state ψ_α, $\psi_{\alpha,t} := \mathrm{e}^{-it\hat{H}}\psi_\alpha$, for the quadratic Hamiltonian $\hat{H}$ obeys the following formula for $0 \le t < T_c$, with $T_c \in]0, \infty]$,*

$$\boxed{\psi_{\alpha,t}(\theta) = \det(A_t)^{\frac{1}{2}}\mathrm{e}^{\alpha_t(\frac{\alpha_t^*}{2} - \theta)}\mathrm{e}^{(\theta - \alpha_t^*)\cdot\frac{\Gamma_t}{2}(\theta - \alpha_t^*)}} \tag{16.147}$$

where $T_c = \inf\{t > 0|\ \det A_t = 0\}$ and

$$\begin{pmatrix} \alpha_t \\ \alpha_t^* \end{pmatrix} = F_t\begin{pmatrix} \alpha \\ \alpha^* \end{pmatrix} = \begin{pmatrix} A_t & B_t \\ \bar{B}_t & \bar{A}_t \end{pmatrix}\begin{pmatrix} \alpha \\ \alpha^* \end{pmatrix}$$

is the pseudoclassical flow and $\Gamma_t = B_t^T (A_t^T)^{-1}$.

Moreover the formula (16.147) extends holomorphically for $t \in \mathbb{C}\backslash\mathscr{Z}$ where $\mathscr{Z} := \{t \in \mathbb{C},\ \det A_t = 0\}$ is a discrete subset of $\mathbb{C}$.

Proof $t \mapsto \psi_{\alpha,t}$ is clearly holomorphic in $\mathbb{C}$. So it is enough to prove the formula (16.147) for $0 < t < T_c$. We have already proved (16.147) when $\alpha = 0$ using that A_t is invertible for t small enough. Now we have

$$\psi_{\alpha,t} = \hat{T}(\alpha_t)\psi_{0,t}.$$

Using that

$$\hat{T}(\gamma) = \mathrm{e}^{\frac{\gamma\cdot\gamma^*}{2}}\mathrm{e}^{a^*\cdot\gamma}\mathrm{e}^{-\gamma^*\cdot a}$$

and $\mathrm{e}^{-\gamma^*\cdot a}f(\theta) = f(\theta-\gamma^*)$ we get (16.147) for any α. □

Remark 108 The solution $\psi_{\alpha,t}$ is everywhere smooth in time $t\in\mathbb{R}$ but its Gaussian shape breaks down when A_t stops to be invertible. We do not analyze here what happens in general when $\det(A_t)$ vanishes, even if it is an interesting question. An important difference with the bosonic case is the sign of the $\frac{1}{2}$-power of the determinant for the prefactor.

Let us consider the particular cases when $L=0$ or $M=0$.

If $L=0$ we find $A_t=\mathrm{e}^{itM/2}$ and $B_t=0$. Then $\Gamma_t=0$ at every time and the shape of the coherent state is constant.

If $M=0$, we have

$$F_t = \begin{pmatrix} \cos(itL) & \sin(itL) \\ -\sin(itL) & \cos(itL) \end{pmatrix}.$$

The spectrum of L is $\{i\nu_j\}_{1\le j\le n}$, $\nu_j\in\mathbb{R}$ hence

$$\det(A_t)=0 \iff t=\frac{(2k+1)\pi}{2\nu_j},\quad k\in\mathbb{Z}.$$

The shape of the coherent states breaks down at these times as we clearly see in the following example.

Assume that $n=2$ and $L=\begin{pmatrix} 0 & 1 \\ -1 & 0 \end{pmatrix} := J$. We find

$$A_t=\cos(itJ),\quad B_t=-\sin(itJ),\quad \Gamma_t=\tan(itJ),$$

where we have used $\cos(itJ)=\cos t(\mathbb{1})$ and $\sin(itJ)=iJ\sin t$. Hence we have

$$\psi_{0,t}(\theta_1,\theta_2)=\cos t\mathrm{e}^{-\frac{i}{2}\theta\cdot\tan(itJ)\theta} = \cos t(1+\tan(itJ)\theta_1\theta_2)$$

For $t=\frac{\pi}{2}$, $\psi_{0,t}(\theta_1,\theta_2)=i\theta_1\theta_2$ which is no more a Gaussian (a Gaussian function of Grassmann variables is always invertible). We can only say that it is a degenerate Gaussian.

In the next paragraph we want to compute the matrix elements $\langle\psi_\beta, U_t\psi_\alpha\rangle$ where $U_t=\mathrm{e}^{-it\hat{H}}$ and deduce from that a computation of the integral kernel of U_t (fermionic analogue of the Mehler formula) hence a formula for the Weyl symbol of U_t (fermionic analogue of the Mehlig–Wilkinson formula).

16.7.4 More on Quadratic Propagators

Proposition 163 *With the notations of the previous section, for every pair of Grassmann generators* $\alpha = (\alpha_1, \ldots, \alpha_n)$, $\beta = (\beta_1, \ldots, \beta_n)$, *we have*

$$\langle \psi_\beta, U_t \psi_\alpha \rangle = \det(A_t)^{\frac{1}{2}} \mathrm{e}^{\frac{1}{2}(\beta\cdot\beta^* + \alpha_t\cdot\alpha_t^*) + \beta^*\cdot\alpha_t} \mathrm{e}^{(\beta^* - \alpha_t^*)\cdot(\frac{\Gamma_t}{2})(\beta^* - \alpha_t^*)} \tag{16.148}$$

for $t \in \mathbb{C}\backslash\mathscr{Z}$.

Proof It is enough to prove (16.148) for $\alpha = 0$ using the formula

$$\langle \psi_\beta, U_t \psi_\alpha \rangle = \mathrm{e}^{\frac{1}{2}(\beta\cdot\alpha_t^* + \beta^*\cdot\alpha_t)} \langle \psi_{\beta-\alpha_t}, U_t \psi_0 \rangle.$$

Let us introduce the notations: $\eta_\gamma(\theta) = \mathrm{e}^{\theta\cdot\gamma}$ and

$$f_t(\gamma^*) = \langle \eta_\gamma, U_t \eta_0 \rangle.$$

Our strategy to compute f_t is to show that it satisfies a quadratic Schrödinger equation, noticing that $f_0(\gamma^*) = 1$.

We have $i\partial_t f_t = \langle \hat{H}\eta_\gamma, U_t \eta_0 \rangle$. An easy computation using $a^*\eta_\gamma(\theta) = \theta\eta_\gamma(\theta)$ and $a\eta_\gamma(\theta) = \gamma\eta_\gamma(\theta)$ gives that

$$\hat{H}\eta_\gamma(\theta) = \frac{1}{2}\left(\theta\cdot M\gamma + \theta\cdot L\theta - \gamma\cdot\bar{L}\gamma - \frac{\mathrm{Tr}M}{2}\right)\eta_\gamma(\theta). \tag{16.149}$$

Plugging this formula in $\langle \hat{H}\eta_\gamma, U_t\eta_0 \rangle$ and the formula

$$\left(\frac{\partial}{\partial\gamma_j} f\right)^* = (-1)^{1+\pi(f)} \frac{\partial}{\partial\gamma_j^*} f$$

we get

$$i\partial_t f_t(\gamma^*) = (\hat{H} f_t)(\gamma^*).$$

Using the previous section main result we get

$$f_t(\gamma^*) = \det(A_t)^{\frac{1}{2}} \mathrm{e}^{\gamma^*\cdot\frac{\Gamma_t}{2}\gamma^*}.$$

This proves (16.148) for $\alpha = 0$ hence for any α. □

Using formula (16.54) we get a kind of fermionic analogue of the Mehler formula for oscillators.

Corollary 56 *The integral kernel* $K_t(\theta, \gamma^*)$ *of the quadratic propagator* U_t *has the following expression, for* $t \in \mathbb{C}\backslash\mathscr{Z}$,

$$\boxed{K_t(\theta, \gamma^*) = \det(A_t)^{\frac{1}{2}} \mathrm{e}^{\theta\cdot\gamma_t^* + \frac{1}{2}(\gamma_t^*\cdot\gamma_t - \gamma^*\cdot\gamma)} \mathrm{e}^{(\theta-\gamma_t)\cdot\frac{\Gamma_t}{2}(\theta-\gamma_t)}}. \tag{16.150}$$

To conclude this section on quadratic propagators for fermions we now compute the contravariant Weyl symbols U_t^w of the quadratic propagator $\hat{U}_t = \mathrm{e}^{-it\hat{H}}$. All the computations can be extended to time dependent Hamiltonians $H(t)$.

Let us introduce the notations: $\mathrm{Sp}\Sigma$ the spectrum of the Hermitian matrix Σ of H and

$$\mathscr{Z}_H = \left\{ \frac{(2k+1)\pi}{\lambda},\ \lambda \in \mathrm{Sp}\Sigma,\ k \in \mathbb{Z} \right\}.$$

Theorem 86 *Let us consider a quadratic fermionic Hamiltonian like (16.134). Then the contravariant Weyl symbol U_t^w of $\mathrm{e}^{-it\hat{H}}$ is given by the formula*

$$\boxed{U_t^w(\xi) = \det\left(\frac{F_t + \mathbb{1}}{2}\right)^{\frac{1}{2}} \exp\left(\begin{pmatrix}\xi^*\\ \xi\end{pmatrix} \cdot (F_t - \mathbb{1})(F_t + \mathbb{1})^{-1}\begin{pmatrix}\xi\\ \xi^*\end{pmatrix}\right)}. \tag{16.151}$$

for t such that $|t| < T_1$ where T_1 is the first time when $\det(F_t + \mathbb{1}) = 0$.

More generally the formula (16.151) is satisfied for every $t \in \mathbb{C}\backslash\mathscr{Z}_H$

Proof First remark that $t \mapsto U_t^w(\xi)$ is holomorphic in $t \in \mathbb{C}$, so it is enough to prove formula (16.151) for $0 \le t < T_1$.

It could be possible to compute the Weyl symbol U_t^w from the kernel of $\hat{U}_t$ using Fourier computation of Gaussians but in this way it seems not easy to have a direct expression with the flow F_t. So we choose a more direct approach, as we did for bosons, making the following ansatz

$$U_t^w(\xi) = f(t)\exp\left(\frac{1}{2}\begin{pmatrix}\xi\\ \xi^*\end{pmatrix} \cdot \Lambda_t^w \begin{pmatrix}\xi\\ \xi^*\end{pmatrix}\right)$$

where $f(t)$ is a complex valued function and Λ_t^w is a $2n \times 2n$ antisymmetric matrix.

Applying the Moyal formula (16.123) to the equation

$$\partial_t U_t^w = H \circledast U_t^w$$

we get

$$\begin{aligned} i(\dot{f}(t) + f(t)\Phi_t^w(\xi))U_t^w(\xi) &= H(\xi)U_t^w(\xi) + \frac{1}{2}H\left(\overleftarrow{\partial_{\xi^*}} \cdot \overrightarrow{\partial_\xi} + \overleftarrow{\partial_\xi} \cdot \overrightarrow{\partial_{\xi^*}}\right)U_t^w \\ &+ \frac{1}{8}H\left(\overleftarrow{\partial_{\xi^*}} \cdot \overrightarrow{\partial_\xi} + \overleftarrow{\partial_\xi} \cdot \overrightarrow{\partial_{\xi^*}}\right)^2 U_t^w \end{aligned} \tag{16.152}$$

where Φ_t^w is the quadratic form associated with the matrix Λ_t^w.

$$\Phi_t^w(\xi) = \frac{1}{2}\xi \cdot E_t\xi + \xi^* \cdot K_t\xi^* + \xi^* \cdot G_t\xi \text{ and } \Lambda_t^w = \begin{pmatrix} E_t & -\frac{G^T}{2} \\ \frac{G}{2} & K_t \end{pmatrix}. \quad (16.153)$$

We also introduce the matrix of the quadratic form H:

$$\Xi = \begin{pmatrix} -\bar{L} & -\frac{M^T}{2} \\ \frac{M}{2} & L \end{pmatrix}.$$

the matrix $\mathscr{J} = \begin{pmatrix} 0 & \mathbb{1} \\ \mathbb{1} & 0 \end{pmatrix}$ and $\Sigma = \mathscr{J}\,\Xi$ the generator of the flow $F_t = e^{it\Sigma}$.

It will be useful to remark that $\mathscr{J}$ is the fermion analogue of the J matrix of the symplectic form and that the flow F_t leaves $\mathscr{J}$ invariant: $F_t^T\,\mathscr{J}\,F_t = \mathscr{J}$.

Now we can compute the right hand side of (16.152) applying carefully the fermionic computation rules we get, with $X = \begin{pmatrix} \xi \\ \xi^* \end{pmatrix}$,

$$H\left(\overleftarrow{\partial_{\xi^*}} \cdot \overrightarrow{\partial_\xi} + \overleftarrow{\partial_\xi} \cdot \overrightarrow{\partial_{\xi^*}}\right) U_t^w = -\mathscr{J}\,\Xi X \cdot \Lambda_t^w X U_t^w, \quad (16.154)$$

$$H\left(\overleftarrow{\partial_{\xi^*}} \cdot \overrightarrow{\partial_\xi} + \overleftarrow{\partial_\xi} \cdot \overrightarrow{\partial_{\xi^*}}\right)^2 U_t^w = \mathrm{Tr}(\mathscr{J}\,\Xi\,\mathscr{J}\,\Lambda_t^w) + \Lambda_t^w X \cdot \mathscr{J}\,\Xi\,\mathscr{J}\,\Lambda_t^w X. \quad (16.155)$$

Plugging this in (16.152) and identifying the quadratic and the constant parts in X we get for Λ_t^w a Riccati equation and for f a Liouville equation as it is expected.

$$i\dot{\Lambda}_t^w = \Xi - \frac{1}{2}(\Lambda\,\mathscr{J}\,\Xi - \Xi\,\mathscr{J}\,\Lambda_t^w) - \frac{1}{4}\Lambda_t^w\,\mathscr{J}\,\Xi\,\mathscr{J}\,\Lambda_t^w. \quad (16.156)$$

It is convenient here to introduce $\Theta_t = \mathscr{J}\,\Lambda_t^w$. So we get

$$i\dot{\Theta}_t = \frac{1}{4}(2 - \Theta_t)\Sigma(2 + \Theta_t), \quad \Theta_0 = 0. \quad (16.157)$$

Equation (16.157) is easily solved by a Cayley transform: $\Theta_t = 2(\mathscr{N}_t - \mathbb{1})(\mathscr{N}_t + \mathbb{1})^{-1}$. So we find $\mathscr{N}_t = F_t^{-1}$ and we get

$$\Lambda_t^w = 2\,\mathscr{J}(F_t - \mathbb{1})(F_t + \mathbb{1})^{-1}. \quad (16.158)$$

Using this result we find for f the Liouville equation

$$\dot{f} = \frac{1}{2}\mathrm{Tr}(\dot{F}_t(F_t + \mathbb{1})^{-1})f(t), \quad f(0) = 1, \quad (16.159)$$

and we get

$$f(t) = \det\left(\frac{F_t + \mathbb{1}}{2}\right)^{\frac{1}{2}}. \quad (16.160)$$

This achieves the proof of (16.151). □

From this result we can compute the covariant Weyl symbol $U_{t,w}$ using the Fourier Gauss formula in Proposition 149.

Corollary 57 *Assume that the quadratic form H is non-degenerate (so the matrices Ξ and Σ are non-degenerate).*

Let T_2 be the smallest time $t > 0$ such that $\det(F_t - \mathbb{1}) = 0$. *Then for* $0 < t < T_2$ *we have the formula*

$$U_{t,w}(\gamma) = f^w(t) \exp\left(\frac{1}{4}\begin{pmatrix}\gamma\\ \gamma^*\end{pmatrix} \cdot \mathscr{J}(F_t + \mathbb{1})(F_t - \mathbb{1})^{-1}\begin{pmatrix}\gamma\\ \gamma^*\end{pmatrix}\right) \tag{16.161}$$

where $f^w(t)^2 = \det(F_t - \mathbb{1})$.

The formula (16.161) is also true for $t \in \mathbb{C}$, $t \neq \frac{2k\pi}{\lambda}$, $k \in \mathbb{Z}$, $\lambda \in$ $\mathrm{Sp}\Sigma$.

Chapter 17
Supercoherent States—An Introduction

Abstract In previous Chapters we have considered coherent states systems for bosons and for fermions separately. Here we introduce super-spaces where it is possible to consider simultaneously bosons and fermions. Our aim is to give a short introduction to this deep and difficult subject by considering some elementary examples where coherent states and quantization are involved.

17.1 Introduction

We have seen before that a general setting for coherent states is related to unitary irreducible representations of Lie-groups. Then we have defined Fermionic coherent states associated with a "group of translations in fermions coordinates". Here we shall consider states of systems mixing bosons and fermions. This is more difficult but very essential to give a mathematical model of supersymmetry. Recall that supersymmetry is a mathematical theory built to describe transformations which exchange bosons and fermions. Up to now it is not known if such transformations exists in Nature.

Anyway it is useful to understand mixed systems with bosons and fermions from a classical and a quantum point of view as well and to compare them.

In brief super-spaces are spaces where it is possible to define classical supersymmetry.

Supercoherent states are a useful tool to build a bridge between the classical world and the quantum world as they do in the more familiar case for systems of a finite number of bosons. So we shall construct here a family of supercoherent states and we shall see that they have many similarities with the coherent states of the harmonic oscillator. In particular they can be obtained in several ways: by translations through a super Weyl–Heisenberg group, as eigenfunctions of a super annihilation operator or by a minimum uncertainty principle.

We do not give here the more general setting, our aim here is to give a first approach and compute some interesting examples, in particular concerning the Bose-Fermi oscillator or super harmonic oscillator.

D. Robert and M. Combescure, *Coherent States and Applications in Mathematical Physics*, Theoretical and Mathematical Physics,
https://doi.org/10.1007/978-3-030-70845-0_17

17.2 Quantum Supersymmetry

At the formal mathematical level quantum supersymmetry can be easily implemented with a unitary involution τ in an Hilbert space $\mathscr{H}$ (see [243, 263] for details). Denote $\mathscr{H}_B = \ker(\tau - \mathbb{1})$ and $\mathscr{H}_F = \ker(\tau + \mathbb{1})$. If $\Pi_\pm = \frac{1}{2}(\mathbb{1} \pm \tau)$ then $\Pi_\pm$ are orthogonal projectors and we have $\mathscr{H}_B = \Pi_+\mathscr{H}$, $\mathscr{H}_F = \Pi_-\mathscr{H}$.

Definition 49 Let $\hat{H}$ be a self-adjoint operator in $\mathscr{H}$ with a τ-invariant domain $D(\hat{H})$.

$\hat{H}$ is said even if $[\hat{H}, \tau] = 0$ ($\hat{H}$ and τ commute).

$\hat{H}$ is said odd if $[\hat{H}, \tau]_+ = 0$ ($\hat{H}$ and τ anti-commute).

Every Hamiltonian $\hat{H}$ in $\mathscr{H}$ have a matrix representation in the decomposition $\mathscr{H} = \mathscr{H}_B \oplus \mathscr{H}_F$,

$$\hat{H} = \begin{pmatrix} \hat{H}_{++} & \hat{H}_{+-} \\ \hat{H}_{-+} & \hat{H}_{--} \end{pmatrix}.$$

So $\hat{H}$ is odd if and only if $\hat{H}_{++} = \hat{H}_{--} = 0$ and $\hat{H}$ is even if and only if $\hat{H}_{-+} = \hat{H}_{+-} = 0$.

A supersymmetry is determined by its generators $\{\hat{Q}_k\}_{1\leq k\leq N}$ where the $\hat{Q}_k$ are self-adjoint and odd operators, so we have $[\hat{Q}_k, \tau]_+ = 0$, and $\hat{Q}_k$ send bosons on fermions and vice-versa.

The Hamiltonian $\hat{H}$ is said supersymmetric if $[\hat{H}, \hat{Q}_k] = 0$ for $1 \leq k \leq N$. But to have better algebraic properties it is assumed that $[\hat{Q}_k, \hat{Q}_j]_+ = 2H\delta_{j,k}, \leq j, k \leq N$. Here for simplicity suppose that $N = 2$. We write the Hamiltonian in the symmetric form

$$\hat{H} = \frac{1}{2}(\hat{Q}_1^2 + \hat{Q}_2^2)$$

or if we note $\hat{Q} = \frac{\hat{Q}_1 + i\hat{Q}_2}{\sqrt{2}}$, we have the complex form:

$$\hat{H} = \frac{1}{2}[\hat{Q}, \hat{Q}^*]_+, \quad \text{with the relations}$$

$$[\hat{H}, \hat{Q}] = [\hat{H}, \hat{Q}^*] = 0, \quad \hat{Q}^2 = (\hat{Q}^*)^2 = 0 \tag{17.1}$$

$\hat{Q}$ is usually called a super-charge for $\hat{H}$.

The main physical problem in supersymmetry is to check if it is broken or preserved. In [263], Witten said:

"The most important question about supersymmetry theory is the question whether there exists in the Hilbert space $\mathscr{H}$ a state $|\Omega\rangle$ which is annihilated by the supersymmetry operators Q_i, $Q_i|\Omega\rangle = 0$."

Assume the spectrum of $\hat{H}$ is purely discrete. So its energy levels are non negative real numbers E. If $E > 0$ the eigenstates for E are pair (ψ_B, ψ_F) where $\psi_F = \hat{Q}\psi_B$.

If $E = 0$ then $\hat{H}\psi = 0$ if and only if $\hat{Q}_1\psi = \hat{Q}_2\psi = 0$ then it is said that supersymmetry is preserved.

But if there exists $\psi \in D(\hat{Q}) \cap D(\hat{Q}^*)$ such that $\hat{Q}\psi = 0$ and $\hat{Q}^*\psi \neq 0$ then we say that supersymmetry is broken. Then the vacuum energy E satisfies $E > 0$.

Let us now consider the well known Witten example of a quantum mechanical system with supersymmetry.

It is a spin $\frac{1}{2}$ exchange model in one degree of freedom.

So $\mathscr{H} = L^2(\mathbb{R}, \mathbb{C}^2)$ with the unitary involution $\sigma_3 = \begin{pmatrix} 1 & 0 \\ 0 & -1 \end{pmatrix}$. The supercharge is

$$\hat{Q} = \frac{1}{i}\left(\frac{d}{dx} + V'(x)\right)\sigma_-, \quad \sigma_- = \begin{pmatrix} 0 & 0 \\ 1 & 0 \end{pmatrix}. \tag{17.2}$$

$$\hat{Q}^* = \frac{1}{i}\left(\frac{d}{dx} - V'(x)\right)\sigma_+, \quad \sigma_+ = \begin{pmatrix} 0 & 1 \\ 0 & 0 \end{pmatrix}. \tag{17.3}$$

An easy computation gives the following Hamiltonian

$$\hat{H} = \frac{1}{2}\left(-\frac{d^2}{dx^2} + V'(x)^2\right)\mathbb{1}_2 - \frac{\sigma_3}{2}V''(x). \tag{17.4}$$

Assume for simplicity that $V(x)$ is a polynomial of even degree: $V(x) = c_0x^{2k} + \cdots c_m x^{2k+m}$ with $c_0 > 0$. Then $\hat{H}$ is a self-adjoint operator with a discrete spectrum.

It is easy to solve equations $\hat{Q}\psi = 0$ and $\hat{Q}^*\psi = 0$. We find

$$\hat{Q}\psi = 0 \iff \psi(x) = \psi(x_0)\mathrm{e}^{V(x_0)-V(x)}. \tag{17.5}$$

$$\hat{Q}^*\psi = 0 \iff \psi(x) = \psi(x_0)\mathrm{e}^{V(x)-V(x_0)}. \tag{17.6}$$

So we see that $\hat{Q}\psi = 0$ has non zero L^2 solution and $\hat{Q}^*\psi = 0$ has no non zero L^2 solution, hence the supersymmetry is broken.

A supersymmetric harmonic oscillator is obtained with $V(x) = \frac{\omega x^2}{2}$, $\omega > 0$. So we get the Hamiltonian

$$\hat{H}_{sos} = \frac{1}{2}\left(-\frac{d^2}{dx^2} + \omega^2 x^2\right)\mathbb{1}_2 - \frac{\omega}{2}\sigma_3$$

It is easy to get the spectral decomposition for $\hat{H}_{sos}$ using the Hermite basis $\{\psi_k\}_{k\in\mathbb{N}}$ (see Chap. 1). Denote $\psi_k(1/2, x) = \begin{pmatrix} \psi_k(x) \\ 0 \end{pmatrix}$ and $\psi_k(-1/2, x) = \begin{pmatrix} 0 \\ \psi_{k-1}(x) \end{pmatrix}$, with $\psi_{-1} = 0$.

For $k \geq 1$, $\{\psi_k(1/2, \cdot), \psi_k(-1/2, \cdot)\}$ is a basis of eigenvectors for the eigenvalue k of multiplicity two and 0 is a non degenerate eigenvalue with eigenvector $\begin{pmatrix} \psi_0(x) \\ 0 \end{pmatrix}$.

Let us remark that besides the position x we have here another degree of freedom $s = 1/2, -1/2$ which is discrete and represents the spin of a fermion.

We can rewrite the supersymmetric harmonic oscillator with creation and annihilation operators for bosons and fermions.

$$a_B = \frac{1}{\sqrt{2\omega_B}}(\omega_B q + \hbar \tfrac{d}{dq}), \quad a_F = \frac{1}{\sqrt{2\omega_F}}\sigma_+ \tag{17.7}$$

$$a_B^* = \frac{1}{\sqrt{2\omega_B}}(\omega_B q - \hbar \tfrac{d}{dq}), \quad a_F^* = \frac{1}{\sqrt{2\omega_F}}\sigma_-. \tag{17.8}$$

We get the super harmonic oscillator,

$$\hat{H} = \omega_B[a_B^*, a_B]_+ + \omega_F[a_F^*, a_F] \tag{17.9}$$

where (a_B^*, a_B) satisfies (CCR) or Weyl–Heisenberg algebra relations and (a_F^*, a_F) satisfies (CAR) or Clifford algebra relations. We shall see that $\hat{H}$ is supersymmetric if $\omega_B = \omega_F$.

It is possible to consider systems with n bosons generators and m fermions generators as well, writing $a_B = (a_{B,1}, \ldots, a_{B,n})$ and $a_F = (a_{F,1}, \ldots, a_{F,m})$ satisfying (CCR) relations for a_B, (CAR) relations for a_F.

The goal of a supersymmetry theory is to put bosons and fermions on the same footing. For example we would like to consider the parameter s for the spin as a classical variable on the same footing as x for the position. In other words the question is to find a kind of classical analogue for the spin. The question seems strange because the spin is purely quantal and disappears in the usual semi-classical limit.

Nevertheless Berezin has invented new spaces where this is possible after quantization, he called them superspaces with different super structures (linear spaces, algebras, groups and manifolds (see [24])). We shall give here an introduction to this wide subject, explaining only enough details to understand some semi-classical properties of supercoherent states. Concerning a more complete presentation we refer to the wealth literature [24, 79, 175, 248].

17.3 Classical Superspaces

The problem considered here is to find a classical analog for the algebra defined in (17.1). In other words we want to construct classical spaces mixing bosons and fermions. We already know classical spaces for bosons: real or complex vector spaces or manifolds and classical spaces for fermions: Grassmann algebras $\mathbb{K}[\theta_1, \ldots, \theta_n]$, $\mathbb{K} = \mathbb{R}$ or $\mathbb{C}$.

It is not obvious to understand in geometrical terms what is a classical space for fermions. We shall come back to this point later. Remark first here that it is not the algebra $\mathbb{K}[\theta]$ but instead a space mysterious B such that the space of "smooth functions" on B with values in the Grassmann algebra $\mathbb{K}$ is $\mathbb{K}[\theta]$. In particular B is a set of dimension 0. This was more or less implicit in the previous sections.

17.3.1 Morphisms and Spaces

The configuration space of a classical system with n bosons and m fermions will be a symbolic space denoted $\mathbb{R}^{n|m}$ such that

$$C^\infty(\mathbb{R}^{n|m}) := C^\infty(\mathbb{R}^n)[\theta_1, \ldots, \theta_n] := C^\infty(\mathbb{R}^n) \otimes \mathcal{G}_m$$

where $\mathcal{G}_m$ is the Grassmann algebra with m generators. In other words $C^\infty(\mathbb{R}^{n|m})$ is a $C^\infty(\mathbb{R}^n)$ module with basis $\{\theta_1, \ldots, \theta_n\}$.

(x, θ) is interpreted as a coordinates system of a point in the symbolic space $\mathbb{R}^{n|m}$; $x = (x_1, \ldots, x_n)$ are the even coordinates and $\theta = (\theta_1, \ldots, \theta_m)$ are the odd coordinates. As for Grassmann algebras, $C^\infty(\mathbb{R}^{n|m})$ is a super-module, with a parity operator $\mathbf{P}$, $C^\infty(\mathbb{R}^n)$-linear. So

$$C^\infty(\mathbb{R}^{n|m}) = [C^\infty(\mathbb{R}^{n|m})]_+ \oplus [C^\infty(\mathbb{R}^{n|m})]_-$$

where $[\bullet]_+$ is the even part and $[\bullet]_-$ is the odd part.

For applications we need to understand transformation map between different superspaces $\mathbb{R}^{n|m}$, $\mathbb{R}^{r|s}$.

A first approach to define a map $\Phi : \mathbb{R}^{n|m} \to \mathbb{R}^{r|s}$ is to use coordinates system: $\Phi(y, \eta) = (x, \theta)$ where x, y are even coordinates, θ, η are odd coordinates. But the symbolic spaces are defined implicitly by the C^∞ functions defined on them. So we interpret Φ as a change of variables in functions $f \in \mathscr{C}^\infty(\mathbb{R}^{r|s})$,

$$\Phi^\star(f)(x, \theta) = f(\Phi(x, \theta)) \tag{17.10}$$

where $f \mapsto \Phi^\star(f)$ is a linear map. It is important to underline here that the meaning of the right hand side in (17.10) is given by the left hand side because superspaces are only symbolic and are defined by their algebra of functions and not by a well identified geometrical object in the usual sense.

To define supermanifolds it is necessary to define local superspaces in an open set U of $\mathbb{R}^n$. Then $U^{n|m}$ is the superspace defined by its algebra of C^∞ functions : $C^\infty(U^{n|m}) := C^\infty(U) \otimes \mathcal{G}_m$. So we have a natural definition for morphisms from $U^{n|m}$ in $V^{s|r}$ where V is an open set of $\mathbb{R}^s$. Let us give now some more formal definitions following [24, 175, 248].

17.3.2 Super Algebra Notions

Definition 50 (*super linear spaces*) A superlinear space V is a vector space on $\mathbb{K} = \mathbb{R}$ or $\mathbb{C}$ with a decomposition ($\mathbb{Z}_2$-grading) $V = V_0 \oplus V_1$. V_0 is the set of even vector, V_1 is the set of odd vectors. Then on V it exists a parity (linear) operator $\mathbf{P}$, defined by $\mathbf{Pv} = \mathbf{v}$ for $v \in V_0$, $\mathbf{Pv} = -\mathbf{v}$ for $v \in V_1$. Elements in $V_0 \cup V_1$ are called homogeneous.

A linear map F from the superlinear space V into the superlinear space W is said superlinear, or even, if it preserves parities, so F sends V_j in W_j, $j = 0, 1$. A linear map from V into W is said odd if $F(V_j) \subseteq W_{j+1}$ ($j \in \mathbb{Z}_2$).

The parity number $\pi(v)$ is defined for homogeneous elements: $\pi(v) = 0$ if $v \in V_0$; $\pi(v) = 1$ if $v \in V_1$

Let us remark that every linear map from V into W is the sum of an even map and an odd map. The following notations will be used: $\mathrm{SHom}(V, W)$ is the linear space of even maps V into W and $\mathrm{Hom}(V, W)$ is the space of all linear maps from V into W. Then $\mathrm{Hom}(V, W)$ is again a super linear space and it even part is the space of even maps from V into W.

$$(\mathrm{Hom}(V, W))_0 = \mathrm{SHom}(V, W).$$

Definition 51 (*superalgebras*) A super algebra is a linear space $\mathscr{A}$ with an associative multiplication with unit 1 $(fg) \mapsto f \cdot g$, such that for every $f, g \in \mathscr{A}_+ \cup \mathscr{A}_-$ we have the parity condition:

$$\pi(f \cdot g) = \pi(f) + \pi(g)$$

The superalgebra $\mathscr{A}$ is said (super)commutative if

$$f \cdot g = (-1)^{\pi(f)\pi(g)} g \cdot f$$

Examples: Grassmann algebras $\mathbb{K}[\theta_1, \ldots, \theta_n]$ are super commutative algebras.

Definition 52 (*superspace*) If B is a superspace (for example defined by a Grassmann algebra) the set of B-points of the superspace $\mathbb{R}^{n|m}$ is the set $\mathbb{R}^{n|m}(B)$ defined by the following equality.

$$\mathbb{R}^{n|m}(B) = \mathrm{Hom}(C^\infty(\mathbb{R}^{n|m}), C^\infty(B)).$$

The interpretation is that $\mathbb{R}^{n|m}(B) = \mathrm{Hom}(B, \mathbb{R}^{n|m})$ so $\mathbb{R}^{n|m}(B)$ can be seen as the "points of $\mathbb{R}^{n|m}$" parametrized by B. An other important trick for a correct interpretation of superspace concerns the product of superspaces. We should like to have for example $\mathbb{R}^{n|m} = \mathbb{R}^{n|0} \times \mathbb{R}^{0|m}$ (of course $\mathbb{R}^{n|0}$ is the usual space $\mathbb{R}^n$ and $\mathbb{R}^{0|m}$ is the space whose C^∞-functions are $\mathbb{R}[\theta_1, \ldots, \theta_m]$).

Let us consider 3 classical super spaces X, Y_1, Y_2. We can understand identification between $\mathrm{Hom}(X, Y_1 \times Y_2)$ and $\mathrm{Hom}(X, Y_1) \times \mathrm{Hom}(X, Y_2)$ through identification between $\mathrm{Hom}(C^\infty(Y_1 \times Y_2), C^\infty(X))$ and $\mathrm{Hom}(C^\infty(Y_1), C^\infty(X)) \times \mathrm{Hom}(C^\infty(Y_2), C^\infty(X))$. This is done as follows.

If Φ_k^* is a morphism from $C^\infty(Y_k)$ into $C^\infty(X)$ then we get a unique morphism Π^* from $C^\infty(Y_1 \times Y_2)$ into $C^\infty(X)$ satisfying

$$\Phi^*(f_1 \otimes f_2) = \Phi_1^*(f_1)\Phi_2^*(f_2), \text{ for } f_k \in \mathbb{C}^\infty(Y_k). \tag{17.11}$$

Conversely if Φ^* is given we get $\Phi_1^*(f_1) = \Phi^*(f_1 \otimes 1)$ and $\Phi_2^*(f_2) = \Phi^*(1 \otimes f_2)$.
So the identification is determined by (17.11).

17.3.3 Examples of Morphisms

We compute here morphism between some super spaces.

(1) Φ is a morphism from $\mathbb{R}^n$ in $\mathbb{R}^r$. It is known that we have $\Phi^\star(f)(x) = f(\Phi(x))$ with Φ smooth function $\mathbb{R}^n \to \mathbb{R}^r$.
(2) Φ is a morphism from $\mathbb{R}^{0|1}$ into $\mathbb{R}$. Using parity we have $\Phi^\star(f) = b_0(f)$ where $b_0(f)$ is real, linear and multiplicative in f. So there exists $x_0 \in \mathbb{R}$ such that $\Phi^\star(f) = f(x_0)$, which means that Φ is constant, as it must be.
(3) Φ is a morphism from $\mathbb{R}^{1|1}$ into $\mathbb{R}$. Using parity, we have $\Phi^\star(f) = c_0(f)$ where c_0 is a morphism from $C^\infty(\mathbb{R})$ in $C^\infty(\mathbb{R})$. Hence there exists a smooth function $\Phi : \mathbb{R} \to \mathbb{R}$ such that $\Phi^\star(f) = f \circ \Phi$
(4) Φ is a morphism from $\mathbb{R}^{1|1}$ into $\mathbb{R}^{0|1}$. Using parity, we have: $\Phi^\star(1) = 1$ and $\Phi^\star(\theta) = c_1\theta$. (1 is the constant function 1). So we have $\Phi^\star(b_0 + b_1\theta) = b_0 + b_1c_1\theta$.
From (3) and (4) we clearly see that (x, θ) are coordinates for $\mathbb{R}^{1|1}$.
(5) Φ is a morphism from $\mathbb{R}^{1|2}$ into $\mathbb{R}^1$. We have

$$\Phi^\star(f) = b_0(f) + b_1(f)\theta_1\theta_2.$$

The super-algebra even morphism conditions give

$$b_0(fg) = b_0(f)b_0(g), \quad f, g \in C^\infty(\mathbb{R}), \quad b_0(1) = 1. \tag{17.12}$$

$$b_1(fg) = b_0(f)b_1(g) + b_1(f)b_0(g). \tag{17.13}$$

From the first condition, there exists $\varphi : \mathbb{R} \to \mathbb{R}$ such that $b_0(f) = f \circ \varphi$. Then the second condition means that b_1 is a derivation at $\varphi(x)$. So there exists $\chi(x, t)$, smooth in a neighborhood of the graph of φ such that:

$$b_1(f)(x) = \chi(x, t)\frac{df}{dt}|_{t=\varphi(x)}. \tag{17.14}$$

17.4 Super-Lie Algebras and Groups

Standard symmetries (at the classical or quantum level as well) are described through group action; Supersymmetries will be described through action of super-groups. Moreover these super-groups are derived by exponentiating super Lie algebras as in the usual case of Lie groups.

17.4.1 Super-Lie Algebras

Definition 53 A super Lie algebra $\mathfrak{s}$ is a super linear space on $\mathbb{K} = \mathbb{R}, \mathbb{C}$ with a bilinear bracket $(A, B) \mapsto [A, B]$ satisfying the following identities for homogeneous elements A, B, C,

$$[A, B] = (-1)^{1+\pi(A)\pi(B)}[B, A] \tag{17.15}$$

$$\pi([A, B]) = \pi(A) + \pi(B) \tag{17.16}$$

$$(-1)^{\pi(A)\pi(C)}[A, [B, C]] + (-1)^{\pi(B)\pi(A)}[B, [C, A]] + (-1)^{\pi(C)\pi(B)}[C, [A, B]] = 0 \tag{17.17}$$

Remark 109 The first equality means that $[A, B]$ is anticommutative excepted if A and B are odd. In this case the bracket is commutative.

The second identity means that the bracket of two even or two odd elements is even and the bracket of an even and odd element is odd.

The third identity is the super-Jacobi identity.

As in the standard case of Lie algebras, many examples of super-Lie-algebra are realized as matrix super-algebras. Let us consider a super-vector space on the field $\mathbb{K} = \mathbb{R}, \mathbb{C}$, $V = V_0 \oplus V_1$ where the even space V_0 has dimension n and the odd space V_1 has dimension m. Any superlinear operator A in V has a matrix representation

$$A = \begin{pmatrix} A_{++} & A_{+-} \\ A_{-+} & A_{--} \end{pmatrix}$$

Denote $\mathrm{End}(V)$ the space of all super linear operators in V. We have seen that $\mathrm{End}(V)$ is a super vector space with parity π. It is a super Lie algebra for the super-bracket:

$$[A, B] = AB - (-1)^{\pi(A)\pi(B)}BA$$

Choosing basis in V_0 and V_1 the super space V is isomorphic to $\mathbb{K}^n \oplus K^m$. This space is denoted $\mathbb{K}^{n,m}$ (not the same meaning as $\mathbb{K}^{n|m}$) and the super-Lie algebra is denoted $\mathfrak{gl}(n|m)$.

We denote $\mathfrak{sl}(n, m)$ the sub-super-Lie-algebra of $\mathfrak{gl}(n|m)$ defined by the super-traceless condition: $\mathrm{Str}A = 0$.

$\mathfrak{su}(n, m)$ is the sub-super-Lie-algebra of matrix A such that:

$$A_{++} = -A^*_{++}, \ A_{-+} = -iA^*_{+-}.$$

For $m = 0$ we recover the classical Lie algebras $\mathfrak{gl}(n)$, $\mathfrak{sl}(n)$, $\mathfrak{su}(n)$.

Definition 54 Let $\mathscr{A}$ be a superalgebra (not necessarily associative). A linear operator D in $\mathscr{A}$ is called a super-derivation if it satisfies, for every homogeneous elements f, g in End$(\mathscr{A})$,

$$D(f \cdot g) = D(f) \cdot g + (-1)^{\pi(D)\pi(f)} f \cdot D(g). \tag{17.18}$$

The basic example is $D = \mathrm{ad}_A$ where $\mathscr{A}$ is a super-Lie algebra and $\mathrm{ad}_A(B) = [A, B]$ (like in standard Lie algebras).

In the super-algebra $C^\infty(\mathbb{R}^{n|m})$, $X_j = \frac{\partial}{\partial_{x_j}}$ and $\Theta_k = \frac{\partial}{\partial_{\theta_k}}$ are superderivations (the first is even, the second is odd). And if $h \in C^\infty(\mathbb{R}^{n|m})$, hX_j and $h\Theta_k$ are again superderivations.

17.4.2 Supermanifolds, A Very Brief Presentation

To define supermanifolds it is necessary to define local superspaces in an open set U of $\mathbb{R}^n$. Then we denote $U^{n|m}$ the superspace defined by its C^∞ functions : $C^\infty(U^{n|m}) := C^\infty(U) \otimes \mathscr{G}_m$. So we have a natural definition for morphisms from $U^{n|m}$ in $V^{s|r}$ where V is an open set of $\mathbb{R}^s$. $U^{n|m}$ is a superspace of dimension $n|m$.

As for a standard manifold, a supermanifold M of dimension $n|m$ is a topological space M_0 (called the underlying space) such that for each point $m \in M_0$ we have an open neighborhood U and a superalgebra $\mathscr{R}_U$ isomorphic to $C^\infty(V^{n|m})$, where V is an open set of $\mathbb{R}^n$. Moreover a gluing condition has to be satisfied (see [248], p. 135). In particular M_0 is a standard manifold of dimension n.

We have discussed here the super analogue of real C^∞ manifolds. It is possible to define real analytic or holomorphic supermanifolds. For that we have to replace $C^\infty(U)$ by the space $C^\omega(U)$ of analytic functions in U. In the complex case U is an open set of $\mathbb{C}^n$ (recall that complex analytic = holomorphic).

Nowadays there exist essentially two kinds of (almost) equivalent definitions of super manifolds. One defines supermanifold by its morphisms [26, 175, 248], it is called the algebro-geometrical approach; the other is closer to the standard definition where a manifold is a set of points but the field of real numbers is replaced by a non commutative and infinite dimensional Banach algebra [76, 226]. Here we shall use the terminology of the algebro-geometrical approach. See [226] for a discussion about comparison of these two definitions.

We only give here a very brief and sketchy introduction to this rich subject. Our goal is only to have a better intuition of what is going on some few examples, in particular concerning the Fermi-Bose (or super) oscillator.

Definition 55 Roughly speaking, a $\mathscr{C}^\infty$-super manifold of dimension $n|m$ is a superspace M defined by an underlying manifold of dimension n such that M is locally isomorph to $U^{n|m}$, where U is a chart (open set in $\mathbb{R}^n$) of the underlying manifold M_0. There are gluing compatibility conditions between the charts as for manifolds where $\mathscr{C}^\infty(U^{n|m})$ replaces $\mathscr{C}^\infty(U)$.

If U is an open set of $\mathbb{R}^n$, the supermanifold $U^{n|m}$ is defined by the underlying space U and with $\mathscr{C}^\infty(U^{n|m}) = \mathscr{C}^\infty(U)[\theta_1, \ldots, \theta_m]$, $\theta_1, \ldots, \theta_m$ are generators of a real Grassmann algebra.

Definition 56 A vector field in the super domain $U^{n|m}$ is a derivation in the super algebra $\mathscr{C}^\infty(U^{n|m})$

As in the standard case the set $\mathscr{V}(U^{n|m})$ of vector fields in $U^{n|m}$ is a super Lie algebra. Note that it is a module over $\mathscr{C}^\infty(U^{n|m})$ with the following basis in coordinates (x, θ):

$$\left\{\frac{\partial}{\partial x_1} \cdots \frac{\partial}{\partial x_n}; \frac{\partial}{\partial \theta_1}, \ldots, \frac{\partial}{\partial \theta_m}\right\}.$$

In a supermanifold M the tangent superspace at "$P \in M$" of coordinates (x, θ) is defined locally in a chart $U^{n|m}$ as the space $\mathscr{V}(U^{n|m})$ restricted at (x, θ) (see [248] for a more precise definition).

17.4.3 Super-Lie Groups

In short, a super-Lie group is a super-manifold and a super group where the group operations are morphisms. It is not the place here to define rigorously all these terms; we refer to [76, 175, 226, 248] for a detailed study of super-manifolds and super groups.

A super Lie-group is defined as a super manifold where the group structure is defined by morphisms

$$\mu : G \times G \to G, \ \iota = G \to G, \quad \nu_0 : \mathbb{R}^{0|0} \to G.$$

These morphisms define respectively the multiplication, the inverse element and the unit element. Of course we need conditions to translate the group properties on morphisms (see [248] for details).

We have defined before the superspace $\mathbb{R}^{n|m}$. It is a commutative Lie group for the "natural" addition with the following morphisms.

$$(x, \theta) + (y, \zeta) = (x + y, \theta + \zeta). \tag{17.19}$$

Addition rule has to be defined in terms of morphisms as follows.

Recall that the space $\mathbb{R}^{n|m}$ is defined by its morphisms and for every superspace B, $\mathbb{R}^{n|m}(B)$ is the set of B points of $\mathbb{R}^{n|m}$, $\mathbb{R}^{n|m}(B) = \mathrm{Hom}(C^\infty(\mathbb{R}^{n|m}), C^\infty(B))$. Addition in $\mathbb{R}^{n|m}(B)$ is defined as follows. It is first remarked that a morphism $\psi^\star \in \mathbb{R}^{n|m}(B)$ is determined by its images on the following generators of $\mathbb{R}^{n|m}$

$$f_j(x,\theta) = x_j, \quad g_k(x,\theta) = \theta_k, \quad 1 \le j \le n, \quad 1 \le k \le m$$

$\psi^\star(f_j)$ and $\psi^\star(g_k)$ determine $\psi^\star$. Hence the additive group $\{\mathbb{R}^{n|m}, +\}$ is defined by the rule, $\psi_1^\star, \psi_2^\star \in \mathbb{R}^{n|m}(B)$,

$$(\psi_1^\star + \psi_2^\star)(f) = \psi_1^\star(f) + \psi_2^\star(f), \text{ for } f = f_j,\ g_k,$$

for any $\psi_1^\star, \psi_2^\star \in \mathbb{R}^{n|m}(B)$ and any superspace B.

For simplicity, assume $n = m = 1$. Practically if $f \in C^\infty(\mathbb{R}^{1|1})$ we have $f(x,\theta) = f_0(x) + f_1(x)\theta$ and the meaning of (17.19) is

$$u^\star(f)(t,s,\theta,\zeta) = f_0(t+s) + f_1(t+s)\theta + f_1(t+s)\zeta.$$

It is easy to write explicitly $\iota^\star$ and $\nu_0^\star$.

As in the standard case, it is very important to have connections between super Lie algebras and super groups. Without going into details of the theory of super Lie groups, (see [26, 248]), let us recall here a definition (up to super-isomorphism) of the super Lie algebra of a super group SG of dimension $n|m$.

Definition 57 The super Lie algebra $\mathfrak{sg}$ of the super group SG is the super linear space of left invariant derivations in the algebra $\mathscr{C}^\infty(U^{n|m})$ where U is a neighborhood 1 of the underlying Lie group G_0.

The easiest example is the super-linear group GL(n, m) corresponding to the super Lie algebra $\mathfrak{gl}(n,m)$. The details can be found in [26].

Elements g of GL(n, m) are superlinear isomorphisms in $\mathbb{R}^{n|m}$. They are parametrized by matrices

$$A(g) = \begin{pmatrix} A_{++} & A_{+-} \\ A_{-+} & A_{--} \end{pmatrix}$$

where the matrices A_{++}, A_{--} have even elements and A_{+-}, A_{-+} have odd elements. The real component of $A(g)$ is

$$A_0(g) = \begin{pmatrix} A_{++} & 0 \\ 0 & A_{--} \end{pmatrix}$$

$A_0(g)$ is invertible if and only if $\det(A_{++})\det(A_{--}) \neq 0$. Denote $G_0 = \mathrm{GL}(n) \times \mathrm{GL}(m)$. G_0 is a standard manifold of dimension $n^2 + m^2$. So we see that the supermanifold GL(n, m) is defined by the sheaf of smooth function

$$\mathscr{C}^\infty(\mathrm{GL(n, m)}) := \mathscr{C}^\infty(G_0) \otimes \mathbb{R}^{2nm}$$

It can be proved that it is a super Lie group [26].

Remark 110 One very useful tool to compute in superanalysis is what Berezin called "the Grassmann analytic continuation principle" [26]. It was sometimes applied implicitly above and we shall often apply it later without more justifications. There exist several ways to explain this principle in a more mathematical rigorous approach (see [79, 138]).

The practical rule is the following. Let f be a real function of n real variables $x_1, \ldots, x_n$ and n nilpotent Grassmann numbers $\xi_1, \ldots, \xi_n$. That means that $\xi_j \in \mathcal{G}_n$ and $\xi_j = \sum_{|\epsilon|>0} a_\epsilon^j \theta^\epsilon$ where $\theta_1, \ldots, \theta_n$ are generators of $\mathcal{G}_n$. Denote $x = (x_1, \ldots, x_n)$ and $\xi = (\xi_1, \ldots, \xi_n)$ (using the usual notation in several variable functions), the Grassmann analytic continuation principle says

$$f(x+\xi) = \sum_\alpha \frac{\xi^\alpha}{\alpha!} \frac{\partial^\alpha}{\partial x^\alpha} f(x). \tag{17.20}$$

Recall that $\alpha = (\alpha_1, \ldots \alpha_n) \in \mathbb{N}^n$, $\xi^\alpha = \xi_1^{\alpha_1} \ldots \xi_n^{\alpha_n}$, hence we have $\xi^\alpha = 0$ for $|\alpha| \geq 2n$ so the Taylor expansion (17.20) is finite.

We consider now more interesting examples when even and odd coordinates are mixed, that give supersymmetries after quantization.

Let us begin by the toy model $\mathbb{R}^{1|1}$ with the law super group:

$$(t, \theta) \cdot (s, \zeta) = (t + s - i\theta\zeta, \theta + \zeta). \tag{17.21}$$

Let us denote SG$(1|1)$ this supergroup (it is easy to prove that it is a supergroup with a non commutative product) and $\mathfrak{sg}(1|1)$ its Lie algebra.

If $f \in C^\infty(\mathbb{R}^{1|1})$, $f(t,\theta) = f_0(t) + \theta f_1(t)$, and (s, η) the coordinates of a supervector v in $\mathbb{R}^{1|1}$, the left translation of f by v is defined by

$$\begin{aligned} {}^\ell\tau_{(s,\eta)} f(t,\theta) &= f(t - s + i\eta\theta, \theta - \eta) \\ &= f_0(t-s) + i\eta\theta f_0'(t-s) + (\theta - \eta) f_1(t-s). \end{aligned} \tag{17.22}$$

Here we have used the Grassmann analytic extension principle. But we have the group property: ${}^\ell\tau_{(s,\eta)} = {}^\ell\tau_{(s,0)} \circ {}^\ell\tau_{(0,\eta)}$ and

$${}^\ell\tau_{(s,0)} f(t,\theta) = f(t-s,\theta), \quad \text{(usual translation by a real number } s\text{)} \tag{17.23}$$

$${}^\ell\tau_{(0,\zeta)} f(t,\theta) = f(t,\theta) + \zeta\left(i\theta\frac{\partial}{\partial t} - \frac{\partial}{\partial\theta}\right) f(t,\theta) \tag{17.24}$$

So we can conclude that the super lie algebra $\mathfrak{sg}(1|1)$, has the two generators $\{\partial_t, \mathcal{D}_\theta\}$, where $\partial_t = \frac{\partial}{\partial t}$ is even, $\partial_\theta = \frac{\partial}{\partial\theta}$, and $\mathcal{D}_\theta = i\theta\partial_t - \partial_\theta$ are odd. We have the commutation rule

$$[\mathcal{D}_\theta, \mathcal{D}_\theta]_+ = -2i\partial_t.$$

By analogous computations we get a basis for right invariant vector fields. Only the odd parts is modified. We get

$$\mathcal{Q}_\theta = i\theta\partial_t + \partial_\theta.$$

Only the last commutation rule is changed:

$$[\mathcal{Q}_\theta, \mathcal{Q}_\theta]_+ = 2i\partial_t. \tag{17.25}$$

Moreover we have the following commutation

$$[\mathcal{Q}, \mathcal{D}] = 0. \tag{17.26}$$

In mechanics the generator of the time translations is the Hamiltonian H. So Eq. (17.25) is a classical analogue for a supersymmetric system.

Consider now a more physical example leading to a classical analogue for the Witten model [263]. This kind of model was considered before by several physicists in the seventies (Wess, Zumino, Salam).

On $\mathbb{R}^{1|2}$ define the following multiplication (non commutative) rule

$$(t, \theta, \theta^*)(s, \zeta, \zeta^*) = (t + s - i(\theta\zeta^* - \zeta\theta^*), \theta + \zeta, \theta^* + \zeta^*). \tag{17.27}$$

For $f \in C^\infty(\mathbb{R}^{1|2})$ we have $f(t, \theta, \theta^*) = f_0(t) + f_1(t)\theta + f_2(t)\theta^* + f_3(t)\theta\theta^*$. The multiplication morphism μ^* is determined by $\mu^*(f_j)$ for $1 \le j \le 3$ (recall that μ^* is a superalgebra morphism). For every $f \in C^\infty(\mathbb{R})$ the meaning of (17.27) is

$$\begin{aligned}\mu^\star(f)(t, s, \theta, \theta^*, \zeta, \zeta^*) &= f(t + s - i(\theta\zeta^* - \zeta\theta^*), \theta + \zeta, \theta^* + \zeta^*)\\ &= f(t+s) - i(\theta\zeta^* - \zeta\theta^*)f'(t+s) - \theta\theta^*\zeta\zeta^* f''(t+s)\end{aligned}$$

$\iota^\star$ and $\nu_0^\star$ are easily computed such that $\{\mathbb{R}^{1|2}, \mu\}$ is a super Lie group that we shall denote SG(1|2).

As we have done for SG(1|1) we can compute generators for the super Lie algebra $\mathfrak{sg}(1|2)$ by identifying left invariant vector fields.

Denote ${}^\ell\tau_{s,\eta,\eta^*}$ the left translation by (s, η, η^*), ${}^\ell\tau_\eta$ the left translation by $(0, \eta, 0)$ and ${}^\ell\tau_{\eta^*}$ the translation by $(0, 0, \eta^*)$. As above we get easily for every $f \in C^\infty(\mathbb{R}^{1|2})$,

$$ {}^\ell\tau_\eta f(t, \theta) = f(t, \eta) + \eta\left(i\theta^*\partial_t - \partial_\theta\right) f(t, \theta) \tag{17.28}$$

$$ {}^\ell\tau_{\eta^*} f(t, \theta) = f(t, \eta) + \eta\left(i\theta\partial_t - \partial_{\theta^*}\right) f(t, \theta) \tag{17.29}$$

So we have got the following basis for $\mathfrak{sg}(1|2)$:

$$\partial_t, \quad \mathcal{D}_\theta = i\theta^*\partial_t - \partial_\theta, \quad \mathcal{D}_{\theta^*} = i\theta\partial_t - \partial_{\theta^*},$$

with one even and two odd generators. The commutation rules of this algebra are

$$\mathscr{D}_\theta{}^2 = \mathscr{D}_{\theta^*}{}^2 = 0 \tag{17.30}$$

$$[\mathscr{D}_\theta, \partial_t] = [\mathscr{D}_{\theta^*}, \partial_t] = 0 \tag{17.31}$$

$$[\mathscr{D}_\theta, \mathscr{D}_{\theta^*}]_+ = -2i\partial_t \tag{17.32}$$

By analogous computations we get a basis for right invariant vector fields. Only the odd parts is modified. We get

$$\mathscr{Q}_\theta = i\theta^*\partial_t + \partial_\theta, \quad \mathscr{Q}_{\theta^*} = i\theta\partial_t - \partial_{\theta^*}.$$

Only the last commutation rule is changed:

$$[\mathscr{Q}_\theta, \mathscr{Q}_{\theta^*}]_+ = 2i\partial_t. \tag{17.33}$$

Moreover we have the commutation

$$[\mathscr{Q}, \mathscr{D}] = 0. \tag{17.34}$$

So Eq. (17.33) is a classical analogue for the quantum supersymmetric system of Witten. This will be more explicit later.

17.5 Classical Supersymmetry

17.5.1 A Short Overview of Classical Mechanics

There are many books concerning this very well known subject. For our purpose we recall here some basic facts concerning Lagrangians and Hamiltonians. We refer for more details to the following books [8, 121, 241].

In physics classical dynamical systems are usually introduced with their Lagrangian $\mathscr{L}$ and their action integral $S = \int dx \mathscr{L}(x)$ if x is a coordinates system for classical paths. For a point particle moving in $\mathbb{R}^n$ with coordinates $q = (q_1, \ldots, q_n)$ we have $S = \int_{t_0}^{t_1} dt \mathscr{L}(q, \dot{q})$, $\dot{q}$ is the time derivative of q. The equations of motion (Euler–Lagrange equations) are deduced from the least action principle

$$\frac{\partial \mathscr{L}}{\partial q} - \frac{d}{dt}\left(\frac{\partial \mathscr{L}}{\partial \dot{q}}\right) = 0 \tag{17.35}$$

An Euler–Lagrange equation may be very difficult to solve. Many informations about its solutions can be obtained studying its symmetries. According the Noether famous theorem, symmetries give integral of motions I, functions of q, $\dot{q}$ conserved along the motion $\frac{d}{dt}I(q, \dot{q}) = 0$. In particular the Hamiltonian energy function H is conserved:

$$H(q,\dot{q}) = \frac{\partial \mathscr{L}}{\partial \dot{q}} \cdot \dot{q} - \mathscr{L}(q,\dot{q}). \tag{17.36}$$

The canonical conjugate momentum p is defined as

$$p = \frac{\partial \mathscr{L}}{\partial \dot{q}}$$

p is a cotangent vector in $(\mathbb{R}^n)^*$. We consider here for simplicity systems without constraints and that the configuration space is the Euclidean space $\mathbb{R}^n$. (see [8] for manifolds).

Let us recall now a statement for the Noether theorem concerning symmetries and integrals of motion.

Theorem 87 *Let* $V(q) = \sum_{1\leq j\leq n} v_j(q)\frac{\partial}{\partial q_j}$ *be a vector field and* ϕ_V^s *its flow:*

$$\frac{d}{ds}\phi_V^s(q_s) = v(q_s), \; q_0 = q, \quad \text{where} \quad v(q) = (v_1(q), \ldots, v_n(q)). \tag{17.37}$$

(NI) Assume that $\mathscr{L}$ *is invariant under* ϕ_V^s *for* s *close to 0. Then*

$$I_V(q,\dot{q}) = v(q) \cdot \frac{\partial \mathscr{L}}{\partial \dot{q}}$$

is an integral of motion: $\frac{dI_V}{dt} = 0$.

(NII) Assume that there exists a smooth function K *on* $\mathbb{R}^n \times \mathbb{R}^n$ *such that*

$$v(q) \cdot \frac{\partial \mathscr{L}}{\partial \dot{q}} + \dot{v}(q)\frac{\partial \mathscr{L}}{\partial q} = \frac{d}{dt}K(q,\dot{q}) \tag{17.38}$$

then $I = v(q) \cdot \frac{\partial \mathscr{L}}{\partial \dot{q}} - K(q,\dot{q})$ *is an integral of motion.*

Remark 111 It is convenient to state Noether's theorem with finite small deformations of size ϵ. Denote $\delta_\epsilon q = \epsilon v(q)$, $\delta_\epsilon \mathscr{L} = \mathscr{L}(q + \delta_\epsilon q, \dot{q} + \delta_\epsilon \dot{q}) - \mathscr{L}(q,\dot{q})$. The invariance assumptions (17.38) means that

$$\delta_\epsilon \mathscr{L} = \epsilon \frac{d}{dt}K(q,\dot{q}) + O(\epsilon^2). \tag{17.39}$$

The energy Hamiltonian H was defined as a function of $(q,\dot{q})$. It is convenient to replace the velocity $\dot{q}$ by the momentum p using the Legendre transform in $\dot{q} \mapsto p$.

$$H(q,p) = p \cdot \dot{q}(q,p) - \mathscr{L}(q,\dot{q}(q,p)). \tag{17.40}$$

If the matrix $\frac{\partial^2 \mathscr{L}}{\partial \dot{q} \partial \dot{q}}$ is non-degenerate, H is a smooth function of (q, p) only (at least locally).

In order to quantize the classical system with Lagrangian $\mathscr{L}$ it is assumed that H is defined globally.

We shall see that for fermions it is not true for interesting examples where the Lagrangian is degenerate: $\dot{q} \mapsto p$ is not onto. But in order to quantize canonically a classical system we have to compute an Hamiltonian and a Poisson bracket. Dirac proposed a method [84] to do that in the degenerate case.

Let us describe roughly the Dirac method which will apply to Fermions. We follow here [147] where the reader can find many details.

When $\dot{q} \mapsto p$ is not onto the range of the mapping $(q, \dot{q}) \mapsto (q, p)$ $(p = \frac{\partial \mathscr{L}}{\partial \dot{q}})$ is a non geometrically trivial part of the phase space. In particular it is not possible to recover $\dot{q}$ from p. In this situation we say that the Lagrangian is singular or degenerate.

Following [84], it will be assumed that this is a smooth manifold $\mathscr{C}$ described by equations $\chi_j(q, p) = 0$, $1 \leq j \leq m$ where the differential $d\chi_j$ are everywhere independent.

The energy Hamiltonian H is always a function of q and p (as in the regular case) but is defined here only under the constraint $(q, p) \in \mathscr{C}$. Dirac assumed that H has an extension to the whole phase space $\mathbb{R}^n \times \mathbb{R}^n$ ($\mathbb{R}^n$ is identified with $(\mathbb{R}^n)^*$) and he computed a modified Poisson bracket $\{\bullet, \bullet\}_{\mathrm{Di}}$ such that for any observable $F \in C^\infty(\mathbb{R}^n \times \mathbb{R}^n)$ the equation of motion is

$$\dot{F} = \{F, H\}_{\mathrm{Di}}, \text{ in particular } \dot{q} = \{q, H\}_{\mathrm{Di}}, \quad \dot{p} = \{p, H\}_{\mathrm{Di}}. \tag{17.41}$$

On $\mathscr{C}$, using definition of p we have

$$dH = \dot{q}dp - \frac{\partial \mathscr{L}}{\partial q}dq = \frac{\partial H}{\partial q}dq + \frac{\partial H}{\partial p}dp \tag{17.42}$$

Then there exists $u_j(q, \dot{q})$, $1 \leq j \leq m$ such that

$$\dot{q} = \frac{\partial H}{\partial p} + \sum_{1 \leq j \leq m} u_j \frac{\partial \chi_j}{\partial p} \tag{17.43}$$

$$\dot{p} = -\frac{\partial H}{\partial q} - \sum_{1 \leq j \leq m} u_j \frac{\partial \chi_j}{\partial q} \tag{17.44}$$

So we have for every observable F

$$\dot{F} = \{F, H\} + \sum_{1 \leq j \leq m} u_j \{F, \chi_j\} \tag{17.45}$$

The constraints χ_j have to be preserved during the time evolution, which gives the conditions

$$\{\chi_k, H\} + \sum_{1\le j\le m} u_j\{\chi_k, \chi_j\} = 0, \text{ for } 1 \le k \le m \tag{17.46}$$

For simplicity assume now $m = 2$ and $\{\chi_1, \chi_2\} = \lambda \neq 0$. From conditions (17.46) we can compute u_1, u_2:

$$u_1 = \frac{\{\chi_2, H\}}{\lambda}, \quad u_2 = -\frac{\{\chi_1, H\}}{\lambda}.$$

Then the Dirac bracket has the following expression:

$$\{F, G\}_{Di} = \{F, G\} + \frac{1}{\lambda}(\{F, \chi_1\}\{\chi_2, G\} - \{F, \chi_2\}\{\chi_1, G\}) \tag{17.47}$$

It is easy to see that $(F, G) \mapsto \{F, G\}_{Di}$ is a Lie bracket on the linear space $C^\infty(\mathbb{R}^n \times \mathbb{R}^n)$.

Let us consider the following example in the configuration space $\mathbb{R}^2$: $\mathscr{L}_M = x\dot{y} - y\dot{x} - V(x, y)$, V is a smooth potential. This Lagrangian is degenerate.

$$p_x := \frac{\partial \mathscr{L}_M}{\partial \dot{x}} = -y, \quad p_y := \frac{\partial \mathscr{L}_M}{\partial \dot{y}} = x \tag{17.48}$$

The energy Hamiltonian is $H = V(x, y)$. We have the two constraints $\chi_1 = p_x + y$, $\chi_2 = p_y - x$ and $\{\chi_1, \chi_2\} = 2$. Then the Dirac bracket is

$$\{F, G\}_{Di} = \frac{1}{2}\{F, G\} - \frac{1}{2}\left(\frac{\partial F}{\partial x}\frac{\partial G}{\partial y} - \frac{\partial F}{\partial y}\frac{\partial G}{\partial x}\right) - \frac{1}{2}\left(\frac{\partial F}{\partial p_x}\frac{\partial G}{\partial p_y} - \frac{\partial F}{\partial p_y}\frac{\partial G}{\partial p_x}\right) \tag{17.49}$$

We can check that the Hamiltonian H and the Dirac bracket (17.49) give the Euler–Lagrange equation for the motion. We find

$$\dot{x} = \{x, V\}_{\text{Di}} = -\frac{1}{2}\frac{\partial V}{\partial y}; \quad \dot{y} = \{y, V\}_{\text{Di}} = \frac{1}{2}\frac{\partial V}{\partial x} \tag{17.50}$$

To prepare a canonical quantization of the Lagrangian $\mathscr{L}_M$ we write the commutation relations

$$\{x, y\}_{\text{Di}} = \{p_x, p_y\}_{\text{Di}} = -\frac{1}{2} \tag{17.51}$$

$$\{x, p_x\}_{\text{Di}} = \{y, p_y\}_{\text{Di}} = \frac{1}{2} \tag{17.52}$$

$$\{y, p_x\}_{\text{Di}} = \{x, p_y\}_{\text{Di}} = 0. \tag{17.53}$$

A Dirac quantization $F \mapsto \hat{F}$ satisfies

$$\{F, G\}_{\text{Di}} \to -i[\hat{F}, \hat{G}]$$

So we get a representation of the Lie algebra (17.51) in $L^2(\mathbb{R}^2)$ satisfying

$$\hat{x} = \frac{1}{2}\left(x + \frac{\partial}{i\partial y}\right), \quad \hat{y} = \frac{1}{2}\left(y - \frac{\partial}{i\partial x}\right)$$
$$\hat{p}_x = -\frac{1}{2}\left(y - \frac{\partial}{i\partial x}\right), \quad \hat{p}_y = \frac{1}{2}\left(x + \frac{\partial}{i\partial y}\right). \tag{17.54}$$

Notice that this representation in $L^2(\mathbb{R}^2)$ is not irreducible because we have the constraints $\hat{p}_x + \hat{y} = 0$, $\hat{p}_y - \hat{x} = 0$ and the relations (17.54) define an Heisenberg Lie algebra of dimension 3. So we can get a quantization equivalent to the Weyl quantization in $L^2(\mathbb{R})$.

17.5.2 Supersymmetric Mechanics

Supermechanics is an extension of classical mechanics. In supermechanics a point has real (or even) coordinates $x = (x_1, \ldots, x_n)$ representing the bosonic degrees of freedom and Grassmann coordinates, $\xi = (\xi_1, \ldots, \xi_M)$, representing the fermionic degrees of freedom. To encode the two kinds of degree of freedom in the same object one introduces a "real super variables" X living in the configuration space of all the system. A super Lagrangian is a function $\mathscr{L}_S$ of X and its derivative $\mathscr{D}X$, with values in a Grassmann algebra. In coordinates we can write $X = (x, \xi)$ where x has real dependent components, ξ have fermionic (real) components.

In coordinates we get a pseudo-classical Lagrangian $\mathscr{L}(x, \dot{x}, \xi, \dot{\xi})$. The conjugate momenta are

$$p = \frac{\partial \mathscr{L}}{\partial \dot{x}}, \quad \pi = \frac{\partial \mathscr{L}}{\partial \dot{\xi}}. \tag{17.55}$$

Assume for a moment that we can define the Hamiltonian H as the Legendre transform of $\mathscr{L}$ in $(\dot{x}, \dot{\xi})$. Let F be an observable on the phase space defined by its coordinates (x, ξ, p, π). Along the motion we want to have as usual

$$\dot{F} = \{F, H\}$$

where F is an extension of the usual Poisson bracket. A direct computation, using equations of motion, gives

$$\dot{F} = \left(\frac{\partial H}{\partial p}\frac{\partial F}{\partial x} - \frac{\partial H}{\partial x}\frac{\partial F}{\partial p}\right) - \left(\frac{\partial H}{\partial \pi}\frac{\partial F}{\partial \xi} + \frac{\partial H}{\partial \xi}\frac{\partial F}{\partial \pi}\right). \tag{17.56}$$

Recall that H and F are in the superalgebra $\mathscr{C}^\infty(\mathbb{R}^{n+n|m+m})$. The right side in (17.56) is the super Poisson bracket $\{F, H\}$, the first term is the usual antisymmetric Poisson bracket in bosonic variables, inside the second parenthesis we have a symmetric form for the contribution of fermionic variables.

For any homogeneous observables F, G the Poisson bracket is a super Lie product in $C^\infty(\mathbb{R}^{2n|2m})$ satisfying

$$\{F, G\} = \left(\frac{\partial F}{\partial x}\frac{\partial G}{\partial p} - \frac{\partial F}{\partial p}\frac{\partial G}{\partial x}\right) + (-1)^{\pi(F)}\left(\frac{\partial F}{\partial \xi}\frac{\partial G}{\partial \pi} + \frac{\partial F}{\partial \pi}\frac{\partial G}{\partial \xi}\right). \tag{17.57}$$

First consider a simple system with one bosonic state and one fermionic state without interaction. Its states are represented by the superfield:

$$X = x + i\theta\xi \tag{17.58}$$

x is a real number depending on time t, ξ is an odd number living in a Grassmann algebra. We want that X is even and "real". We introduce two Grassmann complex conjugate generators $\{\eta, \eta^*\}$ and choose $\xi(t) = \bar{c}(t)\eta + c(t)\eta^*$ where $c(t)$ is a complex number. Then we have $(i\theta\xi)^* = i\theta\xi$ so X is even and real. Remark that X is not a C^∞ function on $\mathbb{R}^{1|1}$ because the coefficient ξ is not a complex number but a Grassmann variable.

As we have seen the superfields X has to be understood as a morphism from $C^\infty(\mathbb{R})$ in $C^\infty(\mathbb{R}^{1|1})$, using the Grassmann analytic extension principle we have:

$$X^*(f)(x, \theta) = f(x) + if'(x)\theta\xi.$$

Let us introduce the super Lagrangian

$$\mathcal{L}_S = \frac{1}{2}\mathcal{D}X \cdot \mathcal{D}(\mathcal{D}X)$$

where $\mathcal{D} = \mathcal{D}_\theta$.

In coordinates we get the pseudo-classical Lagrangian

$$\mathcal{L}_{pcl} = \int d\theta \mathcal{L}_S = \frac{1}{2}(\dot{x}^2 + i\xi\dot{\xi})$$

We can easily solve the Euler–Lagrange equations. We get

$$\ddot{x} = 0, \quad \dot{\xi} = 0.$$

We can consider the same model with 3 superfields $X_j = x_j + i\theta\xi_j$, $1 \le j \le 3$.

$$\mathcal{L}_{pcl} = \frac{1}{2}(\dot{x} \cdot \dot{x} - i\dot{\xi} \cdot \xi).$$

Here we have 3 constraints

$$\chi_j = \pi_j + \frac{i}{2}\xi_j.$$

But we have $\{\chi_j, \chi_k\} = -i\delta_{j,k}$ so the constraints are second order. This system is invariant by any rotation in $\mathbb{R}^3$. Its Dirac canonical quantization gives a spin system [241].

The supergroup SG(1|1) transforms X by right translations, so we have

$$X(t - i\theta\eta, \theta + \eta) = X(t, \theta) + \eta\mathscr{Q}X(t, \theta)$$

where $\mathscr{Q} = \mathscr{Q}_\theta$. So a variation of X is given by $\delta_\eta X = \eta\mathscr{Q}X$, where δ_η is a derivation in $C^\infty(\mathbb{R})[\theta, \eta, \eta^*]$ considered as a $\mathbb{R}(\eta, \eta^*]$ module (coefficients of X are in $\mathbb{R}[\eta, \eta^*]$), $\delta_\eta F = [\eta\mathscr{Q}, F]$.

In coordinates we have $\delta_\eta X = \delta_\eta x + i\theta\delta_\eta\xi$ where

$$\delta_\eta x = i\eta\xi, \quad \delta_\eta\xi = \eta\dot{x}.$$

So we compute the variation of the Lagrangian and the variation of the corresponding action $S = \int dtd\theta\mathscr{L}$. Using that δ_η is a derivation, we get

$$\delta_\eta\mathscr{L}_S = \eta\mathscr{Q}\mathscr{L}_S = \eta(i\theta\partial_t + \partial_\theta)\mathscr{L}. \tag{17.59}$$

Hence we get

$$\partial_\eta S = \int dt\partial_t\left(\int d\theta i\theta\mathscr{L}_S\right). \tag{17.60}$$

By extension of the Noether theorem to the fermionic case we see that the vector field $\mathscr{Q}$ is the generator of a symmetry for the Lagrangian $\mathscr{L}_S$. This symmetry is called supersymmetry because it concerns supervariables mixing usual numbers (real or complex) and Grassmann numbers.

For the corresponding variations of the pseudoclassical Lagrangian $\mathscr{L}_{pcl}$ we have

$$\delta_\eta\mathscr{L}_{pcl} = \frac{3i\eta}{2}\frac{\partial}{\partial t}\xi\dot{x}$$

According to Noether's results we have the conserved charge

$$I_Q = i\xi\frac{\partial\mathscr{L}}{\partial\dot{x}} + \dot{x}\frac{\partial\mathscr{L}}{\partial\xi} - \frac{3i}{2}\frac{\partial}{\partial t}\xi\dot{x} = -2i\xi\dot{x}$$

We can compute the Hamiltonian H by Legendre transform. The phase space here is the super linear space $\mathbb{R}^{2|1}$. We get for the conjugate momentum

$$\pi = \frac{\partial L}{\partial\xi} = -\frac{i}{2}\xi.$$

So this Lagrangian is degenerate, with the constraint $\chi = \pi + \frac{i}{2}\xi$. We get $H = \frac{p^2}{2}$ where p is the conjugate momentum to x.

Now we shall consider a more interesting example involving the superharmonic oscillator. We consider the superfields:

$$X = x + \theta\psi^* + \psi\theta^* + y\theta\theta^*. \tag{17.61}$$

This field may depend on time t, so $x, y, \psi, \psi^\star$ are time dependent, where x, y are real numbers, $\theta, \theta^\star$ are conjugate Grassmann variables, ψ, ψ^* are necessarily odd numbers. As above, to compute these numbers we have to introduce an other pair of Grassmann variables $\{\zeta, \zeta^*\}$, anticommuting with the pair $\{\theta, \theta^*\}$. So we consider that $\psi = c\zeta + d\zeta^*$ where c, d are complex (time dependent) numbers.

As we have already remarked, to avoid mathematical difficulties it is necessary to add extra Grassmann variables.

Recall that the superfield X defined in the formula (17.61) is a morphism from $\mathbb{R}^{1|4}$ in $\mathbb{R}$ which means that it is defined by a morphism $X^\star$ from $C^\infty(\mathbb{R})$ in $\mathbb{C}^\infty(\mathbb{R})[\theta, \theta^*, \zeta, \zeta^*]$ which can be be computed by the the Grassmann analytic continuation principle:

$$(X^\star f)(x, \theta) = f(x) + f'(x)(\psi\theta^* + \theta\psi^*) + f'(x)y\theta\theta^* - f''(x)\theta\theta^*\psi\psi^*. \tag{17.62}$$

There exists other rigorous interpretation about this formula. For a detailed discussion we refer to the paper [138].

Let us introduce a super potential $V(X)$ where $V \in \mathbb{C}^\infty(\mathbb{R})$ and the super Lagrangian

$$\mathscr{L}_S = \frac{1}{2}\mathscr{D}_{\theta^*}X\mathscr{D}_\theta X + V(X). \tag{17.63}$$

Assume that V is polynomial for simplicity. Then the Lagrangian has the supersymmetry defined by the odd generators $\mathscr{Q}_\theta$ and $\mathscr{Q}_{\theta^*}$ defined before. The infinitesimal transformations on fields are $\delta_\eta X = \eta^*\mathscr{Q}_\theta + \eta\mathscr{Q}_{\theta^*}X$ and as above δ_η is a derivation. η, η^* are Grassmann variables, independent of the other Grassmann variables. We also have $\delta_\eta V(X) = \eta^*\mathscr{Q}_\theta + \eta\mathscr{Q}_{\theta^*}V(X)$ and

$$\delta_\eta\mathscr{L}_S = \eta^*\mathscr{Q}_\theta + \eta\mathscr{Q}_{\theta^*}\mathscr{L}_S.$$

As for the above toy model, it results that the Lagrangian $\mathscr{L}_S$ is again supersymmetric with generators $\mathscr{Q}_\theta, \mathscr{Q}_{\theta^*}$.

Now we shall compute in coordinates with the pseudo-classical Lagrangian.

Compute

$$\mathscr{Q}_\theta X = \psi^* + \theta^*(i\dot{x} + y) - i\theta\theta^*\dot{\psi}^* \tag{17.64}$$

$$\mathscr{Q}_{\theta^*}X = -\psi + \theta(i\dot{x} - y) - i\theta\theta^*\dot{\psi}. \tag{17.65}$$

So we get

$$\delta_\eta X = \eta^*\psi^* - \eta\psi + \theta\eta(y - ix) + \eta^*(y + i\dot{x})\theta^*$$
$$- i(\eta^*\dot{\psi}^* + \eta\dot{\psi})\theta\theta^*. \tag{17.66}$$

Hence in coordinates the supersymmetry has the following infinitesimal representation

$$\delta_\eta x = \eta^*\psi^* - \eta\psi \tag{17.67}$$
$$\delta_\eta \psi = \eta^*(y + i\dot{x}) \tag{17.68}$$
$$\delta_\eta \psi^* = \eta(y - ix) \tag{17.69}$$
$$\delta_\eta y = -i(\eta^*\dot{\psi}^* + \eta\dot{\psi}). \tag{17.70}$$

Now we compute the pseudo-classic Lagrangian: $\mathscr{L}_{pcl} := \int d\theta d\theta^* \mathscr{L}_S$. Using the computation rules for Berezin integral we get

$$\mathscr{L}_{pcl} = \frac{1}{2}(\dot{x}^2 + y^2) - V'(x)y + \frac{i}{2}(\psi\dot{\psi}^* - \dot{\psi}\psi^*) + \frac{1}{2}V''(x)\psi\psi^*. \tag{17.71}$$

The real variable can be eliminated because we have $\frac{\partial}{\partial \dot{y}}\mathscr{L}_{pcl} = 0$. From Euler–Lagrange equation we get $\frac{\partial}{\partial y}\mathscr{L}_{pcl} = 0$ so $y = V'(x)$. We get now the following Lagrangian

$$\mathscr{L}_w = \frac{\dot{x}^2}{2} - \frac{V'(x)^2}{2} + \frac{i}{2}(\psi\dot{\psi}^* - \dot{\psi}\psi^*) + \frac{1}{2}V''(x)\psi\psi^*. \tag{17.72}$$

We can compute the two Noether charges associated with the generators $\mathscr{Q}_\theta$, $\mathscr{Q}_{\theta^*}$. The results are

$$Q = (\dot{x} - iV'(x))\psi, \quad Q^* = (\dot{x} + iV'(x))\psi^*. \tag{17.73}$$

An easy exercise is to compute the dynamics for the Fermi oscillator, well known for the Bose (or harmonic) oscillator. The Lagrangian is

$$\mathscr{L}_F = \frac{i}{2}(\psi\dot{\psi}^* - \dot{\psi}\psi^*) + \frac{1}{2}\omega\psi\psi^*. \tag{17.74}$$

The Euler–Lagrange equation gives $\dot{\psi} = -i\omega\psi$ hence we get $\psi(t) = \eta \mathrm{e}^{-i\omega t}$, where η is any Grassmann complex variable.

It is more suggestive to write down the dynamics in real coordinates:

$$\xi_1(t) = \frac{\psi(t) + \psi^*(t)}{\sqrt{2}} = \frac{1}{\sqrt{2}}\left(\eta \mathrm{e}^{-i\omega t} + \eta^* \mathrm{e}^{i\omega t}\right) \tag{17.75}$$
$$\xi_2(t) = \frac{\psi^*(t) - \psi(t)}{i\sqrt{2}} = \frac{1}{i\sqrt{2}}\left(\eta^* \mathrm{e}^{i\omega t} - \eta \mathrm{e}^{-i\omega t}\right). \tag{17.76}$$

17.5.3 Supersymmetric Quantization

The first step is to compute the Hamiltonian H_w for the Lagrangian $\mathscr{L}_w$.

The momenta are defined as usual

$$p = \frac{\partial \mathscr{L}_w}{\partial \dot{x}} = \dot{x} \tag{17.77}$$

$$\pi = \frac{\partial \mathscr{L}_w}{\partial \dot{\psi}} = -\frac{i}{2}\psi^* \tag{17.78}$$

$$\pi^* = \frac{\partial \mathscr{L}_w}{\partial \dot{\psi}^*} = -\frac{i}{2}\psi. \tag{17.79}$$

Hence the Legendre transform is not surjective: the Lagrangian is degenerate. We can extend the Dirac method (see more details in [147]) to fermionic variables with the two constraints

$$\chi_1 = \left(\pi + \frac{i}{2}\psi^*\right), \quad \chi_2 = \left(\pi^* + \frac{i}{2}\psi\right).$$

Recall that here $\{\cdot\}$ is the super-Poisson-bracket. We have $\{\chi_1, \chi_2\} = -i$, so the constraint is of second order. Denote

$$H = \dot{x}p + \dot{\psi}\pi + \dot{\psi}^*\pi^* - \mathscr{L}_w = \frac{1}{2}\left(p^2 + V'(x)^2\right) - V''(x)\psi\psi^*.$$

According to Dirac's method we compute multipliers u_1, u_2 such the evolution of any observable F follows the equation

$$\dot{F} = \{F, H\} + u_1\{F, \chi_1\} + u_2\{F, \chi_2\}.$$

u_1, u_2 are computed with the compatibility conditions $\dot{\chi}_k = 0$ for $k = 1, 2$. We get

$$u_1 = -iV''(x)\psi, \quad u_2 = iV''(x)\psi^*.$$

The total Dirac Hamiltonian is here, after computation,

$$H_{\text{Di}} = H + u_1\chi_1 + u_2\chi_2 = \frac{1}{2}\left(p^2 + V'(x)^2\right) + iV''(x)(\psi^*\pi^* - \psi\pi).$$

To achieve the quantization of our system we define, as in the bosonic case, the Dirac bracket. Let F, G depending only on fermionic variables. Then the Dirac bracket is

$$\{F, G\}_{\text{Di}} = \{F, G\} + \frac{1}{\{\chi_1, \chi_2\}}\left(\{F, \chi_1\}\{\chi_2, G\} + \{F, \chi_2\}\{\chi_1, G\}\right).$$

So we have

$$\{\psi,\psi^*\}_{\mathrm{Di}} = -i,\quad \{\pi,\pi^*\}_{\mathrm{Di}} = \frac{i}{4},\quad \{\psi,\pi\} = \{\psi^*,\pi^*\} = \frac{1}{2}. \tag{17.80}$$

A quantization $F \mapsto \hat{F}$ has to follow the correspondence principle. For $F = x, p$ we consider the usual Weyl–Heisenberg quantization with the commutator rule $[\hat{x},\hat{p}] = i\hbar$. If F and G are fermionic variables then

$$i\hbar\widehat{\{F,G\}} = [\hat{F},\hat{G}] \tag{17.81}$$

where the brackets are symmetric (anticommutators).

In particular we have

$$[\hat{\psi},\hat{\psi}^*] = \hbar,\quad [\hat{\pi},\hat{\pi}^*] = \frac{-\hbar}{4},\quad [\hat{\psi},\hat{\pi}] = [\hat{\psi}^*,\hat{\pi}^*] = \frac{-i\hbar}{2}. \tag{17.82}$$

The Dirac quantum Hamiltonian is

$$\hat{H}_{Di} = \frac{1}{2}\left(\hat{p}^2 + V'(x)^2\right) + iV''(x)(\hat{\psi}^*\hat{\pi}^* - \hat{\psi}\hat{\pi}). \tag{17.83}$$

A realization of the commutation relations is obtained with the Pauli matrices:

$$\hat{\psi} = \sqrt{\hbar}\sigma_-,\quad \hat{\psi}^* = \sqrt{\hbar}\sigma_+ \tag{17.84}$$

$$\hat{\pi} = -\frac{i}{2}\sigma_+,\quad \hat{\pi}^* = -\frac{i}{2}\sigma_-. \tag{17.85}$$

For $\hbar = 1$ we get the Witten supersymmetric Hamiltonian considered at the beginning of this Chapter:

$$\hat{H} = \frac{1}{2}\left(-\frac{d^2}{dx^2} + V'(x)^2\right)\mathbb{1}_2 - \frac{\sigma_3}{2}V''(x). \tag{17.86}$$

We can also remark that the supercharges $\hat{Q}$ and $\hat{Q}^*$ are obtained by quantization of the Noether charges Q, Q^*.

In particular we can consider an harmonic potential $V(x) = \frac{\omega x^2}{2}$. Choosing a quantization such that $\hat{\pi} = -\frac{i}{2}\hat{\psi}^*$ and $\hat{\pi}^* = -\frac{i}{2}\hat{\psi}$ we get the super-harmonic oscillator

$$\hat{H}_{sos} = \frac{1}{2}\left(-\frac{d^2}{dx^2} + \omega^2x^2\right) + \omega\hat{\psi}^*\hat{\psi}. \tag{17.87}$$

We can realize this Hamiltonian with $\hat{\psi} = \theta$ and $\hat{\psi}^* = \partial_\theta$ in the super Hilbert space $\mathscr{H}_S := L^2(\mathbb{R}) \otimes \tilde{\mathscr{H}}^{(2)}$. It is nothing but a Grassmann algebra interpretation of the Witten model introduced before with $\tilde{\mathscr{H}}^{(2)}$ in place of $\mathbb{C}^2$.

The scalar product in the Hilbert space $\mathcal{H}_S$ is defined as

$$\langle F, G\rangle = \int dxd\theta d\theta^* e^{\theta^*\theta} F^*(x,\theta)G(x,\theta), \tag{17.88}$$

where x is a real variable, θ is an holomorphic Grassmann variable.

17.6 Supercoherent States

As canonical coherent states are built on the Heisenberg-Weyl Lie group, supercoherent states are built on the super Heisenberg-Weyl super Lie group. We first consider the simplest case with one boson and one fermion. So we have creation and annihilation operators a_B^*, a_F^*, a_B, a_F for bosons and fermions. They satisfy the super-Lie algebra commutation relations:

$$[a_B, a_B^*] = \mathbb{1}, \quad [a_F, a_F^*]_+ = \mathbb{1}. \tag{17.89}$$

All other relations are trivial.

We have a supersymmetric harmonic oscillator $\hat{H}_{\rm sos}$ with supersymmetry generators Q^*, Q,

$$\hat{H}_{\rm sos} = a_B^* a_B + a_F^* a_F, \quad \hat{Q} = a_B a_F^*, \; \hat{Q}^* = a_F a_B^*. \tag{17.90}$$

This definition of $\hat{H}_{sos}$ may differ of others by a constant.

We have $[\hat{H}_{\rm sos}, \hat{Q}] = 0$ so $(\hat{Q}, \hat{Q}^*)$ generates a global supersymmetry defined by the following unitary operators

$$U_\eta = e^{\eta\hat{Q}^* + \eta^*\hat{Q}}$$

where (η, η^*) are any complex conjugate Grassmann numbers.

Super translations $\hat{T}(z,\gamma)$ are parametrized by (z,γ), z is a complex number, γ is a Grassmann (complex) number

$$\hat{T}(z,\gamma) = \exp(za_B^* - \bar{z}a_B + a_F^*\gamma - \gamma^* a_F). \tag{17.91}$$

Using the Baker–Campbell–Hausdorff formula we have the following useful properties

$$T(z,\gamma) = \exp\left(\frac{1}{2}\gamma^*\gamma - |z|^2\right)\exp(za_B^*)\exp(a_F^*\gamma)\exp(-\bar{z}a_B)\exp(-a_F\gamma^*) \tag{17.92}$$

$$T(z,\gamma)T(u,\delta) = \exp\left(\frac{1}{2}\left(z\bar{u} - \bar{z}u + \delta^*\gamma + \delta\gamma^*\right)\right)T(z+u, \gamma+\delta) \tag{17.93}$$

In particular $T(z,\gamma)^{-1} = T(-z,-\gamma)$ is unitary and

$$T(z,\gamma)^{-1} a_B T(z,\gamma) = a_B + z \tag{17.94}$$
$$T(z,\gamma)^{-1} a_F T(z,\gamma) = a_F + \gamma \tag{17.95}$$

The ground state of $\hat{H}_{sos}$ is the state $\psi_{0,0} := \varphi_0 \otimes \psi_0$ where φ_0 and ψ_0 are the normalized ground states of the bosonic and fermionic oscillators. So we define super-coherent states by displacement of the ground state of the super-harmonic oscillator by super-translations:

$$\psi_{z,\gamma} = T(z,\gamma)\psi_{0,0}. \tag{17.96}$$

Using (17.92) we have

$$\psi_{z,\gamma} = \left(1 + \frac{\gamma\gamma^*}{2}\right)(|z,0\rangle + |z,1\rangle\gamma) \tag{17.97}$$

where we use the notation $|z,\epsilon\rangle = \varphi_z \otimes \theta^\epsilon$, $\epsilon = 0, 1$.

The coherent states family $\psi_{z,\gamma}$ has the following expected properties. The proofs follow easily from results already established for bosonic and fermionic coherent states.

1. (Normalization)
$$\|\psi_{z,\gamma}\|^2 = \langle \psi_{z,\gamma}, \psi_{z,\gamma}\rangle = 1. \tag{17.98}$$

2. (Non orthogonality)
$$\langle \psi_{z,\gamma}, \psi_{u,\delta}\rangle = \exp\left(\gamma^*\delta - \frac{1}{2}(\gamma^*\gamma + \delta^*\delta)\right)\exp\left(-\frac{1}{2}(|z|^2 + |u|^2) + \bar{z}u\right). \tag{17.99}$$

3. (Over-completeness) For every $\psi \in \mathscr{H}_S$ we have
$$\psi(x,\theta) = \int \langle \psi_{z,\gamma}, \psi\rangle \psi_{z,\gamma}(x,\theta)dzd\gamma^*d\gamma. \tag{17.100}$$

4. (Translation property)
$$T(z,\gamma)\psi_{u,\delta} = \exp(\frac{1}{2}\left(z\bar{u} - \bar{z}u + \delta^*\gamma + \delta\gamma^*)\right)\psi_{z+u,\gamma+\delta}. \tag{17.101}$$

5. (Eigenfunctions of annihilation operators)
$$a_B\psi_{z,\gamma} = z\psi_{z,\gamma}, \quad a_F\psi_{z,\gamma} = \gamma\psi_{z,\gamma}. \tag{17.102}$$

In particular we have the averages for energy and supercharges

$$\langle \psi_{z,\gamma}, \hat{H}_{sus} \psi_{z,\gamma} \rangle = |z|^2 + \gamma^* \gamma \quad (17.103)$$

$$\langle \psi_{z,\gamma}, Q \psi_{z,\gamma} \rangle = z\gamma^*, \quad \langle \psi_{z,\gamma}, Q^* \psi_{z,\gamma} \rangle = \bar{z}\gamma. \quad (17.104)$$

Remark 112 Definition and properties of supercoherent states with one boson and one fermion can easily be extended for systems with n bosons and m fermions. Then $z = (z_1, \ldots, z_n) \in \mathbb{C}^n$, $\gamma = (\gamma_1, \ldots, \gamma_m)$ represent a system of m complex Grassmann variables, $a_B = (a_{B,1}, \ldots, a_{B,n})$, $a_F = (a_{F,1}, \ldots, a_{F,m})$, etc. The formulas are the same using the multidimensional notations: $\gamma^* \cdot a_F = \sum_{1 \le j \le m} \gamma_j^* a_{F,j}$. Then the energy Hamiltonian can be $\hat{H} = \omega_B \cdot a_B^* a_B + \omega_F a_F^* \cdot a_F$, $\omega_{B,F}$ are real numbers.

The Hilbert space of this system is $\mathcal{H}_{n,m} = L^2(\mathbb{R}^n) \otimes \tilde{\mathcal{H}}^{(m)}$. To have a supersymmetric system we need that $n = m$ and $\omega_B = \omega_F$.

17.7 Phase Space Representations of Super Operators

As it is well known for bosons and as it was studied before for fermions we can extend to mixed systems bosons+fermions representations formulas for operators in the super-Hilbert space $\mathcal{H}$. Phase space representation means that we are looking for a correspondence between functions on the phase space and operators in some Hilbert space. This correspondence is called quantization in quantum mechanics. Our goal here is to revisit Weyl quantizations for observables mixing bosons and fermions. It could be possible to do it also for anti-Wick quantization.

To have a better analogy between the bosonic and fermionic variables it is nicer to write bosonic Weyl quantization in complex coordinates $\zeta := \frac{q+ip}{\sqrt{2}}$, $\zeta \in \mathbb{C}^n$, $q, p \in \mathbb{R}^n$. The Lebesgue measure in $\mathbb{C}^n$ is $d^2\zeta = |d\zeta \wedge d\zeta^*|$.

Assume for simplicity that $\hat{H}, \hat{G}$ have smooth Schwartz kernels in $\mathcal{S}(\mathbb{R}^{2n})$. We consider their Weyl symbols H_w^w, G_w^w in complex coordinates. Using the same proof as in the fermonic case we have the following Moyal formulas for the Weyl symbols of $\hat{G}\hat{H}$:

$$(G_w \circledast H_w)(\zeta) = \int_{\mathbb{C}^n} d^2\eta G_w(\zeta - \eta) H_w(\eta) e^{1/2(\bar{\zeta}\cdot\eta - \zeta\cdot\bar{\eta})} \quad (17.105)$$

$$(G^w \circledast H^w)(\zeta) = G^w(\zeta) e^{1/2(\overleftarrow{\partial_\zeta} \cdot \overrightarrow{\partial_{\bar{\zeta}}} - \overleftarrow{\partial_{\bar{\zeta}}} \cdot \overrightarrow{\partial_\zeta})} H^w(\zeta). \quad (17.106)$$

The first formula gives the covariant symbol, the second formula the contravariant symbol. We have the following relations

$$G^w(\eta) = \int d^2\zeta e^{\bar{\zeta}\cdot\eta - \zeta\cdot\bar{\eta}} G_w(\zeta). \quad (17.107)$$

$$G_w(\eta) \quad = (2\pi)^{-2n} \int d^2\zeta e^{\zeta\cdot\bar{\eta} - \bar{\zeta}\cdot\eta} Gw(\zeta). \quad (17.108)$$

It is not difficult to get a Moyal formula for classes of operators in the super-space $\mathcal{H}_{n,m} = L^2(\mathbb{R}^n) \otimes \tilde{\mathcal{H}}^{(m)}$. As usual it is enough to establish formulas for smoothing operators then the formulas are extended to suitable classes of symbols. So we shall assume that $\hat{G}$, $\hat{H}$ are linear continuous operators from $\mathcal{S}'(\mathbb{R}^n) \otimes \tilde{\mathcal{H}}^{(m)}$ into $\mathcal{S}(\mathbb{R}^n) \otimes \tilde{\mathcal{H}}^{(m)}$(that means that their Schwartz kernels are in $\mathcal{S}(\mathbb{R}^{2n} \otimes \mathcal{G}_m^c)$). The covariant symbol H_w is defined such that

$$\hat{H} = \int d^2\zeta d^2\gamma H_w(\zeta, \gamma)\hat{T}(-\zeta, -\gamma) \tag{17.109}$$

$$H_w(\zeta, \gamma) = \mathrm{Str}(\hat{H}\hat{T}(\zeta, \gamma)), \tag{17.110}$$

where Str is defined as the usual trace in bosonic variable ζ and the super-trace in the fermionic variable γ. More precisely, for any trace-class operator $\hat{H}$ in $\mathcal{H}_{n,m}$ we have

$$\mathrm{Str}\hat{H} = \mathrm{Tr}(\hat{H}(\mathbb{1} \otimes \chi))$$

where χ is the chirality operator in $\tilde{\mathcal{H}}^{(m)}$ defined in Chap. 16.

The contravariant symbol is the symplectic Fourier transform of the covariant symbol:

$$H^w(\zeta, \gamma) = \int d^2\eta d^2\alpha H_w(\eta, \alpha) \mathrm{e}^{\bar{\eta}\cdot\zeta - \eta\cdot\bar{\zeta} + \alpha^*\cdot\gamma + \alpha\cdot\gamma^*}. \tag{17.111}$$

So we get the following Moyal formula for super symbols

$$(G^w \circledast H^w)(\zeta, \alpha) = G^w(\zeta, \alpha)\mathrm{e}^{1/2\left(\overleftarrow{\partial_\zeta}\cdot\overrightarrow{\partial_{\bar{\zeta}}} - \overleftarrow{\partial_{\bar{\zeta}}}\cdot\overrightarrow{\partial_\zeta} + \overleftarrow{\partial_{\alpha^*}}\cdot\cdot\overrightarrow{\partial_\alpha} + \overleftarrow{\partial_\alpha}\cdot\overrightarrow{\partial_{\alpha^*}}\right)} H^w(\zeta, \alpha). \tag{17.112}$$

Many results explained before for bosons and fermions could be extended to mixed systems. Instead to do that we now discuss a simple application.

17.8 Application to the Dicke Model

This model was studied in [1] as a supersymmetric system. The Hamiltonian for this model is

$$\hat{H} = \Omega\mathbf{a}^*\mathbf{a} + \frac{1}{2}\omega\sigma_3 + g(\mathbf{a}^* + \mathbf{a})\sigma_1 \tag{17.113}$$

It concerns a two-level atom interacting with a monochromatic radiation field. $\mathbf{a}^*$ and $\mathbf{a}$ are one particle creation/annihilation operators, the Pauli matrices $\sigma = (\sigma_1, \sigma_2, \sigma_3)$ denote the radiation field with frequency Ω, ω is the level distance of the states of the atom, g is a coupling constant.

The matrices $\sigma_\pm$ satisfy the (CAR) relation so we can identify them with fermionic creation/annihilation operators. So we can introduce a pair of complex conjugate Grassmann numbers (β, β^*) such that $\sigma_+ \equiv \hat{\beta}^* = \partial_\beta, \sigma_- \equiv \hat{\beta}$ is multiplication by β (see Chap. 16) and $\sigma_3 \equiv 2\hat{\beta}^*\hat{\beta} - \mathbb{1}$. But with this choice $\hat{H}$ is not an even operator. To overcome this problem we add a new Grassmann pair of complex variables (θ, θ^*) and the real Grassmann number $\eta = \theta + \theta^*$. We have $\hat{\eta}^2 = \mathbb{1}$ (In [1] $\hat{\eta}$ is denoted c and is called a Clifford number, see Chap. 16 for more explanations). Hence we have the two pairs of Grassmann variables (β, β^*), (θ, θ^*) to represent the Pauli matrices:

$$\sigma_+ \longleftrightarrow \hat{\eta}\hat{\beta}^*, \quad \sigma_- \longleftrightarrow \hat{\eta}\hat{\beta}$$

where we denote $\hat{\eta} = \theta + \partial_\theta$, $\mathbf{b} = \hat{\beta}$, $\mathbf{b}^* = \partial_\beta$. We have to remark that $(\mathbf{b}\hat{\eta})^*\mathbf{b}\hat{\eta} = \mathbf{b}^*\mathbf{b}$. With this substitution the Hamiltonian $\hat{H}$ is transformed into an even Hamiltonian defined in the space $L^2(\mathbb{R}) \otimes \tilde{\mathscr{H}}^{(2)}$

$$\hat{H}_S = \Omega\mathbf{a}^*\mathbf{a} + \frac{1}{2}\omega(\mathbf{b}^*\mathbf{b} - \mathbb{1}) + g(\mathbf{a}^* + \mathbf{a})(\hat{\eta}\mathbf{b}^* + \mathbf{b}\hat{\eta}) \tag{17.114}$$

Its (contravariant) Weyl symbol is

$$H_S(\zeta, \beta, \theta) = \Omega\left(|\zeta|^2 - \frac{1}{2}\right) + \omega\beta^*\beta + g(\zeta + \bar{\zeta})(\eta\beta^* + \beta\eta).$$

Our aim is to study the dynamics for the Hamiltonian $\hat{H}_S$. It is determined by the von Neumann equation

$$i\frac{\partial\hat{\rho}_t}{\partial t} = [\hat{H}_S, \hat{\rho}_t], \quad \hat{\rho}_{t=0} = \hat{\rho}_0, \tag{17.115}$$

where $\hat{\rho}$ is a density operator (a positive operator of trace 1), ρ_t is the contravariant Weyl symbol of $\hat{\rho}_t$ (also called the Weyl–Wigner Distribution Function).

Let us remark that $\hat{H}_S$ depends only on $\mathbf{a}$ and $\hat{\gamma}$ where $\gamma = \beta\eta$, because we have $\mathbf{b}^*\mathbf{b} = \hat{\gamma}^*\hat{\gamma}$. Moreover we have $[\hat{\gamma}, \hat{\gamma}^*]_+ = \mathbb{1}$.

ρ_t satisfies a Fokker-Planck type equation which can be computed using the Moyal product formula applied to $H_S \circledast \rho_t - \rho_t \circledast H_S$. So we get

$$-\frac{\partial\rho_t}{\partial t} = \overrightarrow{L}\rho_t + \rho_t\overleftarrow{L}^* \tag{17.116}$$

where we find (see also [1])

$$-i\overrightarrow{L} = \Omega(\zeta^* - \frac{1}{2}\overrightarrow{\partial}_{\zeta})(\zeta + \frac{1}{2}\overrightarrow{\partial}_{\zeta^*}) + \omega\left((\beta^* + \frac{1}{2}\overrightarrow{\partial}_{\beta})(\beta + \frac{1}{2}\overrightarrow{\partial}_{\beta^*}) - \frac{1}{2}\right) +$$
$$g\left(\zeta + \zeta^* + \frac{1}{2}(\overrightarrow{\partial}_{\zeta^*} - \overrightarrow{\partial}_{\zeta})\right) + (\theta + \overrightarrow{\partial}_{\theta})\left(\beta^* - \beta + \frac{1}{2}(\overrightarrow{\partial}_{\zeta} - \overrightarrow{\partial}_{\zeta})\right). \tag{17.117}$$

$\hat{H}_S$ being time independent we also have

$$\hat{\rho}_t = \mathrm{e}^{-it\hat{H}_S}\hat{\rho}_0\mathrm{e}^{it\hat{H}_S}.$$

So we start to consider the propagation of the dynamical variables $\mathbf{a}$ and $\hat{\gamma}$. We first compute the commutators

$$[\hat{H}_S, \hat{\gamma}] = -\hat{\gamma} + g(\mathbf{a} + \mathbf{a}^*)(2\hat{\gamma}^*\hat{\gamma} - \mathbb{1}). \tag{17.118}$$
$$[\hat{H}_S, \hat{\gamma}^*] = \hat{\gamma}^* - g(\mathbf{a} + \mathbf{a}^*)(2\hat{\gamma}^*\hat{\gamma} - \mathbb{1}). \tag{17.119}$$
$$[\hat{H}_S, \mathbf{a}] = -g(\hat{\gamma}^* + \hat{\gamma}) - \Omega\mathbf{a}. \tag{17.120}$$
$$[\hat{H}_S, \mathbf{a}^*] = g(\hat{\gamma}^* + \hat{\gamma}) + \Omega\mathbf{a}^*. \tag{17.121}$$

So we find the non-linear differential system

$$\begin{aligned}
i\frac{\partial}{\partial t}\hat{\gamma}_t &= -\hat{\gamma}_t + g(\mathbf{a} + \mathbf{a}^*)(2\hat{\gamma}_t^*\hat{\gamma}_t - \mathbb{1}) \\
i\frac{\partial}{\partial t}\hat{\gamma}_t^* &= \hat{\gamma}_t^* - g(\mathbf{a} + \mathbf{a}^*)(2\hat{\gamma}_t^*\hat{\gamma}_t - \mathbb{1}) \\
i\frac{\partial}{\partial t}\mathbf{a}_t &= -g(\hat{\gamma}_t + \hat{\gamma}_t^*) - \Omega\mathbf{a}_t \\
i\frac{\partial}{\partial t}\mathbf{a}_t^* &= -g(\hat{\gamma} + \hat{\gamma}^*) + \Omega\mathbf{a}_t^*.
\end{aligned}$$

Assume now that the initial Weyl–Wigner distribution $\hat{\rho}$ depends only on $\mathbf{a}$, $\mathbf{a}^*$, $\hat{\gamma}$, $\hat{\gamma}^*$. Then $\hat{\rho}_t$ has the following shape

$$\hat{\rho}_t = \hat{\rho}_t^0(\mathbf{a}_t, \mathbf{a}_t^*) + \hat{\rho}_t^1(\mathbf{a}_t, \mathbf{a}_t^*)\hat{\gamma}_t + \hat{\rho}_t^{1,*}(\mathbf{a}_t, \mathbf{a}_t^*)\hat{\gamma}_t^* + \hat{\rho}_t^2(\mathbf{a}_t, \mathbf{a}_t^*)\hat{\gamma}_t^*\hat{\gamma}_t.$$

Using (17.122) we get that $\mathbf{a}_t$, $\mathbf{a}_t^*$, $\hat{\gamma}_t^*\hat{\gamma}_t$ are functions of t and $\mathbf{a}$, $\mathbf{a}^*$, $\hat{\gamma}$, $\hat{\gamma}^*$. So $\hat{\rho}_t$ can be written as follows

$$\hat{\rho}_t = \hat{\rho}_t^{(0)}(\mathbf{a}, \mathbf{a}^*) + \hat{\rho}_t^{(1)}(\mathbf{a}, \mathbf{a}^*)\hat{\gamma} + \hat{\rho}_t^{(1)*}(\mathbf{a}, \mathbf{a}^*)\hat{\gamma}^* + \hat{\rho}_t^{(2)}(\mathbf{a}, \mathbf{a}^*)\hat{\gamma}^*\hat{\gamma}. \tag{17.122}$$

Now we can give the physical meaning of the bosonic operators $\hat{\rho}_t^{(\bullet)}(\mathbf{a}_t, \mathbf{a}_t^*)$ by taking the average on the fermionic variables (β, θ).

We shall see now that we recover the dynamics generated by the Hamiltonian $\hat{H}$ taking the basis $\{\mathbb{1}_2, \sigma_+, \sigma_-, \sigma_3\}$.

We have to compute the relative trace $\mathrm{Tr}_{\tilde{\mathscr{H}}_2}\left(\hat{\rho}_t(\mathbb{1}\otimes\hat{G})\right)$, where $\hat{G}$ is a fermionic operator in $\tilde{\mathscr{H}}_2$, using the formula

$$\mathrm{Tr}_{\tilde{\mathscr{H}}_2}\left(\hat{\rho}_t(\mathbb{1}\otimes\hat{G})\right)=\frac{1}{4}\int \rho_t(\mathbf{a},\mathbf{a}^*,\beta,\theta)G_w(\beta,\theta)d^2\beta d^2\theta.$$

We compute the covariant Weyl symbols of $\mathbb{1}_2, \hat{\gamma}, \hat{\gamma}^*, \hat{\gamma}^*\hat{\gamma}$.

$$\begin{aligned}
1_w(\beta,\theta) &= \beta^*\beta+\theta^*\theta.\\
\gamma_w(\beta,\theta) &= \beta(\theta-\theta^*).\\
\gamma_w^*(\beta,\theta) &= (\theta^*-\theta)\beta^*.\\
(\gamma^*\gamma_w)(\beta,\theta) &= \theta^*\theta.
\end{aligned}$$

Then we get effective distribution functions ρ_{eff} (for simplicity the time t is not written explicitly nor the contravariant bosonic variable ζ)

$$\begin{aligned}
\rho_{\mathrm{eff}}^{(0)} &= \int \rho\times(\theta^*\theta+\beta^*\beta)d^2\theta d^2\beta=\rho^{(0)}\\
\rho_{\mathrm{eff}}^{(1)} &= \int \rho\times\beta(\theta+\theta^*)(\theta-\theta^*)\beta^*d^2\theta d^2\beta=-2\rho^{(1)}\\
\rho_{\mathrm{eff}}^{(1)*} &= \int \rho\times(\theta+\theta^*)\beta^*\beta(\theta-\theta^*)d^2\theta d^2\beta=-2\rho^{(1)*}\\
\rho_{\mathrm{eff}}^{(2)} &= 2\int \rho\times\theta^*\theta d^2\theta d^2\beta=2\rho^{(2)}.
\end{aligned}$$

Finally after a direct computation we recover the dynamics for the Hamiltonian $\hat{H}$ defined by (17.113) from the dynamics for the super Hamiltonian $\hat{H}_S$ defined by (17.114):

$$\rho_{\mathrm{eff}}=\rho_{\mathrm{eff}}^{(0)}\mathbb{1}_2+\rho_{\mathrm{eff}}^{(1)}\sigma_+ + \rho_{\mathrm{eff}}^{(1),*}\sigma_- + \rho_{\mathrm{eff}}^{(2)}\sigma_3.$$

□

Appendix A
Tools for Integral Computations

A.1 Fourier Transform of Gaussian Functions

This result is the starting point for the stationary phase theorem.

Let M be a complex matrix such that $\Re M$ is positive-definite. We define $[\det M]_*^{1/2}$ the analytic branch of $(\det M)^{1/2}$ such that $(\det M)^{1/2} > 0$ when M is real.

Theorem 88 *Let $\mathscr{A}$ be a symmetric complex symmetric matrix, $m \times m$. We assume that $\Im \mathscr{A}$ is non negative and $\mathscr{A}$ is non degenerate. Then we have the Fourier transform formula for the Gaussian $\mathrm{e}^{i\mathscr{A}x\cdot x/2}$*

$$\int_{\mathbb{R}^m} \mathrm{e}^{i\mathscr{A}x\cdot x/2}\mathrm{e}^{-ix\cdot\xi}d\xi = (2\pi)^{m/2}[\det(-i\mathscr{A})]_*^{-1/2}\mathrm{e}^{(i\mathscr{A})^{-1}\xi\cdot\xi/2}. \qquad (A.1)$$

Proof For $\mathscr{A}$ real formula (A.1) is well known: first we prove it for $m = 1$ then for $m \geq 2$ by diagonalizing $\mathscr{A}$ and using a linear change of variables.

For $\mathscr{A}$ complex (A.1) is obtained by analytic extension of left and right hand side. □

A.2 Sketch of Proof for Theorem 29

Recall that critical set M of the phase f is

$$M = \{x \in \mathscr{O}, \Im f(x) = 0, f'(x) = 0\} \quad .$$

Note that if a is supported outside this set then $J(\omega)$ is $O(\omega^{-\infty})$.

Using a partition of unity, we can assume that $\mathscr{O}$ is small enough that we have normal, geodesic coordinates in a neighborhood of M. So we have a diffeomorphism,

D. Robert and M. Combescure, *Coherent States and Applications in Mathematical Physics*, Theoretical and Mathematical Physics,
https://doi.org/10.1007/978-3-030-70845-0

$$\chi : \mathscr{U} \to \mathscr{O} \quad ,$$

where $\mathscr{U}$ is an open neighborhood of (0, 0) in $\mathbb{R}^k \times \mathbb{R}^{d-k}$, such that

$$\chi(x', x'') \in M \iff x'' = 0$$

and if $m = \chi(x', 0) \in M$ we have

$$\chi'(x', 0)(\mathbb{R}^k_x) = T_m M$$

$$\chi'(x', 0)(\mathbb{R}^{d-k}_{x''}) = N_m M\,, \text{(normal space at } m \in M) \quad .$$

So the change of variables $x = \chi(x', x'')$ gives the integral

$$J(\omega) = \int_{\mathbb{R}^d} e^{i\omega f(\chi(x',x''))} a(x', x'')|\det \chi'(x', x'')| dx' dx'' \quad . \tag{A.2}$$

The phase

$$\tilde{f}(x', x'') := f(\chi(x', x'')).$$

clearly satisfies

$$\{\tilde{f}'_{x''}(x', x'') = 0, \Im\tilde{f}(x', x'') = 0\} \iff x'' = 0$$

Hence, we can apply the stationary phase Theorem 7.7.5 of [153] in the variable x'', to the integral (A.2), where x' is a parameter (the assumptions of [153] are satisfied, uniformly for x' close to 0). We remark that all the coefficients c_j of the expansion can be computed using the above local coordinates and Theorem 7.7.5.

A.3 A Determinant Computation

Here we give the details concerning computations of the determinant (9.33) in Chap. 9.

We write $\alpha = (\alpha', \alpha_4)$ where $\alpha' = \hat{e}_1 + i\cos\gamma\hat{e}_2$ The gradient of the phase (9.28) is $ix + G(p)$ where $G(p)$ is given by

$$G(p) := \frac{2}{(p^2+1)\alpha \cdot w_1(p)} \left(\alpha' + p(\alpha_4 - \alpha \cdot w_1(p)\right) = K(p)M(p)$$

where

$$K(p) = \frac{2}{2\alpha' \cdot p + \alpha_4(p^2 - 1)}$$

and

$$M(p) = \alpha' + \frac{2p}{1+p^2}(\alpha_4 - \alpha' \cdot p)$$

$$K : \mathbb{R}^3 \to \mathbb{C}, \quad M : \mathbb{R}^3 \to \mathbb{C}^3.$$

Since x is a constant the hessian of (9.28) is simply $DG(p)$ where DG (resp. DK, DM) is the first differential of G (resp. K, M). We want to calculate at the critical point p^c with

$$p^c = \left(\frac{\cos\beta}{1 - \sin\beta \sin\gamma}, \frac{\cos\gamma \sin\beta}{1 - \sin\beta \sin\gamma}, 0 \right).$$

Let δp be an arbitrary increase of p. One has:

$$DG(p)(\delta p) = (DK(p) \cdot \delta p) \cdot M(p) + K(p) DM(p) \cdot \delta p$$

We can write;

$$(DK(p) \cdot \delta p)\, M(p) = \left(M(p) \otimes (DK(p))^* \right) \cdot \delta p$$

where $DK(p)^* \in (\mathbb{R}^3)^* + i(\mathbb{R}^3)^* \simeq \mathbb{C}^3$. Using the dual structure the identification of $(\mathbb{R}^3)^* + i(\mathbb{R}^3)^*$ with $\mathbb{C}^3$ is performed via the isomorphism: $u \mapsto (v \mapsto u \cdot v)$. We have:

$$K(p^c) = e^{-i\beta}(1 - \sin\beta \sin\gamma)$$

$$M(p^c) = \alpha' + (i \sin\gamma - e^{i\beta}) p^c.$$

Thus

$$DG(p^c) = e^{-i\beta}(1 - \sin\beta \sin\gamma) DM(p^c) + M(p^c) \otimes (DK(p^c))^*$$

It is convenient to choose as a basis of vectors $(p^c, q^c, \hat{e}_3)$ where

$$q^c = \alpha' + (i \sin\gamma - e^{i\beta}) p^c.$$

The vectors p^c, q^c are $\mathbb{C}$-linearly independent for $\gamma \neq \frac{\pi}{2} + k\pi$. A simple calculus yields

$$DM(p^c) = (i \sin\gamma - e^{i\beta}) \mathbb{1}_{\mathbb{R}^3} - (1 - \sin\beta \sin\gamma) p^c \otimes (q^c)^*$$

$$DK(p^c) = -e^{-2i\beta}(1 - \sin\beta \sin\gamma)^2 (\alpha' + \alpha_4 p^c)$$

Thus we get

$$DG(p^c) = e^{-i\beta}(i \sin\gamma - e^{i\beta})(1 - \sin\beta \sin\gamma) \mathbb{1}_{\mathbb{R}^3} \tag{A.3}$$

$$-e^{-i\beta}(1 - \sin\beta \sin\gamma)^2 p^c \otimes (q^c)^* - e^{-2i\beta}(1 - \sin\beta \sin\gamma)^2 q^c \otimes (q^c + e^{i\beta} p^c)^*$$

Let $H(p)$ be the Hessian matrix in the basis $(p^c, q^c, \hat{e}_3)$ and denote

$$H_1(p) = \frac{H(p)}{1 - \sin\beta \sin\gamma}.$$

The second line of (A.3) yields a matrix in the plane generated by (p^c, q^c) of the form

$$\begin{pmatrix} d_1 & d_2 \\ d_3 & d_4 \end{pmatrix}$$

that we calculate using

$$p^c \otimes (q^c)^* = \begin{pmatrix} q^c \cdot p^c & (q^c)^2 \\ 0 & 0 \end{pmatrix}$$

$$q^c \otimes (q^c)^* = \begin{pmatrix} 0 & 0 \\ q^q \cdot p^c & (q^c)^2 \end{pmatrix}$$

$$q^c \otimes (p^c)^* = \begin{pmatrix} 0 & 0 \\ (p^c)^2 & p^c \cdot q^c \end{pmatrix}.$$

We have

$$(p^c)^2 = \frac{1 + \sin\beta \sin\gamma}{1 - \sin\beta \sin\gamma}$$

$$q^c \cdot p^c = \sin\gamma \frac{i - e^{i\beta} \sin\beta}{1 - \sin\beta \sin\gamma}$$

$$(q^c)^2 = -e^{2i\beta}$$

We get

$$d_1 = -(1 - \sin\beta \sin\gamma)$$

$$d_2 = -e^{-i\beta}(1 - \sin\beta \sin\gamma)$$

$$d_3 = -e^{-i\beta}(1 + i \sin\gamma e^{-i\beta}).$$

$$d_4 = 0.$$

Thus the three-dimensional reduced Hessian matrix H_1 equals

$$H_1 = \begin{pmatrix} d_1 & d_2 & 0 \\ d_3 & 0 & 0 \\ 0 & 0 & d_5 \end{pmatrix}$$

with $d_5 = e^{-i\beta}(i \sin\gamma - e^{i\beta})$. Its determinant equals

$$\det H_1 = -d_2 d_3 d_5.$$

One has

$$|\det H_1|^2 = (1 - \sin\beta \sin\gamma)^2[1 - 2\sin\beta\sin\gamma + \sin^2\gamma][1 + 2\sin\beta\sin\gamma + \sin^2\gamma].$$

Finaly we get the result:

$$|\det H(p^c)| = (1 - \sin\beta\sin\gamma)^4\sqrt{[\sin^2\gamma + 2\sin\beta\sin\gamma + 1][\sin^2\gamma - 2\sin\beta\sin\gamma + 1]}.$$

We see that it doesn't vanishes provided $\sin\beta\sin\gamma \neq 1$.

A.4 The Saddle Point Method

A.4.1 The One Real Variable Case

This result is elementary and very explicit. Let us consider the Laplace integral

$$I(\lambda) = \int_0^a e^{-\lambda\phi(r)} F(r)dr$$

where $a > 0$, ϕ and F are smooth functions on $[0, 1]$ such that $\phi'(0) = \phi(0) = 0$, $\phi''(0) > 0$, $\phi(r) > 0$ if $r \in]0, 1]$. Under these conditions we can perform the change of variables $s = \sqrt{\phi(r)}$ where we denote $r = r(s)$ and $G(s) = F(r(s))r'(s)$. So we have

Proposition 164 *For every $N \geq 1$ we have the asymptotic expansion for $\lambda \to +\infty$,*

$$I(\lambda) = \sum_{0 \leq j \leq N-1} C_j \lambda^{-(j+1)/2} + O(\lambda^{-(N+1)/2}), \tag{A.4}$$

where

$$C_j = \Gamma\left(\frac{j}{2} + 1\right)\frac{G^{(j)}(0)}{2j!}.$$

In particular $C_0 = F(0)\left(\phi''(0)\right)^{-1/2}$.

Proof This is a direct consequence of Taylor expansion applied to G at 0 and using that, for every $b > 0$ and $\varepsilon > 0$,

$$\int_0^b e^{-\lambda s^2} s^j ds = \frac{1}{2}\Gamma\left(\frac{j}{2} + 1\right)\lambda^{-(j+1)/2} + O(e^{-\varepsilon\lambda}).$$

□

A.4.2 The Complex Variables Case

This is an old subject for one complex variable but there are not so many references for several complex variables. Here we recall a presentation given by Sjöstrand [238] or [237].

Let us consider a complex holomorphic phase function in a open neighborhood $\mathscr{A} \times \mathscr{U}$ of (0, 0) in $\mathbb{C}^k \times \mathbb{C}^n$, $\mathscr{A} \times \mathscr{U} \ni (a, u) \mapsto \varphi(a, u) \in \mathbb{C}$.

Assume that

- $\varphi(0, 0) = 0$, $\partial_u\varphi(0, 0) = 0$.
- $\det\partial^2_{(u,u)}\varphi(0, 0) \neq 0$.
- $\Re\varphi \geq 0\ \forall u \in \mathscr{U}$, $\Re\varphi > 0$ for all $u \in \partial\mathscr{U}_{\mathbb{R}}$.

  where $\mathscr{U}_{\mathbb{R}} := \mathscr{U} \cap \mathbb{R}^n$ and $\partial\mathscr{U}_{\mathbb{R}}$ is the boundary in $\mathbb{R}^n$ of $\mathscr{U}_{\mathbb{R}}$.

By the implicit function theorem we can choose $\mathscr{A} \times \mathscr{U}$ small enough such that the equation $\partial_u\varphi(a, z) = 0$ has a unique solution $z(a) \in \mathscr{U}$, $a \mapsto z(a)$ being holomorphic in $\mathscr{A}$. Then we have the following asymptotic result

Theorem 89 *For every holomorphic and bounded function g in $\mathscr{U}$ we have, for $k \to +\infty$,*

$$e^{k\varphi(a,z(a))} \int_{\mathscr{U}_{\mathbb{R}}} e^{-k\varphi(a,\mathbf{r})} g(\mathbf{r}) d\mathbf{r} =$$

$$\left(\frac{2\pi}{k}\right)^{n/2} \left[\det\left(\partial^2_{(u,u)}\varphi(a, z(a))\right)\right]^{-1/2} g(a) \ + \ O(k^{-n/2-1}). \tag{A.5}$$

A.5 Kähler Geometry

Let M be a complex manifold and h an Hermitian form on M:

$$h = \sum_{j,k} h_{j,k} dz_j \otimes dz_k, \quad \overline{h_{j,k}(z)} = h_{k,j}(z).$$

h is a Kähler form if the imaginary part $\omega = \Im h$ is closed ($d\omega = 0$) and its real part $g = \Re h$ is positive definite.

$$\omega = i \sum_{j,k} h_{j,k}(z) dz_j \wedge d\bar{z}_k.$$

(M, h) is said a Kähler manifold if h is a Kähler form on M.

Then on M exists a Riemann metric $g = \Re h$ and symplectic two form ω.

Locally there exists a real -valued function K, called Kähler potential, such that

$$h_{j,k} = \frac{\partial^2}{\partial z_j \partial z_k} K.$$

The Poisson bracket of two smooth functions ϕ, ψ on M is defined as follows

$$\{\phi, \psi\}(m) = \omega(X_\phi, X_\psi)$$

where X_ψ is the Hamiltonian vector field at m defined such that

$$d\psi(v) = \omega(X_\psi, v), \text{ for all v} \in T_m(M).$$

More explicitly:

$$\{\psi, \phi\}(z) = i \sum_{j,k} h^{j,k}(z) \left(\frac{\partial \psi}{\partial z_j} \frac{\partial \phi}{\partial \bar{z}_k} - \frac{\partial \phi}{\partial z_k} \frac{\partial \psi}{\partial \bar{z}_j} \right)$$

where $h^{j,k}(z)$ is the inverse matrix of $h_{j,k}(z)$.

The Laplace-Beltrami operator corresponding to the metric g is

$$\Delta = \sum_{j,k} h^{j,k}(z) \frac{\partial}{\partial z_j} \frac{\partial}{\partial \bar{z}_k}.$$

In particular for the Riemann sphere we have:

$$ds^2 = 4 \frac{d\zeta d\bar{\zeta}}{(1 + |\zeta|^2)^2} \tag{A.6}$$

$$\{\psi, \phi\}(z) = i(1 + |z|^2)^2 \left(\frac{\partial \psi}{\partial z} \frac{\partial \phi}{\partial \bar{z}} - \frac{\partial \phi}{\partial z} \frac{\partial \psi}{\partial \bar{z}} \right) \tag{A.7}$$

$$\Delta = (1 + |z|^2)^2 \frac{\partial^2}{\partial z \partial \bar{z}}. \tag{A.8}$$

For the pseudo-sphere we have:

$$ds^2 = 4 \frac{d\zeta d\bar{\zeta}}{(1 - |\zeta|^2)^2} \tag{A.9}$$

$$\{\psi, \phi\}(z) = i(1 - |z|^2)^2 \left(\frac{\partial \psi}{\partial z} \frac{\partial \phi}{\partial \bar{z}} - \frac{\partial \phi}{\partial z} \frac{\partial \psi}{\partial \bar{z}} \right) \tag{A.10}$$

$$\Delta = (1 - |z|^2)^2 \frac{\partial^2}{\partial z \partial \bar{z}}. \tag{A.11}$$

Appendix B
Remainder Estimate for the Moyal Product

We give here a remainder estimate for the Moyal product for matrix valued symbols. Let us first recall the formal product rule for quantum observables with Weyl quantization. Let $A, B \in \mathscr{S}(\mathbb{R}^{2n}, \mathbb{C}^{m,m})$. The Moyal product $C := A \circledast B$ is the semi-classical observable C such that $\hat{A} \cdot \hat{B} = \hat{C}$. Some computations with the Fourier transform give the following well known formula [153, 217]

$$C(x,\xi) = \exp\left(\frac{i\hbar}{2}\sigma(D_q, D_p; D_{q'}, D_{p'})\right) A(q,p)B(q',p')|_{(q,p)=(q',p')=(x,\xi)}, \quad \text{(B.1)}$$

where σ is the symplectic bilinear form $\sigma((q,p),(q',p')) = p \cdot q' - p' \cdot q$ and $D = i^{-1}\nabla$. By expanding the exponential term, we get

$$C(x,\xi) = \sum_{j\geq 0} \frac{\hbar^j}{j!}(\frac{i}{2}\sigma(D_q, D_p; D_{q'}, D_{p'}))^j A(q,p)B(q',p')|_{(q,p)=(q',p')=(x,\xi)}. \quad \text{(B.2)}$$

So that $C(x, \xi)$ is a formal power series in $\hbar$ with coefficients given by

$$C_j(x,\xi) = \frac{1}{2^j} \sum_{|\alpha+\beta|=j} \frac{(-1)^{|\beta|}}{\alpha!\beta!}(D_x^\beta \partial_\xi^\alpha A).(D_x^\alpha \partial_\xi^\beta B)(x,\xi). \quad \text{(B.3)}$$

It is enough for our purpose to assume that A, B behave like polynomials at infinity. For $\mu \geq 0$ denote $\mathbf{P}_{(m,m)}(\mu)$ the linear space of C^∞ symbols $A : \mathbb{R}^{2n} \to \mathbb{C}^{m,m}$ such that

$$|\partial_X^\gamma A(X)| \leq C_\gamma, \text{ for any } |\gamma| \geq \mu.$$

Assuming that $A \in \mathbf{P}_{(m,m)}(\mu_A)$, $B \in \mathbf{P}_{(m,m)}(\mu_B)$ it is known that, for $\hbar$ fixed, $\hat{A}\hat{B} = \hat{C}$ where $C \in \mathbf{P}_{(m,m)}(\mu_C)$, $\mu_C \geq \max\{\mu_A, \mu_B\}$ (see [153]). Recall the notation $C = A \circledast B$.

D. Robert and M. Combescure, *Coherent States and Applications in Mathematical Physics*, Theoretical and Mathematical Physics,
https://doi.org/10.1007/978-3-030-70845-0

We need here better estimates in small $\hbar > 0$ and a control in the size of the derivatives of A, B.

We introduce the notations $X = (x, \xi) \in \mathbb{R}^n_x \times \mathbb{R}^n_\xi$ and for every $N \geq 1$,

$$A \circledast B(X) - \sum_{0 \leq j \leq N} \hbar^j C_j(X) =: R_N(A, B; X; \hbar). \tag{B.4}$$

The following estimate is a particular case of Theorem A.1 in [40].

Theorem 90 *There exist N_0, N_1 such that for all $N \geq N_0$ and every $\gamma \in \mathbb{N}^{2d}$, there exists a constant $K(N, \gamma)$ such that for any $A \in \mathbf{P}_{(m,m)}(\mu_A)$, $B \in \mathbf{P}_{(m,m)}(\mu_B)$ for every multi-index γ, the following estimate holds, for every $X \in \mathbb{R}^{2n}$,*

$$|\partial_X^\gamma \left(A \circledast B(X) - \sum_{0 \leq j \leq N} \hbar^j C_j(X) \right) | \leq \hbar^{N+1} K(N, \gamma)$$
$$\sum_{N+1 \leq |\alpha|, |\beta| \leq N+N_1+|\gamma|} \|\partial_X^\alpha A\|_{L^\infty} \|\partial_X^\beta B\|_{L^\infty} \tag{B.5}$$

Proof We follow [40] Appendix A. By Fourier transform computations and Taylor formula, we get the following formula

$$R_N(A, B; X; \hbar) = \frac{1}{N!} \left(\frac{i\hbar}{2} \right)^{N+1} \int_0^1 (1-t)^N R_{N,t}(X; \hbar) dt \quad \text{where} \quad R_{N,t}(X; \hbar) =$$
$$(2\pi \hbar t)^{-2n} \int\int_{\mathbb{R}^{2n} \times \mathbb{R}^{2n}} \exp(-\frac{i}{2t\hbar} \sigma(u, v)) \sigma^{N+1}(D_u, D_v) A(u+X) B(v+X) du dv. \tag{B.6}$$

Notice that the integral is an oscillating integral as we shall see below. We shall use the following lemma to estimate $R_{N,t}(X; \hbar)$.

Lemma 109 *Let us consider $F \in \mathscr{S}(\mathbb{R}^{2n} \times \mathbb{R}^{2n})$ and the integral*

$$I(\lambda) = \lambda^{2n} \int\int_{\mathbb{R}^{2n} \times \mathbb{R}^{2n}} \exp[-i\lambda\sigma(u, v)] F(u, v) du dv. \tag{B.7}$$

Then for every real number $m > 2$ there exists $C(m, d) > 0$ such that the following estimate holds

$$|I(\lambda)| \leq C(m, d) \sup_{\substack{u,v \in \mathbb{R}^{2n} \\ |\alpha+\beta| \leq m}} |\partial_u^\alpha \partial_v^\beta F(u, v)|. \tag{B.8}$$

Proof The lemma is proved in a standard way, using cut-off and integration by parts as follows. Let us introduce a cut-off χ_0, C^∞ on $\mathbb{R}$, $\chi_0(x) = 1$ for $|x| \leq 1/2$ and $\chi_0(x) = 0$ for $|x| \geq 1$. We split $I(\lambda)$ into three pieces

$$I_0(\lambda) = \lambda^{2n} \iint_{\mathbb{R}^{2n}\times\mathbb{R}^{2n}} \exp[-i\lambda\sigma(u, v), \chi_0(\lambda(u^2+v^2))F(u, v)dudv,$$

$$I_1(\lambda) = \lambda^{2n} \iint_{\mathbb{R}^{2n}\times\mathbb{R}^{2n}} \exp[-i\lambda\sigma(u, v)](1-\chi_0(\lambda(u^2+v^2))\chi_0(u^2+v^2)F(u, v)dudv,$$

$$I_2(\lambda) = \lambda^{2n} \iint_{\mathbb{R}^{2n}\times\mathbb{R}^{2n}} \exp[-i\lambda\sigma(u, v)](1-\chi_0(u^2+v^2))F(u, v)dudv.$$

For $I_0(\lambda)$, we easily have

$$|I_0(\lambda)| \le \omega_{4d} \sup_{u^2+v^2\le 1} |F(u, v)|, \tag{B.9}$$

where ω_{4d} is the volume of the unit ball in $\mathbb{R}^{4n}$. For $I_1(\lambda)$ and $I_2(\lambda)$, we integrate by parts with the differential operator

$$L = \frac{i}{u^2+v^2}\left(Ju\frac{\partial}{\partial v} - Jv\frac{\partial}{\partial u}\right), \tag{B.10}$$

where J is the matrix associated to the symplectic form ($\sigma(u, v) =< Ju, v >$). Performing $4n$ integrations by parts, we can see that it exists a constant c_n such that

$$|I_1(\lambda)| \le c_n \sup_{\substack{u^2+v^2\le 1\\ |\mu|+|\nu|\le 4n}} |\partial_u^\mu \partial_v^\nu F(u, v)|. \tag{B.11}$$

Similarly, performing $m > nd$ integrations by parts, we get for a constant $c(d, m)$,

$$|I_2(\lambda)| \le c_n \sup_{\substack{u,v\in\mathbb{R}^{2n}\\ |\mu|+|\nu|\le m}} |\partial_u^\mu \partial_v^\nu F(u, v)|. \tag{B.12}$$

□

Now we can complete the proof of the theorem by using Lemma 109, the Leibniz formula and the following elementary estimate, using in $\mathbb{R}^{2n}$ the coordinates $u = (x, \xi)$, $v = (y, \eta)$,

$$\begin{aligned}&|\sigma^N(\partial_x, \partial_\xi; \partial_y, \partial_\eta))A(x, \xi)B(y, \eta)| \le \\ &(2n)^N \sup_{|\alpha|+|\beta|=N} |\partial_x^\alpha \partial_\xi^\beta A(x, \xi)\partial_y^\beta \partial_\eta^\alpha B(y, \eta)|.\end{aligned} \tag{B.13}$$

From the Lemma we get first the Theorem for $A, B \in \mathscr{S}(\mathbb{R}^{2n})$ with universal constants. Then the theorem is proved for any $A, B \in \mathbf{P}_{(m,m)}(\mu)$ by smoothing: $A_\eta(u) = \mathrm{e}^{-\eta u^2}A(u)$ and $B_\eta(v) = \mathrm{e}^{-\eta v^2}B(v)$ for $\eta > 0$ and passing to the limit $\eta \to 0$.

Appendix C
Semi-classical Functional Calculus

C.1 Introduction

In this appendix we give a proof of Theorem 10. There are many papers or books proving that if $\hat{H}$ is a self-adjoint $\hbar$-pseudodifferential operator and f a complex function on $\mathbb{R}$ then $f(\hat{H})$ (defined by the abstract functional calculus) is again a $\hbar$-pseudodifferential operator. Recall that $\hat{H}$ is supposed to be a uniformly elliptic semi-classical Hamiltonian. Moreover it is a self-adjoint operator in $L^2(\mathbb{R}^n)$.

There are several works concerning functional calculus for pseudodifferential operators and each of them have considered different classes of symbols and different classes of functions f.

Notice that if f is holomorphic on a suitable neighborhood of the real axis the problem is solved by proving first that the resolvent $(\hat{H} - z)^{-1}$ is a pseudodifferential operator with a Weyl symbol R_z^H for $z \in \mathbb{C}\backslash\mathbb{R}$ satisfying suitable estimates with a good in $\Im(z)$. Then the functional calculus result follows by using the Cauchy formula on a suitable contour Γ

$$f(\hat{H}) = \frac{1}{2i\pi} \oint_{\Gamma} (z - \hat{H})^{-1} f(z) dz. \tag{C.1}$$

In particular the leading symbol of $f(\hat{H})$ is $f(H(x, \xi))$.

For f only C^∞ smooth on $\mathbb{R}$ the Cauchy formula can be replaced by an almost analytic extension of f, following an idea of Helffer-Sjöstrand [143]. The main motivation is to consider smooth function with compact support allowing cut-off in energies. Here we shall show that the same method can be applied for f satisfying (C.2). f is C^∞-smooth such that there exists $r \in \mathbb{R}$ such that for all $j \in \mathbb{N}$ we have

$$|\frac{d^j f}{dt}(t)| \leq C_j < t >^{r-j}, \quad \forall t \geq 0. \tag{C.2}$$

D. Robert and M. Combescure, *Coherent States and Applications in Mathematical Physics*, Theoretical and Mathematical Physics,
https://doi.org/10.1007/978-3-030-70845-0

C.2 Functional Calculus and Almost Analytic Extension

Let$f$ be such that (C.2) is statisfied with $r < -1$. Notice here that this is not restriction because we can write$f(t) = (t+i)^m g(t)$ and taking m large enough,g satisfies ((C.2) with < -1.

Let $\chi \in C_0^\infty(\mathbb{R})$ be a cut-off function such that $\chi(t) \equiv 1$ in a neighborhood of 0. For a given integer $N \geq 1$ let us define

$$\tilde{f}_N(x+iy) = \sum_{0 \leq k \leq N} f^{(k)}(x) \frac{(iy)^k}{k!} \chi\left(\frac{|y|}{<x>}\right). \tag{C.3}$$

$\tilde{f}$ is an almost analytic extension of$f$ to $\mathbb{C}$. The following property was considered by Hörmander [153]. We use here the presentation given in [71]. $\tilde{f}$ is almost analytic in the following sense. Let us introduce the Cauchy-Riemann operator $\bar{\partial} = \frac{1}{2}\left(\frac{\partial}{\partial x} + i\frac{\partial}{\partial y}\right)$.

Lemma 110 *There exists a constant C_N such that*

$$|\bar{\partial}\tilde{f}_N(x+iy)| \leq C_N <x>^{r-N-1} |y|^N, \quad \forall (x+iy) \in \mathbb{C}. \tag{C.4}$$

with $0 < a < b$.

The proof of this lemma is elementary using (C.2) and (C.3). □

The functional calculus for pseudodifferential operators will be obtained from the following result due to Helffer-Sjöstrand for$f$ with a compact support.

Proposition 165 *Let$f$ satisfying (C.2) with $r < -1$ and$\tilde{f}_N$ an almost analytic extension of$f$ satisfying (C.4) with $N \geq 1$ then we have*

$$f(\hat{H}) = \frac{1}{\pi} \int_{\mathbb{C}} \bar{\partial}_\lambda \tilde{f}_N(\lambda)(\hat{H} - \lambda)^{-1} \mu_L(d\lambda) \tag{C.5}$$

where μ_L is the Lebesgue measure on the complexe plane $\mathbb{C}$.

Proof We follow the proof given in [83, Theorem 8.1].

Remark first that $\frac{1}{\pi\lambda}$ is a fondamental solution for the differential operator $\bar{\partial}_\lambda$, so we have, for every $\lambda \in \mathbb{R}$,

$$f(\nu) = -\frac{1}{\pi} \int_{\mathbb{C}} \bar{\partial}_\lambda \tilde{f}_N(\lambda)(\nu - \lambda)^{-1} \mu_L(d\lambda). \tag{C.6}$$

By the general spectral theorem for self-adjoint operators we have

$$<u, f(\hat{H})v> = \int_{\mathbb{R}} f(\lambda) d(<u, E_\lambda v>), \; u, v \in \mathscr{H}, \tag{C.7}$$

where E_λ is the spectral projector of $\hat{H}$ on $]-\infty, \lambda]$.

Hence plugging formula (C.6) in formula (C.7) and using Fubini's theorem we get the formula (C.5). □

C.2.1 Weyl Calculus and Resolvent Estimates

The main idea is to plug in (C.5) suitable approximations for $(\hat{H} - z)^{-1}$. These approximations (named parametrices) are obtained using the operational calculus (Moyal product) on semi-classical observables and uniform ellipticity of H.

The construction start with

$$b_{\lambda,0}(X) = (H_0(X) - \lambda)^{-1}, \; X = (x, \xi) \in \mathbb{R}^{2n}, \quad \lambda \in \mathbb{C}\backslash\mathbb{R},$$

using Weyl calculus we have

$$(\hat{H} - \lambda)\hat{b}_{\lambda,0} = \mathbb{1} + \hbar\hat{R}_{\lambda,1}$$

with

$$|\partial_X^\gamma R_{\lambda,1}(X)| \leq C\left|\frac{<\lambda>}{\Im\lambda}\right|^{n_0}$$

C, n_0 are independent on $\hbar$.

The last estimate is established using the following elementary estimates in the complex plane. Assume $H_0 > 0$ and $\arg\lambda \in [-\pi/2, \pi/2]$ then we have:

$$|H_0 - \lambda| \geq |\Im\lambda|, \quad |H_0 - \lambda| \geq \left|\frac{\Im\lambda}{\lambda}\right| |H_0|.$$

By induction on N, we get $b_{\lambda,j}$, for any $j \in \mathbb{N}$, such that we have

Theorem 91

$$(\hat{H} - \lambda)\left(\sum_{0\leq j\leq N} \hbar^j\hat{b}_{\lambda,j}\right) = \mathbb{1} + \hbar^{N+1}\hat{R}_{\lambda,N+1}$$

such that

$$\|\hat{R}_{\lambda,N+1}\|_{L^2,L^2} \leq C_N\left|\frac{<\lambda>}{\Im\lambda}\right|^{2N+n_0}$$

Moreover for $j \geq 1$, we have, for the symbol of the parametrix, the formula:

$$b_{\lambda,j} = \sum_{1\leq k\leq 2j-1} d_{j,k}b_{\lambda,0}^{k+1},$$

where $d_{j,k}$ are polynomials (independent on λ) in $\partial_X^\gamma H_\ell$ for $|\gamma| + \ell \leq j$.

Remark 113 The L^2 uniform estimates can be obtained from the constructions of the book [217] (Chap. III, Sect. 3) using improvement in the composition formula obtained in [40].

C.2.2 End of the Proof of Theorem 10

Plugging the parametrix obtained in Theorem 91 in formula (C.5) leads directly to the proof of the Theorem. □

Appendix D
Lie Groups and Coherent States

D.1 Lie Groups and Coherent States

In this appendix we start by a short review of some basic properties of Lie groups and Lie algebras. Then we explain some useful points concerning representation theory of Lie groups and Lie algebras and how they are used to build a general theory of Coherent States systems according to Perelomov [208, 209]. This theory is an extension of the examples already considered in Chaps. 7 and 8.

D.2 On Lie Groups and Lie Algebras

We recall here some basic definitions and properties. More details can be found in [88, 138] or in many other textbooks.

D.2.1 Lie Algebras

A Lie algebra $\mathfrak{g}$ is a vector space equipped with an anti-symmetric bilinear product: $(X, Y) \mapsto [X, Y]$ satisfying $[X, Y] = -[Y, X]$ and the Jacobi identity

$$[[X, Y], Z] + [[Y, Z], X] + [[Z, X], Y] = 0.$$

The map $(\mathrm{ad}\, X)Y = [X, Y]$ is a derivation: $(\mathrm{ad}\, X)[Y, Z] = [(\mathrm{ad}\, X)Y, Z] + [Y, (\mathrm{ad}\, X)Z]$.

If $\mathfrak{g}$ and $\mathfrak{h}$ are Lie algebras a Lie homomorphism is a linear map $\chi : \mathfrak{g} \to \mathfrak{h}$ such that $[\chi X, \chi Y] = [X, Y]$.

A sub-Lie algebra $\mathfrak{h}$ in $\mathfrak{g}$ is a subspace of $\mathfrak{g}$ such that $[\mathfrak{h}, \mathfrak{h}] \subseteq \mathfrak{h}$ $\mathfrak{h}$ is an ideal if furthermore we have $[\mathfrak{h}, \mathfrak{g}] \subseteq \mathfrak{h}$. If χ is a Lie homomorphism then $\ker \chi$ is an ideal.

D. Robert and M. Combescure, *Coherent States and Applications in Mathematical Physics*, Theoretical and Mathematical Physics,
https://doi.org/10.1007/978-3-030-70845-0

In the following we assume for simplicity that the Lie algebras $\mathfrak{g}$ considered are finite-dimensional.

$\mathfrak{g}$ is abelian if $[\mathfrak{g}, \mathfrak{g}] = 0$.

$\mathfrak{g}$ is simple if it is non-abelian and contains only the ideals $\{0\}$ and $\mathfrak{g}$. The centre $Z(\mathfrak{g})$ is defined as

$$Z(\mathfrak{g}) = \{X \in \mathfrak{g},\ [X, Y] = 0,\quad \forall Y \in \mathfrak{g}\}$$

$Z(\mathfrak{g})$ is an abelian ideal.

If $\mathfrak{g}$ has no abelian ideal except $\{0\}$ then $\mathfrak{g}$ is said semi-simple. In particular $Z(\mathfrak{g}) = \{0\}$.

The Killing form on $\mathfrak{g}$ is the symmetric bilinear form $B(X, Y)$ defined as

$$B(X, Y) = \mathrm{Tr}((\mathrm{ad}\, X)(\mathrm{ad}\, X)).$$

$\mathfrak{g}$ is semi-simple if and only if its Killing form is non-degenerate (Cartan's criterion).

D.2.2 Lie Groups

A Lie group G is a group equipped with a multiplication $(x, y) \mapsto x \cdot y$ and equipped with the structure of a smooth connected manifold (we don't recall here definitions and properties concerning manifolds, see [138] for details) such that the group operation $(x, y) \mapsto x \cdot y$ and $x \mapsto x^{-1}$ are smooth maps.

We always assume that Lie groups considered here are analytic.

A useful mapping is the conjugation $C(x)(y) = xyx^{-1}, x, y \in G$. Its tangent mapping at $y = e$ is denoted $\mathrm{Ad}(x)$. $Ad(x) \in GL(\mathfrak{g})$ and $x \mapsto \mathrm{Ad}(x)$ is a homomorphism from G into $GL(\mathfrak{g})$. It is called the adjoint representation of G and $\mathrm{Ad}(G)$ is the adjoint group of G.

We denote ad the tangent map of $x \mapsto \mathrm{Ad}(x)$ at $x = e$.

The Lie algebra $\mathfrak{g}$ associated with the Lie group G is the tangent space $T_e(G)$ at the unit e of G. The Lie bracket on $\mathfrak{g}$ is defined as follows:

Let $X, Y \in \mathfrak{g} = T_e(G)$ and define $[X, Y] = (\mathrm{ad}\, X)(Y)$. $\mathfrak{g}$ is the Lie algebra associated with the Lie group G. A first example of Lie group is $\mathbf{GL}(V)$ the linear group of a finite dimensional linear space V. Here we have $\mathfrak{g} = \mathscr{L}(V, V)$ and the Lie bracket is the commutator $[X, Y] = XY - YX$.

Many informations on Lie groups can be obtained from their Lie algebras through the exponential map exp.

Theorem 92 *Let G be a Lie group with Lie algebra $\mathfrak{g}$. There there exists a unique function* $\exp : \to G$ *such that*

(i) $\exp(0) = e$

(ii) $\frac{d}{dt}\exp(tX)|_{t=0} = X$.

(iii) $\exp((t+s)X) = \exp(tX)\exp(sX)$, *for all* $t, s \in \mathbb{R}$.

(iv) $\mathrm{Ad}(\exp X) = \mathrm{e}^{\mathrm{ad}\, X}$.

There is an open neighborhood U of 0 in $\mathfrak{g}$ and an open neighborhood V of e in G such that the exponential mapping is a diffeomorphism form U onto V. This local diffeomorphism can be extended in a global one if moreover the group G is connected and simply connected.

Other properties of the exponential mapping are given in [88].

A useful tool on Lie group is integration.

Definition 58 Let μ be a Radon measure on the Lie group G. μ is left-invariant (left Haar measure) if $\int f(x)d\mu(x) = \int f(yx)d\mu(x)$ for every $y \in G$ and μ is right -invariant (right Haar measure) if $\int f(x)d\mu(x) = \int f(xy)d\mu(x)$ for every $y \in G$. If μ is left and right invariant we say that μ is a bi-invariant Haar measure.

If there exists on G a bi-invariant Haar measure, G is said unimodular.

Theorem 93 *On every connected Lie group G there exits a left Haar measure μ. This measure is unique up to a multiplicative constant.*

If G is a compact and connected Lie group then a left Haar measure is a right Haar measure and there exists a unique bi-invariant Haar probability measure i.e. every compact Lie group is unimodular.

Remark 114 The affine group $G_{aff} = \{x \mapsto ax + b, \;\; a, b \in \mathbb{R}\}$ is not unimodular but the Heisenberg group $\mathbf{H}_n$ and $SU(1, 1)$ are. This remark is important to understand the differences between the corresponding coherent states associated with these groups. Coherent states associated with the affine group are called wavelets.

In many examples a Lie group G is a closed connected subgroup of the linear group $GL(n, \mathbb{R})$ (or $GL(n, \mathbb{C})$). Then we can compute a left Haar measure as follows (see [230] for details).

Let us consider a smooth system $(x^1, \ldots, x^n)$ on an open set U of G and the matrix of one-forms $\Omega = A^{-1} \sum_{1 \le j \le n} \frac{\partial A}{\partial x^j} dx^j$. Then we have following [230]:

Proposition 166 *Ω is a matrix of left-invariant one-forms in U. Moreover the linear space spanned by the elements of Ω has dimension n. There exist n independent left invariant one-forms $\omega^1, \ldots, \omega^n$ and $\omega^1 \wedge \omega_2 \wedge \cdots \wedge \omega^n$ defines a left Haar measure on G.*

Explicit examples of Haar measures:

(i) On the circle $\mathbb{S}_1 \equiv \mathbb{R}/(2\pi\mathbb{Z})$ the Haar probability measure is $d\mu(x) = \frac{dx}{2\pi}$.

(ii) Haar probability measure on $SU(2)$. Consider the parametrization of $SU(2)$ by the Euler angles (see Chap. 7).

$$g(\theta, \varphi, \psi) = \begin{pmatrix} \cos(\theta/2)e^{-i/2(\varphi+\psi)} & -\sin(\theta/2)e^{i/2(\psi-\varphi)} \\ \sin(\theta/2)e^{-i/2(\psi-\varphi)} & \cos(\theta/2)e^{i/2(\varphi+\psi)} \end{pmatrix} \tag{D.1}$$

A straightforward computation gives

$$2g^{-1}dg = \begin{pmatrix} -i(\cos\theta d\varphi + d\psi) & e^{i\psi}(d\theta - i\sin\theta d\varphi) \\ e^{-i\psi}(d\theta + i\sin\theta d\varphi) & i(\cos\theta d\varphi + d\psi) \end{pmatrix}.$$

So we get, after normalization the Haar probability on $SU(2)$:

$$d\mu(\theta, \varphi, \psi) = \frac{1}{16\pi^2}\sin\theta d\theta d\varphi d\psi.$$

(iii) Haar measure for $SU(1,1)$. The same method as for $SU(2)$ using the parametrization

$$g(\varphi, t, \psi) = \begin{pmatrix} \cosh\frac{t}{2}e^{i(\varphi+\psi)/2} & \sinh\frac{t}{2}e^{i(\varphi-\psi)/2} \\ \sinh\frac{t}{2}e^{i(\psi-\varphi)/2} & \cosh\frac{t}{2}e^{-i(\varphi+\psi)/2} \end{pmatrix}.$$

After computations we get

$$2g^{-1}dg = \begin{pmatrix} i(\cosh td\phi + d\psi) & e^{-i\psi}(dt + i\cosh td\phi) \\ e^{i\psi}(dt - i\cosh td\phi) & i(\cosh td\phi - d\psi) \end{pmatrix}$$

and a left Haar measure:

$$d\mu(t, \phi, \psi) = \cosh tdtd\phi d\psi. \tag{D.2}$$

(iii) Let us consider the Heisenberg group $\mathbf{H}_n$ (see Chap. 1). These group can also be realized as a linear group as follows. Let

$$g(x, y, s) = \begin{pmatrix} 1 & x_1 & x_2 & \cdots & x_n & s \\ 0 & 1 & 0 & \cdots & 0 & y_1 \\ 0 & 0 & 1 & 0\cdots & 0 & y_2 \\ \vdots & \vdots & \vdots & \vdots & \vdots & \vdots \\ 0 & 0 & 0 & \cdots & 1 & y_n \\ 0 & 0 & 0 & \cdots & 0 & 1 \end{pmatrix}$$

where $x = (x_1, \ldots, x_n) \in \mathbb{R}^n, y = (y_1, \ldots, y_n) \in \mathbb{R}^n, s \in \mathbb{R}$. We can easily check that $\{g(x, y, s),\ x, y \in \mathbb{R}^n, s \in \mathbb{R}\}$ is a closed subgroup $\tilde{\mathbf{H}}_n$ of the linear group $GL(2n + 1, \mathbb{R})$. $\tilde{\mathbf{H}}_n$ is isomorphic to the Weyl-Heisenberg, group $\mathbf{H}_n$ by the isomorphism

$$g(x, y, s) \mapsto (s - \frac{x\cdot y}{2}, \frac{x - iy}{\sqrt{2}})$$

The Lebesgue measure $dxdyds$ is a bi-invariant measure on $\tilde{\mathbf{H}}_n$ i.e. the Weyl-Heisenberg group is unimodular.

We shall see that when considering coherent states on a general Lie group G it is useful to consider a left invariant measure on a quotient space G/H where H is a closed subgroup of G; G/H is the set of left coset, it is a smooth analytic manifold with an analytic action of G: for every $x \in G$, $\tau(x)(gH) = xgH$. The following result is proved in [138]

Theorem 94 *There exists a G-invariant measure $d\mu_{G/H}$ on G/H if and only if we have*

$$|\det \mathrm{Ad}_G(h)| = |\det \mathrm{Ad}_H(h)|, \quad \forall h \in H,$$

where Ad_G is the adjoint representation for the group G.

Moreover this measure is unique up to a multiplicative constant and we have for any continuous function f, with compact support in G,

$$\int f(g) d\mu_G(g) = \int_{G/H} \left(\int_H f(gh) d\mu_H(h) \right) d\mu_{G/H}(gH)$$

where $d\mu_G$ and $d\mu_H$ are left Haar measure suitably normalized.

In this book we have considered the three groups $\mathbf{H}_n$, $SU(2)$ and $SU(1, 1)$ and their related coherent states. In each case the isotropy subgroup H is isomorphic to the unit circle $U(1)$ and we have found the quotient spaces: $\mathbf{H}_n/U(1) \equiv \mathbb{R}^{2n}$, $SU(2)/U(1) \equiv \mathbb{S}_2$ and $SU(1, 1)/U(1) \equiv P\mathbb{S}_2$ with their canonical measure. Each of this space is a symplectic space and can be seen as the phase space of classical systems.

D.3 Representations of Lie Groups

The goal of this section is to recall some basic facts.

D.3.1 General Properties of Representations

G denotes an arbitrary connected Lie group, V_1, V_2, V are complex Hilbert spaces, $\mathscr{L}(V_1, V_2)$ the space of linear continuous mapping from V_1 into V_2, $\mathscr{L}(V) = \mathscr{L}(V, V)$, $GL(V)$ the group of invertible mappings in $\mathscr{L}(V)$, $U(V)$ the sub-group of $GL(V)$ of unitary mappings i.e. $A \in U(V)$ if and only if $A^{-1} = A^*$.

Definition 59 A representation of G in V is a group homomorphism $\hat{R}$ from G in $GL(V)$ such that $(g, v) \mapsto \hat{R}(g)v$ is continuous from $G \times V$ into V.

If $\hat{R}(g) \in U(V)$ for every $g \in G$ the representation is said *unitary*.

Definition 60 The subspace $E \subseteq V$ is invariant by the representation $\hat{R}$ if $\hat{R}(g)E \subseteq E$ for every $g \in G$.
The representation $\hat{R}$ is irreducible in V if the only invariant closed subspaces of V are $\{V, \{0\}\}$.

Definition 61 Two representations $(\hat{R}_1, V_1)$ and $(\hat{R}_2, V_2)$ are equivalent if there exists an invertible continuous linear map $A : V_1 \to V_2$ such that

$$R_2(g)A = AR_1(g), \quad \forall g \in G.$$

Irreducible representations are important in physics: they are associated to elementary particles (see [256]).

Let $d\mu$ be a left Haar measure on G. Consider the Hilbert space $L^2(G, d\mu)$ and define $L(g)f(x) = f(g^{-1}x)$ where $g, x \in G, f \in L^2(G, d\mu)$. L is a unitary representation of G called the left regular representation.

The Schur lemma is an efficient tool to study irreducible dimensional representations.

Lemma 111 (Schur) *Suppose $\hat{R}_1$ and $\hat{R}_2$ are finite dimensional irreducible representations of G in V_1 and V_2 respectively. Suppose that we have a linear mapping $A : V_1 \to V_2$ such that $A\hat{R}_1(g) = \hat{R}_2(g)A$ for every $g \in G$. Then or A is bijective or $A = 0$.*

In particular if $V_1 = V_2 = V$ and $A\hat{R}(g) = \hat{R}(g)A$ for all $g \in G$ then $A = \lambda\mathbb{1}$ for some $\lambda \in \mathbb{C}$.

Suppose that $(\hat{R}, V)$ is a unitary representation in the Hilbert space $\mathscr{H}$ then it is irreducible if and only if the only bounded linear operators A in V commuting with $\hat{R}$ ($A\hat{R}(g) = \hat{R}(g)A$ for every $g \in G$) are $A = \lambda\mathbb{1}$, $\lambda \in \mathbb{C}$.

A useful property of a representation $\hat{R}$ is its *square integrability* (see [123] for details).

Definition 62 A vector $v \in V$ is said to be admissible if we have

$$\int_G |\langle \hat{R}(g)v, v\rangle|^2 d\mu(g) < +\infty. \tag{D.3}$$

The representation $\hat{R}$ is said square integrable if $\hat{R}$ is irreducible and there exists at least one admissible vector $v \neq 0$.

If G is compact every irreducible representation is square integrable. The discrete series of $SU(1, 1)$ are square integrable (prove that 1 is admissible using the formula (D.2) for the Haar measure on $SU(1, 1)$).

We have the following result due to Duflo-Moore and Carey (see [123] for a proof).

Theorem 95 *Let $\hat{R}$ be a square integrable representation in V. Then there exists a unique self-adjoint positive operator C in V with a dense domain in V such that:*
(i) The set of admissible vectors is equal to the domain $\mathscr{D}(C)$.
(ii) If v_1, v_2 are two admissible vectors and $w_1, w_2 \in V$ then we have

$$\int_G \overline{\langle \hat{R}(g)v_2, w_2\rangle} \langle \hat{R}(g)v_1, w_1\rangle dg = \langle Cv_1, Cv_2\rangle \langle w_1, w_2\rangle. \tag{D.4}$$

(iii) If G is unimodular then $C = \lambda \mathbb{1}$, $\lambda \in \mathbb{R}$.

Remark 115 If G is unimodular coherent states can be defined as follows. We start from an admissible vector $v_0 \in V$, $\|v_0\| = 1$ and an irreducible representation $\hat{R}$ in V. Define the *coherent state* (or the *analyzing wavelet*) $\varphi_g = \hat{R}(g)v_0$. Then the family $\{\lambda^{-1/2}\varphi_g, |g \in G\}$ is overcomplete in V:

$$\lambda^{-1} \int_G \overline{\langle\varphi_g, \psi_1\rangle}\langle\varphi_g, \psi_2\rangle d\mu(g) = \langle\psi_1, \psi_2\rangle, \quad \psi_1, \psi_2 \in V,$$

where $\lambda = \int_G |\langle\hat{R}(g)v_0, v_0\rangle|^2 d\mu(g)$.

D.3.2 The Compact Case

Representation theory for compact group is well known (for a concise presentation see [170] or for more details [171]). Typical examples are $SU(2)$ and $SO(3)$ considered in Chap. 7. Here G is a compact Lie group. The main facts are the following

1. Every finite dimensional representation is equivalent to a unitary representation.
2. Every irreducible unitary representation of G is finite dimensional and every unitary representation of G is a direct sum of irreducible representations
3. If $\hat{R}_1, \hat{R}_2$ are non equivalent irreducible finite representations of G on V_1 and V_2 then

$$\int_G \overline{\langle\hat{R}_1(g)v_1, w_1\rangle}\langle\hat{R}_2(g)v_2, w_2\rangle = 0, \text{ for all } v_1, w_1 \in V_1, \; v_2, w_2 \in V_2.$$

4. If $\hat{R}$ is an irreducible unitary representation of G then we have

$$(\dim V) \int_G \overline{\langle\hat{R}(g)v_1, w_1\rangle}\langle\hat{R}(g)v_2, w_2\rangle = \overline{\langle v_1, v_2\rangle}\langle w_1, w_2\rangle,$$
$$\text{for all } v_1, w_1 \in V_1, \; v_2, w_2 \in V_2. \tag{D.5}$$

5. (Peter-Weyl Theorem) If we denote $(\hat{R}_\lambda, V_\lambda)$, $\lambda \in \Lambda$, the set of all irreducible representations of G and $M_{\lambda,v,w}(g) = \langle\hat{R}_\lambda(g)v, w\rangle$, then the linear space spanned by $\{M_{\lambda,v,w}(g)|g \in G, v, w \in V_\lambda\}$ is dense in $L^2(G, d\mu)$.

D.3.3 The Non-compact Case

This case is much more difficult than the compact case and there is so far no a general theory of irreducible unitary representations. The typical example is $SU(1,1)$ or equivalently $SL(2,\mathbb{R})$ considered in Chap. 7.

These groups have the following properties

Definition 63 (i) A Lie group G is said *reductive* if G is a closed connected subgroup of $GL(n,\mathbb{R})$ or $GL(n,\mathbb{C})$ stable under inverse conjugate transpose.
(ii) A Lie G is said *linear connected semisimple* if G is reductive with finite center.

Proposition 167 *If G is a linear connected semisimple group its Lie algebra $\mathfrak{g}$ is semisimple.*

It is known that a compact connected Lie group can be realized as a linear connected reductive Lie group ([168], Theorem 1.15).

Let us consider the Lie algebra $\mathfrak{g}$ of G. The differential of the mapping $\Theta(A) = A^{-1,*}$ at $e = \mathbb{1}$ is denoted θ. We have $\theta^2 = \mathbb{1}$ so θ has two eigenvalues ± 1.

So we have the decomposition $\mathfrak{g} = \mathfrak{l} \oplus \mathfrak{p}$ where $\mathfrak{l} = \ker(\theta - 1)$ and $\mathfrak{p} = \ker(\theta + 1)$. Let $K = \{g \in G | \theta g = g\}$. The following result is a generalization of the polar decomposition for matrices or operators in Hilbert spaces.

Proposition 168 (polar Cartan decomposition) *If G is a linear connected reductive group then K is a compact connected group and is a maximal compact subgroup of G. Its Lie algebra is $\mathfrak{l}$ and the map: $(k, X) \mapsto k \exp X$ is a diffeomorphism from $K \times \mathfrak{p}$ onto G.*

D.4 Coherent States According Gilmore-Perelomov

Here we describe a general setting for a theory of coherent states in a arbitrary Lie group from the point of view of Perelomov (for more details see [208, 209]).

We start from an irreducible unitary representation $\hat{R}$ of the Lie group G in the Hilbert space $\mathscr{H}$. Let $\psi_0 \in \mathscr{H}$ be a fixed unit vector ($\|\psi_0\| = 1$) and denote $\psi_g = \hat{R}(g)\psi_0$ for any $g \in G$. In quantum mechanics states in the Hilbert space $\mathscr{H}$ are determined modulo a phase factor so we are mainly interested in the action of G in the projective space $\mathbf{P}(\mathscr{H})$ (space of complex lines in $\mathscr{H}$). We denote by H the isotropy group of ψ_0 in the projective space: $H = \{h \in G | \hat{R}(h)\psi_0 = e^{i\theta}\psi_0\}$.

So the coherent states system $\{\psi_g | g \in G\}$ is parametrized by the space G/H of left coset in G modulo H: if π is the natural projection map: $G \to G/H$. Choosing for each $x \in G/H$ some $g(x) \in G$ we have, with $x = \pi(g)$, $\psi_g = e^{i\theta(g)}\psi_{g(x)}$. Moreover ψ_{g_1} and ψ_{g_2} define the same states if and only if $\pi(g_1) = \psi_{g_2} := x$ hence we have $\psi_{g_1} = e^{i\theta_1}\psi_{g(x)}$ and $\psi_{g_2} = e^{i\theta_2}\psi_{g(x)}$.

So for every $x \in G/H$ we have defined the state $|x\rangle = \{e^{i\alpha}\psi_g\}$ where $x = \pi(g)$. It is convenient to denote $|x\rangle = \psi_{g(x)}$ and $x(g) = \psi(x)$. This parametrization of coherent states by the quotient space G/H has the following nice properties.

We have $\psi_g = e^{i\theta(g)}|x(g)\rangle$ and $\theta(gh) = \theta(g) + \theta(h)$ if $g \in G$ and $h \in H$. The action of G on the coherent state $|x\rangle$ satisfies

$$\hat{R}(g_1)|x\rangle = e^{i\beta(g_1,x)}|g_1.x\rangle \tag{D.6}$$

where $g_1.x$ denotes the natural action of G on G/H and $\beta(g_1, x) = \theta(g_1 g) - \theta(g)$ where $\pi(g) = x$ (β depends only on x, not in g).

Computation of the scalar product of two coherent states gives

$$\langle x_1|x_2\rangle = e^{i(\theta(g_1)-\theta(g_2))}\langle 0|\hat{R}(g_1{}^{-1}g_2)|0\rangle \tag{D.7}$$

where $x_1 = x(g_1)$ and $x_2 = x(g_2)$. Moreover if $x_1 \neq x_2$ we have $|\langle x_1|x_2\rangle| < 1$ and

$$\langle g.x_1|g.x_2\rangle = e^{i(\beta(g,x_1)-\beta(g,x_2))}\langle x_1|x_2\rangle. \tag{D.8}$$

Concerning completeness we have

Proposition 169 *Assume that the Haar measure on G induces a left invariant measure $d\mu(x)$ on G/H (see Theorem 94) and that the following square integrability condition is satisfied:*

$$\int_M |\langle 0|x\rangle|^2 dx < +\infty. \tag{D.9}$$

Then we have the resolution of identity:

$$\frac{1}{d}\int_M \langle x|\psi\rangle\psi_x d\mu(x) = \psi, \quad \forall\psi \in \mathscr{H}. \tag{D.10}$$

where $d = \int_M |\langle 0|x\rangle|^2 dx$. Moreover we have a Plancherel identity

$$\langle\psi|\psi\rangle = \frac{1}{d}\int_M |\langle x|\psi\rangle|^2 d\mu(x). \tag{D.11}$$

Remark 116 (i) As we have seen in Chap. 2, using formula (D.10) and (D.11), we can consider Wick quantization for symbols defined on $M = G/H$.
(ii) When the square integrability condition (D.9) is not fulfilled (the Poincaré group for example) there exists an extended definition of coherent states. This is explained in [3].

Remark 117 When G is a compact semi-simple Lie group and $\hat{R}$ is a unitary irreducible representation of G in a finite dimensional Hilbert space $\mathscr{H}$ then it is possible to choose a state ψ_0 in $\mathscr{H}$ such that if H is the isotropy group of ψ_0 then G/H is a Kähler manifold (see [201]).

Some results concerning coherent states and quantization have also been obtained for non compact semi-simple Lie groups extending results already seen in Chap. 8 for $SU(1, 1)$.

Appendix E
Berezin Quantization and Coherent States

We have seen in Chap. 12 that canonical coherent states are related to Wick and Weyl quantizations. Berezin in the paper [24] has given a general setting to quantize "classical systems".

Let us explain here very briefly the Berezin construction. Let M be a classical phase space i.e. a symplectic manifold with a Poisson bracket denoted $\{\cdot,\cdot\}$ and an Hilbert space $\mathscr{H}$. Assume that for a set of positive numbers $\hbar$, with 0 as limit point, we have a linear mapping $A \mapsto \hat{A}_\hbar$ where A is a smooth function on M and $\hat{A}_\hbar$ is an operator on $\mathscr{H}$. The inverse mapping is denoted $S_\hbar(\hat{A}_\hbar)$. In general it is difficult to describe in details the definition domain and the range of this quantization mapping. Some example are considered in [246].

Nevertheless for a quantization mapping the two following conditions are required, to preserve the Bohr's correspondence condition (semi-classical limit),

$$\lim_{\hbar\to 0} S_\hbar(\hat{A}_\hbar\hat{B}_\hbar)(m) = A(m)B(m)\ \forall m \in M\,, \quad (C1)$$

$$\lim_{\hbar\to 0} \frac{1}{i\hbar} S_\hbar([\hat{A}_\hbar, \hat{B}_\hbar])(m) = \{A, B\}(m)\ \forall m \in M\,, \quad (C2).$$

We have seen in Chap. 1 that these conditions are fulfilled for the Weyl quantization of $\mathbb{R}^{2n}$. In [24] the author have considered the two dimensional sphere and the pseudo-sphere (Lobachevskii plane). In these two examples the Planck constant $\hbar$ is replaced by $\frac{1}{n}$ where n is an integer parameter depending on the considered representation. The semi-classical limit is the limit $n \to +\infty$. For the pseudosphere $n = 2k$, where k is the Bargmann index.

In this section we shall explain some of the Berezin's ideas concerning quantization on the pseudosphere and we shall prove that the Bohr's correspondence principle is satisfied using results taken from Chap. 7.

The same results could be proved for quantization of the sphere [24], using results of Chap. 7.

D. Robert and M. Combescure, *Coherent States and Applications in Mathematical Physics*, Theoretical and Mathematical Physics,
https://doi.org/10.1007/978-3-030-70845-0

Nowadays the quantization problem has been solved in much more general settings in particular for Kähler manifolds (the Poincaré disc $\mathbb{D}$ or the Riemann sphere $\mathbb{S}^2$ are examples of Kähler manifolds) where generalized coherent states are still present. This domain is still very active and is named Geometric Quantization; its study is outside the scope of this book (see the book [264] and the recent review [231]).

We have defined before the coherent states family ψ_ζ, $\zeta \in \mathbb{D}$, for the representations $\mathscr{D}_n^+$ of the group $SU(1,1)$ which is a symmetries group for the Poincaré disc $\mathbb{D}$. Recall that $\{\psi_\zeta\}_{\zeta\in\mathbb{D}}$ is an overcomplete system in $\mathscr{H}_n(\mathbb{D})$ hence the map $\varphi \mapsto \varphi^\sharp$, where $\varphi^\sharp(z) = \langle \psi_z, \varphi\rangle$, is an isometry from $\mathscr{H}_n(\mathbb{D})$ into $L^2(\mathbb{D})$.

Let $\hat{A}$ be a bounded operator in $\mathscr{H}_n(\mathbb{D})$. Its covariant symbol $A_c(z, \bar{w})$ is defined as

$$A_c(z, \bar{w}) = \frac{\langle \psi_z, \hat{A}\psi_w\rangle}{\langle \psi_z, \psi_w\rangle}.$$

It is an holomorphic extension in $(z, \bar{w})$ of the usual covariant symbol $A_c(z, \bar{z})$. Moreover the operator $\hat{A}$ is uniquely determined by its covariant symbol and we have

$$\hat{A}\varphi(z) = \int_{\mathbb{D}} A_c(z, \bar{w})\varphi(w)\langle \psi_z, \psi_w\rangle d\nu_n(w). \tag{E.1}$$

From (E.1) we get a formula for the covariant symbol product of the product of two operators $\hat{A}$, $\hat{B}$. If $(AB)_c$ denotes the covariant symbol of $\hat{A}\hat{B}$ then we have the formula

$$(AB)_c(z, \bar{z}) = \int_{\mathbb{D}} A_c(z, \bar{w})B_c(w, \bar{z})|\langle \psi_z, \psi_w\rangle|^2 d\nu_n(w). \tag{E.2}$$

From our previous computations (Chap. 7) we have

$$|\langle \psi_w, \psi_z\rangle|^2 = \left(\frac{(1-|z|^2)(1-|w|^2)}{|1-\bar{z}w|^2}\right)^n.$$

We have to consider the following operator

$$\mathscr{T}_n F(z, \bar{z}) = \frac{n-1}{4\pi}\int_{\mathbb{D}} F(w, \bar{w})\left(\frac{(1-|z|^2)(1-|w|^2)}{|1-\bar{z}w|^2}\right)^n d\mu(w)$$

for F bounded in $\mathbb{D}$ and C^2-smooth.

Proposition 170 *We have the following asymptotic expansion, for $n \to +\infty$,*

$$\mathscr{T}_n F(z, \bar{z}) = F(z, \bar{z})\left(1 - \frac{n}{(n-2)^2}\right) + \frac{1}{n}(1-|z|^2)^2\frac{\partial^2 F}{\partial z \partial \bar{z}}(z, \bar{z}). \tag{E.3}$$

Proof Using invariance by isometries of $\mathbb{D}$ we show that it is enough to prove formula (E.3) for $z = 0$.

Let us consider the change of variable $w = \frac{\zeta - z}{1-\bar{z}\zeta}$. Denote $G(\zeta, \bar{\zeta}) = F(w, \bar{w})$ then we get $\mathscr{T}_n F(z, \bar{z}) = \mathscr{T}_n G(0, 0)$. A direct computation gives

$$\frac{\partial^2 G}{\partial \zeta \partial \bar{\zeta}}(0, 0) = (1 - |z\bar{z}|)^2 \frac{\partial^2 F}{\partial z \partial \bar{z}}(z, \bar{z}) = \Delta F(z; \bar{z})$$

where Δ is the Laplace-Beltrami operator on $\mathbb{D}$. So we have proved (E.3) for any z if it is proved for $z = 0$.

To prove (E.3) for $z = 0$ we use the real Laplace method for asymptotic expansion of integrals A.4.

We write

$$\mathscr{T}_n F(0, 0) = \frac{n-1}{\pi} \int_0^{2\pi} d\theta \int_0^1 dr F(re^{i\theta}, re^{-i\theta}) re^{-(n-2)\phi(r)} \tag{E.4}$$

where $\phi(r) = \log((1 - r^2)^{-1})$.

Note that $\phi(r) > 0$ for $r \in]0, 1]$ and $\phi(0) = 1$. So method (see Sect. A.4 for a precise statement) to estimate (E.4) with the large parameter $\lambda = n - 2$. Modulo an exponentially small term it is enough to integrate in r in $[0, 1/2]$. Using the change of variable $\phi(r) = s^2$, $s > 0$ we have $r(s) = \sqrt{1 - e^{-s^2}}$ and

$$\mathscr{T}_n F(0, 0) = \frac{n-1}{\pi} \int_0^{2\pi} d\theta \int_0^c ds e^{-\lambda s} K(r(s)) s e^{-s^2} + O(\lambda^{-\infty})$$

where $K(r) = F(re^{i\theta}, re^{-i\theta})$. Now to get the result we have to compute the asymptotic expansion at $s = 0$ for $L(s) = K(r(s)) s e^{-s^2}$. Note that $L(s)$ is periodic in θ and we have to consider only the part of the expansion independent in θ. If $L_0(s)$ is this part, we get after computations

$$L_0(s) = \frac{n-1}{2\pi(n-2)} F(0, 0) + \frac{n-1}{2\pi(n-2)^2} \left(\frac{\partial^2 F}{\partial z \partial \bar{z}}(0, 0) - F(0, 0) \right) + O(\frac{1}{n^2})$$

and formula (E.4) follows. □

It is not difficult, using Proposition 170, to check the correspondence principle (C1) and (C2).

We get (C1) by applying the Proposition to $F_{AB}(w, \bar{w}) = A_c(z, \bar{w}) B_c(w, \bar{z})$. So we have

$$(AB)_c(z, \bar{z}) = \mathscr{T}_n(z, \bar{z}) \underset{n \to +\infty}{\longrightarrow} A_c(z, \bar{z}) B_c(z, \bar{z}).$$

For (C2) we write

$$F_{AB}(z, \bar{z}) = A_c(z, \bar{z}) B_c(z, \bar{z})(1 - \frac{n}{(n-1)^2}) + \frac{1}{n}(1 - |z|^2)^2 \left(\frac{\partial A_c}{\partial \bar{w}}(z, \bar{z}) \frac{\partial B_c}{\partial w}(z, \bar{z}) \right) + O\left(\frac{1}{n^2}\right)$$

So we get

$$n\,(F_{AB}(z,\bar{z})-F_{BA}(z,\bar{z}))=(1-|z|^2)^2\frac{\partial A_c}{\partial \bar{z}}(z,\bar{z})\frac{\partial B_c}{\partial z}(z,\bar{z})-\frac{\partial A_c}{\partial z}(z,\bar{z})\frac{\partial B_c}{\partial \bar{z}}(z,\bar{z})+O\left(\frac{1}{n}\right)$$

But we know that $F_{AB}-F_{BA}$ is the covariant symbol of the commutator $[\hat{A},\hat{B}]$ and the Poisson bracket $\{A,B\}$ is

$$\{A,B\}(z,\bar{z})=i(1-|z|^2)^2\left(\frac{\partial A_c}{\partial \bar{z}}(z,\bar{z})\frac{\partial B_c}{\partial z}(z,\bar{z})-\frac{\partial A_c}{\partial z}(z,\bar{z})\frac{\partial B_c}{\partial \bar{z}}(z,\bar{z})\right)$$

So we get (C2). □

We have seen that the linear symplectic maps and the metaplectic transformations are connected with quantization of the Euclidean space $\mathbb{R}^{2n}$. Here symplectic transformations are replaced by transformations in the group $SU(1,1)$ and metaplectic transformations by the representations $g\mapsto\hat{R}(g)=\mathscr{D}_n^-(g)$. Then we have

Proposition 171 *(i) For any bounded operator $\hat{A}$ in $\mathscr{H}_n(\mathbb{D})$ the covariant symbol A_g of $\hat{R}(g)\hat{A}\hat{R}(g)^{-1}$ is*

$$A_g(\zeta)=A_c(g^{-1}\zeta,\overline{g^{-1}\zeta}). \tag{E.5}$$

(ii) The covariant symbol $R(g)_c$ of $\hat{R}(g)$ is given by the formula

$$R(g)_c(\zeta)=\mathrm{e}^{in\arg(\alpha+\beta\zeta)}\left(\frac{1-|\zeta|^2}{\bar{\alpha}+\bar{\beta}\bar{\zeta}-\beta\zeta-\alpha|\zeta|^2}\right) \tag{E.6}$$

where $g=\begin{pmatrix}\alpha&\beta\\ \bar{\beta}&\bar{\alpha}\end{pmatrix}$.

Proof (i) is a direct consequence of definition of $\mathscr{D}_n^-(g)$ and computations of Chap. 8. For (ii) using formula (8.74) of Chap. 8 in complex variables we get, after computation,

$$\langle\psi_\zeta,\psi_{g^{-1}\zeta}\rangle=\left(\frac{\alpha+\beta z}{|\alpha+\beta z|}\right)^n\left(\frac{1-|\zeta|^2}{\bar{\alpha}+\bar{\beta}\bar{\zeta}-\beta\zeta-\alpha|\zeta|^2}\right)$$

which gives (E.6). □

References

1. Abe, S., Suzuki, N.: Wigner distribution function of a simple optical model: an extended-phase-space approach. Phys. Rev. A **45**, 520–523 (2001)
2. Aharonov, Y., Anandan, J.: Phase change during a cyclic quantum evolution. Phys. Rev. Lett. **58**, 1593–1596 (1987)
3. Ali, S.T., Antoine, J.P., Gazeau, J.P.: Coherent States Wavelets and their Generalizations. Springer, Berlin (2000)
4. Anantharaman, N., Fermanian-Kammerer, C., Maciá, F.: Semiclassical completely integrable systems: long-time dynamics and observability via two-microlocal Wigner measures. Am. J. Math. **137**(3), 577–638 (2015)
5. Aravind, P.: Spin coherent states as anticipators of the geometric phase. Am. J. Phys. **10**(October), 899–904 (1999)
6. Arecchi F.T. Courtens, E.G., R. Thomas, H.: Atomic coherent states in quantum optics. Phys. Rev. A **6**, 2211–2237 (1972)
7. Arnaud, J.A.: Hamiltonian theory of beam mode propagation. In: Wolf, E. (ed.) Progress in Optics XI. North Holland (1973)
8. Arnold, V.: Mathematical Methods of Classical Mechanics. Springer, New-York (1999)
9. Assal, M.: Analyse spectrale des systèmes d'opérateurs h-pseudodifférentiels. Ph.D. thesis, University of Bordeaux (2017)
10. Babich, V.: Eigenfunctions concentrated in a neighborhood of a closed geodesic. Math. Problems in Wave Propagation Theory, V.M. Babich, ed., Sem. Math., V.A. Steklov Math. Inst., 9, Leningrad. Translated by Consultants Bureau, New York (1970)
11. Babich, V., Buldyrev, V.: Asymptotic Methods in Short Wave Diffraction Problems. Nauka Moscow (1972)
12. Balazs, N., Voros, A.: The quantized baker's transformation. Ann. Phys. **190**, 1–31 (1989)
13. Balazs, N.L., Voros, A.: Chaos on the pseudosphere. Phys. Rep. **143**(3), 109–240 (1986)
14. Balian, R., Bloch, C.: Distribution of eigenfrequencies for the wave equation in a finite domain. Ann. Phys. **69**, 76–160 (1972)
15. Balian, R., Bloch, C.: Distribution of eigenfrequencies for the wave equation in a finite domain. Ann. Phys. **85**, 514–545 (1974)
16. Bambusi, D., Grébert, B., Maspero, A., Robert, D.: Growth of Sobolev norms for abstract linear Schrödinger equations. arXiv: 1706.09708v2 [AP] (2018)
17. Bander, M., Itzykson, C.: Group theory and the hydrogen atom. Rev. Mod. Phys. **38**, 330–345 (1966)

D. Robert and M. Combescure, *Coherent States and Applications in Mathematical Physics*, Theoretical and Mathematical Physics,
https://doi.org/10.1007/978-3-030-70845-0

18. Bargmann, V.: Irreducible unitary representations of the Lorentz group. Ann. Math. **2**(48), 568–640 (1947)
19. Bargmann, V.: On unitary ray representations of continuous groups. Ann. Math. **2**(59), 1–46 (1954)
20. Bargmann, V.: On a Hilbert space of analytic functions. Commun. PureAppl. Math **14**, 187–214 (1961)
21. Beals, R.: Characterization of pseudodifferential operators and applications. Duke Math. J. **44**, 45–57 (1977)
22. Beals, R.: Propagation des singularités pour des opérateurs. Journées E.D.P Saint Jean de Monts, conf. No. 19 (1980)
23. Berezin, F.: The Method of Second Quantization. Academic Press, New-York (1966)
24. Berezin, F.: General concept of quantization. Commun. Math. Phys. **40**, 153–174 (1975)
25. Berezin, F.: Models of Gross-Neveu type are quantization of a classical mechanics with nonlinear phase space. Commun. Math. Phys. **63**, 131–153 (1978)
26. Berezin, F.: Introduction to Superanalysis. Reidel Publishing Company, Dordrecht (1987)
27. Berezin, F., Shubin, M.: The Schrödinger equation. Kluwer Academic Publishers (1991)
28. Berry, M.: Quantal phase factors accompanying adiabatic changes. Proc. R. Soc. Lond. **Ser.A 292**, 47–57 (1984)
29. Bertrand, J., Irac-Astaud, M.: Characterization of SU(1,1) coherent states in terms of affine wavelets. J. Phys. A: Math. Gen. **35**, 7347–7357 (2002)
30. Bhatia, R.: Matrix Analysis, vol. 169. Springer, New York (1997)
31. Billingsley, P.: Convergence of Probability Measures. Wiley & Sons, New York - London (1968)
32. Bily, J.M.: Propagation d'états cohérents et applications. Ph.D. thesis, Université de Nantes (2001)
33. Bolte, J., Glaser, R.: Semiclassical propagation of coherent states with spin-orbit interaction. Annales Henri Poincaré **6**, 625–656 (2004)
34. Bonechi, F., De Bièvre, S.: Controlling strong scarring for quantized ergodic toral automorphisms. Duke Math. J. **117**, 571–587 (2003)
35. Bony, J.M.: Caractérisation des opérateurs pseudo-différentiels. Séminaire E. D. P (1996-97), Exposé XXIII (1997)
36. Bony, J.M.: On the characterization of pseudodifferential operators. bonyjm.perso.math.cnrs.fr/IMPbertinoro.pdf (2009)
37. Bouclet, J., De Bièvre, S.: Long time propagation and control on scarring for perturbed quantized hyperbolic toral automorphism. Ann. Henri Poincaré **6**, 885–913 (2005)
38. Bouzouina, A.: Comportement semi-classique de symplectomorphismes du tore quantifiés. Ph.D. thesis, Université de Paris-Dauphine (1997)
39. Bouzouina, A., De Bièvre, S.: Equipartition of the eigenfunctions of quantized ergodic maps on the torus. Commun. Math. Phys. **178**, 83–105 (1996)
40. Bouzouina, A., Robert, D.: Uniform semiclassical estimates for the propagation of quantum observables. Duke Math. J. **111**, 223–252 (2002)
41. Boya, L.J., Perelomov, A.M., Santander, M.: Berry phase in homogeneous Kähler manifolds with linear Hamiltonians. J. Math. Phys. **42**(11), 5130–5142 (2001)
42. Bratteli, O., Robinson, D.: Operator Algebra and Quantum Statistical Mechanics II. Springer, New York (1981)
43. Breuer, H.P., Petruccione, F.: The theory of open quantum systems. Oxford University Press, New York (2002)
44. Brocker, T., Tom Dieck, T.: Representations of compact Lie groups. In: Graduate Texts in Mathematics, vol. 17, pp. 7–87. Springer, New-York (2007)
45. Burq, N.: Mesures semi-clasiques et mesures de défaut. Séminaire N. Bourbaki (1996-1997)
46. Buser, P.: Geometry and spectra of compact Riemann surfaces. Geometry and spectra of compact Riemann surfaces. In: Progress in Mathematics, vol. 106. Birkhäuser, Boston (1992)
47. Cahill, K., Glauber, R.: Density operators for fermions. Phys. Rev. A **59**(2), 1538–1555 (1999)
48. Cassanas, R.: Hamiltoniens quantiques et symétries. Ph.D. thesis, Université de Nantes (2005)

49. Cesari, L.: Asymptotic Behaviour and Stability Problems in Ordinary Differential Equations. Springer, Berlin (1963)
50. Charbonnel, A., Popov, G.: Semiclassical asymptotics for several commuting operators. Commun. Partl. Differ. Equs. **24**, 283–323 (1998)
51. Chazarain, J.: Formule de Poisson pour les variétés riemanniennes. Inven. Math. **24**, 65–82 (1974)
52. Chazarain, J.: Spectre d'un hamiltonien quantique et mécanique classique. Commun. Partl. Differ. Equs. **6**, 595–644 (1980)
53. Chen, L., Oon Lee, J.: Rate of convergence in non linear Hartree dynamics with factorized initial data. J. Math. Phys. **52**, 1–25 (2011)
54. Chodosh, O.: Infinite matrix representations of isotropic pseudodifferential operators. Methods Appl. Anal. **18**, 351–372 (2011)
55. Colin de Verdière, Y.: Spectre du Laplacien et longueurs des géodésiques périodiques I. Commun. Math. Helvetici **27**, 83–106 (1973)
56. Ergodicité et fonctions propres du laplacien: Colin de Verdière, Y. Commun. Math. Phys. **102**, 497–502 (1985)
57. Colin de Verdière, Y.: Bohr-Sommerfeld rules to all orders. Ann. Henri Poincaré **6(5)**, 925–936 (2005)
58. Combes J.M. Duclos, P., Seiler, R.: The born-oppenheimer approximation. In: Rigorous Atomic and Molecular Physics, pp. 185–212. New-York Plenum (1981)
59. Combescure, M.: A quantum particle in a quadrupole radio-frequency trap. Ann. Inst. Henri Poincaré **44**, 293–314 (1986)
60. Combescure, M.: A generalized coherent state approach of the quantum dynamics for suitable time-dependent Hamiltonians. Journées Equations aux Dérivées Partielles, Saint-Jean-de-Monts (1992)
61. Combescure, M.: The squeezed states approach of a semi-classical limit of the time dependent Schrödinger equation. J. Math. Phys. **33**, 3870–3880 (1992)
62. Combescure, M.: Circulant matrices, Gauss sums and mutually unbiased bases, I. The prime number case. CUBO A Math. J. **11**, 73–86 (2009)
63. Combescure, M., Ralston, J., Robert, D.: A proof of the Gutzwiller semi-classical formula using coherent states decomposition. Commun. Math. Phys. **202**, 463–480 (1999)
64. Combescure, M., Robert, D.: Distribution of matrix elements and level spacings for classically chaotic systems. Annales de l'I.H.P. Physique théorique **61**(4), 443–483 (1994)
65. Combescure, M., Robert, D.: Semiclassical spreading of quantum wavepackets an applications near unstable fixed points of the classical flow. Asymptot. Anal. **14**, 377–404 (1997)
66. Combescure, M., Robert, D.: Quadratic quantum Hamiltonians revisited. Cubo A Math. J. **8**(1), 61–86 (2006)
67. Combescure, M., Robert, D.: Coherent States and Applications in Mathematical Physics. Theoretical and Mathematical Physics. Spinger (2012)
68. Cordoba, A., Fefferman, C.: Wave packets and Fourier integral operators. Commun. Partl. Differ. Equs. **3**(11), 979–1005 (1978)
69. Cornfeld, I., Fomin, S.V., Sinai, Y.G.: Ergodic Theory. Springer, Berlin (1982)
70. Courant, R., Hilbert, D.: Methods of Mathematical Physics. Wiley, New York (1966)
71. Davies, E.: The functional calculus. J. Lond. Math. Soc. **52**, 166–176 (1995)
72. Davis, P.: Circulant Matrices. Wiley, New York (1979)
73. De Bièvre, S., Degli Esposti, M., Giachetti, R.: Quantization of a class of piecewise affine transformations of the torus. Commun. Math. Phys. **176**, 73–94 (1995)
74. De Bièvre, S., Faure, F., Nonnenmacher, S.: Scarred eigenstates for quantum cat maps of minimal periods. Commun. Math. Phys. **239**, 449–492 (2003)
75. De Gosson, M.: The Weyl representation of metaplectic operators. Lett. Math. Phys. **72**, 129–142 (2005)
76. De Witt, B.: Supermanifolds. Cambridge University Press, Cambridge (1984)
77. Degli Esposti, M.: Quantization of the orientation preserving automorphism of the torus. Ann. Inst. Henri Poincaré **58**, 323–341 (1993)

78. Degli Esposti, M., Graffi, S., Isola, S.: Stochastic properties of the quantum Arnol'd cat in the classical limit. Commun. Math. Phys. **176**, 73–94 (1995)
79. Deligne, P., Etingof, P., Freed, D., Kazhdan, D., Morgan, J., Morrison, D., Witten, E.: Quantum Fields and Strings: A Course for Mathematicians, vol. 1. AMS, Providence (1999)
80. Delort, J.M.: Growth of Sobolev norms for solutions Schrödinger operators with harmonic potential. Commun. Partl. Differ. Equs. **39**, 1–33 (2014)
81. Derezinski, J., Gérard, C.: Scattering Theory of Classical and Quantum N-Particle Systems. Springer, Berlin (1997)
82. Derezinski, J., Napiorkowski, M., Solovej, J.: On the minimization of Hamiltonians over pure Gaussian states. Lecture Notes Series, Institute for Mathematical Sciences, National University of Singapore Complex Quantum Systems (2013)
83. Dimassi, M., Sjöstrand, J.: Spectral asymptotics in the semi-classical limit. Lecture Note Series, vol. 268. London Mathematical Society (1999)
84. Dirac, P.: Lectures on Quantum Mechanics. Dover Publications, INC., Princeton (2001)
85. Dodonov, V.: Parametric excitation and generation of nonclassical states in linear media. In: Theory of Non classical States of Light. Taylor & Francis, London (2003)
86. Dodonov, V.V., Manko, O., Manko, V.: Quantum non-stationary oscillators: models and applications. J. Russ. Laser Res. **16** (1995)
87. Dozias, S.: Opérateurs h-pseudodifférentiels à flot périodique. Ph.D. thesis, Université de Paris-Nord (1994)
88. Duistermaat, J., Kolk, J.: Lie Groups. Springer, New-York (1999)
89. Duistermaat, J.J., Guillemin, V.: The spectrum of positive elliptic operators and periodic bicharacteristics. Inven. Math. **29**, 39–79 (1975)
90. Egorov, Y.V.: On canonical transformations of pseudodifferential operators. Upeshi Mat. Nauk **25**, 235–236 (1969)
91. Ehrenfest, P.: Bemerkungüber die angenaherte gultigkeit der klassischen mechanik innerhalb der quantenmechanik, remarks on the approximate validity of classical mechanics in quantum mechanics. Z. Phys. **45**, 455–456 (1927)
92. Eisert, J., Plenio, M.B.: Quantum and classical correlations in quantum Brownian motion. Phys. Rev. Lett **89** (2002)
93. Elong, O., Robert, D., Senoussaoui, A.: On semiclassical fourier integral operators, Schrödinger propagators and coherent states. J. Math. Phys. **59**, 1–11 (2018). https://doi.org/10.1063/1.5043363
94. Emmrich, C., Weinstein, A.: Geometry of the transport equation in multicomponent WKB approximations. Commun. Math. Phys. **176**, 701–711 (1996)
95. Erikson, F.: On the measure of solid angles. Math. Mag. **63**, 184–185 (1999)
96. Fedosov, B.: Deformation Quantization and Index Theory. Akademic Verlag, Berlin (1996)
97. Fermanian Kammerer, C.: Semi-classical analysis of a Dirac equation without adiabatic decoupling. Monat. fur Math. **142**, 281–313 (2004)
98. Fermanian Kammerer, C.: Semiclassical analysis of generic codimension 3 crossings. Int. Math. Res. Not. **45**, 2391–2435 (2004)
99. Fermanian Kammerer, C., Gérard, P.: Mesures semi-classiques et croisements de modes. Bull. Soc. Math. Fr. **130**, 145–190 (2002)
100. Fermanian Kammerer, C., Gérard, P.: A landau-zener formula for non-degenerated involutive codimension three crossings. Ann. Henri Poincaré **4**, 513–552 (2003)
101. Fermanian-Kammerer C. Lasser, C., Robert, D.: Propagation of wave packets for systems presenting codimension 1 crossing (2020)
102. Feynman, R.P., Vernon, F.L.: The theory of a general quantum system interacting with a linear dissipative system. Ann. Phys. **24** (1963)
103. Fleming, C., B.L., H., Roura, A.: Exact analytical solutions to the master equation of Brownian motion for a general environment. Ann. Phys. **326** (2011)
104. Folland, G.B.: Harmonic Analysis in the Phase Space, vol. 122. Princeton University Press, Princeton (1989)

105. Folland, G.B.: Quantum Field Theory. Mathematical Surveys Ans Monographs. American Mathematical Society (2008)
106. Ford, G., OĆonnell, R.F.: Exact solution of the Hu–Paz–Zhang master equation. Phys. Rev. D **64** (2001)
107. Galaviz, I., Garcia-Compean, H., Przanowski, M., Turrubiates, F.: Weyl-Wigner-Moyal formalism for Fermi classical systems. Ann. Phys. **323**, 267–290 (2008)
108. Gazeau, J.P.: Coherent States in Quantum Physics. Wiley-VCH Verlag GmbH, Berlin (2009)
109. Gelfand, I., Neumark, M.: Unitary representations of the Lorentz group. Acad. Sci. USSR, J. Phys. **X**(3), 93–94 (1946)
110. Gérard, P.: Microlocal defect measures. Commun. Partl. Differ. Equs. **16**, 1761–1794 (1991)
111. Gérard, P., Leichtnam, E.: Ergodic properties of the eigenfunctions for the Dirichlet problem. Commun. Math. Phys. **71**, 559–607 (1993)
112. Gilmore, R.: Ann. Phys. (NY). Geometry of symmetrized states **74**, 391–463 (1972)
113. Gilmore, R.: On properties of coherent states. Rev. Mex. Fis. **23**, 143–187 (1974)
114. Ginibre, J., Velo, G.: The classical field limit of scattering theory of non relativistic many bosons systems. II. Commun. Math. Phys. **68**, 45–68 (1979)
115. Ginibre, J., Velo, G.: The classical field limit of non relativistic bosons. I. Borel summability for bounded potentials. Ann. Phys. **128–2**, 243–285 (1980)
116. Ginibre, J., Velo, G.: The classical field limit of non relativistic bosons systems, II. asymptotic expansions for general potentials. Ann. I. H. P. Sec. A **33**, 363–394 (1980)
117. Ginibre, J., Velo, G.: On a class of non linear Schrödinger equation with non local interactions. Math. Z **170**, 109–136 (1980)
118. Glauber, R.: Coherent and incoherent states of the radiation field. Phys. Rev. **131**, 2766–2788 (1963)
119. Glauber, R.: The quantum theory of optical coherence. Phys. Rev. **130**, 2529–2539 (1963)
120. Gohberg, I.C., Krein, M.G.: Introduction to the Theory of Linear Nonselfadajoint Operators in Hilbert Space, vol. 18. Translations of Mathematical Monographs. AMS (1969)
121. Goldstein, H.: Classical Mechanics. Addison -Wesley Publishing Company (1980)
122. de Gosson, M.: Maslov Classes, Metaplectic Representation and Lagrangian Quantization. Akademie-Verlag, Berlin (1997)
123. Grossmann A. Morlet, J., Paul, T.: Transforms associated to square integrable group representations. I. General results. J. Math. Phys. **26**, 2473–2479 (1985)
124. Guichardet, A.: Histoire d'un vecteur tricentenaire. SMF-Gazette **117**, 23–33 (2008)
125. Guillemin, V., Sternberg, S.: Geometric asymptotics. In: Mathematical Surveys, vol. 14. AMS (1977)
126. Guillemin, V., Uribe, A.: Circular symmetry and the trace formula. Invent. Math. **96**, 385–423 (1989)
127. Gutzwiller, M.: Periodic orbits and classical quantization conditions. J. Math. Phys. **12**, 343–358 (1971)
128. Gutzwiller, M.: Chaos in Classical and Quantum Mechanics. Springer, Berlin (1991)
129. Hagdorn. G.A., Joye, A.: Recent results on non-adiabatic transitions in quantum mechanics. In: Chernov, N., et al. (eds.) Recent Advances in Differential Equations and Mathematical Physics, vol. 412, pp. 183–198 (2006)
130. Hagedorn, G.: Semiclassical quantum mechanics III. Ann. Phys. **135**, 58–70 (1981)
131. Hagedorn, G.: Semiclassical quantum mechanics IV. Ann. Inst. Henri Poincaré. **42**, 363–374 (1985)
132. Hagedorn, G., Joye, A.: Molecular propagation through small avoided crossings of electron energy levels. Rev. Math. Phys. **11**, 41–101 (1999)
133. Hagedorn, G., Joye, A.: Semiclassical dynamics with exponentially small estimates. Commun. Math. Phys. **207**, 439–465 (1999)
134. Hagedorn, G., Joye, A.: Exponentially accurate semi-classical dynamics: propagation, localization, Ehrenfest times, scattering and more general states. Ann. Henri Poincaré **1–5**, 837–883 (2000)

135. Hall, K., Meyer, G.: Introduction to Hamiltonian Dynamical Systems and the N-body problem. Applied Mathematical Sciences, vol. 90. Springer, Berlin (1991)
136. Halliwell, J.J., Yu, T.: Alternative derivation of the Hu-Paz-Zhang master equation of quantum Brownian motion. Phys. Rev. D **53** (1966)
137. Hannay, J., Berry, M.: Ergodic properties of the eigenfunctions for the Dirichlet problem. Phys. D **1**, 267–291 (1993)
138. Helein, F.: A representation formula for maps on supermanifolds. J. Math. Phys. **49**, 1–19 (2008)
139. Helffer, B.: Théorie spectrale pour des opérateurs globalement elliptiques. Asterisque : No. 112, 1984. Société Mathématique de France (1984)
140. Helffer, B., Martinez, A., Robert, D.: Ergodicité et limite semi-classique. Commun. Math. Phys. **109**, 313–326 (1987)
141. Helffer, B., Robert, D.: Calcul fonctionnel par la transformation de Mellin et opérateurs admissibles. J. Funct. Anal. **53**(3), 246–268 (1983)
142. Helffer, B., Robert, D.: Puits de potentiels généralisés et asymptotique semi-classique. Annales de l'I.H.P. **41**, 291–331 (1984)
143. Helffer, B., Sjöstrand, J.: Multiple wells in the semi-classical limit I. Commun. Partl. Differ. Equs. **9**, 337–408 (1984)
144. Heller, E.J.: Time dependent approach to semiclassical dynamics. J. Chem. Phys. **62**, 1544–1555 (1975)
145. Heller, E.J.: Quantum localization and the rate of exploration of phase space. Phys. Rev. A **35**, 1360–1370 (1987)
146. Helton, J.: An operator algebra approach to partial differential equations. Indiana Univ. Math. J. **26**, 997–1018 (1977)
147. Henneaux, M., Teitelboim, C.: Quantization of Gauge Systems. Princeton University Press, Princeton (1994)
148. Hepp, K.: The classical limit of quantum mechanical correlation functions. Commun. Math. Phys. **35**, 265–277 (1974)
149. Herman, M., Kluk, E.: A semiclassical justification for the use of non-spreading wavepackets in dynamics calculations. J. Chem. Phys. **91**, 27–34 (1984)
150. Hirshfeld, A., Henselder, P.: Deformation quantization for systems with fermions. Ann. Phys. **302**, 59–77 (2002)
151. Hislop, P., Sigal, M.: Introduction to spectral theory. Applied Mathematical Sciences, vol. 113. Springer, Berl (1996)
152. Hörmander, L.: The spectral function of an elliptic operator. Acta Math. **121**, 173–218 (1968)
153. Hörmander, L.: The Analysis of Linear Partial Differential Operators. Grundlehren der mathematischen Wissenschaften, vol. 256. Springer, Berlin (1983)
154. Hörmander, L.: Symplectic classification of quadratic forms and general Melher formulas. Math. Z. **219**, 413–449 (1995)
155. Hornberger, K.: Introduction to decoherence, entangement and decoherence. Lecture Notes in Physics, vol. 768 (2009)
156. Horodecki, M., Horodecki, P., Horodecki, R.: Separability of mixed states: necessary and sufficient conditions. Phys. Lett. A **223** (1966)
157. Howland, J.: Two problems with time-dependent Hamiltonians. Lecture Notes in Physics, vol. 130 (1980)
158. Hu, B.L., Paz, J.P., Zhang, Y.: Quantum brownian motion in a general environment. Phys. Rev. D **45** (1992)
159. Hudson, R.: When is the Wigner quasi-probability non negative? Rep. Math. Phys. **6**, 249–252 (1974)
160. Inönü, E., Wigner, E.: On the contraction of groups and their representations. Proc. Nat. Acad. Sci. **39**, 510–524 (1953)
161. Ivrii, V.: Microlocal Analysis and Precise Spectral Asymptotics. Springer, Berlin (1998)
162. Joos, E., Zeh, H.D.: The emergence of classical properties through interaction with the environment. Z. Phys. B Condens. Matter **59** (1985)

163. Katok, A., Hasselblatt, B.: Introduction to the Modern Theory of Dynamical Systems. Cambridge University Press, Cambridge (1995)
164. Kay, K.: Integral expressions for the semi-classical time dependent propagator. J. Chem. Phys. 4377–4392 (1994)
165. Keller, J., Streifer, W.: Complex rays with an application to Gaussian beams. J. Opt. Soc. Am. **61**, 40–43 (1971)
166. Khuat-Duy, D.: Formule des traces semi-classique pour un énergie critique et construction de quasi-modes à l'aide d'états cohérents. Ph.D. thesis, Université de Paris-Dauphine (1996)
167. Klauder, J., Skagerstam, B.: Coherent States. Worlds Scientific Publishing (1985)
168. Knapp, A.W.: Representation theory of semisimple groups. An overview based on examples. Princeton Mathematical Series, vol. 36. Princeton University Press, Princeton, NJ (1986)
169. Knapp, A.W.: Lie Groups Beyond an Introduction. An overview based on examples. In: Princeton Mathematical Series, vol. 36. Birkhäuser, Boston (1996)
170. Knapp, A.W., Trapa, P.E.: Representations of semisimple Lie groups. In: Representation theory of Lie groups (Park City, UT, 1998), IAS/Park City Math. Ser., vol. 8, pp. 7–87. American Mathematical Society, Providence, RI (2000)
171. Kosmann-Schwarzbach, Y.: Groups and symmetries. Universitext. Springer, New York (2010). From finite groups to Lie groups, Translated from the 2006 French 2nd edition by Stephanie Frank Singer
172. Landau, L., Lifshitz, F.: Quantum Mechanics. Pergamon (1977)
173. Leboeuf, P., Voros, A.: Chaos-revealing multiplicative representation of quantum eigenstates. J. Phys. A **23**, 1165–1174 (1990)
174. Ledoux, M.: The Concentration of Measure Phenomenon. Mathematical Surveys and Monographs, vol.138. American Mathematical Society (2001)
175. Leites, D.: Introduction to the theory of supermanifolds. Russ. Math. Surv. **35**, 3–57 (1980)
176. Leray, J.: Lagrangian Analysis and Quantum Mechanics. MIT (1981)
177. Leray, J.: Lagrangian analysis and quantum mechanics. A mathematical structure related to asymptotic expansions and the Maslov index, (trans. from French). MIT Press, Boston (1981)
178. Lerner, N.: The Wick calculus of pseudo-differential operators and some of its applications. Cubo Mat. Educ. **5**(1), 213–236 (2003)
179. Lieb, E.: Proof of an entropy conjecture of wehrl. Commun. Math. Phys. **62**, 1–13 (1978)
180. Lieb, E.H.: The classical limit of quantum spin systems. Commun. Math. Phys. **31**, 327–340 (1973)
181. Lions, P., Paul, T.: Sur les mesures de Wigner. Revista Mat. Iberoamericana **9**, 553–618 (1993)
182. Littlejohn, R.: The semiclassical evolution of wavepackets. Phys. Rep. **138**, 193–291 (1986)
183. Maas, H.: Siegel's modular forms and Dirichlet series. Lecture Notes in Mathematics, vol. 216. Springer, Berlin (1971)
184. Macià, F.: Some remarks on quantum limits on Zoll manifolds. Commun. Partl. Differ. Equs. **33**, 1137–1146 (2008)
185. Macià, F.: Semiclassical measures and the Schrödinger flow of Riemannian manifolds. Nonlinearity **22**, 1003–1020 (2009)
186. Macià, F.: High-frequency propagation for the Schroödinger equation on the torus. J. Funct. Anal **258–3**, 933–955 (2010)
187. Manè, R.: Ergodic Theory and Differentiable Dynamics. Springer, Berlin (1987)
188. Martinez, A.: An introduction to semiclassical and microlocal analysis. Universitext. Springer, New York, NY (2002)
189. Martinez, A., Sordoni, V.: Twisted pseudodifferential calculus and application to the quantum evolution of molecules. Memoirs of the Mathematical Society, vol. 936. American Mathematical Society (2009)
190. Maspero, A.: Lower bounds on the growth of Sobolev norms in some linear dependent Schrödinger equations. Math. Res. Let **10**, 1–10 (2019)
191. Maspero, A., Robert, D.: On time dependent Schrödinger equations: global well-posedness and growth of Sobolev norms. J. Funct. Anal. **273**, 721–781 (2017)

192. McLachland, N.: Theory and Application of Mathieu Functions. Oxford University Press, New York (1947)
193. Mehlig, B., Wilkinson, M.: Semiclassical trace formula using coherent states. Ann. Phys. (Leipzig) **10**, 501 (2001)
194. Meinrenken, E.: Semi-classical principal symbols and Gutzwiller's trace formula. Rep. Math. Phys. **31**, 279–295 (1992)
195. Moyal, J.E.: Quantummechanics as a statistical theory. Proc. Camb. Philos. Soc. **45**, 99–124 (1959)
196. Aravind, Mukada N., P., Simon, R., : Wigner rotations, Bargmann invariants and geometric phases. J. Phys. A: Math. Gen. **36**, 2347–2370 (2003)
197. Muller, C.: Spherical harmonics. In: Lecture Notes in Mathematics, vol. 17, pp. 7–87. Springer, New-York (1966)
198. urakami, M., Ford, G., OConnell, R.: Decoherence in phase space. Laser Phys.**13**, 180–183 (2003)
199. Nenciu, G.: Linear adiabatic theory. Commun. Math. Phys. **152**, 701–711 (1993)
200. Nielsen, M., Chuang, I.: Quantum Computation and Quantum Information. Cambridge University Press, Cambridge (2000)
201. Onofri, E.: A note on coherent state representations of Lie groups. J. Math. Phys. **16**, 1087–1089 (1975)
202. Pancharatnam, S.: Generalized theory of interference and its applications. Proc. Indian Acad. Sci. Sec. A **44**, 247–262 (1956)
203. Paul, T., Uribe, A.: Sur la formule semi-classique des traces. Note CRAS, Paris **313**(1), 217–222 (1991)
204. Paul, T., Uribe, A.: A construction of quasi-modes using coherent states. Ann. Inst. H. Poincaré Phys. Théor. **59**(4), 357–381 (1993)
205. Paul, T., Uribe, A.: The semiclassical trace formula and propagation of wave packets. Commun. Math. Phys. **132**, 192–242 (1995)
206. Paul, T., Uribe, A.: On the pointwise behavior of semiclassical measures. Commun. Math. Phys. **175**, 229–256 (1996)
207. Pauli, W.: General principles of quantum mechanics. Springer-Verlag, Berlin (1980); Translated from the German by P. Achuthan and K. Venkatesan, With an introduction by Charles P. Enz
208. Perelomov, A.: Generalized coherent states and their applications. Texts and Monographs in Physics. Springer, Berlin (1986)
209. Perelomov, A.M.: Coherent states for arbitrary Lie group. Commun. Math. Phys. **26**, 222–236 (1972)
210. Petkov, V., Popov, G.: Semiclassical trace formula and clustering of eigenvalues for Schrödinger operators. Ann. Inst. Henri Poincaré, sect. Phys.Th. **68**, 17–83 (1998)
211. Petkov, V., Robert, D.: Asymptotique semiclassique du spectre d'hamiltoniens quantiques et trajectoires classiques périodiques. Commun. Partl. Differ. Equs. **10**, 365–390 (1985)
212. Ralston, J.: On the construction of quasimodes associated with stable periodic orbits. Commun. Math. Phys. **51**, 219–242 (1976)
213. Ralston, J.: Gaussian beams and propagation of singularities. In: Studies in Partial Differential Equations, pp. 246–248. MAA Studies Mathematics, vol. 23 (1982)
214. Ramis, J.: Les séries k-sommables et leurs approximations. In: Analysis, Microlocal Calculus and Relativistic Quantum Theory, Proceedings Les Houches 1979, pp. 178–199. Springer, Lecture Notes in Physics, vol. 126 (1980)
215. Reed, M., Simon, B.: Fourier-analysis. Self-adjointness. Academic Press (1979)
216. Rivière, G.: Remarks on quantum ergodicity. J. Mod. Dyn. **7**, 119–133 (2013)
217. Robert, D.: Autour de l'approximation semi-classique. PM 68. Birkhäuser (1987)
218. Robert, D.: Propagation of coherent states in quantum mechanics and applications. In: Wang, X.P. et al. (eds.) Partial differential equations and applications. Proceedings of the CIMPA school, Lanzhou, China, July 2004. Paris: Société Mathématique de France. Séminaires et Congrès, vol. 15, pp. 181–252 (2007)

219. Robert, D.: On the Herman-Kluk semiclassical approximation. Rev. Math. Phys. **22–10**, 1123–1145 (2010)
220. Robert, D.: On some properties of the time evolution of quadratic open quantum systems. St. Petersburg Math. J. **30**, (2019)
221. Robert, D., Tamura, H.: Asymptotic behavior of scattering amplitudes on semi-classical and low energy limits. Ann. Inst. Fourier **39**, 155–192 (1989)
222. Robert, D., Thomann, L.: Random weighted Sobolev inequalities and application to quantum ergodicity. Commun. Math. Phys. **335**, 1181–1209 (2015)
223. Robin, J., Salamon, D.: The Maslov index for path. Topology **32**, 827–844 (1993)
224. Robinson, S.: The semiclassical limit of the quantum dynamics II: Scattering theory. Ann. Inst. Henri Poincaré **48**, 281–296 (1988)
225. Rodnianski, I., Schlein, B.: Quantum fluctuations and rate of convergence towards mean field dynamics. Commun. Math. Phys. **291**, 31–61 (2009)
226. Rogers, A.: Supermanifolds. Theory and Applications. World Publishing Company, Singapore (2007)
227. Rousse, V., Swart, T.: A mathematical justification for the Herman-Kluk propagator. Commun. Math. Phys. **286**, 725–750 (2009)
228. Rudin, W.: Real and Complex Analysis. McGraw-Hill Book Co., New York (1966)
229. Ruelle, D.: Statistical Mechanics. Benjamin, New-York (1969)
230. Sattinger, D.: Lie Groups and Algebras with Applications to Physics, Geometry and Mechanics. Springer, New-York (1986)
231. Schlichenmaier, M.: Berezin-toeplitz quantization for compact kähler manifolds. A review of results. Adv. Math. Phys. **2010**, 1–38 (2010)
232. Schnirelman, A.: Ergodic properties of eigenfunctions. Usp. Math. Nauk **29**, 181–182 (1974)
233. Schrödinger, E.: Der stetige übergang von der Mikro- zur Makromechanik. Naturwissenshaften **14**, 664–666 (1926)
234. Schrödinger, E.: Further studies on solving eigenvalue problems by factorization. Proc. R. Irish Acad. **A46**, 183–206 (1940)
235. Shubin, M., Andersson, S.: Pseudodifferential Operators and Spectral Theory. Springer, Berlin Heidelberg (2001)
236. Simon, R.: Peres-Horodecki separability criterion for continuous variable systems. Phys. Rev. Lett. **84** (2000)
237. Sjöstrand, J.: Propagation of analytic singularities for second order Dirichlet problems. Commun. Part. Differ. Equs. **5**(1), 41–95 (1980)
238. Sjöstrand, J.: Singularités analytiques microlocales. Astérique, 95. Société Mathématique de France (1982)
239. Sokal, A.: An improvement of Watson's theorem on Borel summability. J. Math. Phys. **21**, 261–263 (1980)
240. Soto, F., Claverie, P.: When are the Wigner function of multidimensional system non negative? J. Math. Physics **24**, 197–100 (1983)
241. Takhtajan, L.: Quantum mechanics for mathematicians. In: Quantum Mechanics for Mathematicians. Graduate Studies in Mathematics, vol. 95. AMS, Providence, Rhode Island (2008)
242. Tataru, D.: Phase space transform and microlocal analysis, pp. 505–524. Publ. Cent. Ri. Mat. Ennio Giorgi (2004)
243. Thaller, B.: The Dirac Operator. Springer, Berlin (1990)
244. Thomas, L.E., Villegas-Blas, C.: Asymptotics of Rydberg states for the hydrogen atom. Commun. Math. Phys. **187**, 623–645 (1997)
245. Unterberger, A.: Oscillateur harmonique et opérateurs pseudo-différentiels. Annales de l'Institut Fourier de Grenoble XXXIX **3**, 201–221 (1979)
246. Unterberger, A., Unterberger, J.: Representations of sl(2, r) and symbolic calculi. Integr. Equat. Oper. Th. **18**, 303–334 (1994)
247. Uribe, A.: A symbol calculus for a class of pseudodifferential operators on S^n and band asymptotics. J. Funct. Anal. **59**, 535–556 (1984)

248. Varadarajan, V.: Supersymmetry for Mathematicians: An Introduction. American Mathematical Society, Providence, Rhode Island (2004)
249. Colin de Verdière, Y.: The level crossing problem in semi-classical analysis (I). the symmetric case. Annales de l'Institut Fourier **53**(4), 1023–1054 (2003)
250. Colin de Verdière, Y.: The level crossing problem in semi-classical analysis. II. Ann. Inst. Fourier (Grenoble) **142**, 1423–1441 (2004)
251. Villegas-Blas, C.: The Laplacian on the n-sphere, the hydrogen atom and the Bargmann space representation. Ph.D. thesis, Mathematics Department, University of Virginia (1996)
252. Walters, P.: An Introduction to Ergodic Theory, vol. 79-Graduate Texts in Mathematics. Springer, Berlin (1982)
253. Wang, X.: Time-delay operators in semiclassical limit ii short range potentials. Trans. Am. Math. Soc. **322**, 1509–1563 (1990)
254. Wehrl, A.: On the relation between classical and quantum-mechanical entropy. Rept. Math. Phys. **16**, 353–358 (1979)
255. Weil, A.: Sur certains groupes d'opérateurs unitaires. Acta Math. **111**, 143–211 (1964)
256. Weinberg, S.: The Quantum Theory of Fields, vol. I, II III. Cambridge University Press, Cambridge (1995)
257. Weinstein, A.: Asymptotics of the eigenvalues clusters for the Laplacian plus a potential. Duke Math. J. **44**, 883–892 (1977)
258. Werner, R.F., Wolf, M.: Bound entangled Gaussian states. Proc. Camb. Philos. Soc. **86** (2001)
259. Weyl, H.: The classical groups. Princeton Landmarks in Mathematics. Princeton University Press, Princeton, NJ (1997) Their invariants and representations. Fifteenth printing, Princeton Paperbacks
260. Wigner, E.: On unitary representations of the inhomogeneous Lorentz group. Ann. Math. (2) **40**(1), 149–204 (1939)
261. Wigner, E.: Group Theory and Its Applications to Quantum Mechanics of Atomic Spectra. Academic Press, New-York (1959)
262. Wilkinson, M.: A semi-classical sum rule for matrix elements of classically chaotic systems. J. Phys. A. Math. Gen. **20**, 2415–2423 (1987)
263. Witten, E.: Supersymmetry and Morse theory. J. Differ. Geom. **17**, 661–692 (1982)
264. Woodhouse, N.M.J.: Geometric Quantization. Clarendon, Oxford (1991)
265. Yajima, K.: The quasiclassical limit of quantum scattering theory. Commun. Math. Phys. **69**, 101–129 (1979)
266. Yajima, K.: Existence of solutions for Schrödinger evolution equations. Commun. Math. Phys. **110**, 415–426 (1987)
267. Zelditch, S.: Uniform distribution of the eigenfunctions on compact hyperbolic surfaces. Duke Math. J. **55**, 919–941 (1987)
268. Zelditch, S.: Quantum ergodicity on the sphere. Commun. Math. Phys. **146**, 61–71 (1992)
269. Zinn-Justin, J.: Intégrale de chemin en mécanique quantique: introduction. CNRS Editions/EDP Sciences, Paris (2003)
270. Zurek, W.H.: Pointer basis of quantum apparatus: into what mixture does the wave packet collapse. Phys. Rev. D. **24** (1981)
271. Zworski, M.: Semiclassical Analysis. Graduate Studies in Mathematics, vol. 138. American Mathematical Society (2010)

Index of Concepts

D. Robert and M. Combescure, *Coherent States and Applications in Mathematical Physics*, Theoretical and Mathematical Physics,
https://doi.org/10.1007/978-3-030-70845-0

Index of Symbols

Symbols

D. Robert and M. Combescure, *Coherent States and Applications in Mathematical Physics*, Theoretical and Mathematical Physics,
https://doi.org/10.1007/978-3-030-70845-0

The manufacturer's authorised representative in the EU is Springer Nature Customer Service Centre GmbH, Europaplatz 3, 69115 Heidelberg, Germany. If you have any concerns regarding our products, please contact ProductSafety@springernature.com

Printed and bound by CPI Group (UK) Ltd, Croydon, CR0 4YY
30/09/2026
02225295-0016